D1327640

Chemical Kinetics and Process Dynamics in Aquatic Systems

Patrick L. Brezonik
Professor of Environmental Engineering Science
University of Minnesota

LEWIS PUBLISHERS
Boca Raton   Ann Arbor   London   Tokyo

Library of Congress Cataloging-in-Publication Data

Brezonik, Patrick L.
  Chemical kinetics and process dynamics in aquatic systems /
  Patrick L. Brezonik.
      p.   cm.
  Includes bibliographical references and index.
  ISBN 0-87371-431-8
  1. Water chemistry. 2. Chemical kinetics. I. Title.
GB855.B74   1993
551.48—dc20   92-42114
                                                          CIP

Learning Resources
Centre

This book contains information obtained from authentic and highly regarded sources. Reprinted material is quoted with permission, and sources are indicated. A wide variety of references are listed. Reasonable efforts have been made to publish reliable data and information, but the author and the publisher cannot assume responsibility for the validity of all materials or for the consequences of their use.

Neither this book nor any part may be reproduced or transmitted in any form or by any means, electronic or mechanical, including photocopying, microfilming, and recording, or by any information storage and retrieval system, without prior permission in writing from the publisher.

CRC Press, Inc.'s consent does not extend to copying for general distribution, for promotion, for creating new works, or for resale. Specific permission must be obtained in writing from CRC Press for such copying.

Direct all inquiries to CRC Press, Inc., 2000 Corporate Blvd., N.W., Boca Raton, Florida 33431.

© 1994 by CRC Press, Inc.
Lewis Publishers is an imprint of CRC Press

No claim to original U.S. Government works
International Standard Book Number 0-87371-431-8
Library of Congress Card Number 92-42114
Printed in the United States of America 1 2 3 4 5 6 7 8 9 0
Printed on acid-free paper

To
Carol

PREFACE

This book, like so much that has happened in aquatic chemistry over the past quarter century, had its beginnings in the towns and villages near Zurich, where I spent the summer of 1980 as a visiting faculty guest at EAWAG, the Swiss Federal Institute for Water Research. I decided to write this book primarily to fill a teaching need, there being no text that dealt with kinetic aspects of water chemistry, but the book has evolved as much into a reference as a text. I did not intend to be writing this preface 13 years later. However, if I had finished it in a timely fashion, it would have been a very different book. The field of aquatic chemical kinetics was much less developed in 1980, and my goals for the breadth and depth of treatment were much simpler then.

The intended audience for the book includes graduate students and practicing professionals in many aspects of aquatic sciences and engineering: water chemistry, limnology, aqueous geochemistry, microbial ecology, marine science, environmental and water resources engineering, and related fields. Although the book seeks to describe the current level of sophistication in some fairly advanced aspects of aquatic chemistry, a strong effort was made to create a book that is understandable and useful to nonexperts and those with limited backgrounds in chemistry. Chapters 2 and 3 and the first parts of Chapters 5 and 6 are written in a didactic style and introduce the important concepts and equations of chemical and microbial kinetics for batch systems and continuous-flow reactors. The balance of the book applies these concepts and equations to a variety of chemical and biological processes that occur in and affect the status of natural aquatic environments. Abundant references are included to both the current literature and the key studies of the past that have led to recent developments in aquatic chemical kinetics.

I would like to thank several of my colleagues at the University of Minnesota for reviewing various chapters: Steven Eisenreich, Walter Maier, John Gulliver, and Deb Swackhamer. I appreciate their helpful corrections and suggestions on stating certain concepts correctly, and they introduced me to some important references. The comments of Michael Hoffmann (Cal Tech.) on an early version of the first several chapters were helpful in enhancing the rigor of the text. Alan Stone (Johns Hopkins) read the entire manuscript and made many useful suggestions that improved both the clarity and accuracy of the text. Many graduate students exposed to parts of the manuscript in graduate courses at the University of Minnesota made helpful suggestions and participated in the development of the problems at the end of each chapter. It scarcely needs to be emphasized that any errors and deficiencies that remain are my responsibility.

I owe a special debt of gratitude to Werner Stumm for his hospitality and support at EAWAG, where this book began, and for inviting me to participate in his 1989 workshop on *Aquatic Chemical Kinetics*, which resulted in a book by that name published by John Wiley & Sons. An earlier, shorter version of Chapter 7 appeared in that volume.

I appreciate the support of the University of Minnesota, which provided a stimulating environment, an extensive library system, and the freedom to pursue this venture. The expert assistance of Christine Winter, Project Editor at CRC Press, in converting the manuscript to book form is gratefully acknowledged. Finally, my wife, Carol, and sons, Craig and Nick, have been both understanding and supportive over the long gestation period for this book. The many hours spent away from them while writing this cannot be recovered, but I hope that they and you, the reader, will find that something useful has emerged from that investment.

Patrick L. Brezonik
Minneapolis

THE AUTHOR

Patrick L. Brezonik is a professor of environmental engineering science in the Department of Civil & Mineral Engineering
at the University of Minnesota and director of the University's
Water Resources Research Center. He received a B.S. in
chemistry from Marquette University in 1963 and M.S. (1965)
and Ph.D. (1968) degrees in water chemistry from the University of Wisconsin. Before assuming his present position, Professor Brezonik was on the faculty at the University of Florida
for 15 years. His research interests are focused on surface water

quality and chemistry, especially on the fate and behavior of inorganic contaminants
in lakes and the interactions between chemical and microbial processes in elemental
cycles. He has published widely on such subjects as lake eutrophication, nitrogen
and phosphorus cycling processes, effects of acidic deposition on lakes, and heavy
metal chemistry in surface waters and sediments.

TABLE OF CONTENTS

Chapter 7
Prediction Methods for Reaction Rates and Compound Reactivity 553

SYMBOLS

Author's note: Selection of symbols was based on common usage in the literature. Multiple use of some symbols to denote several different quantities was unavoidable, but in general symbols do not have more than one meaning in a given chapter. For symbols with multiple meanings, definitions restricted to a single chapter or set of chapters are identified in parentheses.

α_i	fraction of an element or compound present as species i
α	Brønsted catalysis coefficient (chapter 7)
α	beam attenuation coefficient (chapter 8)
α_1	hydrogen bond donor acidity (chapter 7)
β	reactivity index (units of concentration) (represents concentration of P at which quenching of $^1O_2$ by P and water are equal) (chapter 8)
β_i	cumulative stability constant for metal with ith ligand
β_{m2}	hydrogen bond acceptor basicity
γ_i	activity coefficient of species i
γ_{O_2}	fraction of encounters of triplet sensitizers with ground state O_2 that yield $^1O_2$ (chapter 8)
δ	thickness or depth of biofilm (chapter 6)
δ_i	atom valence index in molecular connectivity indices
Δ	change in some quantity; displacement from equilibrium
$\Delta\beta$	parameter to quantify A-B character of metal ions
$\Delta E; \Delta E^{\neq}$	internal energy of reaction/activation
$\Delta G; \Delta G^{\neq}$	free energy of reaction/activation
ΔG°	free energy of reaction under standard conditions
$\Delta H; \Delta H^{\neq}$	enthalpy of reaction/activation
$\Delta S; \Delta S^{\neq}$	entropy of reaction/activation
ΔS	change in storage (chapter 5)
$\Delta V^{\neq}$	volume of activation
$\Delta|\delta|_{x-y}$	difference in absolute atomic charge across bond x-y
χ	molecular connectivity index
η	viscosity (g cm^{-1} s^{-1})
η^k	kinematic viscosity (cm^2 s^{-1})
ϵ	molar absorptivity (molar extinction coefficient)
κ	transmission coefficient
λ	wavelength
λ	elementary jump distance of molecules (chapter 3)
λ	eigenvalue (chapter 5)
μ_i	chemical potential of i
μ	specific growth rate (microbial kinetics) (t^{-1})
ν	stoichiometric coefficient
ν	frequency
ν_f	fundamental frequency
π_X	lipophilicity parameter for substituent or compound X
π^*	solvatochromic parameter
Φ	quantum yield
Φ	Thiele modulus (chapters 5,6)

Ψ	eigenvector
ρ_i	density of i
ρ_w	flushing coefficient (t^{-1}) (chapters 5,6)
ρ	Hammett reaction coefficient (chapter 7)
σ	standard deviation
σ	sedimentation coefficient $(year^{-1})$ (chapter 5)
σ	Hammett-type substituent coefficient (chapter 7)
σ_k	HSAB softness parameter (chapter 7)
σ_p	HSAB hardness parameter (chapter 7)
τ	characteristic time (when concentration of reactant in first order reaction reaches 1/e of initial concentration)
τ	average time between elementary jumps of molecules (chapter 3)
τ_i	residence time of i (in a reactor)
θ	empirical temperature coefficient
θ	isokinetic temperature (chapters 6,7)
θ	angle of refracted light (chapter 8)
ξ	extent of reaction
ω	rate of change in mass loading (chapter 5)
ω	electrostatic work (chapter 7)
Ω	degree of $CaCO_3$ saturation
a	absorption coefficient
a	shape parameter (chapter 6)
a_{nar}	activity of compound inducing narcosis
a_s	specific affinity (enzyme kinetics)
A	surface area
A	pre-exponential or frequency factor (in Arrhenius equation)
Ci	curie (unit of radioactive decay); $1\ Ci = 3.7 \times 10^{10}$ disintegrations per second (dps)
dF(t)	residence time distribution function
d_i	diameter of i
D	dilution or flushing rate
D	distribution coefficient (ratio of mean pathlength of light in thin vertical layer divided by thickness of layer) (chapter 8)
D	dielectric constant
D_i	molecular diffusion coefficient for species i
D_L	dispersion coefficient
e	charge on electron
e	eccentricity (chapter 6)
e^-_{aq}	hydrated electron
E	efficiency of substrate use (chapter 6)
E_{act}	activation energy kJ mol^{-1}
E_{bg}	band gap energy
E_f	catalyst effectiveness
$E_o(\lambda)$	scalar irradiance at water surface (photons $cm^{-2}\ s^{-1}$)
E_s	substituent steric parameter (chapter 7)
EC_{50}	median effective concentration (to produce some biological response)
F	flux (areal mass transfer rate)
g	gravitational constant
h	hour

h	Planck's constant (for electromagnetic radiation, the relationship between energy and frequency of radiation)
H	Henry's law constant ($atm \cdot L\ mol^{-1}$ or $mm\ Hg \cdot L\ mol^{-1}$)
I	ionic strength (units of molality)
I_o	incident light intensity (Einsteins $cm^{-2}\ s^{-1}$)
$I_{a,\lambda}$	rate of absorption of light energy at wavelength λ
J	joule
k	rate coefficient
k	Boltzmann constant (gas constant per molecule)
k'	chromatographic retention index
K	equilibrium constant or partition coefficient
K_l	overall mass transfer coefficient for gas or volatile compound between air and water ($cm\ h^{-1}$)
K_R	Robertson light attenuation coefficient
K_S, K_m	half-saturation (Michaelis) constant (enzyme, microbial kinetics)
ℓ	light pathlength (cm or m)
L_A	areal mass loading rate ($g\ m^{-2}\ year^{-1}$)
L_o	ultimate (first stage) BOD; initial concentration of biodegradable organic matter
L_t	mass loading rate ($g\ year^{-1}$)
L_V	volumetric mass loading rate ($g\ m^{-3}\ year^{-1}$)
LC_{50}	median lethal concentration
ln	natural logarithm
M	molarity ($mol\ L^{-1}$)
M	microbial mass (chapter 6)
M_W	molecular weight
n, N	number (of moles); population of organisms
N_o	Avogadro's number
pK	negative logarithm of equilibrium constant
pT	negative logarithm of time in seconds
pε	negative logarithm of relative electron activity ($E_H/0.059$)
P	pressure (atm)
P	Peclet number (chapter 5)
P, P′, P″	dimensionless diffusion control parameters (chapter 6)
P_r	reciprocal Peclet number
q_a	areal water loading rate ($m\ year^{-1}$)
Q	reaction quotient (chapter 2)
Q	partition function (chapter 3)
Q	hydraulic flow ($m^3\ year^{-1}$) (chapters 5, 6)
Q	cell quota (chapter 6)
Q	quenching agent (chapter 8)
Q_{10}	ratio of rate constant at $(T+10)°C$ to rate constant at $T°C$
R	rate of reaction
R	receptor molecule (reacts with sensitizing agent) (chapter 8)
$\hat{R}$	resource requirement
$\mathbf{R_m}$	molar refractivity
R_P	fraction of P loading retained in reactor
R_t	chromatographic retention time of a compound
s	scattering coefficient (chapter 8)

s	shear rate (chapter 6)
S	degree of saturation (chapter 4)
S	substrate in enzymatic reaction (chapter 6)
S	sensitizer molecule (chapter 8)
S_w	aqueous solubility
$t_{1/2}$	half life
T	temperature
u, U	velocity
v	velocity (rate of reaction for enzyme-catalyzed reaction)
v_P	sedimentation velocity for phosphorus
V, V_{max}	maximum velocity for enzyme-catalyzed reactions
V	volume
W_λ	intensity of sun + sky radiation on horizontal plane just below water surface
X_{IP}	reactive photo-intermediate
X_w	mol fraction solubility in water
y	yield coefficient (chapter 6)
Y	maximum yield coefficient
z	depth
Z_λ	light intensity parameter used in calculating near surface photolysis rates
Z_{SD}	Secchi disk transparency (m)
Z_i; z_i	charge on ion i
Z_{ij}	collision frequency between i and j
[i]	molar concentration of i
{i}	activity of i

ACRONYMS

ACT	activated complex theory
AH	aquatic humus
ANC	acid neutralizing capacity
BCF	buffer catalysis factor
BF	bioaccumulation factor
BOD	biochemical oxygen demand
CB	conduction band
CFSTR	continuous-flow, stirred tank reactor
DHLL	Debye-Huckel limiting law
DIE	discrete integrated equation
DOC	dissolved organic carbon (mg L^{-1})
DOM	dissolved organic matter (mg L^{-1})
EDHE	extended Debye-Huckel equation
FA	fulvic acid
FL	foraminiferal lysocline
HA	humic acid
HBA	hydrogen bond acceptor
HBD	hydrogen bond donor

HSAB	(Pearson) hard and soft acid base (theory)	
i/o	inorganic/organic (character of organic compound)	
IP	ion product	
LFER	linear free energy relationship	
LMCT	ligand-to-metal charge transfer	
LSER	linear solvation energy relationship	
MC	molecular connectivity	
NLR	nonlinear regression	
os	outer sphere	
PAR	property-activity relationship	
PCA	principal component analysis	
PES	potential energy surface	
PFR	plug flow reactor	
SP	solubility-related property	
(Q)SAR	(quantitative) structure-activity relationship	
THM	trihalomethane	
TMV	total molecular volume	
TSA	molecular surface area	
TST	transition state theory	
UNIFAC	universal quasi-chemical functional groups activity coefficient	
VB	valence band	

SOME IMPORTANT CONSTANTS

Quantity	Symbol	Value
charge on electron	e	1.602×10^{-19} coulombs
		4.4840×10^{-10} esu
dielectric constant of water	$\mathbf{D}_{H_2O}$	78
Planck's constant	h	6.62×10^{-27} erg s
average time between molecular vibrations (fundamental frequency factor)	v_f	6.2×10^{12} s^{-1}
Avogadro's number	N_o	6.022×10^{23}
Boltzmann constant (gas constant per molecule)	k	1.38×10^{-16} erg K^{-1}
		1.38×10^{-23} J K^{-1}
thermal energy per molecule (at 298°C)	kT	2476 J mol^{-1}
		592 cal mol^{-1}
gas constant	R	8.31 J K^{-1} mol^{-1}
		1.987 cal K^{-1} mol^{-1}
		0.08206 L atm K^{-1} mol^{-1}
molar volume (ideal gas, 0°C, 1 atm)		22.414 L
ice point		273.15 K
gravitational constant	g	980 cm s^{-2}
pi	π	3.14159265
e (base of natural logs)	e	2.718282
Faraday's constant	$\mathscr{F}$	9.648×10^4 coulomb mol^{-1}

SOME USEFUL CONVERSION FACTORS

1 joule (J)	= 1 newton meter
	= 10^7 erg
	= 0.239 calorie
	= 9.9×10^{-3} L atm
	= 6.25×10^{18} electron-volt (eV)
1 calorie (cal)	= 4.184 J
1 eV	= 1.602×10^{-19} J
1 atm	= 760 torr
	= 760 mm Hg
	= 1.013×10^5 Pascal (Pa)
	= 1.013 bar
$ln\mathrm{x}$	= $2.30258 log\mathrm{x}$

"Yes, as everyone knows, meditation and water are wedded forever.... Why did the old Persians hold the sea holy? Why did the Greeks give it a separate deity, and own brother of Jove? Surely all this is not without meaning. And still deeper the meaning of that story of Narcissus, who because he could not grasp the tormenting, mild image he saw in the fountain, plunged into it and was drowned. But that same image, we ourselves see in all rivers and oceans. It is the image of the ungraspable phantom of life; and this is the key to it all."

Moby Dick, chap. 1, Herman Melville

Chemical Kinetics and Process Dynamics in Aquatic Systems

"Iron rusts from disuse, stagnant water loses its purity, and in cold weather becomes frozen; even so does inaction sap the vigors of the mind."

Notebooks, Leonardo da Vinci

CHAPTER 1

Overview

1.1 INTRODUCTION

Quantitative understanding of the chemistry of natural waters involves the two cornerstones of physical chemistry: thermodynamics and kinetics. The principles of thermodynamics define a system's composition at equilibrium — the condition toward which it tends to go in the absence of energy inputs. The rate at which systems approach equilibrium is the domain of kinetics. Aquatic chemists have made major advances over the past two decades in explaining the chemistry of natural waters, via thermodynamic concepts and equilibrium models, and many books have been devoted to these topics.[1] The principles of chemical equilibrium were applied to natural waters most thoroughly by Stumm and Morgan,[2] whose text has served as a conceptual framework for the discipline of aquatic chemistry.

Equilibrium approaches are useful in relating the inorganic composition of lakes, rivers, groundwater, and the oceans to weathering reactions of minerals in watersheds, but the ability of thermodynamics to describe aquatic systems is limited, particularly for organic substances and for waters affected by human activity. Although equilibrium calculations compare favorably with observed concentrations of major and minor ions in some waters, complete chemical equilibrium never occurs in natural waters. Nonequilibrium conditions persist, in part, because many energetically favorable reactions are very slow and, in part, because natural waters are open systems that receive influxes of matter and energy from external sources. The adequacy of equilibrium models thus depends on the relative rates of influxes and reactions for a substance. If influxes are small and reactions are rapid, equilibrium descriptions are adequate; slow reactions and rapid influxes allow nonequilibrium conditions to persist. The fact that the composition of natural waters changes over time also implies that equilibrium descriptions are not always

1

adequate. Moreover, the time-invariant state for open systems is the *steady state* and not the equilibrium state. This implies that kinetic relationships are important in describing aquatic systems.

A major goal of aquatic chemists is to understand the rate-controlling factors for chemical processes in aquatic systems. Kinetics often is considered an empirical science in which reaction rates are measured rather than predicted. Nonetheless, relationships are available to describe the effects of the major state variables (temperature, pressure) on reaction rates and to predict the kinetic behavior of some compounds from the behavior of related compounds. In addition, sophisticated theories are available to explain the mechanisms and energetics of chemical change. These relationships and theories are the ingredients that make kinetics a science rather than a mere collection of facts and rate constants. The search for other unifying principles and predictive relationships is the source of intellectual stimulation to scientists.

Aquatic chemists have placed increasing emphasis on kinetic and process-oriented models of natural waters over the past 20 years.[3] The evolution from static to dynamic models resulted, in part, from concerns about the fate of pollutants in natural waters. As a result, the literature on kinetics in aquatic systems has increased rapidly in recent years. The reactions and processes receiving attention are diverse, ranging from inorganic redox and ligand exchange reactions to organic transformations by biological and photochemical processes, and from reactions in solution to processes at air/water and water/solid interfaces. This literature is scattered widely in journals and books, and a comprehensive treatment of kinetic principles and their application to natural waters has not been available. This book is an effort to fill that void.

The processes of chemical change are studied at three levels in modern chemical kinetics: (1) phenomenological or observational; (2) mechanistic; and (3) statistical mechanical. The first level involves measurement of reaction rates under various physical conditions (e.g., temperature) and interpretation of the data in terms of rate laws based on mass action principles. The second level is concerned with elucidating reaction mechanisms, i.e., the "elementary" steps that comprise a net (stoichiometric) sequence. The third level is concerned with the details of elementary reactions: the ways reactants approach each other and form transition states, mechanisms of bond breaking, and the energetics of these processes. Aquatic chemical kinetics traditionally has focused on phenomenological and, to a lesser extent, mechanistic aspects. As the subject has matured in recent years, chemists have begun to explore the applications of statistical mechanics to understand the detailed mechanisms of some elementary aquatic reactions.

1.2 NATURAL WATERS AS NONEQUILIBRIUM SYSTEMS

Perhaps the most commonly cited example of nonequilibrium in the natural environment is the coexistence of O_2 and N_2 in the atmosphere. If equilibrium prevailed, the atmosphere would be depleted of O_2, and the world's oceans would be dilute solutions of nitric acid or nitrate salts. As illustrated in Example 1-1, this reaction is feasible energetically; nonetheless, its rate is exceedingly slow because of its high activation energy.

Example 1-1. Atmospheric Nonequilibrium

The reaction between N_2 and O_2 to produce nitric acid:

$$N_2 + 2\tfrac{1}{2}O_2 + H_2O \rightarrow 2H^+ + 2NO_3^- \tag{1-1}$$

is slightly endergonic under standard conditions. From compiled free energies of formation for the reactants and products,[2] we find that $\Delta G°$, the Gibbs free energy of reaction under standard conditions, of Equation 1-1 is +14.55 kJ mol^{-1}. Recall that standard conditions are 1-atm pressure, 25°C, and unit concentrations (activities) of all reactants and products. The *actual* free energy of reaction, ΔG, depends on the activities of the reactants and products:

$$\Delta G = \Delta G° + RT \ln Q \tag{1-2a}$$

where Q is the reaction quotient (the product of the activities of the products, each raised to its stoichiometric power, divided by the product of the activities of the reactants, each raised to its stoichiometric power). For reaction 1-1, Equation 1-2a becomes

$$\Delta G = \Delta G° + RT \ln\left(\{H^+\}^2\{NO_3^-\}^2 / P_{N_2}P_{O_2}^{2.5}\right) \tag{1-2b}$$

The standard free energy of reaction 1-1 at pH 7, $\Delta G°_7$, is –65.3 kJ mol^{-1}; i.e., it is exergonic. However, to calculate the actual value of ΔG, we must substitute actual concentrations of the products and reactants into Equation 1-2b. At equilibrium, $\Delta G = 0$, and the reaction quotient is given the symbol K, which is called the equilibrium constant of the reaction. It is apparent from Equation 1-2a that $\Delta G° = -RT \ln K$. The equilibrium constant for reaction 1-1 is

$$K = 10^{-2.6} = \frac{\{H^+\}^2_{aq}\{NO_3^-\}^2_{aq}}{P_{N_2}P_{O_2}^{2.5}} \tag{1-3}$$

where $\{i\}$ denotes the activity of i, and P_i is the partial pressure of i.

Some extreme statements have been made regarding the global equilibrium conditions implied by reaction 1-1. Lewis and Randall[4] stated that if the reaction proceeded to equilibrium, the world's oceans would have a nitric acid concentration greater than 0.1 M, and Hutchinson[5] repeated this conclusion in a widely cited book. Holland[6] estimated that nitrate activity $\{NO_3^-\}$ in the oceans would be $10^{5.7}$ if the reaction were at equilibrium under present conditions of pH, P_{N_2}, and P_{O_2}. None of these statements represents the actual equilibrium situation under the constraints of reactant availability on a global basis.

According to Sillen's model for the origin of the ocean[7] and data on the volumes of the oceans and atmosphere,[6] the geotitration of volatile acids with basic rocks that formed seawater produced ~600 g of sediment and ~3.45 L of air for every liter of seawater. Under present conditions, P_{O_2} is 0.21 atm and P_{N_2} is 0.78 atm; thus, the atmosphere contains 0.0323 mol O_2 and 0.120 mol N_2 per liter of seawater. Clearly, O_2 is the stoichiometrically limiting reactant for reaction 1-1.

Equation 1-3 can be solved readily by trial and error (or computerized numerical methods) for the equilibrium concentrations of H^+ and NO_3^- that could form under the above starting conditions (mass balance constraints). The following values are obtained: $NO_3^- = H^+ = 0.022$ M (pH = 1.65); $P_{O_2} = 0.029$ atm; $P_{N_2} = 0.707$ atm. The equilibrium pH would not be as low as predicted, however, because the large reservoir of sedimentary $CaCO_3$ would neutralize most of the H^+. Consumption of H^+ would increase the oceanic total carbonate concentration more than tenfold to about 2.6×10^{-2} M, and atmospheric P_{CO_2} would increase to $10^{-1.8}$ atm. The equilibrium pH would be near 8.0, which would bring the reaction nearly to completion in terms of O_2 depletion. The equilibrium {NO_3^-} would be 0.027 M. The increase in P_{CO_2} probably would have large effects on global climate, increasing the average temperature, but this was not factored into the calculations. Some kinetic aspects of global carbon models are described in Chapter 5.

The biogenic elements (C, N, P, S) provide other examples of nonequilibrium distributions in natural waters. None of the thousands of natural organic compounds is thermodynamically stable in the presence of O_2. Of course, many compounds are metastable and form more stable substances only at very slow rates. Microorganisms readily metabolize some organic compounds and control the aquatic concentrations of the compounds. Other substances, including many synthetic chemicals, are biologically refractory and are lost from natural waters by physico-chemical processes such as hydrolysis, photolysis, volatilization, and adsorption onto particles.

Photosynthesis is the principal driving force for nonequilibrium conditions in the environment. It produces chemical energy in the form of compounds with carbon, nitrogen, and sulfur in reduced oxidation states. For the average composition of biomass in the biosphere, the photosynthetic equation is[8]

$$1480CO_2 + 1480H_2O + 16NO_3^- + 1.8H_2PO_4^- + SO_4^{2-} + 18H^+$$

$$\xrightarrow{hv, green\ plants} C_{1480}H_{2960}O_{1480}(NH_3)_{16}(H_2PO_4)1.8H_2S + 1514O_2 \tag{1-4}$$

The much higher carbon content of average biospheric biomass compared with the often-cited C:N:P content of algae (106:16:1 — the so-called Redfield ratio) reflects the fact that most of the global biomass is cellulose in woody tissue.

Aerobic respiration returns the elements to equilibrium inorganic states, but when the demands of respiration exceed a system's capacity to provide O_2, anoxic conditions occur, and reduced forms of S, N, Fe, and Mn build up. The reduced species are unstable in the presence of O_2 and are oxidized when mixed into oxygenated water. Oxidation may be microbially mediated or may occur abiotically; in both cases rates can vary greatly, depending on temperature, pH, and other factors. Under some conditions, unstable reduced compounds can persist for extended periods.

Biological processes affect chemical conditions that control equilibria and thus indirectly induce chemical reactions. For example, photosynthesis increases pH, which may cause metal carbonates and hydroxides to precipitate. Conversely, pH decreases resulting from respiration may cause unsaturated conditions and induce mineral dissolution. Biological activity is a pervasive driving force for both organic and inorganic reactions in natural waters.

Changing physical conditions induce nonequilibrium states and act as a driving force for chemical change. The behavior of dissolved gases in water illustrates this. Gas solubility is defined by Henry's Law, which states that the equilibrium aqueous concentration of a gas at a given temperature is proportional to its gas-phase partial pressure. Nonequilibrium conditions can be induced by processes that produce or consume the gas or by changes in temperature, which affect its solubility. Actual concentrations of the dissolved gas depend not only on rates of internal production/ consumption processes, but also on the rate of equilibrium-restoring transfer across the air/water interface.

Rates of chemical reactions also can be limited by physical processes, especially mass transport. Rates of mineral dissolution may be controlled by chemical reactions at the mineral surface or by physical diffusion of reactants to, or products from, the surface. Mass-transport processes are often important in systems with two or more phases, e.g., air/water and sediment/water systems, but rate limitation by mass transport is possible for intrinsically very rapid reactions, even in homogeneous solutions. Mass transport is described mathematically in ways analogous to kinetic descriptions for chemical reactions.

As the above paragraphs suggest, the processes that control the concentrations of compounds in natural waters are complicated. Complete understanding requires information on chemical, photochemical, biological, and physical (transport) processes. For example, decomposition of organic pollutants may be controlled by abiotic reactions (e.g., hydrolysis and photolysis) or by microbial processes. Rates of metal ion oxidation may be controlled by thermochemical, photochemical, or biochemical processes. In addition, physical transport may affect these rates, since diffusion to microbial cells or other reactive surfaces may be the rate-limiting step.

Figure 1-1 illustrates the range of time scales for some processes of interest in aquatic systems and places them in the context of time scales for other important events, on a logarithmic scale of time in seconds. By analogy to the scale for hydrogen ion activity ($pH = -log\{H^+\}$), we can define $pT = -log(\text{time, s})$.[9] The time scales of interest for aquatic processes range over many orders of magnitude. For convenience, we may arbitrarily classify processes into those with "fast kinetics," $pT > 0$, i.e., those that are essentially complete in less than 1 s and may be termed "instantaneous," and those with "slow kinetics," $pT < 0$, which take more than a second to complete. Time scales for fast chemical reactions (like H^+ transfer) extend to the diffusion-controlled limit ($pT \approx 10$), and times for fast submolecular processes extend to about $pT \approx 15$. On the slow end of the scale, processes like the decomposition of organic compounds may require days to years ($pT \approx -5$ to -10), and radioisotope decay may occur over thousands or even millions of years ($pT \approx -14$).

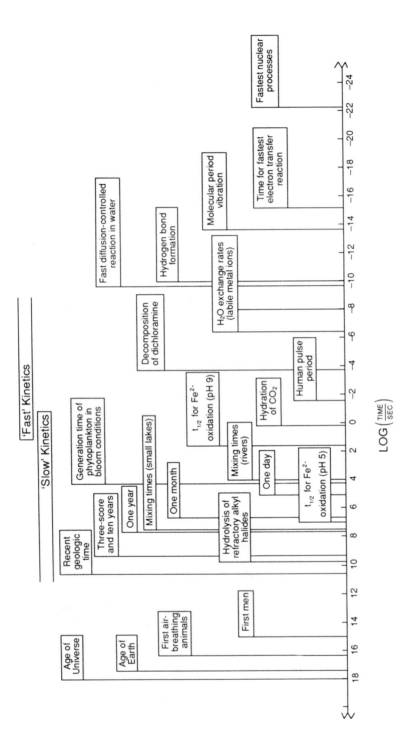

Figure 1-1. A logarithmic time scale for fast and slow processes in aquatic systems, placed in the context of other important events. [Modified from Onwood, D., *J. Chem. Ed.*, 63, 680 (1986).]

The length scale over which environmental processes occur also is of interest, and a crude correlation exists over many orders of magnitude between the characteristic time and length scales for many processes (Figure 1-2a). There is considerable overlap in the range of time scales for mixing in aquatic systems and the time scales for chemical and biological reactions (Figure 1-2b). Where such overlap occurs, the system's behavior can be understood only by use of spatially resolved dynamic models (rather than homogeneous equilibrium models).

Table 1-1 summarizes the functional relationships used to describe rates of important physical, chemical, and biological processes. Much of this book is devoted to deriving and explaining these equations and illustrating their application to kinetic processes in natural waters. Example 1-2 illustrates the use of these expressions to describe the kinetics of important processes in the aquatic chemistry of sulfur. The expressions in Table 1-1 also serve as the building blocks for complex dynamic models that describe the behavior of chemical substances in aquatic environments.

Example 1-2. Kinetic Aspects of the Sulfur Cycle

The sulfur cycle provides a convenient illustration of the kinetic processes formulated in Table 1-1. The purpose of this example is to briefly describe relationships among the major components of the cycle and the dynamic processes involved in these relationships. The illustrations are from recent studies that have improved our quantitative understanding of the cycle's dynamics, but the focus is graphical and descriptive, rather than mathematical and numerical. More detailed treatments are given in later chapters.

The aquatic sulfur cycle involves many chemical species: gases (H_2S, SO_2), ions (SO_4^{2-}, HS^-), solid phases (metal sulfides, pyrites, elemental sulfur), and organic compounds [dimethyl sulfide (DMS), S-containing amino acids, thiols, and sulfate esters]. Species in parentheses are examples; the list is not exhaustive. Many reactions and processes are involved: abiotic redox reactions, some of which are mediated by indirect photochemical processes; hydrolysis and organic substitution reactions involving nucleophilic sulfur species such as HS^-; precipitation and dissolution of solid phases; and microbial redox processes involving the major inorganic sulfur forms, H_2S and SO_4^{2-}. Mass transfer processes, such as diffusion across the sediment-water interface and gas exchange at the air-water interface, provide additional controls on sulfur cycling rates in the environment.

Figure 1-3 illustrates the sources, sinks, and major internal cycling processes for sulfur in a hypothetical lake and its sediments. From the perspective of kinetic processes, many of the sources and sinks of sulfur for lake-sediment systems are not of interest here because they depend primarily on rates of water exchange (stream inflows and outflows) or on processes outside the scope of this book. For example, climatic conditions and anthropogenic emissions of SO_2 are important factors affecting precipitation inputs of sulfur. The dynamics of sulfur cycling in lake-sediment systems involve all the categories of processes mentioned at the end of Section 1.2,[10,11] and nearly all the equations in Table 1-1 are useful in describing these processes.

Inputs and Outflows. By far the major sulfur species entering lakes is sulfate ion, which is dissolved in surface waters, groundwater, and rain. Sulfuric acid is formed in the atmosphere from SO_2 derived from fossil fuel burning; this is a major cause of acid

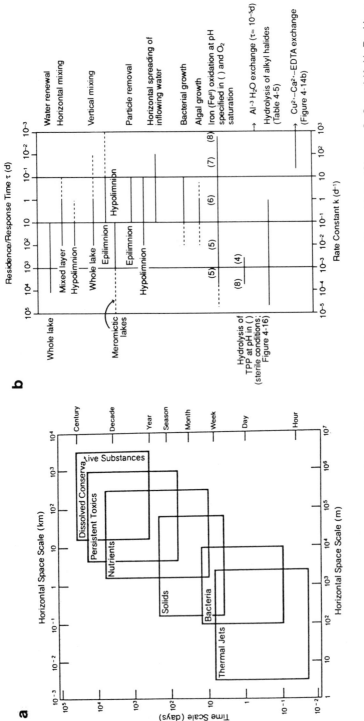

Figure 1-2. (a) Correlation between characteristic time and length scales for important water quality problems (Chapra, S. C. and K. H. Reckhow, *Engineering Approaches for Lake Management*, Vol. 2, Ann Arbor Science/Butterworth, Boston, 1983. With permission.); (b) time scales for mixing processes and some chemical and biological processes in aquatic systems. Solid lines: typical range; dashed lines: less common cases. [Based on Imboden, D. M. and R. P. Schwarzenbach, in *Chemical Processes in Lakes*, W. Stumm (Ed.), Wiley-Interscience, New York, 1985, p. 1.]

precipitation. Conversion of SO_2 (S^{IV}) to sulfuric acid (S^{VI}) in the atmosphere can occur by both gas- and liquid-phase reactions, but the latter are much more important.[12] S^{IV} can be oxidized to S^{VI} by several oxidants, including O_2, O_3, H_2O_2, OH radicals, and nitrous acid in liquid phases (aerosols and cloud droplets). Although present in the atmosphere at much lower concentrations than O_2, O_3 is much more reactive, and SO_2 oxidation by O_3 is a major pathway for H_2SO_4 production in clouds[13] at pH > 5. At lower pH, oxidation by H_2O_2 becomes the predominant mechanism.

S^{IV} exists in three forms as a function of pH: $H_2O \cdot SO_2$ (aquated SO_2), $HO\text{-}SO_2^-$ (bisulfite), and SO_3^{2-} ($pK_1 = 1.9$; $pK_2 = 7.2$). Each form reacts with O_3 by an independent mechanism, and the overall rate expression is the sum of the second-order rate equations for each S^{IV} form. Similar remarks apply to SO_2 oxidation by H_2O_2. The net reaction of S^{IV} with ozone,

$$H_nSO_3^{(n-2)} + O_3 \rightarrow nH^+ + SO_4^{2-} + O_2 \qquad (1\text{-}5)$$

suggests that one oxygen atom is transferred from O_3 to the reactant sulfur species. However, experiments[14] with ^{18}O-labeled O_3 showed that two oxygen atoms are transferred from O_3 to the product sulfate. A mechanism for S^{IV} oxidation by O_3 consistent with the rate equation and isotope-labeling results, is illustrated in Section 4.6; Section 2.5 describes ways to elucidate mechanisms from kinetic information. S^{IV} oxidation in cloudwater is complicated by the fact that the product (H_2SO_4) lowers the pH and affects reactant speciation. Further complications arise from the presence of NH_3, which neutralizes H^+ produced by oxidation and from the presence of dust particles that may contain catalytic quantities of iron and other metals.[12]

Sulfur also enters lakes by dry deposition of sulfate aerosols onto the surface and absorption of SO_2 at the air-water interface. (Minor amounts of sulfur enter lakes from leaves and input of litter from shoreline vegetation.) The main losses of sulfur from most lakes are hydraulic: stream outflow and downward seepage of lakewater through the sediments into regional groundwater aquifers. Volatilization of H_2S (and possibly other sulfur gases, like DMS and COS) may account for minor losses of sulfur from lakes. DMS volatilization from ocean waters is important in the global sulfur cycle,[15] but little information is available on the magnitude of such losses from freshwaters. Rates of gas transfer are described (Equations 1-40,41) as the products of concentration gradients across gas and liquid films at the air-water interface and an overall mass transfer coefficient, K_l. The magnitude of the latter depends on the nature of the gas and on physical conditions like the degree of turbulence at the interface. For highly soluble gases like SO_2, diffusion through the atmospheric boundary layer is the rate-limiting step, but for less-soluble gases, like DMS, diffusion through the liquid film is rate limiting.

In-Lake Processes. Sulfate is reduced to sulfide by microorganisms under anoxic conditions in lake sediments and possibly in bottom waters of lakes.[16] At low sulfate concentrations the rate of microbial reduction is first order (Equation 1-6), but over a broad range of $[SO_4^{2-}]$, the process fits Michaelis-Menten saturation kinetics (Equation 1-31) (Figure 1-4a,b). Growth of the sulfate-reducing (and sulfide-oxidizing) bacteria often is described by the Monod equation (Equation 1-32), which yields a hyperbolic curve similar to Figure 1-4b when the specific growth rate, μ, is plotted vs. substrate concentration. Loss of sulfate by microbial reduction in the sediments induces a gradient in concentration between the overlying water and sediment porewater (Figure 1-4c). This induces a diffusive flux of sulfate across the sediment-water interface, which can be described by Fick's first law

Table 1-1. Functional Expressions for Important Process Rates in Aquatic Systems

Process	Equation	Symbols	Comments	Section*
A. Elementary kinetic expressions				
Zero order	$r = k$ (1-5)	r = rate (mol L^{-1} time^{-1}) k = rate constant	Units of k depend on units of r and reaction order. Zero-order kinetics apply mainly to reactions in two phases	2.2
First order $A \rightarrow P$	$r = -k_1[A]$ (1-6) $[A]_t = [A]_0 \exp\{-k_1 t\}$ (1-7)	$[i]$ = concentration of i at t ($_0$ = initial value), mol L^{-1} (M); A,B, etc. = reactants; P,Q, etc. = products; k_1 has units of t^{-1}	Equation also describes rates of many mechanistically complicated reactions and processes; exponential form is time-integrated equation	2.2
	$t_{1/2} = 0.69/k_1$ (1-8)	$t_{1/2}$ = half-life of reactant	Time required to decrease [A] in half is independent of concentration for first order reactions	
	$\tau = 1/k_1$ (1-9)	τ = characteristic time of reactant	[A] is 1/e of initial [A] at t = τ	
Second order $A + B \rightarrow P$ $2A \rightarrow P$	$r = -k_2[A][B]$ (1-10) $r = -k_2[A]^2$	k_2 has units of M^{-1} t^{-1}	Time-integrated expressions for these two reaction types are given in Section 2.2.3	
Acid/base catalysis $A + H^+ \rightarrow P + H^+$ $A + OH^- \rightarrow P + OH^-$	$r = -k_2[A][H^+]$ (1-11) $r = -k_2[A][OH^-]$ (1-12)		At constant pH, $r = -k'_1[A]$; r is pseudo-first order. Examples assume first-order uncatalyzed reaction	4.3

B. Factors affecting rate constants

Factor	Equation		Variables	Description	*
Temperature	$k = A\exp(-E_{act}/RT)$	(1-13)	A = frequency factor; E_{act} = activation energy (kJ mol⁻¹); R = gas constant (J mol⁻¹ K⁻¹); T = absolute temperature (K)	Arrhenius equation; analogous to Van't Hoff equation for effects of T on equilibria; explanation of this equation spurred development of collision- and transition-state theories for reaction kinetics	3.2
	$k_2/k_1 = \Theta^{(T_2 - T_1)}$	(1-14)	Θ = empirical temperature coefficient	Empirical equation has basis in Arrhenius equation	3.2
	$k = \dfrac{kT}{h}\exp(-\Delta G^{\neq}/RT)$	(1-15)	k = Boltzmann constant (gas constant per molecule, R/N_o, N_o = Avogadro's number; h = Planck's constant; $\Delta G^{\neq}$ = free energy of activation	Fundamental equation derived from transition-state theory	3.5
	$\Delta G^{\neq} = \Delta H^{\neq} - T\Delta S^{\neq}$	(1-16)	$\Delta S^{\neq}$ = entropy of activation	Based on 2nd law of thermodynamics	3.5
	$E_{act} = \Delta H^{\neq} + RT$	(1-17)	$\Delta H^{\neq}$ = enthalpy of activation	Applies to reactions in solution	3.5
Pressure, P	$\ln\left[\dfrac{k_p}{k_1}\right] = \dfrac{-\Delta V^{\neq}}{RT}(P-1)$	(1-18)	P in atm; $\Delta V^{\neq}$ = volume of activation (cm³ mol⁻¹)	Analogous to Planck's equation for effects of P on equilibria	3.6
	$\Delta V^{\neq} = 0.5\Delta S^{\neq}$	(1-19)		Empirical relationship (for $\Delta S^{\neq}$ in cal mol⁻¹ k⁻¹)	3.6
Ionic strength	$\log\dfrac{k}{k_o} = 1.02\,Z_A\,Z_B\,\sqrt{I}$	(1-20)	k_o = rate constant at I = 0; I = ionic strength; Z_i = charge on ion i; $I = \tfrac{1}{2}\Sigma C_i Z_i^2$	The Brønsted equation; for ionic reactions, $A^{\pm n} + B^{\pm m} \rightarrow P$	3.7

* Refers to section of book that describes or uses this equation.

Table 1-1 (continued).

Process	Equation	Symbols	Comments	Section*
C. Diffusion				
Fick's first law	(1-21) $F_x = -D_i d[i]/dx$	F_x = flux (g cm⁻² s⁻¹) in x direction; D_i = diffusivity (diffusion coefficient) of i (cm² s⁻¹)	Gives diffusional flux in one dimension; can be written in three dimensions or spherical coordinates	3.3
Fick's second law	(1-22) $\dfrac{d[i]}{dt} = D_i \dfrac{d^2[i]}{dx^2}$		Gives time rate of change of [i] at point x as function of second derivative of [i] with respect to x	3.3
Stokes-Einstein Equation	(1-23) $D_i = kT/6\pi\eta r_i$	η = viscosity (g cm⁻¹ s⁻¹); r_i = molecular radius	Expresses effect of T and η on D_i	3.3
	(1-24) $[i]_{x,t} = \left\{M/\sqrt{4\pi Dt}\right\}\exp(-x^2/4Dt)$	M = mass	One-dimensional spread by diffusion of slug of i with mass M introduced at t = 0, x = 0	9.5
D. Photochemical reactions				
Direct photolysis	(1-25) $r_\lambda = I_{s_{a,\lambda}}\Phi_d$ (1-26) $= k_{a,\lambda}\Phi_d[S]$ (1-27) $= k_{b,\lambda}[S]$	λ = wavelength; $I_{s_{a,\lambda}}$ = intensity of absorbed light at λ; Φ_d = quantum yield (dimensionless) k_i's = first-order rate constants; S = photolyzing substance; P = product(s)	For general mechanism: $S + h\nu \rightarrow P$ Terms with subscript λ are wavelength dependent	8.2
	(1-28) $k_d = 2.3\Phi_d \displaystyle\int E_o(\lambda)\varepsilon_\lambda \, d\lambda$	$E_o(\lambda)$ = scalar irradiance at water surface (photons cm⁻² s⁻¹); ε_λ = molar absorptivity of S (M⁻¹ cm⁻¹) at λ	Overall rate constant for direct photolysis (overall λ absorbed by S); applied to near-surface conditions	8.2
	(1-29) $r = (k_{obs}/I_t)[S]$	I_t = total solar radiation (μEinsteins cm⁻² s⁻¹)	Empirical equation for photolysis as function of total solar radiation	8.2

Process	Eq.	Equation	Description	Section*
Indirect photolysis	(1-30)	$r = {}^{1s}I_a\,\Phi_{1s}\Phi_{el}\Phi_r$	Φ_{1s} = fraction of absorbed light resulting in triplet states; Φ_{el} = fraction of triplet energy transferred to R; Φ_r = quantum yield for reaction of R; X = photochemical intermediate	For general mechanism: S + hv → X; X + R → P	8.2
E. Biochemical processes					
Enzyme kinetics	(1-31)	$v = \dfrac{V[S]}{K_s + [S]}$	v = r; V = maximum v (at high [S]); S = substrate; K_S = half saturation constant	Michaelis-Menten equation; applies to single-substrate reactions; mathematical form fits many processes	6.2
Microbial growth	(1-32)	$\mu = \dfrac{\mu_{max}[S]}{K_s + [S]}$	μ = growth rate (t⁻¹); S = growth-limiting substrate	Monod equation; analog of Michaelis-Menten equation; derived empirically	6.3
Cell quota model	(1-33)	$\mu = \mu'_{max}\left[\dfrac{Q - Q_o}{Q}\right]$	Q_o = substistence cell quota for growth (limiting [S]) when μ = 0; Q = actual cell quota	Applies mainly to micro-nutrient-limited growth; quota = concentration in cells	6.3
Substrate	(1-34)	$\mu = \dfrac{\mu_{max}[S]}{[S] + K_s + [S]^2/K_i}$	K_i = inhibition constant for S (units of concentration)	For substrates like phenols that inhibit growth at high [S]	6.4
F. Reactor kinetics					
Residence time	(1-35)	$\tau = V/Q = 1/D = 1/\rho_w$	τ = residence time; V = reactor volume; Q = volumetric inflow or outflow rate; D = dilution rate; ρ_w = flushing coefficient		5.1

* Refers to section of book that describes or uses this equation.

Table 1-1 (continued).

Process	Equation		Symbols	Comments	Section*
Plug-flow reactor (PFR)	$\dfrac{V}{Q} = \displaystyle\int_{[A]_i}^{[A]_o} \dfrac{d[A]}{r_A}$	(1-36)	$[A]_o$ = outlet concentration; $[A]_i$ = inlet concentration	Defines τ needed for given effluent[A]; r_A must be integratable; e.g., $r_A = -k[A]$ for first-order reaction	5.2
PFR with first-order reaction	$\dfrac{[A]_o}{[a]_i} = \exp(-k_1\tau)$	(1-37)			
Continuous-flow stirred tank reactor (CFSTR)	$\dfrac{V}{Q} = \dfrac{[A]_o - [A]_i}{r_A}$	(1-38)			
One-dimensional mass transport equation	$D_L \dfrac{\partial^2[A]}{\partial x^2} - u\dfrac{\partial[A]}{\partial x} + \dfrac{\partial[A]}{\partial t} + r_A = 0$	(1-39)	D_L = dispersion coef. (cm² s⁻¹); u = advective velocity (cm s⁻¹)	Includes terms for dispersion advection, reaction, and accumulation in non-steady-state tubular reactor	5.2

G. Physical/heterogeneous processes

Process	Equation		Symbols	Comments	Section*
Gas transfer	$F = k_g\left([C]_g - H[C]_\ell\right)$	(1-40)	K_g, K_ℓ = gas exchange coefficients; [C] = gas concentration in gas (g) or liquid (ℓ); H = Henry's law constant		5.4
	$= k_\ell\left([C]_g / H - [C]_\ell\right)$	(1-41)			
Particle settling	$v = \dfrac{m_2 g(\rho_2 - \rho_1)}{6\pi\eta r_2}$	(1-42)	v = velocity; m_2, ρ_2, r_2 = mass, density, and radius of settling particle; ρ_1 = density of medium; g = gravitational constant; η = viscosity	Stokes Law; assumes spherical particles	5.10

| Nucleation | $J = kC^n$ | (1-43) | J = flux of nuclei; C = concentration of nucleating particles; n = critical number of nucleating molecules | For homogeneous nucleation; rarely occurs in natural waters; n = cluster (nucleus) size above which crystal growth is spontaneous | 4.7 |
| Crystal growth/dissolution | $\dfrac{-d[C]}{dt} = kA\{[C] - [C]_s\}^n$ | (1-44) | A = surface area; $[C]_s$ = solubility of C (M); n = empirical reaction order | If $n > 1$, growth is reaction controlled, not diffusion controlled; If $[C] < [C]_s$, dissolution occurs | 4.7 |

* Refers to section of book that describes or uses this equation.

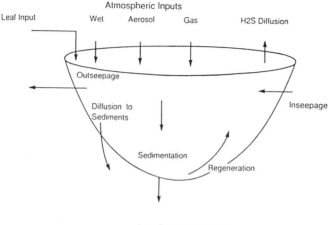

Long-Term Accumulation

Figure 1-3. Schematic diagram of sulfur mass balance in a hypothetical seepage lake, illustrating important sources, sinks, and internal losses. [Adapted from Baker, L. A., P. L. Brezonik, and N. Urban, in *Biogenic Sulfur in the Environment*, E. S. Saltzman and W. J. Cooper (Eds.), ACS Symp. Ser. 393, Am. Chemical Society, Washington, D.C., 1989, p. 79.]

(Equation 1-21) or by diffusion-reaction (diagenetic) models.[17] Some microbially reduced sulfur is sequestered by metals, especially iron, and stored in the sediments as iron monosulfides and pyrite. A significant fraction of the sulfide is converted into organic forms that remain in the sediments, but a large fraction of the reduced sulfur (perhaps more than 50%) is reoxidized to sulfate at the oxic-anoxic interface.[11] This reaction can be mediated by certain bacteria and also proceeds abiotically.

Sulfide and other inorganic ions containing sulfur in intermediate oxidation states are relatively strong nucleophiles (Lewis bases); i.e., they are "electron rich" and react by donating a pair of electrons to various electrophiles (Lewis acids). For example, HS^- is a "soft" nucleophile — in the sense of Pearson's hard and soft acid-base (HSAB) concept. As such, it reacts with soft electrophiles such as saturated carbon atoms to which highly electronegative elements (Br, Cl) are attached. The net effect is a nucleophilic substitution reaction in which -SH displaces a halide ion on alkyl halides (RX), producing a thiol (RSH) or thiolate ion, RS^-.[18] Such reactions are analogous to hydrolysis reactions in which H_2O acts as the nucleophilic agent. Although alkyl halides are not found naturally in lakes and sediment porewaters, they may occur in these environments at trace quantities as a result of spills or movement of contaminated groundwater. In some cases reaction of these compounds with HS^- is more important than hydrolysis. Rates of nucleophilic substitution reactions involving reduced sulfur compounds and alkyl halides are discussed in Sections 4.4.4 and 7.5.1.

Chemical oxidation of sulfide by O_2 (called autoxidation) is mechanistically complicated,[19] but follows second-order kinetics (Equation 1-10). The reaction is catalyzed by some heavy metals and is strongly affected by pH,[20] increasing rapidly near pK_1 (~7.0) (Figure 1-4d). These results suggest that HS^- reacts much more rapidly with O_2 than does the totally protonated species, H_2S. This trend is followed for autoxidation rates of several other elements (e.g., Fe^{II}, Cu^I). Strong oxidants like ozone rapidly oxidize sulfide to sulfate or to

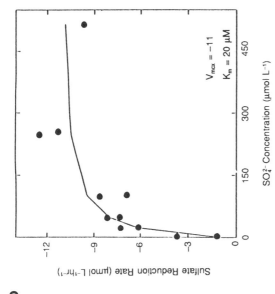

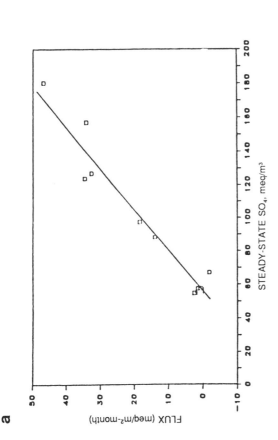

Figure 1-4. (a) Rate of sulfate loss from the water column in littoral enclosures of a low-alkalinity (primarily the result of microbial sulfate reduction in sediments) is first order in water column sulfate concentration over a range of [SO_4^{2-}] typical of such lakes (From T. E. Perry, M.S. thesis, University of Minnesota, 1986), but (b) over a broad range of [SO_4^{2-}] in laboratory cultures, sulfate reduction rates fit a Michaelis-Menten relationship (saturation kinetics) (N. Urban and P. Brezonik, unpublished data). (c) Typical gradients of [SO_4^{2-}] vs. depth in the region of the sediment-water interface (represented by horizontal dashed line) in lakes where sediment organic matter exerts sufficient O_2 demand to deplete dissolved O_2 near the interface (C. J. Sampson and P. L. Brezonik, unpublished data from Little Rock Lake, WI). (d) Effect of pH on abiotic oxidation of sulfide species (H_2S and HS^-) [Redrawn from Millero, F. J., et al., *Environ. Sci. Technol.*, 21, 439 (1987). With permission.] (e) Arrhenius plot of dissolution rate vs. $1/T$ for tetragonal FeS (mackinawite) to obtain apparent activation energy, E_{act} [Redrawn from Pankow, J. F. and J. J. Morgan, *Environ. Sci. Technol.*, 13, 1248 (1979). With permission.]

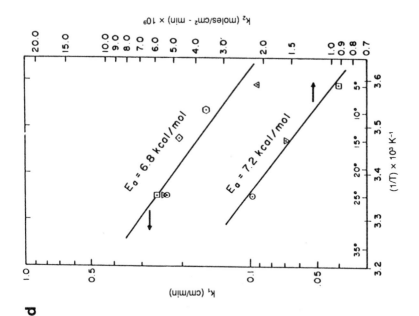

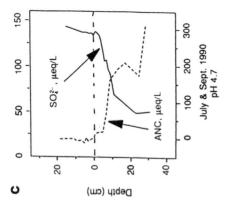

Figure 1-4 c and d.

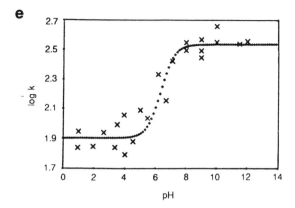

Figure 1-4 e.

nontoxic (and nonodorous) intermediate oxidation states, a fact of importance for drinking-water treatment. The effect of temperature on rates of abiotic sulfur redox reactions is described by the Arrhenius equation (Equation 1-13) and its fitted parameter E_{act}, the activation energy, which expresses the sensitivity of a reaction rate to changes in temperature. Over narrow temperature ranges, biotic reactions also can be described by this equation or by Equation 1-14.

Metal sulfides are unstable in the presence of oxygen. Oxidation of iron sulfides and pyrites results in the formation of a highly acidic solution and is responsible for the problem of acid mine drainage. Like other multiphase (heterogeneous) reactions, oxidation of metal sulfide is mechanistically complex, and the rate-limiting step may involve chemical reaction at the mineral surface or diffusion of a reactant to, or product from, the surface. The effect of temperature on rates of such processes sometimes can be used to distinguish diffusion control from chemical (surface reaction) control. The former are characterized by low activation energies (2 to 5 kcal mol^{-1} or ~8 to 21 kJ mol^{-1}) and relatively small temperature effects (rate increases of ~1.3- to 1.5-fold for a 10°C increase in temperature. The latter have larger E_{act} values (>7 kcal mol^{-1} or ~29 kJ mol^{-1}) and correspondingly larger temperature coefficients (typically two times or larger for a 10°C increase. The nonoxidative dissolution of mackinawite, a crystalline form of FeS, is catalyzed by H$^+$ at pH < 4.5, but is independent of H$^+$ at higher pH.[21] E_{act} for both the catalyzed and uncatalyzed dissolution is about 6.7 kcal mol^{-1} (Figure 1-4e); this value is at the low end of chemical control and suggests that diffusion limitation also may play a role.

Sulfate is removed from lake waters by planktonic assimilation and subsequent gravitational settling of seston (suspended matter, including nonmotile plankton). Under quiescent conditions the settling rate is described by Stokes Law (Equation 1-42). Seston deposition is not a major loss mechanism for sulfate in most lakes, where sulfate concentrations are greatly in excess of algal needs, but it can be important in dilute systems (e.g., acid-sensitive softwater lakes) with low sulfate levels (< ~50 to 80 µeq L^{-1}).[22] Moreover, one recent study[23] found that putrefaction of sedimentary protein (presumably derived from sedimenting seston) accounts for about one fourth of the H$_2$S in the sediments of eutrophic lakes.

The dynamic mass balance of sulfate in a lake can be described by completely mixed reactor models (Equation 1-38), which treat the lake as a homogeneous chemical reactor. Such models[24] can be used to evaluate the responses of acid-sensitive lakes to changes in atmospheric deposition of acid sulfate. Typically they include a source or input term, a hydraulic outflow term, and terms for any internal loss or generation processes. Internal losses by microbial sulfate reduction and seston deposition are significant in acid-sensitive lakes with moderate-to-long water residence times (> ~1 year), and these internal sinks can be portrayed as simple first-order reaction terms in the model.

1.3 SCOPE OF THIS BOOK

This book treats the dynamics of processes that affect the composition of natural waters. Because these processes include biochemical and physical processes as well as chemical reactions, the book's scope is broader than that of conventional chemical kinetics, but its emphasis is on chemical substances and their transformations. The rate equations used to describe biochemical and physical processes are based on the same principles used in chemical kinetics. Phenomenological aspects of reaction kinetics (i.e., quantitative expression of their rates) are covered in detail in Chapter 2. As the equations in part (a) of Table 1-1 indicate, rates are expressed as differential equations in which the rate of reactant loss or product formation is proportional to the concentration of one or more reactants. The proportionality coefficient or rate constant, k, for a reaction varies with physical conditions — temperature, pressure, and ionic strength; the units of k depend on the reaction order, i.e., the sum of exponents on concentration terms in the rate equation. For first-order equations, k has units of reciprocal time. Rate equations for elementary (one-step) reactions follow the law of mass action: rates are proportional to the concentrations of reactants in the stoichiometric expression. Selecting the rate equation that best fits a set of data and developing accurate estimates of kinetic parameters are important phenomenological facets of kinetics.

With few exceptions, the fundamental processes of chemical change involve only one or two molecules (or ions) at a time. Most reactions thus proceed by a sequence of mono- or bimolecular steps and involve transient intermediates that are difficult to measure. Rate equations for multistep reactions cannot be predicted from reaction stoichiometry. The overall rate may be a complicated function of the rate expressions for the elementary steps, but many such reactions fit simple equations, over some range of experimental conditions. Rate equations provide important information to elucidate reaction mechanisms, but several mechanisms may fit a set of kinetic data (Section 2.7). Consequently, mechanisms can be inferred from rate equations, but not proven unequivocally.

Several chapters describe the factors that affect values of k for reactions in solution. Aside from temperature, pressure, and ionic strength (Table 1-1b, Chapter 3), these include light (Table 1-1d, Chapter 8) and catalysis by acids, bases, metal ions, and surfaces (Chapter 4). The Arrhenius equation, which describes the effect of temperature on rate constants, was developed by analogy to Van't Hoff's equation, which relates the effect of temperature on chemical equilibrium. In turn, explanation of the Arrhenius equation led to the modern theories of reaction rates (collision- and transition-state theories). These concepts are developed in Chapter

3, along with a description of the way reactants in solution encounter each other by diffusional processes (Table 1-1c).

Chemical reactions usually are studied in closed systems in the laboratory, but natural waters and engineered systems for water and wastewater treatment usually behave as flow-through chemical reactors. The two most common types of flow-through reactors, plug-flow (tubular) reactors and completely mixed tank reactors (Table 1-1f), can be used to describe processes in rivers and lakes, respectively. The kinetic properties of these systems have been developed extensively by chemical engineers (see Chapter 5).

The Michaelis-Menten and Monod equations (Table 1-1e) are the starting points for kinetic treatment of enzymatic and microbial processes, but as Chapter 6 shows, biological kinetics is a much more complicated subject than implied by these simple equations. Correlations between the bioactivity of molecules and their physicochemical properties or molecular structure are of interest to aquatic scientists faced with predicting the ecological and toxicological effects of thousands of synthetic organic contaminants. These property-activity and structure-activity relationships are analogous to linear free-energy relationships (LFERs) that exist between the equilibrium and kinetic properties of many series of chemical reactions. Principles and applications of these relationships to natural water systems are described in Chapter 7.

Comprehensive information on the kinetics of chemical and biological processes that control the behavior of substances in natural waters is still rather sparse, but the database is increasing rapidly. Chapters 4 and 8 summarize such information for a variety of natural solutes and contaminants introduced to natural waters by human activity. Examples are provided in other chapters to illustrate the principles covered in this book.

The bibliography that follows the references for this chapter lists general texts and monographs on chemical and biochemical kinetics, as well as reviews and books on the kinetics of aquatic processes. The list includes the major references on these subjects, but is not exhaustive. Several of the texts were important sources for material in this book, and some treat theoretical aspects in greater depth than aimed for here. Serious students of aquatic kinetics should be familiar with these books and papers.

REFERENCES

1.   Sillen, L.G., in *Oceanography*, M. Sears (Ed.), Am. Assoc. Adv. Sci., Washington, D.C., 1961, pp. 549–581; Garrels, R.M. and M.E. Thompson, *Am. J. Sci.* 26D, 57 (1962); Garrels, R.M. and C.L. Christ, *Solutions, Minerals, and Equilibria,* Harper and Row, New York, 1965; Stumm, W. (Ed.), *Equilibrium Concepts in Natural Water Systems,* Adv. Chem. Ser. 67, Am. Chem. Soc., Washington, D.C., 1967; Jenne, E. (Ed.), *Chemical Models*, Adv. Chem. Ser. 103, Am. Chem. Soc., Washington, D.C., 1979.

2.   Stumm, W. and J.J. Morgan, *Aquatic Chemistry*, 2nd ed., Wiley-Interscience, New York, 1981.

3. Stumm, W. (Ed.), *Chemical Processes in Lakes*, Wiley-Interscience, New York, 1985; Hites, R. and S.J. Eisenreich (Eds.), *Chemical Pollutants in Lakes*, Adv. Chem. Ser. 216, Am. Chem. Soc., Washington, D.C., 1987.

4. Lewis, G.N. and M. Randall, *Thermodynamics*, McGraw-Hill, New York, 1923; 2nd ed. (revised by K.S. Pitzer and L. Brewer), 1961.

5. Hutchinson, G.E., *A Treatise on Limnology*, Vol. I, John Wiley & Sons, New York, 1957.

6. Holland, H.E., *The Chemistry of the Oceans and Atmosphere*, Wiley-Interscience, New York, 1978.

7. Sillen, L.G., in *Oceanography*, M. Sears (Ed.), Publ. No. 67, Am. Assoc. Adv. Science, Washington, D.C., 1960, pp. 549–581.

8. Deevey, E.S., Jr., *Sci. Am.*, 223, 148 (1970).

9. Onwood, D., *J. Chem. Ed.*, 63, 680 (1986).

10. Nriagu, J. (Ed.), *Sulfur in the Environment*, John Wiley & Sons, New York, 1978.

11. Baker, L.A., P.L. Brezonik, and N. Urban, in *Biogenic Sulfur in the Environment*, E.S. Saltzman and W.J. Cooper (Eds.), ACS Symp. Ser. 393, Am. Chem. Soc., Washington, D.C., 1989, pp. 79–100.

12. Hoffmann, M.R. and J.G. Calvert, Chemical transformation modules for Eulerian acid deposition models. Vol. 2. The aqueous phase chemistry. Off. Res. Dev., U.S. EPA, Washington, D.C., Rept. EPA/600/3-85/036, 1985. (NTIS No. PBS85-198653).

13. Hoffmann, M.R., *Atmos. Environ.*, 20, 1145 (1986).

14. Espenson, J.H. and H. Taube, *Inorg. Chem.*, 4, 704 (1965).

15. Saltzman, E.S. and W.J. Cooper (Eds.), *Biogenic Sulfur in the Environment*, ACS Symp. Ser. 393, Am. Chem. Soc., Washington, D.C., 1989; especially see Chaps. 1,2,9–13.

16. Brezonik, P.L., L.A. Baker, and T.E. Perry, in *Chemistry of Aquatic Pollutants*, Hites, R. and S.J. Eisenreich (Eds.), Am. Chem. Soc., Adv. Chem. Ser. 216, Washington, D.C., 1987, pp. 229–260.

17. Berner, R.A., *Early Diagenesis, a Theoretical Approach*, Princeton University Press, Princeton, NJ, 1980.

18. Barbash, J.E. and M. Reinhard, pp. 101–138, in reference 15.

19. Chen, K.Y. and J.C. Morris, *Environ. Sci. Technol.*, 6, 529 (1972); Millero, F.J. and J.P. Hershey, pp. 282–313 in reference 15.

20. Millero, F.J., S. Hubinger, M. Fernandez, and S. Garnett, *Environ. Sci. Technol.*, 21, 439 (1987).

21. Pankow, J.F. and J.J. Morgan, *Environ. Sci. Technol.*, 13, 1248 (1979).

22. Baker, L.A., J.E. Tacconi, and P.L. Brezonik, *Verh. Int. Verein. Limnol.*, 23, 346 (1988).

23. Dunnette, D.A., pp. 72–78 in reference 15.

24. Baker, L.A. and P.L. Brezonik, *Water Resour. Res.*, 24, 65 (1988).

BIBLIOGRAPHY

General References on Chemical Kinetics

Amis, E.S., *Solvent Effects on Reaction Rates and Mechanisms*, Academic Press, New York, 1966.

Bamford, C.H. and C.F.H. Tipper (Eds.), *Comprehensive Chemical Kinetics*, Elsevier, Amsterdam (24 volumes to date, beginning in 1969).

Benson, S.W., *The Foundations of Chemical Kinetics*, McGraw-Hill, New York, 1960.

Benson, S.W., *Thermochemical Kinetics*, 2nd ed., McGraw-Hill, New York, 1976.

Entelis, S.G. and R.P. Tiger, *Reaction Kinetics in the Liquid Phase*, Halsted Press, New York, 1976.

Eyring, H.S., S.H. Lin, and S.M. Lin, *Basic Chemical Kinetics*, Wiley- Interscience, New York, 1980.

Frost, A.A. and R.A. Pearson, *Kinetics and Mechanism*, 2nd ed., John Wiley & Sons, New York, 1961.

Laidler, K.J., *Chemical Kinetics*, McGraw-Hill, New York, 1965.

Laidler, K.J., *Reaction Kinetics*, Vol. I and II, Pergamon Press, Oxford, 1963.

Moore, J. and R.A. Pearson, *Kinetics and Mechanism*, 3rd ed., John Wiley & Sons, New York, 1981.

Weston, R.E., Jr. and H.A. Schwarz, *Chemical Kinetics*, Prentice-Hall, Englewood Cliffs, NJ, 1972.

Biochemical Kinetics and Microbial Growth

Bailey, J.E. and D.F. Ollis, *Biochemical Engineering Fundamentals*, McGraw-Hill, New York, 1977.

Button, D.K., Kinetics of microbial growth and nutrient transport, *Microb. Rev.,* 49, 270 (1985).

Cornish-Bowden, A. *Principles of Enzyme Kinetics*, Butterworths, London, 1976.

Jacquez, J.A., *Compartmental Analysis in Biology and Medicine*, Elsevier, New York, 1972.

Segel, I.H., *Enzyme Kinetics*, Wiley-Interscience, New York, 1975.

Chemical Engineering and Reactor Kinetics

Hill, C.G., Jr., *An Introduction to Chemical Engineering Kinetics and Reactor Design,* John Wiley & Sons, New York, 1977.

Levenspiel, O., *Chemical Reaction Engineering*, John Wiley & Sons, New York, 1962.

Geochemical and Aquatic Kinetics

Aagaard, P. and H.C. Helgeson, Thermodynamic and kinetic constraints on reaction rates among minerals and aqueous solutions. I. Theoretical considerations, *Am. J. Sci.*, 282, 237 (1982).

Berner, R.A. and J.W. Morse, Dissolution kinetics of calcium carbonate in sea water. IV. Theory of calcite dissolution, *Amer. J. Sci.*, 274, 108 (1974).

Brezonik, P. L., Chemical kinetics and dynamics in natural water systems, in *Water and Water Pollution Handbook*, L. Ciaccio (Ed.), Marcel Dekker, New York, 1972, pp. 831–913.

Hoffmann, M. R., Trace metal catalysis in aquatic systems, *Environ. Sci. Technol.*, 14, 1061 (1980).

Hoffmann, M. R., Thermodynamic, kinetic, and extrathermodynamic considerations in the development of equilibrium models for aquatic systems, *Environ. Sci. Technol.*, 15, 345 (1981).

Lasaga, A.C. The kinetic treatment of geochemical cycles, *Geochim. Cosmochim. Acta,* 44, 815 (1980).

Lasaga, A.C. and R.J. Kirkpatrick (Eds.), *Kinetics of Geochemical Processes*, Mineralogical Society of America, Washington, D.C., 1981.

Liss, P.S. and P.G. Slater, Flux of gases across the air-sea interface, *Nature,* 247, 181 (1974).

Mabey, W. and T. Mill, Critical review of hydrolysis of organic compounds in water under environmental conditions, *J. Phys. Chem. Ref. Data*, 7, 383 (1978).

Pankow, J.F. and J.J. Morgan, Kinetics for the aquatic environment, *Environ. Sci. Technol.*, 15, 1155,1306 (1981).

Rimstidt, J.D. and H.L. Barnes, The kinetics of silica-water reactions, *Geochim. Cosmochim. Acta,* 44, 1683 (1980).

Stumm, W. (Ed.), *Chemical Processes in Lakes*, Wiley-Interscience, New York, 1985; especially see D.M. Imboden and R.P. Schwarzenbach, Spatial and temporal distribution of chemical substances in lakes: modeling concepts, pp. 1–30; and J.J. Morgan and A.T. Stone, Kinetics of chemical processes of importance in lacustrine environments, pp. 389–426.

Stumm, W. (Ed.), *Aquatic Chemical Kinetics*, Wiley-Interscience, New York, 1990.

Zafiriou, O.C., J. Joussot-Dubien, R.G. Zepp, and R.G. Zika, Photochemistry of natural waters, *Environ. Sci. Technol.*, 18, 358A (1984).

Zepp, R.G., Assessing the photochemistry of organic pollutants in aquatic environments in *Dynamics, Exposure, and Hazard Assessment of Toxic Chemicals,* R. Haque (Ed.), Ann Arbor Science Publ., Ann Arbor, MI, 1980, pp. 69–110.

Zepp, R.G. and D.M. Cline, Rates of direct photolysis in aquatic environments, *Environ. Sci. Technol.*, 11, 359 (1977).

"It's as if a horse walks into a church. You can convince yourself there's no problem using scientific evidence — if you analyze the smell, take dirt samples from the floor, note shadows on the wall. But all this obscures the obvious. The problem is, there's a horse in the church!"

Mikhail Grachev, Lake Baikal Limnological Institute
(Quoted in *National Geographic*, 1992)

CHAPTER 2

Rate Expressions for Chemical Reactions

2.1 INTRODUCTION

This chapter describes the equations used to define rates of chemical reactions. After first defining some terms and concepts, we describe rate equations for simple one-step reactions that proceed to completion in closed systems, and then proceed to more complicated cases important in natural waters: reversible, consecutive, and chain reactions. Problems involved in estimating kinetic parameters (rate constants) from experimental data and fitting rate equations to data are discussed next. Methods to derive rate equations for reaction mechanisms are described along with general principles used to infer mechanisms from kinetic data. Finally, some analytical aspects of kinetic studies are discussed. The chapter is not an exhaustive treatment of the topic; rate equations for catalyzed reactions, chemical reactors, and biological processes are presented in Chapters 4, 5, and 6, respectively. Several texts on kinetics,[1,2] provide comprehensive treatments of these facets of chemical rate processes.

2.1.1 Extent of Reaction, ξ

Rates of chemical reactions depend on the nature of the reactants, their concentrations, and physical conditions such as temperature. In general, the rate of a chemical reaction can be written as

$$rate = d\xi / Vdt \qquad (2\text{-}1)$$

V is the volume of the reacting system, ξ denotes the extent of reaction, and t is time. Extent of reaction is a simple concept developed for irreversible thermodynamic systems (by de Donder in 1920) and defined in terms of changes in numbers of moles of the species in a reaction. Consider the general reaction

$$aA + bB + \ldots \rightleftarrows pP + qQ + \ldots \qquad (2\text{-}2)$$

Symbols a, b, p, and q are stoichiometric coefficients of reactant and product species A, B, P, and Q, respectively. Stoichiometric coefficients are symbolized as v_i, and Equation 2-2 thus can be written as

$$v_A A + v_B B + \ldots + v_P P + v_Q Q + \ldots = 0 \qquad (2\text{-}3)$$

where $v_A = -a$, $v_B = -b$, $v_P = p$, and $v_Q = q$. By convention v is negative for reactants and positive for products. Equation 2-3 can be generalized to

$$\sum_i v_i M_i = 0 \qquad (2\text{-}4)$$

where M_i is the molecular weight of the ith species; the sum is over all reactants and products. Equation 2-4 follows from the law of mass conservation.

The extent of reaction (ξ) is the change in the number of moles of a substance participating in a reaction divided by its stoichiometric coefficient:

$$\xi = (n_i - n_{io}) / v_i \qquad (2\text{-}5)$$

n_i is the number of moles of i present at a given time, and n_{io} is the initial number of moles. Differentiating Equation 2-5 with respect to time yields

$$d\xi / dt = (1 / v_i) dn_i / dt \qquad (2\text{-}6)$$

and substituting Equation 2-6 into Equation 2-1 gives

$$rate = dn_i / v_i V dt = d[C_i] / v_i dt \qquad (2\text{-}7)$$

for a constant volume system (because molar concentration [C] = n/V). The rate of a reaction thus is equal to the rate of change in concentration of a reacting species (product or reactant) divided by its stoichiometric coefficient.

2.1.2 Order, Stoichiometry, and Molecularity

With few exceptions that are only apparent contradictions of the rule, increasing reactant concentration increases the reaction rate. In a closed system, reactant concentrations decrease as the reaction proceeds. Consequently, reaction rates decrease with time, asymptotically approaching zero. Chemical reactions can be grouped into two categories based on complexity of the pathways by which they proceed. *Elementary* reactions proceed in one step, and product(s) are formed directly when reactants collide with the proper orientation and sufficient energy. No intermediates are involved, and the reaction rate follows the law of mass action exactly (with few exceptions, as explained below). Only simple transformations can take place in a single step or collision: transfer of an electron or electron pair, bond formation between species with electrons and empty orbitals available for sharing, and breaking of a bond between two atoms. Many chemical transformations involve much more extensive changes in molecular structure than can take place in a single collision. Such *multistep* reactions proceed by a series of elementary steps that together define the reaction mechanism. The algebraic sum of the elementary steps gives the overall reaction stoichiometry. Although rates of multistep reactions are functions of (some) reactant concentrations, overall stoichiometry is not a good predictor of the form of the rate equation.

The first step in analyzing the kinetics of a reaction is to fit the data to an empirical rate equation. The sum of the exponents in this equation defines the reaction order. Reactions that fit the rate equation $-d[A]/dt = k[A]$ are first order, regardless of the number of reactants in the stoichiometric equation. Reactions that fit the equation $-d[A]/dt = k[A][B]$ are second order overall and first order in A and B. The order of multistep reactions need not be an integer, but may be fractional or even zero. Order is determined by the fit of data to a rate equation and need not agree with reaction stoichiometry.

The number of reactant species in an elementary reaction is known as its "molecularity"; most elementary reactions are uni- or bimolecular. With two exceptions,* order, stoichiometry, and molecularity are synonymous terms for elementary reactions. For reactions that involve several steps, we may refer only to the molecularity of each step. Definition of molecularity thus requires knowledge of the reaction mechanism. In multistep reactions one step often is much slower than the rest and limits the overall rate of reaction. In many cases the stoichiometry of this step determines overall reaction order. For example, if this step is bimolecular, the reaction likely will follow second-order kinetics, at least over some range of concentrations. Mechanisms ultimately determine the overall rate equation and reaction order, but the converse is not true. As described in Section 2.7, several mechanisms may lead to the same overall rate equation and same reaction order. Elucidation of mechanisms is a complicated process that requires various types of information. Many texts are available on this subject (see references at the end of Chapter 1). Mechanisms of some aquatic reactions are described in Chapters 4 and 8.

---

* In dissociation of diatomic molecules, collision with another molecule is needed to provide the energy for dissociation, but for reasons described in Section 3.4.5, the reaction may fit first-order kinetics. Recombination of two atoms into diatomic molecules requires that a third body (inert molecule) collide simultaneously with the atoms to absorb some of the collisional energy; otherwise they will fly apart again. Such reactions are termolecular and third order ($A + A + M \rightarrow A_2 + M$), although the net stoichiometry is $A + A \rightarrow A_2$.

2.1.3 Differential and Integrated Rate Expressions

Rates of most reactions can be described by equations of the general form:

$$rate = d[A]/dt = kC^aC^b \cdots \qquad (2\text{-}8)$$

The superscripts (reaction orders for each reactant) are small integers (1 or 2 for elementary reactions), and k is the rate constant. The value of k depends on temperature and other physical conditions, but not on time.[*] The units of k depend on the order of the rate equation; in general, k has units of $[C]^{1-n} t^{-1}$, where n is the overall order of the rate equation. Solution of the rate equation for a given reaction requires knowledge of the rate constant and reactant orders. It often is convenient to use integrated forms of the rate equations that relate reactant concentrations as functions of time. The integrated expressions can be used to solve kinetic problems, e.g., compute the concentration of a reactant remaining (or product formed) for any desired reaction time. For simple elementary reactions, standard integrated forms are available, as discussed in the following section. Rate equations for more complicated reactions may have several terms and/or denominators with reactant terms, and these often cannot be solved analytically. In such cases numerical methods may be used to determine concentrations as a function of time, or simplifying assumptions may be made to obtain a pseudoanalytical solution (see Sections 2.3 and 2.5).

Integrated expressions for elementary reactions can be written so that rate constants can be evaluated graphically from a plot of a function of [C] vs. a function of time; the nature of the functions depends on reaction order. These plots are the most common way to evaluate rate constants, but caution should be observed in using them, because the possibility of interferences from competing and side reactions increases during the time course of a reaction. Further information on the estimation of rate constants is given in Section 2.4.

2.2 ELEMENTARY RATE EQUATIONS

2.2.1 First-Order Kinetics

A reaction whose rate depends linearly on the concentration of a single reactant follows first-order kinetics. If the initial concentration of reactant A is $[A]_o$, and the amount reacted per unit volume at any time is x, then $([A]_o - x) = [A]$, the concentration remaining at time t, and the rate is:

---

[*]   This is true in normal kinetics, which apply to systems that are well mixed in three dimensions, such as aqueous solutions and gas-phase reactions. In contrast, diffusion-limited reactions with geometrical constraints, such as occur in solid, viscous, or porous media or at interfaces, may exhibit "fractal kinetics", in which the rate coefficient varies with time: $k_{obs} = kt^{-h}$, $0 < h < 1$ (t > 1), where k_{obs} is the instantaneous (measured) rate constant and k does not vary with time.[3] These anomalies derive from nonrandom distribution of reactants. We will not treat fractal kinetics further in this book, and rate coefficients will be assumed to be time invariant.

$$-d[A]/dt = k_1[A] = k_1([A]_o - x) = dx/dt \qquad (2\text{-}9)$$

Equation 2-9 is valid if the reaction is irreversible, i.e., it proceeds only in a forward direction, with no tendency for product to revert to reactant. The integrated form of Equation 2-9 is

$$[A]/[A]_o = \exp(-k_1 t) \qquad (2\text{-}10)$$

or
$$\ln[A]/[A]_o = \ln\big([A]_o - x\big)/[A]_o = -k_1 t \qquad (2\text{-}11)$$

The reaction time course can be followed by measuring either the reactant concentration remaining or the amount of product formed at various times, whichever is more convenient. (If the reaction stoichiometry is known, the amount of product formed can be translated into the amount of reactant consumed; if $[A]_o$ is known, $[A]$ can be calculated.) The left side of Equation 2-11 is a dimensionless ratio, and the right side also must be dimensionless. The units of k_1 thus are reciprocal time. The preferred unit is s^{-1}, but min^{-1}, h^{-1}, and other time units may be used for slow reactions. The units in which reactant concentrations are expressed thus do not affect the value of k_1. Because rates vary with temperature, the temperature at which k_1 was measured must be specified.

For graphical solution the log_{10} of Equation 2-10 can be taken as

$$\log[A] - \log[A]_o = \log(e^{-k_1 t}) \qquad (2\text{-}12\text{a})$$

or
$$\log[A] = -0.434 k_1 t + \log[A]_o \qquad (2\text{-}12\text{b})$$

A plot of $log[A]$ vs. t yields a straight line for reactions following first-order kinetics (Figure 2-1). The y-intercept is $log[A]_o$, and k_1 is obtained from the slope, $k_1 = -2.3(\text{slope})$.

A useful kinetic concept for first order reactions is the half-life $t_{1/2}$ — the time required for half of the reactant to react. From Equation 2-11,

$$t_{1/2} = \frac{1}{k_1} \ln \frac{[A]_o}{0.5[A]_o} = \ln(2)/k_1 = 0.693/k_1 \qquad (2\text{-}13)$$

Equation 2-13 shows that the half-life of a first-order reaction is independent of initial concentration. It follows that a constant fraction of reactant is lost in successive time periods of equal length. Half the initial concentration remains at $t = t_{1/2}$; one-fourth remains at $t = 2t_{1/2}$; one-eighth at $t = 3t_{1/2}$, and so on. Dimensionless plots of $log[A]/[A]_o$ vs. time expressed in units of $t_{1/2}$ thus yield straight lines (Figure 2-1).

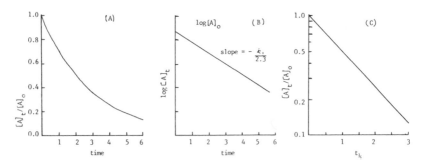

Figure 2-1. First-order kinetic plots: (A) arithmetic dimensionless plot of fraction initial A remaining vs. time in multiples of $t_{1/2}$; (B) semilog plot of log[A] vs. t linearizes data; (C) dimensionless plot of log fraction [A] remaining vs. time in multiples of $t_{1/2}$.

Other fractional lives may be calculated readily from Equation 2-11. In general, if $t_{1/i}$ is the time required for A to decrease to $1/i$th of its initial value,

$$t_{1/i} = -(1/k_1)\ln(1/i) = \ln(i)/k_1 \qquad (2\text{-}14)$$

and

$$[A]/[A]_o = (1/i)^n \qquad (2\text{-}15)$$

n is dimensionless time expressed in multiples of $t_{1/i}$. When $1/i = 1/e$, $t_{1/i}$ is given the symbol τ, and Equation 2-14 becomes $\tau = 1/k_1$, since $ln(1/e) = -1$. τ is known as the characteristic time or mean lifetime of a reaction and is analogous to τ_w, the hydraulic residence time, and τ_s, the substance residence time, in completely mixed reactors (Chapter 5). τ represents the average time before a molecule reacts. From the nature of the exponential function, the concentration of A is $1/e$th (0.37) of $[A]_o$ at $t = \tau$. From Equation 2-14 and the fact that $k = 1/\tau$, we see that $t_{1/2} = 0.69\tau$.

Decay of radioisotopes follows first-order kinetics, and decay rates usually are expressed in half-lives. Rates of radioisotope decay cannot be modified by chemical or normal physical means and depend only on the amount of isotope remaining at any time. Microbial die-off also follows first-order kinetics, and the disappearance of pathogens or indicator organisms at a constant concentration of disinfectant usually is assumed to be first order in microbe concentration. A common term in disinfection kinetics is t_{99} (or $t_{.01}$), the time required for 99% die-off. From Equation 2-14

$$t_{99} = -(1/k_1)\ln(1/100) = 4.605/k_1 \qquad (2\text{-}16)$$

Many reactions of interest in natural waters follow first-order kinetics under some circumstances, even though they are mechanistically and stoichiometrically complex. Examples are given throughout the book. For example, loss rates of nutrients and trace metals from lakes often follow first-order kinetics, although several mechanisms, including biological uptake, chemical precipitation, and adsorption onto suspended solids, probably are involved. Chemical residence times of substances in lakes and oceans generally are calculated from chemical reactor theory (Chapter 5), with first-order kinetics assumed. The growth of microorganisms and the exertion of biochemical oxygen demand also follow first-order kinetics under many circumstances.

Reactions whose rates depend on the concentrations of several reactants can be made pseudo-first order by holding constant the concentrations of all but one reactant. For reactions involving dissolved gases, concentrations can be fixed at a desired level by sparging the solution with a gas mixture containing a fixed partial pressure of the reactant gas. Reactions in which the solvent acts as a reactant often are pseudo-first order; hydrolysis reactions fit into this category. For example, hydrolysis of esters, amides, and condensed phosphates (Chapter 4) are first order (ignoring effects of acid or base catalysis) because water is both a solvent and reactant and stays at constant concentration during the reaction.

2.2.2 Second-Order Kinetics

Two types of elementary reactions exhibit second-order kinetics, and many reactions with complex mechanisms also fit second-order expressions, at least under limited experimental conditions. Elementary bimolecular reactions of the types $A + B \rightarrow$ products and $2A \rightarrow$ products are second order. In the former case, if $[A]_o$ and $[B]_o$ are the initial concentrations of A and B, x is the amount of each that has reacted at time t, and the reaction is irreversible, the rate equation is

$$dx / dt = k_2 [A][B] = k_2 ([A]_o - x)([B]_o - x) \qquad (2\text{-}17)$$

If $x = 0$ at $t = 0$ and $[A]_o \neq [B]_o$, the integrated form of this equation is

$$\frac{1}{[A]_o - [B]_o} \ln \frac{[B]_o([A]_o - x)}{[A]_o([B]_o - x)} = \frac{1}{[A]_o - [B]_o} \ln \frac{[B]_o[A]}{[A]_o[B]} = k_2 t \qquad (2\text{-}18)$$

Rate constant k_2 has units of t^{-1} conc^{-1} and is numerically equal to the reaction rate when [A] and [B] are unity. The magnitude of k_2 depends on the nature of the reaction, physical factors like temperature, and the units of t and concentration. Concentrations in gas-phase reactions usually are expressed in atm, and the preferred unit for k_2 is atm^{-1} s^{-1}. Solution concentrations are usually in mol L^{-1} (M^{-1}), and k_2 commonly is in M^{-1} s^{-1}.

Graphical solutions are obtained by rearranging Equation 2-18:

$$\log[A]/[B] = 0.43 k_2 ([A]_o - [B]_o)t - \log[B]_o / [A]_o \qquad (2\text{-}19)$$

Since $[A]_o$ and $[B]_o$ are constant in a given experiment, a plot of $\log$[A]/[B] vs. t gives a straight line for second-order reactions, and k_2 is obtained from the slope of the line (Figure 2-2a):

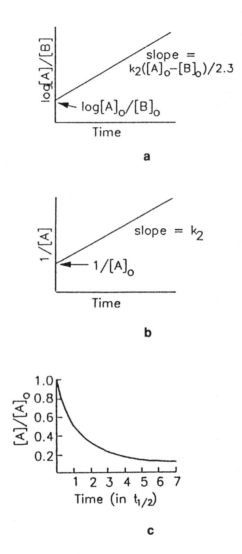

Figure 2-2. Second-order kinetic plots: (a) general second-order reaction, A + B → P, showing slope and intercept values; (b) plot for 2A → P or A + B → P when $[A]_o$ = $[B]_o$; (c) fraction of A remaining vs. time in multiples of $t_{1/2}$ for second-order reaction, 2A → P, showing increasing length of fractional times.

$$k_2 = 2.303(slope)/([A]_o - [B]_o) \qquad (2\text{-}20)$$

In reactions of the type $2A \rightarrow$ products or where $[A]_o = [B]_o$, Equations 2-18 to 2-20 are indeterminate. The rate expression for reactions of this type is

$$dx/dt = k_2[A]^2 = k_2([A]_o - x)^2 \qquad (2\text{-}21)$$

which integrates to

$$1/[A] = 1/([A]_o - x) = k_2 t + 1/[A]_o \qquad (2\text{-}22)$$

Plots of 1/[A] vs. t are linear with a slope of k_2 (Figure 2-2b).

In studying second-order reactions, experimentalists may hold the concentration of one reactant constant by making its concentration much higher than that of the other reactant. The apparent rate constant thus obtained is pseudo-first order, since Equation 2-17 then integrates to a first-order-like equation:

$$-d[A]/dt = k_2[A][B] = k'[A] \qquad (2\text{-}23a)$$

or
$$\ln[A]/[A]_o = k't \qquad (2\text{-}23b)$$

where $k' = k_2[B]$, and B is the fixed reactant. The value of the apparent first-order rate constant k' depends on the concentration of the fixed reactant. Under certain conditions, Equation 2-23 also applies to reactions where B is a catalyst. Rate constants obtained when $[A]_o = [B]_o$ are true second-order rate constants, as dimensional analysis of k_2 in Equation 2-22 demonstrates. Plots of Equation 2-22 are simpler than those of Equation 2-19, and k_2 is obtained directly as the slope of the former equation. Setting initial reactant concentrations equal thus is a convenient way to simplify analysis of second-order reactions.

The half-life of a second-order reaction is an ambiguous concept if concentrations of A and B are not equal, since they then will have different times for a half-reaction. In reactions where the concentrations are the same or the two reacting species are identical, the half-life is obtained from Equation 2-22:

$$t_{1/2} = 1/k_2[A]_o \qquad (2\text{-}24)$$

Thus $t_{1/2}$ is inversely proportional to initial concentration, and the time for successive decreases in [A] by a factor of one half doubles for each period. Thus $t_{1/4} = t_{1/2} + 2t_{1/2} = 3t_{1/2} = 3/k_2[A]_o$; and $t_{1/8} = t_{1/2} + 2t_{1/2} + 4t_{1/2} = 7t_{1/2}$, and so on (Figure 2-2c). For second-order reactions where one reactant (B) is present in fixed concentration (pseudo-first order), the half-life is given by

$$t_{1/2} = \ln 2 / k_2 [B]_o \qquad (2\text{-}25)$$

and the half-life is independent of variable reactant A.

It is often convenient to write rate equations in terms of stoichiometric coefficients[2,4] (see Section 2-1). For general reaction

$$v_a A + v_b B + \ldots \rightarrow v_p P + v_q Q + \ldots$$

the reaction rate can be defined as

$$Rate = \frac{-1d[A]}{v_a dt} = \frac{-1d[B]}{v_b dt} = \frac{1d[P]}{v_p dt} = \frac{1d[Q]}{v_q dt} = \ldots \qquad (2\text{-}26)$$

where the v_i are stoichiometric coefficients. According to this convention, second-order reactions of the type $2A \rightarrow$ products have the rate expression

$$-d[A]/dt = 2k_2 [A]^2 \qquad (2\text{-}27)$$

which integrates to

$$2k_2 t = 1/[A] - 1/[A]_o \qquad (2\text{-}28)$$

and the half-life is given by

$$t_{1/2} = 1/2k_2 [A]_o \qquad (2\text{-}29)$$

Caution must be used in taking rate constants and rates from the literature, since both the above convention and the one used for Equations 2-21 and 2-22 are used in the literature. Use of stoichiometric coefficients yields second-order rate constants half as large as those obtained without the coefficients.

Many reactions important in aqueous environmental chemistry are second order; examples include oxidation of sulfide and various chlorination reactions. In most cases, however, the reactions are complicated by consecutive or competing reactions or catalytic effects, and simple second-order formulations do not fully describe the kinetics. Several examples are presented in Chapter 4.

2.2.3 Third-Order Kinetics

Elementary reactions whose rates depend on the concentration of three reactants are not too common, except where catalysis is involved. This is understandable given that reactions occur when molecules collide; the probability of simultaneous collision of three or more reactants is much smaller than the probability of bimolecular collisions. Several types of third-order reactions can be considered. In the most general case the three reactants are different species at different initial concentrations, e.g., A + B + C → products, and the rate expression is

$$dx / dt = k_3([A]_o - x)([B]_o - x)([C]_o - x) \qquad (2\text{-}30)$$

An integrated form of Equation 2-30 is available,[2] but it is quite complicated. Conditions where it applies are uncommon and should be avoided. More commonly, two reactants are identical, e.g., 2A + B → products. Concentrations of A and B at any time are ($[A]_o - 2x$) and ($[B]_o - x$); note that stoichiometric coefficients are essential in this case. The rate equation is

$$dx / dt = k_3[A]^2[B] = k_3([A]_o - 2x)^2([B]_o - x) \qquad (2\text{-}31)$$

which can be integrated to

$$\left[\frac{([A]_o - [A])(2[B]_o - [A]_o)}{[A]_o[A]} + \ln \frac{[B]_o[A]}{[A]_o[B]} \right] = (2[B]_o - [A]_o)^2 k_3 t \quad (2\text{-}32\text{a})$$

or in terms of stoichiometric coefficients:

$$\frac{1}{[A]} - \frac{1}{[A]_o} + \frac{1}{[A]_o - [B]_o v_A / v_B} \ln \left[\frac{[B][A]_o}{[A][B]_o} \right] = ([A]_o v_B - [B]_o v_A) k_3 t \quad (2\text{-}32\text{b})$$

The simplest third-order reaction is 3A → products; (A + B + C → products, with initial concentrations all equal, is equivalent). The rate equation is

$$(1 / v_A) d[A] / dt = k_3[A]^3 \qquad (2\text{-}33)$$

which integrates to

$$-2v_A k_3 t = 1 / [A]^2 - 1 / [A]_o^2 \qquad (2\text{-}34)$$

The left sides of Equations 2-32 and 2-34 can be graphed vs. t to obtain k_3.

Pseudo-second-order expressions arise when one of the reactants has a fixed concentration (e.g., one is a catalyst). The integrated expression then is

$$\frac{1}{[A]_o - [B]_o} \ln \frac{[B]_o[A]}{[A]_o[B]} = k_2[C]t \qquad (2\text{-}35)$$

where C is the fixed reactant or catalyst.

2.2.4 The Anomaly of Zero-Order Reactions

At the beginning of this chapter we stated that reaction rates depend on the concentration of reactants, as required by the law of mass action. Nevertheless, zero-order reactions, whose rates are independent of concentration, are common. Rate expressions of such reactions may be considered to contain concentration to the zero power:

$$-d[A]/dt = k_o[A]^o = k_o \tag{2-36}$$

which integrates to

$$[A]_o - [A] = k_o t \tag{2-37}$$

Thus, a simple arithmetic plot of concentration vs. time is linear.

The violation of mass action principles is only apparent in these cases. Zero-order kinetics usually implies that heterogeneous or homogeneous catalysis is involved in the reaction. The catalyst may be a surface with a fixed number of reactive sites (heterogeneous catalysis) or a dissolved catalyst (e.g., some enzymes). For example, the reactants may be gases that react only when adsorbed to a solid surface that acts as a catalyst. If the adsorption tendency is strong, adsorption/reaction sites on the surface will be completely covered over a broad range of reactant pressures (P), and the reaction rate will be independent of P as long as the surface is covered. Surface coverage will depend on P at some low value, but this may not be attained in a given experiment. Analogous conditions can occur in liquid/solid systems where solutes react only on a surface, and even in homogeneous solution where solutes react only when bound to a catalytic site on some dissolved molecule. A single elementary reaction cannot exhibit zero-order kinetics.

Some reactions may seem to be zero order simply because reactant concentrations have not changed sufficiently during an observation period to cause curvature in a plot of concentration vs. time. A reaction may occur at a constant rate in an open system, because continuous influx of reactant from external sources maintains steady-state concentrations.

2.3 RATE EQUATIONS FOR MORE COMPLICATED REACTIONS

Most reactions of interest in aquatic systems are not elementary (one-step) processes, and in many cases their rate equations are more complicated than those described in the preceding section. This section deals with rate equations for some of the most common situations that complicate kinetic descriptions: reversible, consecutive, and chain reactions.

2.3.1 Reversible Reactions

The reactions treated above were assumed to proceed only in the direction indicated, and products were assumed not to "back-react" to form the original reactants.[*] This assumption is valid only for reactions where equilibrium lies far to the right. If appreciable concentrations of both products and reactants exist at equilibrium, the reaction is reversible and the back reaction will begin once the products accumulate. Equilibrium constants for reactions are related to rate constants for forward and back reactions by the ratio $K_{eq} = k_{for}/k_{rev}$; this follows from the law of microscopic reversibility.[5] For reactions where K_{eq} is small ($\sim 10^{-2}$ to 10^2), the forward and reverse rate constants will be of fairly comparable magnitude. Of course, both may be very small or large, since only the ratio of the k's is defined by K_{eq}.

Kinetic expressions for reversible reactions are complicated by the addition of a term for the back reaction. Consider a simple reversible first-order reaction, $A \rightleftarrows P$. If we take the back reaction into account, the net rate of change in [A] is $-d[A]/dt$ = $rate_{for} - rate_{rev}$, or

$$-d[A]/dt = k_f[A] - k_r[P] \qquad (2\text{-}38)$$

k_f and k_r are first-order rate constants for the forward and reverse reactions. With x as the net concentration of A that has reacted, Equation 2-38 becomes

$$dx/dt = k_f\left([A]_o - x\right) - k_r([P]_o + x) \qquad (2\text{-}39)$$

Equation 2-39 has the general form dx/(ax + b)=dt, where a =$-(k_f+k_r)$ and b=$k_f[A]_o$ + $k_r[P]_o$. The indefinite integral of dx/(ax + b) is a standard form: $(1/a)ln|ax + b|$ + C; evaluating this between t = 0 and t = t, substituting for a and b, and rearranging yields

$$\ln\frac{(k_f[A]_o - k_r[P]_o)}{k_f([A]_o - x) - k_r([P]_o + x)} = (k_f + k_r)t \qquad (2\text{-}40)$$

If $[P]_o = 0$, then $[P] = [A]_o - [A]$, and Equation 2-38 becomes

$$-d[A]/dt = k_f[A] - k_r([A]_o - [A]) \qquad (2\text{-}41)$$

which integrates to

$$\ln\{k_f[A]_o/(k[A] - k_r[A]_o)\} = k^*t \qquad (2\text{-}42a)$$

---

[*]   By definition, *elementary* reactions in kinetics are unidirectional; the reverse reaction is considered to be a separate elementary reaction.

or
$$[A] = (1/k^*)[A]_o\{k_r + k_f \exp(-k^*t)\} \qquad (2\text{-}42b)$$

where $k^* = k_f + k_r$.

Equation 2-42b can be written in terms of the equilibrium constant for the reaction, $K = k_f/k_r = [P]_{eq}/[A]_{eq}$. If $[P]_o = 0$, then $[P]_{eq} = [A]_o - [A]_{eq}$, and

$$[A]_{eq}/[A]_o = [A]_{eq}/\{[A]_{eq} + [P]_{eq}\} = 1/(1 + K) = k_r/(k_f + k_r) \quad (2\text{-}42c)$$

If we divide both sides of Equation 2-42b by $[A]_{eq}$ and use the identities defined in Equation 2-42c to simplify the expression, we obtain

$$[A]/[A]_{eq} = 1 + K \exp(-k^*t) \qquad (2\text{-}43a)$$

or
$$\frac{([A] - [A]_{eq})}{([A]_o - [A]_{eq})} = \exp(-k^*t) \qquad (2\text{-}43b)$$

Equation 2-43b is comparable to the simple first-order equation (2-11) for irreversible reactions, except that $[A]$ is corrected for equilibrium concentration $[A]_{eq}$, and the reaction approaches equilibrium with a rate constant k^* equal to the **sum** of the forward and back rate constants. Equation 2-43b applies to transfer rates of O_2 and other gases to unsaturated solutions (Chapter 4) and also to the kinetics of two-box compartment models and CFSTRs in which a step change in input has occurred (Chapter 5). Plots of the natural *log* of the left side of Equation 2-43b vs. t for reversible first order reactions thus are straight lines with slopes of $-k^* = -(k_f + k_r)$. Since $[A]_{eq}/[A]_o = k_r/(k_f + k_r) = 1/(1 + K)$, rate data from a single experiment are sufficient to calculate both rate constants, provided that the reaction reaches equilibrium so that $[A]_{eq}$ can be measured. Equation 2-43a can be reformulated[5] to illustrate how the magnitude of K affects the shape of first-order plots (Figure 2-3). As K decreases from infinity (for irreversible reactions), the ratio $[A]_{eq}/[A]_o$ increases. All curves in the figure have the same value of k_f and the same initial slope; decreasing K means that k_r increases. As a result, the curves begin to deviate from the irreversible first-order curve more quickly.

Integrated expressions for higher-order reversible reactions are much more complicated. Second-order reversible expressions can be simplified by writing the rate law in terms of equilibrium concentrations and Δ, the displacement of reactant or product concentration from equilibrium ($\Delta = [i]_{eq} - [i]$).[6] Table 2-1 lists differential and integrated expressions for several second-order reaction types. Sorption of *E. coli* T_2 bacteriophage onto activated carbon was described as a reversible second-order reaction.[7] The complications of higher-order equations can be avoided by working under conditions where a reaction becomes pseudo-first order. Reversible kinetics often can be avoided altogether by starting with $[P]_o = 0$ (for all products) and measuring the initial forward velocity. k_r can be determined similarly or calculated from k_f and K_{eq} (at least for elementary reactions; see Section 2.5.3).

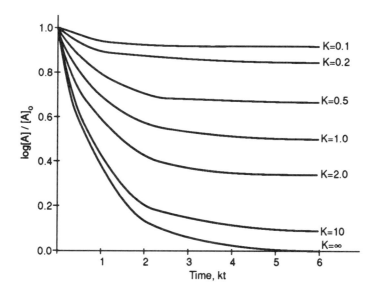

Figure 2-3. Effect of equilibrium constant K (= k_f/k_r) on shape of reactant disappearance curves for reversible first-order reactions. Dimensionless time in units of kt; k = $k_f + k_r$; k_f fixed at 10 t^{-1}; k_r varied to give K values shown by curves. Graph computed from Equation 2-43a reformulated to $[A]/[A]_o = 1/(1 + K) + \{K/(K + 1)\}\exp(-kt)$.

Table 2-1. Integrated Rate Equations for Some Reversible Reactions

---

1. $A + B \rightleftharpoons C + D$; $-d[A]/dt = k_f[A][B] - k_r[C][D]$

$$\ln\frac{\Delta}{\Delta(1-1/K)+[A]_e +[B]_e +\left([C]_e +[D]_e\right)/K} = -k_f\{[A]_e + [B]_e + ([C]_e + [D]_e)/K\}t + \text{Const}$$

2. $A + B \rightleftharpoons 2C$; $-d[A]/dt = k_f[A][B] - k_r[C]^2$

$$\ln\frac{\Delta}{\Delta(1-4/K)+[A]_e +[B]_e +4[C]_e /K} = -k_f\{[A]_e + [B]_e + 4[C]_e/K\}t + \text{Const}$$

3. $A + B \rightleftharpoons C$; $-d[A]/dt = k_f[A][B] - k_r[C]$

$$\ln\frac{\Delta}{\Delta+[A]_e +[B]_e +1/K} = -k_f\{[A]_e + [B]_e + 1/K\}t + \text{Const}$$

Note: Δ = displacement of concentration of reactant A from equilibrium; i.e., $\Delta = [A] - [A]_e$, where $_e$ indicates equilibrium concentration. K = equilibrium constant for the reaction. Const = constant of integration, which can be evaluated by plotting the left side of each equation vs. t. Const = y-intercept value. [From King, E. L., *Int. J. Chem. Kinetics*, 14, 1285 (1982).]

2.3.2 Consecutive Reactions

Many processes important in natural waters do not proceed by a single reaction, but involve several steps and intermediate products before forming the final product. Such reactions as

$$A \xrightarrow{k_i} B \xrightarrow{k_{ii}} C \qquad (2\text{-}44)$$

are termed consecutive reactions. In any sequence of reactions of differing speeds, the slowest step determines the rate of the overall process, since each reaction depends on the previous one for its reactants. In sequence 2-44, conversion of A to B determines the rate of final product formation if $k_{ii} \gg k_i$, but if $k_i \gg k_{ii}$, conversion of B to C controls the rate of final product formation. When k_i and k_{ii} are comparable in magnitude, both reactions must be considered. In this section we consider consecutive reactions where the rate constants of individual steps are such that all steps must be considered. We are interested in the time-dependent concentrations of the intermediate products, as well as the final product. This feature differentiates consecutive reactions from that of deriving rate equations for reactions with multistep mechanisms in which intermediates are formed that then react to form the observed products. In these cases the intermediates do not build up appreciable concentrations in the reaction medium, and we need rate equations for product formation that do not include unmeasured intermediate species.

Two-Step First-Order Sequence

Reaction sequence 2-44 yields the following rate expressions:

$$-d[A]/dt = k_i[A] \qquad (2\text{-}45a)$$

$$d[B]/dt = k_i[A] - k_{ii}[B] \qquad (2\text{-}45b)$$

$$d[C]/dt = k_{ii}[B] \qquad (2\text{-}45c)$$

Integration of Equation 2-45a yields the familiar expression for first-order reactions, $[A] = [A]_o \exp(-k_i t)$, but integration of Equation 2-45b is less straightforward. Substitution of Equation 2-10 for [A] into Equation 2-45b yields

$$d[B]/dt = k_i[A]_o \exp(-k_i t) - k_{ii}[B] \qquad (2\text{-}45d)$$

which is not integratable directly, but can be solved by the change of variable or integrating factor techniques described in basic calculus texts. The former method is based on the fact that for any variable $U = f(X)$, $dU = f'(X)dX$.* If we let $U = [B]\exp(k_i t)$, then $[B] = U\exp(-k_i t)$, and

$$d[B]/dt = (dU/dt)\exp(-k_i t) - Uk_i \exp(-k_i t)$$

Substitution of these equations for [B] and d[B]/dt into Equation 2-45d introduces the factor $\exp(-k_i t)$ into every term, and it cancels out, leaving

---

* For use of the integrating factor approach, see Section 5.2.3.

$$dU \, / \, dt = k_i[A]_o + (k_i - k_{ii})U \qquad (2\text{-}45e)$$

which is a standard form that integrates to

$$\frac{1}{k_i - k_{ii}} \ln\{k_i[A]_o + (k_i - k_{ii})U\} = t + \text{constant} \qquad (2\text{-}45f)$$

Substitution of [B] exp(k_it) for U into Equation 2-45f and evaluating the constant of integration by setting [B] = 0 at t = 0 yields

$$[B] = \frac{k_i[A]_o}{k_{ii} - k_i} \{\exp(-k_i t) \; - \; \exp(-k_{ii} t)\} \qquad (2\text{-}46)$$

The integrated expression for [C] can be obtained from a materials balance (assuming $[C]_o = [B]_o = 0$): $[C] = [A]_o - [A] - [B]$. Substitution of Equations 2-10 and 2-46 for [A] and [B] and rearranging leads to

$$[C] = \frac{[A]_o}{k_{ii} - k_i} \{k_{ii}(1 - \exp(-k_i t)) - (1 - \exp(-k_{ii} t))\} \qquad (2\text{-}47)$$

Equations 2-46 and 2-47 are indeterminate when $k_i = k_{ii}$. In that unlikely case, d[B]/dt = k_i([A] − [B]), which integrates to

$$[B] = k_i t[A]_o \exp(-k_i t) \qquad (2\text{-}48)$$

Figure 2-4 shows the time course of [A], [B], and [C] for several ratios of k_i/k_{ii}. In each case, [B] reaches a maximum at some intermediate time and then decreases, gradually approaching 0. Both the time for [B] to reach its maximum, t_{maxB}, and $[B]_{max}$ depend on the relative sizes of k_i and k_{ii}. Since d[B]/dt = 0 at the maximum, it follows from Equation 2-45b that $k_i[A] = k_{ii}[B]$, or $[B] = [A]k_i/k_{ii}$. Substituting Equation 2-10 for [A] into this expression yields

$$[B]_{max} = (k_i \, / \, k_{ii})[A]_o \exp(-k_i t_{max B}) \qquad (2\text{-}49)$$

By equating Equations 2-46 and 2-49 at t = t_{maxB}, we can eliminate [B] and solve for t_{maxB}:

$$t_{max B} = \frac{1}{k_i - k_{ii}} \ln(k_i \, / \, k_{ii}) \qquad (2\text{-}50)$$

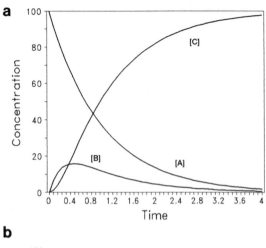

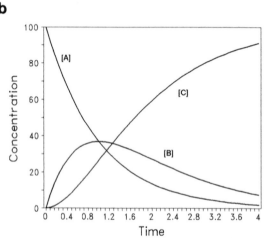

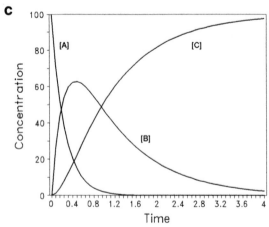

Figure 2-4. Concentration vs. time for consecutive first-order reactions, $A \rightarrow B \rightarrow C$: (a) $k_1 = 1\ h^{-1}$, $k_2 = 4\ h^{-1}$; (b) $k_1 = k_2 = 1\ h^{-1}$; (c) $k_1 = 4\ h^{-1}$, $k_2 = 1\ h^{-1}$. Simulations performed by *Acuchem*.[20]

For the cases in Figure 2-4a and c, where $k_{ii} = 4k_i = 1$ h^{-1} and $k_i = 4k_{ii} = 1$ h^{-1}, respectively, we find $t_{maxB} = 0.46$ h. For $k_i = k_{ii}$, Equation 2-50 is indeterminate, but since $d[B]/dt = k_i([A] - [B])$, and since $d[B]/dt = 0$ at t_{maxB}, it is apparent that t_{maxB} occurs where $[A] = [B]$. It is unlikely that k_i and k_{ii} are exactly equal in real systems, and for practical purposes Equations 2-46, 2-49, and 2-50 are always applicable. Also apparent in the graphs of Figure 2-4 is an initial lag before [C] begins to increase. The duration of the lag (called the induction period) depends on the relative magnitudes of k_i and k_{ii}. Because $d[C]/dt = k_{ii}[B]$, formation of C is most rapid at $[B] = [B]_{max}$.

Secular Equilibrium

Consecutive first-order reactions with $k_{ii} > k_i$ exhibit the interesting feature of secular or transient equilibrium, whereby the ratio [B]/[A] approaches a constant value at $t \gg t_{maxB}$. (Secular equilibrium is a steady-state phenomenon of kinetic processes; it should not be confused with thermodynamic equilibrium.) If we solve for $[A]_o$ using the first-order equation $[A] = [A]_o \exp(-k_1 t)$, and substitute the result into Equation 2-46, we obtain

$$\frac{[B]}{[A]} = \frac{k_i}{k_{ii} - k_i} \{1 - \exp(-(k_{ii} - k_i)t)\} \qquad (2\text{-}51)$$

If $k_{ii} > k_i$, the exponential term approaches zero as t becomes large, and thus [B]/[A] approaches a constant value:

$$\lim[B]/[A] = k_i / \left(k_i - k_{ii}\right) \qquad (2\text{-}52)$$

If $k_i > k_{ii}$, the exponent in Equation 2-51 is positive, the exponential term does not approach zero at large t, and secular equilibrium does not occur. Instead, [B]/[A] continues to increase with time (if the reaction is irreversible).

Example 2-1. Secular Equilibrium in Radionuclide Decay

Secular equilibrium is common in radionuclide decay. For example, radium-226 decays slowly ($t_{1/2} = 1620$ year, $k_i = 4.28 \times 10^{-4}$ year^{-1}) to radon-222, which decays rapidly ($t_{1/2} = 3.8$ days, $k_{ii} = 66$ year^{-1}) to polonium-218:

$$^{226}Ra \xrightarrow{\;k_i\;} {}^{222}Rn \xrightarrow{\;k_{ii}\;} {}^{218}Po$$

^{222}Rn is a useful tracer for vertical mixing in lakes and coastal waters.[8] Its use in this regard relies on the fact that ^{226}Ra occurs in much greater quantities in sediments than in overlying waters, and sediments act as a boundary source of ^{222}Rn. For diffusion calculations it is necessary to determine the total ^{222}Rn activity in water samples and the fraction that arises from secular equilibrium with ^{226}Ra present in the water column. This is done by purging a water sample with an inert gas and collecting the ^{222}Rn in a liquid N_2 trap. The Rn activity (Rn) is determined by alpha spectroscopy. The purged sample is set aside for several weeks to allow ^{226}Ra in the water to reach secular equilibrium with ^{222}Rn. From Equation 2-51 we compute that (Rn)/(Ra) = 92% of the limiting value after 14 days and 98% after 21 days. Purging after 21 days and measuring the ^{222}Rn thus yields the (Rn) in secular equilibrium with ^{226}Ra in the water. Subtraction of this activity from the initial activity gives the (Rn) arising from sediment-water exchange processes (diffusion, gas ebullition). Similar techniques are used to determine levels of ^{90}Sr, ^{226}Ra, and ^{210}Pb in environmental samples[9] because the daughters (^{90}Y, ^{222}Rn, and ^{210}Po, respectively) can be isolated and concentrated conveniently from interfering matrices.

The mass concentration of ^{226}Ra could be calculated from Equation 2-52 after secular equilibrium is attained, but radioisotope levels are expressed in terms of activity (disintegrations per second (dps) or curies (Ci); 1 Ci = 3.7×10^{10} dps). At secular equilibrium the activity of ^{222}Rn (in dps L^{-1}) is equal to that of the parent ^{226}Ra. This is apparent from Equation 2-52. If $k_{ii} \gg k_i$, as is true for ^{226}Ra decay, Equation 2-52 simplifies to $k_i[A] \approx k_{ii}[B]$; i.e., the rate of the first step is equal to the rate of the second step. Therefore, at secular equilibrium the rate of decay (or activity) of ^{226}Ra is equal to that of ^{222}Rn.

More Complicated Reaction Sequences

Analytical solutions are known for rate equations of some more complicated sequences, including (1) three-step sequence of first-order reactions, A $\rightarrow$ B $\rightarrow$ C $\rightarrow$ D;[10] (2) five-step sequence of first-order reactions, A $\rightarrow$ B $\rightarrow$ C $\rightarrow$ D $\rightarrow$ E $\rightarrow$ F;[11] (3) three-step sequence of first-order reactions with a competing pathway, A $\rightarrow$ B $\rightarrow$ (C or D) $\rightarrow$ E;[12] and (4) competing first- and second-order reactions of the type, A $\rightarrow$ B, 2A $\rightarrow$ C.[13] Sequences much more complicated than these lead to systems of differential equations that cannot be solved analytically. If the rate constants for individual reactions in a sequence are greatly different, the reactions can be treated separately, as mentioned earlier. The more general approach is to simulate the system of coupled differential equations numerically, provided initial concentrations for the reactants and products and rate constants are known for each reaction. (Methods to determine k values for each reaction are described in Section 2.4.) Concentrations of all reactants and products then are computed as a function of time.

A wide range of methods are available to integrate coupled differential equations numerically, including simple Eulerian approaches like initial slope and trapezoidal methods and more complicated techniques like Runge-Kutta routines.[14] The simplest routines are sufficient to simulate sequences involving a few reactants and products when the reactions do not have greatly different charcteristic times.[15] Table 2-2 describes several algorithms that can be used to solve simple first-order reaction sequences (such as (1) and (2) in the preceding paragraph) and competing first-order and second-order reactions such as sequence (4) above. Because analytical solutions are available for these sequences, exact values can be obtained to compare with the results of numerical integration.

Sets of rate equations for some reaction sequences are difficult to solve numerically because of an instability called "stiffness". This problem arises whenever greatly differing characteristic times (or rate constants) exist in a reaction

Table 2-2. Some Numerical Methods for Solving Rate Equations[a]

General form	Solution for first-order rate equation

A. Euler-Cauchy (initial slope method)

$$C_{t+\Delta t} = C_t + \left(\frac{dC}{dt}\right)\Delta t \qquad\qquad C_{t+\Delta t} = C_t - kC_t\Delta t$$

Δt is the integration time step

B. Runge-Kutta method

$C_{t+\Delta t} = C_t + \phi\Delta t$
where Δt is the integration time step
and $\phi = [r_1 + 2r_2 + 2r_3 + r_4]/6$

The r_i are values of dC/dt (symbolized below as $C'(t,C)$) calculated at various values of t and C:

$r_1 = C'(t,C_t)$

$r_2 = C'(t + \frac{1}{2}\Delta t, C_t + \frac{1}{2}\Delta t r_1)$

$r_3 = C'(t + \frac{1}{2}\Delta t, C_t + \frac{1}{2}\Delta t r_2)$
$r_4 = C'(t + \Delta t, C_t + \Delta t r_3)$

$r_1 = -kC_t$

$r_2 = -k\{C_t + \frac{1}{2}\Delta t(-kC_t)\}$

$r_3 = -k\{C_t + \frac{1}{2}\Delta t(-kC_t + \frac{1}{2}k^2 C_t\Delta t)\}$

$r_4 = -k\{C_t + \Delta t(-kC_t + \frac{1}{2}k^2 C_t\Delta t - \frac{1}{4}k^3 C_t[\Delta t]^2)\}$

Combining r_1-r_4 into F yields:

$$\phi = kC_t\left\{-1 + \tfrac{1}{2}k\Delta t - \tfrac{1}{6}k^2[\Delta t]^2 + \tfrac{1}{24}k^3[\Delta t]^3\right\}$$

C. Comparison of results for numerical methods with analytical solution

Consider a first-order reaction with $C_t = 10.0$ and $k = 1$ (h^{-1}). The analytical solution for the rate equation is $C_{t+\Delta t} = C_t e^{-k\Delta t}$, where Δt is the integration time step. The following values of $C_{t+1\Delta t}$ are obtained for $\Delta t = 0.1$ and 1.0 h:

Integration time step, Δt:	0.1 h	1.0 h
Analytical Solution	9.048	3.679
Initial Slope	9.00	0
Runge-Kutta	9.048	3.74

The superiority of the Runge-Kutta method over the initial slope method is especially evident when the time-step of integration is large, but it yields more accurate values even when Δt is small (relative to k).

[a] See reference 14 for a more detailed description of these methods.

sequence. This situation results in development of steady-state concentrations for one or more intermediates. One would expect that short computational time steps would be required when concentrations are changing rapidly during the initial stages of a simulated reaction sequence, and that time steps could increase as the reaction slows down and/or some intermediates reach steady state. Just the opposite occurs, however, and this leads to very inefficient integration by traditional methods. This seemingly anomalous behavior can be explained in terms of a negative feedback inherent in steady-state chemical systems;[16] slight displacement from steady state induces a rapid return, and the integration time step must be shorter than the characteristic time of the restoring force to maintain stability and prevent the error from propagating from step to step.[17] A "backward differencing" method proposed by Curtiss and Hirschfelder[16] overcomes the stability constraint on the integration time step, and Gear[18] developed an efficient algorithm based on this idea to solve stiff equations. Many programs used to simulate the kinetics of chemical reactions are based on this algorithm.[19] For example, *Acuchem*[20] is a Gear-based program for microcomputers, capable of simulating the kinetics of systems of moderate complexity (up to 80 reactions). The Gear algorithm automatically increases the time step when rapidly reacting species achieve steady states, and it solves a predictor-corrector equation involving the Jacobian matrix J of partial derivatives for each species. Elements of J are

$$J_{ij} = \frac{\partial}{\partial [X_j]} \frac{\partial [X_i]}{\partial t}$$

(2-53)

Figure 2-5 contrasts the performance of a Gear algorithm and a nonstiff differential equation solver in simulating a system with steady-state behavior. The latter is constrained to very short time steps during the steady-state period and as a result required 75 times as many steps and 50X longer execution time.

When a large number of reactions are involved in a simulation, conversion of the reaction equations into the set of differential equations for all reacting species is tedious and subject to error. Simple compilers have been developed to do this automatically in some simulation packages (e.g., *Acuchem*), for chemical reactions. Consider a sequence of m elementary steps of the form

$$v_A A + v_B B + \cdots \rightarrow v_P P + v_Q Q + \cdots$$

(2-54)

The rate of the *i*th reaction can be expressed in terms of the rate of change of its reactants or products:

$$R_i = \pm k_i [A]^{v_A} [B]^{v_B} \cdots = \pm k_i \prod_{j=1}^{n} [X_j]^{v_j}$$

(2-55)

where each step is assumed to follow mass-action kinetics, and n is the order of the step ($n_{max} = 3$ for elementary reactions). The time rate of change of any species X_g (a reactant, intermediate, or product) can be written,

$$\frac{d[X_g]}{dt} = \sum_{i=1}^{m} v_{gi} R_i = \sum_{i=1}^{m} \pm k_i v_{gi} \prod_{j=1}^{n} [X_{ij}]^{v_{ij}}$$

(2-56)

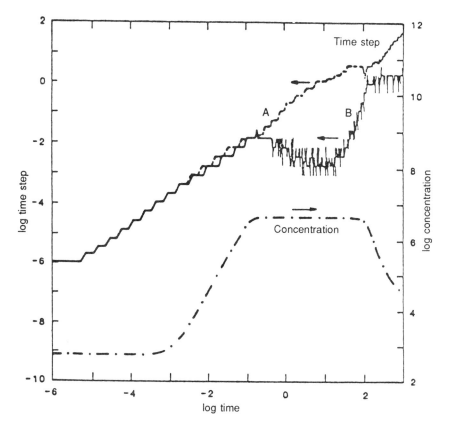

Figure 2-5. Performance comparison of an algorithm designed to solve stiff differential equations (line A) and a nonstiff equation solver (line B) for a kinetic simulation involving steady-state behavior (broken line). Note small time steps required by program B during steady-state period. [From Edelson, D., *Science*, 214, 981 (1981). With permission.]

Elements of the Jacobian matrix (Equation 2-53) can be expressed in similar manner:

$$J_{ij} = \frac{\partial}{\partial[X_r]}\frac{\partial[X_g]}{\partial t} = \sum_{i=1}^{m} k_i \nu_{gi} \nu_{ri}[X_{ri}]^{(\nu_{ri}-1)} \prod_{j=1 \neq r}^{n}[X_{ij}^2]^{\nu_{ij}} \qquad (2\text{-}57)$$

Programs based on Equations 2-56 and 2-57 allow the user to define the chemical reactions and species to be considered. The differential equations and Jacobian matrix then are assembled by the program.

Consecutive-reaction kinetics are very common in atmospheric chemistry. The process of smog formation includes over 100 reactions, and the reaction system that maintains ozone levels in the stratosphere is similarly complex. Simulation of such reaction sequences is complicated by branching or competing pathways and by widely disparate time constants for various reactions. Early simulation models for atmospheric reactions assumed a quasi-steady state (QSS) for species with large rate constants and eliminated them from consideration in the rest of the simulation. However, this approximation can lead to results that differ greatly from those obtained by complete numerical integration using Gear routines,[21] and QSS approximations are not recommended. Consequently, most programs for atmospheric reaction processes[22] use the Gear algorithm.

Several important aqueous reactions involve consecutive kinetics, but compared with atmospheric processes, most aqueous sequences are relatively simple. Hydrolysis of condensed phosphates to orthophosphate is a consecutive process in which an orthophosphate ion is released from a condensed phosphate compound in each step (Section 4.4.3). This process is of interest relative to the eutrophication of surface waters. Chlorination of phenol involves consecutive and competing reactions that produce several intermediates and products that are of interest in drinking-water treatment because of their very low threshold odor concentrations (Section 4.5.2). Example 2-2 illustrates the use of *Acuchem* to simulate the sequential chlorination of phenol.[23] Chlorination of ammonia and amines is a consecutive process that is second order overall and first order in each reactant, and the kinetics of breakpoint chlorination, in which chloramines are oxidized to N_2 by chlorine, has been simulated by computer models (see Section 4.5.1). Mineralization and oxidation of organic nitrogen, an important microbial process in soils and aquatic systems, can be modeled as a three-reaction sequence. Finally, photodegradation of many organic contaminants involves complicated sequences and competing reactions (Chapter 8). In most cases, insufficient information is available on rates of individual steps and exact reaction pathways to allow kinetic simulations.

Example 2-2. Simulation of Second-Order Sequence: Chlorination of Phenol

Phenol, a common contaminant in wastewaters, reacts readily with chlorine to form a series of chlorinated phenols that have low threshold odor concentrations and are more toxic than the parent compound. The reactions are second order overall (first order in each reactant) and follow the sequence

$$P \begin{array}{c} \nearrow 2CP \rightarrow 24DCP \searrow \\ \searrow 4CP \rightarrow 26DCP \nearrow \end{array} 246TCP \rightarrow\rightarrow X \qquad (1)$$

where P = phenol, C = chloro-, D = di-, T = tri-, numbers indicate position of substitution (e.g., 24DCP = 2,4-dichlorophenol), and X = unspecified oxidation products. Rates vary with pH because both reactants are weak acids and the actual reactants are HOCl and the phenate anions. Rates of the reactions in Equation (1) were measured by Lee and Morris[23] over the pH range 6 to 10, and further details on the sequence are given in Section 4.5.3. Table 4-8 summarizes the rate constant data (M^{-1} min^{-1}) for rate equations expressed in terms of total concentrations of $[Cl]_t$ ($[HOCl] + [OCl^-]$) and $[P]_t$ ($[phenol] + [phenate]$).

It is easy to simulate the kinetics of the above sequence by *Acuchem*. No knowledge of programming is required, and aside from a few simple lines of input defining integration times and tolerances, the only information needed for the input file is a list of the reactions, rate constants, and initial concentrations of reactants, as shown in the following format:

;ID	Reaction	Rate Constant, M^{-1} min^{-1}, pH = 7.0
1,	P + C = 2CP	1.12E3
2,	P + C = 4CP	1.12E3
3,	2CP + C = 24DCP	1.58E3
4,	2CP + C = 26DCP	1.58E3
5,	4CP + C = 24DCP	8.93E2
6,	24DCP + C = 246TCP	1.76E3
7,	26DCP + C = 246TCP	4.95E3
8,	246TCP + C = X	5.16E2
end		

;Reactant	Initial Concentration, M
P,	1.0E-5
C,	1.0E-4
end	

Lines beginning with a semicolon are comments and are not read by the program. ID is an identification number for each reaction. The equal sign in each reaction is equivalent to an arrow ($\rightarrow$) in chemical reactions. All reactions are unidirectional, in spite of the equal sign; reverse reactions must be written explicitly and assigned a rate constant. Only zero-, first-, and second-order reactions are allowed, but higher-order reactions can be treated as a series of bimolecular steps. For example, the reaction A + B + C $\rightarrow$ P, can be written as A + B $\rightarrow$ D, k_1; D + C $\rightarrow$ P, k_2; where D is a dummy variable and k_1 is assigned a very large value.[21] Rate constants for steps 1 and 2 above are set equal to half the value for reaction of P (Table 4-8) under the assumption that products 2CP and 4CP have equal probabilities of formation. The same is true for steps 3 and 4. Figure 2-6 illustrates output for the above sequence, as plotted by a companion program *Acuplot*. The output from *Acuchem* is easily transferred to spreadsheets and other programs for statistical analysis and more refined graphical presentation.

2.3.3 Chain Reactions

This multistep reaction mechanism occurs when the reactants form intermediates that react with more reactant to yield product plus more intermediate; e.g.,

$$i \qquad A_2 \rightleftharpoons 2A$$

$$ii \qquad A + B_2 \rightleftharpoons P + B$$

$$iii \qquad B + A_2 \rightleftharpoons P + A \qquad\qquad (2\text{-}58)$$

Overall $\qquad\qquad A_2 + B_2 \rightarrow 2P$

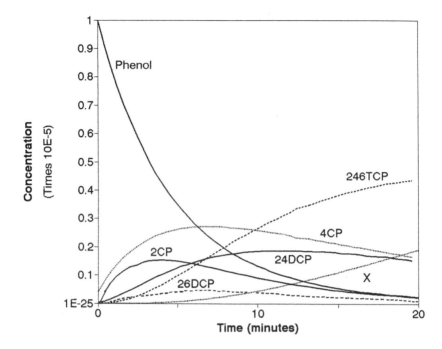

Figure 2-6. Simulation of chlorination of phenol at pH 7 by *Acuchem*.[20] Initial concentrations: phenol [P] = 1×10^{-5} *M* ; [HOCl] = 1×10^{-4} *M*. Product abbreviations explained in Example 2-2. Rate constants from Table 4-6.

All chain reactions include three stages: initiation, propagation, and termination. In sequence 2-58, the forward reaction of step 1 represents initiation, and its reverse represents termination. Steps 2 and 3 are chain propagation steps. Species that propagate the chain typically are atoms or free radicals, which are highly reactive because they have an unpaired electron.

Chain reactions are common in organic chemistry and oxidation processes. For example, hydrocarbons and carbon monoxide are oxidized by chain mechanisms. Photochemical smog formation and processes involved in maintaining ozone in the stratosphere include free-radical chain reactions. The oxidation of H_2S (sulfite) and FeS_2 (iron pyrite) by O_2 in aquatic systems are thought to be chain reactions. Some ozonation reactions involving organic compounds also occur by chain mechanisms with free-radical intermediates (Chapter 8).

The H_2-Br_2 Reaction

A classic example of a chain reaction is the hydrogen-bromine reaction, $H_2 + Br_2 \rightarrow 2HBr$, which follows the mechanism in Equation 2-58, with $A_2 = Br_2$, $B_2 = H_2$, and P = HBr. The main sequence and derivation of the rate equation is given in Example 2-3. The kinetics of this reaction were first studied by Bodenstein and Lind in 1906,[24] who fit their data to an empirical equation:

$$d[HBr]/dt = \frac{k[H_2][Br_2]^{1/2}}{1 + k'[HBr]/[Br_2]} \qquad (2-59)$$

Herzfeld, Christiansen, and Polanyi in 1919 independently explained Equation 2-59 in terms of a chain mechanism.[25] Although this reaction occurs in the gas phase and has no environmental interest, it is a convenient example because its mechanism is so well known. Many chain reactions in natural aquatic systems are so complicated that derivation of an overall rate equation is impractical (see Example 2-7 for an exception). As the first chain reaction analyzed mechanistically, the H_2-Br_2 reaction also has historical significance. The chain is initiated by thermal dissociation or photolysis of Br_2 to $2Br$. Chain-propagating steps form two HBr and regenerate a Br atom in each cycle. The reaction continues until the chain is terminated by recombination of two Br to Br_2. Side reactions complicate the sequence when water vapor and O_2 are present. The first propagation step is reversible, but the second is not. Both chain initiation and chain termination involve another molecule (any "third-body" species M) to supply energy for dissociation or absorb energy from recombination (see footnote, Section 2.1.2). M cancels out in deriving the rate equation and is not considered in the example. The H_2-Cl_2 reaction follows a similar but more complicated scheme with more side reactions.[2]

Example 2-3. Derivation of the Rate Equation for the H_2-Br_2 Chain Reaction

A. Mechanism (elementary reaction sequence):

(1)   Chain initiation (a)

$$Br_2 \xrightarrow{\;k_1\;} 2Br$$

Chain propagation (b)

$$Br + H_2 \underset{k_3}{\overset{k_2}{\rightleftharpoons}} HBr + H$$

(c)

$$H + Br_2 \xrightarrow{\;k_4\;} HBr + Br$$

(2)   Chain termination (d)

$$2Br \xrightarrow{\;k_5\;} Br_2$$

B. The rate of product formation for the mechanism is

$$d[HBr]/dt = k_2[H_2][Br] + k_4[Br_2][H] - k_3[HBr][H]$$

C. A steady-state assumption can be made (Section 2.5.1) to eliminate unmeasurable reactive intermediates, H and Br:

(3a) $d[Br]/dt = 0 = 2k_1[Br_2] + k_3[HBr][H] + k_4[Br_2][H]$

$$-k_2[H_2][Br] - 2k_5[Br]^2$$

(3b) $d[H]/dt = 0 = k_2[H_2][Br] - k_3[HBr][H] - k_4[Br_2][H]$

From Equation 3b we find

(4) $$[H] = \frac{k_2[H_2][Br]}{k_3[HBr] + k_4[Br_2]}$$

The addition of Equations 3a and 3b eliminates terms containing k_2, k_3, and k_4, and yields

(5) $$[Br] = \{k_1[Br_2]/k_5\}^{0.5} = K^{0.5}[Br_2]^{0.5}$$

K is the equilibrium constant for dissociation of Br_2. Substituting Equation 5 into 4 yields an equation for [H] with no intermediates on the right side:

(6) $$[H] = k_2 K^{0.5}[H_2][Br_2]^{0.5} / (k_3[HBr] + k_4[Br_2])$$

D. The final rate equation is obtained by substituting Equations 4 and 5 into Equation 2, rearranging, and simplifying:

(7) $$\frac{d[HBr]}{dt} = \frac{2k_2 K^{0.5}[H_2][Br_2]^{0.5}}{1 + (k_3[HBr]/k_4[Br_2])}$$

Equation 5 of the H_2-Br_2 derivation expresses the fact that at steady state the rates of the initiation and termination reactions are equal, which implies that $[Br]_{ss}$ is approximately equal to the equilibrium concentration for the dissociation-recombination reaction (Equation 1a). The rate of product formation (Equation 7) is identical to empirical Equation 2-59, where $k = 2k_2 K^{1/2}$ and $k' = k_3/k_4$. Note that the reaction is first order in H_2, but fractional order in Br_2. Chain reactions characteristically have complicated kinetic expressions, often with fractional orders. The presence of product HBr in the denominator of the rate equation also is notable. Constant $k' = k_3/k_4$ is called the inhibition constant, since it indicates the degree of inhibition of the reaction by HBr. Several other seemingly reasonable elementary reactions involving components of the H_2-Br_2 reaction can be written, but they can be eliminated based on low collision probabilities for the reactants or high energies of bond dissociation.[2]

Experimental Evidence for Chain Reactions

Evidence that a reaction has a chain mechanism sometimes can be obtained directly by detecting electronic spectra of free-radical intermediates, although not all reactions with free radicals have chain mechanisms. A number of other phenomena are characteristic of chain mechanisms. An induction period before product formation begins often occurs in chain reactions (see discussion of sulfide oxidation in Section 4.6.2). Chain termination occurs readily on the walls of reaction vessels (where radicals adsorb and recombine), and rates of chain reactions often decrease with increasing surface-area-to-volume ratio of the reaction vessel. Many substances inhibit chain reactions by removing chain-carrying radicals. Gas-phase inhibitors include nitric oxide (NO) and alkenes; Lewis bases (e.g., amines and alcohols) act as chain breakers in solution by forming less-active radicals or nonradicals. On the other hand, transition metals and substances that readily form free radicals accelerate chain reactions by initiating chains (usually the rate-limiting step; see Section 4.2.4). As mentioned previously, chain reactions usually have complicated rate equations with fractional orders for some reactants. Finally, the appearance of a product in the denominator of the rate equation is strong evidence for a chain mechanism.[26] Any given chain reaction may not exhibit all these characteristics, but the occurrence of several is good evidence for a chain mechanism.

2.4 DETERMINATION OF RATE EQUATIONS AND RATE PARAMETERS FROM EXPERIMENTAL DATA

Analysis and interpretation of kinetic data involves two basic questions: what equation best fits the rate data, and what are the best estimates of the rate coefficients for the equation that can be obtained from the data? This section describes methods to fit data to rate equations and estimate kinetic parameters from such data. Fitting data to a rate equation establishes the reaction order for each reactant and allows one to predict reaction rates for given experimental conditions. Knowledge of the rate equation is the first step in evaluating a reaction's mechanism, and thus this is a subject of considerable importance.

2.4.1 Fitting Data to Rate Equations

To derive the rate equation that fits a set of data, we must determine the order for each reactant and evaluate the rate constant(s). Two general approaches are used: **differential** and **integral** methods. The former use reaction rates directly in plots or calculations to determine the rate equation and rate constant; the latter use plots of data according to integrated rate equations.

By definition, the rate of a reaction is equal to the decrease in reactant concentration or increase in product concentration per unit time. Instantaneous rates can be determined as tangents to curves of reactant or product concentrations vs. time (Figure 2-7A). Plots of such instantaneous rates vs. various functions (first, second order) of reactant concentrations constitute a differential method of reaction rate analysis. The plot yielding a straight line indicates the order of the reaction for a given reactant. Graphical fitting of tangents to curves is not very accurate, and this is not recommended for careful work. Example 2-4 describes a more accurate differential method.

Example 2-4. Finite Difference Method to Fit n and k[27]

This method evaluates n and k for nth order reactions that fit the equation

(1) $$-d[A]/dt = k[A]^n$$

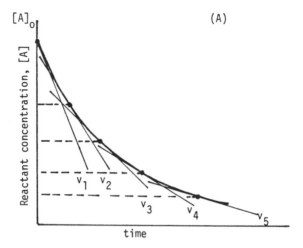

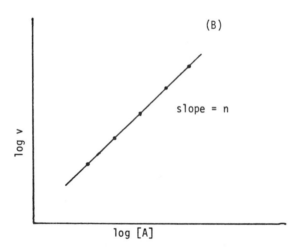

Figure 2-7. Differential method for determining reaction order of rate equation: (A) on plot of reactant concentration remaining vs. time, draw tangents to curve at various points and determine slopes, v_i (instantaneous reaction rates), and $[A]_i$, the corresponding reactant concentration; (B) plot $log v_i$ vs. $log[A]_i$. Slope of line $= n$, the reaction order. (Adapted from Hill, C. G., Jr., *Chemical Engineering Kinetics and Reactor Design*, John Wiley & Sons, New York, 1977.)

Reactions having several reactants can be fit to Equation 1 by fixing the concentrations of all but one reactant (A) at much higher values than the variable reactant. Concentrations of the fixed reactants then become imbedded in the computed value of k (see Equation 2-60 below).

For n ≠ 1, (2)

$$1/[A]^{n-1} - 1/[A]_o^{n-1} = (n-1)kt$$

or (3)

$$[A]^{n-1} = \{(n-1)kt + 1/[A]_o^{n-1}\}^{-1}$$

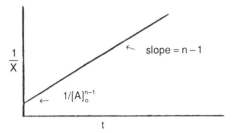

Figure 2-8. Plot of 1/X (from Equation 6, Example 2-4) vs. t to obtain reaction order n (from slope) and rate constant k (from y-intercept).

or (4)
$$[A]^n = [A]\{(n-1)kt + 1/[A]_o^{n-1}\}^{-1}$$

From Equations (1) and (4):

(5)
$$\frac{-d[A]/dt}{[A]} = X = k\left[(n-1)kt + \frac{1}{[A]_o^{n-1}}\right]^{-1}$$

or

(6)
$$1/X = (n-1)t + 1/k[A]_o^{n-1}$$

X can be obtained by finite differences as $X \approx (1/[A])(\Delta[A]/\Delta t)$. For three successive values $[A]_{n-1}$, $[A]_n$, $[A]_{n+1}$, at times t_{n-1}, t_n, t_{n+1}, we define

(7)
$$\Delta_1 = [A]_{n-1} - [A]_n$$

(8)
$$\Delta_2 = [A]_n - [A]_{n+1}$$

(9)
$$X_n = \left[\frac{\Delta_1 + \Delta_2}{t_{n+1} - t_{n-1}}\right]$$

For accurate results, $[A]_i$ should be determined at close, uniform intervals.

Kinetic parameters commonly are evaluated from time course data for a reaction and plots of observed concentrations vs. time according to the integrated rate expressions given in Section 2.3. If the data fit a straight line on such a plot, the order of the reaction is considered to be determined, and the rate constant is estimated from the slope of the plot. However, data often do not fit any rate expression exactly. In some cases data may fit several expressions (e.g., first and second order) equally well (or poorly), and it may be difficult to select the correct order. A lack of fit has many causes: (1) random noise in analytical measurements; (2) solution inhomogeneities and sampling errors; (3) contaminants that cause competing reactions; (4) branching pathways in complicated reactions; (5) pH variations caused by H^+ consumption or

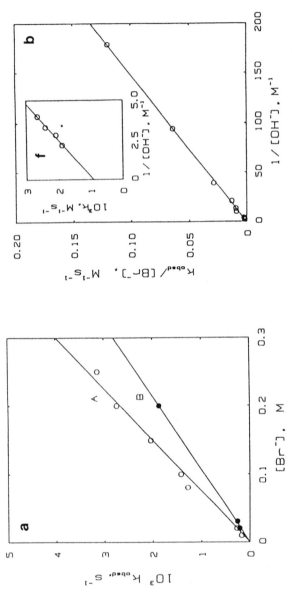

Figure 2-9. The rate equation for the oxidation of bromide by hypochlorite was determined by following the loss of OCl⁻ from its absorbance at 292 nm at large excess of Br⁻ ([OCl⁻]$_o$ = 6 × 10⁻⁴ M; [Br⁻] = 0.10–0.25 M) and several fixed values of pH. These conditions gave good first-order plots, indicating the reaction is first order in OCl⁻ and yielding pseudo-first-order rate constants, k_{obs}. (a) k_{obs} (25°C, I = 0.5) is a linearly related to [Br⁻], indicating the rate equation is first order in Br⁻, and k_{obs} normalized to [Br⁻] is inversely related to [OH⁻]. The small intercept at high [OH⁻] indicates the reaction has a small hydroxide-independent term and gives the rate constant, k_o, for this term. The slope of the line yields the rate constant, k_1, for an [OH⁻]-dependent term. These results lead to the rate equation: $-d[OCl⁻]/dt = k_o + (k_1/[OH⁻])[Br⁻][OCl⁻]$, where k_o = 0.9 × 10⁻³ $M⁻¹$ s⁻¹ and k_1 = 6.6 × 10⁻⁴ s⁻¹. [From Kumar, K. and D. W. Margerum, *Inorg. Chem.*, 26, 2706 (1987). With permission.]

production in poorly buffered solutions; (6) product inhibition; and (7) back reaction if the reaction is reversible.

Consequently, it often is desirable to measure initial velocities, i.e., rates during a period when only a small fraction of the reactants are consumed. By design, product concentrations can be kept low, and opportunities for side reactions are reduced. The measurement of initial rates usually entails analysis of small amounts of product in the presence of large amounts of reactants. Determination of reaction order by this method is done by fixing the concentrations of all but one reactant and measuring the reaction rate in a series of vessels, each with a different concentration of the variable reactant. The process is repeated for each reactant. To avoid changes in concentrations of the fixed reactants during reaction, they are added at much higher values than is the variable reactant (e.g., see Figure 2-9). The rate then simplifies (usually) to pseudo-first order. If the reaction is allowed to occur for only a short time, concentrations of the variable reactant do not change much, and the initial rate can be plotted vs. the initial concentration of the variable reactant. Initial rates can be measured as the slope of a recorder trace of reactant concentration. If the time interval for the reaction is small, $d[A]/dt$ can be approximated by $\Delta[A]/\Delta t$, where $\Delta[A] = [A]_o - [A]_t$.

To determine the order for a reactant, initial-rate data are plotted on a *log-log* graph (Figure 2-7b). The order is equal to the slope of the fitted line, and the y-intercept is the product of the intrinsic rate constant and concentrations of the fixed reactants:

$$\log(rate) = a \log[A] + \log(k[C_{fixed}]) \tag{2-60}$$

This equation was obtained by taking logs of both sides of the general rate equation (2-8). If the reactants are supplied in the exact ratio of their appearance in the stoichiometric reaction, the slope of a *log-log* plot of rate vs. concentration of any reactant is equal to the overall reaction order.[4]

The above approaches are adequate to derive rate expressions for reactions that fit equations of the type: rate $=k[A]^a[B]^b[C]^c\cdots$. If a variable order is found for some reactant, the rate equation includes denominator terms and involves a multistep mechanism. In such cases one can fit observed data to an empirical equation, but one may not be able to determine microscopic (elementary) rate constants from the overall reaction rate. For example, if a reaction is first order in A at low [A], but zero order in A at high [A], we can fit the data to an equation of the type

$$rate = \frac{k[A]\cdots}{1+k'[A]} \tag{2-61}$$

Values of k and k' can be estimated by nonlinear regression (Section 2.4.3). The form of the rate equation indicates that the reaction has at least two steps, one of which probably is a rapid-equilibrium step, and the overall reaction involves at least three elementary rate constants (cf. reaction sequence 2-82 and corresponding rate equations in Section 2.5.1). The constants in Equation 2-61 do not correspond to elementary rate constants; these must be determined by studying the elementary steps separately. For another example, compare empirical Equation 2-59 for the H_2-Br_2 reaction with Equation 7 derived from the mechanism in Example 2-3.

2.4.2 Parameter Estimation for Elementary Rate Equations

The methods described in the preceding section are not equally reliable in providing estimates of kinetic parameters. To obtain precise and accurate parameter estimates, one must pay careful attention not only to gathering data, but also to experimental design and statistical analysis of data. This section describes treatment of data for optimal estimation of kinetic parameters, particularly rate coefficients when the form of the rate equation is known. Estimation of reaction order, a small integral number, requires much less attention to statistics. We focus on integral methods because they are used more widely by chemists, but the principles described here apply generally to parameter estimation methods. The statistics of parameter estimation for the Michaelis-Menten equation, a differential expression for enzyme kinetics, are described in Section 6.2.1, and an entire book has been written on this subject.[28]

Determinate or nonrandom error introduces bias into computed kinetic parameters, but does not introduce uncertainty per se. We will assume that bias is not a problem or can be accounted for and will focus on random error, which is quantified by the standard deviation, σ, or variance, σ^2. Concentration data may exhibit three kinds of random error:

Error type	Characteristic of σ
E equal (constant)	σ^2 constant (independent of [A])
C counting	σ^2 proportional to [A]
P proportional	σ^2 proportional to $[A]^2$ (σ proportional to [A])

In some cases more than one type of error may be present. Type-E errors are independent of the size of measurements; uncertainty in detecting a titration end point introduces a constant error. Type-C errors are associated with measurements of radiotracer activity. Type-P errors increase as measurement size increases; for example, uncertainty in titrant strength introduces a random error proportional to the amount of titrant consumed (hence, proportional to concentration). The coefficient of variation or relative error, σ/X, is constant when we have proportional error (X is the size of a measurement).

The simplest way to estimate rate constant values is to use rate expressions integrated over discrete time intervals (hereafter called discrete integrated equations, DIEs). Equations 2-11, 2-18, and 2-22 are examples of DIEs. They require data for only two time periods: an initial concentration and a measurement of reactant remaining, $[A]_t$, after some reaction time. These equations yield imprecise estimates of k when based on unreplicated measurements, and they are more appropriate for estimating $[A]_t$ at various reaction times, once k is estimated by other methods. The reliability of k's estimated from DIEs can be improved by replication, i.e., by setting up identical reaction flasks and measuring $[A]_t$ after a given time; measurement of $[A]_o$ also must be replicated. Mean values of $[A]_o$ and $[A]_t$ yield better estimates of k than do single pairs. If measurement error is normally distributed, the reliability of mean values increases with $n^{1/2}$, where n = number of replicates; in theory one can obtain highly accurate estimates of k by using large numbers of replicates. Quantitative measures of precision for k's computed from DIEs are given in Section 2.4.4. Replication does not compensate for one major deficiency of DIEs, however: they do not allow one to determine whether a reaction is deviating from a given rate form over time because of side reactions or build-up of inhibitors. In this respect, graphical and initial velocity methods are preferable.

Because rate data are temporally nonlinear (except for zero-order reactions), data on $[A]_t$ must be transformed to obtain linear plots. The advantage of such plots is that they allow use of simple statistics (linear regression) to obtain the line of best fit and kinetic parameters. For example, k is obtained from the slope and $[A]_o$ from the y-intercept of *log*-transformed concentration data for first-order reactions: $log[A]_t = log[A]_o - kt$. For statistically optimal results, the "error structure" of the data needs to be considered before data are transformed because transforming data may cause relatively more weight to be placed on unreliable data in computing a regression than the data merit.

If the data for a first-order reaction exhibit purely proportional (P) error, *log* transformation to Equation 2-11 yields the appropriate weighting,[29] and this yields the optimal statistical fit for k by linear regression (LR). In contrast, if data exhibit only constant (E) error, proper weighting in regression is achieved by fitting the *untransformed* data to an equation. For first-order reactions the appropriate equation is $[A]_t = [A]_o exp(-kt)$. This is nonlinear in k, however, and requires a fit by a nonlinear regression (NLR) model. Note: an equation may be nonlinear in one parameter and linear in another. The first-order exponential form is linear in $[A]_o$, but nonlinear in k. In general, function F is nonlinear in x_i if $\partial F/\partial x_i$ contains x_i.

If data with type-E error are analyzed by *log* transformation and LR, they are no longer weighted according to their reliability, and the regression produces less-precise parameter estimates; that is, the variances of the estimated parameters are not minimized, and the procedure is not statistically efficient. This statement also applies to data with type-P error that is analyzed by NLR. In both cases the lost efficiency can be regained by weighting the transformed data according to their uncertainty; this is done by multiplying each datum by the reciprocal of its variance, $1/\sigma^2$. However, the error structure of data may not be known in advance, and unless an experiment is properly designed, one may not be able to extract this information after the fact.

The appropriateness of linear vs. nonlinear models reflects a fundamental requirement of regression analysis:[30] the dependent variable Y must exhibit equal variance over the range of independent variable X; this is called homoskedasticity and represents the situation for constant error. However, if $[A]_t$ exhibits type-P error, we have $\sigma(y) \equiv dy = cy$, where c is the constant coefficient of variation; this is called heteroskedasticity. Type-P error becomes constant error (homoskedastic) in *log*-transformed space: $d\ln y = (dy)/y = cy/y = c$.[31] Other assumptions of regression analysis are[30]

1. **Normality** — measured values of Y at a given X are distributed normally with mean $\mu_{Y/X}$ and variance σ^2
2. **Independence** — a given observation of Y is not influenced by other observations
3. **Linearity** — mean values $\mu_{Y/X}$ fall on a straight line (i.e., the linear model is valid)

Because parameter reliability depends on the validity of these assumptions, it is important to evaluate data relative to these assumptions before applying statistical procedures.

Of the main assumptions in regression analysis, the ones most likely to be violated are equality of variance and linearity. These assumptions can be checked by examining the residuals of the regression, i.e., the differences between observed values and those predicted by the regression:

$$e_i = \left| Y_i - \hat{Y}_i \right| \tag{2-62}$$

Y_i is a measured value at a given X, and $\hat{Y}_i$ is the predicted value for a given X. Linearity can be examined by plotting e_i vs. $\hat{Y}_i$. If the assumption of linearity holds, there should be no relationship between residual values and predicted values. A simple scatter plot of Y vs. X (or $log[A]_t$ vs. t for first-order data) also shows whether the data fit a straight line.

Several methods are useful in deciding whether error is proportional or constant.[32] In the residuals plot approach, one plots the absolute value of the residuals vs. $\hat{Y}_i$. If the error is constant, the plotted values will produce a horizontal band (zero slope), but if the error is proportional, the residuals will increase with increasing $\hat{Y}_i$ (nonzero slope) (Figure 2-10). The statistical significance of the slope can be determined by a t test using the estimated standard error of the slope.[30] The F_k method computes the ratio

$$F_k = \frac{e_n^2 + e_{n-1}^2 + \cdots + e_{n-k}^2}{e_1^2 + e_2^2 + \cdots + e_k^2} \tag{2-63}$$

where the measured or predicted values of Y (e.g., $[A]_t$) are ranked in increasing order (n is largest), and e_i is the associated residual. If error is constant, $F_k \approx 1$, and if error is proportional, $F_k > 1$.[32] F_k follows the F distribution with k,k degrees of freedom. The optimum value of k is arbitrary and depends on the magnitude of n; if $n \approx 10$, $k = 3$ is reasonable. Finally, repeated measurements of [A] at two values about a decade apart to yield two estimates of σ provide a simple way to determine error structure. If the two values of σ are about the same, we are dealing with type-E error; if the values are about a decade apart, the error is proportional.[33]

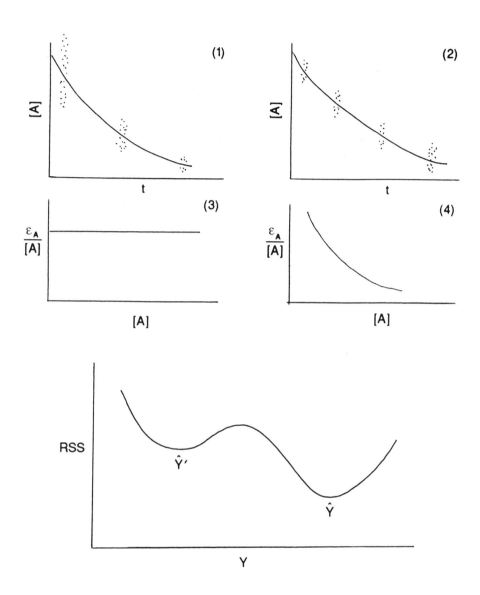

Figure 2-10. (a) Sketches showing effects of (1) proportional and (2) constant error on hypothetical reactant vs. time curves; (3) relative error is constant fraction of [A] for proportional error, but (4) constant error leads to increasing relative error as [A] decreases. (b) Locally optimal solution ($\hat{Y}'$) and true optimum ($\hat{Y}$) in regression analysis; RSS = root mean sum of squares.

Table 2-3.   **Efficiencies (%) of Unweighted Least-Squares Analyses of Linearized Data for y = Aexp(–kt)**

Range of y (decades)	C (counting)		E (equal or constant)	
	k	A	k	A
0.5	92.0	94.0	70.0	71.0
1.0	68.0	69.0	29.0	21.0
1.5	47.0	42.0	8.8	3.2
2.0	28.0	20.0	2.1	0.4
3.0	9.0	3.0	0.07	0.005
4.0	2.0	0.4	0.002	0.00005

Data from Kalantar, A. H., *Int. J. Chem. Kinetics,* 19, 923 (1987).

Note: Efficiency was calculated from the variance of extracted parameters using proper weights divided by the variance obtained without weights. Data were spaced at 12 equal intervals of time beginning at t = 0 that yielded the ranges of y listed in column 1. For proportional (P) errors, efficiencies averaged $100 \pm 1.9\%$ for k and $100 \pm 1.1\%$ for A.

How important is correct weighting for accurate estimation of kinetic parameters? No simple answer can be given, but in general, the importance of correct weighting increases as the reliability (precision) of the data decreases and as the range of values for $[A]_t$ expands. These issues have been addressed in several recent studies.[31,33,34] If the data are very precise, lack of weighting has little effect on the reliability of estimated parameters, and both fitting approaches (linear fit to *log* transform, and direct nonlinear fit to exponential form) give unbiased estimates. If the range of measured values for [A] is small, all values are about equally reliable (whether we are dealing with E or P error). However, at long reaction times, $[A]_t$ is much smaller than $[A]_o$, and if we are dealing with E errors, the later data will be much less reliable than the early data. Table 2-3 shows the dramatic effect of increasing data range on the efficiency of parameter estimation by unweighted linear regression of *log*-transformed first-order data with type-E and -C errors. When the data span one order of magnitude, the efficiency is less than 30% of weighted LR (or unweighted NLR), both of which are assumed to give the best possible estimate. This is equivalent to ignoring 70% of the data and treating the remainder with proper relative weighting. The situation is even more discouraging for larger data ranges. On the other hand, when the data span only half an order of magnitude (typical of some kinetic experiments), the efficiency of unweighted LR is 70% of weighted LR (or NLR), and weighting is not so critical.

Rathbun and Tai[34] compared the two fitting approaches for kinetic data on volatilization of two chlorinated ethanes and absorption of O_2 by water. The former data fit simple first-order equations (2-10, 11); the latter fit a reversible first-order equation similar to Equation 2-43a and its *log* transform. Ratios of coefficients for 77 different experiments fitted by the two methods ranged from 0.955 to 1.05, with a mean of 1.01. This indicates that essentially no bias exists in the two methods and may reflect high precision in the measurements. Root-mean-square (RMS) errors of prediction were smaller for NLR fits than for LR fits of *log*-transformed data, and based on this, the authors suggested that the gas exchange data exhibited E, rather than P, error.

However, simulations with synthetic data and computer-generated random E and P errors[31] indicated that the situation is more complicated than suggested above. According to this analysis, RMS error of prediction cannot be used to decide the error structure of kinetic data because unweighted NLR always produces smaller RMS values, whether the errors are type E or P. Another statistical parameter, the fit-predicted standard deviation of k, $\sigma(k)$, did yield the correct trends: lower $\sigma(k)$ with NLR for E-type error and lower $\sigma(k)$ for LR on *log*-transformed data for type-P error. $\sigma(k)$ could serve as a criterion for deciding on the error structure of kinetic data.

Optimal statistical treatment of kinetic data is more complicated when the error structure is mixed or cannot be determined (e.g., on data from the literature). In general, these circumstances favor the use of robust statistical methods (techniques that do not depend on the distributional characteristics of the data). Robust methods for the Michaelis-Menten equation are described in Section 6.2.1. The use of DIEs with replication to improve estimates of $[A]_0$ and $[A]_t$ is a robust approach. If one is confident that the measured values of $[A]_t$ are not affected by competing reactions, build-up of interferences, and other problems that cause deviation from the assumed kinetic form, this approach can yield reliable results. Potential problems can be evaluated by plotting data from a preliminary experiment and selecting a time interval in the linear portion of the graph for a carefully replicated set of measurements.

2.4.3 Linear and Nonlinear Regression Methods

Conventional regression analysis is a "least squares" technique; i.e., it minimizes the sum of squares (SS) of differences between measured $[A]_t$ values and values predicted for the same t from the line of best fit. The linear regression model has an exact analytical solution[30] and is straightforward to calculate. Nonlinear regression is not so well behaved. There are no analytical solutions, and minimization of the SS must be done iteratively. Many algorithms and minimization criteria have been developed for this purpose, but until recently the methods were unreliable and consumed large amounts of computer time. Efficient algorithms to fit data to nonlinear equations are available.[35] These programs use Marquardt's rapidly converging gradient-expansion algorithm,[36] which is generally considered to be the fastest and most reliable technique. Nonlinear regression methods are sensitive to starting estimates; the closer these are to the true values, the faster convergence is achieved, and the more likely the calculations will converge to the true parameter values. If starting estimates are poor, the regression may converge to a local optimum (Figure 2-10) rather than to the true values. These problems are not so severe when there is only one fittable parameter, such as a first-order rate constant, but they become more serious as the number of fittable parameters increases.

Although one may obtain close fit of data to an equation with many parameters, the fitted coefficients may lack physical meaning. Ultimately, such analyses become curve-fitting exercises. This may be satisfactory for predictive purposes, but is not satisfactory for making mechanistic inferences.

2.4.4 Effects of Analytical and Physical Uncertainties on Parameter Estimates

Rates of reactions depend on physical conditions as well as on reactant concentrations, and these conditions must be controlled if valid data are to be obtained. The most important physical factors for aqueous reactions are temperature, ionic strength, and pH; light intensity is a critical variable in photochemical reactions. As discussed in Chapter 3, reaction rates increase with temperature, and increases of two times (or more) over a 10°C range are common. If the rate doubles for a 10° rise, a variation of ±1°C will cause the rate to vary by ±7% and introduce an uncertainty of this magnitude into the estimated rate constant. Thus, for accurate work careful control and measurement of temperature is essential. Kinetic experiments commonly are done in thermostatted systems where the temperature can be controlled to within ±0.1 or even ±0.01°C.

Temperature affects equilibria as well as directly affecting reaction rates. Recognition of this fact can be important if a reactant's concentration is calculated from an equilibrium relationship rather than measured; equilibrium constants applicable to the temperature of interest must be used. For example, K_w (the ion product of water) is 0.185×10^{-14} at 5°C, 1.00×10^{-14} at 25°C, and 1.469×10^{-14} at 30°C. Since $K_w = \{H^+\}\{OH^-\}$, where $\{i\}$ denotes activity of i, it follows that at constant pH of 7.00, $\{OH^-\}$ is 1.00×10^{-7} at 25°C, but only 1.85×10^{-8} at 5°C, and 1.47×10^{-7} at 30°C. Similarly, solubilities of gases decrease nonlinearly as temperature increases. When temperature effects on equilibria (and thus on actual reactant concentrations) were taken into account, several researchers[37] found that the rate constant for oxidation of Fe^{2+} by O_2 was nearly constant over the temperature range 5 to 30°C (see Section 4.6.5).

The effects of pH are important in reactions catalyzed by acids or bases (Chapter 4) and in reactions where H^+ or OH^- are reactants or products. If a reaction is catalyzed by H^+ and the rate is first order in $\{H^+\}$, an uncertainty in pH of ±0.1 will cause an uncertainty of about ±25% in the rate constant. An uncertainty of 1 pH unit leads to a tenfold uncertainty in k. Careful measurement and control of pH is very important in the kinetics of oxidation of Fe^{II} and Mn^{II}, which are second order in $\{OH^-\}$ (Section 4.6). In these cases a pH uncertainty of 0.1 units leads to an uncertainty in k of $(10^{\pm0.1})^2$, or -37 to $+58\%$. Because these reactions consume OH^-, it is especially important to buffer solutions to prevent pH change during reaction. As noted by Davison and Seed,[38] an uncertainty of ±0.2 pH units could account for the approximately sixfold range in reported rate constants for Fe^{II} oxidation: 1 to 6×10^{13} M^{-2} atm^{-1} min^{-1} in the pH range 6.5 to 7.4 at 25°C. They proposed that the true value is close to 2×10^{13}.

For nonionic reactions, small shifts in ionic strength (I) generally have negligible effects on reaction rates. For ionic reactions, $\log k$ varies in proportion to $I^{1/2}$, as a result of variations in ion activity with I (see Section 3.7). Indirect effects of increasing I are caused by changes in equilibrium properties of ionic species. In addition to the "nonspecific" effects of I on reaction rates, large effects occur when the added electrolyte interacts specifically with ionic reactants. For example, large changes in reaction rates of metal ions can be caused by adding ions that form complexes having different reactivities than the uncomplexed metal ions have. Reactions between ions and solid surfaces also are affected by I. The electric double layer surrounding solid particles is compressed as I increases, and this affects transport of ions to and from the surface, allowing like-charged ions and particles to approach more closely before being influenced by electrostatic repulsion.

Random errors in all measurements of reactant concentrations, as well as uncertainties in the time at which measurements are made, must be considered jointly to properly evaluate the reliability of computed rate constants. The expected error in a rate constant can be computed by standard procedures of propagated error analysis.[39] In general, for any dependent variable $y = f(x_i)$,

$$\sigma_y^2 = \sum_i \left\{ \partial f(x) / \partial x_i \right\}^2 \sigma_{xi}^2 \qquad (2\text{-}64)$$

The variance in y thus is the vector sum of the variances of each component variable in y times a weighting factor that expresses the sensitivity of y to changes in x_i. The weighting factor is the square of the partial derivative of the function (y) with respect to x_i. Equation 2-64 arises from a Taylor expansion of the variance truncated above first-order terms to eliminate consideration of covariances, skewness, and kurtosis. We can generalize Equation 2-64 by substituting Δy and Δx_i as more general measures of the uncertainty in y and x_i in place of the statistical measure σ. For example, Δx_i may be the measurement error associated with a given experimental variable. In addition, it is convenient to consider relative errors, $\Delta y/y$ and $\Delta x_i/x_i$, rather than absolute errors in each variable. This is done by dividing each term in Equation 2-64 by y^2 and then multiplying and dividing each term by the square of the independent variable in that term:[1]

$$\left(\frac{\Delta y}{y} \right)^2 = \sum \left(\frac{x_i}{y} \frac{\partial \Delta y}{\partial x_i} \right)^2 \left(\frac{\Delta x_i}{x_i} \right)^2 \qquad (2\text{-}65)$$

This is equivalent to a logarithmic partial derivative expression:

$$\left(\frac{\Delta y}{y} \right)^2 = \sum \left(\frac{\partial \ln f(x)}{\partial \ln(x_i)} \right)^2 \left(\frac{\Delta x_i}{x_i} \right)^2 \qquad (2\text{-}66)$$

Table 2-4. Equations to Estimate Relative Error in Some Kinetic Parameters

For an nth-order reaction, $-d[A]/dt = k[A]^n$, integration from t_1 to t_2 (for $n \neq 1$) yields:

$$\frac{[1]}{\left[[A]_2^{n-1}\right]} - \frac{[1]}{\left[[A]_1^{n-1}\right]} = -k(1-n)(t_2 - t_1) \tag{1}$$

or

$$k = \frac{[A]_1^{n-1} - [A_2^{n-1}]}{(n-1)(t_2 - t_1)[A]_1^{n-1}[A]_2^{n-1}} \tag{2}$$

(If $t_1 = 0$, $[A]_1 = [A]_o$). If the random errors in t_1, t_2, $[A]_1$, and $[A]_2$ are independent and n is known, the relative error or uncertainty $\Delta k/k$ for an nth-order reaction can be obtained from Equation 2-66 in the text and Equation 2 (above):

$$\left(\frac{\Delta k}{k}\right)^2 = \left(\frac{\partial \ln k}{\partial \ln t_1}\right)^2\left(\frac{\partial t_1^2}{t_1}\right) + \left(\frac{\partial \ln k}{\partial \ln t_2}\right)^2\left(\frac{\partial t_2}{t_2}\right)^2 + \left(\frac{\partial \ln k}{\partial \ln[A]_1}\right)^2\left(\frac{\partial[A]_1}{[A]_1}\right)^2$$

$$+ \left(\frac{\partial \ln k}{\partial \ln[A]_2}\right)^2\left(\frac{\partial[A]_2}{[A]_2}\right)^2 \tag{3}$$

i.e.,

$$\left(\frac{\partial k}{k}\right)^2 = \left(\frac{\partial t_1}{t_2 - t_1}\right)^2 + \left(\frac{\partial t_2}{t_2 - t_1}\right)^2 + \left(\frac{(n-1)[A]_2^{n-1}}{[A]_1^{n-1} - [A_2^{n-1}]}\right)^2\left(\frac{\partial[A]_1}{[A]_1}\right)^2$$

$$+ \left(\frac{(1-n)[A]_1^{n-1}}{[A]_1^{n-1} - [A_2^{n-1}]}\right)^2\left(\frac{\partial[A]_2}{[A]_2}\right)^2 \tag{4}$$

The relative error of first-order reactions can be obtained by solving Equation 2-11 for k and applying Equation 2-66:

$$k = \frac{1}{[t_2 - t_1]}\ln\frac{[A]_1}{[A]_2} \tag{5}$$

$$\left(\frac{\partial k}{k}\right)^2 = \left(\frac{\partial t_1}{t_2 - t_1}\right)^2 + \left(\frac{\partial t_2}{t_2 - t_1}\right)^2 + \left(\ln\frac{[A]_2}{[A]_1}\right)^2\left(\frac{\partial[A]_1}{[A]_1}\right)^2 + \left(\ln\frac{[A]_2}{[A]_1}\right)^2\left(\frac{\partial[A_2]}{[A_2]}\right)^2 \tag{6}$$

Material derived from Benson, S. W., *Foundations of Chemical Kinetics*, McGraw-Hill, New York, 1960.

Equation 2-66 can be used to estimate the precision of common kinetic parameters like k's and activation energies. In such applications of Equation 2-66, k's are estimated from discrete time-integrated expressions (based on reactant or product concentrations at two measurement times), and activation energy is estimated from the discrete form of the Arrhenius equation. Table 2-4 presents equations to estimate relative error in rate constants for reactions of various orders; relative errors in E_{act} estimated from rate constants measured at two temperatures are discussed in Section 3.2.2.

If measurement times are known to much greater precision than concentrations, terms in the equations of Table 2-4 that involve Δt can be neglected. It is interesting to note that for a given set of measurement conditions, the relative error in k is independent of reaction order for n = 0–4 and depends only on the extent of reaction ξ over the two measurement times and on analytical precision.[1] Consequently, large uncertainties in k occur when reactant concentrations are small, unless analytical precision is very high. For example, to determine k to within ±10%, one must attain an analytical precision of ±2.0% for 30% conversion of reactants to products, ±0.7% for 10% conversion, and better than ±0.1% at 1% conversion. This reflects the difficulty in measuring small differences in two large values. The accuracy of k can be improved by measuring product concentrations during the early stage of reaction.

It is more common to estimate rate constants graphically from a set of measurements of [A] over time rather than from measurements at only two times. The precision of k's estimated from the slopes of appropriate plots is determined by regression analysis as the standard error of estimate for the slope of the line. For such estimates to be statistically reliable, the assumptions made in regression analysis must be valid. As described earlier, the assumption that σ^2 for [A] is constant probably causes the most problems in this regard.

2.4.5 Further Complications in Parameter Estimation For First- and Second-Order Reactions

Two types of complications can arise in estimating rate parameters for first-order or pseudo-first-order reactions of environmental interest. First, progress in some reactions can be measured only in terms of the amount of product formed. The amount of reactant consumed is readily calculated, but $[A]_t$, the concentration remaining, cannot be calculated if $[A]_o$ is unknown, which sometimes is the case. The second complication arises when reactions of mixtures of related reactants are monitored by nonspecific analytical methods that yield a common response for a series of related compounds, or when the mixture yields a common product that is monitored to follow reaction progress.

Reactions in Which $[A]_o$ and $[A]_t$ Cannot Be Measured

(A) **First-Order Cases**. Consider a first-order reaction with $[P]_o = 0$ and $v_A = -v_P$. Assume the reaction is irreversible and proceeds to completion, $[A]_o$ is unknown, and $[A]_t$ cannot be measured directly. Reaction progress is evaluated by measuring $[P]_t$. For the above conditions it is clear that $[P]_t = [A]_o - [A]_t = x$, where x is the concentration of A that has reacted at time t. Moreover, $[P]_\infty = [A]_o$, because the reaction proceeds to completion at long incubation times, and $[A]_t \to 0$ as $t \to \infty$. Consequently, if we let the reaction proceed to completion and measure [P], we have determined $[A]_o$, and from each measured $[P]_t$ value we can compute $[A]_t$. In this case, we can apply the calculated values of $[A]_o$ and $[A]_t$ to the conventional first-order equation (Equation 2-10) and determine the rate constant.

Sometimes it is not possible to measure $[P]_\infty$; thus, $[A]_t$ cannot be calculated, and Equation 2-10 cannot be used directly. From the above relationships we see that $[A]_t = [A]_o - x$. Substituting for $[A]_t$ in Equation 2-10 and rearranging, we obtain

$$\{[A]_o - x\} / [A]_o = \exp(-k_1 t),$$

or $$x = [A]_o \left\{1 - \exp\left(-k_1 t\right)\right\} \tag{2-67}$$

The dilemma is clearly illustrated by Equation 2-67; we have one equation and two unknowns, $[A]_o$ and k_1, for any series of x and t values. Furthermore, plots of x vs. t are curvilinear (Figure 2-1a), making it difficult to extrapolate $[A]_o$ accurately from the asymptote. The best way to obtain $[A]_o$ and k_1 is to fit the x(t) data to Equation 2-67 by nonlinear least squares regression.[35]

Example 2-5. Sedimentary Production of Perylene from an Unknown Precursor

The problem of an unmeasured (and unknown) reactant applies to the production of perylene, a polynuclear aromatic hydrocarbon, in recent marine sediments. Gschwend et al.[40] found perylene levels increased with depth in a core from Buzzards Bay, MA, and attributed the increase to formation from unknown precursors under anoxic conditions. It is unlikely that deposition rates to the sediment were higher in the past than at present. The core was dated by ^{210}Pb methods so that a profile of perylene concentration vs. depth could be converted to concentration vs. time since burial, i.e., approximately the time of reaction (Figure 2-11). Because the precursor was unknown and thus not measurable, neither the initial reactant concentration at time of burial nor the reactant concentration at later times (greater depths) was known, and the data could not be fit directly to Equation 2-10. The authors used a nonlinear least-squares procedure and found a good fit to Equation 2-67. An equally good fit was obtained, however, to a second-order model (Equation 2-22).

Equation 2-67 and the problem of unmeasurable $[A]_o$ and $[A]_t$ apply to the kinetics of BOD exertion (BOD = biochemical oxygen demand[9]). The standard BOD test is designed so that the rate of O_2 uptake depends (first order) on the concentration of biodegradable organic matter L, but L is not directly measurable. In effect, the product of biodegradation is the consumption of O_2. In the context of Equation 2-67, the amount of O_2 consumed at any time, x(t), measures the extent of reaction: $x(t) = L_o - L_t$, where L_o, the ultimate first-stage BOD, represents the initial concentration of biodegradable organic matter, and L_t is the concentration remaining at time t, expressed in O_2 equivalents. If the reaction proceeds to completion, the

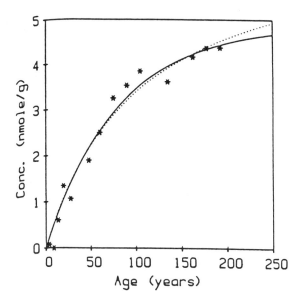

Figure 2-11. Concentration of perylene vs. age of deposition in a core from Mountain Pond, ME and curves of fit to a first-order model (solid line) and second-order model (dotted line). [From Gschwend, P. M., P. H. Chen, and R. H. Hites, *Geochim. Cosmochim. Acta*, 47, 2115 (1983). With permission.]

amount of O_2 depletion (x_∞), in theory, is equivalent to L_0, but for several reasons it is not practical to determine L_0 this way. First, BOD exertion is a slow process, and 20 to 30 days may be needed for complete oxidation. Second, because the reaction slows down as biodegradable organic matter becomes depleted, it is difficult to determine when the reaction is complete, and the uncertainties are compounded by the inherent variability of biological processes. Finally, a nitrogenous BOD caused by oxidation of ammonium to nitrite and nitrate (nitrification) often begins before the carbonaceous BOD is completely exerted. Inhibitors can be used to prevent nitrification,[9] but the other problems preclude direct determination of L_0.

Prior to the availability of nonlinear least-squares fitting routines, environmental engineers developed several calculation methods to estimate L_0 and k_1 from BOD data (i.e., amount of O_2 consumed during various incubation periods). A least-squares fit by trial and error was proposed,[41] but the method is tedious and never was widely used. Both of the most common fitting methods were developed by Thomas: the "method of moments"[42] and the Thomas graphical procedure.[43] Both are simple to use and provide reasonably reliable estimates of k and L. If greater accuracy is desired, their results can be used as initial estimates for a NLR fit by a computer program (also see reference 44).

Method of moments. This procedure assumes that because of sampling and analytical errors, observed BOD data are randomly scattered about the theoretical first-order curve.[42] The first two moments of the observed BOD values are set equal to the respective moments of the fitted theoretical curve. If x_i is the observed value at t_i, and $\hat{x}$ is the curve-fitted value at t_i, we have

I. $$\sum_{i=0}^{n} x_i = \sum_{i=0}^{n} \hat{x} = \sum_{i=0}^{n} L\left\{1 - \exp(-k_1 t_i)\right\} \qquad \text{(zeroth moment)}$$

or $$\sum_{i=0}^{n} x_i = (n+1)L - L\sum_{i=0}^{n} \exp(-k_1 t_i) \qquad (2\text{-}68)$$

II. $$\sum_{i=0}^{n} t_i x_i = \sum_{i=0}^{n} t_i \hat{x} = \sum_{i=0}^{n} t_i L\left\{1 - \exp(-k_1 t_i)\right\} \qquad \text{(first moment)}$$

or $$\sum_{i=0}^{n} t_i x_i = L\sum_{i=0}^{n} t_i - L\sum_{i=0}^{n} t_i \exp(-k_1 t_i) \qquad (2\text{-}69)$$

Note that L can be eliminated by dividing Equation 2-68 by 2-69:

$$\sum x_i / \sum t_i x_i = \left\{n + 1 - \sum \exp(-k_1 t_i)\right\} / \left\{\sum t_i - \sum t_i \exp(-k_1 t_i)\right\} \quad (2\text{-}70)$$

The left side of Equation 2-70 is a function only of experimentally measured values, and the right side is a function only of k_1 and t_i. For any specified temporal sequence of measurements, the value of the right-hand function can be determined for various values of k_1, and the results can be plotted as in Figure 2-12a, where the abscissa is given as the left side of Equation 2-70, which consists of measured terms only. Because $\hat{x} = 0$ at $t = 0$, Equation 2-68 can be rewritten as

$$\sum_{i=1}^{n} x_i = nL_o - L_o\sum_{i=1}^{n} \exp(-k_1 t_i)$$

or $$\frac{1}{nL_o}\sum_{i=1}^{n} x_i = 1 + \frac{1}{n}\sum_{i=1}^{n} \exp(-k_1 t_i) \qquad (2\text{-}71)$$

Again, the right side of Equation 2-71 is a function only of k_1 for a given time sequence, and its reciprocal, equal to $nL_o/\Sigma x_i$, is plotted vs. k_1 for several time sequences in Figure 2-12. Once k_1 is obtained from the appropriate plot of $\Sigma x_i/\Sigma t_i x_i$ (Figure 2-12a), it can be used to estimate L_o from Figure 2-12b.

Graphical method. Often called the "Method of Thomas," this procedure relies on the similarity of the functions: $F_1 = (1 - 10^{-kt})$ and $F_2 = (2.3kt)\{1 + (2.3/6)kt\}^{-3}$.[43] Readers can verify this similarity by computing F_1 and F_2 for arbitrary values of k and t. If $kt < {\sim}1$, F_1 and F_2 yield similar values. Because of this similarity, we can replace F_1 with F_2 in Equation 2-67:

$$x \approx L_o(2.3kt)\{1 + (2.3/6)kt\}^{-3} \qquad (2\text{-}72)$$

Rearranging terms and taking cube roots leads to the expression

$$(t/x)^{1/3} = 1/(2.3kL_o)^{1/3} + \{(2.3k)^{2/3}/6L_o^{1/3}\}t \qquad (2\text{-}73)$$

A plot of $(t/x)^{1/3}$ vs. t thus yields a straight line whose y-intercept A and slope B define k_1 and L_o: $A = 1/(2.3kL)$; $B = (2.3k)^{2/3}/6L_o^{1/3}$. By algebraic manipulation we find that

$$k_1 = 2.61B/A; \; L_o = 1/(A^2B) \qquad (2\text{-}74)$$

The subjectivity in fitting a straight line to a plot of $(t/x)^{1/3}$ vs. t can be avoided by using calculators with regression programs.

(B) Second-Order Cases. These are encountered much less frequently than the first-order cases described above. Livesey[45] described a simple graphical method to linearize second-order data for reactions of the type $C + D \rightarrow A$, where concentrations of reactants are equal and the reaction is followed by measuring the appearance of A rather than the loss of C or D. Such cases could occur, for example, in rapid reactions where the product can be monitored by spectroscopic methods, but neither reactant has suitable spectral characteristics.

Competing and Concurrent Reactions

(A) Concurrent First-Order Reactions. If a reactant can undergo several first-order reactions, its overall reaction rate is the sum of the individual rates:

$$-d[A]/dt = k_1[A] + k_2[A] + \cdots = \left\{\sum k_i\right\}[A] = k'[A] \qquad (2\text{-}75)$$

The reaction thus follows a first-order equation with a rate constant equal to the sum of the constants for the individual reactions. This case also applies to more complicated reactions if they can be made pseudo-first order in A.

(B) Competing Second-Order Reactions. If a series of related compounds react with a common reagent, the relative rate constants can be determined without measuring individual rates.[4] Consider the second-order reactions of related compounds A and B with common reagent C to form products P and Q, respectively:

$$A + C \xrightarrow{k_A} P; \; -d[A]/dt = k_A[A][C] \qquad (2\text{-}76a)$$

$$B + C \xrightarrow{k_B} Q; \; -d[B]/dt = k_B[B][C] \qquad (2\text{-}76b)$$

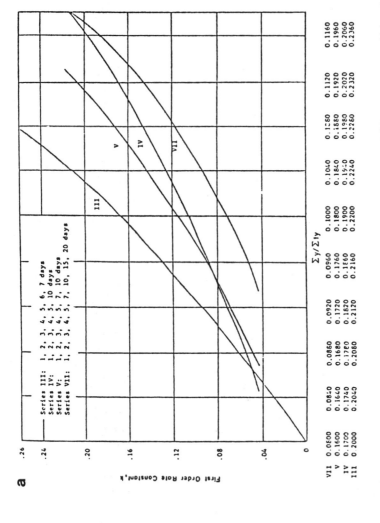

Figure 2-12. Method of moments for BOD evaluation. (a) Plot of k vs. Σx/Σtx for four time series of BOD measurements. Values below abscissa are scale for Σx/Σtx for the different time series. (b) nL/Σx vs. k for same four time series. [From Thomas, H. A., Jr., *Sewage Works J.*, 12, 504 (1940). With permission.]

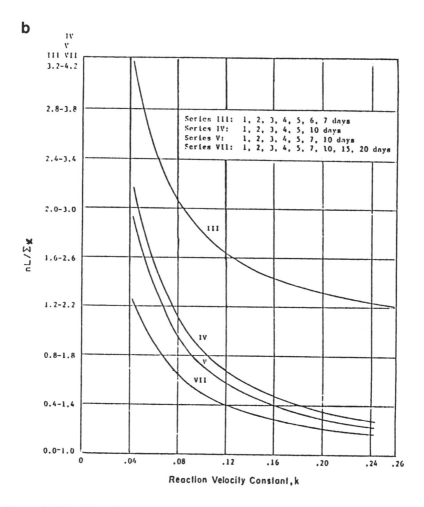

Figure 2-12 (continued).

Dividing Equation 2-76a by 2-76b yields an equation independent of time and [C]:

$$d[A]/d[B] = k_A[A]/k_B[B] \qquad (2\text{-}77a)$$

which is integrated to

$$\ln[A]/[A]_o = \{k_A/k_B\}\ln[B]/[B]_o \qquad (2\text{-}77b)$$

Because $[A]=[A]_o-[P]$, and $[B]=[B]_o-[Q]$ (assuming $[P]_o$ and $[Q]_o=0$), Equation 2-77a can be rearranged to

$$\ln\{1-[P]/[A]_o\} = \{k_A/k_B\}\ln\{1-[Q]/[B]_o\} \qquad (2\text{-}77c)$$

If [P] and [Q] are small compared with $[A]_o$ and $[B]_o$, Equation 2-77c can be simplified to the approximate expression:[4]

$$k_A / k_B \approx [P][B]_o / [Q][A]_o \tag{2-77d}$$

Relative rate constants for a series of reactions with a common reactant thus can be determined without measuring individual rates from ratios of initial reactant concentrations and either reactant concentrations (Equation 2-77b) or product concentrations (Equation 2-77d) at any time.

(C) Curve Peeling. A common complication in evaluating rate constants for simple reactions involves mixtures of related compounds that react to form a common product or a series of analytically indistinguishable products. If the reaction progress is followed by measuring product concentrations, one is faced with a problem of resolving concentration(s) of the individual reactants as a function of time. If the reaction goes to completion, the ultimate product concentration, $[P]_\infty$, gives the total initial concentration of reactants, and $[P]_\infty - [P]_t$ gives the sum of reactant concentrations remaining at any time. However, because the reactants react at different rates (with different rate constants), a plot of $log(\Sigma[A_i])$ vs. t will be curvilinear.

In simple cases where relatively large differences exist in rate constants, a graphical method called curve peeling[46] can be used to resolve the system. Consider two reactants, A' and A'', with first-order rate constants $k' \gg k''$ and $[A']_o \approx [A'']_o$. At long reaction times the faster-reacting substance, A', has disappeared, and the asymptotic slope of the plot as $t \to \infty$ yields k'' (Figure 2-13). Extrapolation of the asymptotic line to $t = 0$ yields $[A'']_o$, which allows one to compute $[A'']$ for any reaction time. Subtraction of computed $[A'']_t$ from the values of $[A_{obs}]_t$ not on the asymptotic line yields values of $[A']_t$ that do fit a straight line, the slope and y-intercept of which are k' and $[A']_o$, respectively. The technique can be applied to more complex mixtures if the rate constants are suitably separated. Hydrolysis reactions of mixtures of organic compounds (e.g., alkyl halides, phosphate esters) can be monitored easily by measuring the common inorganic product or by measuring changes in conductivity. Such data can be resolved by the above method if the mixture is simple and the rate constants are sufficiently different. The accuracy of estimates for the coefficients decreases with each successive pair of $[A_i]_o$ and k_i extracted by graphical curve peeling. Nonlinear least-squares routines that extract all the coefficients simultaneously are available in computer packages[47] to deconvolve curves. These procedures yield coefficients of equal reliability and are preferred over graphical methods. Deconvolution of curvilinear data is important in analyzing tracer data by compartment models (Chapter 5), as well as in analyzing depth profiles of contaminants in sediments.

(D) Kinetic Spectrum Analysis. In still more complicated cases, one may be faced with mixtures of many closely related compounds and a near continuum of rate constants. Dissociation reactions of metal complexes with natural aquatic macromolecules, such as aquatic humus (AH), may be in this category, for example. AH is a complex mixture of subunits and functional groups arranged in nearly limitless ways (see Section 8.4.3), and the concept of a single molecular entity called AH is not meaningful. AH has many functional groups that complex metal ions, and

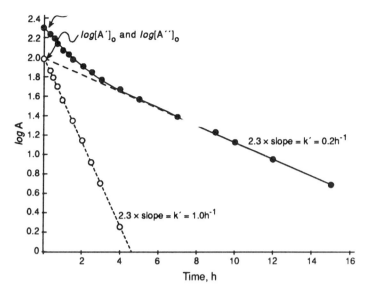

Figure 2-13. Graphical analysis of mixture of two analytically indistinguishable reactants disappearing by first-order kinetics. $[A']_o = [A'']_o = 100$; $k' = 1.0$ h^{-1}; $k'' = 0.2$ h^{-1}. Closed circles: observed (total) concentrations of reactants; solid curve: observed disappearance curve. Dashed line: extrapolated line for first-order loss of slow reacting A''; k'' computed from slope. Open circles; $log[A'] = log\{[A_{obs}] - [A'']\}$, where $[A'']$ is computed from extrapolated $[A'']_o$ and k'' for each data point not lying on dashed line. Note that in this case, where k' is only $5 \times k''$, the initial slope of the observed loss curve is affected by loss of both A' and A'' and does not yield an accurate estimate of k'. At long reaction times (> 5 h), the observed loss rate = loss rate for A'', which allows accurate estimate of k''. Subtracting computed $[A'']$ for early reaction times from $[A_{obs}]$ yields accurate estimates of $[A']$ and k'.

numerous studies have been performed to quantify metal-AH complexation equilibria. The subject is of interest because complexation affects metal transport in aquatic systems, as well as metal toxicity and availability to aquatic organisms.[48] Many analytical methods used to study metal binding by AH assume that metal-AH complexes are nonlabile, i.e., once formed, complexes dissociate slowly. Controversies have developed over the validity of these assumptions.[49] The kinetic lability of complexes also may affect the bioavailability and toxicity of metal ions. Kinetic studies on dissociation rates of metal-AH complexes are complicated by the heterogeneous nature of AH, which has binding sites of varying strength and lability. Concentrations of individual binding sites cannot be measured directly, and dissociation rates are followed by measuring the common product, free metal ion by ion-selective electrode or by complexing the dissociated metal with a colorimetric reagent (see Section 4.4.2).

A "kinetic spectrum" method was proposed to resolve rate constants for mixtures of first-order reactions[50] and was used to analyze Cu^{II}-AH dissociation rates. In general, for a mixture of reacting components, the total reactant concentration at time t is

$$[C]_t = \sum_{i=1}^{n} [C_i]_o \exp(-k_i t) \qquad (2\text{-}78)$$

where $[C_i]_o$ is the initial concentration of the ith component (e.g., the Cu^{II} complex with the ith-AH binding site). If the rate constant is the variable of integration, the summation can be replaced by the integral

$$[C]_{k,t} = \int F(k,t) \exp(-kt) dk \qquad (2\text{-}79)$$

F(k,t) is a continuous decay function that at a given initial concentration and time of measurement depends only on the rate constants. $[C]_{k,t}$ is the Laplace transform of F(k,t). Olson and Shuman[50] evaluated this expression and derived a useful distribution function from it:

$$H(k,t) = \frac{\partial^2 [C]_{k,t}}{\partial (\ln t)^2} - \frac{\partial [C]_{k,t}}{\partial \ln t} = \sum_{i=1}^{n} [C_i]_o k_i^2 t^2 \exp(-k_i t) \qquad (2\text{-}80)$$

H(k,t) is obtained by numerical differentiation of [C] vs. $\ln$ t data. The function is symmetric in k and t such that $\partial H(k,t)/\partial \ln k = \partial H(k,t)/\partial \ln t$, and

$$\int H(k,t) d(\ln k) = \int H(k,t) d(\ln t) = \sum_{i=1}^{n} [C_i]_o \qquad (2\text{-}81)$$

Consequently, the function is the same whether plotted vs. $\ln$ k or $\ln$ t, and the area under the curve is equal to the initial concentration of all components. Plots of H(k,t) vs. $\ln$ k yield curves with a series of peaks corresponding to individual components (Figure 2-14); peak maxima occur at abscissa values corresponding to $\ln k_i$. Although $\ln$ k data are not directly available from experimental data, it happens that the value of $\ln k_i$ at a peak maximum equals 2/t, where t is the measurement time associated with the peak in H(k,t). This can be shown by setting $\partial H(k,t)/\partial(\ln k) = 0$ (where the peak is maximum) and solving for t in terms of k_i. An example of the kinetic spectrum method for a two-component system is shown in Figure 2-14. Although complete resolution of the peaks was obtained, it is apparent that large differences (~43-fold[50]) must exist in rate constants of individual components to achieve this.

The kinetic spectrum method is analogous to an affinity spectrum method[51] to obtain stability constants associated with a continuum of binding-site energies in polymeric ligands like proteins and AH. The latter method was criticized,[52] however, as very sensitive to the presence of random errors in data. In addition, the method was unable to reproduce binding constants for a synthetic data set and produced spectra that lacked physical realism. A similar analysis of the kinetic spectrum method has not been undertaken.

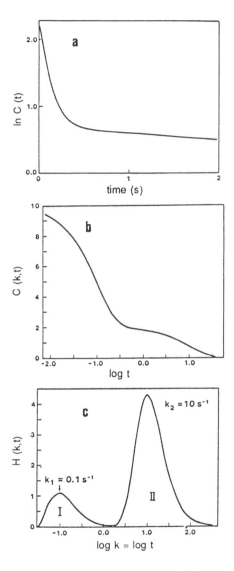

Figure 2-14. Simple example of kinetic spectrum method. (a) Conventional first-order plot for composite function $[C](t) = [C]^{\circ}_1\exp(-k_1t) + [C]^{\circ}_2\exp(-k_2t) = 2\exp(-0.1\ t) + 8\exp(-10\ t)$. (b) Data from $[C](t)$ function in (a) applied to Equation 2-79 to obtain $[C]_{k,t}$ and plotted vs. $\log t$. (c) Kinetic spectrum of $[C](t)$ function; $2.3 \times$ (area I) $= [C]^{\circ}_1$, and $2.3 \times$ (area II) $+ [C]^{\circ}_2$. Note that peak maxima correspond to $\log k_1$ and $\log k_2$. [From Olson, D. L. and M. S. Shuman, *Anal. Chem.*, 55, 1103 (1983).]

2.5 RELATIONSHIPS BETWEEN RATE EQUATIONS AND REACTION MECHANISMS

In single-step reactions, mechanisms and stoichiometries are identical. Reactions having more than two reactants seldom proceed in such a manner, however, because the probability of simultaneous collision of three or more molecules is small compared with the probability of bimolecular collisions. Many reactions with only one or two reactants also proceed by multistep sequences that belie their stoichiometric simplicity. Rate equations for multistep reactions often are different from those expected based on stoichiometry. Relationships between mechanisms and rate equations are described in this section.

2.5.1 Derivation of Rate Equations for Known Mechanisms

If we know the mechanism for a reaction — which is true *a priori* only for hypothetical reactions — we can (in principle) determine overall rates of product formation or reactant loss by following a few simple rules. In practice, the amount of algebraic manipulation required to eliminate intermediate terms may make such derivations infeasible for complicated mechanisms. The exact form of the rate equation for multistep reactions depends on the reactant or product being considered. In deriving the rate equation for a mechanism, we can rely on the tenet that elementary steps follow the law of mass action; i.e., the rate equations are consistent with their stoichiometry. As mentioned earlier, one step in multistep reactions usually is inherently slower than the rest and is rate limiting. For example, if the mechanism consists of a series of elementary second-order steps, the step with the lowest rate constant is the rate-limiting step. At steady state in such a sequence of elementary steps, the actual rate of the rate-limiting step is equal to the rates of the steps following it.

Using the law of mass action and the elementary steps contained in a mechanism, we can readily write a series of equations that provide reaction rates as a function of various reactants, intermediates, and products. In general, the rate of product formation depends (ultimately) on concentrations of reactants in the rate-limiting step. One or more of these often is an intermediate product; because these occur at low concentrations and are short-lived species, measurement of their concentrations is difficult. Consequently, the presence of intermediates in the final rate equation is undesirable. Our objective is to develop an overall rate equation in terms of the quantities accessible to us (concentrations we can measure) or quantities that are most important to us, such as concentrations of the original reactants and final products.

Two procedures are used to eliminate intermediates from rate expressions: the equilibrium constant and steady-state approaches. The former assumes that rapid elementary steps reach equilibrium such that reactant and (intermediate) product concentrations can be related by an equilibrium constant. The latter assumes that intermediates reach a steady concentration (rates of formation and loss balance each other) so that the rate equation for an intermediate becomes an easily solved algebraic equation. The following paragraphs derive rate expressions for the same mechanism by these procedures.

Equilibrium Constant Approach

Many reactions have mechanisms involving one or more rapid, reversible step(s) that occur on characteristic times ($1/k_{step}$) far shorter than the characteristic time of the rate-limiting step ($1/k_{rls}$). These rapid, reversible steps may occur before or after the rate-limiting step and are referred to as pseudo-equilibrium steps, leading to the formation of one (or more) intermediate(s) that then react slowly (in a rate-limiting step) to form intermediate or final products. A simple example is

$$A + B \quad \underset{k_2}{\overset{k_1}{\rightleftharpoons}} \quad I \qquad rapid \qquad (2\text{-}82a)$$

$$I + B \quad \xrightarrow{k_3} \quad C \qquad slow \qquad (2\text{-}82b)$$

Net: $\quad A + 2B \quad \rightarrow \quad C \qquad\qquad\qquad (2\text{-}82c)$

In this case, the pseudo-equilibrium step precedes the rate-limiting step and is called a "pre-equilibrium" step; the second step is rate determining and also yields the final product.

Sequence 2-82 is stoichiometrically analogous to the consecutive reaction sequences described in Section 2.3.2; the difference between consecutive reactions and multistep mechanisms is one of degree. In consecutive reactions the intermediates are assumed to be relatively stable species that build up to measurable levels during the reaction. In the above example, we assume that rates of the reverse of the first step and the slow second step together are such that [I] does not reach levels we can measure conveniently. From a macroscopic viewpoint only the net reaction is observed. Stated another way, a mass balance involving only A, B, and C in sequence 2-82 would be complete within experimental error at all times, because the amount of unmeasured I always is small. This is not true for consecutive reactions (cf. Figure 2-4).

The rate of product formation for reaction sequence 2-82 is

$$d[C]/dt = k_3[I][B] \qquad\qquad (2\text{-}83)$$

For the aforementioned reasons, we wish to eliminate [I] from this expression. If the first step is rapid and reaches equilibrium, the forward and reverse rates can be equated: $k_1[A][B] = k_2[I]$, or

$$[I] = (k_1/k_2)[A][B] = K[A][B] \qquad\qquad (2\text{-}84)$$

since $K_{eq} = k_1/k_2$. Substitution of Equation 2-84 into Equation 2-83 yields an equation in terms of measurable reactants only:

$$d[C]/dt = \left(k_1 k_3 / k_2 \right)[A][B]^2 = k_3 K[A][B]^2 = k_{obs}[A][B]^2 \qquad (2\text{-}85)$$

This expression is identical to the rate equation for a one-step reaction of the same stoichiometry. Thus, one cannot distinguish between a one-step reaction and the two-step mechanism, based only on the observed rate equation. The equilibrium constant approach is not completely accurate, however, because I is removed in the second step; thus, [I] may be lower than the equilibrium value defined by Equation 2-84. The magnitude of the difference depends on the relative rates of the two steps and relative concentrations of the reactants.

Steady-State Approach

A generally more accurate approach to deriving rate equations assumes that [I] rapidly reaches steady state (but not necessarily an equilibrium value). The rate of change in [I] for mechanism 2-82 is

$$d[I]/dt = k_1[A][B] - k_2[I] - k_3[I][B] \qquad (2\text{-}86a)$$

If [I] never builds up appreciably in the medium, its rate of change must be much less than that for any of the reactants. It is likely that [I] builds up quickly to a low value that remains relatively constant but slowly declines as reactants are consumed. The steady-state approximation, $d[I]/dt = 0$, transforms Equation 2-86a into an algebraic equation that can be solved for [I]:

$$[I] = \frac{k_1[A][B]}{k_2 + k_3[B]} \qquad (2\text{-}86b)$$

Substitution of Equation 2-86b into Equation 2-83 gives an expression for product formation in terms only of constants and reactants:

$$d[C]/dt = \frac{k_1 k_3 [A][B]^2}{k_2 + k_3[B]} \qquad (2\text{-}87a)$$

Depending on the relative magnitudes of k_2 and $k_3[B]$, two limiting conditions may occur. If $k_3[B] \gg k_2$, we can ignore k_2 in the denominator, and Equation 2-87a simplifies to a second-order expression:

$$d[C]/dt \approx k_1[A][B] \qquad (2\text{-}87b)$$

Consequently, at high concentrations of B, the reaction follows rate expression 2-87b. On the other hand, if $k_2 >> k_3[B]$,

$$d[C]/dt \approx k'[A][B]^2 \tag{2-87c}$$

and the rate is third order overall, as in Equation 2-85 and the termolecular (one-step) reaction.

Table 2-5 lists several simple reaction mechanisms and shows how rate expressions are derived by both the equilibrium constant and steady-state approaches, and Example 2-6 uses the steady-state approach to derive the rate equation for a chain reaction. The table shows that several mechanisms can yield the same rate expressions and that rate expressions do not unequivocally define a mechanism. In summary, rate equations for multistep mechanisms are derived from mass-action rate expressions for the elementary steps by assuming that either (1) equilibrium conditions exist for intermediates formed in rapid steps, or (2) they rapidly achieve steady-state concentrations. With either assumption one can eliminate terms involving intermediates from rate expressions for reactions with a few steps and few intermediates. The steady-state approach is more robust and yields equations that are applicable over a broader range of conditions. Rate equations derived by these methods often decompose to simpler expressions at high or low concentrations of one or more reactants; the limiting equations may help decide which of several mechanisms is most likely.

Example 2-6. Mechanism for Ozonation of Cyanide

Ozonation is a common method for removal of cyanide from waste streams. The reaction fits the rate equation: $R = k_{CN}[O_3][CN^-]^{1/2}$, and the product is cyanate. The fractional order for CN^- rules out direct reaction of CN^- and O_3 and suggests a chain mechanism. A chain mechanism similar to that for decomposition of O_3 was proposed as follows:[53]

(1) $\quad O_3 + OH^- \rightarrow O_3^- + OH \cdot$ ⎫
                                                                ⎬ initiation (slow)
(2) $\quad O_3^- \rightarrow O_2 + O^-$ ⎭

(3) $\quad OH \cdot + OH^- \rightleftharpoons H_2O + O^-$ ⎫
                                                                    ⎬ very fast (equilibria)
(8) $\quad CN^- + H_2O \rightleftharpoons HCN + OH^-$ ⎭

(4) $\quad O_3 + O^- \rightarrow O_2 + O_2^-$ ⎫
                                                          ⎬ propagation (fast)
(7) $\quad HCN + O_2^- \rightarrow HCNO + O$ ⎭

(6) $\quad O_2^- + O^- + H_2O \rightarrow O_2 + 2OH^-$ $\quad$ termination

Table 2-5. Derivations of Rate Equations for Three Simple Mechanisms by Equilibrium and Steady-State Assumptions

Mechanism	Rate equations[a]	Equilibrium assumption	Overall rate of product formation	Steady-state assumption	Overall rate of product formation	Limiting rate equations
I. A → C						
(1) $A \underset{k_2}{\overset{k_1}{\rightleftharpoons}} I$ (fast) $I + A \xrightarrow{k_3} C$ (slow)	$\dot{A} = k_2 I - k_1 A - k_3 AI$ $\dot{I} = k_1 A - k_2 I - k_3 AI$ $\dot{C} = k_3 AI$	$k_2 I = k_1 A$ $I = k_1 A/k_2$ $I = KA$	$\dot{C} = k_3 KA^2$ $= k'A^2$	$\dot{I} = 0$ $I = \dfrac{k_1 A}{k_2 + k_3 A}$	$\dot{C} = \dfrac{k_1 k_3 A^2}{k_2 + k_3 A}$	$k_2 \gg k_3, \ \dot{C} = k'A^2$ $k_3 \gg k_2, \ \dot{C} = k_1 A$ $k_1 \approx k_2, \ \dot{C} = \dfrac{k_1 A^2}{k'' + A}$
(2) $A + A \underset{k_2}{\overset{k_1}{\rightleftharpoons}} I$ (fast) $I \xrightarrow{k_3} C$ (slow)	$\dot{A} = k_2 I - k_1 A^2$ $\dot{I} = k_1 A^2 - k_2 I - k_3 I$ $\dot{C} = k_3 I$	$k_2 I = k_1 A^2$ $I = \dfrac{k_1}{k_2} A^2$ $I = KA^2$	$\dot{C} = k_3 KA^2$ $= k'A^2$	$\dot{I} = 0$ $I = \dfrac{k_1 A^2}{k_2 + k_3}$	$\dot{C} = \dfrac{k_1 k_3 A^2}{k_2 + k_3}$	$k_2 \gg k_3, \ \dot{C} = k'A^2$ $k_3 \gg k_2, \ \dot{C} = k_1 A^2$ $k_2 \approx k_3, \ \dot{C} = k''A^2$
II. 2A + B → C						
(1) $A + B \underset{k_2}{\overset{k_1}{\rightleftharpoons}} I$ (fast) $I + A \xrightarrow{k_3} C$ (slow)	$\dot{A} = k_2 I - k_1 AB - k_3 AI$ $\dot{I} = k_1 AB - k_2 I - k_3 AI$ $\dot{B} = k_2 I - k_1 AB$ $\dot{C} = k_3 AI$	$k_2 I = k_1 AB$ $I = \dfrac{k_1}{k_2} AB$ $I = KAB$	$\dot{C} = k_3 KA^2 B$ $= k'A^2 B$	$\dot{I} = 0$ $I = \dfrac{k_1 AB}{k_2 + k_3 A}$	$\dot{C} = \dfrac{k_1 k_3 A^2 B}{k_2 + k_3 A}$	$k_2 \gg k_3, \ \dot{C} = k'A^2 B$ $k_3 \gg k_2, \ \dot{C} = k_1 AB$ $k_2 \approx k_3, \ \dot{C} = \dfrac{k'A^2 B}{1 + k''A}$

(2) $A + A \underset{k_2}{\overset{k_1}{\rightleftharpoons}} I$ (fast)

$I + B \xrightarrow{k_3} C$ (slow)

$\dot{A} = k_2 I - k_1 A^2$ $k_2 I = k_1 A^2$ $\dot{C} = k_3 KA^2 B$ $\dot{I} = 0$ $\dot{C} = \dfrac{k_1 k_3 A^2 B}{k_2 + k_3 B}$ $k_2 >> k_3,\ \dot{C} = k' A^2 B$

$\dot{I} = k_1 A^2 - k_2 I - k_3 BI$ $I = \dfrac{k_1}{k_2} A^2$ $= k' A^2 B$ $I = \dfrac{k_1 A^2}{k_2 + k_3 B}$ $k_3 >> k_2,\ \dot{C} = k_1 A^2$

$\dot{B} = -k_3 BI$ $I = KA^2$ $k_2 \approx k_3,\ \dot{C} = \dfrac{k' A^2 B}{1 + k'' A}$

$\dot{C} = k_3 BI$

III. A + B → C + D

(1) $A \underset{k_2}{\overset{k_1}{\rightleftharpoons}} I + C$ (fast)

$I + B \xrightarrow{k_3} D$ (slow)

$\dot{A} = k_2 IC - k_1 A$ $k_2 IC = k_1 A$ $\dot{D} = k_1 k_3 AB/C$ $\dot{I} = 0$ $\dot{C} = \dfrac{k_1 k_3 AB}{k_2 C + k_3 B}$

$\dot{I} = k_1 A - k_2 IC - k_3 IB$ $I = k_1 A/k_2 C$ $= k' AB/C$ $I = \dfrac{k_1 A}{k_2 C + k_3 B}$ $\dot{C} = \dfrac{k'' AB}{k' + k_3 B}$ [b]

$\dot{D} = k_3 IB$ $I = KA/C$ $= k' AB$ [b]

(2) $A + B \underset{k_2}{\overset{k_1}{\rightleftharpoons}} I$

$I \xrightarrow{k_3} C + D$

$\dot{A} = \dot{B} = k_2 I - k_1 AB$ $k_2 I = k_1 AB$ $\dot{C} = k_3 KAB$ $\dot{I} = 0$ $\dot{C} = \dfrac{k_1 k_3 AB}{k_2 + k_3}$

$\dot{I} = k_1 AB - k_2 I - k_3 I$ $I = \dfrac{k_1 AB}{k_2}$ $I = \dfrac{k_1 AB}{k_2 + k_3}$ $\dot{C} = k' AB$

$\dot{C} = k_3 I$ $I = KAB$

[a] $\dot{A} = d[A]/dt$.

[b] If $C = H_2O$, as in Co^{II} exchange (Equation 2-90), it is constant and subsumed into k'.

Steps 3 and 8 are very rapid and treated as equilibria. Rate constant subscripts given below correspond to the equation numbers (as defined by reference 53); step 3 is ignored in the following rate equation derivation. Based on the above equations, the rate of reaction is given by step 7:

$$(9) \qquad R = k_7[HCN][O_2^-]$$

O_2^-, O^-, and O_3^- are intermediates. We make the same steady-state assumption for them as we made for the intermediates in Example 2-2:

$$(10) \quad d[O_2^-]/dt = 0 = k_4[O_3][O^-] - k_7[HCN][O_2^-] - k_6[O_2^-][O^-]$$

$$(11) \quad d[O^-]/dt = 0 = k_2[O_3^-] + k_7[HCN][O_2^-] - k_4[O_3][O^-] - k_6[O_2^-][O^-]$$

$$(12) \quad d[O_3^-]/dt = 0 = k_1[O_3][OH^-] - k_2[O_3^-]$$

Solving 12 for $[O_3^-]$ yields:

$$(13) \quad [O_3^-] = (k_1/k_2)[O_3][OH^-]$$

Substituting 13 into 11 and then adding the result to 10 yields:

$$(14a) \qquad 2k_6[O_2^-][O^-] = k_1[O_3][OH^-]$$

or $(14b) \qquad [O_2^-] = k_1[O_3][OH^-]/2k_6[O^-]$

Substituting 14b into 10 and simplifying yields:

$$(15) \quad k_1[OH^-]\{k_7[HCN] + k_6[O^-]\} = 2k_4k_6[O^-]^2$$

If we assume that $k_7[HCN] \gg k_6[O^-]$, we can solve 15 for $[O^-]$:

$$(16) \quad [O^-] = \{k_1k_7[OH^-][HCN]/2k_4k_6\}^{1/2}$$

This is reasonable in that the propagation steps 4 and 7 are assumed to be rapid compared with initiation and termination. Also, O^- is a highly unstable species, and its concentration should be much lower than that of HCN, even allowing for dissociation of HCN at high pH. Substituting 16 into 14b yields

$$(17) \quad [O_2^-] = [O_3]\{2k_6k_7[HCN]/k_1k_4[OH^-]\}^{-1/2}$$

Finally, from ${}^cK_a = [H^+][CN^-]/[HCN]$ and ${}^cK_w = [H^+][OH^-]$, we can write $[HCN] = {}^cK_w[CN^-]/{}^cK_a[OH^-]$. Substituting this and 17 into 9 and simplifying yields

$$R = d[HCNO]/dt = \frac{\left[k_1 k_4 k_7^c K_w\right]^{1/2}}{\left[2k_6^c K_a\right]}\left[O_3\right]\left[CN^-\right]^{1/2} = k_{CN}\left[O_3\right]\left[CN^-\right]^{1/2}$$

Reaction between O_2^- and CN^- (instead of HCN, as in Equation 7) was rejected initially because collisions between two like-charged ions are improbable. However, over the pH range 9.4 to 11.5 used in the study, the ratio $[CN^-]/[HCN]$ is $\sim 10^{1.4}$ to $10^{3.5}$ ($pK_a = 8$). This should compensate for the unfavorable electrostatic status of CN^- and suggests that CN^- could be a reactant with O_2^-. However, k_{CN} was constant over the pH range of the study; if CN^- were a reactant, k_{CN} should vary with $[OH^-]^{1/2}$.

2.5.2 Inferring Mechanisms from Rate Equations

Determination of the mechanism for a given reaction usually must be done iteratively. Initial experiments to determine rate constants, reactant order, and an empirical rate equation can suggest one or more mechanisms that fit the data. As demonstrated in Table 2-5, proof that a reaction occurs by a certain mechanism is not possible from rate equations alone, and several mechanisms often can be written that conform to the same equation. As a second step, reaction conditions (e.g., reactant concentrations, pH, ionic strength, temperature) are varied, and effects on the rate equation and rate constant are observed. The nature of these effects may rule out or support various hypothesized mechanisms. When experimental methods have been exhausted and a choice still exists, Ockham's razor (*"Entia non sunt multiplicanda praeter necessitatem"*)* may be the deciding factor; i.e., the simplest mechanism consistent with the experimental evidence is accepted as the most likely one.

Differences in rate expressions under limiting conditions of reactant concentration sometimes permit us to select among alternative mechanisms or may suggest further experiments to test hypothesized mechanisms.[55] For example, the reaction used in the previous section (A + 2B → C) could occur by the mechanism in Equation 2-82 or by the following mechanism:

$$B + B \underset{k_2}{\overset{k_1}{\rightleftharpoons}} I$$

$$I + A \xrightarrow{k_3} C$$

(2-88a)

The rate equation for the mechanism in Equation 2-88a is

---

* William of Ockham, a 14th century English philosopher, theologian, and scholastic thinker, based scientific knowledge on experience gained by the senses and on logical propositions resulting from "self-evident truths." His writings stressed the principle that "entities must not be multiplied beyond what is necessary", meaning that one should accept the simplest theory consistent with available facts. A related maxim attributed to Ockham expresses similar sentiments:[54] "*...quia frustra fit per plura quod potest per pauciora*" (because it is vain to do by more what can be done by fewer).

$$d[C]/dt = \frac{k_1 k_3 [A][B]^2}{k_2 + k_3 [A]} \qquad (2\text{-}88b)$$

If $k_3[A]$ is large compared with k_2 (i.e., at sufficiently high $[A]$), the rate equation becomes independent of $[A]$ and second order in $[B]$:

$$d[C]/dt \approx k_1 [B]^2 \qquad (2\text{-}88c)$$

In contrast, the mechanism in Equation 2-82 is first order in A under all conditions and first order in B at high $[B]$; the mechanism in Equation 2-88a always is second order in B. It is not always possible to detect these distinctions because limiting conditions may not be experimentally achievable. If k_2 is large and k_3 small, it may not be possible to reach conditions where $k_3[A] \gg k_2$.

Deduction of mechanisms from rate equations requires insight and intuition, both of which are products of experience. Some rules that help deduce mechanisms and judge their reasonableness are given in Table 2-6. A few examples of the use of rate data and chemical insight to select among several mechanisms are given below.

1. The redox reaction, $2Fe^{III} + U^{IV} \rightarrow 2Fe^{II} + U^{VI}$, obeys a simple second-order rate law:[56]

$$d[U^{VI}]/dt = k[Fe^{III}][U^{IV}] \qquad (2\text{-}89a)$$

This shows that the reaction proceeds by a multistep sequence (because the one-step reaction would be third order), but does not permit us to choose among several possible sequences, such as

$$Fe^{III} + U^{IV} \rightarrow Fe^{II} + U^{V} \qquad \text{(rate determining)}$$

$$Fe^{III} + U^{V} \rightarrow Fe^{II} + U^{VI} \qquad (2\text{-}89b)$$

or $$Fe^{III} + U^{IV} \rightarrow Fe^{I} + U^{VI} \qquad \text{(rate determining)}$$

$$Fe^{III} + Fe^{I} \rightarrow 2Fe^{II} \qquad (2\text{-}89c)$$

The redox chemistry of the metals suggests the first mechanism is more likely, however. U^{V} is moderately stable in weakly acidic solutions, but Fe^{I} is not.

2. A more informative rate equation is that for the ligand substitution reaction, $Co(CN)_5(H_2O)^{2-} + I^- \rightarrow Co(CN)_5I^{3-} + H_2O$, which has the form:[57]

$$d[Co(CN)_5 I^{3-}]/dt = \frac{k_1 k_3 [Co(CN)_5(H_2O)^{2-}][I^-]}{k_2 + k_3 [I^-]} \qquad (2\text{-}90a)$$

Table 2-6. Guidelines for Developing Reaction Mechanisms

1. Complete analysis of reaction products over time may provide information on intermediates.

2. Atomic and electronic structures of reactants and products may suggest the nature of intermediates.

3. All elementary reactions must be feasible from a bond energy viewpoint. Extensive bond rearrangements do not occur in elementary steps.

4. All intermediates produced in elementary steps must be consumed in other steps.

5. Most elementary steps are bimolecular; the rest are unimolecular or termolecular. Reactions with four or more species must occur in several steps.

6. A postulated mechanism for a forward reaction must hold for the reverse reaction. Thus, reverse reactions must have molecularity ≤ 3. Also, the rate-limiting step for the forward reaction is the rate-limiting step for the reverse reaction. These concepts follow from the principle of microscopic reversibility.

7. Highly reactive intermediates do not react exclusively with each other because their concentrations are low and they are much more likely to encounter stable molecules.

8. When the overall order is greater than 3, the reaction probably has one or more equilibrium and intermediate(s) prior to the rate-limiting step.

9. Inverse orders arise from rapid equilibria before the rate-limiting steps.

10. Rate laws with noninteger orders arise from multistep mechanisms, often involving free radicals and chain mechanisms.

11. If the reaction order increases with the concentration of (a) reactant(s), the reaction may have two or more parallel paths.

12. If $v_i > n_i$ (order for species i), one or more intermediates and reactions occur after the rate-limiting step.

Modified from Hill, C. G., Jr., *Chemical Engineering Kinetics and Reactor Design,* John Wiley & Sons, New York, 1977, and from Edwards, J. O., E. F. Greene, and J. Ross, *J. Chem Ed.,* 45, 381 (1968).

This form was determined from the fact that the reaction rate is second order overall and first order in I^- at low $[I^-]$, but first order overall (in the cobalt complex) and independent of $[I^-]$ at high $[I^-]$. This supports the hypothesis that the reaction proceeds by a two-step dissociative mechanism with an intermediate five-coordinate cobalt species that is probable short-lived (cf. Section 4.4.2); see guideline 9, Table 2-6:

$$Co(CN)_5(H_2O)^{2-} \underset{k_2}{\overset{k_1}{\rightleftharpoons}} Co(CN)_5^{2-} + H_2O$$

$$Co(CN)_5^{2-} + I^- \overset{k_3}{\longrightarrow} Co(CN)_5 I^{3-} \tag{2-90b}$$

Readers can verify that this mechanism yields the proper rate equation. Alternative one- or two-step mechanisms yield simple second-order rate equations for all concentrations of the reactants, as illustrated in Table 2-5 (reaction 3).

3. The chlorination of secondary amines like dimethyl-, diethyl-, methylethanol, and diethanolamine by N-chlorosuccinimide (NSCl) is a reversible reaction for which two mechanisms have been postulated: (1) a direct transfer mechanism and (2) a two-step process in which NSCl hydrolyzes to form HOCl, which chlorinates the amines in a second step:[58]

1.
$$\begin{array}{c} CH_2CO \\ \diagdown \\ NCl + RR'NH \\ CH_2CO \diagup \end{array} \rightleftharpoons \begin{array}{c} CH_2CO \\ \diagdown \\ NH + RR'NCl \\ CH_2CO \diagup \end{array}$$

2.
$$\begin{array}{c} CH_2CO \\ \diagdown \\ NCl + H_2O \\ CH_2CO \diagup \end{array} \rightleftharpoons \begin{array}{c} CH_2CO \\ \diagdown \\ NH + HOCl \\ CH_2CO \diagup \end{array} \quad \text{slow}$$

$$HOCl + RR'NH \rightleftharpoons RR'NCl + H_2O \quad \text{fast}$$

The rate-limiting step for the above mechanism is the hydrolytic step, and the steady-state assumption for [HOCl] leads to the following rate equation:

$$d[RR'NCl]/dt = k_1k_3[RR'NCl][NSCl]/\{k_2[NSH] + k_3[RR'NH]\}$$

In the presence of excess RR'NH, the reaction rate should be independent of the concentration or nature of RR'NH (guideline 12 of Table 2-6.) The rate of the direct transfer mechanism depends on the concentration and nature of both reactants (either direction) regardless of concentration. Experiments over a wide range of concentration ratios for NSCl and various secondary amines[58] yielded rate equations that always were first order in both reactants (in both directions). These results are consistent with the direct transfer mechanism.

2.5.3 The Principle of Microscopic Reversibility

First formulated by Tolman in 1924,[59] this principle states that at equilibrium any molecular process and the reverse of that process occur at the same rate. If a reaction occurs by several elementary pathways, at equilibrium the forward and reverse rates are equal for each pathway. This principle is an application of the second law of thermodynamics and the laws of statistical mechanics. Moreover, the reaction pathway established as most probable for the forward direction also is the most probable pathway for the reverse process. The latter statement is known as the principle of detailed balancing, and it has important implications regarding the relationship between rate constants and equilibrium constants. For a one-step reaction $A \rightleftharpoons B$, the forward and reverse rates are equal at equilibrium, $k_f[A] = k_r[B]$, and $[B]/[A] = k_f/k_r = K_{eq}$; i.e., the equilibrium constant is equal to the ratio of forward and reverse rate constants of the elementary reactions. The principle of detailed balancing also provides a means to determine whether a mechanism is reasonable or not. Postulated elementary steps that involve more than three products can be ruled out because the reverse reaction is very unlikely. This principle can be useful in analyzing reactions of geochemical interest.[60]

Two cautions must be observed in using the principle of detailed balancing to relate equilibrium and rate constants. First, rate constants and equilibrium constants must be expressed in similar concentration units. Second, the principle applies strictly only to elementary reactions. For multistep reactions, k_f/k_r is equal to *some* equilibrium constant, but not necessarily K_{eq} for the reaction as written. In general, $k_f/k_r = K_{eq}^n$, where n is the number of times each elementary reaction must occur to yield the overall reaction stoichiometry; usually $n = 1, 2,$ or $1/2$. If the forward and reverse rate laws are known for a multistep reaction, one can compute K_{eq} for the reaction quotient obtained from the two rate laws, but one generally cannot compute one rate constant from the other rate constant and an independently determined K_{eq} unless n is known. Reasons for this are explained by example elsewhere.[5,61]

2.6 EXPERIMENTAL ASPECTS OF KINETIC STUDIES

The experimental techniques used to study the kinetics of reactions are extremely diverse, and comprehensive treatment of this subject is beyond our scope. This section serves to introduce the subject and describe analytical procedures in a general way. Experimental approaches and cautions that should be observed in kinetic studies are emphasized here.

2.6.1 Analysis of Reacting Solutions for Reactants and Products

Kinetic studies deal with changing concentrations of reactants and products in inherently unstable situations. This poses certain requirements for analytical procedures. Most important, the time for measurement of reactant or product concentrations must be small compared with the time of reaction. For reactions that approach completion on time scales of hours to days, no special analytical techniques are needed, since reactions used in chemical analysis are complete in seconds to minutes. If the reaction being investigated is heterogeneous, it can be stopped by separating the two phases by filtration or by centrifugation. For example, in mineral dissolution studies a sample of the solution can be filtered and stored for later analysis of soluble reactants or products. Similarly, in studies involving microorganisms, reactions can be stopped by separating the microbial biomass from the solution. Standard analytical procedures can be used to measure reactants or products in solution, or if isotope tracers are used, microbial uptake of labeled material can be measured directly.

The reactant or product being measured also can be changed rapidly into a nonreacting form, or an aliquot of the reaction medium can be quenched at appropriate time intervals by chemical methods. For example, the analysis of iodide by a kinetic method[10] is based on its catalysis of the ceric-arsenite redox reaction. The reaction is quenched after a specified time by adding thiosulfate, which oxidizes arsenite essentially instantaneously. The remaining Ce^{IV} then can be determined at leisure from its absorbance at 410 nm or by titration with ferrous ammonium sulfate. Chlorination reactions can be stopped by adding arsenite, which immediately reduces chlorine species to chloride, and oxidation of Fe^{II} can be stopped by acidifying the solution.

Nonspecific (physical) methods often are used to follow reactions. Changes in UV or visible absorbance caused by disappearance of reactant or appearance of product are easily measured, e.g., in reactions involving chlorine species (see Section 4.5). Similarly, changes in pH can be used for reactions that produce or consume protons. Hydrolysis and dissolution reactions can be followed by measuring conductivity. Some reactions can be followed by measuring volume changes. This technique is useful for reactions that do not involve optical changes (e.g., polymerization reactions) or have reactants that are difficult to measure by other analytical methods. The technique relies on the fact that products and reactants usually have different partial molal volumes, which causes the solution volume to change during the reaction. Although the volume changes are small, they can be measured accurately with volumetric flasks called dilatomers, which have a calibrated capillary tube at the top. Because solution volume changes with temperature, the reaction vessel must be carefully thermostatted. In general, one needs to be cautious to avoid faulty interpretations caused by interferences when using nonspecific methods.

2.6.2 Continuous Measurement Methods

Physical parameters are especially well suited for continuous monitoring of reactions. Changes in absorbance can be monitored essentially instantaneously by standard spectrophotometric methods. Concentrations of some species can be monitored continuously by electrometric methods. Rates of O_2 uptake by microorganisms often are followed by continuous monitoring with amperometric oxygen electrodes. Reactions that produce or consume protons can be monitored with pH electrodes.[61] This is especially convenient when other reactants and products are difficult to analyze by conventional means or when removal of sample from the reaction vessel may introduce contaminants.

Example 2-7. Continuous Measurement of Dissolution of FeS by Monitoring pH

The anoxic dissolution of mackinawite (tetragonal ferrous sulfide, FeS) was studied by monitoring pH and computing the extent of dissolution.[62] This enabled the authors to monitor the reaction in a sealed vessel and avoid oxidation of Fe^{II} and S^{-II}. Because S^{2-} is a strong base and Fe^{2+} is a weak acid, dissolution causes pH to increase. The rate equation is derived below. The reactions are

(1) $FeS_{(s)} \rightarrow Fe^{2+} + S^{2-}$

(2) $S^{2-} + H^+ \rightleftharpoons HS^-$ $K_2^{-1}; \ pK_2 \approx 14$

(3) $HS^- + H^+ \rightleftharpoons H_2S$ $K_1^{-1}; \ pK_1 = 7.02$

(4) $Fe^{2+} + H_2O \rightleftharpoons FeOH^+ + H^+$ $^*K_1; \ p^*K_1 = 9.50$

Reactions 2–4 are very rapid and can be treated as equilibria. The change in [H⁺] for any time during the reaction can be obtained from a proton balance equation. If we neglect [OH⁻], which is very small, this equation is

$$(5) \quad -\Delta\left[H^+\right] = -\left\{\left[H^+\right] - \left[H^+\right]_o\right\} = \left[HS^-\right] + 2\left[H_2S\right] - \left[FeOH^+\right]$$

Initial values of species on the right side of Equation 5 are assumed to be zero; from Equations 2–4 we see that formation of HS⁻ and FeOH⁺ from the initial products of dissolution each consumes one H⁺; formation of H₂S consumes two H⁺. In an unbuffered system of low ionic strength, $[S]_T = [H_2S] + [HS^-]$; $[S^{2-}]$ is negligible except at very high pH; and $[Fe]_T = [Fe^{2+}] + [FeOH^+]$. We also can define $[H_2S] = \alpha_o[S]_T$; $[HS^-] = \alpha_1[S]_T$; and $[FeOH^+] = \alpha_{FeOH}[Fe]_T$, where α_i is the fraction of the element present as species i. The α's are functions of [H⁺] and equilibrium constants:

$$(6) \quad \alpha_o = \left\{1 + K_1/\left[H^+\right] + K_1K_2/\left[H^+\right]^2\right\}^{-1}$$

$$(7) \quad \alpha_1 = \left\{1 + \left[H^+\right]/K_1 + K_2/\left[H^+\right]\right\}^{-1}$$

$$(8) \quad \alpha_2 = \left\{1 + \left[H^+\right]/K_2 + \left[H^+\right]^2/K_1K_2\right\}^{-1}$$

Derivation of α's is given in aquatic chemistry texts,[63] and α values at any pH are readily calculated. Since $[S]_T = [Fe]_T$ from the stoichiometry of dissolution, we can substitute the α relationships into Equation 5 and solve for $[S]_T$:

$$(9) \quad [S]_T = -\Delta\left[H^+\right]\left\{2\alpha_o + \alpha_1 - \alpha_{FeOH}\right\}^{-1}$$

The dissolution rate was found to be first order in [H⁺] in unbuffered solutions at pH < 4.3 (Figure 2-15); i.e.,

$$(10) \quad d\left[H^+\right]/dt = -k_{obs}\left[H^+\right]$$

Because $\alpha_o \approx 1$ and α_1 and $\alpha_{FeOH} \approx 0$ under these conditions, Equation 9 becomes

$$2[S]_T = -\Delta\left[H^+\right] = -\left(\left[H^+\right] - \left[H^+\right]_o\right)$$

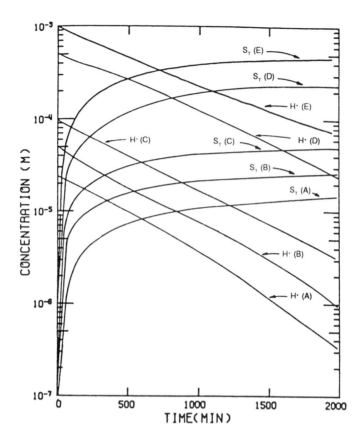

Figure 2-15. Measured changes in (H+) and calculated dissolution of mackinawite (FeS) in terms of total soluble sulfide [S^{-II}]$_t$ from Equation 9 of Example 2-7 for experiments with HCl ranging from (A) 2.5×10^{-5} M HCl to (E) 1×10^{-5} M HCl in 0.05 M NaCl.

Differentiating with respect to t, we obtain

$$(11) \quad d[S]_T / dt = -1/2 \, d[H^+]/ dt = -1/2 \, k_{obs}[H^+]$$

Finally, $k_{obs} = k_1 A/V$; i.e., the observed rate constant is equal to an intrinsic rate constant corrected for the surface area of dissolving mineral and volume of solution.

Ion-selective electrodes for cations and anions, such as F$^-$, Cl$^-$, CN$^-$, NO$_3^-$, Ca^{2+}, and some heavy metals (Cd, Cu), can be used to monitor the kinetics of reactions involving these ions. However, the response times for some of these electrodes are relatively slow ($\sim$ minutes), and their use is limited to slow reactions, such as some ligand-exchange and dissolution/precipitation reactions and some adsorption reactions in which intraparticle diffusion is rate limiting (for an exception, see Example

4-2). One way to circumvent the problem of long response times is to conduct studies in flow-through reactors (Chapter 5) and monitor steady-state concentrations of a reactant or product. The residence time of the reacting fluid can be changed by varying flow rates or reactor volume. Attention must be paid to the possible introduction of artifacts such as streaming potentials caused by varying flow rates past a potentiometric electrode.

When continuous monitoring is used to follow a reaction rate, instrument response time may become an important factor. Response times depend on several physical and electronic factors. In optical instruments, response time depends primarily on electronic signal dampers, since light transmission is instantaneous, and photomultiplier detectors also respond very rapidly. Response times of some electrometric instruments (e.g., conductivity meters) depend on electronic factors, while others depend on the time required to equilibrate electrode membranes. Response times of potentiometric electrodes range from a few seconds to several minutes. The upper value generally applies to electrodes near their detection limits. Response times of O_2 electrodes depend on diffusion rates of O_2 through a semipermeable membrane and are on the order of tens of seconds.

The response characteristics of an instrument can be determined by observing its output to a known input signal such as a step function. For example, an O_2 electrode might be equilibrated in a solution of sodium sulfite at zero dissolved O_2 and then rapidly moved into an air-saturated solution. If instrument response is first order (which is common), the approach to a new steady state after a step change in input from $[C]_1$ to $[C]_2$ is given by

$$([C] - [C]_1) / ([C]_2 - [C]_1) = 1 - \exp(-t/t_o) \qquad \text{(2-91a)}$$

or
$$B(\tau) = 1 - \exp(-\tau) \qquad \text{(2-91b)}$$

t_o is the instrument time constant; $B(\tau)$ is the dimensionless output concentration caused by a change in input; and τ is dimensionless output time.

Instrument response characteristics are not of concern in most kinetic studies, which use conventional laboratory instruments to analyze discrete samples from an experiment. Nor are they typically of concern when spectrophotometers and other instruments with rapid response times are used for continuous monitoring. However, when monitoring is done with complicated flow systems, such as auto-analyzers[64] or slowly responding O_2 electrodes,[65] response or lag times can cause differences between the observed concentration-time profile and the actual one. A simple equation to compute the difference (C_{lag}) between observed and actual dissolved O_2 concentrations in respiring microbial systems is[65]

$$[C]_{lag} = at_o \qquad \text{(2-92)}$$

where a is the (zero-order) rate of oxygen consumption, and t_o is the electrode time constant. The effect of electrode response time was negligible ($[C]_{lag} < 0.1$ mg L^{-1}) for

O_2 consumption rates of activated sludge suspensions.[65] Significant transient differences between actual and observed concentrations were found when a step change in glucose was applied to a microbial reactor and glucose concentration was monitored colorimetrically with an autoanalyzer, but the asymptotic lag concentration after the system had adjusted to the step input change was negligible (< 0.1 mg L^{-1}).[64]

2.6.3 Rapid Reactions

If a reaction takes more than several seconds to complete, the conventional analytical methods just described are adequate for kinetic studies. Many reactions of interest in aquatic studies belong in this category. Reactions that are complete in less than a few seconds are too rapid to be followed by such methods. These reactions often are considered "instantaneous" and are treated by equilibrium models (Section 5.7). In some respects this viewpoint is satisfactory because the time scale of primary interest in environmental studies is that in which we experience changes — fractions of minutes or longer periods. Kinetic studies also are used, however, to evaluate reaction mechanisms, and these may be of environmental interest even for very rapid reactions. For example, the kinetics of rapid redox and ligand exchange reactions are important in understanding the availability of metal ions, such as iron, to phytoplankton. Knowledge of redox and complex dissociation kinetics for trace metals is important in interpreting results of analytical methods like anodic stripping voltammetry.[49] Differences in rates of very rapid steps, such as water exchange rates of hydrated metal ions, also lead to major differences in overall rates of ligand exchange processes, which are many orders of magnitude slower (Section 4.4.2).

A variety of methods and special instruments are available to study the rates of rapid reactions.[5,66,67] For reactions that proceed to completion in time scales of ~10^{-3} to a few seconds, (pT ≈ –1 to ~3), rapid continuous-flow and stopped-flow methods can be used. In continuous-flow devices the reactants are rapidly mixed in a small cell and pumped through small-bore tubing of varying length, past an observation device such as a photomultiplier tube. A change in absorbance in the UV-visible range must accompany the reaction. The length of the tubing determines the reaction time at which observations are made; for a given length and flow rate, a single reaction time is measured. In stopped-flow devices, pumping is stopped momentarily, and the reaction time course is displayed by directing the photomultiplier output to an oscilloscope. Stopped-flow devices can measure reactions that are complete in a few milliseconds and have the advantage (over continuous-flow systems) of allowing the time course of reaction to be viewed and recorded. Commercially available rapid-flow systems are widely used to study enzyme kinetics. They also are used in kinetic analyses,[68] wherein concentrations are determined by measuring reaction rates rather than equilibrium amounts of product. A basic assumption in kinetic analysis is that the rate is proportional to the concentration of the measured analyte.

Until recently, rapid-flow systems were limited to reactions that take at least several milliseconds (ms) to be completed ($t_{1/2}$ > ~3 ms, first order k < 200 s^{-1}), because they required complete mixing of the reagent streams before measurements were made, and it is not possible to mix two streams homogeneously in shorter time. A data treatment technique was described recently[69] that permits accurate determination of pseudo-first-order rate constants by stopped-flow spectroscopy up to at least 2000 s^{-1} ($t_{1/2} > 0.3$ ms). The technique takes into account the influence of mixing rates on the reaction profile after flow has stopped; the net effect of mixing is to make the observed rate constant (k_{obs}) smaller than the true constant (k_r). Rates of mixing and chemical reaction can be treated as sequential first-order processes, leading to the relationship $1/k_{obs} = 1/k_r + 1/k_{mix}$. For a given stopped-flow system k_{mix} can be evaluated by measuring k_{obs} for a reaction whose rate constant is known.

"Pulsed-accelerated flow spectrophotometry" (PAFS) extends the applicability of flow methods to even faster reactions and can measure pseudo-first-order rate constants as large as 12,000 s^{-1} ($t_{1/2} \approx 60$ μs).[70] Because mixing is proportional to mean flow velocity, the mixing and reaction steps can be resolved by varying the reagent flow rate during an injection into an observation cell from a special syringe. The method has been used to measure rate constants for rapid reactions of chloramines, hypochlorite, and iodine (Section 4.5).

Reactions that are complete in much less than ~10^{-4} s must be studied with the reactants already mixed. Obviously, if the reaction is that fast, the system will be at equilibrium. The key to very rapid kinetic techniques is to apply a rapid perturbation and follow the approach to a new equilibrium. Shock-tube methods rely on very rapid heating of gas samples (in a few μs). The temperature jump method of Eigen et al.[71] is an analogous technique for rapid reactions in solution. In this technique a large capacitor is charged to a high voltage and then discharged across electrodes placed in solution. The discharge lasts 10 to 20 μs, and ohmic heating causes the solution temperature to rise a few degrees within the period of discharge.[5] The increase perturbs the system so that it is no longer at equilibrium. The shift to a new equilibrium can be followed photometrically with a high-speed oscilloscope displaying the photomultiplier output.

The temperature-jump method belongs to a class of procedures known as relaxation methods, which involve rapid physical perturbation of equilibrium systems and monitoring of some physical attribute (usually an optical characteristic) as the system "relaxes" to a new equilibrium. If the perturbation is small, relaxation follows first-order kinetics regardless of the overall reaction order. Proof of this statement is found in many texts.[3,5,26] Relaxation methods have been used to study the kinetics of the dehydration step in metal-ligand complexation reactions, including complexation by oxide surfaces.

Some types of rapid liquid-phase reactions can be studied by their effects on interphase transfer rates and application of mass transfer theory. For example, a wetted-wall short-column method[72] has been used to measure rates of Cl_2 hydrolysis in water. The method is based on the enhancement of Cl_2 transfer across a gas-liquid interface caused by rapid chemical reaction of Cl_2 in the liquid film (Section 5.4), and it should be applicable to other gases that react in solution (e.g., SO_2).

Rotating disk and ring-disk electrodes are used to study the kinetics of rapid complexation,[73] dissociation,[74] and ligand-exchange reactions in aqueous solutions. For example, rotating-disk mercury film electrodes (RDE) have been used to distinguish between kinetically labile and nonlabile metal complexes and measure dissociation rate constants.[74] The rate of solution transport across the face of the RDE depends on the rotation rate. The time available for a complex to dissociate as it moves across the RDE depends on its rotation rate, which can vary from ~50 to 3000 rpm. The electrode must be maintained at a potential where the free metal ion is reduced to the metallic state, but the complex is not. This criterion may limit use of this method in studying metal complexation with natural dissolved organic matter (DOM), because reduction potentials of free metal ions and metals complexed with DOM seem to be similar. The amount of metal deposited in a given time is determined by integrating the oxidation current in an anodic stripping step. This current is a function of the dissociation rate constant, ratio of uncomplexed to complexed metal, and rotation rate.[73] Use of mercury RDEs is limited to electroactive metals like Cd, Cu, Pb, and Zn. Rotating disks of crystalline salts have been used to study mineral dissolution kinetics[75] and to determine whether a dissolution is limited by diffusion from the surface or chemical reactions at the surface. The mathematics of diffusion near rotating disks was treated by Levich.[76]

2.6.4 Photochemical Reactions

The main experimental differences between photochemical and thermochemical reactions concern the control and measurement of light. By definition, photochemical reactions require light; rates of direct photolysis are directly proportional to light intensity. Quantification of photochemical reaction rates thus requires knowledge of light intensity. Because reactants absorb light at specific wavelengths, spectral distribution also must be measured.

Light Sources

For reactions of environmental interest, the natural light source is solar radiation. Natural light is used in photoreaction kinetics, but its intensity varies temporally because of changing cloud cover and the Earth's rotation. These variations also cause significant changes in spectral quality, particularly in the UV-B range (280 to 320 nm). The sun thus is a nonconstant light source, and this is especially problematic for slow reactions (characteristic times of hours) and for reactions induced by UV-B light (Section 8.2.3). Sunlight is useful when one is interested in relative reaction rates for several compounds or in the nature of photochemical products (e.g., "screening" studies on photodegradation of petroleum residues). This approach provides realistic results, but it is difficult to extrapolate them to other environmental conditions. Chapter 8 and Zepp[77] give details on calibrating photochemical rate constants in terms of total solar radiation.

Careful photokinetic studies to measure rate constants are done in the laboratory with artificial light sources. Mercury vapor lamps have been widely used, but because they emit light at wavelengths (λ) below 280 nm, they are inappropriate to study reactions occurring in natural environments. The stratospheric ozone layer absorbs an increasing fraction of incoming solar radiation at $\lambda < 320$ nm (in the UV-

B range), and essentially no radiation less than 280 nm penetrates to the Earth's surface. Some organic compounds absorb only at $\lambda < 280$ nm. Thus they are not capable of direct photolysis in natural environments, but they may react if exposed to light from a mercury lamp. Low-pressure Hg vapor lamps emit almost all their radiation in a narrow band near 253 nm and are especially inappropriate; medium- and high- pressure Hg lamps have broader spectral outputs, but still emit light in the environmentally undesirable region below 280 nm.[78] The latter sources can be used with filters to remove $\lambda < 300$ nm, but the spectral quality of the light still does not match that of the sun. Borosilicate glass (Pyrex) absorbs light below 305 nm, and the use of such glass for reaction vessels eliminates problems of spurious reactions induced by low-wavelength light. On the other hand, some aquatic contaminants absorb light and photolyze only in the UV-B range or lower. In these cases quartz vessels, which are transparent to lower wavelengths, should be used along with a light source that does not produce emissions below 280 nm.

Ordinary incandescent lamps have the opposite problem. They emit too much light at $\lambda > 500$ nm, which is not photochemically useful, and not enough in the photoactive UV-A and blue regions. Fluorescent light has a flatter spectral output, and "Grolux" fluorescent lamps are even more like sunlight in spectral quality. Xenon arc lamps have a broad, strong, and even spectral output similar to the sun's and are widely used to simulate sunlight in laboratory photochemical studies. Xenon or medium-high-pressure Hg lamps often are used with filters and mono-chromators when narrow band (monochromatic) light is desired as a light source (e.g., reference 79; see references 78 and 80 for details). Their high outputs (500 to 2000 watts) are needed to obtain sufficient intensity in the narrow band to induce photochemical reactions at measurable rates. In contrast, ordinary fluorescent or incandescent lamps have very low outputs over narrow spectral ranges.

Although continuous light sources are desired for rate studies, pulse or flash sources are useful to study photophysical processes and reaction mechanisms. Until the advent of lasers, light sources for flash photolysis were limited by a tradeoff between relatively long pulses (milliseconds to, at best, a few microseconds) or small energy output. Developments in lasers over the past 20 years have greatly decreased pulse length, while still providing adequate energy. Commercially available devices can provide pulses of 0.1 to 1.0 J in 10 to 20 ns or less — enough to photolyze a large fraction of the sample or put a large fraction of the reactant molecules into the same excited electronic state simultaneously. Laser flash photolysis to measure photochemical reaction kinetics in ns to µs time domains is now fairly routine (see Section 8.3.5 for applications to phototransient production from aquatic humus). Recent developments in laser technology permit observations at picosecond (10^{-12} s) and even femtosecond (10^{-15} s) scales — pulses short enough to resolve an individual molecular vibration or measure the time of bond breaking in a reaction.[81]

A variety of lasers are available.[78] Ruby lasers emit at 694.3 nm, but with a nonlinear crystal the frequency can be doubled to 347.2 nm. Neodymium lasers emit at 1065 nm, but this can be doubled to 532 nm, tripled to 355 nm, or quadrupled to 266 nm. Tunable lasers made with dyes and a monochromator provide a narrow spectral output over a range of wavelengths. Continuous lasers are available that provide very-high-intensity output in a narrow spectrum.

Because lamps give off heat (large amounts for high-intensity sources), precautions need to be be taken to prevent heating of samples. To obtain a broad area of evenly distributed light, the source lamp often is housed in a unit containing mirrors and collimating lenses (and perhaps a monochromator) that produce parallel light beams. Care must be taken to keep the light system and reaction vessels optically clean (avoid fingerprints and dust). Usually, one is interested in irradiation of multiple samples, and special care must be observed to insure homogeneous sample irradiation. This often is achieved with a "merry-go-round reactor" (MGRR), which has a light source at the center. The samples (at constant distance from the center) are rotated continuously.[82]

Measurement of Light Intensity

Light intensity is measured in units of energy per unit area per time or photons per area per time (Table 2-7). The energy of a photon depends on the wavelength; thus, the relationship between the two types of units depends on λ. Thermopiles and radiometers are the primary intruments for measuring light intensity, and these are often used in gas-phase studies. Because the vessels used to contain liquids inevitably reflect, absorb, and/or scatter some of the incident light, it is much more difficult to accurately measure the light intensity received by photolyzing solution, and most studies in aqueous photochemistry use a secondary method called chemical actinometry. This approach relies on the fact that quantum yields (Φ) for some chemical reactions are insensitive to temperature, reactant concentration, light intensity, and wavelength of absorbed light. Φ is the number of molecules undergoing a particular photochemical process per number of photons absorbed by the molecule. Φ values have been determined very accurately for a few chemical reactions having the above characteristics, and light intensity can be measured accurately by chemical analysis of the amount of product formed upon irradiation.

The most widely used chemical actinometer is the ferrioxalate system.[80,83] Simple to use, accurate, and sensitive over a wide range of wavelengths, it is based on the simultaneous reduction of Fe^{III} to Fe^{2+} and the oxidation of oxalate to CO_2, which occurs by a charge-transfer mechanism (see Section 8.4.2). The products of reaction do not absorb light, but the Fe^{2+} formed during irradiation can be measured by a sensitive colorimetric procedure with phenanthroline. Solutions of 0.006 M $K_3Fe(C_2O_4)_3$ in 0.05 M H_2SO_4 absorb more than 99% of incident light in a 1-cm path for λ up to 390 nm; 0.15-M solutions absorb more than 99% of incident light in 1 cm up to 436 nm. Φ for the reaction decreases slowly from 1.24 at 302 nm to 1.14 at 405 nm and 1.01 at 436 nm.

The uranyl oxalate actinometer has an almost constant Φ of 0.57 to 0.58 over the range 302 to 436 nm.[80,84] The system is based on uranyl ion-photosensitized oxidation of oxalic acid, loss of which is measured titrimetrically with $KMnO_4$. Although widely used in the past, it is not nearly so sensitive as the ferrioxalate system. Actinometers based on photoionization of the leucocyanides of crystal

Table 2-7. Relationships Among Units of Light Intensity

Energy units

    1 joule (J) cm^{-2} s^{-1} = 107 ergs cm^{-2} s^{-1} = 1 watt cm^{-2}

    1 lux = 1 lumen m^{-2} = 1.61 × 10^{-7} watts cm^{-2}

    1 foot-candle = 1 lumen ft^{-2} = 1.73 × 10^{-6} watts cm^{-2}

    1 langely = 1 cal cm^{-2} s^{-1} = 4.19 watts cm^{-2}

    1 eV[a] = 23.06 × 10^{3} cal = 9.66 × 10^{4} J

Photon units

    1 Einstein (Ei) = 6.024 × 10^{23} photons = 1 mol of photons

Photon-energy conversions

    From the Planck-Einstein equation: E = hv = hc/λ, where h = Planck's constant
    (6.6256 × 10^{-27} erg s), v = frequency of light, λ = wavelength of light, and c = speed
    of light (3 × 10^{10} cm s^{-1}), we can show that

        1 watt cm^{-2} = 8.358 × 10^{-9} λ Ei cm^{-2} s^{-1}, where λ is in nm

At 300 nm, 1 watt cm^{-2} = 2.5 μEi cm^{-2} s^{-1}, and 1 Ei = 95.3 kcal mol^{-1} = 398 kJ mol^{-1}

[a] eV, a molecular energy unit, is related here to cal and J on a molar basis.

---

violet and malachite green in ethanol to form the respective dyes are about ten times as sensitive as the ferrioxalate system and have $\Phi = 1.0$ over the range 248 to 330 nm.[80] Unfortunately, the product dyes absorb strongly in the UV, thus acting as internal filters and leading to erroneously low results. Actinometers based on benzophenone-*cis*-1,3-pentadiene,[85] o-nitrobenzaldehyde,[86] and p-nitroanisole/pyridine[87] have been used to measure solar intensity in photolysis studies on aquatic pollutants.

Quantum Yields (Φ)

Determination of Φ for photochemical reactions involves two kinds of measurements: (1) the amount of product formed and (2) the amount of light absorbed. The former is straightforward and can be achieved by any appropriate analytical procedure. The latter is complicated, especially for compounds absorbing only in the UV-B range and below. The simplest and most accurate measurements of Φ are done with parallel monochromatic light,[88] but polychromatic light in the near-UV range can be used to determine Φ for prediction of environmental photolysis rates,[89] since similar data are needed in both cases.

Φ values for direct photolysis of most compounds in solution are independent of λ over the range in which the compound absorbs light. In general,

$$\Phi = \frac{r}{(A/V)I_o(1 - I_t/I_o)} = \frac{r}{(A/V)I_oF_s} = \frac{r}{(A/V)\int I_a(\lambda)d\lambda} \quad (2\text{-}93)$$

r is the reaction rate mol L^{-1} s^{-1}; A is the area exposed to light; V is solution volume; I_o is incident light intensity in Einsteins cm^{-2} s^{-1}; I_t is light intensity transmitted through the reacting solution; F_s is the fraction of incident light absorbed by the reacting system; and I_a is the absorbed light intensity (a function of λ). Integration is over the wavelength range of incident light. Procedures to obtain Φ for direct photolysis of organic pollutants in aqueous solution were outlined by Draper[89] and Zepp.[90]

If monochromatic light is used and light is weakly absorbed by the system ($F_s < 0.1$), the general rate equation for direct photolysis simplifies to a first-order expression (see reference 90 for a derivation):

$$-d[A]/dt = r = 2.303 I_{o\lambda} (A/V) \, \varepsilon_\lambda \ell \Phi[A] \qquad (2\text{-}94)$$

Subscript λ refers to the wavelength of incident (monochromatic) light; ε is molar absorptivity of A; and ℓ is light pathlength. Thus, Φ can be determined from the slope of a first-order plot of ln[A] vs. t:

$$\Phi = -slope / 2.303 I_{o\lambda} (A/V) \, \varepsilon_\lambda \ell \qquad (2\text{-}95)$$

$I_{o\lambda}$ is determined by actinometry, using conditions used for the pollutant. If all incident light is absorbed by the actinometer, $I_{o\lambda}(A/V) = r_{act}/\Phi_{act}$ (subscript "act" refers to the actinometer solution). ε_λ of A is determined by spectrophotometry. This should be done in water (ε_λ varies among solvents), but because many organic compounds are poorly soluble in water, this may not be possible. Alternatively, ε_λ should be determined in a solvent with a refractive index close to that of water (e.g., acetonitrile, methanol) or a mixture of water and such a solvent. In theory the absorption bands of nonpolar compounds are the same in different solvents having the same refractive index.[91]

Finally, ℓ, the average pathlength of the photolysis cell, is determined experimentally. At sufficiently high [A], light absorption by solvent is negligible compared with that by A, and the reaction rate becomes[90]

$$r = \Phi I_{o\lambda} (A/V) (1 - 10^{-\varepsilon \ell [A]}) \qquad (2\text{-}96)$$

When $\varepsilon_\lambda \ell [A] > 2$, essentially all light is absorbed, and Equation 2-96 simplifies to

$$r_{max} = \Phi I_{o\lambda} (A/V) \qquad (2\text{-}97)$$

The ratio of Equation 2-96 to Equation 2-97 is

$$r / r_{max} = X = 1 - 10^{-\varepsilon \ell [A]} \qquad (2\text{-}98)$$

and a plot of $-log(1 - X)$ vs. ε_λ[A] is a straight line with slope = ℓ.

2.6.5 Experimental Methods to Elucidate Mechanisms

Numerous experimental techniques are used in mechanistic studies, and this discussion is only a brief introduction. Effects of temperature, pH, ionic strength, light, and pressure on rates provide valuable information on the rate-limiting step, nature of the transition state, and overall mechanism. The basis for interpreting these variables is given in Chapters 3 and 8, and examples of their use in mechanistic studies are presented in Chapters 4 and 8. Although the intermediates in multistep reactions are unstable, they often can be detected by physico-chemical methods. For example, spectroscopic techniques, like electron paramagnetic resonance spectroscopy (ESR), are used to detect free radicals in chain and photochemical processes. In some cases intermediates can be isolated in pure form or as alternate products by adding substances that compete with the normal step involving the intermediate. The role of hydroxyl radicals ·OH in photochemical reactions is determined in this way.

Radioisotopes are widely used as tracers to measure reaction rates. They have the advantages of convenience, ease of measurement, and sensitivity, and also can provide mechanistic information. Both stable and radioisotopes of carbon (^{14}C, ^{13}C), oxygen (^{18}O), and hydrogen (^{2}D, ^{3}T) are used to elucidate mechanisms of organic reactions. By labeling the atoms at a site in a reacting molecule and determining where the labeled atoms appear in products, one can determine where bonds are broken and/or formed. The sequence of products in a series of consecutive reactions can be determined by measuring the distribution and amount of isotopic label in the products as a function of time. This technique is used widely in metabolic studies. One of the first major successes of this technique was in elucidating the pathway of carbon fixation in photosynthesis.

Important mechanistic information can be obtained from the *kinetic isotope effect*. This refers to rate effects caused by differences in isotope masses. Heavier isotopes have greater inertia and move more slowly than lighter isotopes. This causes small differences in dissociation energies for the bonds between a given element and two isotopes of another element (e.g., C-H and C-D) and can be used to obtain information on the rate-limiting steps. For example, if the rate-limiting step involves dissociation of hydrogen from a carbon atom, substitution of deuterium for hydrogen will result in a lower dissociation rate because the bond dissociation energy is higher for C-D than C-H.[5] The isotope effect has been used widely in mechanistic studies of organic reactions.[2] Because H and D have a greater relative difference in mass than isotope pairs of all other elements, the isotope effect is most easily measured by substituting D for H at the suspected reactive site in a reactant.

Small isotope effects are observed in the geocycling of many elements, including C, N, O, and S. Analysis of stable isotope ratios (e.g., $^{13}C/^{12}C$; $^{15}N/^{14}N$; $^{18}O/^{16}O$; and $^{34}S/^{32}S$) is an important branch of geochemistry.[92] In general, heavier isotopes react more slowly, leading to enrichment of lighter isotopes in products, but the fractionation ratio for a given element varies among chemical and biological reactions. Information on isotope abundances has been used to infer the provenance of solutes in aquatic systems. For example, $^{15}N/^{14}N$ ratios have been used to infer the relative importance of fertilizer vs. soil organic N[93] or municipal wastewater vs. nonpoint agricultural drainage[94] as sources of nitrate in runoff and groundwater. Quantitative interpretation of isotope ratio data for such purposes is difficult, however, and sometimes controversial[93] because numerous processes can influence the isotope ratios of an element in a given chemical species and sample.

PROBLEMS

1. A glass vial contains 0.5% potassium by weight. The natural abundance of ^{40}K, which decays by β^- emission, is 0.00118%, and its half-life is 1.28×10^9 years. If the vial weighs 15 g, calculate the expected rate of β^- production by the vial (in disintegrations per minute, dpm).

2. O'Brien and Birkner [*Env. Sci. Technol.*, 11, 1114 (1977)] reported the following data for oxygenation of S^{II} in aqueous solution. Determine the order of the reaction with respect to $[O_2]$ and $[S^{-II}]$ and evaluate the second-order rate constant for the reaction:

Initial rate			Initial rate		
(*M*/min)	$[S^{-II}]$, *M*	$[O_2]$, *M*	(*M*/min)	$[S^{-II}]$, *M*	$[O_2]$, *M*
0.29×10^{-4}	0.22×10^{-4}	10^{-3}	0.41×10^{-3}	10^{-3}	0.22×10^{-3}
1.03×10^{-4}	1.1×10^{-4}	10^{-3}	1.2×10^{-4}	10^{-3}	0.57×10^{-3}
6.80×10^{-4}	5.2×10^{-4}	10^{-3}	1.3×10^{-4}	10^{-3}	0.90×10^{-3}
13.0×10^{-4}	12.1×10^{-4}	10^{-3}	1.4×10^{-4}	10^{-3}	1.0×10^{-3}

3. The decomposition of ozone in water fits the general rate equation:

$$-d[O_3]/dt = k_d[O_3]^a[OH^-]^b$$

Staehelin and Hoigne [*Env. Sci. Technol.*, 16, 678 (1982)] reported the following data for decrease in O_3 versus time all at the same fixed pH. Determine the order of the reaction with respect to O_3 by evaluating k_{OH}, the apparent rate constant at fixed pH (hence fixed OH⁻). From the reported values of k_{OH} at different pH values (last two columns), determine the overall reaction order, write the rate equation, and evaluate k_d.

$[O_3]$	Time	$[O_3]$	Time	$[O_3]$	Time		
M	s	*M*	s	*M*	s	k_{OH}	pH
12.0	0	3.00	0	0.300	0	2.0×10^{-4}	8.0
8.76	15	2.49	10	0.261	10	6.3×10^{-4}	8.5
6.36	30	1.77	30	0.225	20	1.9×10^{-3}	9.0
5.04	45	1.50	40	0.180	30	6.0×10^{-3}	9.5
4.08	60	1.23	50	0.159	40	1.9×10^{-2}	10.0
3.24	75	1.02	60	0.123	50		
2.16	90	0.90	70	0.114	60		
		0.75	80	0.075	80		
		0.63	90				

4. The reaction, $2CH_3NHCl \rightarrow CH_3NCl_2 + CH_3NH_2$, may proceed either directly (as written) or by the mechanism:

$$CH_3NHCl + H_2O \rightarrow CH_3NH_2 + HOCl \qquad (slow)$$

$$CH_3NHCl + HOCl \rightarrow CH_3NCl_2 + H_2O \qquad (fast)$$

The first mechanism leads to a second-order reaction; the second to a first-order one. In an experiment at pH 3.5 with $[CH_3NHCl] = 5.5 \times 10^{-4} M$, the following results were obtained:

Time (min)	0	10	20	30	40
$[CH_3NHCl]$ $(10^{-4} M)$	5.0	3.12	2.28	1.79	1.47

Determine the reaction order (and thus apparent mechanism) of the reaction. Also determine the reaction rate constant and $t_{1/2}$ under the above conditions.

5.   What is the ratio of k_A/k_B necessary to have 99% of A react on the same time that only 1% of B reacts (both by first-order reactions)?

6.   Microbial nitrification (the oxidation of ammonium ion to nitrate) can be modeled as a two-step, consecutive, first-order process:

$$NH_4^+ + 1.5O_2 \rightarrow NO_2^- + H_2O + 2H^+$$

$$NO_2^- + 0.5O_2 \rightarrow NO_3^-$$

Model the time course of the three nitrogen forms for the following conditions: $[NH_4^+]_0$ = 2.0 mM, $[NO_2^-]_0 = [NO_3^-]_0 = 0$; $k_{NH_4^+} = 0.2$ day^{-1}; $k_{NO_2^-} = 1.0$ day^{-1}. Assume O_2 is supplied continuously from the atmosphere and its concentration does not change. *Note:* this sequence can be solved manually or simulated readily by *Acuchem* or other differential equation solvers.

7.   The following data were obtained from a kinetic study on the hydrolysis of a mixture of alkyl chlorides. Because of analytical limitations, the concentrations of individual compounds (RCl) in the mixture could not be determined. Instead, concentrations of chloride released as a product of hydrolysis ($RCl + H_2O \rightarrow ROH + H^+ + Cl^-$) were measured by ion chromatography. Resolve the data by the method of curve peeling: i.e., determine the number of component alkyl chlorides in the mixture, their initial concentrations, and their first-order hydrolysis rate constants.

Time (h)	0	0.25	0.50	1.0	2.0	3.0	4.0	5.0	7.0	9.0	10	12	16	18	25
$[Cl^-]$ (mg/L)	0	27	51	89	144	178	199	214	231	239	241	245	247	249	250

REFERENCES

1.   Benson, S.W., *Foundations of Chemical Kinetics*, McGraw-Hill, New York, 1960.
2.   Hill, C.G., Jr., *Chemical Engineering Kinetics and Reactor Design*, John Wiley & Sons, New York, 1977.
3.   Kopelman, R., *Science*, 241, 1620 (1988).
4.   Weston, R.E., Jr. and H.A. Schwarz, *Chemical Kinetics*, Prentice-Hall, Englewood Cliffs, NJ, 1972.

5. Adamson, A.W., *A Textbook of Physical Chemistry*, 2nd ed., Academic Press, New York, 1979.

6. King, E.L., *Int. J. Chem. Kinetics*, 14, 1285 (1982).

7. Cookson, J.T. and W.J. North, *Environ. Sci. Technol.*, 1, 46 (1967).

8. Imboden, D. and Th. Joller, *Limnol. Oceanogr.*, 29, 831 (1984); Martens, C.S., G.W. Kipphut and J.V. Klump, *Science*, 208, 285 (1980); Broecker, W.S., p. 116 in: *Diffusion in Oceans and Fresh Waters*, T. Ichiye (Ed.), Lamont Geol. Lab., Palisades, New York, 1965, p. 116. For limitations of the radon flux method to determine sediment-water transport rates see Gruebel, K.A. and C.S. Martens, *Limnol. Oceanogr.*, 29, 587 (1984).

9. American Public Health Association, American Water Works Association, Water Pollution Control Federation, *Standard Methods for the Examination of Water and Wastewater*, 16th ed., Washington, D.C., 1985; Eakins, J.D. and R.T. Morrison, *Int. J. Appl. Radiat. Isotopes,* 29, 531 (1978); Krishnaswamy, S. and D. Lal, in *Lakes: Chemistry, Geology, and Physics*, A. Lerman (Ed.), Wiley- Interscience, New York, pp. 153–177.

10. Farooqi, J. and P.H. Gore, *Tetrahedron Lett.*, 34, 2983 (1977).

11. Ja'far, A., P.H. Gore, E.F. Saad, D.N. Waters, and G.F. Moxon, *Int. J. Chem. Kinetics,* 15, 697 (1983).

12. Gore, P.H., A.M.G. Nassar, D.N. Waters, and G.F. Moxon, *Int. J. Chem. Kinetics,* 12, 107 (1980).

13. Leitich, J., *Int. J. Chem. Kinetics*, 11, 1249 (1979).

14. Chapra, S.C. and K.H. Reckhow, *Engineering Approaches for Lake Management, Vol. 2: Mechanistic Modeling*, Butterworth (Ann Arbor Science), Boston, 1983.

15. DeTar, D.F., *J. Chem. Ed.*, 44, 191 (1967).

16. Curtiss, C.F. and J.O. Hirschfelder, *Proc. Nat. Acad. Sci. U.S.A.*, 38, 235 (1952).

17. Edelson, D., *Science,* 214, 981 (1981).

18. Gear, C.W., *Comm. ACM*, 14, 176 (1971).

19. Field, R.J., *J. Chem. Ed.*, 58, 408 (1981); tech. rept. NDRL 2121, Notre Dame Radiation Laboratory, University Notre Dame, Notre Dame, IN, 1980; Edelson, D., *Comput. Chem.*, 1, 29 (1976); Stabler, R.N. and J.P. Chesick, *Int. J. Chem Kinetics*, 10, 461 (1978); Carver, M.B. and A.W. Boyd, *Int. J. Chem. Kinetics*, 11, 1097 (1979).

20. Braun, W., J.T. Herron, and D.K. Kahaner, *Int. J. Chem. Kinetics*, 20, 51 (1988).

21. Farrow, L.H. and D. Edelson, *Int. J. Chem. Kinetics*, 6, 787 (1974).

22. Hecht, T.A., J.H. Seinfeld, and M.C. Dodge, *Env. Sci. Technol.*, 8, 327 (1974); Falls, A.H. and J.H. Seinfeld, *Env. Sci. Technol.*, 12, 1398 (1978).

23. Lee, G.F., in *Principles and Applications of Water Chemistry*, S.D. Faust and J.V Hunter (Eds.), John Wiley & Sons, New York, 1967, pp. 54–72; Lee, G.F. and J.C. Morris, *Int. J. Air Water Poll.*, 6, 419 (1962).

24. Bodenstein, M. and S.C. Lind, *Z. Physik. Chem.*, 57, 168 (1907).

25. Herzfeld, K.F., *Ann. Physik*, 59, 635 (1919); Christiansen, J.A., *Kgl. Danske Videnskab. Selskab. Mat.-Fys. Medd.,* 1, 14 (1919); Polanyi, M., *Z. Electrochem.,* 26, 50 (1920).

26. Laidler, K.J., *Chemical Kinetics*, McGraw-Hill, New York, 1965.

27. Ross, S.D., *Int. J. Chem. Kinetics*, 14, 535 (1982).

28. Endrenyi, L. (Ed.), *Kinetic Data Analysis*, Plenum Press, New York, 1981.

29. Boyle, W.C., P.M. Berthouex, and T.C. Rooney, *J. Environ. Eng. Div. Am. Soc. Civil Eng.,* 100, 391 (1974); Cornish-Bowden, A., p. 105 in ref. 28.

30. Watts, D.G., p. 1 in reference 28. Principles of regression analysis are described in many books on statistics; some good texts include: Box, G.E.P., W.G. Hunter, and J.S. Hunter, *Statistics for Experimenters*, Wiley, New York, 1978; Draper, R.N. and H. Smith, *Applied Regression Analysis*, 2nd ed., Freeman, New York, 1981.

31. Kalantar, A.H., *Chemosphere*, 16, 79 (1987).

32. Endrenyi, L. and F.H.F. Kwong, pp.89–105 in reference 28; Neter, J. and W. Wasserman, *Applied Linear Statistical Models*, Irwin Publishers, Homewood, IL, 1974.

33. Kalantar, A.H., *Int. J. Chem. Kinetics*, 19, 923 (1987); *J. Phys. Chem.*, 90, 6301 (1986).

34. Rathbun, R.E. and D.Y. Tai, *Chemosphere*, 13, 715 (1984).

35. Marquardt, D.W., *SIAM J. Appl. Math.*, 11, 432 (1963); Bard, Y., *Nonlinear Parameter Estimation,* Academic Press, New York, 1974; Bevington, P.R., *Data Reduction and Error Analysis for the Physical Sciences,* McGraw-Hill, New York, 1969. Several large statistical packages such as SAS and SPSS have NLR routines.

36. Schreiner, M., M. Kramer, S. Krischer, and Y. Langsam, *PC Tech. J.*, May, p. 170 (1985).

37. Stumm, W. and G.F. Lee, *Ind. Eng. Chem.*, 53, 143 (1961); Sung, W. and J.J. Morgan, *Environ. Sci. Technol.*, 14, 561 (1980).

38. Davison, W. and G. Seed, *Geochim. Cosmochim. Acta*, 47, 67 (1983).

39. Shoemaker, D.P. and C.W. Garland, *Experiments in Physical Chemistry*, McGraw-Hill, New York, 1974, chap. 2.

40. Gschwend, P.M., P.H. Chen, and R.A. Hites, *Geochim. Cosmochim. Acta*, 47, 2115 (1983).

41. Reed, L.J. and E.J. Therioult, *J. Phys. Chem.*, 35, 673,950 (1931).

42. Thomas, H.A., Jr., *Sewage Works J.*, 12, 504 (1940).

43. Thomas, H.A., Jr., *Water Sewage Works, 97*, 123 (1950).

44. Several other approaches to solving this type of problem by linearized plots have been described. The most popular is that of Guggenheim (*Phil. Mag.*, 2, 538 [1926]), which requires data taken at a fixed time interval. Best results are obtained when many observations are available; therefore, it is most suited for reactions that can be monitored continuously (e.g., by optical methods). For a computerized version of the Guggenheim method, see Hobey, W.D., W.-H. Shen, and D.A. Gouin, *Am. Lab.,* March, 97 (1986); the statistical reliability of this method was evaluated by McKinnon, GH., C.J. Backhouse, and A.H. Kalantar, *Int. J. Chem. Kinetics*, 17, 655 (1985).

45. Livesey, D.L., *Int. J. Chem. Kinetics*, 18, 281 (1986).

46. Atkins, G.L., *Multicompartment Models for Biological Systems*, Methuen, London, 1969.

47. Berman, M. and M.F. Weiss, *SAMM User's Manual*, Nat. Inst. of Health, Bethesda, 1974; Gomeni, R. and C. Gomeni, *Comput. Biol. Med.*, 9, 39 (1979); Estreicher, J., C. Revillard, and J.-R. Scherrer, *Comput. Biol. Med.*, 9, 49 (1979).

48. Brezonik, P.L., C.E. Mach, and S. King. The influence of water chemistry on metal bioaccumulation and toxicity, in *Ecotoxicology of Metals: Current Concepts and Applications,* M. Newman and A. MacIntosh (Eds.), Lewis Publishers, Chelsea, MI, 1991, pp. 1–29.

49. For example, see Tuschall, J.R. and P.L. Brezonik, *Anal. Chem.*, 53, 1986 (1981) and ensuing discussions (54, 1000, 2116 [1982]) regarding the effect of complex lability on the validity of metal-AH binding constants determined by anodic stripping voltammetry.

50.   Olson, D.L. and M.S. Shuman, *Anal. Chem.*, 55, 1103 (1983); *Geochim. Cosmochim. Acta,* 49, 1371 (1985).

51.   Ninomiya, K. and J.D. Ferry, *J. Coll. Sci.*, 14, 36 (1959); Thakur, A.K., P.J. Munson, D.L. Hunston, and D. Rodbard, *Anal. Biochem.*, 103, 240 (1980).

52.   Turner, D.R., M.S. Varney, M. Whitfield, R.F.C. Mantoura, and J.P.Riley, *Geochim. Cosmochim. Acta*, 50, 289 (1986).

53.   Zeevalkink, J.A., D.C. Visser, P. Arnoldy, and C. Boelhouwer, *Water Res.*, 14, 1375 (1980).

54.   Hutchinson, G.E., in *Introduction to Population Ecology*, Yale University Press, New Haven, CT, 1978, p. 2.

55.   Edwards, J.O., E.F. Greene, and J. Ross, *J. Chem. Ed.*, 45, 381 (1968).

56.   Halpern, J., *J. Chem. Ed.,* 45, 372 (1968); Betts, R.H., *Can. J. Chem.*, 33, 1780 (1955).

57.   Haim, A. and W.K. Wilmarth, *Inorg. Chem.*, 1, 573, 585 (1962).

58.   Antelo, J.M., F. Arce, J. Franco, M.C. Carcia Lopez, M. Sanchez, and A. Varela, *Int. J. Chem. Kinetics*, 20, 297 (1988).

59.   Tolman, R.C., *Phys. Rev.*, 23, 699 (1924); *The Principles of Statistical Mechanics,* Clarendon Press, Oxford, 1938.

60.   Lasaga, A.C., in *Kinetics of Geochemical Processes*, A.C. Lasaga and R.J. Kirkpatrick (Eds.), Mineralogical Society of America, Washington, D.C., 1981, pp. 1–68.

61.   Morgan, J.J. and A.T. Stone, in *Chemical Processes in Lakes*, W. Stumm (Ed.), Wiley-Interscience, New York, 1985, pp. 389–426.

62.   Pankow, J.F. and J.J. Morgan, *Environ. Sci. Technol.*, 13, 1248 (1979).

63.   Snoeyink, V. and D. Jenkins, *Water Chemistry*, John Wiley & Sons, New York, 1980; Butler, J.N., *Ionic Equilibrium; a Mathematical Approach*, Addison-Wesley, Reading, MA, 1964.

64.   Baillod, C.R. and W.C. Boyle, Jr., *Environ. Sci. Technol.*, 3, 1205 (1969).

65.   Mueller, J.A., W.C. Boyle, Jr., and E.N. Lightfoot, *Appl. Microbiol.*, 15, 674 (1967).

66.   Caldin, E.F., *Fast Reactions in Solution*, John Wiley, New York, 1964; Mark, H.B. and G.A. Rechnitz, *Kinetics in Analytical Chemistry*, Wiley-Interscience, New York, 1968.

67.   Onwood, D., *J. Chem. Ed.*, 63, 680 (1986).

68.   Brezonik, P.L., in *Water and Water Pollution Handbook*, Vol. 3, L. Ciaccio, (Ed.), Marcel Dekker, New York, 1972, pp. 831–913.

69.   Dickson, P.N. and D.W. Margerum, *Anal. Chem.*, 58, 3153 (1986).

70.   Jacobs, S.A., M.T. Nemeth, G.W. Kramer, T.Y. Ridley, and D.W. Margerum, *Anal. Chem.*, 56, 1058 (1984).

71.   Eigen, M., W. Kruse, G. Maass, and L. De Maeyer, *Progr. React. Kinetics*, 2, 285 (1964); Finholt, J.E., *J. Chem. Ed.*, 45, 394 (1968).

72.   Aieta, E.M. and P.V. Roberts, *Environ. Sci. Technol.*, 20, 44 (1986).

73.   Shuman, M.S. and L.C. Michael, in Proceedings Int. Conf. Heavy Metals in the Environment, Vol. I, Toronto, Canada, 1975, pp. 227–248.

74.   Shuman, M.S. and L.C. Michael, *Environ. Sci. Technol.*, 12, 1069 (1978).

75.   Zutic, V. and W. Stumm, *Geochim. Cosmochim., Acta*, 48, 1493 (1984).

76.   Levich, V.G., *Physicochemical Hydrodynamics*, Prentice Hall, Englewood Cliffs, N.J., 1962.

77.   Zepp, R.G., in *Dynamics, Exposure and Hazard Assessment of Toxic Chemicals*, R. Haque (Ed.), Ann Arbor Science Publ., Ann Arbor, MI, pp. 69–110.

78.   Arnold, D.R., N.C. Baird, J.R. Bolton, J.C.D. Brand, P.W.M. Mayo, and W.R. Ware, *Photochemistry: An Introduction*, Academic Press, New York, 1974.

79. Zepp, R.G., G.L. Baughman, and P.F. Schlotzhauer, *Chemosphere*, 10, 119 (1981).

80. Calvert, J.G. and J.N. Pitts, Jr., *Photochemistry*, John Wiley & Sons, New York, 1966.

81. Shank, C.V., *Science*, 233, 1276 (1986); Zewail, A.H., *Science*, 242, 1645 (1988).

82. Moses, F.G., R.S.H. Liu, and B.M. Monroe, *Mol. Photochem.*, 1, 245 (1969).

83. Hatchard, C.G. and C.A. Parker, *Proc. Roy. Soc. (London)*, A235, 518 (1956).

84. Discher, C.A., P.F. Smith, I. Lippman, and R. Turse, *J. Phys. Chem.*, 67, 2501 (1963).

85. Lamola, A.A. and G.S. Hammond, *J. Chem. Phys.*, 41, 2129 (1965).

86. Pitts, J.N., Jr., J.K. Wan, and E.A. Schuck, *J. Am. Chem. Soc.*, 86, 3606 (1964).

87. Dullin, D. and T. Mill, *Environ. Sci. Technol.*, 16, 815 (1982).

88. ECETOC Task Force Photodegradation, An assessment of test methods for photodegradation of chemicals in the environment, Tech. Report No. 3, Ecol. & Toxicol. Cent., Europ. Chem. Ind., Ave. Louise 250, B63, Brussels, Belgium.

89. Draper, W.M., *Chemosphere*, 14, 1195 (1985).

90. Zepp, R.G., *Environ. Sci. Technol.*, 12, 327 (1978).

91. Strickler, S.J. and R.A. Berg, *J. Chem. Phys.*, 37, 814 (1962).

92. Fritz, P. and J.C. Fontes (Eds.), *Handbook of Environmental Isotopic Geochemistry*, Elsevier, Amsterdam, 1980; Hoefs, J., *Stable Isotope Geochemistry*, 3rd ed., (Minerals and Rocks: 9), Springer-Verlag, New York, 1987.

93. Kohl, D.H., G.B. Shearer, and B. Commoner, *Science*, 174, 1331 (1971); Hauck, R.D., *J. Env. Qual.*, 2, 317 (1973); Meints, V.W., G.B. Shearer, D.H. Kohl, and L.T. Kurtz, *Soil Sci.*, 119, 421 (1975).

94. Showers, W.J., D.M. Eisenstein, H. Paerl, and J. Rudek, Stable isotope tracers of nitrogen sources to the Neuse River, NC, Report 253, Water Resources Research Institute, University of North Carolina, Chapel Hill, 1990.

95. Kumar, K. and D.W. Margerum, *Inorg. Chem.*, 26, 2706 (1987).

"The fascination of a growing science lies in the work of the pioneers at the very borderland of the unknown, but to reach this frontier one must pass over well travelled roads … ."

G. N. Lewis and M. Randall, Thermodynamics, 1923

CHAPTER 3

Theoretical Aspects of Kinetics and Effects of Physical Conditions on Reaction Rates

3.1 INTRODUCTION

Kinetics is the study of change and is concerned especially with **rates** of change. From a metaphysical perspective, this definition was disquieting before the development of modern scientific knowledge. For a long time natural philosophers wondered, "How can we have knowledge about something that is changing into something else?" Ancient Greek philosophers believed that only changeless things could be the subject of scientific study,[1] and they had difficulty comprehending how a thing could cease to exist and become something else. Modern science provides partial answers to the puzzle. The most fundamental chemical entities, atoms, do not change during chemical reaction; only their positions change relative to each other. The driving force for change is the difference in energy between reactants and products. The second law of thermodynamics tells us that the criterion for spontaneous reaction is a decrease in a system's free energy and an increase in the entropy of the universe. Thermodynamics does not predict reaction rates, however; it predicts only whether a reaction can occur (without input of energy) and the extent to which it can occur. Although the physical world is a dynamic, rather than a static, environment, time is not a thermodynamic variable. To this extent kinetics is a more fundamental science than thermodynamics.

It is interesting to note that kinetic principles were used in the early development of equilibrium concepts. In the mid-1800s chemical equilibrium was shown to involve a balance between forward and reverse rates of reaction rather than a static condition. The law of mass action expresses reaction rates as proportional to reactant concentrations, and the concept of equilibrium constants was derived from kinetic concepts of mass action rather than vice versa. van't Hoff later showed that the equilibrium constant for a reaction is equal to the ratio of rate constants for the forward and reverse reactions, $K_{eq} = k_f/k_r$. In turn, modern kineticists borrowed from thermodynamic theories to explain reaction rates. For example, transition-state theory expresses reaction rates as a function of the energy difference between reactants and a transition state, i.e., the activation energy, E_{act}. The reactants and transition state are considered to be in quasiequilibrium, so that thermodynamic relationships apply. Many theories have been developed during the past 100 years to explain how reactions occur and why they occur at a given rate, but the philosophical aspects of kinetics remain somewhat elusive. Even the most advanced theories leave some fundamental questions unanswered (see Sections 3.3.4–5 and 3.4.3).

This chapter describes relationships between thermodynamics and kinetics, in the context of explaining the physical factors affecting aqueous reaction rates. Based on observed relationships between temperature (the most important physical variable) and reaction rate, Arrhenius developed the concept of activation energy in the late 19th century. This concept spawned the modern theories of kinetics, collision, and transition-state theory, which are developed for aqueous reactions after short discussions on the structure of water, forces between reactants, and diffusion in water. This theoretical base then is used to explain the effects of pressure and ionic strength on reaction rates.

3.2 EFFECTS OF TEMPERATURE ON REACTION RATES: THE ARRHENIUS EQUATION

It is well known that rates of chemical reactions increase with increasing temperature. Within physiological limits, increasing temperature also increases rates of biological processes, and physical processes such as diffusion are temperature dependent, as well. The sensitivity of process rates to changes in temperature varies widely. Physical processes generally increase by a factor of 1.1 to 1.4 for a 10°C rise, but increases of 1.5 times to 3.0 times are common for chemical reactions over a 10°C rise. Biological process rates often are assumed to increase by two times per 10°C increase, but this is a large simplification.

3.2.1 Effect of Temperature on Chemical Equilibria: The van't Hoff Equation

The effect of temperature on chemical equilibria is expressed quantitatively by the van't Hoff equation, which relates the value of the equilibrium constant to temperature as a function of the enthalpy of reaction ($\Delta H°$):

$$d \ln K_{eq} / dT = \Delta H° / RT^2 \qquad (3\text{-}1a)$$

or since $dT/T^2 = -d(1/T)$,

$$d \ln K_{eq}/d(1/T) = -\Delta H^{\circ}/R \qquad (3\text{-}1b)$$

R is the gas constant (J deg^{-1} mol^{-1}), and T is absolute temperature in K. If ΔH° is constant over a given temperature range, Equation 3-1b can be integrated to

$$\ln K_2/K_1 = \Delta H^{\circ}(T_2 - T_1)/RT_1T_2 \qquad (3\text{-}2)$$

More complicated equations result if ΔH° varies with temperature.[2]

3.2.2 The Arrhenius Equation for Temperature Effects on Kinetics

Based on the idea that the equilibrium constant of a reaction is equal to the ratio of the rate constants for the forward and reverse steps, Arrhenius suggested in 1889 that an equation similar to that of van't Hoff would describe the relationship between rate constants and temperature:

$$d \ln k/dT = E_{act}/RT^2$$

or
$$d \ln k/d(1/T) = -E_{act}/R \qquad (3\text{-}3)$$

E_{act}, the activation energy of the reaction, is considered to be the energy that reactants must absorb to react. For Equation 3-3 to be useful as a means of predicting the effect of T on k, E_{act} must be constant (or a simple function of temperature). If E_{act} is constant, Equation 3-3 integrates to

$$\ln k = -E_{act}/RT + \ln A \qquad (3\text{-}4)$$

where lnA is a constant of integration, or

$$k = A\exp(-E_{act}/RT) \qquad (3\text{-}5)$$

A is called the pre-exponential or frequency factor. Between limits of T_1 and T_2, Equation 3-4 becomes

$$\ln k_2/k_1 = \frac{(T_2 - T_1)E_{act}}{RT_1T_2} \qquad (3\text{-}6)$$

which is similar to Equation 3-2 for chemical equilibria. Equations 3-3 to 3-6 are forms of the Arrhenius equation. According to Equation 3-4, a plot of lnk vs. 1/T gives a straight line (Figure 3-1) if the Arrhenius equation is obeyed. This provides a convenient graphical method to evaluate E_{act} (the slope is E_{act}/R) and estimate k at any temperature. Many studies have shown that E_{act} is approximately constant

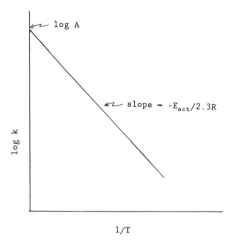

Figure 3-1. Arrhenius plot of log k vs. 1/T, according to Equation 3-4.

over wide temperature ranges for numerous reactions. We shall show later that E_{act} is closely related to the enthalpy of activation, $\Delta H^{\neq}$.

3.2.3 Variations in E_{act} with Temperature

Although the concept of E_{act} was developed to explain the effect of temperature on single-step chemical reactions, it has been used widely to quantify the effects of temperature on chemical reactions with multistep mechanisms and various physical and biological processes as well. When applied to such processes, E_{act} should be referred to as an "apparent activation energy". The apparent E_{act} for many physical processes does not vary with temperature. However, several aqueous transport processes, including the self-diffusion of water, viscous flow, and ion mobility (as measured by electrical conductivity), exhibit apparent activation energies that vary with temperature (Figure 3-2). These variations have been interpreted as resulting from temperature-induced changes in the "flickering cluster" structure of water[3] (Section 3.3).

A change in activation energy with temperature for chemical reactions usually is considered to indicate a change in reaction mechanism. Either the rate-limiting step changes, or a different reaction sequence altogether may occur. In some cases a change in E_{act} signifies a change from a heterogeneous to a homogeneous process.[4] The former mechanisms usually have lower activation energies and are favored at low temperature. An example is the half-calcination reaction of dolomite, which is important in preparing half-calcined dolomite for absorption of SO_2 in pressurized fluid-bed combustion chambers:

$$CaMg(CO_3)_2 \xrightarrow{heat} CaCO_3 + MgO + CO_2 \qquad (3\text{-}7)$$

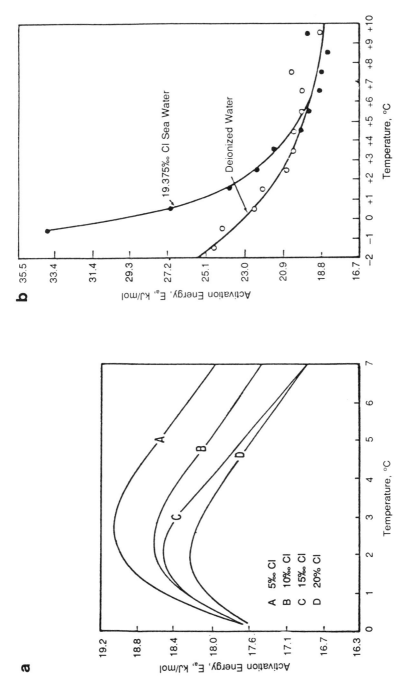

Figure 3-2. Temperature dependency of E_{act} for some transport processes in aqueous solutions of varying salinity: (a) conductivity; (b) viscous flow.[(a)Redrawn from R.A. Horne and R.A. Courant, *J. Geophys. Res.*, 69, 1152 (1964), copyright by the American Geophysical Union; (b) redrawn from R.A. Horne, et al., *J. Phys. Chem.*, 69, 3988 (1965). With permission of the American Chemical Society.]

The value of E_{act} over the temperature range 700 to 800°C is 284 kJ mol^{-1} and only 146 kJ mol^{-1} over the range 640 to 700°C.[5] These changes have been related to morphological changes in calcite, one of the products. At low temperatures the crystals are large and have "relic structures" of the original dolomite, but they become smaller and more randomly oriented at high temperatures. More recent studies[6] suggest that these effects may be related to shifts in active sites for growth of calcite from grain boundaries to edges with increasing temperature. Interpretation of E_{act} for related heterogeneous reactions of SO_2 with $CaCO_3$ and $CaSO_3$ is complicated by their multistep nature.[7]

3.2.4 Physical Interpretation of Activation Energy

According to the concept of activation energy, reactants do not pass directly into products, but first go to an activated intermediate or transition state (Figure 3-3). In an exothermic reaction the reactants have a higher enthalpy than the products; the enthalpy difference, ΔH, is the enthalpy of reaction. The transition state has a higher energy than the reactants, thus requiring absorption of a certain amount of energy, E_{act}. In most cases the energy is obtained from collision between two molecules. Once the transition state is formed, it proceeds spontaneously to form products and release an amount of energy equal to $E_{act} + \Delta H$.[*] From the concept of microscopic reversibility, which states that reactions proceed by the same pathway in both directions, the activation energy of the reverse reaction, $E_{act,r}$ must be $E_{act,f} + \Delta H_r$, and $\Delta H_r = -\Delta H_f$.

The enthalpy of reaction can be either positive (endothermic) or negative (exothermic), but activation energies of elementary reactions always are positive (or zero in some cases). If E_{act} is negative, it can be assumed that the reaction is a composite of several steps and that the observed (apparent) E_{act} is the sum of a negative ΔH for a rapid pre-equilibrium first step and a smaller positive E_{act} for a later step. For example, E_{act} is negative for the atmospheric reaction $2NO + O_2 \rightarrow NO_2$, the mechanism for which is thought to involve two steps:[4]

$$2NO \rightleftharpoons N_2O_2 \qquad \text{(rapid)}$$
$$N_2O_2 + O_2 \rightarrow 2NO_2 \quad \text{(slow)} \tag{3-8}$$

The overall rate constant k_{net} is the product of equilibrium constant K_1 for the rapid first step and rate constant k_2 for the rate-limiting second step (see Section 2.7.1). $E_{act,net}$ is the sum of ΔH_1 (which is negative and expresses the effect of

---

[*] Strictly speaking, the energy difference between products and reactants is ΔE, the change in internal energy, and $\Delta H = \Delta E + \Delta(PV)$. For reactions in solution, $\Delta(PV)$ is negligible, and $\Delta H \approx \Delta E$. For gases, $\Delta H = \Delta E$ only if there is no net change in the number of moles of gases during reaction; i.e., $\Delta n = 0$, where Δn = moles of products minus moles of reactants. When $\Delta n \neq 0$, $\Delta H = \Delta E + P\Delta V \approx \Delta E + (\Delta n)PV \approx \Delta E + (\Delta n)RT$. Similar considerations apply to energy differences between reactants and transition states. The difference in internal energy is $\Delta E^{\neq}$, which is related to the enthalpy of activation by $\Delta E^{\neq} = \Delta H^{\neq} - \Delta n^{\neq}RT$. The observed activation energy, $E_{act} = \Delta E^{\neq} + RT$ (see Section 3.5.3).

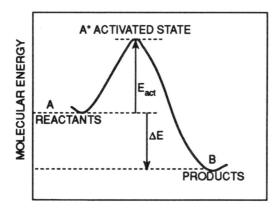

Figure 3-3. Time-energy course (reaction profile) for an exergonic reaction, showing the concept of activation energy and the activated (transition) state.

temperature on the equilibrium status of this step) and E_{act} (which is positive) for the second step.

Apparent activation energies for biochemical processes are positive only within physiological limits; increasing denaturation of enzymes at higher temperatures causes reaction rates to decrease, thus producing negative apparent E_{act} values above the temperature optimum.

3.2.5 Parameter Estimation for the Arrhenius Equation

The Arrhenius equation has the same basic exponential form as the integrated first-order expression: $y = a\exp(-bx)$, where y and x are variables, and a and b are fitted parameters:

	Variables		**Parameters**	
Equation	x	y	a	b
first order	t	$[A]_t$	$[A]_o$	k
Arrhenius	1/T	k	A	E_{act}/R

The comments on parameter estimation in Section 2.4.2 thus apply as well to the Arrhenius equation. Usually, $[A]_o$ is known in first-order reactions, so that only k must be determined. The Arrhenius pre-exponential factor (A) often is not estimated explicitly, since it is not needed for the primary application of the equation (to compute temperature effects on k). Once E_{act} is estimated, one can use the discrete integrated form, Equation 3-6, which does not contain A.

As Equation 3-6 shows, E_{act} can be estimated from k values at two temperatures. (The use of this equation does not allow one to determine whether E_{act} is constant over the two temperatures, but for most aqueous reactions this is a reasonable assumption.) By propagated error analysis (Equation 2-62), we can show that the relative error in E_{act} for rate data at two temperatures is

$$\left[\frac{\Delta E_{act}}{E_{act}}\right]^2 = \left[\frac{T_2}{T_2 - T_1}\right]^2 \left[\frac{\Delta T_1}{T_1}\right]^2 + \left[\frac{T_1}{T_2 - T_1}\right]^2 \left[\frac{\Delta T_2}{T_2}\right]^2$$

$$+ \left[\frac{1}{\ln(k_2/k_1)}\right]^2 \left[\left[\frac{\Delta k_1}{k_1}\right]^2 + \left[\frac{\Delta k_2}{k_2}\right]^2\right] \tag{3-9}$$

The precision of E_{act} thus depends on the size of the temperature interval used for the two measurements; the denominators of all three terms on the right side of Equation 3-9 decrease as the temperature interval decreases. The precision of the estimate can be improved by making replicate measurements of k at the two temperatures (to decrease Δk_1 and Δk_2, the standard deviations of k at T_1 and T_2).

The usual approach to improving the precision of E_{act} is to measure k at several temperatures and plot the data as in Figure 3-1. If the data are precise, determination of the slope is simple, but for accurate results a line should be fit by least-square regression. The same problems arise as described earlier for estimating parameters of rate equations, i.e., to achieve statistical efficiency we must know the error structure of the values of k at the various temperatures.[8] If the error in k is constant over the temperature range used, the appropriate form is the exponential (Equation 3-5) fit by nonlinear regression (NLR) without weighting of the data. Data should be fit to the *log*-transformed equation by unweighted linear regression only in the unlikely case that the error in k is proportional to temperature, i.e., the relative error, $\Delta k/k$, is independent of temperature. If Δk varies point by point, the proper weighting should be used for each point ($w_i = 1/\Delta k_i^2$ for Equation 3-5; $w_i = k_i^2/\Delta k_i^2$ for the *log*-transformed equation).

The importance of correct weighting has been demonstrated by computer simulations with synthetic data to which a normally distributed random error was added.[8] Incorrect weighting not only increased the variance of the estimated parameters (A and E_{act}), but also changed their average values. In addition, a close correlation between the fitted parameters (A and E_{act}) was found in the computer simulations. As a result, it is necessary to compute the covariance matrix in order to accurately compute the confidence interval of k values interpolated or extrapolated from the Arrhenius equation. Finally, as pointed out in Section 2.4.3, the inefficiency of incorrect weighting increases as the range of the data increases. For a decade range in k (roughly a 35°C range around room temperature if $E_{act} = 50$ kJ mol^{-1}), incorrect weighting results in an efficiency of only 30% (Table 2-3). In terms of precision of the estimated parameters, this is equivalent to throwing out 70% of the available data and basing the estimates on the remaining 30%.[9]

3.2.6 Other Empirical Expressions for Temperature Effects

A simple mathematical expression often is used to express the effect of temperature on rates of biological processes:

$$k_2/k_1 = \theta^{(T_2-T_1)} \tag{3-10a}$$

k_1 and k_2 are rate constants at temperatures (in °C) T_1 and T_2, respectively, and θ is an empirical constant that typically has a value in the range 1.03 to 1.10. θ can be estimated from k values at two temperatures:

$$\log k_2 - \log k_1 = (T_2 - T_1)\log \theta \tag{3-l0b}$$

Over narrow temperature limits, such as those of interest for biological processes, Equation 3-10a,b is an approximation of the Arrhenius equation. If E_{act} is large (as in biological processes), then $E_{act}/2.3RT_1T_2$ is approximately constant over narrow temperature limits and can be given the value $log\theta$. Thus,

$$\log \frac{k_2}{k_1} = \frac{E_{act}(T_2-T_1)}{2.3RT_1T_2} \approx (T_2-T_1)\log \theta \tag{3-11}$$

θ is not truly constant with temperature, and for a given set of rate constants at various temperatures, the value of θ obtained from Equation 3-11 depends slightly on the temperature range used to compute θ. Consider a reaction with $E_{act} = 12.0$ kcal mol^{-1} (~50 kJ mol^{-1}), a typical value for biological processes. Let the rate constant at 30°C be 1.00 arbitrary units. From the Arrhenius equation we compute that rate constants at 20 and 10°C would be 0.507 and 0.245, respectively. Using Equation 3-11, we compute $\theta = 1.0705$ over the range 20 to 30°C. Substituting θ into Equation 3-11 we extrapolate k at 10°C = 0.256, about a 4% error compared with the Arrhenius value. Similarly, over the range 10 to 20°C the rate constants obtained from the Arrhenius equation yield a θ of 1.0755, and substituting this value into Equation 3-11 yields a calculated k at 30°C of 1.05 (a 5% error).

Typical values of θ for important biological processes are listed in Table 6-2. A range of values exists for most processes, and the effect of temperature on process rates cannot be predicted accurately by indiscriminant use of literature values. For example, a range of θ from 1.05 to 1.13 has been reported[10,11] for denitrification in a wide range of environmental habitats, including marine and freshwater sediments, soils, and sewage sludges.

3.3 PROPERTIES OF WATER AND REACTANTS IN AQUEOUS SOLUTION

3.3.1 The Structure of Liquid Water

Water is by far the most common liquid on the Earth's surface, and its unique properties (Table 3-1) enable life to exist as we know it. Water is the medium for all reactions discussed in this book; a description of its structure and properties thus is in order. Relative to its low molecular weight, water has a very high boiling point

Table 3-1. Properties of Water

Property	Value	Comparison with normal liquids	Environmental significance
State (room temperature and pressure)		Liquid rather than gas like H_2S, H_2Se	Provides medium for life
Heat capacity (specific heat)	1.0 cal $g^{-1}°C$	Very high	Moderates climate
Latent heat of fusion	79 cal g^{-1}	Very high	Moderating effect; stabilizes temperature
Latent heat of evaporation	540 cal g^{-1}	Very high	Moderating effect; important in precipitation-evaporation balance
Density	1.0 g cm^{-3} (at 4°C)	High; anomalous maximum at 4°C for pure water	Freezing from surface; controls temperature distribution and water circulation
Surface tension	73 dyne cm^{-1} (at 20°C)	Very high	Affects adsorption, wetting, membrane transport
Dielectric constant	78	Very high	Makes water a good solvent for ions; shields electric fields of ions
Dipole moment	1.84×10^{-18} esu	High compared with organic solvents	Polarity important cause of above characteristics and the "solvent" properties of water
Viscosity	0.01 g $cm^{-1} s^{-1}$ (1.0 centipoise) at 20°C	Somewhat higher than organic solvents	Slows movement of solutes (reactants)
Transparency		High, especially in mid-visible range	Allows thick zone for photosynthesis and photochemical reactions
Heat conduction		Very high	Important in heat transfer in stagnant systems

Adapted from Horne, R. A., *Marine Chemistry,* Wiley-Interscience, New York, 1969.

and freezing point. Among common liquids it has the highest heat capacity, heat conduction, heats of vaporization and fusion, and dielectric constant. The latter parameter, which measures the attenuation rate of coulombic forces in a solvent compared to attenuation in a vacuum, is especially important in the dissolution of salts in water. The high dielectric constant of water permits similarly charged ions to approach each other much more closely before repulsive coulombic forces become important than is the case for solvents with low dielectric constants.

The unique properties of liquid water reflect the fact that the water molecules do not behave independently, but are attracted to each other and to many solutes by moderately strong hydrogen bonds. The strength of hydrogen bonds in ice or liquid water is about 23 kJ mol^{-1},[12] which is much weaker than the O–H bond of water (464 kJ mol^{-1}), but stronger than London-van der Waals dispersion forces (< ~4 kJ mol^{-1}). The angle between the two O–H bonds (105°) in water molecules is greater than the 90° expected for perpendicular p orbitals (Figure 3-4a). This is caused by repulsion between the two hydrogens and indicates some s-p hybridization in the oxygen. Oxygen's four remaining valence electrons occupy orbitals opposite from the hydrogens, in a distorted cube-like arrangement; this explains the molecule's large dipole moment. The unshared electron pairs attract hydrogen atoms of adjacent water molecules, forming moderately strong hydrogen bonds with lengths of 1.74 Å (measured in ice by X-ray diffraction), and leading to a three-dimensional structure.

The structure of ice is known with great accuracy (Figure 3-4b). Ice-I, the form that occurs under environmental conditions, has an open, hexagonal ring structure. Each ring contains six water molecules, each of which is surrounded by the oxygen atoms of four adjacent water molecules, in a tetrahedral arrangement. The density of ice calculated from measured bond lengths and the structure in Figure 3-4b is in good agreement with the measured density of ice.[3]

The degree of structure in liquid water is intermediate between the crystalline arrangement of ice and the absence of structure of gaseous water, where each molecule behaves independently. Models of the structure of water must account for the properties listed in Table 3-2, which suggest that water is a structured medium, and the variations in these properties with temperature. The heat of fusion of ice (330 J g^{-1}) indicates that only about 15% of the hydrogen bonding in ice is lost upon melting. The higher density of water than ice and the density maximum at 4°C also must be explained. The structure of water has been a subject of interest for many years, and many models have been proposed, including relic ice and clathrate structures.[3] The flickering-cluster model of Frank and Wen[13,14] now is accepted as reasonable if not exact.

According to the flickering-cluster model, water forms large clusters of hydrogen-bonded molecules of indeterminate structure (Figure 3-5a). The clusters have very short lives (~10^{-10} s) and are constantly forming and disintegrating with thermal fluctuations at the microregion.[3,12] Although the lifetime of individual clusters may seem trivially short, it is about a thousand times longer than the time between molecular vibrations (~10^{-13} s), and clusters thus can be said to have a real albeit highly ephemeral existence. The formation of hydrogen bonds is a cooperative phenomenon; formation of one bond facilitates formation of the next, because of dipole effects. Consequently, rather large clusters are formed. The average cluster has 65 water molecules at 0°C, but only 12 at 100°C.[14] However, because the number of clusters increases with temperature, the fraction of molecules not in clusters decreases much more slowly. Unclustered water is considered to be "free".

Clusters have been described by some as icelike, but Frank and Wen eschewed any specific structural interpretation. According to Stillinger,[12] water consists of a random, macroscopic network of hydrogen bonds. The anomalous properties of water are said to arise from the competition between relatively bulky ways of

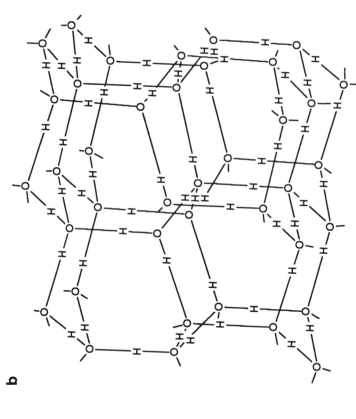

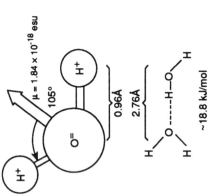

Figure 3-4. (a) Structure of water molecule; (b) arrangement of water molecules in ice I lattice to form hexagonal rings; and (c) detailed view of tetrahedral arrangement of water molecules surrounding each water molecule in ice I. [(a), (c) From Horne, R. A., *Marine Chemistry*, Wiley-Interscience, New York, 1969. With permission; (b) from G. Nemethy and H.A. Scheraga, *J. Chem. Phys.*, 36, 3382 (1962). Copyright, American Institute of Physics.]

c

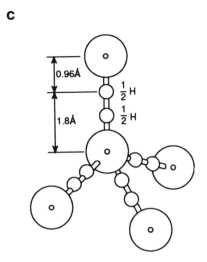

Figure 3-4 (continued).

connecting molecules into local patterns by strong bonds and tetrahedral angles and more compact arrangements that have more bond strain and broken bonds. These alternatives constitute a continuum of structural possibilities rather than a pair of options. If the clusters were relic ice structures, the polygons formed by hydrogen bonding would reflect the hexagonal structure of ice and contain an even number of molecules (6, 10, 14, etc.). Breakage of a hydrogen bond in the three-dimensional hexagonal ring structure of ice forms a 10-membered ring; loss of a second hydrogen bond yields a 14-membered ring (see Figure 3-4b). The frequency distribution of polygon size in water (Figure 3-5b) shows no evidence for relic ice structures, however; and there is no way that five-member polygons (the modal size) could result from ice-I.

Solvation of small ions produces structures incompatible with the general liquid structure, and the net effect is an unstructured transition zone between the hydrated ion and the rest of the solvent (Figure 3-5c). Changes in viscosity upon addition of salts indicate changes in the degree of solvent structure.[3] Some salts, like NaCl, are overall structure makers that increase viscosity, while others, like KCl, and especially 2:2 electrolytes like $CaSO_4$, are structure breakers that decrease viscosity. Nonpolar solutes prefer the interstices between clusters and promote structure in liquid water. As a result, they have high negative entropies of solution.

3.3.2 Forces Between Atoms, Molecules, and Ions

The attractive and repulsive forces between ions and/or neutral molecules are important in solution-phase kinetics because the reactants must approach each other closely if chemical reaction is to occur. Molecular (or ionic) interactions can be

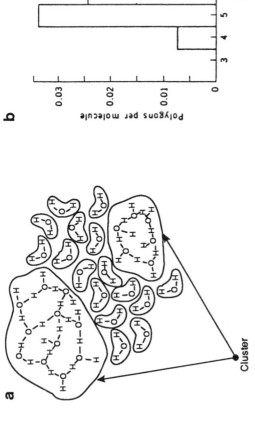

Figure 3-5. (a) Frank-Wen flickering cluster model for liquid water. (b) Frequency distribution of polygon sizes formed by hydrogen bond in water at 10°C. (c) Fractions of water molecules involved in various number of hydrogen bonds at 10°C. (d) "Two-zone" model for effects of structure-breaking solutes (K^+, Li^+) on structure of water: zone A, structurally-enhanced, electrostatically-enhanced, electrostricted hydration sphere; zone B, region of disrupted structure; zone C, region of "normal" water. (e) "Three-zone" model for structure-enhancing solute ions; zones $D_1 + D_2$ represent solvated ion in Frank-Wen cluster. (f) Orientation preference for water molecules next to nonpolar solutes preserves maximum number of hydrogen bonds. [(a) from Nemethy, G. and H.A. Scheraga, *J. Chem. Phys.*, 36, 3382 (1962). Copyright Amer. Inst. Physics; (b,c,f) from Stillinger, F. H., *Science*, 209, 451 (1980). Copyright 1980 by AAAS; (d) from Frank, H.S. and W. Y. Wen, *Disc. Faraday Soc.*, 24, 133 (1957). Copyright, Royal Society of Chemistry, London; (e) from Horne, R.A., *Surv. Progr. Chem.*, 4, 1 (1968). Copyright, Academic Press.]

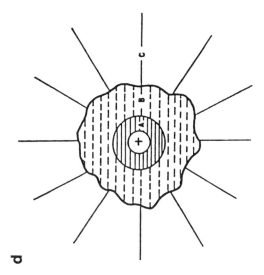

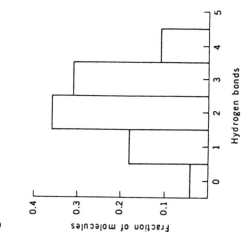

Figure 3-5 (continued).

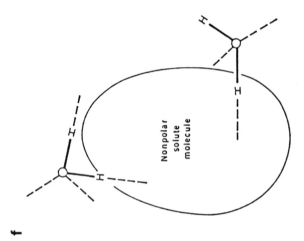

f

Nonpolar
solute
molecule

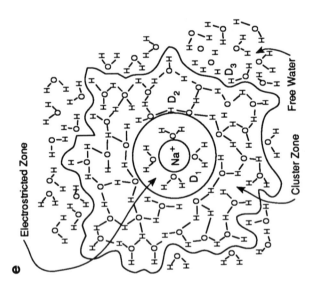

e

Electrostricted Zone

Cluster Zone   Free Water

Figure 3-5 (continued).

divided into short-range and long-range forces. The former forces occur when the electronic orbitals of molecules overlap. The force is attractive when overlap leads to bond formation and highly repulsive when bonding interactions cannot occur. Long-range attractive or repulsive forces extend beyond the electronic orbitals of molecules; their magnitude varies inversely as a power function of intermolecular distance.

It is convenient to express intermolecular interactions in terms of the potential energy (V) of interaction.[15] Force and potential are related by the formula: $F(r) = -dV(r)/dr$; r is the internuclear distance. Long-range potentials are essentially electrostatic and include ionic and dipole interactions.

The potential (in ergs) between two ions with charges $z_A e$ and $z_B e$ separated by a distance r_{AB} (cm) is given by Coulomb's Law:

$$V(r) = z_A z_B e^2 / D r_{AB} \qquad (3\text{-}12)$$

D is the dielectric constant of the solvent, and e is the charge on an electron (4.80×10^{-10} esu). V is positive for ions of the same sign and negative for ions of opposite sign. Positive V implies repulsive forces (higher energy); negative V implies attractive forces. As mentioned earlier, bulk water has a high dielectric constant ($D = 78$). However, water bound in the first coordination sphere of metal ions has a much lower D. Consequently, the appropriate value of D to calculate potentials when reactants are separated by only a few water molecules is uncertain. Because D appears in all equations for long-range interactions, this problem also extends to potentials between nonionic species.

For interactions between an ion and a polar molecule, the potential is attractive and is given by

$$V(r) = -\frac{z_A e \mu_B \cos \theta}{D r_{AB}^2} \qquad (3\text{-}13)$$

μ_B is the dipole moment of the polar molecule (in esu cm or 10^{-18} debyes), and θ is the angle between the direction of the dipole and r_{AB}.[15,16] If the ion is aligned with the direction of the dipole ($\theta = 0$), Equation 3-13 becomes $V(r) = -z_A e \mu_B / r_{AB}^2$. Statistical averaging of V(r) over all orientations causes V(r) to be proportional to r_{AB}^{-4}.

Electric fields (from ionic solutes) can induce dipoles in nonpolar molecules, and the potential for ion-induced dipole interactions is

$$V(r) = -\frac{z_A^2 e^2 \alpha_B}{2 D r_{AB}^2} \qquad (3\text{-}14)$$

α_B is the polarizability (cm^3) of the nonpolar molecule and relates the size of the induced dipole to the size of the electric field:

$$\mu = \frac{\alpha_B z_A e}{D r_{AB}^2} \qquad (3\text{-}15)$$

Attractive interactions between neutral molecules (excluding hydrogen bonding) are known as van der Waals forces. Several types exist, but in each V(r) is proportional to r_{AB}^{-6}. For end-on dipole-dipole interactions, V(r) is given by

$$V(r) = \frac{-2\mu_A\mu_B}{Dr_{AB}^3} \qquad (3\text{-}16a)$$

and the average interaction potential over all orientations is

$$V(r) = -\frac{2\mu_A^2\mu_B^2}{3kTDr_{AB}^6} \qquad (3\text{-}16b)$$

where **k** is the Boltzmann constant (gas constant per molecule).

Polar molecules induce dipoles in nonpolar molecules, and the potential for dipole-induced dipole interactions is given by

$$V(r) = -\frac{(\alpha_A\mu_B^2 + \alpha_B^2)}{D^2 r_{AB}^6} \qquad (3\text{-}17)$$

Finally, even nonpolar molecules can induce dipoles in nonpolar molecules because of ephemeral fluctuations in the electron distribution around any molecule. Induced dipole-induced dipole interactions, called London dispersion forces, yield the following potentials for molecules of the same kind:

$$V(r) = -\frac{3\alpha^2 h\nu_o}{4Dr^6} \qquad (3\text{-}18)$$

where $h\nu_o$ is the ionization energy (ergs). Assumptions inherent in the equations for V(r) and their limitations are discussed elsewhere.[15,16]

Development of an accurate theoretical model for molecular interaction potentials that includes both short- and long-range interactions has proven elusive, but several empirical expressions are available. The best-known and most-realistic model is the Lennard-Jones [6–12] potential (Figure 3-6):

$$V(r) = 4e\left[(\sigma/r)^{12} - (\sigma/r)^6\right] \qquad (3\text{-}19)$$

e is the minimum potential or depth of the "potential well", and σ is the value of r at which V = 0. The equation consists of an attractive term proportional to r^{-6}, representing London dispersion forces, and an empirical, rapidly increasing short-range repulsive potential proportional to r^{-12}.

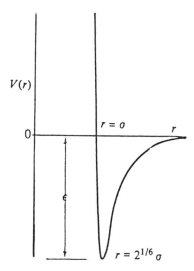

Figure 3-6. Lennard-Jones [6–12] model (Equation 3-19) for intermolecular potentials. (From Weston, R. E., Jr. and H. A. Schwarz, *Chemical Kinetics,* Prentice-Hall, Englewood Cliffs, NJ, 1972. With permission.)

3.3.3 Diffusion of Solutes in Water: Fick's Laws

In the absence of mechanical agitation, reactants in solution become mixed and encounter each other by diffusion. This process is caused by the kinetic motion of molecules and is characterized by small, random movements often described as a random walk. Molecules in any fluid are constantly moving and colliding with each other. The direction of movement for a molecule changes with each collision, and the distance a molecule travels between collisions (the mean free path) depends on its size and the structure of the solvent. As described above, the exact structure of water is unknown and constantly changing, but the mean free path or elementary jump distance is thought to be about the same as the diameter of water molecules themselves.

The net effect of the random molecular movements in liquids is to homogenize solutions, i.e., eliminate regions of high or low solute concentrations and the gradients between such regions. The basis for the decrease in concentration gradients, which is a fundamental characteristic of diffusion, can be understood on probabilistic grounds by a simple two-box example of a one-dimensional concentration gradient, as decribed by Fischer et al.[17] Suppose we have two small boxes of equal size separated by a permeable membrane. The left box contains 100 solute molecules; the right contains 50. If the solutes move randomly, the number of molecules crossing the membrane should be proportional to the number of molecules near it. Thus, the number passing from left to right is proportional to the number on the left, and the number passing from right to left is proportional to the number on the right. Suppose that there is a probability of 0.1 that a solute molecule in either box crosses the membrane during time period Δt. At the end of Δt the left box will have lost ten of its original solute molecules and gained five from the right box, for a net loss of five. The right box will lose five of its original molecules, but gain ten from the left, for a net gain of five. The boxes now have 95 and 55 molecules, respectively, and the concentration difference has decreased from 50 to 40. The flux F of molecules across a surface like the membrane in our example is defined as the net rate of mass transfer per unit area of the surface and per unit time. Thus,

$$F = P_t(M_1 - M_r)/A \qquad (3\text{-}20a)$$

P_t (units of time^{-1}) is the probability of transfer per unit time; M is mass of solute; and A is the membrane surface area. Defining $[C]_1 = M_1/A\Delta x$ and $[C]_r = M_r/A\Delta x$ as solute concentrations leads to

$$F = P_t\Delta x\left([C]_1 - [C]_r\right) \qquad (3\text{-}20b)$$

Taking the limit as $\Delta x \to 0$ gives $\partial[C]/\partial x = ([C]_r - [C]_1)/\Delta x$, and substituting into Equation 3-20b yields

$$F = -P_t(\Delta x)^2 \partial[C]/\partial x \qquad (3\text{-}20c)$$

Net transport thus is always down gradient, from high to low concentration. In the two-box example, transfer probability is a function of molecular motion, which depends on temperature and solvent viscosity, and box size. (The bigger the box, the fewer the molecules close to the membrane.) As Fischer et al.[17] point out, mass transfer should not depend on an arbitrarily defined box size. The only way to avoid having F depend on Δx is for $P_t(\Delta x)^2$ to be constant (for a given temperature, solute, and solvent). This term has units of (length)^2/time, e.g., cm^2 s^{-1}, and is called the diffusion coefficient, D. This leads directly to Fick's first law of diffusion:

$$F = -D\partial[C]/\partial x \qquad (3\text{-}20d)$$

Equation 3-20d was first formulated as an empirical relationship by Adolph Fick in 1855[18] in analogy to earlier laws by Fourier for heat flow* and Ohm for conductance of electrical current. For diffusion in three dimensions, Fick's first law can be written as

$$F_{xyz} = -D(\partial[C]/\partial x + \partial[C]/\partial y + \partial[C]/\partial z) = -D\nabla C \qquad (3\text{-}20e)$$

F_{xyz} is the mass flux vector with components F_x, F_y, and F_z.

---

* The equations governing heat flow and molecular diffusion are mathematically identical, and solutions of many common diffusion and heat flux problems are available in several texts.[19,20]

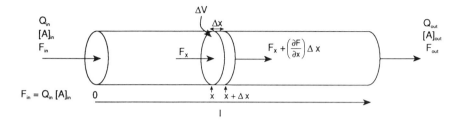

Figure 3-7. Diffusive flux in control volume of tubular reactor.

Diffusion coefficients of gases are roughly between 0.1 and 1.0 cm² s⁻¹ under environmental conditions, but for solutes in water they are much lower — roughly 10^{-6} to 10^{-5} cm² s⁻¹. Diffusion coefficients vary with temperature, viscosity, and solute size according to the Stokes-Einstein equation:

$$D_i = \mathbf{k}T/6\pi\eta r_i \tag{3-21}$$

where $\mathbf{k}$ is Boltzmann's constant, η is viscosity (g cm⁻¹ s⁻¹) and r is molecular radius. Viscosity varies with temperature, and the variation in D_i with T primarily reflects the change in viscosity with T. According to Walden's rule, $D\eta \approx$ constant.

A second important diffusion relationship can be derived from Fick's first law and the principle of mass conservation. Consider the diagram of fluxes for the control volume between x and x + Δx in Figure 3-7. If the flux at x is F_x, the flux at x + Δx is $F_x + (\partial F/\partial x)\Delta x$, and the difference simply is $(\partial F/\partial x)\Delta x$. If the diffusing species is conservative (i.e., it is neither formed nor reacts within the system), then the change in flux across the control volume must be equal to the rate of change in concentration within the control volume:

$$\Delta[C]/\Delta t = -\Delta F/\Delta x$$

or as Δt and Δx → 0, $(\partial[C]/\partial t)_x = -(\partial F/\partial x)_t$. Substituting Fick's first law for F yields:

$$\frac{\partial[C]}{\partial t} = \frac{\partial}{\partial x}\left[D\frac{\partial[C]}{\partial x}\right] = D\left[\frac{\partial^2[C]}{\partial x^2}\right] \tag{3-22a}$$

This equation is known as Fick's second law and it relates the time rate of change in concentration at a given point to its second derivative with respect to distance from the point. In three dimensions, Fick's second law is

$$\partial[C]/\partial t = D\nabla^2[C] \tag{3-22b}$$

3.4 ENCOUNTER THEORY FOR REACTIONS IN SOLUTION

3.4.1 Diffusion and Encounter of Uncharged Solutes

Quantitative descriptions of reaction rates in solution must consider the fact that reactants are surrounded by solvent molecules and that collisions between reactants require displacement of the solvent. The standard physical model for solution-phase reactions portrays the reactants in a solvent "cage" (Figure 3-8a). The molecules vibrate against the walls (i.e., the solvent molecules) of this loose, low-energy cage and occasionally escape to an adjacent position. The frequency with which reactant molecules encounter each other in the same solvent cage has been derived from principles of random diffusion. A simple model described by Adamson[4] is summarized below.

Random diffusion of molecules in liquids is assumed to occur by elementary jumps of distance $\lambda = 2r$, where r is the molecular radius. From the Einstein-Smoluchowski equation, which defines diffusion coefficients in terms of λ, we can write,

$$D = \lambda^2/2\tau; \quad \text{or} \quad \tau = \lambda^2/2D \qquad (3\text{-}23a)$$

D is the diffusion coefficient (cm^2 s^{-1}), and τ is the average time between jumps. Equation 3-23a assumes a continuous medium. For a liquid with a semicrystalline structure, such as water, a more accurate relationship is[4]

$$D = \lambda^2/6\tau; \quad \text{or} \quad \tau = \lambda^2/6D \qquad (3\text{-}23b)$$

The self-diffusion coefficient of water is $\sim 1 \times 10^{-5}$ cm^2 s^{-1}, and λ for water molecules is about 4×10^{-8} cm (4 A). The length of time a water molecule spends in a given solvent cage, τ, thus is $\sim 2.5 \times 10^{-11}$ s. We can compare this to the molecular vibration frequency based on thermal energy:

$$\nu_f = \mathbf{k}T/h \qquad (3\text{-}23c)$$

where ν_f is frequency (s^{-1}), and h is Planck's constant. At 25°C, $\nu_f = 6.2 \times 10^{12}$ s^{-1}; i.e., the average time between vibrations is $\sim 1.5 \times 10^{-13}$ s. Consequently, the average molecule vibrates about $2.5 \times 10^{-11}/1.5 \times 10^{-13}$ or 150 times in its solvent cage before jumping to a new position.

A similar analysis can be done for solutes. The encounter frequency for two solute molecules by random diffusion (Figure 3-8) is of special interest. Because the structure of liquid water is poorly known and because the shapes of solute molecules affect the probability of encounter, the analysis necessarily is simplified. If we assume the solute molecules A and B are spherical and about the same size as water, then each solute molecule will have 12 nearest neighbors, and on each diffusional jump it will encounter six new ones. The probability that one of the new molecules

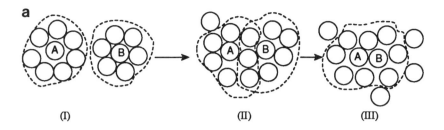

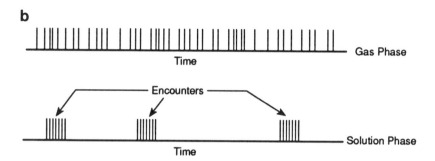

Figure 3-8. (a) Encounter of reactants A and B in solution: (I) reactants in separate cages; (II) cages of A and B overlap; and (III) A and B in same cage. (b) Frequency of collisions in gas phase and of encounters and collisions in solution (time scale not exact). (Redrawn from Adamson, A. W., *A Textbook of Physical Chemistry*, 2nd ed., Academic Press, New York, 1979.)

encountered by A is a molecule of B is related to the mole fraction of B: $X_B \approx$ (molecules of B cm^{-3})/(molecules of solvent cm^{-3}) $= n_B/(1/\gamma\lambda^3)$, where γ is a geometric packing factor, and λ^3 is the molecular volume. The probability that A will encounter a molecule of B on any diffusional jump is $6X_B$. Substitution into Equation 3-23b gives

$$1/\tau_{AB} = 6D\left(6n_B\gamma\lambda^3\right)/\lambda^2 = 36\gamma\lambda n_B D \tag{3-24a}$$

For water, $\gamma \approx 0.74$, and the effective diffusion coefficient $D_{AB} = D_A + D_B$. λ is the sum of molecular radii, r_{AB}. Equation 3-24a thus becomes

$$1/\tau \approx 25r_{AB}D_{AB}n_B \tag{3-24b}$$

The number of encounters per centimeter^3 per second is given by Equation 3-24b multiplied by n_A, the number of molecules of A per cubic centimeter. If [A] and [B] are expressed in mol L^{-1} (*M*), we multiply by Avogadro's number/1000, and the encounter frequency becomes (*M* s^{-1}):

$$Z_{e,AB} = 2.5 \times 10^{-2} r_{AB} D_{AB} N_o [A][B] \qquad (3\text{-}25)$$

Dividing both sides of Equation 3-25 by [A][B] gives the frequency factor, A_e, or encounter frequency per unit concentration of A and B, with units of $M^{-1}\,s^{-1}$. A_e has the units of a second-order rate constant and is equivalent to the rate constant (k_D) for a diffusion-controlled reaction (one that occurs on every encounter). $Z_{e,AB}$, the rate of such a reaction, is equal to A_e when reactant concentrations are 1 M. The value of A_e depends on the diffusion coefficients and radii of the reactants; for small molecules it is in the range $\sim 10^9$ to $10^{10}\,M^{-1}\,s^{-1}$. For example, if $D_A = D_B = 1 \times 10^{-5}\,cm^2\,s^{-1}$ and $r_{AB} = 4 \times 10^{-8}$ cm, $A_e = 1.2 \times 10^{10}\,M^{-1}\,s^{-1}$; and $Z_{e,AB} = 1.2 \times 10^{10}$ $M\,s^{-1}$ when [A] and [B] = 1 M.

A more rigorous derivation of the diffusion-encounter rate constant[15,16] is based on diffusion toward the surface of a sphere around the reacting molecule and leads to values of k_D (or A_e) that are about half those given by Equation 3-25:

$$k_D = 4\pi D_{AB} r N_o / 1000 \qquad (3\text{-}26a)$$

or $\qquad\qquad\qquad k_D = 4\pi D_A r N_o / 1000 \qquad (3\text{-}26b)$

if the reactants are the same species. r is the radius at which A and B are in contact; for small ions $r \approx 2\text{-}5 \times 10^{-8}$ cm. Given the assumptions involved in both derivations, the differences in predicted values from Equations 3-25 and 3-26 are not too important.

3.4.2 Assumptions and Implications of the Encounter Model

Equation 3-26 was first derived by Smoluchowski[21] for encounter rates of co-agulating colloids. Several difficulties arise in applying this model to encounter rates of ions and molecules in aqueous solution:[22-24]

1. The model neglects long-range forces between reactants.
2. It assumes that diffusive displacements are small compared with the dimensions of the particles, but this is not true for small molecules.
3. It assumes that each reactant acts as a stationary sink around which a concentration gradient of the other reactant is established, but in reality all molecules are moving simultaneously.
4. It treats the solvent as a structureless continuum, but aqueous solutions are structured.

Modifications that have been proposed to eliminate these assumptions[15,16,22-24] generally result in much more complicated models. As described below, coulombic interactions between ions can be accounted for, but quantifying long-range forces between nonionic reactants is still problematic. Inaccuracies from the second assumption have been minimized by adding a final jump-like step to the diffusion model.[23] Motion up to the second coordination sphere is treated as diffusion with infinitely small displacements, but the transition from the second to the first coordination sphere is treated as a discrete jump with a certain probability. Assumption (3) is a problem primarily in solvents with very low viscosity and

Table 3-2. Rate Constants for Some Diffusion-Limited Elementary Reactions

Reaction	k, 10^{10} M^{-1} s^{-1}	
	Measured	Calculated
$OH + OH \rightarrow H_2O_2$	0.5	0.6
$OH + C_6H_6 \rightarrow C_6H_6OH$	0.33	0.8
$H^+ + NH_3 \rightarrow NH_4^+$	4.3	4.0
$H^+ + SO_4^{2-} \rightarrow HSO_4^-$	10.0	6.5
$H^+ + OH^- \rightarrow H_2O$	14.0	11.0
$H^+ + HS^- \rightarrow H_2S$	7.5	9.0
$e^-_{aq} + NO_3^- \rightarrow NO_3^{2-} (\rightarrow NO_2^- + \cdot O^-)$	0.85	0.97
$e^-_{aq} + Co(NH_3)_6^{3+} \rightarrow Co(NH_3)_6^{2+}$	8.2	9.3

Summarized from Weston, R. E., Jr. and H. A. Schwarz, *Chemical Kinetics*, Prentice-Hall, Englewood Cliffs, NJ, 1972.

reactions with slow chemical steps,[24] but in the latter case diffusion-encounter is not the rate-controlling step anyway. Perhaps the most serious difficulty in improving diffusion-encounter models is that of describing the effects of discrete solvent structure on reacting molecules as they approach each other.

When electrostatic forces of attraction or repulsion between reacting ions are taken into account, Equation 3-26a becomes[15]

$$k_D = \frac{-4\pi D_{AB} z_A z_B r_o N_o}{1000\left[1 - \exp(z_A z_B r_o / R)\right]} \tag{3-27}$$

$r_o = 7.1 \times 10^{-8}$ cm at 25°C is a constant with units of length ($r_o = e^2/DkT$) see reference 15 for a derivation). Values of k_D computed for diffusion-limited reactions between ions are in good agreement with observed rate constants (Table 3-2).

The diffusion-encounter model for reactions in solution differs from the collision model for gas-phase reactions. Collision frequencies of gas-phase reactants can be calculated from the kinetic theory of gases (Section 3.4.4). If colliding gas molecules do not react, they immediately separate and move on independent paths. The frequency factor for gas-phase collisions is about 10^{11} M^{-1} s^{-1}. For the same concentrations, reactants in solution encounter each other less frequently than they do in the gas-phase (Figure 3-8b), but because of the solvent-cage effect, solutes remain together as an encounter complex for about 2.5×10^{-11} s and have ~150 collisions with each other. When activation energies are high, the probability that a collision leads to a reaction is very small, and the difference in collision patterns between gas- and solution-phase reactions is not of consequence in predicting reaction rates. When activation energies are low or zero, a reaction will occur after a few collisions (or even one), and every encounter will lead to a reaction. Such reactions are diffusion controlled, and their rates are given by the rate of encounters. Rates of diffusion-controlled encounters are moderately temperature dependent and have an E_{act} of 12 to 17 kJ mol^{-1}, the apparent activation energy for viscous flow.

The solvent-cage phenomenon has important implications for free radical formation by photoactivation in solution. Because radicals are highly reactive and have little or no activation energy for recombination, a pair formed in a solvent cage may recombine before the radicals can diffuse apart.[25]

3.4.3 Slow Reactions of Ions and Molecules in Solution

The diffusion-encounter model presented above does not consider chemical transformation once reactants encounter each other. The overall encounter-reaction process can be considered as a two-step sequence with two rate constants: k_D for diffusion-encounter, and k_c for chemical reaction. When $k_D \gg k_c$, the chemical step controls the overall reaction, and the overall rate constant (k_r) approaches k_c. Conversely, when $k_c \gg k_D$, diffusion controls, and k_r approaches k_D. When k_D and k_c are of comparable size, both must be considered. In analogy to the law governing electrical resistances in series, the total resistance for the reaction is equal to the sum of the resistances for the separate steps. Resistance is the inverse of conductance, and rate constants are analogous to conductivities. Thus, we can write,

$$1/k_r = 1/k_D + 1/k_c \qquad (3\text{--}28a)$$

or
$$k_r = k_D k_c /(k_D + k_c) \qquad (3\text{--}28b)$$

Substitution of Equation 3-27 into 3-28b yields an expression for the overall rate constant of rapid ionic reactions.

For slow reactions ($k_c \ll k_D$), the observed rate constant k_r can be considered equal to the product of an intrinsic rate constant k_R and an exponential function of the interaction potential energy:[15]

$$k_r = k_R \exp(-V(r)/\mathbf{k}T) \qquad (3\text{-}29a)$$

Substitution of Equation 3-12 for V(r) of ionic interactions leads to

$$k_r = k_R \exp\left\{-z_A z_B e^2 / D r_{AB} \mathbf{k}T\right\} \qquad (3\text{-}29b)$$

Unfortunately, there is no way to make *a priori* estimates of k_R for slow reactions, and Equation 3-29b has no predictive value. Nonetheless, it does show that the rate depends on the dielectric constant of the medium as well as on the product of ionic charges. Similar substitution of Equation 3-13, which defines V(r) for ion-dipole interactions, into Equation 3-29a yields an equation that expresses the effect of dielectric constant on rates of ion-dipole reactions.[15] However, because ion-dipole-solvent interactions are more complicated than predicted by simple electrostatic theory, this equation has limited utility. Formation of ion-dipole encounter complexes involves substitution of a polar reactant for a polar water molecule in an ion hydration sphere. Consequently, specific water-ion interactions may be as important as ion-dipole interactions in affecting the reaction rate. Complicated models have been developed to describe the effects of the dielectric constant on rates of ion-dipole and dipole-dipole reactions in various solvents,[26] but these models are not useful in predicting effects of ionic strength on rates of such reactions.

Electrostatic factors like the dielectric constant of the solvent are not important in reactions between neutral molecules. In fact, for reactions of neutral molecules that occur in the gas phase as well as in solution, the solvent tends to play a minor role as space filler. Rate constants for such reactions often are similar in a variety of solvents and in the gas phase.[25]

3.4.4 Collision Model for Gases

Arrhenius originally proposed that activation energy represents the minimum energy level reactants must attain for reaction. The exponential factor in his equation represents the fraction of molecules having this energy. This factor is the Maxwell-Boltzmann distribution:

$$n/N = \exp(-E/RT) \tag{3-30}$$

n is the number of molecules having an energy of E or greater at temperature T, and N is the total number of molecules in the system. When E is large, n/N is very small; increasing T at constant E increases the fraction of molecules with energy greater than E. If molecules require a certain energy to react, increasing T will increase the fraction that can react and will increase the rate of reaction.

Further development of collision theory came early in the 20th century, when the pre-exponential factor in the Arrhenius equation was quantified as the frequency of collisions between reactants. The exponential factor became regarded as the fraction of colliding pairs with $E > E_{act}$. Rates of bimolecular reactions thus were expressed as

$$\text{Rate} = \begin{bmatrix} \text{number of collisions of} \\ \text{reactants per unit time} \end{bmatrix} \begin{bmatrix} \text{fraction of collisions whose} \\ \text{impact pairs have } E > E_{act} \end{bmatrix} \tag{3-31}$$

The collision frequency of gases (Z_{AB}) can be computed from the kinetic theory of gases, which is derived from classical mechanics:

$$Z_{AB} = d_{AB}^2 \{8\pi kT/m_{AB}\}^{1/2} n_1 n_2 = (\text{Constant})T^{1/2}[A][B] \tag{3-32}$$

Z_{AB} is the number of collisions per cubic centimeter per second between two kinds of molecules; d_{AB} is the average collision diameter [$(d_A + d_B)/2$]; m_{AB} is the reduced mass [$m_A m_B/(m_A + m_B)$] — a term derived from classical mechanics;[15] k is the Boltzmann constant (R/N_o, where N_o is Avogadro's number); n_A and n_B are concentrations in molecules/cm³; and [A] and [B] are in mol/L. For gases at standard conditions there are 10^{10} to 10^{11} collisions per molecule per second or about 10^{28} collisions cm⁻³ s⁻¹. Substituting Equation 3-32 and the Maxwell-Boltzmann term into Equation 3-31, we obtain for bimolecular gas-phase reactions[27]

$$\text{Rate} = (\text{Constant})T^{1/2}\exp\left(-E_{act}/RT\right)[A][B] \qquad (3\text{-}33)$$

The rate constant ($M^{-1}\,s^{-1}$) is

$$k = A'T^{1/2}\exp\left(-E_{act}/RT\right) \qquad (3\text{-}34a)$$

where $\qquad A' = d_{AB}^2\{8\pi k/m_{AB}\}^{1/2}\,N_o/1000 \qquad (3\text{-}34b)$

A' is reduced by a factor of two for collisions between like molecules. Equation 3-34a implies that the Arrhenius pre-exponential factor, A, varies with temperature; $A = A'T^{1/2}$. The Arrhenius equation assumes A is constant and only approximates the temperature dependency of rate constants. Plots of lnk vs. 1/T thus are not perfectly linear. However, the factor $T^{1/2}$ is small compared with the exponential term, and deviations from linearity are small.

Collision theory yields accurate predictions of reaction rates for a few simple bimolecular gas-phase reactions, but measured rates for most reactions are lower than predicted by factors of 10 to 10^6. This difficulty was recognized early, and Equation 3-5 was modified by adding an empirical steric factor ($P = A_{obs}/A_{calc}$). Simple collision theory assumes that only activation energy and a rigid sphere collision frequency determine the rate of reaction. It is obvious, however, that not all collisions between complicated molecules lead to reaction; the reactant molecules must have the proper spatial orientation at collision so that energy and group transfer can take place. The steric factor represents the fraction of energetically favorable collisions that also have the proper spatial orientation. Steric factors of 10^{-1} to 10^{-2} are understandable on this basis, but factors as low as 10^{-6} are difficult to explain on the basis of steric effects. Classical collision theory, which treats reacting molecules as structureless, hard spheres, thus is not adequate to predict reaction rates from first principles because it does not take molecular complexity into account, except in terms of the empirical steric factor. Vibrational and rotational energy states for reacting molecules must be considered for a more accurate rate model, and this requires quantum or statistical mechanical approaches.

3.4.5 Reconciliation of First-Order Reactions with Collision Theory: The Lindeman Model

The fact that many reactions are first order would seem to pose major difficulties for classical collision theory. The activation energy should come from the kinetic energy provided by collision, but first-order kinetics implies that the rate does not depend on collision frequency. Lindeman[28] explained this by postulating a mechanism with a time lag between activation and reaction:

$$A + A \underset{k_{-1}}{\overset{k_1}{\rightleftharpoons}} A + A^* \qquad (3\text{-}35a)$$

$$A^* \overset{k_2}{\rightarrow} P \qquad (3\text{-}35b)$$

Some fraction of collisions of A with other molecules of A (or any other molecule M) results in energy gain to an activated state A^*. Although A^* has the energy for reaction, it may not be focused in the appropriate bond, and the reaction must await vibrational redistribution into that bond. A^* can collide with a ground-state molecule during this period, with the probable result that A^* will be deactivated. Steady-state treatment of A^* leads to the expression:

$$d[P]/dt = \frac{k_1 k_2 [A]^2}{k_2 + k_{-1}[A]}$$

(3-36a)

If deactivation is much faster than reaction, then $k_{-1}[A] \gg k_2$, and Equation 3-36a reduces to a first-order expression:

$$d[P]/dt = k_2 K_1 [A] = k_{app}[A]$$

(3-36b)

$K_1 = k_1/k_{-1}$ is the equilibrium constant for reaction 3-33a, and $K_1[A]$ is the equilibrium concentration of A^*. At sufficiently low $[A]$, $k_{-1}[A] < k_2$, and $d[P]/dt \approx k_1[A]^2$. Reactions following this mechanism thus become second order at low reactant concentrations.[29]

Major advances have been made in collision theory for first-order reactions since the Lindeman model was formulated. The Rice-Ramsberger-Kassel (RRK) model showed that the potential energy of the activated state could come from both the kinetic energy of collision and a molecule's internal energy by vibrational redistribution. This explains the high-frequency factors ($> \mathbf{k}T/h$) (see Section 3.5.2) of some first-order reactions. The Rice-Ramsberger-Kassel-Marcus (RRKM) model[15] uses statistical mechanics to predict rates at which different forms of the excited state A^* (molecules with sufficient energy to react) are converted to $A^{\ddagger}$, the excited species with the proper configuration to react. This model is a blend of quantized RRK collision theory and transition-state theory, which is described in the following section.

3.5 TRANSITION-STATE/ACTIVATED-COMPLEX THEORY

3.5.1 Nature of the Transition State

Deficiencies in collision theory for gas-phase reactions led chemists to develop a more fundamental theory of reaction kinetics, called the absolute reaction rate or transition-state theory (TST). Many individuals contributed to the theory, but its prime developer was Henry Eyring, who articulated it in 1935.[30-33] The theory is based on quantum and statistical mechanics, and it uses partition functions to quantize energy states of reactants and intermediates.

The transition state is a somewhat artificial concept. It has been described as the "reactant configuration of no return";[34] i.e., once the reactants reach this configuration, they must proceed to form products. It is artificial in the sense that the reactants are continuously changing rather than in a static configuration. Even if we ignore this problem, the transition state is so short lived that problems arise in

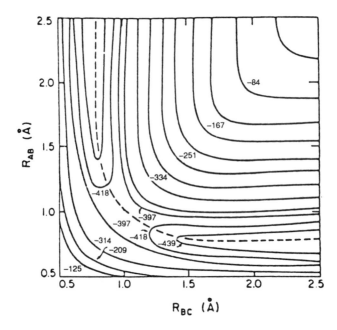

Figure 3-9. Potential energy surface for the reaction H + H$_2$ → H$_2$ + H, based on London-Eyring-Polanyi valence bond method.[31] R = interatomic distance (Å); dashed line represents reaction coordinate. Transition state occurs at "saddlepoint", where R$_{AB}$ = R$_{BC}$ ≈ 0.9 Å. Energies in kcal mol^{-1} relative to zero energy at infinite separation of reactants. Saddlepoint energy is ~8 kcal mol^{-1} higher than that of reactants or products. (From Eysing, H., S.H. Lin, and S.M. Lin, *Basic Chemical Kinetics*, Wiley-Interscience, New York, 1980. With permission.)

attributing thermodynamic (equilibrium) properties to it. Theoretical calculations place the lifetime of such transition states in the subpicosecond (<10^{-12} s) region. Nonetheless, recently developed ultrashort-pulse laser techniques have made it possible to observe transition-state configurations and lifetimes directly for simple gas-phase reactants by "real-time femtosecond transition-state spectroscopy".[34] These methods are an important complement to theoretical calculations of transition state conditions by the "*ab initio*" methods described briefly below.

The nature of the transition state can be understood by examining the potential energy surface (PES) (Figure 3-9) of the reaction: H' + H–H → H'–H + H.[4,33] This was the first reaction to be quantized by TST techniques.[35] H is a hydrogen atom; the symbol ' is used for tracking purposes. The reactants and products exist at the bottom of potential energy wells; i.e., they are stable species. As the diagram shows, the potential energy of the system increases as the reactants approach each other. An infinite number of pathways is possible, but most involve large increases in energy. The path with the smallest energy increase is preferred, and the maximum point on the minimum-energy path occurs at the "saddle point" of the PES. At this point, called the transition state, the surface is at a maximum in the direction of the reaction coordinate, but it increases in all other directions. In the H′ + H–H reaction, the amount of extra energy at the transition state (i.e., E_{act}) is 33.4 kJ/mol, and it occurs when the atoms are each 0.9 Å apart (compared with a bond length of 0.74 Å in H_2). If the reaction continues, products form spontaneously and the potential energy is released as kinetic energy. A plot of the energy a system must have to continue reacting vs. extent of reaction is called the reaction profile (see Figure 3-3).

Statistical mechanics quantifies the energy states of molecules and defines the probability that a molecule is in a given energy state (including the transition state). According to Boltzmann's theory, energy is distributed among molecules by $\exp(-E_j/kT)$; $\mathbf{k}$, the Boltzmann constant, has a value of 1.38×10^{-16} erg K^{-1}. The probability P_j that a molecule is in a state with energy E_j is

$$P_j = \{\exp(-E_j/\mathbf{k}T)\} / \Sigma \exp(-E_i/\mathbf{k}T) = \{\exp(-E_j/\mathbf{k}T)\}/Q \qquad (3\text{-}37)$$

The denominator (Q) of Equation 3-37 is called the partition function and is the sum of the Boltzmann distribution over all allowed energy states. $\Sigma P_j = 1$ by definition. Q is a key parameter in predicting thermodynamic functions, and the field of statistical mechanics is concerned with evaluating Q from rotational, translational, and vibrational properties of molecules. This subject is covered in physical chemistry texts, and an introduction for geochemists also is available.[36]

Until recently, exact quantum mechanical (*ab initio*) calculations of the potential energies of reactants along possible reaction pathways were feasible only for very simple molecules, such as H_2. The tremendous expansion of computing power over the past two decades and the development of efficient algorithms now allows numerical estimation of PES's for systems with three to six atoms (e.g., H + CO and OH + H_2) by solving the Schrodinger wave equation, using basis set techniques.[37] The PES actually is a hypersurface of dimension $3N - 6$, where N is the number of atoms in the systems. By use of simplifying assumptions, approximate solutions also can be developed for more complicated molecules, including reactants in solutions.[38] Solution-phase reactions are considerably more complicated than gas-phase reactions, because changes in solvation, as well as changes in bonding, affect the system's potential energy.

Progress has been reported recently in applying these techniques to reactions of interest in aqueous geochemistry. For example, Lasaga and co-workers[39,40] have calculated the shape and interatomic distances of the transition states for water- and ion-induced dissolution of quartz and related silica surfaces (Figure 3-10). Programs that run on microcomputers are now available for such calculations. (Execution times are relatively long with such systems, and runs often are made overnight.) A critical assumption in making these calculations is that only atoms within a few atoms distance of the reaction center affect the potential energy of the transition state.[39] This simplifies the problem to a reasonable number of atoms so that the effects of changes in interatomic distances on the system's total potential energy can be estimated. This simplification is not thought to be too restrictive to allow the analysis of transition-state conditions for reactions of interest in aquatic environ-

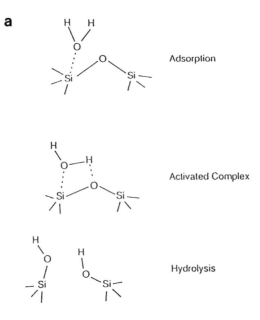

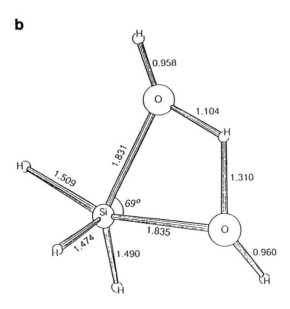

Figure 3-10. (a) Elementary steps in hydrolysis of silica, and (b) structure of transition state for hydrolysis reaction, as calculated by the linear synchronous transit method. [From Lasaga, A. C. and G. V. Gibbs, in *Aquatic Chemical Kinetics,* W. Stumm (Ed.), Wiley-Interscience, New York, 1990, p. 259. With permission.]

ments; however, it still is not possible to compute the global PES or detailed reaction pathways for solution-phase reactions by standard *ab initio* methods.[*]

3.5.2 Activated Complex Theory: A Thermodynamic Simplification of TST

A descriptive approach to TST based on thermodynamic formalisms can be derived readily and is easy to understand. Because simplifying assumptions in this approach violate some tenets of TST, some authors refer to it as "activated complex theory" (ACT). ACT is useful in describing the factors affecting reaction rates, and we will rely on it in later discussions. A more detailed version of the following derivation is given in texts on physical chemistry.[4] Building on Lindemann's model for first-order reactions, we can describe the activated intermediate or transition state $A^{\neq}$ or $AB^{\neq}$ as a species in equilibrium with the reactants, e.g., for bimolecular reactions:

$$A + B = AB^{\neq}, \quad K^{\neq} = \left[AB^{\neq}\right]/[A][B] \tag{3-38a}$$

or

$$\left[AB^{\neq}\right] = K^{\neq}[A][B] \tag{3-38b}$$

The reaction rate is controlled by dissociation of the intermediate and thus is proportional to both its concentration and its frequency of dissociation to products, $AB^{\neq} \rightarrow$ products:

$$\text{Forward rate} = \begin{bmatrix} \text{Concentration of} \\ \text{activated complex} \end{bmatrix} \begin{bmatrix} \text{Frequency of its} \\ \text{dissociation to product} \end{bmatrix} \tag{3-39}$$

The complex breaks down to products when the potential energy of its reacting bond becomes equal to the bond's vibrational energy. $AB^{\neq}$ is not just the energized molecule of collision theory, although both have the same energy, E_{act}. The transition state is a special species in which one vibrational degree of freedom[**] corresponds to movement along the reaction coordinate so that the reactants are converted into products with the release of the extra energy of the transition state (E_{act}) plus the energy of reaction.

Planck's law relates vibrational energy to the frequency of vibration:

$$E_{vib} = h\nu \tag{3-40a}$$

---

[*] Even a "simple" four-body reaction of $OH + H_2$ requires calculations at 10^6 geometries;[37] calculations for more complicated systems must be limited to critical regions of the PES, as determined from experimental data or other estimation methods.

[**] A degree of freedom represents a mode of molecular motion. Molecules have three types of degrees of freedom: translational, rotational, and vibrational. The energy of a molecule is the sum of these three kinds of energy.[33]

where h is Planck's constant (6.62×10^{-27} erg s), and v is frequency (s^{-1}). The bond being broken in the activated complex is weak compared with the ground-state molecule (note the increase in bond length in Figure 3-9). By definition, it will break on the next vibration, without further input of energy. Thermodynamics does not enable us to calculate the energy of the bond exactly, but it must have at least the thermal distribution. Thus,

$$E_{pot} \approx kT \qquad (3\text{-}40b)$$

When $E_{vib} = E_{pot}$, $hv = kT$, or

$$v_f = kT/h \qquad (3\text{-}41)$$

Equation 3-41 gives the frequency of transition-state dissociation to form products. At 25°C, $kT/h = 6.2 \times 10^{12}$ s^{-1}. This factor is called the fundamental frequency, and according to this simplified treatment, it is the same for all reactions except outersphere electron transfer reactions (see Section 7.4.1). The reciprocal of v_f, ~10^{-13} s, is a measure of the lifetime of the transition state. This is the length of one molecular vibration period at 25°C.

We now can express Equation 3-39 in mathematical form. The concentration of transition-state complex $AB^\neq$ is given by Equation 3-38b, and its dissociation frequency is given by Equation 3-41. Thus,

$$\text{Rate} = \left[AB^\neq\right]\left[v_f\right] = K^\neq[A][B](kT/h) \qquad (3\text{-}42)$$

The conventional second-order equation is rate = $-d[A]/dt = k[A][B]$. Thus,

$$k = \frac{\text{Rate}}{[A][B]} = \frac{kT}{h} K^\neq \qquad (3\text{-}43)$$

More rigorous derivations of Equation 3-43, based on statistical mechanics, are found in texts on kinetics.[31-33,36] The above derivation is inaccurate in several respects.[41] First, $K^\neq$ is not a true "thermodynamic" equilibrium constant for the reaction A + B $\rightleftharpoons$ $AB^\neq$ because a factor corresponding to the reaction coordinate "normal mode oscillation" was factored out. This is not apparent in the above derivation, but is explicit in derivations based on statistical mechanics. Second, because Equation 3-40b is an approximation (because the right side is an indeterminate quantity), v_f is **not** a universal frequency of activated complex decomposition. Finally, a correction factor κ, the "transmission coefficient," should be added to Equation 3-43 ($k = \kappa v_f K^\neq$) to account for processes like quantum tunneling, barrier recrossing, and solvent frictional effects that are not covered by the theory. κ is thought to range from about 0.1 to 1 for solution-phase reactions at ambient temperatures.[41]

3.5.3 Expression of Activation Energy as Thermodynamic Functions

Equation 3-43 is the fundamental relationship of ACT, but by itself it is not very useful. $K^{\neq}$, the equilibrium constant between the transition state and reactants, cannot be measured by conventional equilibrium methods, because $AB^{\neq}$ is so ephemeral. From thermodynamics, however, we know that

$$\Delta G^{\circ} = -RT \ln K \qquad (3\text{-}44a)$$

or
$$K = \exp\left(\Delta G^{\circ} / RT\right) \qquad (3\text{-}44b)$$

and
$$\Delta G^{\circ} = \Delta H^{\circ} - T\Delta S^{\circ} \qquad (3\text{-}45)$$

Thus
$$K = \exp\left(\Delta S^{\circ} / R\right) \exp\left(-\Delta H^{\circ} / RT\right) \qquad (3\text{-}46)$$

where ΔG°, ΔH°, and ΔS° are the standard free energy, enthalpy, and entropy of reaction. Identifying $K^{\neq}$ with K, changing the corresponding thermodynamic functions, and substituting into Equation 3-43 yields

$$k = \frac{kT}{h} \exp(-\Delta G^{\neq} / RT) = \frac{kT}{h} \exp(\Delta S^{\neq} / R) \exp(-\Delta H^{\neq} / RT) \qquad (3\text{-}47)$$

$\Delta G^{\neq}$, $\Delta S^{\neq}$, and $\Delta H^{\neq}$, the free energy, entropy, and enthalpy of activation, represent the difference in the respective thermodynamic variable between the transition state and the reactants, when all are in their standard states (i.e., at unit concentration). The thermodynamic functions associated with transition states actually are more numerous and complicated than described here.[42]

It is apparent from Equation 3-47 that the rate constant for a reaction is related exponentially to the free energy of activation. The larger $\Delta G^{\neq}$ is, the slower the reaction will be. The experimental activation energy, E_{act}, is related to the internal energy of activation $\Delta E^{\neq}$ and enthalpy of activation $\Delta H^{\neq}$, as follows. Taking logs of Equation 3-43 and differentiating with respect to T, we obtain

$$\ln k = \ln\left(k/h\right) + \ln T + \ln K^{\neq}$$
$$\qquad (3\text{-}48)$$

or
$$d \ln k / dT = 1/T + d \ln K^{\neq} / dT$$

From the Arrhenius equation, the left side of Equation 3-48 is equal to E_{act}/RT^2, and according to the Clapeyron equation, $d \ln K^{\neq}/dT = \Delta E^{\neq}/RT^2$. Consequently,

$$E_{act} = \Delta E^{\neq} + RT \qquad (3\text{-}49a)$$

In general, $\Delta E = \Delta H - P\Delta V = \Delta H - \Delta n RT$, where Δn is the change in the number of moles for a reaction. Glasstone et al.[31] assumed that the volume of activation, $\Delta V^{\neq}$, was nearly 0 in solution (see Section 3.6.1). Thus, $\Delta E^{\neq} \approx \Delta H^{\neq}$, and

$$E_{act} = \Delta H^{\neq} = RT \qquad (3\text{-}49b)$$

for solution-phase reactions. Although $\Delta V^{\neq}$ is not exactly zero in solution, the volume changes are much smaller (by ~10^{-3}) than those for gas-phase reactions, and the assumption is reasonable. In general, $\Delta G^{\neq} = E_{act} - RT - T\Delta S^{\neq}$ for reactions in solution.

For gas-phase reactions

$$E_{act} = \Delta H^{\neq} + RT(1 - \Delta n^{\neq}) \qquad (3\text{-}49c)$$

where $\Delta n^{\neq}$ is the change in the number of molecules in going from reactants to activated complex. For bimolecular gas-phase reactions, $\Delta n^{\neq} = -1$ (two molecules form one activated complex), and $E_{act} = \Delta H^{\neq} + 2RT$ (see Reference 32).

Substitution of Equation 3-49b into Equation 3-47 leads to the following expressions for rate constants of solution-phase reactions:

$$k = (\mathbf{k}T/h)\exp(\Delta S^{\neq}/R)\exp(-E_{act}/RT)\exp(RT/RT)$$

i.e.,
$$k = e(\mathbf{k}T/h)\exp(\Delta S^{\neq}/R)\exp(-E_{act}/RT) \qquad (3\text{-}50)$$

3.5.4 Structural Interpretation of $\Delta S^{\neq}$

Comparing Equation 3-50 and the Arrhenius equation (Equation 3-5), we see that the pre-exponential factor A of the latter is related to the frequency of transition-state dissociation and the entropy of activation:

$$A = (e\mathbf{k}T/h)\exp(\Delta S^{\neq}/R) \qquad (3\text{-}51)$$

Entropy of activation is analogous to the concept of entropy in equilibrium systems; i.e., it is a measure of disorder in the transition state compared with the reactant state. At first thought one might expect that $\Delta S^{\neq}$ is always negative in bimolecular reactions, because the joining of two reactants to form an activated complex decreases the randomness of the system. These simple predictions do not consider the contributions of solvent structure to entropy, however. $\Delta S^{\neq}$ often is negative for associative reactions, but it is not always so. If solvent molecules are released in forming the transition state, entropy increases. In many cases this more than offsets the decrease caused by reactant association.

Solvent effects on $\Delta S^{\neq}$ are especially important for aqueous reactions of ions and polar molecules. Ions of like sign produce activated complexes with a greater charge density than that of the reactants, causing tighter binding of nearby solvent molecules, especially in the hydration sphere where the solvent has a low dielectric constant. This "electrostriction" process represents a loss in entropy. Ions of opposite sign yield activated complexes with lower charge densities, which loosens solvent molecules and increases entropy. Similar comments apply to polar molecules: formation of a complex more polar than the reactants leads to a more negative $\Delta S^{\neq}$, and formation of a less-polar complex causes an increase in $\Delta S^{\neq}$. For example, hydrolysis of esters results in more polar-activated states because the carbonyl groups become ionized. Such reactions have negative $\Delta S^{\neq}$ and low-frequency factors. Because charges are not fully developed in dipoles, effects on $\Delta S^{\neq}$ are less than those for reactions of ions. Also, because transition states represent intermediate states between reactants and products, $\Delta S^{\neq}$ normally is less than ΔS of reaction. The ratio $\Delta S^{\neq}/\Delta S^{\circ}$ is a measure of the extent to which transition states resemble reactants or products.

3.5.5 Calculation and Use of Activation Energy Functions: Some Cautions

Determination of $\Delta G^{\neq}$, $\Delta H^{\neq}$, and $\Delta S^{\neq}$ from kinetic measurements is straightforward (Example 3-1). As Equation 3-47 indicates, $\Delta G^{\neq}$ is exponentially related to k and can be computed from k at any temperature. As Equation 3-45 indicates, the effect of temperature on $\Delta G^{\neq}$ depends on the value of $\Delta S^{\neq}$. Because the enthalpy of activation is closely related to the experimental activation energy (Equations 3-49b,c), $\Delta H^{\neq}$ can be computed directly from the E_{act} obtained from an Arrhenius plot. $\Delta S^{\neq}$ then can be obtained from Equations 3-45 or 3-47. The three "thermodynamic" functions of activation energy can be evaluated from rate measurements at a minimum of two temperatures by Equations 3-6, 3-47, or 3-50, but more accurate results are obtained by using k values for a larger number of temperatures.

Experimental activation energies (E_{act}) can be determined for any reaction or reaction sequence, and thus the functions $\Delta G^{\neq}$, $\Delta H^{\neq}$, and $\Delta S^{\neq}$ can be computed. As we saw in Section 3.2, the apparent E_{act} for multistep reactions may be a composite of the E_{act} for a rate-limiting step and ΔE° of reaction for earlier pre-equilibrium steps. In such cases, E_{act} (hence, $\Delta H^{\neq}$) has no physical meaning other than to express the temperature dependency of the overall reaction rate. Similarly, the other functions of activation energy lose their physical significance for multistep reaction sequences. Caution should be observed in interpreting such computed values as measures of free energy or entropy differences between reactant and transition states.

Example 3-1. Calculation of Kinetic Parameters for Hydration of CO_2

Johnson[43] reported that the first-order rate constant for hydration of CO_2 at circumneutral pH and 25°C is $k_h{}' = 3.7 \times 10^{-2}$ s^{-1}. The reaction is bimolecular ($CO_2 + H_2O \rightleftharpoons H_2CO_3$) and only *pseudo*-first order ($[H_2O]$ is constant). Thus, it is appropriate to convert to a second-order rate constant, $k_h = 6.67 \times 10^{-4}$ M^{-1} s^{-1} (assuming $[H_2O] = 55.5$ M). Corresponding values at 5°C are 3.66×10^{-3} s^{-1} and 6.60×10^{-5} M^{-1} s^{-1}. Values of the activation parameters can be estimated from equations given previously, as follows. In the absence of information to the contrary, we will assume that the transmission coefficient (κ) = 1, and that $\nu_f = \mathbf{k}T/h$ applies.

From Equation 3-6,

$$E_{act} = RT_1T_2 \ln\{k_{h,2}/k_{h,1}\}/(T_2 - T_1)$$

$$= 8.31(278)(298)\ln\{6.67\times10^{-4}/6.60\times10^{-5}\}/20$$

$$= 79,622\ J\ mol^{-1}$$

From Equation 3-5,

$$A = k_h/\exp(-E_{act}/RT)$$

$$= 6.67\times10^{-4}/\{\exp(-79,622/8.31\times298)\}$$

$$= 6.67\times10^{-4}/(1.09\times10^{-14}) = 6.13\times10^{10}\ M^{-1}s^{-1}$$

From Equation 3-47,

$$\Delta G_{25°C}^{\ddagger} = RT\{\ln(v_f) - \ln(k)\}$$

$$= 8.31(298)\{\ln(6.2\times10^{12}) - \ln(6.67\times10^{-4})\}$$

$$= 91,050\ J\ mol^{-1}$$

$$= 77.1\ kJ\ mol^{-1}$$

From Equation 3-49b,

$$\Delta H^{\ddagger} = E_{act} - RT$$

$$= 79,622 - 8.31(298) = 77,146\ J\ mol^{-1}\ or\ 77.1\ kJ\ mol^{-1}$$

From Equation 3-50,

$$\Delta S^{\ddagger} = E_{act}/T + R\ln k - R\ln e - R\ln(v_f)$$

$$= 79,622/298 + 8.31(\ln 6.67\times10^{-4}) - 8.31 - 8.31\ln(6.2\times10^{12})$$

$$= 267.19 - 60.77 - 8.31 - 244.78$$

$$= -46.67\ J\ K^{-1}\ mol^{-1}$$

Or from Equation 3-45,

$$\Delta S^{\neq} = \left(\Delta H^{\neq} - \Delta G^{\neq}\right)/T$$

$$= (77.1 - 91.1)/298 = -47.0 \ J \ K^{-1} \ mol^{-1}$$

Also, from Equation 3-51,

$$A = \left(ev_f\right)\exp\left(\Delta S^{\neq}/R\right)$$

$$= \left(2.72 \times 6.2 \times 10^{12}\right)\exp\left(-46.7/8.31\right) = 6.11 \times 10^{10} \ M^{-1}s^{-1}$$

The value of k_h is ~5 × 10^{-14} as large as that for diffusion-controlled reactions. The frequency factor A has a value near the normal range (10^{11} M^{-1} s^{-1}), which is to be expected for a reaction between two uncharged reactants. The small negative value of $\Delta S^{\neq}$ indicates the transition state is slightly more ordered than the reactant state and probably can be accounted for in terms of the two reactants forming one transition-state complex. If E_{act} were 0, the given value of k would require A to be 5 × 10^{-14} lower than the normal value, since A is directly proportional to A (at constant E_{act}). In this case, $\Delta S^{\neq}$ would have a large negative value because A is exponentially related to $\Delta S^{\neq}$ (Equation 3-51). That would imply a much greater decrease in the randomness of the transition state than expected for a reaction of two uncharged molecules.

Under alkaline conditions the reaction of $CO_2 + OH^- \rightarrow HCO_3^-$ becomes competitive with the hydration reaction. Johnson[43] reported the following values for the reaction with OH^-: k_{OH} = 7.1 × 10^3 M^{-1} s^{-1} (25°C) and 7.0 × 10^2 M^{-1} s^{-1} (5°C). Using the equations given above for the CO_2 hydration reaction, we calculate the following activation parameter values:

$$E_{act} = 79,713 \ J \ mol^{-1}$$

$$A = 6.9 \times 10^{17} \ M^{-1}s^{-1}$$

$$\Delta H^{\neq} = 77,246 \ J \ mol^{-1}$$

$$\Delta G^{\neq} = 50,954 \ J \ mol^{-1} \ at \ 25°C$$

$$\Delta S^{\neq} = 88.2 \ J \ mol^{-1} \ K^{-1}$$

Although the activation energies are almost identical for the reactions of CO_2 with H_2O and OH^-, k_{OH} is ~10^7 × k_h, and $\Delta G^{\neq}$ is 40 kJ mol^{-1} lower (at 25°C) for the reaction with OH^-. In contrast to the negative entropy of activation for the reaction of CO_2 with H_2O, the reaction with OH^- has a large positive $\Delta S^{\neq}$ and an abnormally high pre-exponential factor, signifying a large loosening of solvent water molecules in the transition state.

3.5.6 Comparison of Collision and Transition State Theories

Transition-state (or activated-complex) theory (TST/ACT) is not a substitute for collision theory, but complements it. The collision-encounter theory for reactions in solution is well founded, and much evidence exists for the cage model and for diffusion-controlled limits on reaction rates. ACT provides further detail on reaction mechanisms at the moment of bond formation and rupture. The energized complex in collision theory may be anywhere along the reaction coordinate, but the transition state has the configuration corresponding to the saddle point (Figure 3-9). ACT treats the transition state $AB^{\neq}$ as a chemical species in thermal equilibrium with its surroundings — its vibrations and rotations correspond to those at the ambient temperature. In contrast, collision theory treats $AB^{\neq}$ as a set of vibrationally excited states of the reactants. Which viewpoint is correct is somewhat a philosophical issue. Because transition states are so ephemeral (mean lifetime of $\sim 10^{-13}$ s), their free energy and other equilibrium properties cannot be measured directly. Whether they can be regarded as definite chemical entities in specific thermodynamic states remains uncertain,[4] detection of some transition states by femtosecond spectroscopy[37] not withstanding. Nonetheless, ACT is a convenient formalism. It provides a reasonable explanation for linear free-energy relationships (Chapter 7), and its derived function $\Delta S^{\neq}$ yields useful insights on the structures of reaction intermediates.

3.5.7 Irreversible Thermodynamics and Rates of Chemical Change

For systems far from equilibrium, the overall reaction rate is simply proportional to the concentration of activated complex, as described in Section 3.5.1, but for systems near thermodynamic equilibrium, the rate is also proportional to the chemical affinity (**A**) of the overall reaction, as demonstrated in this section. **A** is related to Gibbs free energy (G) and extent of reaction (ξ) as follows:

$$A = -(\partial G / \partial \xi)_{P,T} = \Sigma \mu_i \nu_i = RT \ln(K/Q) \tag{3-52}$$

μ_i is the chemical potential of the ith species in the reaction; ν_i is its stoichiometric coefficient (positive for products, negative for reactants; see Section 2.1.1); K is the equilibrium constant for the reaction; and Q is the reaction quotient, i.e., the ratio of the product concentrations to the product of the reactant concentrations at a given ξ.

For any reaction, $A + \cdots \rightleftharpoons B + \cdots$, the net reaction rate $r_n = r_f - r_r$, and $r_f = \alpha \nu_f C^{\neq}$, where α is a coefficient representing the probability of the complex to decompose to products rather than reactants; ν_f is the frequency factor of activated-complex decomposition; and $C^{\neq}$ is the concentration of the activated complex. According to Laidler,[25] $\alpha_f = \alpha_r$ in most cases. For a one-step reaction, the law of microscopic reversibility requires that the same pathway applies to both directions. The activated complex thus is the same in both directions. If the activated complex is in equilibrium with reactants for each direction, as required by ACT, it can be shown that

$$r_f / r_r = K/Q = \exp(A/RT) \tag{3-53}$$

The net reaction rate can be rewritten as: $r_n = r_f - r_r = r_f(1 - r_r/r_f)$. Substituting Equation 3-53 for r_r/r_f, we obtain:

$$r_n = r_f \{1 - \exp(-A/RT)\} \qquad (3\text{-}54)$$

Equation 3-54 relates the rate of reaction to chemical affinity and thus represents a formal link between kinetics and thermodynamics. For reactions far from equilibrium, A is large, and the exponential term in Equation 3-54 becomes negligible. However, as a reaction approaches equilibrium, A becomes more important in controlling the net reaction rate. The exponential term can be written as a polynomial expansion:[44]

$$\exp(-A/RT) = 1 + \sum_m \frac{(-A/RT)^m}{m!} \qquad (3\text{-}55)$$

Truncating Equation 3-55 after $m = 1$ introduces negligible error for reactions close to equilibrium, and substitution into Equation 3-54 yields

$$r_n = r_f^* (A/RT) \qquad (3\text{-}56)$$

r_f^* is the forward reaction rate at equilibrium. Thus, near equilibrium, r_n is proportional to the chemical affinity of the overall reaction. It is important to realize that most reactions of environmental interest occur far from equilibrium conditions, and Equations 3-54 and 3-56 have more theoretical interest than practical importance. Dissolution of minerals like calcite in near-saturated solutions may fit Equation 3-56.

3.6 EFFECTS OF PRESSURE ON RATES OF REACTIONS IN SOLUTION

Pressure (P) affects reaction rates in solution in several ways. Its direct effects on rate constants are related to a parameter known as the volume of activation, which is the difference between the molar volumes of the transition and reactant states. In addition, pressure effects on equilibria may change the forms and concentrations of reactants (relative to 1 atm). Equations describing effects of P on rate constants are analogous to those for equilibria. In both cases the effects are important only at high P, and in natural systems they need be considered only in the depths of the oceans or deep groundwater.

3.6.1 Mathematics of Pressure Effects

The effects of pressure on chemical equilibria were described by Planck in 1887. He related the change in equilibrium constants with P to volume changes between products and reactants:

$$(\partial \ln K / \partial P)_T = -\Delta V / RT \tag{3-57}$$

ΔV is the difference in partial molal volumes between products and reactants $(\Sigma V^\circ_{prod} - \Sigma V^\circ_{react})$. From the relationship between Gibbs free energy and equilibrium constants we can write,

$$\Delta V = (\partial \Delta G / \partial P)_T \tag{3-58}$$

If a reaction produces an increase in volume, ΔG_r increases with P, and K_{eq} decreases with P. Conversely, in a reaction with $\Delta V < 0$, K increases with P, and the reaction proceeds further with increasing P. These trends are a simple extension of Le Chatelier's principle.

The effects of pressure on reaction rates can be derived easily from Equations 3-43 and 3-57 and ACT. According to Equation 3-43, the rate constant is proportional to $K^{\neq}$. Thus from Equation 3-57 we can write,

$$(\partial \ln K^{\neq} / \partial P)_T = -\Delta V^{\neq} / RT = (\partial \ln k / \partial P)_T \tag{3-59}$$

where $\Delta V^{\neq}$ is the volume of activation. Equation 3-59 was derived by van't Hoff in a way analogous to that for the Arrhenius equation. If the transition state is larger than the reactant state ($\Delta V^{\neq} > 0$), the rate decreases with increasing P. Equation 3-59 implies that tangents to curves of $\ln$ k vs. P are equal to $-\Delta V^{\neq}/RT$ at any P. If the plots are straight lines, the slope is constant and $\Delta V^{\neq}$ is independent of P. In this case Equation 3-59 can be integrated to

$$\ln \frac{k_P}{k_{1\,atm}} = \frac{-\Delta V^{\neq}(P-1)}{RT} \tag{3-60}$$

where k_P and k_{1atm} are the rate constants at pressures P and 1 atm, respectively. Plots of $\ln$ k vs. P rarely are linear, however, and $\Delta V^{\neq}$ thus varies with P.[45] This implies that the reactant and transition states (usually) have different molar compressibilities. Interpretation of $\Delta V^{\neq}$ usually is done for values extrapolated to P = 0 ($\Delta V^{\circ\neq}$). These values are 20 to 70% larger than corresponding values at P = 1000 atm.[45] Values of $\Delta V^{\neq}$ are known for only a few reactions of environmental interest, but reported values generally are between −20 and +20 cm^3 mol^{-1}.[45,46]

3.6.2 Components of $\Delta V^{\neq}$ and Relationships Between $\Delta V^{\neq}$ and $\Delta S^{\neq}$

Values of $\Delta V^{\neq}$ reflect two separate processes in forming transition states: one dealing with the structure of the reactant molecules, the other with the surrounding solvent:

$$\Delta V^{\neq} = \Delta V^{\neq}_{str} + \Delta V^{\neq}_{solv} \tag{3-61}$$

Conversion of reactants into a transition state causes a volume decrease for bimolecular processes and an increase for unimolecular reactions (assuming a bond is being broken). Moderately accurate calculations of $\Delta V^{\neq}_{str}$ can be made based on the geometry of reacting molecules;[15] the main uncertainty in this appoach lies in estimating the extent of elongation of the bond being broken. In a unimolecular reaction the bond can be modeled as a cylinder with volume $\pi r^2 l$, where r is the van der Waal's radius (~2 Å), and l is the bond length. If the bond length in the transition state increases by 1 Å, $\Delta V^{\neq}_{str}$ increases by about 12 Å^3 per molecule, or about 7 cm^3 per mol^{-1}.[15]

Reorganization of solvent molecules may cause either positive or negative volume changes. In reactions of ions or highly polar molecules, solvent changes are more important than structural changes. Solvent volume changes are closely related to solvent entropy changes. A reaction that results in tighter binding of solvent molecules results in negative values of both $\Delta S^{\neq}$ and $\Delta V^{\neq}$. The converse is true of reactions that cause looser binding of solvent. Reaction between ions of like sign causes a higher charge density in the transition state, resulting in increased "electrostriction" and a decrease in volume and entropy. These reactions tend to be slow because of the coulombic energy required to bring like charges together. Reactions between oppositely charged ions or dipoles yield transition states with a lower net charge, which releases some bound solvent, increasing both volume and entropy. Unimolecular decomposition of a neutral molecule to yield two ions has $\Delta V^{\neq} < 0$ because the transition state is partially ionized, which increases solvent binding.

Laidler and Chen[46] found a correlation between $\Delta V^{\neq}$ and $\Delta S^{\neq}$ for hydrolysis of amides and esters, and this relationship is useful in estimating pressure effects on reactions for which $\Delta V^{\neq}$ is unavailable. Figure 3-11a shows that $\Delta V^{\neq} \approx 0.5 \Delta S^{\neq}$ when $\Delta V^{\neq}$ is in cm^3 mol^{-1} and $\Delta S^{\neq}$ is in cal K^{-1} mol^{-1}. It is often more convenient to vary temperature and measure $\Delta S^{\neq}$ than to vary P, but the general validity of the relationship in Figure 3-11a has not been evaluated for other reactions of environmental interest. Reactions with $\Delta S^{\neq} < 0$ have abnormally low frequency factors (A $<< 10^{11}$ M^{-1} s^{-1} for bimolecular reactions). Because $\Delta V^{\neq}$ also is less than 0, these reactions are accelerated by increases in P. Reactions with $\Delta S^{\neq} \approx 0$ have normal frequency factors (~10^{11} M^{-1} s^{-1}). Because $\Delta V^{\neq} \approx 0$, pressure has little effect on rates of these reactions. Finally, reactions with $\Delta S^{\neq}$ and $\Delta V^{\neq} > 0$ have abnormally high frequency factors, and their rates decrease with increasing P. These relationships are illustrated in Figure 3-11b, c.

Some chemists[47] consider $\Delta V^{\neq}$ more useful than $\Delta S^{\neq}$ in determining reaction mechanisms of metal-ligand complexes, because the latter term involves both energetic and geometric factors, while $\Delta V^{\neq}$ depends only on positions of substituents. Precise determination of $\Delta V^{\neq}$ now is readily achieved, and a growing body of such data is available to interpret mechanisms of ligand exchange, hydrolysis, isomerization, and redox reactions for metal-ligand complexes.

3.6.3 Examples of Pressure Effects on Reactions in Natural Waters

Few studies have considered the effects of pressure on reaction rates in natural waters, but Equation 3-60 can be used to predict the magnitude of such effects. Pressure increases by 1 atm for every ~10 m increase in water depth. The average

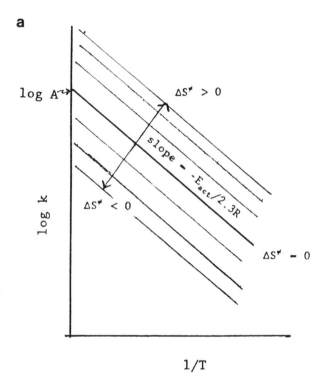

Figure 3-11. Relationships among activation parameters: (a) effect of varying entropy of activation, $\Delta S^{\neq}$ on plots of $\ln k$ vs. $1/T$ for constant E_{act}. Dark line is for $\Delta S^{\neq} = 0$ ("normal" reactions, $\log A \approx 11$). For $\log A \gg 11$, $\Delta S^{\neq} > 0$ (fast reactions); for $\log A \ll 11$, $\Delta S^{\neq} < 0$ (slow reactions); (b) correlation between $\Delta V^{\neq}$ and $\Delta S^{\neq}$. [From Laidler, K. J. and D. T. Y. Chen, *Can. J. Chem.*, 37, 599 (1959). With permission.] (c) Effects of pressure on k for reactions with positive, negative, and zero entropies of activation.

depth z_{av} of the oceans is about 4 km, and in the deepest areas $z_{max} \approx 10.5$ km. Thus, at oceanic z_{av}, P is 0.4 katm (10^3 atm), and at z_{max} it is 1.05 katm. The deepest lake in the world is Baikal (Siberia) with z_{max} of 1741 m (P = 0.17 katm). The deepest lakes in North America are Great Slave (Northwest Territories) and Crater, both with z_{max} of 610 m (P = 61 atm). These depths and pressures are convenient values to examine pressure effects on reaction rates.

Assuming $\Delta V^{\neq}$ is independent of P and that deep water is 4°C, we can plot the ratio k_P/k_{1atm} vs. $\Delta V^{\neq}$ for the above pressures (Figure 3-12). As mentioned earlier, reported values of $\Delta V^{\neq}$ are mostly between ±20 cm³ mol⁻¹, and this range is reasonable for estimating maximum effects. From the figure we see that for $\Delta V^{\neq} = +20$ cm³ mol⁻¹, the rate constant would decrease by a factor of 0.42 at a pressure of 1.0 katm and would increase by a factor of 2.41 if $\Delta V^{\neq}$ were −20 cm³ mol⁻¹. Changes in rate constants are much less pronounced in lakes. Corresponding ratios of $k_P/k_{1\ atm}$ for $\Delta V^{\neq} = \pm20$ cm³ mol⁻¹ are 0.86 and 1.16 for Lake Baikal and 0.95 and 1.05 for the deepest North American lakes. Given the measurement uncertainty in most rate

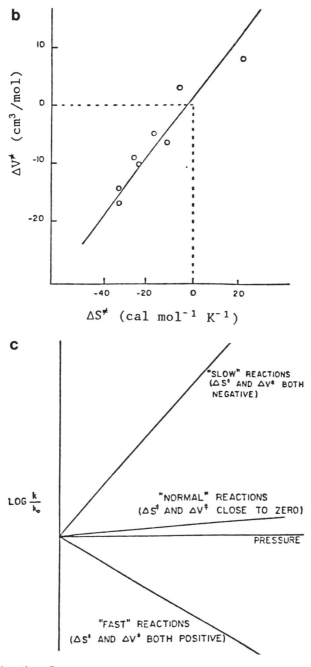

Figure 3-11 (continued).

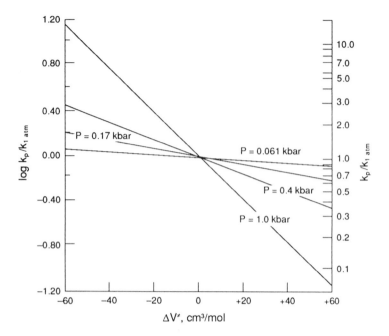

Figure 3-12. Relationship between $k_P/k_{1\ atm}$ and volume of activation, $\Delta V^{\neq}$, for several pressures at T = 277 K.

constants, pressure effects on reaction rate can be safely ignored in freshwaters, but the correction may be of some significance in the deep sea.

Adams[48] evaluated the effects of pressure on the hydrolysis rate of the nerve gas GB (isopropylmethylphosphonofluoridate), which has been disposed of by burial in steel canisters in the deep ocean. Adams estimated that $t_{1/2}$ for hydrolysis of GB to the less toxic product isopropylmethylphosphonate (see Figure 4-1c) would increase by ~25% at a depth of 4 km and by ~69% at 10 km. He estimated $\Delta V^{\neq}$ for alkaline hydrolysis to be –12 cm³ mol⁻¹ from a reported value of $\Delta S^{\neq}$ and the relationship in Figure 3-11a. The half-life of GB at 1 atm and 25°C is 52 min;[48] at 4 km and 25°C it is about 65 min. Based on these values and the fact that water movement at the ocean bottom is slow, it is reasonable to conclude that leakage of GB from canisters on the ocean floor should not lead to widespread contamination by the parent compound. Of course, this says nothing about the persistence of the reaction product.

3.7 EFFECTS OF IONIC STRENGTH ON REACTION RATES

Ionic strength effects are important in dealing with rates of ionic reactions in aqueous solution.[49,50] The activity rate theory of Bronsted[51] and Bjerrum[52] was the first adequate treatment for such reactions. This theory is based on ACT and assumes that reactions between ions proceed through an activated complex in equilibrium with the individual ions:

$$A^{zA} + B^{zB} \rightleftharpoons (A-B)^{z(A+B)} \rightarrow \text{products} \tag{3-62}$$

The overall rate is proportional to the concentration of the activated complex, in agreement with ACT. The complex and reactants are related by an equilibrium expression in which the ion concentrations are expressed in terms of thermodynamic activities. Thus,

$$K^{\ne} = \frac{a^{\ne}}{a_A a_B} = \frac{C^{\ne}}{C_A C_B} \frac{\gamma^{\ne}}{\gamma_A \gamma_B} \tag{3-63}$$

$K^{\ne}$ is the formation constant for the transition state complex; the a_i are activities, C_i are molar concentrations, and γ_i are activity coefficients, which are functions of ionic strength (I), as described by various forms of the Debye-Huckel equation; e.g.,

Extended Debye-Huckel Equation (EDHE):

$$-\log \gamma_i = Z_i^2 A I^{1/2} / (1 + B a_i I^{1/2}) \tag{3-64a}$$

Guntelberg approximation:

$$-\log \gamma_i = Z_i^2 A I^{1/2} / (1 + I^{1/2}) \tag{3-64b}$$

Debye-Huckel Limiting Law (DHLL) :

$$-\log \gamma_i = 0.51 Z_i^2 I^{1/2} \tag{3-64c}$$

A and B are constants characteristic of the solvent at a specified temperature and pressure; a_i is a fitted parameter interpreted as the effective ionic diameter (cm); I $= 0.5 \Sigma m_i z_i^2$. For water at 25°C and P = 1 atm, A = 0.5085, and B = 0.3281 × 10^{-8}. Values of a_i for most ions are in the range 2 to 6 × 10^{-8} cm (see Reference 2 for a tabulation). The EDHE applies at I < 0.1, the Guntelberg equation at I < 0.01, and the DHLL at I < 0.005, but the equations often are used with lower accuracy above these limits.

For a given I (and thus a given set of γ_i), the reaction rate for Equation 3-62 is given by the bimolecular mass action expression:

$$-dC_A / dt = k_2 C_A C_B \tag{3-65}$$

As implied by Equation 3-62, the rate also is proportional to the concentration of the activated complex: $-dC_A/dt = (kT/h)C^{\neq}$. When Equation 3-63 is used to replace $C^{\neq}$, we obtain

$$-dC_A/dt = k_2^\circ C_A C_B \gamma_A \gamma_B / \gamma^{\neq} \tag{3-66}$$

$k_2^\circ = (kT/h)K^{\neq}$, the rate constant at $I = 0$, can be obtained experimentally.

Thus, k_2 in Equation 3-65 is the product of intrinsic constant k_2° and the ion activity coefficients. These cannot be measured directly for transition-state complexes, but can be estimated by analogy between the structure of the transition state and stable compounds. By equating Equations 3-65 and 3-66, simplifying, and taking logs, we find

$$\log k_2 = \log k_2^\circ = \log(\gamma_A \gamma_B / \gamma^{\neq}) \tag{3-67}$$

Using the EDHE to replace the activity coefficients, we obtain

$$\log k_2 = \log k_2^\circ - \frac{Z_A^2 A I^{1/2}}{1 + B a_A I^{1/2}} - \frac{Z_B^2 A I^{1/2}}{1 + B a_B I^{1/2}} + \frac{(Z_A + Z_B)^2 A I^{1/2}}{1 + B a_{\neq} I^{1/2}} \tag{3-68a}$$

Equation 3-68a can be simplified at low I by using the Guntelberg or Debye-Huckel limiting equations. In the latter case we obtain

$$\log k_2 = \log k_2^\circ + \left\{-0.51 Z_A^2 - 0.51 Z_B^2 + 0.51(Z_A + Z_B)^2\right\} I^{1/2}$$

or $\qquad \log k_2 = \log k_2^\circ + 1.02 Z_A Z_B I^{1/2} \tag{3-68b}$

and with the Guntelberg equation we obtain

$$\log k_2 = \log k_2^\circ + 1.02 Z_A Z_B I^{1/2} / (1 + I^{1/2}) \tag{3-68c}$$

Equation 3-68b, c is called the Bronsted equation, and it predicts that plots of $\log k_2$ vs. $I^{1/2}$ are linear, with slopes nearly equal to the product of reactant ionic charges, $Z_A Z_B$. If the charges on both ions have the same sign, the product is positive, and k_2 increases with I. If the charges have opposite signs, the product is negative, and k_2 decreases with I. The physical interpretation of Equations 3-68a-c is that increasing I shields reactant charges, thus decreasing the distance over which they exert coulombic attraction or repulsion.

Figure 3-13 shows that Equation 3-68 holds for a variety of aqueous ionic reactions. Not only are the directions of change correct, but the slopes are equal to the products of the reactant charges. All the rate constants in Figure 3-13 are

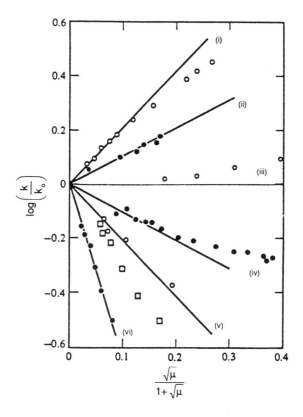

Figure 3-13. Effect of ionic strength on rate constants: (i) $BrCH_2COO^- + S_2O_3^{2-}$; (ii) $e^-_{aq} + NO_2^-$; (iii) $H_3O^+ + C_{12}H_{22}O_{11}$ (inversion of sucrose); (iv) $H_3O^+ + Br^- + H_2O_2$; (v) $OH^- + Co(NH_3)_5Br^{2+}$ (ionic strength varied with NaBr [circles] and Na_2SO_4 [squares]); (vi) $Fe(H_2O)_6^{2+} + Co(C_2O_4)_3^{3-}$. (From Weston, R. E., Jr. and H. A. Schwarz, *Chemical Kinetics*, Prentice-Hall, Englewood Cliffs, NJ, 1972. With permission.)

normalized to the rate at $I = 0$ so that they pass through a common origin, but reactions with negative slopes (those for reactants of opposite sign) actually have much higher rate constants than those with positive slopes. In the former case, electrostatic attraction accelerates collisions; in the latter, electrostatic repulsion inhibits collisions.

The limits of the Debye-Huckel equations should be recognized in using the Bronsted equation, and deviations are to be expected at moderate I. The deviation for reaction of $Co(NH_3)_5Br^{2+}$ with OH^- when Na_2SO_4 is used to vary I (line 5, Figure 3-13) results from interaction of SO_4^{2-} with the cobalt complex to form an uncharged ion pair that reacts less readily with OH^-.[15,49]

If one reactant is uncharged, the product $Z_A Z_B$ is zero, and the Bronsted equation predicts that the rate constant is unaffected by I. This is approximately true for some ion-molecule reactions, but reaction rates for an ion and a polar molecule or two polar molecules are affected by I, although to a lesser extent than occurs with two ions (see plot for inversion of sucrose, a highly polar molecule, in Figure 3-13).

Similarly, the dissociation of HCO_3^- to form $CO_2 + OH^-$ (the reverse of the CO_2 hydration reaction at high pH) is nearly independent of salinity.[43]

Ionic strength affects activities of nonelectrolytes slightly, and activity coefficients of nonelectrolytes are given by

$$\log \gamma_o = b_o I \tag{3-69}$$

b_o is an empirical constant. For values of b_o and I up to 0.1, $\log \gamma_o \approx 0$; i.e., $\gamma_o \approx 1.0$. For reactions between two nonelectrolytes in aqueous solution, Equation 3-69 implies that $\log k_2$ is a linear function of I:

$$\log k_2 = \log k_2^\circ + \left(b_A + b_B - b_{AB\neq}\right) I \tag{3-70}$$

This equation also applies if one reactant is an ion and the other is a neutral molecule.[32] Until recently, activity coefficients for nonelectrolytes were determined empirically, but some predictive techniques are now available (see Section 7.8). Activity coefficients of transition-state complexes are even more problematic. Consequently, it is difficult to predict the effects of I on reaction rates of nonelectrolytes, but they generally are small compared with the effects on electrolytes. For example, k for the hydration of CO_2 is independent of salinity up to 36 parts per thousand (roughly the value for seawater; i.e., $I \approx 0.7$).[43]

3.8 RANGES OF VALUES FOR KINETIC PARAMETERS: SUMMARY AND PERSPECTIVE

Preceding sections of this chapter quantified relationships among important kinetic parameters, including rate constants, energy of activation (E_{act}), free energy and entropy of activation ($\Delta G^\neq$ and $\Delta S^\neq$), and volume of activation ($\Delta V^\neq$). This section presents limiting values for these parameters and summarizes typical values, where such exist.

A huge range exists for reaction rates and rate constants in the natural environment. Table 3-3 ranks first-order rate constants and corresponding half-lives for a variety of chemical reactions in aqueous systems, and Table 3-4 is a similar ranking for second-order reactions. Among the slowest "reactions" of interest is the decay of certain radionuclides,* which usually is expressed in terms of half-lives (the decay being first order). For example, $t_{1/2}$ for ^{14}C is 5750 years, and the corresponding k is 4×10^{-13} s^{-1}. ^{238}U has a much longer $t_{1/2}$, 4.5×10^9 years, and smaller k, 5×10^{-19} s^{-1}. Perhaps the slowest process in the universe is the decay of the proton (if it is not a stable particle), which has an estimated half-life greater than 10^{33} years. Decomposition of recalcitrant organic compounds such as some PCBs and chlorinated pesticides are among the slowest chemical reactions in natural waters. Half-lives for the degradation

---

* Radionuclide decay is not a chemical reaction, because it involves nuclear, rather than electronic (bond), transformations.

Table 3-3.   Rate Constants and Associated Half-Lives for Some First-Order Reactions of Interest in Aquatic Systems

k_1, s^{-1}	$t_{1/2}$	Reaction	Section[a]
5E8	1.4E-9 s	$Cu(H_2O)_4^{2+} \cdot SO_4^{2-} \rightarrow Cu(H_2O)_3SO_4 + H_2O$	Table 4-2
3.2E6	2.2E-7 s	$Fe(H_2O)_6^{2+} \cdot SO_4^{2-} \rightarrow Fe(H_2O)_5SO_4^+ + H_2O$	Table 4-2
2.5E5	4E-6 s	$^1O_2 + H_2O \rightarrow O_2 + H_2O$ (quenching of singlet O_2)	8.3.4
1E5	7E-6 s	$CH_3COOH + H_2O \rightarrow H_3O^+ + CH_3COO^-$	4.4.1
1-2E4	7 – 3.5E-5 s	$Ni(H_2O)_6^{2+} \cdot SO_4^{2-} \rightarrow Ni(H_2O)_5SO_4^\circ + H_2O$	4.4.2
4E2	1.7E-3 s	$H_2S + H_2O \rightarrow H_3O^+ + HS^-$	4.4.1
28	0.02 s	$Cl_2 + H_2O \rightarrow HOCl + HCl$ (I = 0.5)	4.5.1
10	0.07 s	$Al(H_2O)_6^{3+} \cdot SO_4^{2-} \rightarrow Al(H_2O)_5SO_4^+ + H_2O$	4.4.2
0.5-16	1.4 – 0.04 s	Al^{3+}-H_2O exchange (measured in various ways)	4.4.2
3.7E-2	18.6 s	$CO_2 + H_2O \rightarrow H_2CO_3$	Ex. 3-1; 4.4.1
5.8E-4	20 min	$Fe(desferal)^{3+} + 6H_2O \rightarrow Fe^{3+} + desferal$; $1 M\, H^+$	4.4.2
2.9E-4	40 min	direct photolysis of benzo[a]pyrene	Table 8-9
1.8E-4	64 min	$HCO_3^- \rightarrow CO_2 + OH^-$	4.4.1
5.5E-5	3.5 h	decomposition of H_2O_2 in lake water (in dark)	8.3.3
4E-6	48 h	GB (nerve gas) + $H_2O \rightarrow$ nontoxic product	3.6.3; 4.4.6
1E-6	192 h	$P_2O_7^{4-} + H_2O \rightarrow 2HPO_4^{2-}$; pH 8.3 (sterile)	4.4.3
8E-7	10 days	direct photolysis of parathion	Table 8-9
5.3E-8	—	production of $\cdot OH$ from NO_3^- photolysis (annual average surface value at 40° latitude)	Example 8-2
2.3E-8	350 days	CH_3CCl_3 (TCE) + $H_2O \rightarrow CH_3CCl_2OH$	Table 4-4
5.8E-10	38 years	$CH_2BrCHBrCH_2Cl + H_2O \rightarrow CH_2{=}CBrCH_2OH$	Table 4-4
4E-13	5750 years	$^{14}C \rightarrow {}^{14}N + \beta^-$ (radioactive decay)	

Adapted from Pankow, J. F. and J. J. Morgan, *Environ. Sci. Technol.*, 11, 1155 (1981).

[a] Indicates location in text where reaction is discussed.

of the most chlorinated (and most stable) compounds may be on the order of hundreds of years ($k < 10^{-11}$ s^{-1}). The predominant loss mechanism for these compounds is sorption onto solids and burial in the sediments. Figure 1-1 shows other examples of slow reactions.

Characteristic times (and $t_{1/2}$) for second-order reactions are inversely proportional to initial reactant concentration. (Recall that the concept of $t_{1/2}$ is meaningful only when the two reactants have the same initial concentrations.) If $k_2 = 1$ M^{-1} s^{-1}, typical of a moderately slow reaction, $t_{1/2} = 10^3$ s (16.7 min) when the reactants are at millimolar levels, and $t_{1/2} = 10^6$ s (11.6 days) when the reactants are at micromolar levels.

Reactions complete in a few seconds ($k_1 > {\sim}1$ s^{-1}; $k_2 > 10^6$ M^{-1} s^{-1} at micromolar reactant concentrations) often are viewed as instantaneous in environmental studies, and equilibrium rather than kinetic models may suffice for them (see Section 5.6). However, the kinetics of rapid reactions are of interest for many reasons, and it is interesting to consider the upper limit for reaction rates. From the Arrhenius equation we know that k approaches a maximum equal to the pre-exponential factor A, as T becomes large compared with E_{act}. This may happen either because T is large or E_{act} is small; in natural aquatic systems the latter applies. According to ACT, k $\rightarrow v_f$ as $\Delta G^\neq \rightarrow 0$. Thus, $k_{max} = 6 \times 10^{12}$ s^{-1} is the maximum first-order rate constant for any chemical transformation. The reciprocal of k_{max} is the time for one molecular vibration. Reactions in aqueous solution are considerably slower than k_{max} because

Table 3-4. Rate Constants and Associated Half-Lives for Some Second-Order Reactions of Interest in Aquatic Systems

k_2 $M^{-1}s^{-1}$		$t_{1/2}{}^a$ (s)	Reaction	Section[b]
7E10	1 μs	1.4E-6	$H^+ + HS^- \rightarrow H_2S$	Table 3-2
4E10		2.5E-6	$H^+ + NH_3 \rightarrow NH_4^+$	Table 3-2
7E9		1.4E-5	$C_6H_6 + \cdot OH \rightarrow C_6H_5OH$ (phenol)	Table 8-6
6E9		1.7E-5	$Ca^{2+} + Y^{-4} \rightarrow CaY^{-2}$; Y=EDTA	Table 4-3
1E9		1.0E-4	$O_3 + Ph^- \rightarrow$ oxid. products (Ph^- = phenate ion)	Table 8-7
3E8		3.3E-4	$HOCl + CH_3NH_2 \rightarrow CH_3NHCl + H_2O$	Table 4-7
3E8		3.3E-4	$Fe^{2+} + \cdot OH \rightarrow FeOH^{2+}$	Table 8-6
2E8		5.0E-4	$Cu^{2+} + F^- \rightarrow CuF^+$	Table 4-2
2E8		5.0E-4	$Ph^- + {}^1O_2 \rightarrow$ oxid. products (Ph^- = phenate ion)	Table 8-6
2E7		5.0E-3	$(C_2H_5)_2S + {}^1O_2 \rightarrow (C_2H_5)_2SO$	8.3.4; 8.5.2
4E6		2.5E-2	$HOCl + NH_3 \rightarrow NH_2Cl + H_2O$	4.5.1
2E6		5.0E-2	$Co^{2+} + gly \rightarrow Co(gly)^+ + H^+$	Table 4-3
1E6		1.0E-1	$Fe^{2+} + F^- \rightarrow FeF^+$	Table 4-2
3E5	1 s	—	${}^1O_2 + H_2O \rightarrow O_2 + H_2O$ (quenching of singlet oxygen)	8.3.4
1E4		1.0E1	$HOCl + ClO_2^- \rightarrow ClO_3^- + H^+ + Cl^-$	4.5.1
8E3		1.3E1	$Ni^{2+} + F^- \rightarrow NiF^+$	Table 4-2
7E3		1.4E1	$CO_2 + OH^- \rightarrow HCO_3^-$	Example 3-1
2E3	1 min	5.0E1	$HOCl + Ph \rightarrow$ 2-ClPh; (pH 7), Ph = phenol	Table 4-6
1E3		1.0E2	$O_3 + Ph \rightarrow$ oxid. products	Table 8-7
1E3		1.0E2	$HOCl + Br^- \rightarrow HOBr + Cl^-$	4.5.1
1E3		1.0E2	$Ni^{2+} + Y^{-4} \rightarrow NiY^{-2}$; Y = EDTA	Table 4-3
3E2	1 hour	3.3E2	$NH_2Cl + HOCl \rightarrow NHCl_2 + H_2O$	4.5.1
2.5E2		4.0E2	$Fe^{2+} + O_2 \rightarrow Fe^{III}$ (pH 8)	4.6.2
1E2		1.0E3	$Cu^I + H_2O_2 \rightarrow Cu^{II} + \cdot OH + OH^-$	4.6.3
9E1		1.1E3	o-xylene + $O_3 \rightarrow$ oxid. products	Table 8-7
8E0		1.3E4	$SO_3^- + NH_2Cl + H_2O \rightarrow SO_4^{2-} + NH_4^+ + Cl^-$	4.5.1
2.5E0		4.0E4	$Fe^{2+} + O_2 \rightarrow Fe^{III}$ (pH 7)	4.6.2
1E-1	1 day	1.0E6	$O_3 + TCE \rightarrow$ oxid. products; TCE = tetrachloroethylene	Table 8-7
6E-2		1.7E6	$2NH_2Cl \rightarrow NHCl_2 + NH_3$	4.5.1
2.5E-2		4.0E6	$Fe^{2+} + O_2 \rightarrow Fe^{III}$ (pH 6)	4.6.2
1E-2		1.0E7	$H_2S + O_2 \rightarrow S_{ox} + 2H^+$ $S_{ox} = S, S_2O_3^{2-}, SO_4^{2-}$	4.6.1
9E-4	1 year	1.1E8	$OCl^- + Br^- \rightarrow BrO^- + Cl^-$	4.5.1
2.5E-4		4.0E8	$Fe^{2+} + O_2 \rightarrow Fe^{III}$ (pH 5)	4.6.2

Format adapted from Pankow, J. F. and J. J. Morgan, *Environ. Sci. Technol.*, 11, 1155 (1981).

[a] Calculated from $t_{1/2} = 1/k_2[A]_o$; $[A]_o$ assumed = 10^{-5} M for both reactants.
[b] Table or section of text where reaction is described.

the rate at which reactants diffuse together to form an encounter complex in a solvent cage is much slower than v_f. The encounter frequency of neutral molecules in water at 25°C is about 10^9 to 10^{10} M^{-1} s^{-1}; precise estimates depend on the diffusion coefficients and radii of the molecules, as described by Equation 3-25. Second-order rate constants for diffusion-controlled reactions between neutral molecules are defined by Equation 3-26. Electrostatic forces affect encounter frequencies for ions (Equation 3-27), and rate constants for diffusion-controlled reactions of oppositely charged ions range up to about 10^{12} M^{-1} s^{-1}. Among the most rapid reactions of ions in water are simple ion-association reactions (ion-pair and complex formation), some of which are near the diffusion-controlled limit, and acid-recombination reactions ($H^+ + A^- \rightarrow HA$), which generally are diffusion controlled. It is hazardous to generalize about the speed of major categories of reactions, however. Rate constants for water exchange in the primary hydration sphere of metal ions have a range of $\sim 10^{16}$ (Section 4.4.2), and hydrolysis rate constants for alkyl halides range over over a factor of $\sim 10^9$ (Section 4.4.4).

Gas-phase recombinations of free radicals and some free radical-neutral molecule reactions have activation energies of zero. Observed activation energies for the most rapid solution-phase reactions (diffusion-controlled processes) are not zero, as one might expect, but have small E_{act} values ~ 10 to 17 kJ mol^{-1}, reflecting the small temperature dependence of the rate-controlling step — diffusion. The increase in D with T actually reflects the decrease in viscosity, η, with increasing T. According to Walden's rule, ηD is a constant independent of T. Typical values of E_{act} for other elementary chemical reactions are in the range of ~ 21 to 80 kJ mol^{-1} or more. For a normal frequency factor of 5×10^{10} M^{-1} s^{-1}, an activation energy of 80 kJ mol^{-1} leads to a second-order rate constant $k_2 = 5 \times 10^{-4}$ M^{-1} s^{-1} at 25°C. At millimolar concentrations for the reactants, this is equivalent to a reaction rate of $\sim 5 \times 10^{-10} M$ s^{-1} ($1.6 \times 10^{-2} M$ year^{-1}; $t_{1/2} \sim 23$ days). Some multistep processes have very large activation energies. The denaturation of some proteins ($E_{act} \approx 543$ kJ mol^{-1}) is an extreme case; this value implies the rate doubles for every 1°C increase. Activation energies for multistep reactions are difficult to interpret because they may reflect energy changes (both E_{act} and ΔH) for several elementary reactions.

Decreases in rate constants from diffusion-controlled values are caused by increases in E_{act}, decreases in pre-exponential factor A (equivalent to decreases in $\Delta S^{\neq}$) or both (Figure 3-11c). If we assume $E_{act} = 12$ kJ mol^{-1} (reasonable for diffusion), $\Delta S^{\neq} = 0$ and T = 298 K, we obtain a value of $k = 1.3 \times 10^{11}$ M^{-1} s^{-1} from Equation 3-50, which is the general range of the encounter frequencies described above. Keeping $\Delta S^{\neq}$ (hence A) constant, we find that k decreases by a factor of 3.1×10^{-4} for every 20-kJ mol^{-1} increase in E_{act}. For $k \approx 1$ M^{-1} s^{-1}, $E_{act} = 75.2$ kJ mol^{-1}, if $\Delta S^{\neq}$ and A are unchanged.[*] Because of the exponential relationship between k and E_{act}, large changes in k are associated with small changes in E_{act} (at constant A). For example, if $E_{act} = 40$ kJ mol^{-1} and T = 298 K, a doubling of k changes E_{act} to 38.3 kJ mol^{-1}, only a 4% change. E_{act} is calculated from the change in k with temperature rather than from the absolute value of k, and the above remarks do not have direct implications on the accuracy with which E_{act} can be determined.

Entropies of activation may be positive, negative, or near zero, and as noted earlier, the sign and magnitude of $\Delta S^{\neq}$ are reflected in the magnitude of A. A decrease in $\Delta S^{\neq}$ of 40 J K^{-1} mol^{-1} decreases k by a factor of 8.1×10^{-3}, and an increase

---

[*]   A and E_{act} often do not behave independently, at least for series of related reactions. Correlations between the two parameters lead to the so-called "compensation effect", an important phenomenon in linear free-energy relationships (see Section 7.2). In addition, statistical simulations[8] indicate a high covariance (i.e., lack of independence) between the two parameters.

of that size increases k by a factor of 125. The range of $\Delta S^{\neq}$ (ca. ±170 J K^{-1} mol^{-1}) can account for a range of about 10^{17} in A (~10^4 to 10^{22} M^{-1} s^{-1}) and k, since k is directly proportional to A. For example, entropy effects account for all of the ~$10^{5.5}$-fold difference in water exchange rates for Al^{3+} and Cr^{3+}:[54]

	$log\ k'_1$ (s^{-1})	$log\ k_2$ (M^{-1} s^{-1})	E_{act} (kJ mol^{-1})	$\Delta S^{\neq}$ (J mol^{-1} K^{-1})
Al^{3+}	−0.06	−1.8	115.8	99.1
Cr^{3+}	−5.56	−7.3	112.0	−18.0

k'_1 is a pseudo-first-order rate constant. The difference in $\Delta S^{\neq}$ is sufficient to lower k_2 by $10^{-6.1}$. The highly favorable $\Delta S^{\neq}$ for the Al^{3+} exchange suggests a dissociative activation step,[54] whereas the small negative $\Delta S^{\neq}$ for Cr^{3+} exchange suggests a transition state in which the net binding of entering and leaving water molecules is tighter than in the reactant state.

If E_{act} remained at the diffusion-control value and $\Delta S^{\neq}$ decreased to the low end of its range, −170 J K^{-1} mol^{-1}, k would decrease to ~1.7×10^2 M^{-1} s^{-1}. Very slow second-order reactions (k < 1 M^{-1} s^{-1}) thus must have high values of E_{act}, as well as unfavorable entropies of activation. $\Delta S^{\neq}$ usually is lower in absolute value than ΔS_r, and the ratio $\Delta S^{\neq}:\Delta S_r$ is a measure of the extent to which the transition state resembles the reactants or the products.

Entropies and volumes of activation are positively correlated, and as a rule of thumb, $\Delta V^{\neq} = 0.5\Delta S^{\neq}$ when the latter is expressed in cal K^{-1} mol^{-1}. ($\Delta V^{\neq} \approx 0.12\Delta S^{\neq}$ when the latter is expressed in J K^{-1} mol^{-1}.) Relatively few data are available for $\Delta V^{\neq}$, but reported values are within ±20 cm^3 mol^{-1}.

PROBLEMS

1.   The dechlorination of HOCl by sulfite (SO_3^{2-}) is a two-step process. In the first step, which is very rapid, Cl$^+$ transfer to SO_3^{2-} results in formation of chlorosulfate ion ($ClSO_3^-$) and OH$^-$. The second step is the hydrolysis of chlorosulfate: $ClSO_3^- + H_2O \rightarrow Cl^- + SO_4^{2-} + 2H^+$. This reaction also is rapid but can be followed by stop-flow techniques by monitoring the increase in H$^+$ using the acid-base indicator, dinitrophenol, which absorbs light at 410 nm. Yiin and Margerum [*Inorg. Chem.*, 27, 1670 (1988)] reported the following first-order rate constants for hydrolysis of chlorosulfate ion. Determine E_{act} and the activation parameters $\Delta G^{\neq}$ (25°C), $\Delta H^{\neq}$, and $\Delta S^{\neq}$.

T (K)	275	278	283	288	293	298	298
k (s^{-1})	48	63	92	124	211	254	258

2.   Vogel and Reinhard [*Env. Sci. Technol.*, 20 992 (1986)] measured hydrolysis rate constants for several halogenated aliphatic compounds. Experiments were conducted

at elevated temperatures to obtain data in a reasonable time period. In each case, estimate the first-order rate constant and $t_{1/2}$ at 25°C.

		k_{obs} (10^{-5} s^{-1})		
	pH	81°C	61°C	
1,2-dibromopropane	7	1.3	0.21	
(1.2-DBP)	9	1.3	0.21	
	11	3.7	0.50	
	pH	76°C	61°C	45°C
1,3-dibromopropane	7	4.4	0.69	0.10
(1,3-DBP)	9	4.4	0.69	0.14
	11	5.5	0.86	0.14
	pH	90°C	76°C	50°C
1-bromo-3-phenylpropane	7	11	3.3	0.078
(1-BPP)	9	10	3.1	0.078
	11	11	3.3	0.14

3.   In experiments with tritiated naphthalene, Alexander et al. [*Geochim. Cosmochim. Acta,* 46, 219 (1982)] found that the exchange of H atoms in the 1-position on this aromatic molecule with H atoms in water was catalyzed by clays such as bentonite. Given data obtained at various temperatures, estimate the half-life of hydrogen atoms in the 1-position on naphthalene at 30, 50, and 100°C. This process is of interest relative to interpretation of D/H (deuterium/hydrogen) ratios on aromatic compounds in petroleum, which have been proposed as markers for the source material of petroleum products. What do your results suggest about the usefulness of this idea?

Ionic form of bentonite	Temperature (°C)	k_{ex} (10^{-3} h^{-1})
sodium	275	147
sodium	254	41
sodium	222	11.1
sodium	209	9.0
sodium	190	1.84
sodium	177	0.845
hydrogen	120	164

4.   Data on the thermal decarboxylation of acetic acid to CH_4 and CO_2 were reported by Kharaka et al. [*Geochim. Cosmochim, Acta,* 47, 397 (1982)], who suggested that this process could account for an important fraction of natural gas in hydrothermal environments. Given the data tabulated below for several experiments, determine the reaction order, rate constants at 200 and 300°C, the apparent E_{act}, and $t_{1/2}$ for acetic acid at 100, 200, and 300°C. Discuss the apparent effects of pH on this reaction.

<table>
<thead>
<tr><th colspan="2">300°C, pH 2.55</th><th colspan="2">200°C, pH 4.55</th></tr>
<tr>
<th>Reaction time (hours)</th>
<th>Conc. of unreacted acetate (mg/L)</th>
<th>Reaction time (hours)</th>
<th>Conc. of unreacted acetate (mg/L)</th>
</tr>
</thead>
<tbody>
<tr><td>0</td><td>31,150</td><td>0</td><td>31,000</td></tr>
<tr><td>193</td><td>31,000</td><td>479</td><td>30,000</td></tr>
<tr><td>620</td><td>27,900</td><td>645</td><td>29,500</td></tr>
<tr><td>1221</td><td>24,000</td><td>1146</td><td>29,000</td></tr>
<tr><td>1537</td><td>22,500</td><td>4150</td><td>25,200</td></tr>
<tr><td>2330</td><td>18,300</td><td></td><td></td></tr>
<tr><td>3192</td><td>14,500</td><td></td><td></td></tr>
<tr><td>3693</td><td>12,200</td><td></td><td></td></tr>
<tr><td>4393</td><td>10,500</td><td></td><td></td></tr>
</tbody>
</table>

<table>
<thead>
<tr><th colspan="2">300°C, pH 7.01</th><th colspan="2">300°C, pH 4.55</th></tr>
<tr>
<th>Reaction time (hours)</th>
<th>Conc. of unreacted acetate (mg/L)</th>
<th>Reaction time (hours)</th>
<th>Conc. of unreacted acetate (mg/L)</th>
</tr>
</thead>
<tbody>
<tr><td>0</td><td>31,500</td><td>0</td><td>31,000</td></tr>
<tr><td>332</td><td>30,900</td><td>979</td><td>23,000</td></tr>
<tr><td>503</td><td>31,300</td><td>1345</td><td>19,250</td></tr>
<tr><td>638</td><td>31,600</td><td>2878</td><td>14,000</td></tr>
<tr><td>1118</td><td>32,250</td><td>3212</td><td>13,700</td></tr>
<tr><td>1430</td><td>30,250</td><td></td><td></td></tr>
</tbody>
</table>

5. The hydrolysis of pyrophosphate, $H_2P_2O_7^{2-} + H_2O \rightarrow 2H_2PO_4^-$, is a pseudo-first-order reaction at constant pH. The half-life of pyrophosphate at 75°C is 140 h; at 100°C it is 13 h. Compute E_{act} for the reaction and estimate the time required for 99% hydrolysis of a pyrophosphate solution at 20°C.

6. The rate expression for oxygenation of ferrous iron is generally reported as: $-d[Fe^{II}]/dt = k[OH^-]^2PO_2[Fe^{II}]$. Given the following data, find the relationship between ionic strength (I) and the rate constant k (i.e., find the slope of the Brønsted relationship for this reaction).

Ionic strength (I)	0.009	0.012	0.020	0.040	0.060	0.110
$k(10^{13}\ M^{-2}\ atm^{-1}\ min^{-1})$	4.0	3.1	2.9	2.2	1.8	1.2

Note: see Chapter 4 (Section 4.6.5) for a discussion of the need for caution in using these findings to make mechanistic inferences about this reaction.

7. Predict the effect of (A) decreasing the dielectric constant of the solution, (B) increasing the ionic strength, and (C) increasing pressure on rates of the following reactions:

(1) $S_2O_8^{2-} + 2I^- \rightarrow I_2 + 2SO_4^{2-}$
(2) $CO(NH_3)_5Br^{2+} + NO_2^- \rightarrow CO(NH_3)_5NO_2^+ + Br^-$
(3) $Cu^{2+} + HCO_3^- \rightarrow CuHCO_3^+$
(4) $Cu^{2+} + NH_3 \rightarrow CuNH_3^{2+}$

Estimate the sign of $\Delta S^{\neq}$ in each case.
The following data were obtained for reaction (1) at different ionic strengths. Do the data fit the Brønsted equation? What is the value of $z_A z_B$?

I	2.45	3.65	4.45	6.45	8.45	12.45(all $\times 10^{-3}$)
k (M^{-1} s^{-1})	1.05	1.12	1.16	1.18	1.26	1.39

REFERENCES

1.   Denbigh, K.G., *The Principles of Chemical Equilibrium*, 2nd ed., Cambridge University Press, Cambridge, 1966.
2.   Stumm, W. and J.J. Morgan, *Aquatic Chemistry*, 2nd ed., Wiley-Interscience, New York, 1981.
3.   Horne, R.A., *Marine Chemistry*, Wiley-Interscience, New York, 1969.
4.   Adamson, A.W., *A Textbook of Physical Chemistry*, 2nd ed., Academic Press, New York, 1979.
5.   Siegel, S., L.H. Fuchs, B.R. Hubble, and E.L. Nielsen, *Environ. Sci. Technol.*, 12, 1411 (1978).
6.   Steen, C.L., K. Li, and F.H. Rogan, *Environ. Sci. Technol.*, 14, 588 (1980).
7.   Van Houte, G., L. Rodrique, M. Genet, and B. Delmon, *Environ. Sci. Technol.*, 15, 327 (1981).
8.   Heberger, K., S. Kemeny, and T. Vidoczy, *Int. J. Chem. Kinetics*, 19, 171 (1987).
9.   Kalantar, A.H., *Int. J. Chem. Kinetics*, 19, 923 (1987).
10.   Messer, J.J. and P.L. Brezonik, *Ecol. Modeling*, 21, 277 (1984).
11.   Lewandowski, Z., *Water Res.*, 16, 19 (1982).
12.   Stillinger, F.H., *Science*, 209, 451 (1980).
13.   Frank, H.S. and W.Y. Wen, *Disc. Faraday Soc.*, 24, 133 (1957); Eisenberg, D. and W. Kauzmann, *The Structure and Properties of Water*, Oxford University Press, Oxford, 1969.
14.   Nemethy, G. and H.A. Sheraga, *J. Chem. Phys.*, 36, 3382 (1962).
15.   Weston, R.E., Jr. and H.A. Schwarz, *Chemical Kinetics*, Prentice-Hall, Englewood Cliffs, NJ, 1972.
16.   Entelis, S.G. and R.P. Tiger, *Reaction Kinetics in the Liquid Phase*, Halsted Press, New York, 1976.
17.   Fischer, H.B., E.J.List, R.C.Y.Koh, J.Imberger, and N.H.Brooks, *Mixing in Inland and Coastal Waters*, Academic Press, New York, 1979.
18.   For a description of Fick's experiments, see R.A. Berner, *Early Diagenesis, a Theoretical Approach*, Princeton University Press, Princeton, NJ.
19.   Crank, J., *The Mathematics of Diffusion*, 2nd ed., Clarendon Press, Oxford, 1975.
20.   Carslaw, H.S. and J.C. Jaeger, *Conduction of Heat in Solids*, Clarendon Press, Oxford, 1959.
21.   Smoluchowski, M., *Z. Physik. Chem.*, 92, 129 (1917).

22. Noyes, R.M., *Progr. React. Kinetics*, 1, 129 (1961).
23. Burshtein, A.I. and B.I. Yakobson, *Int. J. Chem. Kinetics*, 12, 261 (1980).
24. Sitarski, M., *Int. J. Chem. Kinetics*, 13, 125 (1981).
25. Laidler, K.J., *Chemical Kinetics*, McGraw-Hill, New York, 1965.
26. Amis, E.S., *Solvent Effects on Reaction Rates and Mechanisms*, Academic Press, New York, 1966.
27. Frost, A.A. and R.A. Pearson, *Kinetics and Mechanism*, John Wiley & Sons, New York, 1961.
28. Lindeman, F.A., *Trans. Faraday Soc.*, 17, 598 (1922).
29. Hinshelwood, C.N., *Proc. R. Soc. (London)*, A113, 230 (1926).
30. Eyring, H. *J. Chem. Phys.*, 3, 107 (1935); *Chem. Rev.*, 17, 65 (1935).
31. Glasstone, S., K.J. Laidler, and H. Eyring, *The Theory of Rate Processes*, McGraw-Hill, New York, 1941.
32. Benson, S.W., *Thermochemical Kinetics*, 2nd ed., John Wiley & Sons, New York, 1976.
33. Eyring, H., S.H. Lin, and S.M. Lin, *Basic Chemical Kinetics*, Wiley-Interscience, New York, 1980.
34. Zewail, A.H., *Science*, 242, 1645 (1988).
35. Hirschfelder, J.O., H. Eyring, and B. Topley, *J. Chem. Phys.*, 4, 170 (1936).
36. Lasaga, A.C., in *Kinetics of Geochemical Processes*, A.C. Lasaga and R.J. Kirkpatrick (Eds.), Mineral. Soc. Am., Washington, D.C., 1981, chap. 4.
37. Dunning, T.H., Jr., L.B. Harding, A.F. Wagner, G.C. Schatz, and J.M. Bowman, *Science*, 240, 453 (1988).
38. Lasaga, A.C. and G.V. Gibbs, in *Aquatic Chemical Kinetics*, W. Stumm (Ed.), Wiley-Interscience, New York, 1990, pp. 259–289.
39. Lasaga, A.C. and G.V. Gibbs, *Phys. Chem. Min.*, 14, 107–117 (1987); *Phys. Chem. Min.*, 16, 29–41 (1988).
40. Halgren, T.A. and W.N. Lipscomb, *Chem. Phys. Lett.*, 49, 225–232 (1977); Hehre, W.J., L. Radom, P.R. Schleyer, and J.A. Pople, *AB INITIO Molecular Orbital Theory*, Wiley, New York, 1986.
41. Pacey, P.D., *J. Chem. Ed.*, 58, 812 (1981).
42. Kraut, J., *Science*, 242, 533 (1988).
43. Johnson, K.S., *Limnol. Oceanogr.*, 27, 849 (1982).
44. Aagaard, P. and H.C. Helgeson, *Am. J. Sci.*, 282, 237 (1982).
45. Kohnstam, G., *Progr. React. Kinetics*, 5, 335 (1970).
46. Laidler, K.J. and D.T.Y. Chen, *Can. J. Chem.*, 37, 599 (1959).
47. Lawrance, G.A. and D.R. Stranks, *Acct. Chem. Res.*, 12, 403 (1979).
48. Adams, W.A., *Environ. Sci. Technol.*, 6, 928 (1972).
49. Perlmutter-Hayman, B., *Progr. React. Kinetics*, 6, 239 (1971).
50. Davies, C.W., *Progr. React. Kinetics*, 1, 163 (1961).
51. Bronsted, J.N., *Z. Phys. Chem.*, 102, 169 (1922).
52. Bjerrum, N., *Z. Phys. Chem.*, 108, 82 (1924).
53. Pankow, J.F. and J.J. Morgan, *Environ. Sci. Technol.*, 11, 1155 (1981).
54. Morgan, J.J. and A.T. Stone, in *Chemical Processes in Lakes*, W. Stumm (Ed.), Wiley-Interscience, New York, 1985, pp. 389–426.

"Do you not see that even drops of water falling upon a stone in the long run beat a way through the stone?"

De Rerum Natura, Book 4, Lucretius

CHAPTER **4**

Kinetics of Chemical Reactions in Aquatic Systems: From Homogeneous Catalysis to Reactions at Interfaces

OVERVIEW

This chapter describes rate equations and mechanisms of important chemical reactions in aqueous solutions, with emphasis on reactions in natural aquatic systems. Many solution-phase reactions are affected by homogeneous catalysts, and the first two parts of this chapter are devoted to such reactions. Part I describes the principles of homogeneous catalysis (mechanisms and rate equations), and Part II applies this knowledge to important examples in aquatic systems: acid/base-catalyzed hydrolysis and chlorination reactions and metal-catalyzed oxidation and oxygenation reactions. Not all chemical reactions in aquatic systems occur in solution, however; the last part of the chapter deals with reactions at air-water and solid-water interfaces. Transfer of gases and volatile compounds across the air-water interface is primarily a phase change rather than a chemical reaction, but some gases, such as CO_2 and SO_2, do react chemically with water as soon as they cross the air-water interface. Moreover, rates of gas transfer are described by kinetic equations analogous to those for chemical reactions. Mineral dissolution and precipitation reactions are of vital importance in regulating the ionic composition of natural waters. Complexation and redox reactions at solid-solution interfaces also affect the composition of natural waters.

PART I. HOMOGENEOUS CATALYSIS

4.1 INTRODUCTION

4.1.1 Characteristics of Catalysts

Catalysts usually are defined as substances that change reaction rates, but do not undergo chemical change themselves. Homogeneous catalysts are in the same phase as the reactants; heterogeneous catalysts are solid surfaces that accelerate reactions that are much slower in gas or liquid phases. The above definition of a catalyst is not completely accurate, as the following remarks indicate. First, there are many examples of products that act as catalysts. This is called autocatalysis, and it is important in the oxidation of manganese and possibly iron. Second, a substance cannot affect a process without somehow becoming involved. Thus, it is more accurate to state that (normally) there is no *net* change in the amount of the catalyst during reaction. Catalysts have been defined as substances whose concentrations appear in rate expressions to higher powers than one would predict from stoichiometric equations.[1]

The "requirement" of no net change is violated in some heterogeneous reactions where the catalyst is used up during a reaction. Also, chain reactions are catalyzed by substances that decompose to yield free radicals, and the original compound may not be regenerated. Chain initiation usually is the rate-limiting step, and radical formers facilitate this step. These substances often are called *initiators*. Although small amounts of catalysts usually have large effects on reaction rates, this is not an essential property of catalysts. The concept of "catalytic" quantities derives from the fact that most catalysts are recycled during reactions; they combine with reactants to form intermediates and are released later in the reaction sequence. Thus, they can be used again and again during a reaction. A definition that takes all these characteristics and exceptions into account would be quite complicated.

Except for heterogeneous and chain-initiation reactions, where catalysts may not be recovered in original form, catalysts are not changed during a reaction. This implies that they impart no energy to the system. Therefore, catalysis has no effect on the position of equilibrium. Because equilibrium constants are ratios of forward- and reverse-rate constants, it follows that catalysts influence forward and reverse rates in the same proportion. This fact can be derived from the principle of microscopic reversibility.

4.1.2 Catalysis as a Change in Reaction Mechanism

Some books state that catalysts work by lowering the activation energy of a reaction. Although this is true, it is more accurate to state that catalysts change reaction mechanisms to ones with lower activation energy. A change in mechanism is a key feature of all catalytic action. In many cases catalysts replace a slow one-step reaction with a series of faster reactions. That a series of reactions can be faster than a single reaction when the net result is the same (i.e., the thermodynamic driving force is constant) may be puzzling. However, reaction rates do not depend on the free energy of reaction, but on the free energy of activation. The $\Delta G^{\neq}$ for each step in a catalyzed series thus is less than $\Delta G^{\neq}$ for the single-step reaction.

In many cases the effectiveness of catalysts can be explained in terms of the probability of reactant collisions (i.e., effects on the frequency factor, A). Substitution of a series of more probable bimolecular steps for a single improbable termolecular step can lead to an overall increase in reaction rate. Consider a redox reaction of the type

$$2A^{2+} + B^{+} \rightarrow 2A^{+} + B^{3+} \tag{4-1a}$$

If intermediate oxidation state B^{2+} does not exist, the uncatalyzed reaction must proceed by a single termolecular step. However, if an ionic catalyst C has an intermediate oxidation state, the following mechanism can take place:

$$A^{2+} + C^{+} \rightarrow A^{+} + C^{2+}$$

$$A^{2+} + C^{2+} \rightarrow A^{+} + C^{3+}$$

$$\frac{C^{3+} + B^{+} \rightarrow C^{+} + B^{3+}}{\text{Net:} \quad 2A^{2+} + B^{+} \rightarrow 2A^{+} + B^{3+}} \tag{4-1b}$$

All three steps in sequence 4-1b are bimolecular. Collision frequencies for bimolecular reactions are about 100 times greater than those for termolecular reactions.[2] Thus, if energy factors are equal, the catalyzed sequence should be faster than the one-step reaction.

In other cases the change in mechanism is more subtle. For example, in acid-catalyzed hydrolysis of esters (Section 4.2.1), the mechanism involves protonation of the reacting species to form a positively charged species (an electrophile) that is more susceptible to attack by a nucleophilic water molecule than is the unprotonated reactant.

4.2 TYPES AND MECHANISMS OF HOMOGENEOUS CATALYSTS

Three classes of homogeneous catalysts are important for reactions in aqueous solutions: (1) acids and bases that facilitate group transfer and cleavage reactions; (2) substances with multiple oxidation states, such as transition metals, that facilitate electron transfer; and (3) substances that initiate free-radical formation in chain reactions. The most common and best understood homogeneous catalysts are acids and bases.

4.2.1 Mechanisms of Acid-Base Catalysis

Many types of acid-base catalysis occur, reflecting the variety of reactions influenced by acids and/or bases and the diversity of acid-base definitions. Examples range from the transfer of Arrhenius acids and bases (H^+ and OH^-) to or from a hydrolyzing reactant molecule to the catalysis of hydrolysis and redox reactions by Bronsted acids or bases and to the electrophilic interactions of metal ions (Lewis acids) with nucleophilic reactants (Lewis bases). Catalysis by H^+ (or H_3O^+) is called *specific* acid catalysis; catalysis by OH^- is specific base catalysis. *General* acid or base catalysis, also known as buffer catalysis, refers to catalysis by undissociated Bronsted acids and bases. Structural and energetic explanations for the effectiveness of acid and base catalysis depend on the nature of the reactants and type of reaction being catalyzed. Examples of mechanisms for acid-base catalysis are shown in Table 4-1 and Figure 4-1.

Arrhenius and Bronsted Acid-Base Catalysis

Frost and Pearson[3] described eight general mechanisms for acid-base catalysis, some of which are important only for organic reactions in nonaqueous solutions. Laidler[4] differentiated mechanisms depending on whether the proton acceptor in the product formation step was a solvent (water) or solute molecule. If the proton acceptor is H_2O, the mechanism is termed protolytic (cases I and II in Table 4-1). If the second step is slow (case I), the rate depends on $\{H^+\}$ (actually $\{H_3O^+\}$), yielding specific acid catalysis, but if the second step is rapid (case II), the reaction is subject to general acid catalysis. Mechanisms in which the proton acceptor in the second step is a solute molecule are called prototropic and result in general acid catalysis (case III) regardless of the relative rates of the first and second steps. Similar mechanisms occur for base catalysis.

Acid catalysis involves an increase in the electrophilic character of the reaction site. Addition of H^+ to an electronegative atom (N or O) in an organic molecule tends to withdraw electrons from a central carbon atom, making it more attractive to nucleophiles. For example, protonation of the carbonyl oxygen on an ester induces a positive charge on the carbonyl carbon, promoting nucleophilic attack by H_2O and catalyzing hydrolysis of the ester (Figure 4-1a).

Base catalysis may remove a proton from a reactant, forming a more nucleophilic anion (a better electron donor) than the parent molecule, but complete removal of the proton is not necessary. For example, in the hydrolysis of esters and amides, the base may interact with the proton of an attacking water molecule, promoting a partial negative charge on the attacking oxygen atom and making it more nucleophilic (Figure 4-1b). Similarly, the OH^--catalyzed hydrolysis of the nerve gas GB (also known as sarin), an organofluorophosphonate (Figure 4-1c), may occur either by direct attack of OH^- on the fluorophosphate or by OH^- interaction with a proton on an attacking H_2O molecule. Both mechanisms yield the same rate equation and cannot be distinguished on kinetic grounds. Hydrolysis of octahedral metal ion complexes containing amine ligands (NH_3, RNH_2) is much faster under basic conditions than at neutral or acidic pH. Specific base catalysis in these cases involves OH^- extraction of a proton from an amine N group to create the conjugate base of the original complex in a rapid preequilibrium step (Figure 4-1d). Why this "internal conjugate base" (ICB) mechanism enhances the rate is uncertain, but it

Table 4-1. Mechanisms of Acid-Base Catalysis

Type	Mechanism	Rate expression	Comments
I. Specific H^+	$S + HA \underset{k_2}{\overset{k_1}{\rightleftharpoons}} SH^+ + A^-$ $SH^+ + H_2O \underset{slow}{\overset{k_3}{\rightarrow}} P + H_3O^+$	$P = k_1 k_3 [S][HA]/k_2[A^-]$ $= (k_1 k_3/k_2 K_a)[S][H^+]$ where $K_a = [H^+][A^-]/[HA]$	For protolytic case, expression applies when $k_3 \ll k_2[A^-]$ whether initial H^+ transfer is from Bronsted acid (HA) or H_3O^+.
II. General acid	$S + HA \underset{k_2}{\overset{k_1}{\rightleftharpoons}} SH^+ + A^-$ $SH^+ + H_2O \underset{fast}{\overset{k_3}{\rightarrow}} P + H_3O^+$	$P = [S]\{\sum k_i [HA]_i\}$	Expression applies when $k_3 \gg k_2[A^-]$; rate-controlling step is formation of intermediate SH^+. P written for presence of several Bronsted acids in system.
III. General acid	$S + HA \underset{k_2}{\overset{k_1}{\rightleftharpoons}} SH^+ + A^-$ $SH^+ + A^- \overset{k_3}{\rightarrow} P + HA$	$P = \dfrac{k_1 k_3 [S][HA]}{(k_2 + k_3)}$ or $P = k'[S][HA]$	Prototropic mechanism yields general acid catalysis regardless of relative sizes of k_2 and k_3.
IV. Specific OH^-	$HS + B \underset{k_2}{\overset{k_1}{\rightleftharpoons}} S^- + BH^+$ $S^- + H_2O \underset{slow}{\overset{k_3}{\rightarrow}} P + OH^-$	$P = k_1 k_3 [S^-][B]/k_2[BH^+]$ $= (k_1 k_3/k_2 K_B)[S^-][OH^-]$	For protolytic case, expression applies when $k_3 \ll k_2[BH^+]$ regardless of nature of proton acceptor in first step.
V. General base	$HS + B \underset{k_2}{\overset{k_1}{\rightleftharpoons}} S^- + BH^+$ $S^- + H_2O \underset{fast}{\overset{k_3}{\rightarrow}} P + OH^-$	$P = k[HS][B]$ $P = [HS]\{\sum k_i [B_i]\}$	Expression applies when $k_3 \gg k_2[BH^+]$; rate-controlling step is formation of S^-; P written for presence of several Bronsted bases.
VI. General base	$HS + B \underset{k_2}{\overset{k_1}{\rightleftharpoons}} S^- + BH^+$ $S^- + BH^+ \overset{k_3}{\rightarrow} P + B$	$P = \dfrac{k_1 k_3 [S][HA]}{(k_2 + k_3)}$ or $P = k'[S][HA]$	Prototropic case yields general base catalysis regardless of relative sizes of k_2 and k_3.

Adapted from Laidler, K. J., *Chemical Kinetics*, McGraw-Hill, New York, 1965.

a Acid-catalyzed hydrolysis of esters (and amides)

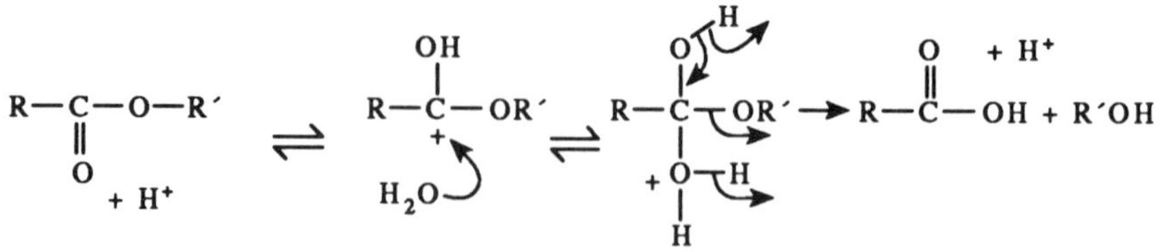

b Base-catalyzed hydrolysis of amides

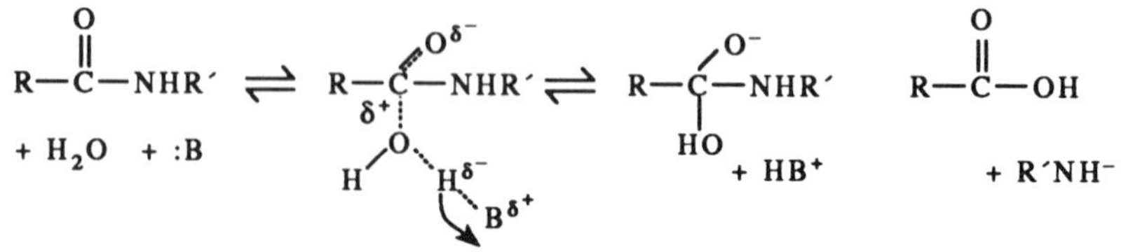

c

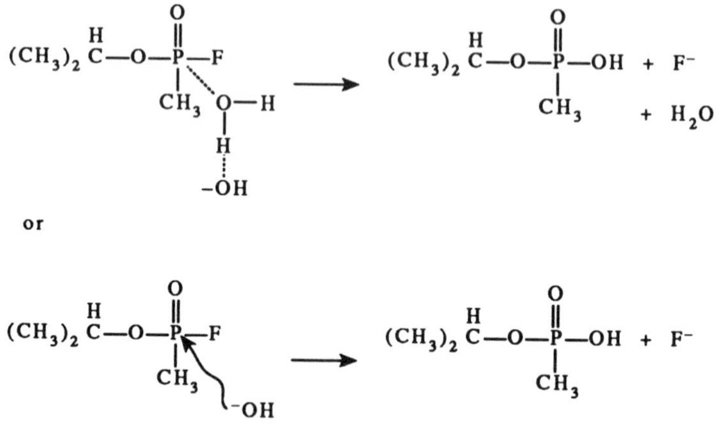

or

Figure 4-1. Mechanisms of acid-, base-, and metal-catalyzed reactions of various compounds and classes of compounds: (a) H^+, esters; (b) Bronsted base, amides; (c) OH^-, sarin (a neurotoxin); (d) OH^-, octahedral metal-amine complexes; (e) heavy metals, amides, peptides; (f) intramolecular Co^{III}-assisted, glycylglycine; (g) intramolecular M^{II}-assisted, imidazole methyl esters; (h) pH-independent, metal-catalyzed, 2-pyridylmethyl hydrogen phthalate ester; (i) Cu^{II}, parathion; (j) transition metals, oxaloacetate decarboxylation; (k) M^{II}, adenosine triphosphate (ATP). All but (j) are hydrolysis reactions.

d

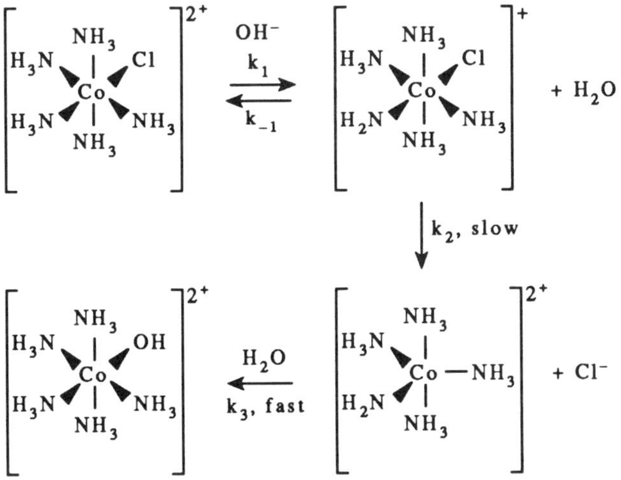

e Cu^{2+}-catalyzed hydrolysis of amino acid esters

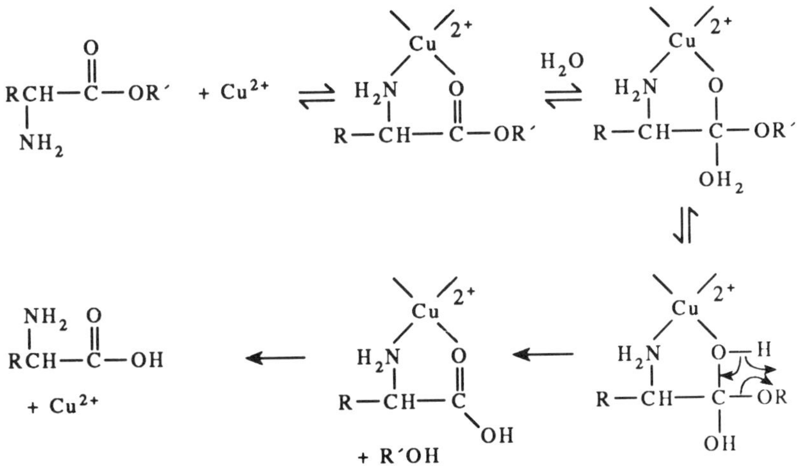

Figure 4-1 (continued).

f

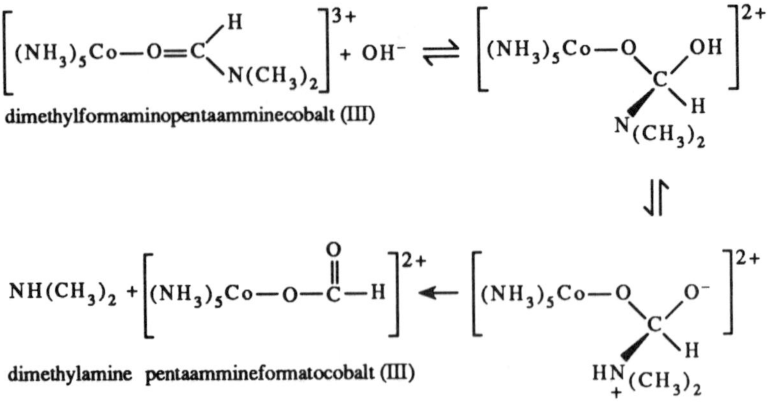

$$\left[(NH_3)_5Co-O=C\begin{smallmatrix}H\\N(CH_3)_2\end{smallmatrix}\right]^{3+} + OH^- \rightleftharpoons \left[(NH_3)_5Co-O\begin{smallmatrix}OH\\C\\H\\N(CH_3)_2\end{smallmatrix}\right]^{2+}$$

dimethylformaminopentaamminecobalt (III)

$$NH(CH_3)_2 + \left[(NH_3)_5Co-O-\overset{O}{\overset{\|}{C}}-H\right]^{2+} \leftarrow \left[(NH_3)_5Co-O\begin{smallmatrix}O^-\\C\\H\\HN(CH_3)_2\\+\end{smallmatrix}\right]^{2+}$$

dimethylamine pentaammineformatocobalt (III)

g Metal ion assisted intramolecular hydrolysis of glycylglycine

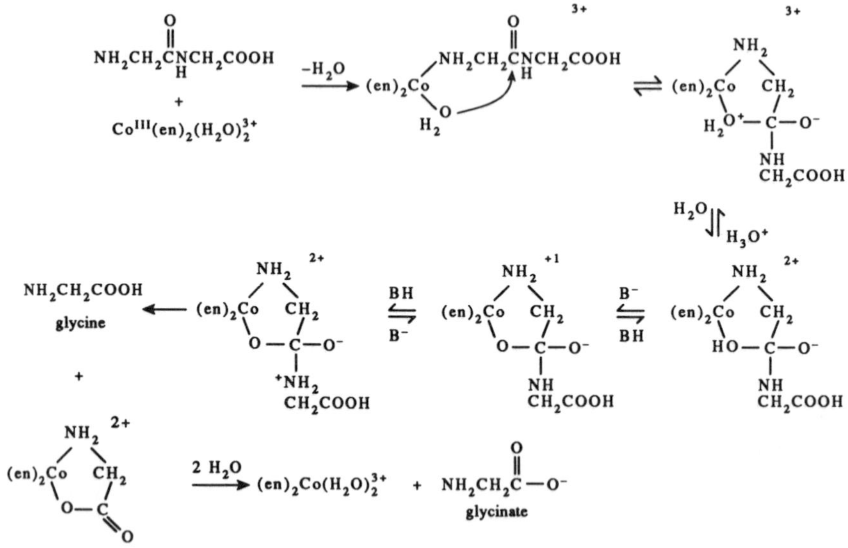

Figure 4-1 (continued).

h Metal ion assisted hydrolysis of a methyl imidazole

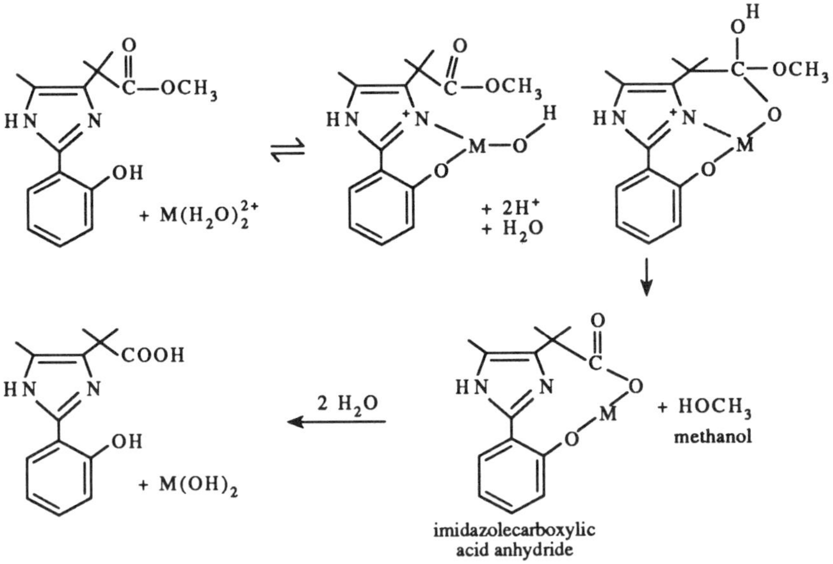

i Ester hydrolysis by metal ion stabilization of transition state

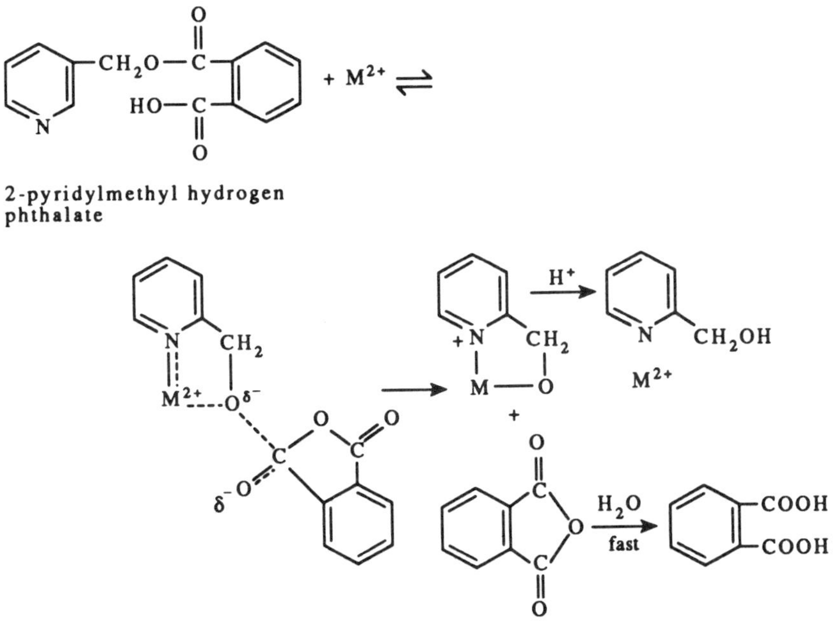

2-pyridylmethyl hydrogen
phthalate

Figure 4-1 (continued).

j Cu(II) assisted hydrolysis of parathion

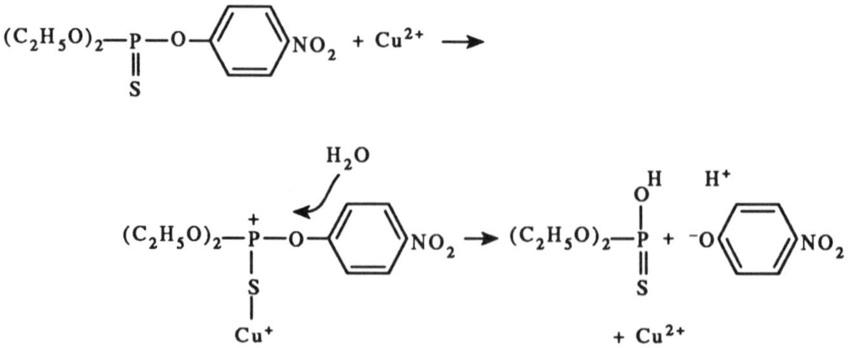

k Decarboxylation of oxaloacetate

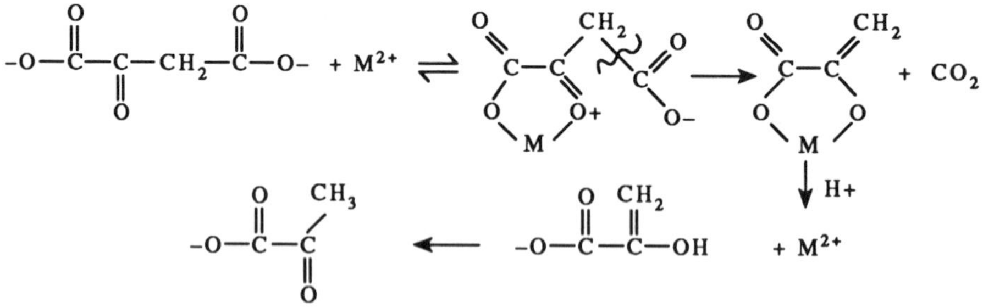

l Metal ion-catalyzed hydrolysis of ATP

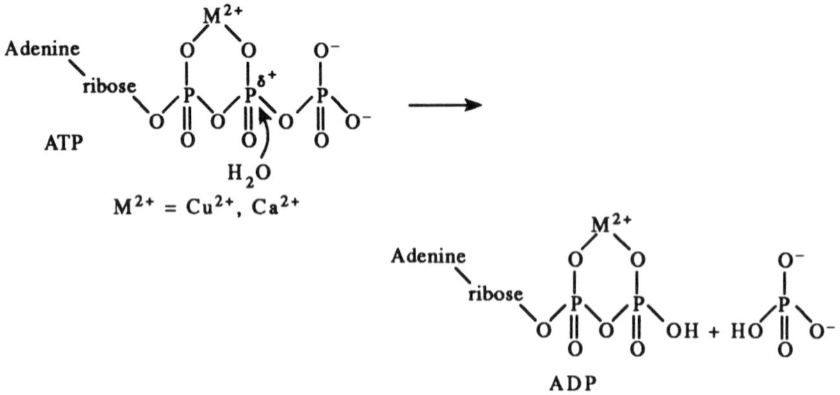

ADP

Figure 4-1 (continued).

appears that the amido group ($-NR_2$) stabilizes the five-coordinate (trigonal bipyrimidal) transition state of the rate-limiting step in the mechanism.[5]

As the reactions in Figure 4-1a–c show, hydrolysis of a given compound may occur by acid catalysis at low pH and base catalysis at high pH. At intermediate pH, where both [H_3O^+] and [OH^-] are low, hydrolysis may proceed by attack by H_2O. This usually is regarded as an uncatalyzed reaction, but H_2O is amphiprotic and can act as both an acid and a base catalyst.

Acid-base catalysis is not restricted to hydrolysis reactions. Many redox reactions of chloramines, hypochlorite, and sulfite involve general acid or base catalysis. For example, disproportionation of monochloramine proceeds by general acid catalysis:[6]

$$NH_2Cl + HA = NH_3Cl^+ + A^- \qquad (4\text{-}2a)$$

$$NH_3Cl^+ + NH_2Cl \rightarrow NHCl_2 + NH_3 + H^+ \qquad (4\text{-}2b)$$

On the other hand, reaction of hypochlorous acid with dichloramine is general base catalyzed:

$$NHCl_2 + HOCl + B \rightarrow NCl_3 + OH^- + HB^+ \qquad (4\text{-}3)$$

The base assists proton removal from $NHCl_2$ as the nitrogen attacks the Cl of HOCl.[7] The kinetics and mechanisms of reactions involving oxidants used in water treatment (such as chloramines and HOCl) are described in Section 4.5.

Lewis Acid Catalysis

Metal cations, which act as Lewis acids and thus behave as electrophiles, can catalyze reactions that are catalyzed by H^+ and Bronsted acids. For example, hydrolysis of esters can be catalyzed by divalent metal ions, but H^+ is a much more effective catalyst because of its higher charge density. Metal ions are effective catalysts for hydrolysis of amino acid esters, amides, and peptides, however, because coordination of the metal ions with basic N and O atoms in these molecules lowers electron density and promotes nucleophilic attack by water. Figure 4-1 portrays several mechanisms for metal-catalyzed hydrolysis reactions of such compounds. Where structures are suitable, the metal ion and organic compound form a five- or six-membered ring, with the O and N atoms of the amino acid serving as bonding ligands[8-10] (Figure 4-1e). Hydrolysis of metal-coordinated amides can proceed by *intermolecular* mechanisms in which the attacking OH^- or H_2O is independent (as in Figure 4-1e), or by *intramolecular* mechanisms in which the attacking OH^- is coordinated to the metal ion. Examples of intramolecular mechanisms include the Co^{III}-assisted hydrolysis of glycylglycine[9] (Figure 4-1f) and the hydrolysis of methyl esters of substituted imidazoles[11] (Figure 4-1g). Co^{II} and Ni^{II} are effective catalysts of the latter reaction in the μM range, but Mg, Ca, and Mn are not. Zn forms a catalytically inactive dimeric complex.

Rate enhancements of 10^4 to 10^7 are common for the hydrolysis of metal complexes of amino acids compared with uncomplexed species,[8,9] and some reactions are catalyzed further by added base. An enhancement of 10^{10} was reported for intramolecular hydrolysis of Co^{III}-coordinated propylglycylglycine in the presence of phosphate, compared with the uncatalyzed hydrolysis of glycylglycine.[9] Such base-catalyzed reactions of amides and esters are called metal-ion-activated reactions, and the rate-limiting step involves nucleophilic attack on the carbonyl carbon by the base. This is the most common mechanism involving metal ions in amide and ester hydrolysis. The effectiveness of activation is correlated with the binding strength of metal ions, e.g., $Cu^{2+} > Co^{2+} > Mn^{2+} > Ca^{2+}$;[12] $Cu^{2+} > Ni^{2+} > Zn^{2+}$;[13] these correlations represent linear free energy relationships (see Chapter 7). Metal enhancement of base-catalyzed amide hydrolysis affects $\Delta S^{\neq}$ primarily rather than $\Delta H^{\neq}$ and produces a temperature-independent acceleration.

Some pH-independent metal-catalyzed reactions also are known. In these cases the metal ion binds weakly with the reactant, and catalytic effectiveness is not related to binding strength with the reactant. The metal catalyzes these reactions by binding to the leaving group oxygen in the transition state.[14] This is important for poor leaving groups like alcohols with high pK_as. For example, the metal ion neutralizes the strongly nucleophilic alcoholate anion in Figure 4-1h, thus stabilizing it and the transition state and lowering E_{act}. In contrast, metal catalysis is not important in the intramolecular nucleophilic hydrolysis of phenolic esters[13,15] because phenols are good leaving groups. However, metal-activated OH^--catalyzed hydrolysis of these esters occurs at pH > 6, with $Zn^{2+} > Cu^{2+} > Co^{2+}$ in effectiveness.

The formation of five-membered chelates is important in the hydrolysis of aminothiol esters;[16] Cu^{II}, which forms especially strong bonds with reduced sulfur, is a much more effective promoter than Ni^{II}. Similarly, Cu^{II} is a moderately effective catalyst for the hydrolysis of parathion, a phosphorothionate pesticide (Figure 4-1i); a 20-fold enhancement of the rate was observed at pH 8.5 and $[Cu^{II}] = 3 \times 10^{-7} M$.[17]

Decarboxylation of malonic acid (acetonedicarboxylic acid) is catalyzed by various metal ions,[18] and the effectiveness of catalysis is correlated with the dissociation constant of the metal malonate complex (Figure 4-2). Catalytic activity follows the Irving-Williams order, $Mn < Co < Ni < Cu > Zn$, which agrees with the binding abilities of the metals; this is another example of a linear free-energy relationship. Similar results have been reported for metal-enhanced decarboxylation of oxaloacetate,[19] which probably occurs via the formation of a five-membered ring (Figure 4-1j).

Activation by metal ions is important in the OH^--catalyzed hydrolysis of condensed phosphates,[10,12,20,21] and the degree of enhancement is related to the complexing ability of the metals. For example, trimetaphosphate hydrolysis is ~360 times faster when $10^{-4} M$ $CaCl_2$ is added to a reaction medium containing $10^{-3} M$ $NaOH$.[21] In general, hydrolysis rates of trimetaphosphate species decrease with increasing charge as follows: $M^{II}P_3O_9^- > M^IP_3O_9^{2-} > P_3O_9^{3-}$; the divalent metal complexes hydrolyze several orders of magnitude faster than $P_3O_9^{3-}$ alone. If the rate-limiting step is hydrolysis of the $M^{II}P_3O_9^-$ complex and if formation-dissociation of the complex is a rapid preequilibrium step, the reaction mechanism is similar to example VI in Table 2-5. The overall rate constant at a given pH is the product of the equilibrium formation constant for the complex, the free metal ion concen-

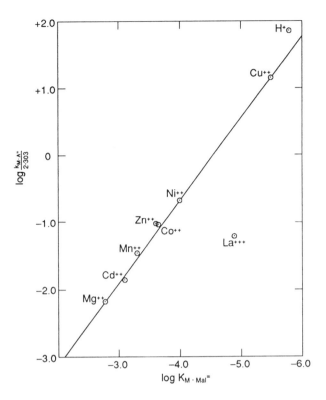

Figure 4-2. The rate constant, k, for metal-catalyzed decarboxylation of malonic acid (acetonedicarboxylic acid) is correlated with K_{MMa}, the dissociation constant for corresponding metal-malonate complex — a linear free-energy relationship. [From Prue, J.E., *J. Chem. Soc.*, II, 2331 (1952). With permission.]

tration, and the intrinsic rate constant for hydrolysis of the metal complex ($k_{obs} = k_o K_f [M^{2+}]$). Hydrolysis of the biochemical intermediate ATP similarly is catalyzed by Mg^{2+}, Ca^{2+}, and Cu^{2+} (Figure 4-1k).

The role of metal ions in catalyzing reactions of bio-organic compounds has received much attention in efforts to explain the functions of metal coenzymes.[22-24] Many of these studies, including some discussed above, involve model organic compounds not found in the environment and metal concentrations much higher than found in natural waters, e.g., >$10^{-3} M$ for transition metals;[14,15,18,19] their applicability to natural conditions thus is doubtful. Electrophilic/nucleophilic interactions involving metal ions also play catalytic roles in organic synthesis reactions such as nitration, addition, condensation, elimination, and group transfer processes,[3,12,23,24] but they typically require extreme conditions (strong acid, nonaqueous solvents) and are unlikely to occur under environmental conditions.

4.2.2 Redox Reactions

Many redox reactions are subject to homogeneous catalysis, and several types of catalytic mechanisms are possible, including recyclable metal ions with several oxidation states, ligand transfer processes, and acid-base catalysis, as mentioned for sulfite and chlorine species in the previous section. Many authors have suggested that metal ions are important in catalyzing redox reactions in natural waters, but very few studies have reported evidence for such reactions under *in situ* conditions. The examples described below involve metal ions and/ or reactants that are unlikely to be important in natural aquatic systems, but some are potentially important for reactors treating industrial wastewaters.

Catalysis by Ions with Intermediate Oxidation States

If an element has two stable oxidation states that differ by two electrons, the intermediate oxidation state is unstable with respect to disproportionation. Some intermediate states have such low stability that they never are important, but mildly unstable states can be important in catalyzing otherwise slow redox reactions. For example, the mechanism in Equation 4-1b accounts for catalysis of the following reaction by Mn^{2+}:[33]

$$2Ce^{4+} + Tl^+ \rightarrow 2Ce^{3+} + Tl^{3+} \tag{4-4}$$

Manganese has a moderately stable intermediate oxidation state, Mn^{III}, but Tl^{2+} is very unstable.

Metal ions are catalysts for reactions involving strong oxidants such as persulfate and ozone. For example, oxidation of Cr^{III}, V^{IV}, Ce^{III}, and other cations by persulfate is catalyzed by Ag^+. Though highly exergonic, these reactions are very slow in the absence of catalysts. The reaction rates are independent of the metal ion concentrations and first order in Ag^+ and $S_2O_8^{2-}$.[25] Rate constants for the reactions are identical, and the rate-limiting step thus involves the same species — probably $AgS_2O_8^-$.[10,25,26] The breakdown products of this complex oxidize the metal ions rapidly, thus accounting the independence of overall rates on metal concentration. One proposed mechanism involves Ag^{3+} and SO_4^{2-} as products of the rate-limiting step:[27]

$$S_2O_8^{2-} + Ag^+ \rightarrow\rightarrow 2SO_4^{2-} + Ag^{3+} \tag{4-5}$$

An alternative mechanism involves Ag^{2+} and the sulfate radical ion:[26]

$$S_2O_8^{2-} + Ag^+ \rightarrow\rightarrow SO_4^{2-} + SO_4^- + Ag^{2+} \tag{4-6}$$

Equilibrium is established rapidly between Ag^{2+} and Ag^{3+}, and the form of the rate equation is of no help in distinguishing between the mechanisms. Other metal ions with two or more oxidation states, notably Cu, also catalyze redox reactions of persulfate. Cu^{3+} and SO_4^- may be products of the initial step, in a mechanism analogous to Equation 4-6. Ozone rapidly oxidizes Ag^+ to Ag^{3+}, and Ag^+ may be a good catalyst for reactions in which ozone is the oxidant.[26] For example, Ag^+ has been used to catalyze the ozonation of Mn^{2+}.[28]

Catalysis of the reaction $2Ce^{4+} + As^{III} \rightarrow 2Ce^{3+} + As^{V}$ by iodide may proceed by a mechanism similar to Equation 4-1b:[29]

$$I^- + Ce^{4+} \rightarrow Ce^{3+} + I^0$$

$$I^0 + Ce^{4+} \rightarrow Ce^{3+} + I^+ \qquad (4\text{-}7)$$

$$I^+ + As^{III} \rightarrow I^- + As^{V}$$

This reaction is the basis for the measurement of iodide in natural waters[30] by kinetic analysis. (The reaction rate is proportional to [I⁻].) The exact mechanism is more complicated than that in sequence 4-7, but it is reasonable to assume that I⁻ catalyzes the reaction by facilitating a sequence in which the molecularity of each step is less than that for a one-step reaction.

Some organic compounds behave as analogs of the inorganic ions described above. Quinones and hydroquinones are stable compounds that differ in oxidation state by two electrons; semiquinones are unstable intermediates:

$$Q + H^+ + e^- \rightarrow QH$$

$$QH + H^+ + e^- \rightarrow QH_2 \qquad (4\text{-}8)$$

Overall $\qquad Q + 2H^+ + 2e^- \rightarrow QH_2 \quad E^0 = 0.699\ V$

Q stands for quinone ($O=C_6H_4=O$). Many electroactive-substituted quinones exist.[31] Such compounds can participate in two-electron redox reactions like that in Equation 4-1b or in one-electron redox cycles, where the quinone or hydroquinone is converted to the semiquinone and back again:

$$A + H^+ + Q \rightleftharpoons QH + A_{ox}$$

$$\underline{B_{ox} + QH \rightleftharpoons B + Q + H^+}$$

Net $\qquad A + B_{ox} \rightleftharpoons A_{ox} + B \qquad (4\text{-}9)$

Some quinones and semiquinones are important biochemical molecules, and these structures also are found in humic substances. As such, they may be important in the redox chemistry of iron and manganese in humic-rich waters.[32,33]

Metal-Initiated Chain Reactions

Chain reactions are initiated by substances that form free radicals, which then induce chain-propagating steps. The chain length may be relatively long (10^2 to 10^3 product molecules from a single chain) if chain termination is slow. The substance initiating the chain may not be regenerated, but small amounts of initiator still may produce large amounts of product.

A metal-initiated chain reaction that has been well studied under laboratory conditions is the oxidation of oxalic acid by Cl_2.[26,34] The reaction is slow at room temperature and is accelerated by several metals. For example, Fe^{2+} reduces Cl_2 to Cl_2^- and initiates a chain reaction:

Chain initiation: (a) $Fe^{2+} + Cl_2 \rightarrow Fe^{3+} + Cl_2^-$

Chain propagation: (b) $Cl_2^- + HC_2O_4^- \rightarrow H^+ + 2Cl^- + C_2O_4^-$

                        (c) $C_2O_4^- + Cl_2 \rightarrow Cl_2^- + 2CO_2$

                                    (4-10)

Chain termination: (d) $Cl_2^- + C_2O_4^- \rightarrow 2Cl^- + 2CO_2$

                        (e) $Cl_2^- + Cl_2^- \rightarrow 2Cl^- + Cl_2$

Mn^{3+} induces the reaction by reacting with oxalate to produce $C_2O_4^-$, which starts the chain by step 4-10c. Cu^{2+} and Co^{2+} accelerate the reaction by facilitating step 4-10b; V^{IV}, Mn^{2+}, and NH_4^+ inhibit the reaction by breaking the chain.

Some important redox reactions in natural waters (e.g., oxidation of pyrite by O_2) are thought to involve metal-catalyzed chain reactions, but the mechanisms are not well understood. Because radicals are such unstable species, they tend to react with whatever chemicals they encounter. In laboratory systems the list of participating chemicals can be kept small so that chain propagation along the pathway of interest is favored. In contrast, natural systems contain a great variety of solutes. Consequently, chain-propagation steps for a given reaction may be terminated rapidly, and a wide variety of products may be produced.

Autoxidation

Oxidation of compounds by ground-state molecular oxygen (O_2) is called autoxidation or oxygenation. Autoxidation is important in the spoilage of food, especially by the development of rancidity via formation of fatty acids from unsaturated fats, and the environmental degradation of organic pigments and paints. Reduced inorganic forms such as Fe^{II}, Mn^{II}, and sulfide are subject to chemical oxidation by O_2 in natural water; the kinetics of these reactions are described later in this chapter. Reaction rates for autoxidation of inorganic species vary widely, depending on pH and the exact chemical form (i.e., degree of protonation, ligand binding) of the ion being oxidized.

Autoxidation of organic compounds generally requires photochemical activation and/or induction by metal ions. In the absence of light or metal catalysts, the reactions typically are very slow, despite favorable energetics. Antioxidants retard autoxidation reactions further by scavenging radical intermediates and terminating chains. Photoactivation seems not to be important in the autoxidation of inorganic species (S^{-II}, Fe^{II}, Mn^{II}), but very few studies have examined this carefully. Photoreduction of ferric and manganic oxyhydroxides is important, however.

Ground-state O_2 is unusual among simple molecules in having two orbital electrons with unpaired spins, i.e., it is in a triplet state $(^3\Sigma_g^-)$. In singlet oxygen, $^1O_2$, which is the first excited state of O_2, the electron spins are paired (see Figure 8-2). $^1O_2$ is generated photochemically in natural waters by sensitizer molecules such as aquatic humus (Section 8.3.4). The reactivities of ground-state and singlet O_2 toward various classes of compounds are quite different, and their mechanisms of oxidation also are dissimilar. Reactions with ground-state O_2 proceed by free-radical mechanisms that may be chain-like. Organic radicals can be induced by heavy metals via one-electron transfers or photochemically through hydrogen extraction by photoexcited molecules (so-called photosensitizers). Singlet oxygen is a more selective reagent and reacts directly by addition to electron-rich organic compounds such as olefins, organic sulfides, and some aromatic compounds. $^1O_2$ thus is an electrophile. Photochemical reactions of O_2 are treated in Chapter 8.

Reduction of O_2 in autoxidation reactions proceeds via one-electron transfers called the Haber-Weiss mechanism[35,36] to form the following radical and ionic intermediates: $O2^-$ (superoxide radical ion), HO_2 (hydrogen superoxide radical), HO_2^- (hydroperoxide ion), H_2O_2 (hydrogen peroxide), and $\cdot OH$ (hydroxyl radical). Reaction sequence 4-11 is a general mechanism for the autoxidation of a reducing agent (A^+) by a series of one-electron steps involving induction by a metal ion (M^{2+}) and the various oxygen-containing intermediates:

Chain initiation:	(a)	$A^+ + M^{2+} \rightarrow A^{2+} + M^+$
Chain propagation:	(b)	$A^{2+} + O_2 \rightarrow O_2^- + A^{3+}$
	(c)	$(H^+ + O_2^- \rightleftharpoons HO_2)$
	(d)	$O_2^- + A^+ \rightarrow O_2^{2-} + A^{2+}$
or		$HO_2 + A^+ \rightarrow HO_2^- + A^{2+}$ (4-11)
	(e)	$(O_2^{2-} + 2H^+ \rightleftharpoons H_2O_2)$
Chain termination:	(f)	$M^{2+} + A^{2+} \rightarrow A^{3+} + M^+$

H_2O_2 is further reduced to H_2O in two separate, relatively rapid steps:

	(g)	$H^+ + H_2O_2 + A^+ \rightarrow A^{2+} + \cdot OH + H_2O$
	(h)	$H^+ + A^+ + \cdot OH \rightarrow A^{2+} + H_2O$
or	(i)	$H^+ + A^{2+} + \cdot OH \rightarrow A^{3+} + H_2O$
Overall	(a – i):	$2A^+ + 4H^+ + O_2 \rightarrow 2A^{3+} + 2H_2O$

Catalyst M^{2+} also serves as a chain breaker by reacting with the product of the chain initiation step (A^{2+}) to produce product A^{3+}. Steps b and d and g–i are the main chain-propagating steps, and c and e are rapid acid-base equilibria.

Anion Catalysis of Redox Reactions between Two Cations

Reactions between cations tend to be slow because charge repulsion suppresses collisions between the reactants. The presence of certain anions accelerates many of these reactions. The simplest explanation for such findings is that the anions form complexes with one or both of the reacting cations, thus reducing their net charge and facilitating encounters. This accounts for some of the catalytic effect of Cl^- on the oxidation of Sn^{2+} by Fe^{3+}, but it does not account for the entire effect. The reaction order with respect to Cl^- is more than three in the [Cl^-] range 0.04 to 0.48 M,[37] which implies that more than three Cl^- are associated with the activated complex. This is greater than the extent of Cl^- complexation with the cations in this concentration range; only 1 to 2 Cl^- are complexed with Fe^{III} and Sn^{II} in this range. A possible explanation for the high order of Cl^- involves stabilization of the intermediate Sn^{3+}, which forms complexes with a larger number of Cl^- ligands. The complexed Sn^{3+} then reacts rapidly with Fe^{3+} to form product Sn^{IV}.

More typically, Cl^- catalysis of electron exchange reactions between simple cations is first order.[38] The mechanism likely involves transfer of a Cl atom rather than simple electron transfer. The activated complex contains a Cl bridge between the two metal ions, and a Cl atom originally associated with the oxidant (e.g., Fe^{III}) stays with the reductant (e.g., Sn^{II}) on rupture of the complex, leaving an electron behind and yielding Fe^{II} and Sn^{III} (Figure 4-3).

Electron transfer between metal complexes may occur by either of two general mechanisms: inner-sphere or outer-sphere exchange. Atom transfer is an example of an inner-sphere mechanism, i.e., electron transfer between reactants that share a ligand in their primary coordination spheres. The electron is transferred via the bridging group. Although atom or ligand transfer is common in inner-sphere redox processes, it is not required. Whether the bridging group transfers to the reducing agent, stays with the oxidizing agent or transfers from reducing agent to oxidizing agent depends on the substitution labilities of the reactants and products.[39] In outer-sphere mechanisms, the primary coordination spheres of the reactants remain intact. An outer-sphere mechanism can be inferred if a redox reaction between two substitutionally inert complexes is rapid.[5] Section 7.4 discusses outer-sphere electron transfer processes in greater detail and describes a linear free-energy relationship for such reactions.

Support for the atom transfer mechanism is provided by applying the principle of microscopic reversibility to electron exchange between two oxidation states of the same element.[26] Suppose the exchange between Fe^{2+} and a monochloro Fe^{3+} complex occurred by simple electron transfer: $FeCl^{2+} + Fe^{2+} = FeCl^+ + Fe^{3+}$. According to the principle of microscopic reversibility, the forward and reverse reactions proceed by the same pathway. At equilibrium the forward and reverse rates are equal, and this implies that exchange is equally likely to occur by collision of $FeCl^+$ and Fe^{3+} as by $FeCl^{2+}$ and Fe^{2+}. The former process is unlikely, however,

(a) $^*Fe^{II}(H_2O)_6^{2+} + Fe^{III}Cl(H_2O)_5^{2+} \rightleftharpoons$

$$(H_2O)_5\,^*Fe^{2+}\overset{\frown}{\cdots Cl}Fe(H_2O)_5^{2+} \rightarrow (H_2O)^*Fe^{III}Cl^{2+} + Fe^{II}(H_2O)_6^{2+}$$

$$H_2\ddot{O} \downarrow \qquad H_2\ddot{O}$$

*added to distinguish between the two Fe atoms

(b) $(NH_3)_5Co^{III}Cl^{2+} + Cr^{II}(H_2O)_6^{2+} \rightleftharpoons \{(NH_3)_5Co^{III} - Cl\cdots Cr^{II}(H_2O)_5\}^{4+}$

"precursor" complex

$$\{(NH_3)_5Co^{III}\cdots Cl\cdots Cr^{II}(H_2O)_5\}^{4+} \rightarrow \{(NH_3)_5Co^{III}\cdots Cl - Cr^{II}(H_2O)_5\}^{4+}$$

activated complex   "successor" complex

$$H_3O^+$$
$$\{(NH_3)_5Co^{III}\cdots Cl - Cr^{II}(H_2O)_5\}^{4+} \rightarrow \rightarrow Co^{II}(H_2O)_6^{2+} + Cr^{III}Cl(H_2O)_5^{2+} + 5NH_4^+$$

Figure 4-3. Mechanism of inner-sphere electron exchange reaction: (a) Fe^{II}-Fe^{III} with chlorine atom transfer; (b) reduction of Co^{III} complexes by Cr^{II} with chlorine atom transfer.

because Fe^{2+} has little tendency to form Cl^- complexes. The dilemma can be resolved by postulating electron exchange by Cl transfer from $FeCl^{2+}$ to Fe^{2+} (Figure 4-3); the products are Fe^{2+} and $FeCl^{2+}$!

Evidence for inner-sphere electron transfer with Cl^- as the bridging ligand was obtained by Taube and co-workers[40,41] in a classic experiment on the reduction of Co^{III} complexes by Cr^{II} (see reference 39 for a review). Reduction of $Co(NH_3)_6^{3+}$ by $Cr(H_2O)_6^{2+}$ is slow (k = 10^{-3} M^{-1} s^{-1}) and occurs by an outer-sphere mechanism. Substitution of Cl^- for a NH_3 on Co^{III} greatly accelerates the reaction (k = 6×10^5 M^{-1} s^{-1}, a factor of 10^8 increase):

$$Co(NH_3)_5Cl^{2+} + Cr(H_2O)_6^{2+} \xrightarrow{H^+}$$
$$Co(H_2O)_6^{2+} + Cr(H_2O)_5Cl^{2+} + 5NH_4^+$$

(4-12)

Both chloro complexes are substitution inert and have half-lives greater than 1 min, but rate constants for exchange of H_2O in the aquo complexes are very large (> 10^9 s^{-1} for Cr^{II} and 2×10^5 s^{-1} for Co^{III}). The only way Cl^- could be transferred between the metals at such a high rate is by direct attack of $CoCl^{2+}$ on Cr^{2+}, as shown in Figure 4-3. Co^{II} complexes are highly labile, and the NH_3 groups exchange rapidly with H_2O.

4.3 RATE EQUATIONS FOR CATALYZED REACTIONS

Development of rate expressions for catalyzed homogeneous reactions is treated in many kinetics texts.[1-4,26,42] This section is limited to some general principles, but more examples are found in other sections of this chapter. Rates of catalyzed reactions usually are proportional to the concentration of the catalyst:

$$\text{rate} = f[C] \tag{4-13a}$$

where f is a function of the rate constant and reactant concentrations. This implies the rate is zero in the absence of catalyst, which often is effectively so. In some cases, a second term must be added for the uncatalyzed rate:

$$\text{rate} = f[C] + f' \tag{4-13b}$$

Rate equations for catalyzed reactions can be derived by using steady-state or equilibrium constant assumptions if the mechanism is known. Conversely, the nature of the rate expression under varying reactant and catalyst concentrations is useful in elucidating mechanisms. Rates of catalyzed reactions often are independent of the concentration of one or more reactants. For example, if the catalyst concentration is small compared with the other reactants, the rate equation for sequence 4-1b is $d[B^{3+}]/dt \approx k[C]_T[B^+]$, when the last step is much slower than the first two, and $d[B^{3+}]/dt \approx k'[C]_T[A^{2+}]$, when the last step is much faster than the first two. Under some circumstances, rates of reactions having a single reactant depend only on catalyst concentration. Rate expressions for redox reactions catalyzed by trace metals often are first order in oxidant and zero order in reductant.[10]

4.3.1 Rate Equation for a Recycled Catalyst Mechanism

The simplest example of a reaction with a recycled catalyst involves a single reactant in the following mechanism:

$$A + C \underset{k_2}{\overset{k_1}{\rightleftharpoons}} X \xrightarrow{k_3} P + C \tag{4-14}$$

The net reaction is $A \rightarrow P$, C is the catalyst, and X is an intermediate compound or complex. Rate equations for sequence 4-14 are

$$-d[A]/dt = k_1[A][C] - k_2[X] \tag{4-15a}$$

$$d[C]/dt = (k_2 + k_3)[X] - k_1[A][C] \tag{4-15b}$$

$$d[X]/dt = k_1[A][C] - (k_2 + k_3)[X] \tag{4-15c}$$

$$d[P]/dt = k_3[X] \tag{4-15d}$$

An analytical solution is not possible for these simultaneous equations, but they can be solved by invoking the steady-state assumption for [X]:[43]

$$d[X]/dt = 0 = k_1[A][C] - (k_2 + k_3)[X]$$

$$X = k_1[A][C]/(k_2 + k_3)$$

(4-15e)

Equation 4-15e contains [C], the concentration of free catalyst, which we may not be able to measure. A mass balance can be written for the catalyst:

$$[C]_T = [C] + [X]$$

(4-15f)

where $_T$ denotes the total amount of catalyst in the system. Solving Equation 4-15f for [C], substituting into Equation 4-15e and solving for [X], we obtain

$$[X] = \frac{k_1[A][C]_T}{k_1[A] + k_2 + k_3}$$

(4-15g)

Substitution of Equation 4-15g into Equation 4-15d yields the rate of product formation:

$$d[P]/dt = \frac{k_1 k_3[A][C]_T}{k_1[A] + k_2 + k_3} = \frac{k_3[A][C]_T}{(k_2 + k_3)/k_1 + [A]}$$

(4-15h)

This equation can be simplified to

$$v = \frac{V[A]}{K_A + [A]}$$

(4-15i)

where v is the rate of product formation, $K_A = (k_2 + k_3)/k_1$, and $V = k_3[C]_T$.

Figure 4-4a shows the time course for concentrations of A, C, X, and P in this mechanism, and demonstrates that the steady-state assumption used to derive Equation 4-15i is valid (at least when $k_1 = k_2 = k_3$), except at the beginning of the reaction. Figure 4-4b shows the variation in reaction rate vs. [A] at constant $[C]_T$ and demonstrates that reaction of a single substance with a recycled catalyst is a mixed-order (hyperbolic) process. Two limiting cases can be distinguished. When [A] $\gg$ K_A, the reaction is zero order in [A], and v = V. Mechanistically, this occurs when all the catalyst is bound as intermediate X. When $K_A \gg$ [A], the equation simplifies to a first-order expression (at constant $[C]_T$). Equation 4-15i is equivalent to the Michaelis-Menten equation for enzyme-catalyzed reactions; see Chapter 6.

4.3.2 Rate Equations for Autocatalyzed Reactions

Catalysis by the product of a reaction is called autocatalysis and is characterized by an s-shaped time-course for product formation (Figure 4-5a). The initial rate is lower than that at intermediate times when the catalyst (P) has built up. Eventually,

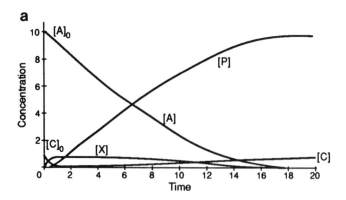

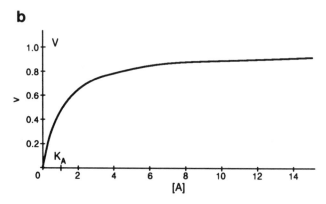

Figure 4-4. Kinetics of irreversible reaction of single reactant A with recycling catalyst C (Equation 4-14): (a) time course for all species for $k_1 = k_2 = k_3 = 1$; (b) normalized rate of reaction v/V vs. reactant concentration, expressed in terms of K_A. Note: K_A has units of concentration because k_2 and k_3 are in time^{-1} and k_1 is in conc.$^{-1}$ time^{-1}.

the rate slows down as the system runs out of reactant. The rate equation for a first-order autocatalytic reaction is

$$d[P] / dt = k_1 [A] + k_2 [P][A] \tag{4-16}$$

k_1 is the rate constant for the uncatalyzed first-order reaction, and k_2 is a second-order autocatalytic rate constant. If $[P]_o = 0$, $[A] = [A]_o - [P]$, and Equation 4-16 can be integrated to

$$(k_1 + k_2[A]_o)t = \ln \frac{[A]_o (k_2[P] - k_1)}{k_2([A]_o - [P])} \tag{4-17}$$

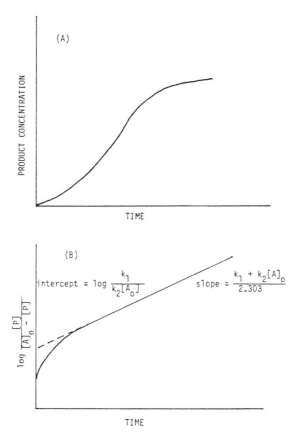

Figure 4-5. (A) S-shaped plot of [P] vs. time for autocatalyzed reaction; (B) linearized plot of autocatalytic reaction rate data.

Equations 4-16 and 4-17 can be written in terms of $[A]_o$ and $[A]$, assuming $[P]_o = 0$ and $[A] = [A]_o - [P]$:[44]

$$-d[A]/dt = k_1[A] + k_2[A]([A]_o - [A]) \tag{4-18a}$$

and
$$(k_1 + k_2[A]_o)t = \ln\{[A]_o(k([A]_o - [A]) + k_1)/k_1[A]\} \tag{4-18b}$$

or
$$[A] = \frac{[A]_o(k_1 + k_2[A]_o)}{k_2[A]_o + k_1\exp\{k_1 + k_2[A]_o t\}} \tag{4-18c}$$

Plots of $ln[P]/([A]_o - [P])$ or $ln[A]/([A]_o - [A])$ vs. time are curvilinear, but approach a straight line when $k_2[P]$ or $k_2([A]_o - [A]) >> k_1$ (Figure 4-5b). Extrapolation of this line to the y-intercept yields $-lnk_2[A]_o/k_1$, and the slope of the line is $k_1 + k_2[A]_o$.[44] Values of k_1 and k_2 thus can be estimated from the intercept and slope. Alternatively, one can estimate k_1 from initial rate measurements at $[P]_o = 0$ and adjust k_2 iteratively until a good fit is obtained between predicted and measured values.

Homogeneous autocatalysis occurs in the hydrolysis of esters in poorly buffered systems, because one of the products (the organic acid) produces H^+, which catalyzes the hydrolysis. Heterogeneous autocatalysis is important in Mn^{2+} and possibly Fe^{2+} oxidation in natural waters (Section 4.6.5). Rates of biological processes such as the general population growth of microorganisms can be described by autocatalytic kinetic equations.

4.3.3 Rate Expressions for Acid-Base Catalysis

If all combinations of specific and general acid-base catalysis are considered, the apparent rate constant for a reaction is expressed by

$$k_{obs} = k_o + k_H\{H^+\} + k_{OH}\{OH^-\} + \sum_i k_{HX(i)}[HX_i] + \sum_j k_{x(j)}[X_j] \quad (4\text{-}19)$$

HX_i and X_j represent all other acids and bases in solution besides H^+ (H_3O^+) and OH^-. The first term is the uncatalyzed aqueous rate constant; it can be written more precisely as $k_{H_2O}[H_2O]$. The next two terms reflect specific acid and base catalysis, and the last two reflect general acid-base or buffer catalysis.

In most cases some types of catalysis are unimportant, and the corresponding terms in Equation 4-19 can be ignored. All terms but k_o are functions of pH, and plots of k_{obs} vs. pH are used to determine the types of acid-base catalysis that influence a reaction and to evaluate the rate constants. For example, if a reaction is subject only to specific acid-base catalysis, Equation 4-19 becomes

$$k_w = k_o + k_H\{H^+\} + k_{OH}K_w / \{H^+\} \quad (4\text{-}20)$$

since $K_w = \{H^+\}\{OH^-\}$. If k_H and k_{OH} are similar in magnitude, the second term will be much greater than the third term, in strong acid solution. At pH 2 the third term is less than 1% of the second term, unless k_{OH}/k_H is 10^8 or more. Such differences in rate constants normally are not encountered. In the region where acid catalysis predominates, a plot of $log\ k_w$ vs. pH is linear and has a slope of -1 (assuming the reaction is first order in H^+). The value of k_H can be determined by extrapolating the line to the y-intercept (pH = 0). At high pH the second term is negligible compared to the third, and a plot of $log\ k_w$ vs. pH has a slope of $+1$. The value of k_{OH} is obtained from the y-intercept at pH 14.

Figure 4-6 illustrates the possible types of pH-dependent profiles for acid-base-catalyzed reactions. Each graph in the figure represents the pH response of k_{obs} for

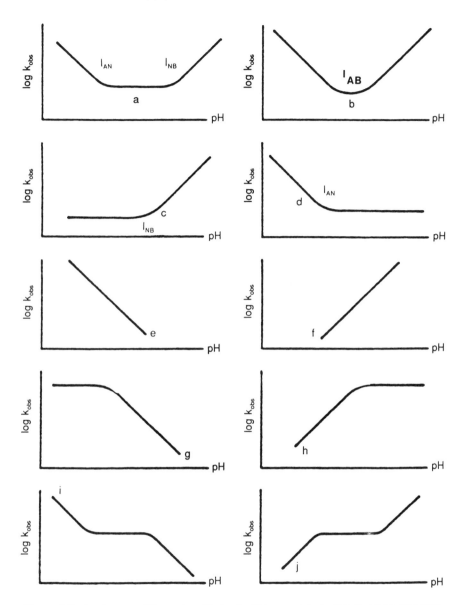

Figure 4-6. *Log* k_{obs} vs. pH for various kinds of acid-base catalysis. See text for explanation of curves. I_{ij} = pH where two types of reactions occur at same rate; A = H^+-catalyzed, N = neutral, B = OH^--catalyzed. (From Hill, C.G. Jr., *An Introduction to Chemical Engineering Kinetics and Reactor Design*, John Wiley & Sons, New York, 1977. With permission.)

different combinations of acid and base catalysis superimposed on an uncatalyzed rate constant.[42] The pH values at which two processes contribute equally to the observed rate are denoted as subscripted I values on the curves. For example, I_{AN}

represents the intersection of an acid-catalyzed rate with an uncatalyzed rate. Because both rates contribute equally to the observed rate at this point, k_{obs} has a value two times that of the point of intersection. I values are useful in tabulating data on hydrolysis rates[45] because they indicate the pH at which a process becomes important.

Curve **a** in Figure 4-6 is a combination of specific acid catalysis at low pH, specific base catalysis at high pH, and an uncatalyzed reaction at intermediate pH, where the rate is pH independent. At pH $< I_{AN}$, $k_{obs} \approx k_H\{H^+\}$, or $log\ k_{obs} = log\ k_H -$ pH. At pH $> I_{NB}$, $k_{obs} \approx k_{OH}\{OH^-\} \approx k_{OH}K_w/\{H^+\}$, or $log\ k_{obs} = log\ k' + $ pH. At intermediate pH, $k_{obs} = k_o$. Curves **b** to **f** represent reactions where one or more of the terms in Equation 4-20 are unimportant. In curve **b** the uncatalyzed reaction never predominates. Curve **c** represents reactions where specific base catalysis occurs at high pH, the uncatalyzed reaction predominates at low pH, and specific acid catalysis never becomes important. Curve **d** represents a combination of specific acid catalysis and the uncatalyzed reaction. In curves **e** and **f** the uncatalyzed reactions also are unimportant. Curves **g** and **h** illustrate general acid and general base catalysis, respectively, and curves **i** and **j** are combinations of general and specific acid-base catalysis. Curve **g** represents catalysis by an undissociated acid, and curve **h** catalysis by an unprotonated base. In both cases specific acid-base catalysis is presumed absent, and the intersection of the horizontal and diagonal lines occurs at the pK of the acid or base. Regarding curve **g**, at a given C_T of Bronsted acid the concentration of undissociated form HX is constant at pH $<$ pK; thus, k_{obs} is constant. At pH $>$ pK, log[HX] decreases with pH, and k_{obs} decreases correspondingly.

Interpretation of curves for general acid-base catalysis and mixtures of general and specific catalysis must be done in a stepwise fashion.[42] First, k_o, k_H, and k_{OH} are determined from rate measurements in solutions containing only strong acid or base. Strong acids and bases whose corresponding cations and anions do not act as catalysts must be used as the sources of H^+ and OH^-. Rate constants for general acid and base catalysis then can be determined from measurements with buffer solutions of the weak acid and its conjugate base. If either acid or base catalysis is absent, the rate constant for the other process can be obtained from the horizontal part of the plot; for example, $k_{obs} \approx k_{HX}$[HX] $\approx k_{HX}C_T$ at pH $<<$ pK in Figure 4-6g. If neither type of catalysis can be ruled out, the experiments and computations become complicated.

4.3.4 Metal-Promoted Acid-Base-Catalyzed Hydrolysis

Many variations are possible here. In general, rate equations are derived by considering the equilibrium speciation of the hydrolyzing compound with respect to complex formation by various metal ions that promote hydrolysis. The rate equation is written in terms of total analytical concentrations of the hydrolyzing compound, total metal ion concentrations, and equilibrium (complex formation) constants relating the metal and ligand. A simple example involving the hydrolysis of trimetaphosphate is given below to illustrate the approach.

Example 4-1. Metal-Promoted Hydrolysis of Trimetaphosphate

Trimetaphosphate, $P_3O_9^{3-}$, (symbolized P_3) undergoes acid- and base-catalyzed hydrolysis. Complexation by cations enhances the rate. Under alkaline conditions and in the presence of Na^+, the observed rate is[21]

(1) $$\text{Rate} = k_{obs}[P_3]_T[OH^-] = k_{OH}[P_3][OH^-] + k_{Na}[NaP_3][OH^-]$$

Dividing both sides of Equation 1 by $[P_3]_T[OH^-]$ yields an expression for k_{obs}:

(2) $$k_{obs} = k_{OH}[P_3]/[P_3]_T + k_{Na}[NaP_3]/[P_3]_T$$

$[P_3]_T = [P_3] + [NaP_3]$ is the total analytical concentration of trimetaphosphate, k_{OH} is the base-catalyzed hydrolysis constant of uncomplexed P_3 anion, and k_{Na} is the base-catalyzed hydrolysis constant of the Na complex. Because concentrations of the P_3 species are not easily measured, it is desirable to write Equation 2 in terms of $[P_3]_T$ and measurable reactants. The equilibrium dissociation constant for NaP_3 is $K_{Na} = [Na][P_3]/[NaP_3]$. ($K_{Na}$ is a concentration-based constant valid at a given ionic strength.) Solving for $[NaP_3]$ and substituting into the equation for $[P_3]_T$, we obtain

$$[P_3]_T = [P_3]\{1 + [Na]/K_{Na}\}$$

or (3) $$[P_3]/[P_3]_T = K_{Na}/\{K_{Na} + [Na]\}$$

Introducing Equation 3 into Equation 2 yields

$$k_{obs} = k_{OH}K_{Na}/\{K_{Na} + [Na]\} + k_{Na}K_{Na}[NaP_3]/[P_3]\{K_{Na} + [Na]\}$$

or (4) $$k_{obs} = \{k_{OH}K_{Na} + k_{Na}[Na]\}/\{K_{Na} + [Na]\}$$

If $[P_3]_T$ is small compared with $[Na]_T$, $[Na] \approx [Na]_T$, and the observed rate constant is expressed in terms of measurable concentrations, intrinsic rate constants, and the measured complex dissociation constant.

Introduction of another complexing cation into the medium (e.g., Ca^{2+}) adds additional terms to the expressions for k_{obs} and $[P_3]_T$:

(5) $$k_{obs} = k_o[P_3]/[P_3]_T + k_{Na}[NaP_3] + k_{Ca}[CaP_3]$$

(6) $$[P_3]_T = [P_3] + [NaP_3] + [CaP_3] = [P_3]\{1 + [Na]/K_{Na} + [Ca]/K_{Ca}\}$$

(7) $$k_{obs} = \frac{k_o K_{Na}K_{Ca} + k_{Na}[Na]K_{Ca} + k_{Ca}[Ca]/K_{Na}}{K_{Na}K_{Ca} + K_{Ca}[Na] + K_{Na}[Ca]}$$

Equation 7 is readily derived from Equations 5 and 6 and the expressions for K_{Na} and K_{Ca} by algebraic manipulation.

PART II. KINETICS OF CHEMICAL REACTIONS
IN AQUEOUS SOLUTION

4.4 DISSOCIATION AND HYDROLYSIS REACTIONS IN NATURAL WATERS

4.4.1 Formation and Dissociation of Acids

Dissociation of simple acids normally is considered to be a very rapid elementary reaction, but rate constants actually are proportional to the strength of the acid. This fact derives from the law of microscopic reversibility, which leads to the relationship, $K_a = k_d/k_r$, and from the fact that rates of the reverse reaction ($H^+ + A^- \rightarrow HA$) are very rapid and independent of the nature of the conjugate base A^-. In fact, the rate of recombination is diffusion limited, and $k_r \approx 4 \times 10^9 \ M^{-1} \ s^{-1}$ (Section 3.4.1). The above considerations lead to the simple relationship, $k_d = \alpha K_a$, where the coefficient of proportionality equals k_r. This is another example of a linear free-energy relationship. Dissociation rate constants for some common acids are $\sim 10^5 \ s^{-1}$ for acetic acid ($pK_a = 4.7$), $10^{2.6} \ s^{-1}$ for H_2S ($pK_1 = 7.0$), and $\sim 5 \ s^{-1}$ for NH_4^+ ($pK_a = 9.3$).

The carbonic acid system is of special interest because it is so important in buffering the pH of natural waters. Based on the above analysis, the dissociation of bicarbonate (HCO_3^-; $pK_2 = 10.33$ at 25°C) is fairly slow: $k_d \approx 0.2 \ s^{-1}$. The rate of dissociation of carbonic acid cannot be obtained in this manner, however, because pK_1 (6.35 at 25°C) is a composite equilibrium constant based on hydration of aqueous CO_2 and dissociation of carbonic acid, H_2CO_3:

$$K_1 = \frac{\{H^+\}\{HCO_3^-\}}{\{H_2CO_3^*\}} = 4.46 \times 10^{-7} \quad (25°C) \qquad (4\text{-}21a)$$

where
$$[H_2CO_3^*] = [CO_2]_{aq} + [H_2CO_3] \qquad (4\text{-}21b)$$

and $CO_{2aq} + H_2O = H_2CO_3; \quad K_h = 1.5 \times 10^{-3} = \{H_2CO_3\}/\{CO_2\}_{aq} \qquad (4\text{-}21c)$

It is easy to show that $K_1 = K_a/(1 + K_h^{-1})$, where K_a is the dissociation constant of true carbonic acid (H_2CO_3). $K_a \approx 1 \times 10^{-4}$; thus, H_2CO_3 actually is a relatively strong acid, e.g., stronger than acetic acid. Furthermore, it is apparent from the value of K_h that CO_{2aq} is the dominant species contributing to $H_2CO_3^*$. With a pseudo-first-order rate constant $k_h = 0.037 \ s^{-1}$ (25°C) and E_{act} of 79 kJ mol^{-1}, the hydration of CO_2 is relatively slow (see Example 3-1).

According to Johnson,[46] the hydration of CO_2 is the sum of two reactions:

$$CO_2 + H_2O \rightarrow H_2CO_3, \quad k_i \qquad (4\text{-}21d)$$

and $CO_2 + H_2O \rightarrow H^+ + HCO_3^-, \quad k_{ii} \qquad (4\text{-}21e)$

H_2CO_3 and HCO_3^- are in very rapid equilibrium with each other so that the two reactions cannot be separated, and $k_h = k_i + k_{ii}$. Similarly, the dehydration reaction is the sum of the reverse reactions for Equations 4-21d and 4-21e, and the dehydration rate constant $k_d = k_{-i}/K_a + k_{-ii}$, where K_a is the dissociation constant for H_2CO_3, and k_{-i} and k_{-ii} are the rate constants for the reverse reactions of Equations 4-21d, e. At 25°C, $k_d = 7.6 \times 10^4 \, M^{-1} \, s^{-1}$.[46] Hydration of CO_2 is sufficiently slow that it can be demonstrated easily in the laboratory,[47] and the slow rate may limit primary production (Example 6-10). The reaction is subject to general base catalysis. The enzyme carbonic anhydrase is a strong catalyst of CO_2 hydration in organisms. At one time it was thought to play a role in CO_2 dynamics in surface waters, but experiments by Goldman and Dennett[48] indicated this was not the case. At pH > 8, the direct reaction of CO_2 with OH^- to produce HCO_3^- becomes competitive with the two-step hydration/dissociation pathway: $CO_2 + OH^- \rightarrow HCO_3^-$; $k = 7.1 \times 10^3$ $M^{-1} \, s^{-1}$ (25°C). The reverse reaction, $HCO_3^- \rightarrow CO_2 + OH^-$, has a rate constant of $1.8 \times 10^{-4} \, s^{-1}$ at 25°C.[46]

4.4.2 Complex Formation and Ligand Exchange

Complexation reactions are important regulators of metal ion speciation in water; in turn, speciation affects metal reactivity and toxicity.[49,50] The literature on rates and mechanisms of complexation reactions is voluminous, and several comprehensive reviews are available.[51,52] We will examine five types of complexation reactions:

1. Water exchange in the primary coordination sphere
2. Formation of complexes from aquated metal ions and monodentate ligands
3. Hydrolysis of complexes with monodentate ligands
4. Formation of complexes with multidentate ligands
5. Multidentate ligand-exchange reactions

Reactions in the first two categories are typically very fast, and special techniques, such as relaxation methods, are required to observe them.

Metal ions exist in water as hydrated species (aquo complexes); water molecules act as ligands and donate a pair of electrons from their oxygen atoms to empty orbitals of the metal ion. The number of ligand atoms bound to a metal ion is the coordination number. This can range from 2 to 8, but by far the most common coordination number is six. The structure of such complexes is octahedral, with the metal ion at the center. For metals such as Cu^{II} that have coordination numbers of four, the complexes are either square planar or tetrahedral. Ag^I is unusual among heavy metal ions in having a coordination number of two.

Water Exchange Rates

Water exchange rates are known for all metal ions of interest in aquatic systems (Figure 4-7). Rate constants for H_2O exchange (k_{-H_2O}) span more than 16 orders of magnitude, from >$10^9 \, s^{-1}$ for Cu^{2+} and Pb^{2+} to <$10^{-7} \, s^{-1}$ for Rh^{3+}. It is apparent, however, that only a few ions have rate constants slower than $1 \, s^{-1}$. Slow exchange rates are measured by isotope dilution, and rapid rates usually are measured by

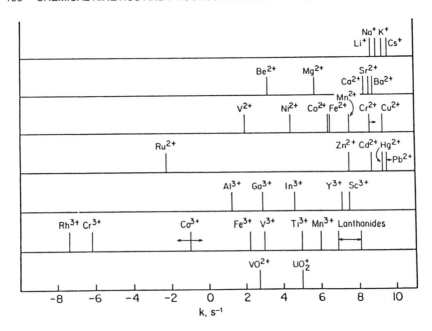

Figure 4-7. Water exchange rate constants (k_1, s^{-1}, 25°C) for aquo metal ions. [From Margerum, D.W., et al., in *Coordination Chemistry*, Vol. 2, A.E. Martell (Ed.), ACS Monograph 174, American Chemical Society, Washington, D.C., 1978, p. 1. With permission.]

NMR methods, which are adaptable to a wide range of rates.[51,53] The order of the rate constants in the first two rows of Figure 4-7 suggests the values are related to ion size, at least for Group IA and IIA ions. Indeed, over a wide range of metal ions, k_{-H_2O} is correlated crudely with Z/d^3,[51,54] where Z is the charge of the metal, and d is one half the sum of the radii of the metal and oxygen atoms (Figure 4-8). Among transition metals, exchange rates are related to d electron configuration and can be estimated from the change in d orbital energy between ground and transition states by an offshoot of molecular orbital theory called the angular overlap model.[5]

Metal ions can be grouped into four classes based on their water exchange rates:[5]

1. $k_{-H_2O} > 10^8$ s^{-1}, i.e., rates are diffusion controlled, or nearly so. This includes most Group IA and IIA ions except Be^{2+} and Mg^{2+}, as well as Cd^{2+}, Cu^{2+}, Hg^{2+} and Pb^{2+}.
2. $k_{-H_2O} = 10^4-10^8$ s^{-1}. This includes most first-row transition metal divalent ions and Mg^{2+}.
3. $k_{-H_2O} = 1-10^3$ s^{-1}: Al^{3+} and Be^{2+}.
4. $k_{-H_2O} = 10^{-7}-10^{-3}$ s^{-1}: Cr^{3+} and Co^{3+}.

Complexes with half-lives toward substitution greater than 1 min (class 4) are called kinetically inert, and complexes with shorter half-lives (classes 1–3) are called labile. Obviously, a wide range of reactivity exists in the labile class. A fair amount of uncertainty exists in k_{-H_2O} values of some metal ions, and values depend on the measurement method. Reported values[51] for Al^{3+} range from 0.5 s^{-1} (measured by sulfate exchange and pressure-jump relaxation) to 16 s^{-1} (measured by NMR).

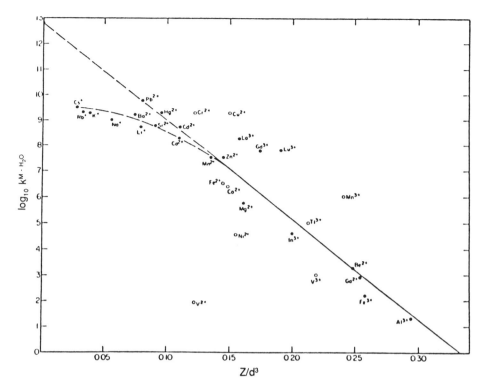

Figure 4-8. Correlation between water exchange rate constants and Z/d^3 from an electro-static ion-dipole model. Open circles: ions expected to deviate from model because of ligand-field or Jahn-Teller effects. [From Margerum, D.W., et al., in *Coordination Chemistry*, Vol. 2, A.E. Martell (Ed.), ACS Monograph 174, American Chemical Society, Washington, D.C., 1978, p. 1. With permission.]

The difference between Cr^{2+} ($k_{-H_2O} = 10^{+8.7}$ s^{-1}) and Cr^{3+} ($k_{-H_2O} = 10^{-6}$ s^{-1}) is astounding, considering that we are comparing two forms of the same element that differ by just one electron. Cr^{3+} has three d electrons, each occupying one of the three degenerate t_{2g} orbitals. This gives rise to a symmetric octahedral structure for $Cr(H_2O)_6^{3+}$, which has a very high stability. In contrast, Cr^{2+} has four d electrons, each in a separate orbital ($3t_{2g}$, $1e_g$). The other e_g orbital is empty, causing a nonsymmetric orbital configuration and leading to an elongated octahedral structure for $Cr(H_2O)_6^{2+}$ that is much more labile.

Complex Formation Rates with Monodentate Ligands

Rates of reaction of aquated metal ions with monodentate ligands to form 1:1 complexes generally are fast and roughly comparable to the water exchange rates described above. Formation of true complexes is not an addition reaction, but a substitution: $M(H_2O)_n^{m+} + L^- = ML(H_2O)_{n-1}^{m-1} + H_2O$. The forward reaction is called anation if L is an anion; the reverse reaction is called hydrolysis. Relaxation methods suggest that anation is a two or perhaps three step process:

$$M\left(H_2O\right)_n^{m+} + L^- \underset{k_r}{\overset{k_f}{\rightleftharpoons}} \left\{M\left(H_2O\right)n \cdot L\right\}^{m-1} \tag{4-22a}$$

$$\left\{M(H_2O)_n \cdot L\right\}^{m-1} \xrightarrow{k_{-H_2O}} ML(H_2O)_{n-1}^{m-1} + H_2O \tag{4-22b}$$

This sequence is called the Eigen mechanism. The first step involves diffusion of ligand into the metal ion's outer hydration sphere to form an outer-sphere (ion pair) complex. This rapid, diffusion-controlled step is readily reversible and can be treated as a pre-equilibrium step with equilibrium constant $K_{os} = k_f/k_r$. The second (rate-controlling) step involves loss of a water molecule from the inner coordination sphere of the metal ion and its replacement by the ligand. Measured values of K_{os} for nonlabile complexes range from about 10 M^{-1} for 3+,1– ions to $10^{3.5}$ for 3+,2– ions.[51] Fuoss[55] and Eigen[56] developed a theoretical model to estimate K_{os} for labile complexes, for which direct measurements are difficult. A simplified expression is[51]

$$K_{os} = (4/3)(\pi Na^3 e^{-b})10^{-3} M^{-1} \tag{4-23}$$

N is Avogadro's number; a is the center-to-center distance between M and L (estimated as 4–5×10^{-8} cm); and $b = Z_M Z_L e_o^2/aDkT$. e_o is the electronic charge, 4.8 $\times 10^{-10}$ esu; D is the dielectric constant (78 for water); and kT is thermal energy per molecule, 4.11×10^{-14} erg at 298 K. Equation 4-23 leads to estimates of $K_{os} \approx 1.0$ for 1+,1– complexes and ~2×10^2 for 2+,2– complexes. For outer-sphere complexes with uncharged ligands, Equation 4-23 yields $K_{os} \approx 0.1$.

Whether the second part of the anation process—replacement of a primary water of hydration by a ligand (Equation 4-22b) — is one step or two independent steps has been a controversial subject. Present consensus is for a mechanism that actually is somewhere in between. Three mechanisms can be hypothesized for Equation 4-22b:

1. An associative pathway (A) in which the ligand enters the coordination sphere before the water molecule leaves (temporarily increasing the coordination number of the complex)
2. A dissociative pathway (D), in which the water molecule leaves the coordination sphere completely before the ligand enters (yielding an intermediate with a lower coordination number)
3. An interchange pathway (I) in which the leaving of H_2O and entering of L occur more or less simultaneously. This mechanism could be predominantly associative (I_a) or dissociative (I_d), depending on the importance of bond making or breaking in forming the transition state

Considerable evidence has been amassed to show that I_d mechanisms are involved in anation reactions of metal ions that form octahedral complexes. This includes cationic complexes of most metal ions. Associative mechanisms are unlikely for octahedral complexes for steric reasons — additional coordination positions are not available. Moreover, rates of anation reactions involving a given metal ion and a series of ligands of the same charge do not vary with ligand basicity or

Table 4-2. Rate Constants for Anation and Water Exchange Reactions

Metal	Ligand	k_f (M^{-1} s^{-1})	k_{-H_2O} (s^{-1})	k_f/k_{-H_2O} (M^{-1})[a]
Be^{2+}	F^-	7.2E2	3.0E3	0.24
Mn^{2+}	F^-	2.7E6	3.1E7	0.09
Fe^{2+}	F^-	1.4E6	3.2E6	0.4
Co^{2+}	F^-	1.8E5	2.0E6	0.09
Ni^{2+}	F^-	8.4E3	3.0E4	0.3
Cu^{2+}	F^-	2.2E8	5.0E9	0.044
Co^{2+}	imidazole	1.3E5	2.0E6	0.065
Cu^{2+}	imidazole	5.7E8	5.0E9	0.11
Ni^{2+}	imidazole	4.9E3	3.0E4	0.16
Ni^{2+}	NH_3	4.5E3	3.0E4	0.15
Ni^{2+}	pyridine	3.4E3	3.0E4	0.11
Ni^{2+}	HF	3.1E3	3.0E4	0.10
Ni^{2+}	N_2H_4	2.5E3	3.0E4	0.08

Values tabulated from Margerum, D.W., et al., in *Coordination Chemistry,* Vol. 2, A.E. Martell (Ed.), ACS Monograph 174, American Chemical Society, Washington, D.C., 1978, p. 1.

[a]   Since $k_f \approx K_{os}k_{-H_2O}$, this ratio is equivalent to K_{os}.

nucleophilicity, and the nature of the metal ion influences the rate more than the nature of the ligand (Table 4-2). This implies that bond formation with the ligand is unimportant in the rate-determining step and rules out A and I_a mechanisms. There is no evidence for intermediates of lower coordination number in anation reactions of *cationic* complexes, which tends to rule out D pathways, but relatively stable five-coordinate intermediates have been detected in anation reactions of *anionic* octahedral complexes; e.g.,[57]

$$Co(CN)_5 H_2O^{2-} \rightleftharpoons Co(CN)_5^{2-} + H_2O$$
$$Co(CN)_5^{2-} + X^- \rightarrow Co(CN)_5 X^{3-}$$

(4-24)

Although release of H_2O ligands has not been shown to be a separate, rate-determining step in anation reactions, their rates still are highly correlated with water exchange rates of metal ions (Table 4-2). This is the expected situation for I_d mechanisms in which bond breaking plays a more important role in forming transition states than does bond formation. In general, replacement of water by monodentate ligands increases the exchange rate of the remaining coordinated waters[51] in transition metal complexes. The same is true for multidentate ligands that occupy only some of a metal's coordination sites.

From the equilibrium constant approach, one can derive a second-order rate equation for the anation mechanism in Equation 4-22: $d[ML]/dt = k_f[M][L]$, where $k_f = K_{os}k_{-H_2O}$. If k_f and k_{-H_2O} can be determined independently, one can estimate K_{os} from their ratio. Example 4-2 illustrates the above principles in terms of the kinetics of Al^{3+} complexation by F^- ion.[58] This reaction is of interest because Al^{3+} and its soluble hydroxy complexes, primarily $AlOH^{2+}$, can reach levels toxic to fish and other aquatic organisms in acidic surface waters (pH $< \sim 5$). Complexation by F^- or organic ligands reduces the toxicity of aqueous Al.

Example 4-2. Kinetics of Aluminum Fluoride Complex Formation

Plankey et al.[58] studied the kinetics of AlF^{2+} formation by measuring the loss of free F^- with an ion-selective electrode. The electrode responds only to the activity of free F^- and has a response time less than 1 s. Initial rate ($\mathbf{R}$) measurements showed that AlF^{2+} formation was always first order in $[Al^{3+}]$, but $\mathbf{R}/[F^-]$ increased linearly with $[F^-]$ (Figure 4-9a), suggesting that the reaction fits a rate equation of form:

(1) $\mathbf{R} = (d[AlF^{2+}]/dt)_{t=o} = K_1[Al^{3+}][F^-] + K_2[Al^{3+}][F^-]^2$

Because acid-base speciation of both Al^{III} and F^- changes within the pH range of interest, four separate formation pathways must be considered:

I.   (2a) $\qquad (H_2O)_6\,Al^{3+} + F^- \rightleftharpoons [(H_2O)_5\,Al(H_2O),F]^{2+} \qquad K_{os}^1$

(2b) $\qquad [(H_2O)_5\,Al(H_2O),F]^{2+} \underset{k_{21}}{\overset{k_{12}}{\rightleftharpoons}} (H_2O)_5\,AlF^{2+}$

II. (3a) $\qquad (H_2O)_5\,AlOH^{2+} + F^- \rightleftharpoons [(H_2O)_5\,AlOH,F]^+ \qquad K_{os}^2$

(3b) $\qquad [(H_2O)_5\,AlOH,F]^+ \underset{k_{32}}{\overset{k_{23}}{\rightleftharpoons}} (H_2O)_4\,AlOHF^+ + H_2O$

(3c) $\qquad (H_2O)_4\,AlOHF^+ + H^+ \underset{k_{43}}{\overset{k_{34}}{\rightleftharpoons}} (H_2O)_5\,AlF^{2+}$

III. (4a) $\qquad (H_2O)_6\,Al^{3+} + HF \rightleftharpoons [(H_2O)_6\,Al,HF]^{+3} \qquad K_{os}^3$

(4b) $\qquad [(H_2O)_6\,Al,HF]^{3+} \underset{k_{54}}{\overset{k_{45}}{\rightleftharpoons}} (H_2O)_5\,AlF^{2+} + H_3O^+$

IV. (5a) $\qquad (H_2O)_5\,AlOH^{2+} + HF \rightleftharpoons [(H_2O)_5\,AlOH,HF]^{2+} \qquad K_{os}^4$

(5b) $\qquad [(H_2O)_5\,AlOH,HF]^{2+} \underset{k_{65}}{\overset{k_{56}}{\rightleftharpoons}} (H_2O)_4\,AlOHF^+ + H_3O^+$

(5c) $\qquad (H_2O)_4\,AlOHF^+ + H^+ \underset{k_{43}}{\overset{k_{34}}{\rightleftharpoons}} (H_2O)_5\,AlF^{2+}$

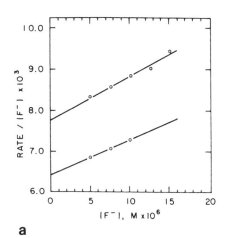

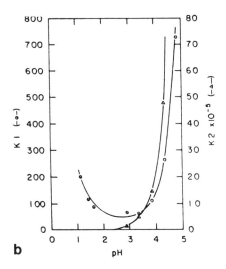

a   b

Figure 4-9. Kinetics of AlF²⁺ complex formation: (a) Rate/[F⁻] vs. [F⁻] at 25°C and pH 4.35; upper line, [Al³⁺] = 2.82 × 10⁻⁵ M; lower line, [Al³⁺] = 2.35 × 10⁻⁵ M. (b) Effect of pH on observed K_1 and K_2 (Equation 1 of Example 4-1). Note: observed K_2 increases more rapidly at pH > ~4.4 than predicted by Equation 12 (Example 4-2), probably because hydrolyzed species of Al^{III} provide more reaction pathways than those considered in the model. [From Plankey, B.J., H.H. Patterson, and C.S. Cronan, *Environ. Sci. Technol.*, 20, 160 (1986). With permission.]

The K_{os} are association constants for outer-sphere complexes, values for which were calculated from the Fuoss model,[55] which is similar to Equation 4-23. If initial conditions are considered, we can neglect reverse reaction terms, and the initial rate of AlF²⁺ formation from the four pathways becomes:

$$(6) \quad (d[AlF^{2+}]/dt)_{t=0} = k_{12}K_{os}^1[Al^{3+}][F^-] + k_{23}K_{os}^2[AlOH^{2+}][F^-]$$

$$+ k_{45}K_{os}^3[HF][Al^{3+}] + k_{56}K_{os}^4[AlOH^{2+}][HF]$$

Coordinated waters are eliminated for simplicity. Al³⁺ and AlOH²⁺ are related by two rapid protolysis reactions:

$$(7) \qquad Al^{3+} + F^- \underset{k_6}{\overset{k_5}{\rightleftharpoons}} AlOH^{2+} + HF$$

$$(8) \qquad Al^{3+} \underset{k_8}{\overset{k_7}{\rightleftharpoons}} AlOH^{2+} + H^+$$

Equations 7 and 8 can be solved by the steady-state approximation to give [AlOH²⁺]; assuming $k_8 \gg k_6$, we obtain:

$$(9) \qquad [AlOH^{2+}]_{ss} = \{k_5[Al^{3+}][F^-] + k_7[Al^{3+}]\} / k_8\{H^+\}$$

Rate Equation 6 for appearance of AlF^{2+} can be simplified by substituting Equation 9 for $[AlOH^{2+}]$, and $\{H^+\}[F^-]/K_{HF}$ for [HF] and collecting terms:

$$(10) \qquad ([AlF^{2+}]/dt)_{t=0} = K_1[Al^{3+}][F^-] + K_2[Al^{3+}][F^-]^2$$

where (11) $\qquad K_1 = k_{12}K_{os}^1 + \dfrac{k_7}{k_8}k_{56}K_{os}^4 K_{HF} + \dfrac{k_7 k_{23}}{k_8\{H^+\}}K_{os}^2 + k_{45}K_{os}^3 K_{HF}\{H^+\}$

and (12) $\qquad K_2 = k_5 k_{56}K_{os}^4 K_{HF}/k_8 + k_5 k_{23}K_{os}^2/k_8\{H^+\}$

Equation 10 has the same form as Equation 1, which was based on measured data. Plots of measured values of K_1 and K_2 vs. $1/\{H^+\}$ (Figure 4-10a,b) allow us to extract the microscopic rate constants for the four pathways (from the slopes and intercepts of the plots). For example, the y-intercept in the plot of K_2 vs. $1/\{H^+\}$ is near zero (Figure 4-10b), and thus the first (pH-independent) term in Equation 12, which represents pathway IV, must be zero or very small. The first two terms in Equation 11 are pH independent and constitute the y-intercept in the plot of K_1 vs. $1/\{H^+\}$ (Figure 4-10a). Because the second term in Equation 11 represents pathway IV (it is almost the same as the first term in Equation 12), which we just showed to be negligible, it is apparent that the y-intercept in Figure 4-10a is approximately $k_{12}K_{os}^1$. Because K_{os}^1 can be calculated from the Fuoss model, the y-intercept yields a measure of k_{12}.

According to Equation 11, K_1 is proportional to $1/\{H^+\}$ at high pH, where pathway II and the third term of Equation 11 predominate; the plot of measured K_1 vs. $1/\{H^+\}$ agrees with this prediction. The slope of this plot provides a means of estimating k_{23}, since K_{os}^2 can be calculated independently, and the ratio k_7/k_8 is the acid dissociation (equilibrium) constant for Al^{3+}, which also is known independently. In contrast, K_1 is proportional to $\{H^+\}$ at low pH (Figure 4-10c), where pathway III >> pathway II and the last term in Equation 11 >> the third term. This allows estimation of k_{45}. Constant k_{56} was assumed to be equal to k_{23} by postulating a dissociative mechanism of water replacement for the hydrolyzed cation, $(H_2O)_5AlOH^{2+}$.[58] Finally, the reverse rate constant values ($k_{21}, k_{32}, k_{54}, k_{65}$) were calculated from the ratio of the equilibrium constant values for each reaction (obtained from the literature) and the forward rate constants. Table 4-3 summarizes the measured and calculated constants for the reactions.

Overall, the rate of AlF^{2+} formation is quite sensitive to pH and temperature. The rate decreases rapidly with decreasing pH in the range ~3 to 5 and increases more slowly at pH < 3 (Figure 4-9b). Equilibration times of several hours may be required under "worst-case" conditions (low temperature and pH ~3). However, under conditions more typical of acidic lakes and streams, reaction half-lives of seconds to minutes are likely.

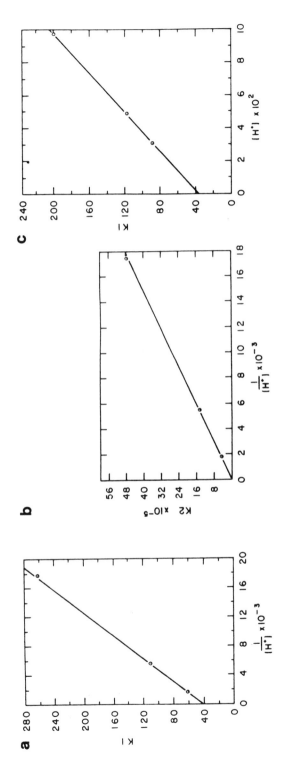

Figure 4-10. Plots of observed K_1 and K_2 to obtain microscopic rate constants for Al-F complexation: (a) K_1 vs. $1/\{H^+\}$; (b) K_2 vs. $1/\{H^+\}$; (c) K_1 vs. $\{H^+\}$ at low pH (see Example 4-2 for derivation of k's from these plots). [From Plankey, B.J., H.H. Patterson, and C.S. Cronan, *Environ. Sci. Technol.*, 20, 160 (1986). With permission.]

Table 4-3.   Rate Constants for Aluminum Fluoride Complexation[a]

A. Observed rate constants[b]

pH		4.85	4.35	3.85	3.36	2.88	1.62	1.42	1.12
K_1,	$(M^{-1}s^{-1})$	726	262	110	61	65	88	117	199
K_2,	$(M^{-2}\,s^{-1})$	2.74E8	4.77E6	1.45E6	4.69E5				

B. Derived microscopic constants (25°C)[c]

Path	1 $Al^{3+} + F^-$ $k_{12}K_{os}^{1}$	2 $AlOH^{2+} + F^-$ $k_{23}K_{os}^{2}$	3 $Al^{3+} + HF$ $k_{45}K_{os}^{3}$	4 $AlOH^{2+} + HF$ $k_{56}K_{os}^{4}$
K_{os} (M^{-1})	5.7	2.0	0.96	0.61
k_{xy} (s^{-1})	6.5	1.8E3	1.5	1.8E3
k_{yx} (s^{-1})	1.4E-5	6.2E-3		
k_{yx} $(M^{-1}\,s^{-1})$			7.3E-4	2.27

[a]   See Example 4-2; from Plankey, B.J., H.H. Patterson, and C.S. Cronan, *Environ. Sci. Technol.*, 20, 160 (1986).
[b] K_1 and K_2 are observed rate constants in the two-term rate expression Equation 1 of Example 4-2.
[c]   The k_{xy} are the forward rate constants for the four paths; the k_{yx} are the corresponding reverse rate constants.

Hydrolysis of 1:1 Complexes

By the principle of microscopic reversibility, hydrolysis and anation occur along the same reaction coordinate in opposite directions:

$$M(H_2O)_{n-1}L^{m-1} + H_2O \underset{\text{anation, } k_f}{\overset{\text{hydrolysis, } k_{hy}}{\rightleftharpoons}} M(H_2O)_n^{m+} + L^- \tag{4-25}$$

An interesting consequence of this principle is that whereas the rate of anation is independent of the nature of L^-, the rate of hydrolysis does depend on the nature of L^-. This can be understood by considering the equilibrium constant for reaction 4-25: $K_{eq} = k_{hy}/k_f$. For a given metal ion and a series of like-charged ligands, k_f is roughly constant. Therefore, k_{hy} is proportional to K_{eq} (this is a linear free-energy relationship). Unless K_{eq} is constant (which generally is not the case for related series of inner-sphere complexes), k_{hy} depends on the nature of L^-.

If we consider k_f for complex formation to be approximately equal to $K_{os}k_{-H_2O}$, we can estimate rates of 1:1 complex dissociation from formation (stability) constants in a manner analogous to that described above for acid-base reactions. For $M + L = ML + H_2O$, $K_f = \{ML\}/\{M\}\{L\} = k_f/k_{hy}$. Thus, $k_{hy} \approx K_{os}k_{-H_2O}/K_f$. Stability constants for monodentate ligands with most metal ions range from $<10^1$ to $\sim10^3$, and $k_{-H_2O} > 10^6\,s^{-1}$ for most metals of interest in natural waters. K_{os} can be estimated from Equation 4-23 and generally is ~1 for monovalent cations and less than 10 for divalent ions. Dissociation rates of simple ML complexes thus are rapid ($k > \sim10^3$

Table 4-4. Rate Constants for Formation of Some Chelates[a]

M	L	k_f	M	L	k_f	M	L	k_f
EDTA (10°C)			Nitrilotriacetate			Glycine oligomers[b]		
Ca^{2+}	HY	3.7E7	Cd^{2+}	HNTA	2.1E5	Co^{2+}	gly	1.5E6
Ca^{2+}	Y	6.1E9	Cd^{2+}	NTA	4.0E9	Co^{2+}	digly	4.6E5
Mn^{2+}	Y	9.0E6	Ni^{2+}	HNTA	7.5	Co^{2+}	trigly	3.1E5
Co^{2+}	Y	1.5E5	Ni^{2+}	NTA	>5.0E4	Ni^{2+}	gly	4.1E4
Ni^{2+}	Y	1.0E3	Cu^{2+}	HNTA	1.1E5	Ni^{2+}	digly	2.1E4
Cu^{2+}	Y	1.2E8	Cu^{2+}	NTA	2.0E8	Ni^{2+}	trigly	8.0E3
Pb^{2+}	Y	>5.0E8	Pb^{2+}	HNTA	4.4E6	Ni^{2+}	tetragly	4.2E3
La^{3+}	Y	8.6E7	Pb^{2+}	NTA	1.5E11			
Amino acids			Organic acids			Miscellaneous		
Co^{2+}	proline	3.5E5	Ni^{2+}	oxalate	7.4E4	Zn^{2+}	porphyrin	0.2
Co^{2+}	leucine	1.3E6	Ni^{2+}	lactate	2.6E4	Mg^{2+}	ATP	1.3E7
Co^{2+}	serine	2.0E6	Ni^{2+}	phthalate	6.3E5	Fe^{2+}	bipyridyl	1.6E5
Co^{2+}	alanine	6.0E5	Ni^{2+}	tartrate	1.8E5	Fe^{2+}	phenanthroline	5.6E4
Co^{2+}	tyrosine	1.3E5	Co^{2+}	salicylate	1.3E5	Ca^{2+}	murexide	>6.0E7

[a] Values tabulated from Margerum, D.W., et al., in *Coordination Chemistry*, Vol. 2, A.E. Martell (Ed.), ACS Monograph 174, American Chemical Society, Washington, D.C., 1978, p.1; values at 25°C, except for EDTA.

[b] Formation rates for the glycine oligomers are even slower than suggested by these numbers, since the rates are inversely proportional to $(H^+)^2$.

to 10^4 s^{-1}). Dissociation rates for complexes of Al^{3+} ($k_{-H_2O} \approx 10$ s^{-1}) and Fe^{3+} ($k_{-H_2O} \approx 10^2$ s^{-1}) are substantially slower.

Chelate Formation And Dissociation

Formation rates of complexes with multidentate ligands (chelates) are more variable than those for monodentate ligands. Table 4-4 lists rate constants for some complexes of interest in aquatic systems; Margerum et al.[51] give a more comprehensive listing. Factors like steric hindrance, electrostatic repulsion, and proton transfer from partially protonated multidentate ligands tend to decrease formation rates; electrostatic attraction between metal ions and highly negative multidentate ligands enhances rates. Electrostatic attraction and repulsion can greatly affect K_{os} and k_f (Figure 4-11). The range in k_f for reactions of Ni^{2+} with ligands ranging in charge from +3 to −4 is about 10^6.

However, formation rate constants for many chelates are about equal to $K_{os}k_{-H_2O}$, (the same as expected for monodentate ligands), and the rate-limiting step is formation of a singly bound intermediate. If the chelate formation rate is smaller than that for a monodentate ligand (after correcting for electrostatic effects), ring closure probably is the rate-limiting step. Once the first ring is formed, formation of additional rings usually is not rate limiting, but the need to transfer protons from partially protonated ligands can cause exceptions, as occurs in the formation metal-peptide complexes with Pb^{2+}, Ni^{2+}, and Cu^{2+}.[59] Oligomers of glycine form square

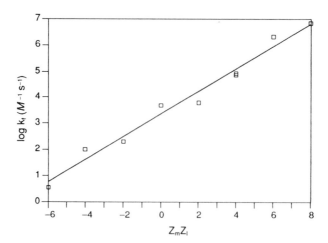

Figure 4-11. Effect of ionic charge on formation rate constants of some Ni^{2+} complexes: $H_3tetren^{3+}$ (–6), H_2trien^{2+} (–4), Hen^+ (–2), NH_3 (0), NCS^- (+2), $C_2O_4^{2-}$ (+4), IDA^{2-} (+4), NTA^{3-} (+6), $HP_3O_{10}^{4-}$ (+8). Numbers in parentheses are product of ionic charges for Ni-L complex. [k_f values from Margerum, D.W., et al., in *Coordination Chemistry*, Vol. 2, A.E. Martell (Ed.), ACS Monograph 174, American Chemical Society, Washington, D.C., 1978, p. 1.] *Note:* en = ethylenediamine.

planar complexes with these metals, and binding occurs with the amide N atoms, which lose their proton in the process. Fully deprotonated amide N does not exist as a free species in water. Formation rates for these complexes are limited by the deprotonation step and are slower than rates of complexation with glycine itself (Table 4-4).

Fully protonated ligands are quite unreactive toward complex formation (HF and HCN are exceptions among monodentate ligands). Among multidentate polyaminocarboxylate ligands (e.g., EDTA and NTA, which are common industrial chemicals that find their way into natural waters via municipal and industrial waste effluents), the less-protonated forms react faster than the more-protonated forms, with few exceptions, and $k_{f,L}/k_{f,HL}$ is in the range of 10 to 300 for reactions of divalent metals with various polyamines, dicarboxylic acids, polypyridines, and EDTA.[51] Even larger enhancements (10^4 to 10^6) are observed for NTA, some amino acids (glycine, serine), and some other aminocarboxylates. These have been explained by postulating ring closure as the rate-limiting step or by an ICB mechanism (Figure 4-1d) that catalyzes reaction of the deprotonated form, but is not possible for protonated forms.

Dissociation rates of chelates are expectedly much slower than dissociation rates of complexes with monodentate ligands. Formation constants for chelates are large — K_f often much greater than 10^8; chelates are thermodynamically stable compounds. Thus, even if the rate of chelate formation is very rapid, dissociation will be slow. Chelate dissociation reactions are generally acid catalyzed. In the absence of assistance from protons, polydentate complexes can have half-lives of many years for dissociation. Even in 1-M acid, the half-life for complete dissociation of the Fe^{3+}-complexing siderophore desferrioxamine B (desferal) is ~20 minutes,[60] even though initial steps in the unraveling of the chelate are much faster at this pH (Figure 4-12a).

Metal-Ligand Exchange Reactions

Rates of metal-metal exchange with a given chelating agent, $ML + M' \rightarrow M'L + M$, tend to be slow, but they often are much faster than one would expect for a mechanism that involves complete dissociation of ML before M'L begins to form. These reactions are acid catalyzed, and direct attack by the exchanging metal ions affects the kinetics in some cases. Two general pathways exist: *adjunctive* mechanisms, in which dinuclear intermediates M-L-M', are important; and *disjunctive* mechanisms, in which ML dissociates stepwise, perhaps catalyzed by acid, to form free L (or a protonated or partially protonated form) before M'L begins to form. Rates of exchange occurring by the adjunctive mechanism are independent of [M], but rates for disjunctive mechanisms are inversely proportional to [M], which promotes reformation of the complex.[61] Stepwise transfer of inflexible macrocyclic ligands (such as porphyrins and some Fe-complexing siderophores) between metal ions should be much more difficult than stepwise transfers of flexible ligands, like polyaminocarboxylates, and linear trihydroxamates, like desferal. Data on macrocyclic ligands are limited, but their dissociation rates are thought to be very slow.

Numerous rate measurements have been made on metal-metal exchange rates of EDTA and other aminocarboxylate complexes, and the mechanisms of exchange, which are complicated, have been studied in detail.[51,62-64] The rates often conform to the expression: $d[M'Y]/dt = k_{M'}^{MY}[M'][MY]$, where $Y = EDTA$, but the reaction order may vary as metal concentrations change.[62] The second-order expression suggests that the rate-limiting step involves a dinuclear M-Y-M' intermediate in which one nitrogen of EDTA is bonded to M and the other to M' (the adjunctive mechanism). In fact, values of $k_{M'}^{NiY}$ have been correlated with stability constant values of M'IDA, where IDA = iminodiacetate (half an EDTA molecule).[63] The general mechanism is illustrated in Figure 4-12b. Detailed studies on the Cu^{2+}-NiY^{2-} exchange[62] have shown that a Cu-assisted pathway is important only at very high Cu^{2+} concentrations ($>10^{-2} M$), and a proton-assisted pathway occurs only at pH < 2 and low Cu (Figure 4-12c). Similar pathways apply to the exchange of Pb^{2+}, Ni^{2+}, and Co^{2+} with $Ce^{3+}EDTA$ complexes,[64] but proton-assisted dissociation is important up to pH ~6 in this case. Exchange rates are proportional to water exchange rates of the metal ions, even though the overall exchange rate is many orders of magnitude slower.

Complexation of Cd^{2+} and Cu^{2+} by EDTA in seawater is a relatively slow process with a time scale of tens of minutes to hours (Figure 4-13).[61,65] The slowness of the process reflects kinetic competition by much higher concentrations of Ca^{2+} and Mg^{2+} in seawater rather than intrinsically slow CdY and CuY formation reactions. CaY and MgY complexes form very rapidly (and preferentially over the Cd or Cu complexes, for mass-action reasons). Formation of CdY or CuY thus must proceed as a metal-metal exchange reaction rather than a simple anation mechanism. In addition, free Ca^{2+} in solution competes with Cu^{2+} for any free Y formed by dissociation of CaY. Although Mg concentrations are higher than Ca concentrations in seawater (5×10^{-2} vs.

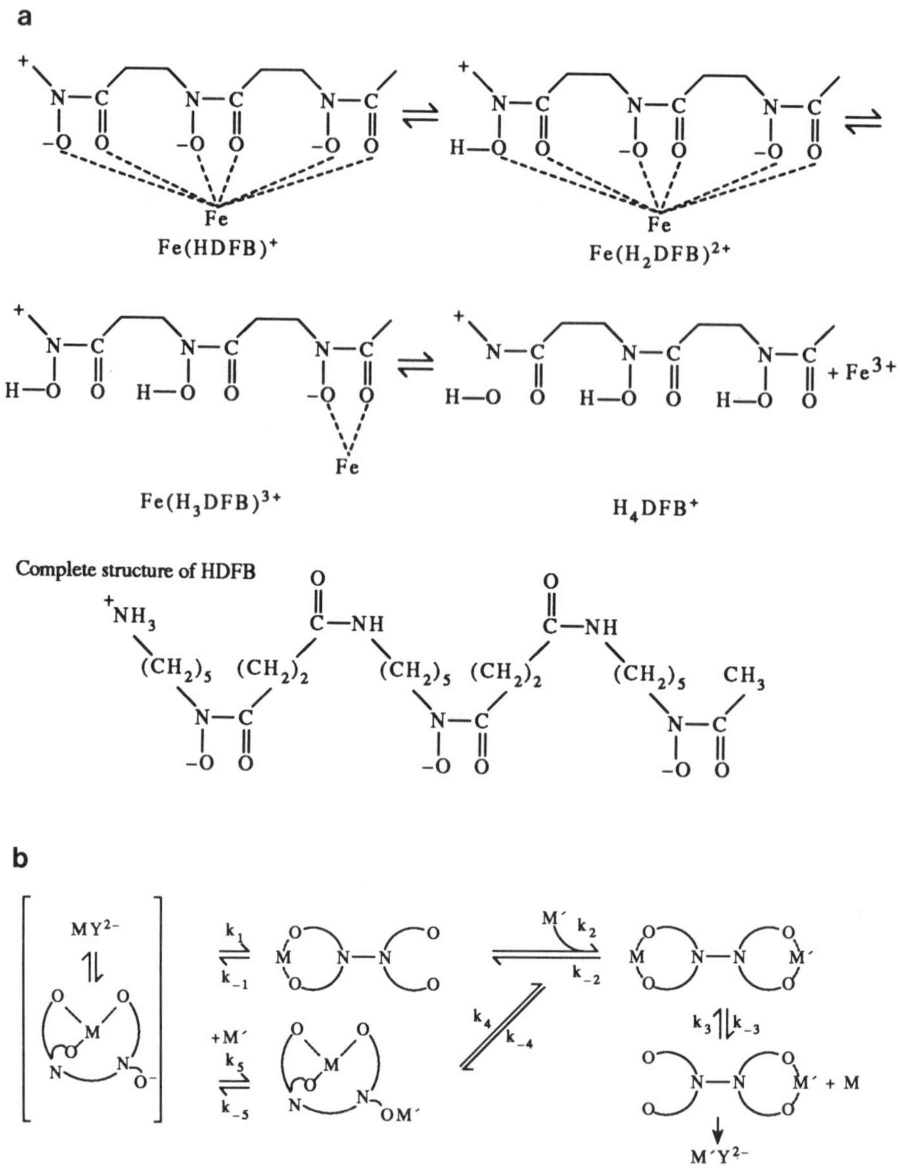
a
Fe
Fe(HDFB)+
Fe
Fe(H2DFB)2+
Fe
Fe(H3DFB)3+
H4DFB+
+ Fe3+
Complete structure of HDFB
b
MY2−
k1
k−1
M′
k2
k−2
k4
k−4
+M′
k5
k−5
OM′
k3
k−3
M′ + M
M′Y2−

Figure 4-12. Reaction mechanisms of some metal ions and chelating agents: (a) acid dissociation of the desferrioxamine B complex of Fe^{3+} (Fe-desferal) [from Biruš, M., et al., *Inorg. Chem.*, 26, 1000 (1987)]; (b) general mechanism of M-M′ EDTA exchange [from Margerum, D.W. and G.R. Dukes, in *Metal Ions in Biological Systems*, Vol. 1, H. Sigel (Ed.), Marcel Dekker, New York, 1973]; (c) detailed mechanism for Cu-Ni EDTA exchange (only species 1, 4, 6 and protonated forms of 3 are observable intermediates) [from Margerum, D.W., D.L. Janes, and H.M. Rosen, *J. Am. Chem. Soc.*, 87, 4463 (1965)]; (d) instrinsic mechanisms for Al^{3+} complexation by fulvic acid (FA) [according to Plankey, B.J. and H.H. Patterson, *Environ. Sci. Technol.*, 21, 595 (1987)].

c

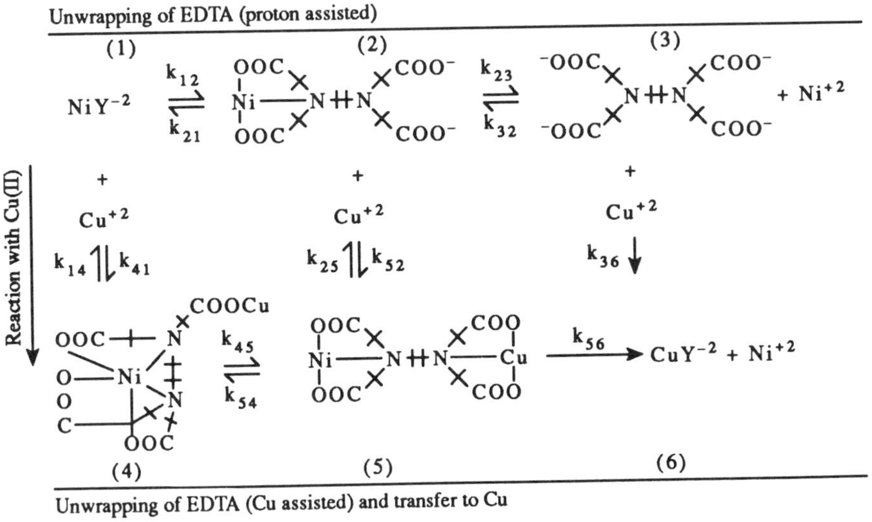
Unwrapping of EDTA (proton assisted)
Reaction with Cu(II)
Unwrapping of EDTA (Cu assisted) and transfer to Cu

d

$$Al^{3+} + FA^- \underset{k_{-1}}{\overset{k_1}{\rightleftharpoons}} Al(FA)^{2+}$$

$$AlOH^{2+} + FA^- \underset{k_{-2}}{\overset{k_2}{\rightleftharpoons}} Al(FA)OH^+$$

$$Al^{3+} + HFA \underset{k_{-3}}{\overset{k_3}{\rightleftharpoons}} Al(FA)^{2+} + H^+$$

$$AlOH^{2+} + HFA \underset{k_{-4}}{\overset{k_4}{\rightleftharpoons}} Al(FA)OH^+ + H^+$$

Figure 4-12 (continued).

1×10^{-2} *M*), Ca forms stronger complexes with EDTA and is more important in hindering the formation of CuEDTA complexes. As Figure 4-14a shows, Cu-CaY and Cu-MgY exchange rates are independent of both [Ca] and [Mg] at seawater concentrations, and Cu-MgY rates are about ten times faster. However, the second-order rate constants for both ions increase at lower alkaline earth concentrations, implying a shift from an adjunctive pathway to a disjunctive one. The latter pathway is acid catalyzed,

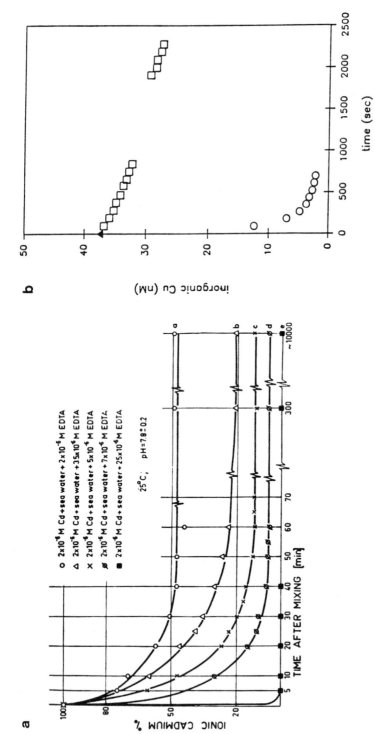

Figure 4-13. (a) Rate of Cd²⁺ disappearance from seawater at various concentrations of EDTA; Cd²⁺ measured polarographically. [From Maljković, D. and M. Branica, *Limnol. Oceanogr.*, 16, 779 (1971). With permission.] (b) Rate of inorganic Cu disappearance from laboratory solutions containing 125 nM CaEDTA, 0.5 M NaCl, 0.002 M NaHCO₃, and 10⁻⁵ M Caₜ (circles) or 10⁻² M Caₜ (squares). Cu measured amperometrically; initial concentration = 37.5 nM. [From Hering, J.G. and F.M.M. Morel, *Environ. Sci. Technol.*, 22, 1469 (1988). With permission.]

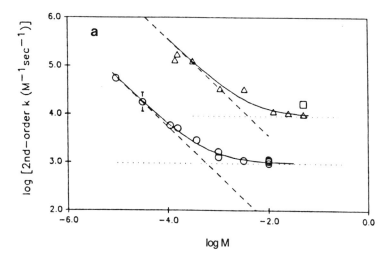

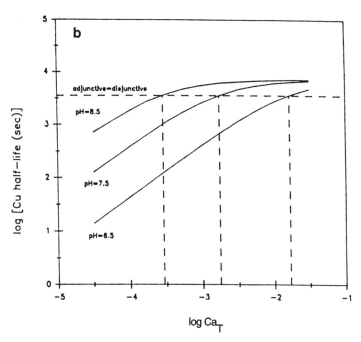

Figure 4-14.   (a) Dependence of second-order rate constant for $-d[Cu]/dt = k[EDTA]_T[Cu]$ on $log[Mg]$ (triangles) or $log[Ca]$ (circles); (b) log of pseudo-first-order $t_{1/2}$ for Cu with respect to reaction with 10^{-7} M CaEDTA as function of $[Ca]_T$ for three pH values; vertical dashed lines show $[Ca]_T$ at which adjunctive and disjunctive mechanisms contribute equally to reaction of Cu with CaEDTA; (c) predicted values of $logt_{1/2}$ for Cu in seawater for adjunctive-type reaction of Cu with Ca-bound ligands at 10^{-7} M L_T, where L represents an aminopolycarboxylate chelating agent (1 = NTA; 5 = EDTA; other ligands identified in Hering and Morel[61]); (d) predicted $t_{1/2}$ of various metals relative to $t_{1/2}$ of Cu for metal-exchange reactions. [All from Hering, J.G. and F.M.M. Morel, *Environ. Sci. Technol.*, 22, 1469 (1988). With permission.]

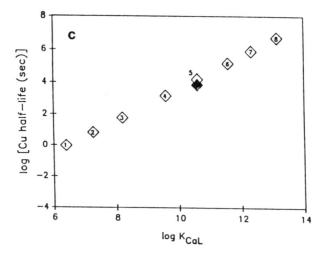

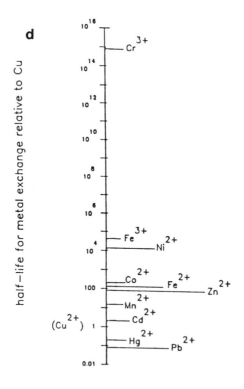

Figure 4-14 (continued).

and pH is an increasingly important factor in the half-life of free Cu as Ca decreases below 10^{-2} M (Figure 4-14b). For disjunctive pathways, it can be shown that the ratio of rate constants, $k\,_{Cu}^{MgY}/k\,_{Cu}^{CaY}$ is given by the ratio of the CaY and MgY stability constants, K_{CaY}/K_{MgY}, which is ~60.[61]

Overall, the rate of CuY formation from CaY is given by

$$\frac{-d[Cu]}{dt} = \left\{ \left[k_{Cu}^{Y} + \frac{k_{Cu}^{HY} K_{HY} \{H^{+}\}}{K_{CaY}} \right] \frac{1}{[Ca]} + k_{Cu}^{CaY} \right\} [CaY][Cu] \qquad (4\text{-}26)$$

Values of the constants are: $k\,_{Cu}^{Y} = 2.3 \times 10^{9}$ M^{-1} s^{-1}; $k\,_{Cu}^{HY} = 5.7 \times 10^{7}$ M^{-1} s^{-1}; $k\,_{Cu}^{CaY} = 1 \times 10^{3}$ M^{-1} s^{-1}; $K_{HY} = 6.3 \times 10^{9}$; $K_{CaY} = 4.1 \times 10^{10}$ (at ionic strength = 0.1). It is apparent from Figure 4-14b that hindrance of CuY formation is less important in freshwater, where pH often is lower and [Ca] usually is less than $10^{-3.3}$ M. However, half-lives still are much longer than 1 min at typical Ca levels and neutral pH. In contrast, natural water concentrations of Ca and Mg have little effect on formation of CuNTA (NTA = nitrilotriacetate), which is virtually complete in seawater within the measurement time of 2 min.[61] NTA is a much weaker chelating agent (K_{Ca} = ~$10^{6.3}$), and theoretical calculations suggest that $t_{1/2}$ is less than 1 s for such comparatively weak ligands (Figure 4-14c).

The slowness of the above complexation reactions has important biological implications. Free metal ions often are toxic to aquatic organisms at low concentrations, whereas complexed metals are not. For example, slow equilibration of Cu added to algal cultures containing artificial chelators resulted in high toxicity.[66] As mentioned previously, exchange rates of the type ML + M′ $\rightleftharpoons$ M′L + M are proportional to the water exchange rates of M′, even though overall exchange is many orders of magnitude slower than water exchange. Consequently, we can predict that exchange rates of many trace metals with Ca complexes of strong ligands will be even slower than those for Cu, since Cu^{2+} has one of the fastest water exchange rates of all metal ions ($k_{-H_2O} \approx 5 \times 10^{8}$ s^{-1}) (Figure 4-7). Among the trace metals of interest in natural waters, only Pb^{2+} (7×10^{9}) and Hg^{2+} (2×10^{9}) have higher k_{-H_2O} values. Exchange rates for Co^{2+}, Fe^{2+}, and Zn^{2+} should be ~100 times slower, while rates for Ni^{2+} and Fe^{3+} should be more than 10^{4} slower. Clearly, the potential exists for long equilibration times (days to months) under some environmentally relevant conditions: neutral to alkaline pH, moderate to high concentrations of Ca, presence of strong ligands like EDTA. Equilibration times may be even longer when both M and M′ are transition metals.[67]

Ligand-ligand exchange reactions of the type MY + Y′ → MY′ + Y are surprisingly rapid when MY′ is the thermodynamically stable complex. These reactions often are much faster than metal-metal exchange rates and much faster than solvent dissociation rates of the MY complex. It can be expected that such reactions will be much faster when Y does not satisfy all coordination sites of M than when it does. In the former case, Y′ can easily displace coordinated water molecules and gain a "foothold" for further reaction. Exchange rate constants have been reported for numerous metal-aminocarboxylate complexes,[51] and values in the range 10^{3} to 10^{5} M^{-1} s^{-1} are common. However, few data are available for compounds found in natural waters.

Ligand Exchange Reactions Involving Aquatic Humic Matter

Aquatic humus, a mixture of polyfunctional, aromatic, and aliphatic macromolecules derived from decomposing vegetation, is responsible for the yellow-brown
color of many surface waters. Most of the humic matter in surface waters is
classified as **fulvic acid** (FA) rather than **humic acid** (HA). This classification is
operationally defined, based on solubility of humic substances in acid and base
solutions (FA is soluble in acids, but HA is not; both are soluble in bases). Recent
studies suggest that the number-weighted average molecular weight of aquatic FA
is about 800 to 1000. Structural aspects of AH are described in Section 8.4.
Complexation reactions involving trace metal ions and natural polyelectrolyte
ligands such as aquatic humus (AH) have received much study over the past 20
years, and substantial progress has been made in quantifying the equilibrium
relationships for these complicated substances. In contrast, rates of dissociation and
exchange reactions for metal-AH complexes have received little attention until
recently. Because AH is structurally heterogeneous, we can expect complicated
kinetics. As recent studies have shown, binding sites on AH exhibit a range of
intrinsic binding strengths,[68] and a wide range of dissociation rates has been
observed for metal-AH complexes.

Two approaches have been used to deconvolve experimental data from such
cases: curve peeling and kinetic spectrum analysis. Choppin and Nash[69] used
xylenol orange as a competing ligand to study the dissociation of Th^{IV}-AH
complexes. They resolved their rate data graphically into three parallel first order
reactions (Figure 4-15a), each of which had a different pH dependency. An
empirical rate expression was fit to the data:

$$\text{Rate} = k_1[ThAH] + k_2[ThAH]\{H^+\}^{0.3} + k_3[ThAH]\{H^+\}^{0.7} \qquad (4\text{-}27)$$

Characteristic times computed from the pseudo-first-order rate constants associated with the lines in Figure 4-15a are on the order of a few minutes to one hour,
but a large fraction (50 to 85%) of the AH-bound Th^{IV} was not dissociated after 24
h. Th^{IV} thus appears to be bound to at least four kinds of AH sites, corresponding
to the three deconvolved rate expressions and the kinetically inert sites. The lack of
pH dependency for k_1 (over the range 4.3 to 6.2) suggests that sites associated with
this constant are completely ionized at pH 4.3. Fractional orders in $\{H^+\}$ were
associated with sites 2 and 3, and these cannot be interpreted mechanistically. This
suggests that the curve peeling essentially was a curve-fitting exercise and that
interpretation of the results in terms of four physically distinct binding sites
probably is too simple.

Studies on Fe and Al dissociation from fulvic acid (FA) found only two
kinetically distinguishable components.[70] Similarly, Al-FA complex formation
experiments over the pH range 3.0 to 4.5 showed evidence for two kinetic
components,[71] which were interpreted as representing two types of binding sites.
The faster-reacting site had rate and equilibrium constants similar to Al-salicylic
acid; FA is thought to contain binding sites of this general type. Overall, Al binding

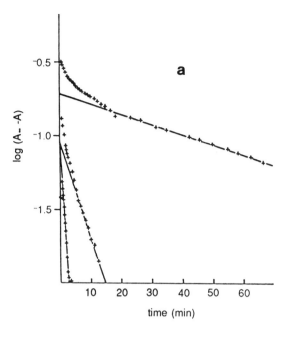

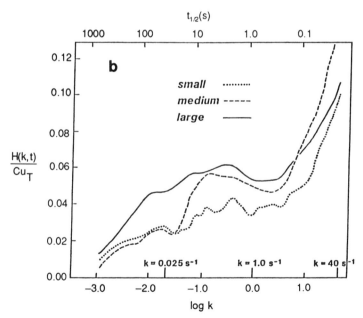

Figure 4-15. (a) Resolution of time plot for dissociation of thorium humate by curve peeling ([ThHu] = 5.77 × 10⁻⁶ *M*; [Hu] = 1.2 × 10⁻⁴ *M*; pH = 4.33). [From Choppin. G.R. and K.L. Nash, *J. Inorg. Nucl. Chem.*, 43, 357 (1981). With permission.] (b) Application of kinetic spectrum method to dissociation of Cu\ :sup:`II` complexes with three size fractions of organic extracts from the Pamlico River, NC. Distribution function H(k,t) is normalized to Cu_T (see Section 2.4.4). [From Olson, D.L. and M.S. Shuman, *Geochim. Cosmochim. Acta*, 49, 1371 (1985). With permission.]

by FA under environmental conditions is fairly fast, albeit not instantaneous, and equilibrium should be approached in a few minutes when Al- and FA-containing waters are mixed. Because the acid-base speciation of Al and FA changes within the pH range of interest, four elementary formation reactions must be considered for each binding site (Figure 4-12d). This is similar to the complexation of Al by F^- (Example 4-2), and the two reactions have about the same rate at 25°C. However, Al-FA complexation rates are only slightly affected by temperature, but Al-F^- complexation rates decrease rapidly with temperature. On this basis, Plankey and Patterson[71] hypothesized that FA complexation of Al should be favored at low temperature. They found, however, that the rate of Al-F^- complex formation is *enhanced* in the presence of FA and concluded that this was due to the formation of mixed-ligand complexes, such as $Al(FA)F^{2+}$ and especially $Al(FA)OHF^+$ (the charge on FA is ignored in these species). Apparently, $Al(FA)^{3+}$ and $Al(FA)OH^{2+}$ react with F^- more rapidly than do free Al^{3+} and $AlOH^{2+}$. In addition, it appears that FA enhances the protolysis of Al^{3+} to $AlOH^{2+}$; the latter species reacts with F^- more rapidly than does Al^{3+}.

The kinetic spectrum method described in Section 2.4.4 has been applied to dissociation kinetics of Cu^{2+}-AH complexes[72] (Figure 4-15b). Reaction progress was determined by a competing ligand method involving formation of $Cu(PAR)_2^{2+}$ ($\lambda_{max} = 508$ nm; PAR = 4-(2-pyridylazo)resorcinol). When excess PAR is present, formation of the 1:2 complex is pseudo-first order in Cu^{2+}. Formation of the Cu-PAR complex was complete in ~20 ms in the absence of AH, but 50 ms was the practical lower limit of measurements. Since $k = 2/t$ (see Figure 2-11), this corresponds to an upper limit of 40 s^{-1} for estimated k values. About 40% of the Cu-AH dissociated with rate constants greater than 40 s^{-1}, and 60 to 65% had rate constants greater than 1 s^{-1}. Most of the Cu complexed with AH thus is labile, but a broad spectrum of rate constants for slower-dissociating sites was found in AH from a river-estuary system. Caution should be observed in interpreting metal-AH dissociation rates estimated from competing-ligand methods because the competing ligand may attack the AH complex and accelerate its dissociation.

No kinetic hindrance by Ca was observed in Cu-FA complex formation.[61] This could be explained by the fact that AH binding sites have relatively low affinity for Ca (effective stability constants, K'_{CaAH}, are in the range ~10^2 to 10^4; cf. Figure 4-14c). It also is possible that Ca and trace metals like Cu do not actually compete for the same binding sites on AH; various studies have found that Ca has little or no effect on equilibrium binding of Cu by AH,[73] but competitive effects were observed for Cu and Cd.[74]

4.4.3 Hydrolysis of Condensed Phosphates

Hydrolysis of condensed phosphates is of interest because the product, orthophosphate, is the most common limiting nutrient for planktonic growth. Condensed phosphates are used in great quantities as "builders" in detergents; in this capacity they chelate Ca^{2+} and Mg^{2+}, buffer pH and ionic strength, and generally increase the cleansing effectiveness of surfactants. Condensed phosphates are not themselves available for plant growth, but many algae have phosphatase enzymes that catalyze their hydrolysis to orthophosphate.

The simplest condensed phosphate is pyrophosphate, $P_2O_7^{4-}$ (PP*), which hydrolyzes to two orthophosphate ions. The reaction is catalyzed by H^+, OH^-, and phosphatase enzymes. Because water is a reactant, the hydrolysis is pseudo-first order. The rate constant (k_{PP}) for chemical hydrolysis of PP (in the absence of organisms) was found to be 1.0×10^{-6} s^{-1} at pH 8.3 (25°C) in lake water and 1.0×10^{-7} s^{-1} in sterile algal growth medium.[75] The disparity may reflect homogeneous catalysis by some solutes in the lake water. Hydrolysis of PP is promoted by complexation with metal ions (Sections 4.2.1 and 4.3.4). $[Ca^{2+}]$ was higher in the lake water than in the growth medium (30 vs. 10 mg L^{-1}), but other ionic differences also may have contributed to the difference in k.

The most effective and widely used condensed phosphate compound in detergents is tripolyphosphate, $P_3O_{10}^{5-}$ (TPP), which hydrolyzes in two steps:

$$TPP \rightarrow PP + P \qquad (4\text{-}28a)$$

$$PP \rightarrow P + P \qquad (4\text{-}28b)$$

Prior to the development of ion chromatography, orthophosphate (P) was the only easily measured product in this sequence. Because P is a product of both steps, k_{TPP} cannot be determined from temporal data on [P] and conventional first-order plots. Clesciri and Lee[75] derived an analytical solution for [P] from the rate equations for TPP, PP, and P:

$$d[TPP]/dt = -k_{TPP}[TPP] \qquad (4\text{-}29a)$$

$$d[P]/dt = k_{TPP}[TPP] + 2k_{PP}[PP] \qquad (4\text{-}29b)$$

$$d[P]/dt = k_{TPP}[TPP] + 2k_{PP}[PP] \qquad (4\text{-}29c)$$

i.e., $[P] = [P]_o + [TPP]_o \left[1 + \dfrac{2}{k_{PP} - k_{TPP}} (k_{PP} + Ck_{TPP}) \right] -$

$$\qquad (4\text{-}29d)$$

$[TPP]_o [\exp(-k_{TPP}t) + \dfrac{2}{k_{PP} - k_{TPP}} (k_{PP} \exp(-k_{TPP}t) + Ck_{TPP} \exp(-k_{PP}t))]$

where
$$C = \dfrac{[PP]_o (k_{PP} - k_{TPP})}{k_{TPP}[TPP]_o} - 1$$

---

* Linear condensed phosphates (e.g., pyrophosphate) are weak acids and are incompletely deprotonated in natural waters. Kinetic studies have not distinguished among the protonated species, and for simplicity we ignore the protons in writing reactions of these compounds. In contrast, trimetaphosphate ($P_3O_9^{3-}$), a six-membered ring, is a strong acid that is completely dissociated at pH 1.

Equation 4-29d can be simplified by making $[P]_o$ and $[PP]_o = 0$. Because k_{PP} can be determined independently, the equation has only one fitted parameter, k_{TPP}. Clesciri and Lee fit TPP hydrolysis data to this equation and derived values of k_{TPP} (sterile conditions, 25°C) of $2.3 \times 10^{-6} s^{-1}$ (lake water) and $3.2 \times 10^{-7} s^{-1}$ (algal medium). The higher lake water value probably reflects catalysis by Ca^{2+} and other cations.

Within the pH range of interest in natural waters, chemical hydrolysis of TPP is an acid-catalyzed reaction.[76] Available data are scattered, but it is apparent that the plot of k_{obs} vs. pH is nonlinear (Figure 4-16a). The reaction is first order in H^+ only at pH < ~2 and is fractional order (<1) over the pH range of interest in natural waters. A neutral hydrolysis mechanism (or perhaps a base-catalyzed reaction) may explain the decrease in slope at pH > ~5, but additional data are needed to verify this. Even more important than the effect of pH on TPP hydrolysis is the presence of organisms, i.e., catalysis by phosphatase enzymes. As Figure 4-16b shows, the rate constant can vary by at least 10^4 at a given pH and temperature, depending on the nature and population density of microorganisms. In fact, hydrolysis of TPP in most surface waters probably is biologically controlled, and the relatively slow chemical hydrolysis is a minor process. Hydrolysis of TPP in wastewater is rapid ($t_{1/2} \approx 30$ min) because of high microbial concentrations. E_{act} for TPP hydrolysis has been measured many times; the mean value for sterile conditions is 105 kJ mol^{-1}.[76] Not surprisingly, E_{act} is lower in the presence of organisms (≈ 46 kJ mol^{-1}), but a wide range of values has been reported (~25 to 75 kJ mol^{-1}).

Hydrolysis of trimetaphosphate (TMP) is catalyzed by both H^+ and OH^-, but acid catalysis is much more important: $k_H \approx 2.5 \times 10^{-2} M^{-1} s^{-1}$; $k_{OH} \approx 5 \times 10^{-5} M^{-1} s^{-1}$.[21] Neutral hydrolysis of TMP is not important at any pH, and k_w has a minimum at pH 7.7. The situation is complicated by the fact that TMP hydrolysis is promoted by complex-forming cations,[21,77] including Na^+ and Ca^{2+}. E_{act} for the acid-catalyzed reaction is ~96 kJ mol^{-1}; that for the base-catalyzed reaction is surprisingly lower (~68.6 kJ mol^{-1}). The Arrhenius preexponential factor for k_H is larger (1.3×10^{16}) than that for k_{OH} (1.6×10^{12}); thus, $\Delta S^{\neq}$ is larger for the acid-catalyzed reaction.

4.4.4 Specific Acid-Base-Catalyzed Hydrolysis of Organic Compounds

Chemical hydrolysis is a major pathway for the degradation of organic compounds in aquatic environments. Hydrolysis involves the substitution of an –OH group for some other functional group or organic fragment. The products of hydrolysis are more polar and, in general, more easily degraded by microorganisms:

Halides: $R-X + H_2O \rightarrow ROH + HX$ (4-30a)

Esters, amides: $RCO-Y + H_2O \rightarrow RCOOH + HY$ (4-30b)

Epoxides: $R_1R_2\underset{\underset{O}{\diagdown\diagup}}{C}-CR_3R_4 + H_2O \rightarrow R_1R_2\underset{HO}{C}-\underset{OH}{C}R_3R_4$ (4-30c)

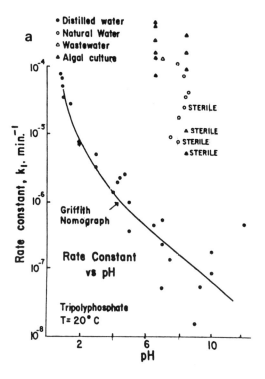

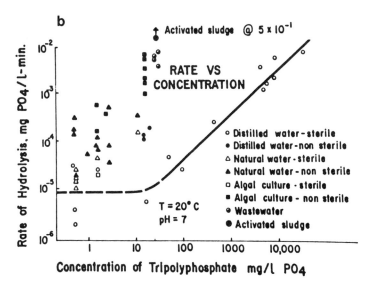

Figure 4-16.   (a) Effect of pH on pseudo-first-order rate constant for hydrolysis of tripolyphosphate (TPP); (b) effect of organism concentration and [TPP] on TPP hydrolysis rate. [From Heinke, G.W., Proc. 12th Conf. Great Lakes Res., International Association of Great Lakes Research, 1969, p. 766. With permission.]

Y stands for an alcoholic (RO–) or amino (–NR$_2$, R may = H) group for esters and amides, respectively. These reactions commonly are catalyzed by H$^+$ and/or OH$^-$.

Rate constants for uncatalyzed and acid- and base-catalyzed hydrolysis of numerous organic compounds in aqueous solution were compiled by Mabey and Mill;[45] Schwarzenbach and Gschwend[78] recently treated mechanistic aspects of these reactions for aquatic contaminants. Generally consistent patterns of H$^+$ and/ or OH$^-$ catalysis are displayed within classes of compounds (Figure 4-17). For example, aliphatic and allylic halides are hydrolyzed by uncatalyzed and base-catalyzed reactions, but not by H$^+$-catalyzed reactions. I_{NB} is above 11 for many alkyl halides (Figure 4-17a), and base-catalyzed hydrolysis is not significant for most of these compounds in natural waters. An exception to this generalization is 1,1,2,2-tetrachloroethane (Table 4-5); also, I_{NB} for the soil fumigant 1,2-dibromo-3-chloropropane is about 5.5. This compound decomposes slowly ($t_{1/2}$ = 141 year at pH 7 and 15°C) by HBr or HCl elimination to allylic intermediates, which hydrolyze rapidly to 2-bromoallyl alcohol.[79] Half-lives for other haloalkanes at pH 7 range from a few days to a few years, with a few notable exceptions (Table 4-5).

Epoxides hydrolyze to diols by neutral, acid-, and base-catalyzed reactions, but acid-base catalysis generally is not important in controlling the persistence of these compounds in natural waters ($I_{AN} < 5$ in most cases, and $I_{NB} > 11$; Figure 4-17b). However, hydrolysis of styrene oxide is acid-catalyzed below pH 7.[80] In contrast to haloalkanes, for which adsorption and surface catalysis were found to be unimportant, hydrolysis of styrene oxide was four times faster in saturated sandy subsoil than in buffered solutions.

Aliphatic and aromatic esters and amides undergo both acid- and base-catalyzed hydrolysis, and in many cases the uncatalyzed reaction is not competitive at any pH. Simple esters are slow to hydrolyze, and base catalysis predominates even at pH 7; halogenated esters are more reactive ($t_{1/2}$ of minutes to hours) and have comparable uncatalyzed and base-catalyzed rates at pH 7 (Figure 4-17c). Most amides are very slow to hydrolyze at pH 7. Because I_{AB} usually is between pH 6 and 7, the acid- and base-catalyzed reactions are competitive under natural water conditions (Figure 4-17d).

Carbamates (R$_1$OC(O)NR$_2$R$_3$) hydrolyze to yield an alcohol, an amine, and CO$_2$ by acid- and base-catalyzed and uncatalyzed reactions. Limited data tabulated for these compounds[45] suggests that base catalysis predominates under environmental conditions (Figure 4-17e). Dialkyl alkylphosphonates {R$_1$P(O)(OR$_2$)$_2$} are very slow to hydrolyze at pH 7 and 25°C ($t_{1/2}$ of years to centuries), and in all cases base-catalyzed reactions predominate ($I_{AB} \approx$ pH 3). In contrast, phosphoric acid esters {(RO)$_3$PO} hydrolyze 100 times or more faster than phosphonates, and at pH 7 the uncatalyzed reactions predominate (Figure 4-17f).

Hydrolysis is an important mechanism for degradation of pesticides in aquatic systems. The reactions are pH dependent, with base catalysis predominant. For example, degradation of malathion is slow and unimportant under acidic conditions, but alkaline degradation is fast enough to be competitive with photolysis.[84] In contrast, hydrolysis of methoxychlor is independent of pH[85] and its $t_{1/2}$ in water is about 1 year. Esters of 2,4-D hydrolyze rapidly in basic conditions, and this is the

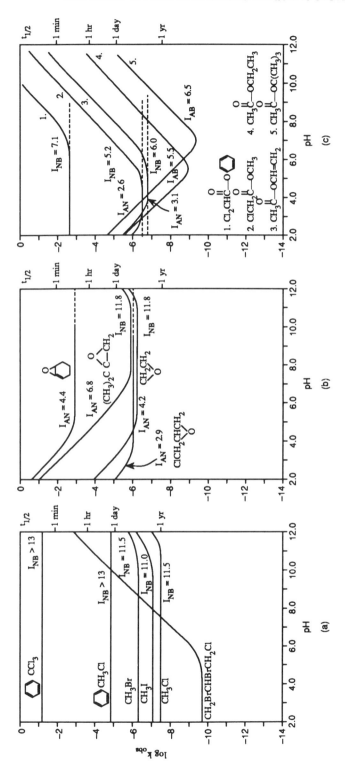

Figure 4-17. $Log\,k_{obs}$ vs. pH for six classes of organic compounds: (a) alkyl halides; (b) epoxides; (c) esters; (d) amides; (e) carbamates; (f) alkyl phosphonates and phosphoric acid esters. [Data from Mabey, W. and T. Mill, *J. Phys. Chem. Ref. Data*, 7, 383 (1978).]

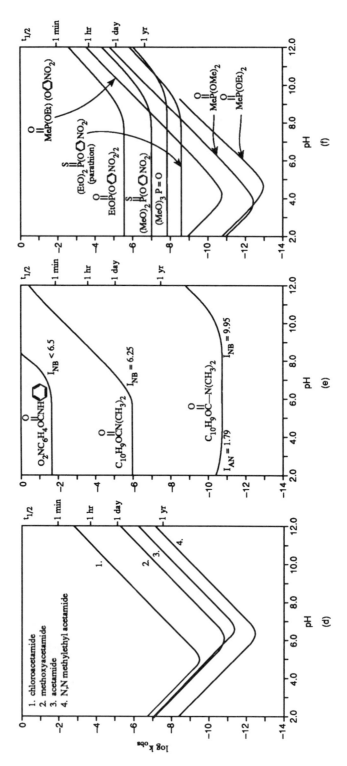

Figure 4-17 (continued).

Table 4-5. Reactivities of Alkyl Halides Toward Hydrolysis

Compound	k_1[a] s^{-1}	E_{act} kJ mol^{-1}	A[a]	$t_{1/2}$ days	Ref.
1,1,1-trichloroethane	2.7E-8	118	1.1E13	350	80
1,1,2,2-tetrachloroethane	1.8E-7[b]	—	—	45	80
dichloromethane	1.5E-8	—	—	550	82
trichloroethane	1.7E-8	—	—	475	82,83
tetrachloromethane	~3E-13	—	—	(7000 year)	82,83
bromomethane	4.1E-7	110	8.1E12	20	45,81
bromoethane	2.6E-7	111	8.6E12	30	45,81
isopropyl bromide	3.7E-6	107	2.2E13	2.1	45,80,81
1-bromopropane	3.0E-7	111	9.9E12	36	45,81
2-bromopropane	3.9E-6	104	6.0E12	2.1	45,81
1-bromo-3-phenylpropane	2.8E-8	103	3.1E10	290	81
1,2-dibromoethane[c]	8.9E-9	103	9.9E9	910	81
1,2-dibromoethane[c]	5.5E-9	107	3.3E10	1500	80
1,2-dibromopropane	2.5E-8	106	9.1E10	320	81
1,2-dibromopropane (pH 11)	6.9E-8	98	1.2E10	(116)	81
1,3-dibromopropane	1.7E-7	94	5.5E9	48	81
1,2-dibromo-3-chloropropane	5.8E-10	94	2.9E9	(38 year)	79

[a] k_1 at pH 7 and 25°C; A = Arrhenius preexponential factor.
[b] Base-catalyzed at least to pH 6; pseudo-first-order value for pH 7.
[c] Commonly known as ethylene dibromide.

major loss mechanism for these compounds in many waters.[86] Base hydrolysis is more important than acid hydrolysis for aldicarb (AS) and its oxidation products, aldicarb sulfoxide (ASO) and aldicarb sulfone (ASO$_2$).[87,88] Aldicarb, an important thiocarbamate pesticide, is a common groundwater contaminant beneath sandy agricultural soils in the U.S. Hydrolysis products of each compound include the corresponding oxime, CO_2, and methylamine, and the hydrolysis rates decrease in order: ASO$_2$ > ASO > AS. Published rate constants for hydrolysis of AS vary widely ($t_{1/2}$ of months to years), making prediction of its environmental persistence imprecise.

4.4.5 Nucleophilic Substitution Reactions of Reduced Sulfur Compounds

Sulfide and other ions containing sulfur in intermediate oxidation states react with certain organic compounds in ways that are analogous to hydrolysis. Reduced sulfur species are relatively strong nucleophiles; in hydrolysis reactions, H_2O acts as a nucleophilic agent. (Recall that nucleophiles are "electron-rich" compounds that donate electrons to an electrophile to form a bond.) Thiosulfate, HS$^-$, and polysulfides (S_x^{2-}) are "soft" nucleophiles — in the sense of Pearson's hard and soft acid-base (HSAB) concept (see Section 7.10.3). As such, they tend to react with soft electrophiles such as saturated carbon atoms to which highly electronegative elements such as Br and Cl are attached, e.g., alkyl halides, RX. The net effect is a nucleophilic substitution (S_N2) reaction in which the sulfur nucleophile displaces a halide ion on RX. For example, reaction of HS$^-$ produces a thiol (RSH) or thiolate ion, RS$^-$. The latter are also reactive nucleophiles, and further reactions may lead to formation of thioethers, RSR.[89,90] Reaction of alkyl halides with HS⁻ can be more important than hydrolysis in some cases. This is most likely to be true for primary alkyl bromides at high pH in anoxic environments with high S⁻ᴵᴵ levels. For example, in contaminated groundwater around landfills, S^{-II}_T may reach 500 μM. A variety of alkyl thiols, such as ethanethiol, and disulfides such as diethyl disulfide, have been detected in such waters (see reference 90).

In general, the basic forms of nucleophiles that are Bronsted acids are much more reactive (2 to 4 orders of magnitude) in S_N2 reactions than their conjugate acids. Reaction rates thus increase markedly in the region around pK_a of the nucleophile. Only limited information is available to verify this for sulfur nucleophiles,[90] but Haag and Mill[91] showed that HS⁻ is much more reactive toward *n*-hexyl bromide than is H_2S.

Water is by far the most abundant nucleophile in aqueous solutions. Whether a given sulfur nucleophile, S_{nu}, is competitive with hydrolysis can be predicted from the ratio $R_{S/h} = log(k_S/k_{H_2O})_{SN}$, where subscript SN stands for "substitution nucleo-philic." If $R_{S/h} > log\{[H_2O]/[S_{nu}]\}$, reaction with S_{nu} exceeds the rate of the hydrolysis reaction. Based on order-of-magnitude estimates for concentrations of sulfur nucleophiles in aquatic systems, Barbash and Reinhard[90] concluded that some brominated compounds could be competitive with hydrolysis (Table 4-6). Alkyl chlorides react more slowly. Some of the halogenated sulfur compounds produced by such reactions are quite toxic.

4.4.6 General Acid-Base (Buffer) Catalysis

According to an analysis by Perdue and Wolfe,[92] general acid-base (or buffer) catalysis is not likely to be important at buffer concentrations less than $10^{-3}\,M$. Buffer catalysts seldom exceed this concentration in natural waters. Perdue and Wolfe derived the following equations to support this conclusion:

$$k_{obs}/k_w = 1 + C_B(BCF) \tag{4-31}$$

k_{obs} and k_w are defined by Equations 4-19 and 4-20 as pseudo-first-order rate constants at a given pH for the sum of all hydrolysis processes (k_{obs}) and the sum of all processes except general acid-base catalysis (k_w); C_B is the total concentration of buffer ($\Sigma[H_iB]$); and BCF is the buffer catalysis factor:

$$BCF = \frac{\sum_{i=1}^{n}\left[\Theta a_i A_i^{\alpha} + (1-\Theta)b_i B_i^{\beta}\right]}{C_w + \Theta\{H^+\}D^{\alpha} + (1-\Theta)[OH^-]D^{\beta}} \tag{4-32}$$

$A_i = K_iC_w/K_w$; $B_i = C_w/K_i$; C_w = molar concentration of water (55.3 M at 25°C); K_i is the acid dissociation constant for the *i*th acid; $D = C_w^2/K_w$; Θ is the fraction of k_{H_2O} attributable to acid catalysis by H_2O; $(1-\Theta)$ is the fraction of k_{H_2O} attributable to base catalysis by H_2O; a_i and b_i are mole fractions of the *i*th acid and its conjugate base, respectively; and α and β are fitted parameters that characterize the reactant's

Table 4-6. Compounds that may React with Sulfur Nucleophiles at Rates Comparable to Reaction with Water (Hydrolysis)[a]

Compound	Hydrolysis $log\ k_{H_2O}$ $(M^{-1}\ s^{-1})$	Sulfur nucleophile	Nucl. subst. $log\ k_{S,SN}$ $(M^{-1}\ s^{-1})$	Concentration of nucleophile[b] μM
$BrCH_2CH_2Br$	-10	HS^-	-3.2	1000
$BrCH_2CH_2Br$		$S_2O_3^{2-}$	-4.0	100
$BrCH_2CH_2Br$		SO_3^{2-}	-3.6	100
$ClCH_2CH_2Cl$	-11	HS^-	-5.2	1000
n-hexBr	-8.5	HS^-	-3	1000
CH_3Br	-8.1	$S_2O_3^{2-}$	-1.05	100
n-pentBr	-8.5	$S_2O_3^{2-}$	-3.3	100

[a] Information derived from Barbash, J.E. and M. Reinhard, in *Biogenic Sulfur in the Environment*, E.S. Saltzman and W.J. Cooper (Eds.), ACS Symp. Ser. 393, American Chemical Society, Washington, D.C., 1989, p. 101.

[b] Maximum concentration expected in sediment porewater or contaminated groundwater. At this concentration, hydrolysis of the compound in column 1 is slower than reaction with the sulfur nucleophile.

sensitivity to acid or base strength. In evaluating conditions that maximize BCF, Perdue and Wolfe assumed $\Theta = 0.5$, but they showed that results are independent of Θ.

For a given buffer system and given C_B, only pH, α, and β are variables in Equation 4-32. Perdue and Wolfe used numerical iteration and analytical methods to show that the values of α and β maximizing BCF depend primarily on pH:

$$\alpha_{opt} = (\text{intercept}) + (\text{slope})\text{pH}; \quad \beta_{opt} = 1 - \alpha_{opt} \qquad (4\text{-}33)$$

The intercept varies slightly depending on K_i of the catalyst; the range of the intercept reported for ten buffers is 0.087 (silicic acid) to 0.125 (phthalic acid). The slope is a constant with a value of 0.057. It is apparent from Equation 4-33 that the sum of α_{opt} and β_{opt} is unity, and $\alpha_{opt} \approx \beta_{opt} \approx 0.5$ at pH 7.

From the optimum values of α and β for a given buffer and pH, corresponding optimum values of BCF can be calculated from Equation 4-32. Results are plotted in Figure 4-18 as isopleths of BCF_{opt} over the pK_i and pH ranges 2 to 12. BCF_{opt} is greater than 100 only in a limited domain near pH and $pK_i \approx 7$, and $BCF_{max} = 106$ at $pK_i = pH = 7.00$. Substituting an approximate upper limit of 100 for BCF in Equation 4-31, we can compute the C_B necessary to increase the hydrolysis rate by any given factor compared with that from the uncatalyzed reaction plus specific acid-base catalysis. For example, at $k_{obs}/k_w = 1.10$, buffer catalysis is 10% of the combined kinetic contributions included in k_w, and $C_B \approx 10^{-3}\ M$. Thus, buffers enhance hydrolysis rates by 10% or more (compared with uncatalyzed plus specific acid-base catalyzed rates) only when $C_B > 10^{-3}\ M$. Such conditions are uncommon in natural waters, but may be encountered in laboratory studies where buffers are added to control pH. The hydrolysis rate of aldicarb sulfoxide (a major metabolite of aldicarb) increased 12% in 0.02 M borate compared with unbuffered solutions at pH 8.5.[88] However, the BCF reported by Perdue and Wolfe for borate at this pH is 50, which should induce a doubling of the hydrolysis rate. Reasons for the discrepancy are unknown.

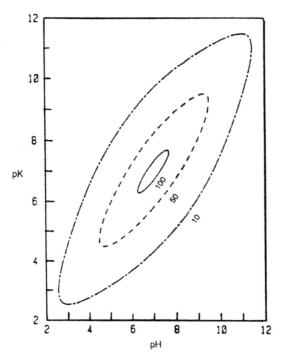

Figure 4-18. Isopleths of BCF (buffer catalysis factor) as function of pH and buffer pK_a. [From Perdue, E.M. and N.L. Wolfe, *Environ. Sci. Technol.*, 17, 635 (1983). With permission.]

General acid (buffer) catalysis by natural fulvic acids (FA) has been hypothesized several times, but evidence for this is weak. Khan[93] reported that hydrolysis of the herbicide atrazine to a nonphytotoxic hydroxy derivative is catalyzed by FA. The apparent first-order rate constant for atrazine hydrolysis increases with [FA] and decreases in a slight nonlinear fashion as pH increases from 3 to 7. The carboxylic acid groups of FA are associated with a wide range of electron-withdrawing and -donating conditions, and FA does not have a single acid dissociation constant. Instead, it exhibits increasing deprotonation over a broad pH range. The trend noted above thus suggests that hydrolysis of atrazine is catalyzed by protonated sites on FA, the number of which decreases slowly over the observed pH range. On the other hand, general acid catalysis by humic and fulvic acids was not observed in the hydrolysis of the 1-octyl ester of 2,4-D.[94] The presence of humic substances actually caused a decrease in hydrolysis rate, apparently because association of the hydrophobic ester with humic material (by sorption or partitioning) protected the ester from OH^--catalyzed hydrolysis. Some evidence for FA catalysis of $Al-F^-$ complex formation was described earlier (Section 4.4.2; see reference 71).

4.4.7 Metal-Promoted Hydrolysis

Although catalysis by trace metals in natural waters has been hypothesized frequently, little direct information is available to support this conjecture. Mabey and Mill[45] concluded that metal-ion catalysis of hydrolysis reactions is not significant at typical concentrations of metal ions in natural waters, but catalysis by metal ions may be more important in soils and sediments.[95] Nonetheless, a few examples of metal-promoted hydrolysis reactions in aquatic environments have been reported. Hydrolysis of the organophosphorus pesticide, parathion, is catalyzed by Cu^{2+} at submicromolar concentrations.[17] The nerve gas GB, isopropyl methylphosphonofluoridate, decomposes by acid-, base-, and uncatalyzed hydrolysis to nontoxic isopropyl methylphosphonic acid:

$$(CH_3)_2 HCO - \underset{F}{\overset{CH_3}{P}} = O + H_2O \rightarrow (CH_3)_2 HCO - \underset{OH}{\overset{CH_3}{P}} = O + HF \qquad (4\text{-}34)$$

The pH profile for k_{obs} in distilled water has the shape of Figure 4-6a. Epstein[96] found that hydroxo-metal complexes significantly enhance the hydrolysis rate of GB at neutral and slightly alkaline pH. For example, addition of 1 mg L^{-1} of Cu^{2+} at pH 6.5 and 25°C decreased $t_{1/2}$ from 175 h to 2 h. The main cations enhancing GB hydrolysis in seawater are Ca^{2+} and Mg^{2+}. From the pH dependency of enhancement (Figure 4-19), it appears that monohydroxo complexes ($CaOH^+$ and $MgOH^+$) are involved rather than the free ions. At pH > 6.5, hydrolysis is enhanced by ~50-fold in seawater compared with distilled water.

4.5 CHLORINATION REACTIONS

4.5.1 Introduction

All compounds containing chlorine in oxidation states greater than –I are unstable with respect to redox equilibria in aqueous solution:

State	Species	Name	Comments
–I	Cl^-	chloride	stable in aqueous solution
0	Cl_2	chlorine	disproportionates to +I and –I states
+I	$HOCl$	hypochlorous acid	weak acid; common disinfecting agents
	OCl^-	hypochlorite	good oxidants
+III	$HOClO$	chlorous acid	acid highly unstable; ion used
	ClO_2^-	chlorite	as bleaching agent
+IV	ClO_2	chlorine dioxide	disinfectant; oxidizing agent
+V	$HOClO_2$	chloric acid	strong acid and oxidant
	ClO_3^-	chlorate	
+VII	$HOClO_3$	perchloric acid	very strong acid; powerful oxidant
	ClO_4^-	perchlorate	explosive in concentrated form

In general, these compounds act as oxidizing agents for organic matter and reduced inorganic ions of Fe, Mn, N, and S. The high reactivity of oxidized Cl forms

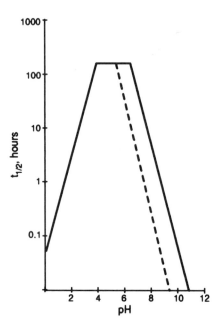

Figure 4-19. Half-life of nerve gas GB in distilled water (solid line) and seawater (dashed line) at 25°C. Decomposition by hydrolysis according to Equation 4-41. [From Epstein, J., *Science,* 170, 1396 (1970). With permission.]

and their commercial availability at low to moderate costs makes them useful in treating water. Chlorine (Cl_2) and salts of hypochlorite [NaOCl and $Ca(OCl)_2$] are the most common disinfectants used in municipal water treatment, but chlorination also is practiced to oxidize reduced inorganic ions (Fe^{2+}, Mn^{2+}, H_2S, and HS^-) and natural, color-causing organic compounds (humates and fulvates), which cause water quality problems to consumers.

Chlorine (Cl_2) is a soluble gas that disproportionates in water to the $-I$ (Cl^-) and $+I$ (HOCl, OCl^-) states. The equilibria among these species are rapid and pH dependent; collectively they are known as "free chlorine":[97,98]

$$Cl_2 + H_2O \rightleftharpoons HOCl + H^+ + Cl^- \quad K = 1.0 \times 10^{-3} \quad (98) \qquad (4\text{-}35)$$

$$HOCl \rightleftharpoons H^+ + OCl^- \qquad K_1 = 3.2 \times 10^{-8}, pK_1 = 7.50 \qquad (4\text{-}36)$$

At freshwater concentrations of Cl^- and pH > 5, the hydrolysis of Cl_2 is nearly complete, and Cl_2 generally is not a significant species in municipal water treatment (Table 4-7). HOCl is the predominant species at pH < 7.5, and OCl^-, which is a much less effective disinfectant, predominates at higher pH. Cl_2 and H_2OCl^+ occur under acidic conditions and may be involved in acid-catalyzed reactions of HOCl. Cl^+ probably does not exist in aqueous solution, except as a highly ephemeral reaction intermediate, but is important in nonaqueous systems.

Table 4-7. Distribution of Free Chlorine Species vs. pH at 15°C[a]

pH	Cl_2	HOCl	OCl^-
5	3.6E-4	0.997	0.003
6	3.6E-5	0.975	0.025
7	2.9E-6	0.797	0.203
8	1.0E-7	0.280	0.720
9	1.0E-9	0.038	0.962

From Morris, J.C., in *Water Chlorination,* R.L. Jolley (Ed.), Ann Arbor Science, Ann Arbor, MI, 1978, p. 21.

[a] Fraction of total free chlorine at $[Cl^-] = 350$ mg L^{-1} (10^{-2} *M*).

The advantages of the above compounds for water treatment include high effectiveness (rapid rates of microbial kill), low cost, and ability to maintain a residual, ensuring disinfection within distribution systems. Disadvantages arise from the high and nonselective reactivity of the free chlorine species. Maintenance of a free residual is difficult in the presence of organic matter, but this problem can be circumvented by adding ammonium, with which free chlorine reacts rapidly to form *combined* chlorine forms (chloramines such as NH_2Cl and $NHCl_2$) that are less reactive. Aside from acting as oxidants, free Cl participates in substitution reactions, yielding organochlorine compounds such as chlorinated phenols and trihalomethanes (e.g., chloroform, $CHCl_3$). ClO_2, which acts only as an oxidant, is preferred for water disinfection in some countries.

Because of the water quality and human health significance of chloramines and organochlorine compounds, the kinetics of their formation and reaction have been studied in detail over the past several decades. The literature on this topic is extensive, but much of it has been summarized in a series of conference proceedings.[99] Compared with other aquatic redox reactions, the rates of chlorination reactions typically are fast (many occur on time scales of subseconds to seconds) and mechanistically simple (often occurring in one or two elementary steps). The reactants and products generally have strong absorption bands in the UV range, which facilitates measurement of rapid reaction rates.

4.5.2 Inorganic Reactions

Hydrolysis of Cl_2

Hydrolysis of Cl_2 (Equation 4-35, a disproportionation reaction) is rapid and essentially complete at neutral pH. Numerous measurements have been made of the reaction rate (see reference 100 for a review), and a wide range has been reported for the first-order hydrolysis rate constant. The "best estimate" reported by Aieta and Roberts[100] is 12.2 s^{-1} at 20°C (95% CI = 10.7 to 14.0 s^{-1}). Margerum et al.[98] reported a somewhat higher value, 28.6 s^{-1} at 25°C and I = 0.5. The half-life of Cl_2 in water thus is only ~0.06 s. Several techniques have been used to measure k for this fast reaction, including temperature-jump relaxation, stopped-flow spectrophotometry, and gas-liquid mass transfer.[100] The reaction has an E_{act} of 60.2 kJ

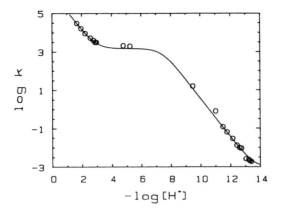

Figure 4-20. Variation with pH (in absence of buffer catalysts) of second-order rate constant (M^{-1} s^{-1}) for oxidation of Br⁻ by free chlorine, expressed in terms of $[OCl^-]_T$ and $[Br^-]$ (25°C; I = 0.50). [From Kumar, K. and D.W. Margerum, *Inorg. Chem.*, 26, 2706 (1987). With permission.]

mol⁻¹ (95% CI = 49 to 71 kJ mol⁻¹) and an Arrhenius pre-exponential factor of 6.8 × 10¹¹, which is in the "normal" range.

Reactions with Other Halogen Compounds

If Br⁻ is present in a water treated with chlorine, it is oxidized rapidly to HOBr. Over a broad range of pH, the rate of this reaction varies widely (Figure 4-20), but it is almost constant over the pH range 3 to 8, which includes most natural waters. It is apparent from the shape of Figure 4-20 that HOCl is much more reactive than OCl⁻. The second-order rate constant for OCl⁻ as the oxidant is k = 0.9(±0.1) × 10⁻³ M^{-1} s^{-1} (see Figure 2-9), and the second order rate constant for HOCl as oxidant is 1.3 × 10³ M^{-1} s^{-1}.[101] HOCl thus is more than 10⁶ more reactive with Br⁻ than is OCl⁻. The increase in k at pH < 3 in Figure 4-20 reflects specific acid (H⁺) catalysis of the HOCl + Br⁻ reaction. The uncatalyzed reaction with HOCl predominates from pH 3 to 13, and reaction with OCl⁻ becomes important only above pH 13. Rates of Br⁻ oxidation increase with increasing buffer concentration at fixed pH under both acidic and basic conditions, and the following mechanisms were proposed by Kumar and Margerum:[101]

$$HA + OCl^- + Br^- \rightarrow A^- + OH^- + BrCl \quad \text{(slow)} \quad \text{(general acid catalysis)} \quad \text{(4-37a)}$$

$$BrCl + 2OH^- \rightleftharpoons OBr^- = Cl^- = H_2O \quad \text{(fast)} \quad \text{(4-37b)}$$

$$HA + HOCl + Br^- \rightarrow A^- + H_2O + BrCl \quad \text{(slow)} \quad \text{(general acid assisted)} \quad \text{(4-38a)}$$

$$BrCl + Br^- \rightleftharpoons Br_2 + Cl^- \quad \text{(fast)} \quad \text{(4-38b)}$$

$$Br_2 + Br^- \rightleftharpoons Br_3^- \quad \text{(fast)} \quad \text{(4-38c)}$$

These mechanisms postulate the transfer of electrophilic Cl^+ to Br^-. In contrast, the direct transfer mechanism previously thought to apply to this reaction ($HOCl + Br^- \rightarrow HOBr + Cl^-$);[102] involves nucleophilic attack by Br^- at the oxygen in $HOCl$, which is equivalent to OH^+ transfer from Cl^- to Br^-.[101] This is considered less likely because the transition states would involve expanding the coordination of oxygen, which is difficult compared with expansion of Cl:

$$A - H - \overset{Br^-}{\underset{Cl^-}{O}} \qquad or \qquad A - H - \overset{Br^- H}{\underset{Cl^-}{O}}$$

Free chlorine and hypochlorous acid react rapidly with chlorite to form ClO_2 and ClO_3^- ($k = 1.3$ to $1.7 \times 10^4 M^{-1} s^{-1}$ at 20°C).[100] The former is favored under acidic conditions where Cl_2 predominates, and the latter is favored at neutral and alkaline conditions where $HOCl$ predominates. When Cl_2 gas is contacted with chlorite solutions, $Cl_{2(aq)}$ reacts with ClO_2^- faster than it hydrolyzes,[103] and formation of ClO_2 is favored. Although the stoichiometric reaction is $Cl_2 + 2ClO_3^- \rightarrow 2ClO_2 + 2Cl^-$, the reaction is second order overall (first order in both Cl_2 and ClO_3^-). A mechanism that satisfies this information is:[104]

$$*Cl_2 + ClO_3^- \rightarrow (*Cl - ClO_2) + *Cl^- \qquad \text{(slow)} \qquad \text{(4-39a)}$$

$$2\{*Cl - ClO_2\} \rightarrow 2ClO_2 + Cl_2 \qquad \text{(fast)} \qquad \text{(4-39b)}$$

The asterisk is used to track Cl atoms derived from Cl_2, and the compound in parentheses represents an unstable intermediate.

Formation and Reactions of Chloramines

The kinetics of chlorine reactions with ammonium to form chloramines and subsequent reactions of chloramines were studied over a long period of time by Morris and students[105] (see references 97,106–109 for reviews) and more recently by Margerum and co-workers.[7,110] Reaction rates and product distribution depend on pH, relative concentrations of $HOCl$ and NH_4^+, time, and temperature. The best-studied reaction (and the simplest one to measure) is that of $HOCl$ and NH_4^+ to form monochloramine, NH_2Cl. This is the major product formed under normal conditions of water chlorination in the presence of NH_4^+. When the rate is expressed in terms of total analytical concentrations of reactants, k_{obs} varies with pH, decreasing under both acidic and basic conditions (maximum rate at pH 8.5), consistent with $HOCl$ and NH_3 as the actual reactants. A wide range in k is found among reported studies, but all agree that the reaction is second order and yield k's within an order of magnitude. The best estimate for k when reactants are expressed in terms of the neutral molecules is $4.2 \times 10^6 M^{-1} s^{-1}$ at 25°C; E_{act} is only 13 kJ mol^{-1},[107] i.e., in the range of diffusion-controlled processes. The reaction thus is quite rapid, being more than 99% complete in much less than 1 min under typical conditions for water chlorination.

The kinetics of reactions involving dichloramine (either as reactant or product)

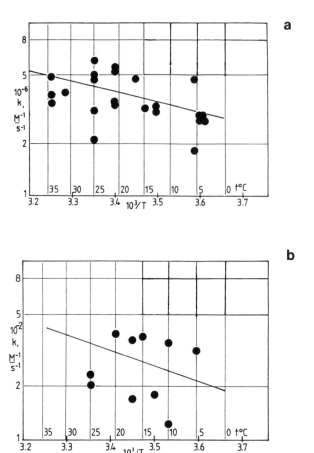

Figure 4-21.   Arrhenius plots of rate constants for formation of (a) NH$_2$Cl and (b) NHCl$_2$. [From Morris, J.C. and R.A. Isaac, in *Water Chlorination,* Vol. 4, R. L. Jolley et al. (Eds.), Ann Arbor Science, Ann Arbor, MI, 1983, p. 49. With permission.]

are difficult to study because other competing reactions interfere with measurements. Most workers have estimated the rate constant for NHCl$_2$ formation from NH$_2$Cl and HOCl indirectly from the final ratio of products (NH$_2$Cl: NHCl$_2$) formed by HOCl in the presence of excess NH$_4^+$.[107] The rate constant for NHCl$_2$ formation can be calculated from this ratio and the known value of k for NH$_2$Cl formation from an equation for the distribution of products in consecutive second-order reaction schemes.[106] Direct measurements of the reaction in stopped-flow apparatus have yielded k values in the same range as the indirect calculations.[107]

The formation of NHCl$_2$ from NH$_2$Cl and HOCl is much slower (~10^{-4} times) than the formation of NH$_2$Cl from NH$_3$ and HOCl, because NH$_2$Cl is a much weaker base (poorer nucleophile) than NH$_3$. Based on a best-fit analysis of published data from five groups of authors, Morris and Isaac[107] reported k = 3.5 × 10^2 M^{-1} s^{-1} at 25°C when the reaction is expressed in terms of neutral reactants (Figure 4-21). The composite data yielded a surprisingly low E$_{act}$ of 17 kJ mol^{-1}, but the plot shows that there is considerable uncertainty in both k and E_{act}. Morris[106] earlier reported $E_{act} = 29$ kJ mol^{-1} for this reaction; given that the reaction is relatively slow, this value seems more reasonable. In contrast to the reaction of NH_3 with HOCl, the reaction of NH_2Cl with HOCl is subject to specific- and general-acid catalysis. According to Morris, k for $NHCl_2$ formation can be expressed as: $k = 3.4 \times 10^2\{1 + 5 \times 10^4 \times 10^{-pH} + 2 \times 10^2[HAc]\}$, where HAc stands for acetic acid (the only Bronsted acid studied). The factor before [HAc] would change for other Bronsted acids.

Because the reactions producing NH_2Cl and $NHCl_2$ have different sensitivities to pH, the distribution of chloramines resulting from the reaction of free chlorine with NH_3 varies with pH. At near neutral pH, NH_2Cl predominates (when Cl:N < 1), but at pH < 5.5, formation of $NHCl_2$ becomes competitive. For equimolar concentrations of chlorine and ammonium at 25°C, the following distributions occur:[108]

pH	% NH_2Cl	% $NHCl_2$
5	13	87
6	57	43
7	88	12
8	97	3

Dichloramine also is formed by disproportionation of monochloramine: $2NH_2Cl \rightarrow NHCl_2 + NH_3$. This can be viewed as the chlorination of one monochloramine by a second monochloramine molecule. Since NH_2Cl is a weaker chlorinating agent (electrophile) than HOCl, the rate of reaction would be expected to be slower than that between NH_2Cl and HOCl, and the estimated k of 5.6×10^{-2} M^{-1} s^{-1} agrees with this prediction; i.e., the specific rate is $\sim 10^{-4}$ as fast. This reaction also is specific- and general-acid catalyzed, and the rate constant can be expressed as $k = 5.6 \times 10^{-2}\{1 + 1.3 \times 10^5 10^{-pH} + 35[HAc]\}$.[106]

A second mechanism complicates the formation of $NHCl_2$ from NH_2Cl: slow, first-order hydrolysis of NH_2Cl to HOCl and NH_3, followed by reaction of the HOCl with another molecule of NH_2Cl. Granstrom[111] evaluated the hydrolysis reaction $(NH_2Cl + H_2O \rightarrow HOCl + NH_3)$ and found $k = 8.7 \times 10^7 exp(-17,000/RT)$ (s^{-1}).

Dichloramine is an unstable compound in aqueous solution and decomposes by several mechanisms. Most important is an oxidation mechanism producing Cl$^-$ and N_2, resulting in a loss of active chlorine. This complicated process, called breakpoint chlorination (Figure 4-22), is a common practice in water treatment. Loss of $NHCl_2$ from solution is an autocatalytic process because HOCl and NCl_3 formed in decomposition steps accelerate further reactions.[7] The presence of even small quantities of NH_4^+ greatly increases the stability of $NHCl_2$ (by a factor of $\sim 10^5$). Losses of less than 10% in 24 h were noted in solutions of $NHCl_2$ prepared by disproportionation of NH_2Cl at pH 3.5 to 4 (which yields $NHCl_2$ and NH_4^+ as products), but substantial loss of $NHCl_2$ occurred at pH 4 in less than 1 h when NH_4^+ was removed from solutions by ion exchange.[7] Inhibition of $NHCl_2$ decomposition by NH_4^+ is explained by its scavenging of HOCl, which reacts rapidly with $NHCl_2$, accelerating its decomposition.

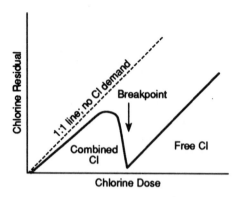

Figure 4-22. General scheme of breakpoint chlorination: difference between total residual Cl and chlorine dose reflects chlorine demand, primarily from ammonium and amines. Before breakpoint most Cl is in combined forms, primarily mono- and dichloramine; after the breakpoint, combined residual consists of slow-reacting organic chloramines. Added Cl remains in free form after the breakpoint. Sharpness of breakpoint and minimum observed Cl concentration depend on pH, temperature, and time of reaction. Loss of residual Cl at breakpoint is caused by oxidation of di- and trichloramines to N_2 according to reaction 4-40a, b and other reactions. Isaac and Morris[109] described a kinetic model for breakpoint chlorination.

Under neutral and basic conditions, decomposition of $NHCl_2$ is thought to proceed by coupled reactions:

$$NHCl_2 + HOCl + B \rightarrow NCl_3 + BH^+ + OH^- \qquad (4\text{-}40a)$$

$$NHCl_2 + NCl_3 + OH^- \rightarrow N_2 + 2HOCl + 3Cl^- + H_2O \qquad (4\text{-}40b)$$

When [HOCl] is low, reaction 4-40a is slower than reaction 4-40b, and [NCl_3] remains low.[7] Reaction 4-40a is general-base catalyzed, and values of the third-order rate constant k_B (M^{-1} s^{-1}) at I = 0.5 and 25°C are 1.6×10^4 (HPO_4^{2-}), 1×10^5 (OCl^-), 6×10^6 (CO_3^{2-}), and 3.3×10^9 (OH^-).[7]

Trichloramine is a fairly stable substance in dilute solution, but it is explosive in pure form. NCl_3 is formed from $NHCl_2$ and HOCl under acidic conditions and is an important component of breakpoint chlorination reactions. According to Kumar et al.,[110] the equilibrium constant for the reaction

$$NHCl_2 + HOCl \rightleftharpoons NCl_3 + H_2O \qquad (4\text{-}41)$$

is 1.6×10^8 M^{-1} (based on the ratio of forward to reverse rate constants). This is substantially higher than a value reported earlier by Morris and Isaac[107] and implies that NCl_3 is rather stable. However, the stability implied by this value is misleading in that NCl_3 reacts further with $NHCl_2$ in the breakpoint process (Equation 4-40b), and therefore NCl_3 and $NHCl_2$ cannot reach a true equilibrium in Equation 4-41. Saguinsin and Morris[112] reported a forward rate constant for Equation 4-41 of 3.4 M^{-1} s^{-1}, but Hand and Margerum[7] found a significantly different rate constant and mechanism, as described above. The decomposition of NCl_3 is base catalyzed and fits the rate equation:[110]

$$-d[NCl_3]/dt = 2(k_0 + k_1\{OH^-\} + k_{HB}[HB]\{OH^-\} + k_2\{OH^-\}^2)[NCl_3] \qquad (4\text{-}42)$$

Values of the rate constants at $I = 0.5$ and $25°C$ are $k_0 = 1.6 \times 10^{-6}$ s^{-1}; $k_1 = 8$ M^{-1} s^{-1}; $k_2 = 890$ M^{-1} s^{-1}; k_{HB} $(M^{-1}$ $s^{-1}) = 2.1 \times 10^3$ $(H_2PO_4^-)$, 7.6×10^2 $(B(OH)_3)$, 128 (HPO_4^{2-}), and 65 (HCO_3^-). The overall stoichiometry of the reaction is

$$2NCl_3 + 6OH^- \rightarrow N_2 + 3OCl^- + 3Cl^- + 3H_2O \qquad (4\text{-}43)$$

but it proceeds by two pathways, (1) a specific-base/general-acid catalyzed decomposition to $NHCl_2$, i.e., the reverse of the NCl_3 formation reaction (Equation 4-40a), and (2) a rapid reaction of NCl_3 with dichloramine to yield N_2:

$$NCl_3 + BH^+ + OH^- \rightarrow NHCl_2 + HOCl + B \qquad (4\text{-}44a)$$

$$NHCl_2 + NCl_3 + 3OH^- \rightarrow N_2 + 2HOCl + 3Cl^- + H_2O \qquad (4\text{-}44b)$$

Cl_2NClOH^- was hypothesized to be an intermediate in both reactions.[110] The mechanism of N_2 formation is still unknown, and it is obvious from the complicated stoichiometry of Equation 4-44b that the reaction proceeds via several steps.

Dechlorination Kinetics

Active chlorine is lost from solution by a variety of oxidation reactions, including the breakpoint process described above, oxidation of organic compounds (described in the following section), and oxidation of reduced inorganic ions of Fe, Mn, and S. The latter reactions tend to be very fast; oxidation rates of organic compounds vary widely. Many studies have been reported on rates of active chlorine loss from natural and drinking waters. Rates of dechlorination generally fit first-order kinetics (proportional to the amount of chlorine residual remaining), but rate constants depend on temperature, pH, chemical forms comprising the chlorine residual, and the nature and concentrations of oxidizable organic and inorganic species.[113,114] Consequently, no general statements can be given, and models of dechlorination rates are inexact at best.

In addition to the normal dechlorination rates caused by the presence of oxidizable compounds, active chlorine can be removed from waters more rapidly by adding specific dechlorinating agents. The most common is sulfite, which reacts very rapidly with free chlorine forms and most chloramines; e.g.,

$$HSO_3^- + HOCl \rightarrow SO_4^{2-} + 2H^+ + Cl^- \qquad (4\text{-}45)$$

The oxidation of sulfite by monochloramine is general-acid assisted and fits the rate expression $-d[NH_2Cl]/dt = k_{HA}[HA][SO_3^{2-}]_T[NH_2Cl]$, where HA is the general acid and $[SO_3^{2-}]_T$ is the sum of sulfite and bisulfite ions.[115] The value of k_{HA} for the uncatalyzed reaction (HA = H_2O) is $7.7\ M^{-1}\ s^{-1}$. Values of k_{HA} increase with the acid strength of HA: H_3O^+ (8×10^{10}), $H_2PO_4^-$ (1.3×10^6), $B(OH)_3$ (5.8×10^3), NH_4^+ (1.7×10^2), all in $M^{-2}\ s^{-1}$.

Finally, chlorine oxidizes nitrite to nitrate:[116]

$$NO_2^- + HOCl \rightarrow NO_3^- + H^+ + Cl^-; \quad k = 7.2\ M^{-1}s^{-1} \qquad (4\text{-}46)$$

This reaction is rapid at neutral pH, but decreases at higher pH, as HOCl dissociates to OCl⁻.[108] Nitrite also accelerates the decay of monochloramine (pseudo-first-order $k = 5.9 \times 10^{-4}\ s^{-1}$ at $[NO_2^-] = 18.8$ mM, pH = 7.5), but the mechanism and products of this reaction are unknown.[117]

4.5.3 Reactions with Organic Compounds

Introduction

Although studies on these reactions date back to the 1920s, much of our information is quite recent. Research on this topic was stimulated by reports in the mid-1970s[118] on the widespread presence of chlorinated organic contaminants in drinking water and the finding that chlorination of water containing natural organic matter and synthetic organic contaminants leads to the formation of these compounds.[119,120] Three categories of reactions occur between organic compounds and chlorine species:

1. Oxidation of organic functional groups
2. Addition to double bonds
3. Electrophilic substitution

In the first category, chlorine species are reduced to chloride ion, and chlorinated organic compounds are not formed. The time scale of chlorine oxidation reactions varies widely depending on the nature of the organic compound. Reactions with organic sulfur and some aromatic and nitrogen heterocyclic rings are very fast ($t_{1/2} < 1$ s). Oxidation of amines occurs on time scales of minutes to hours, and nonselective oxidation of organic matter occurs over minutes to days.[113]

HOCl can add to double bonds to form vicinal halides:[97]

$$>C=C< +ClOH \rightarrow >C_{\underset{Cl^+}{\diagdown}} \; C< +OH^- \rightarrow -\underset{Cl}{C} - \underset{OH}{C} - \qquad (4\text{-}47)$$

This could be significant in waters containing highly unsaturated plant pigments (carotenoids, xanthophylls), but addition reactions are slow unless the double bond

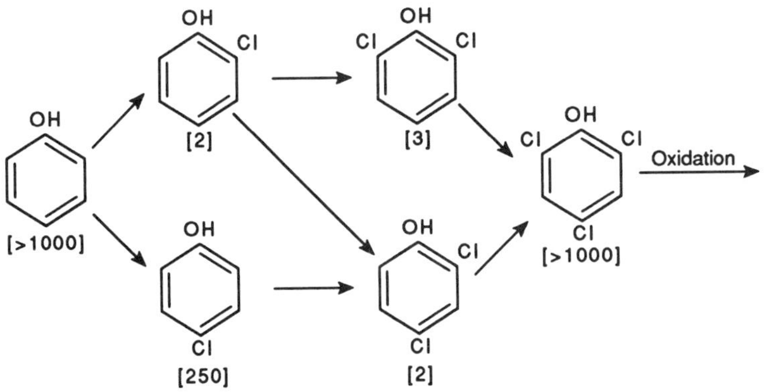

Figure 4-23. General scheme for chlorination of phenol. Numbers next to phenols are their threshold odor concentrations (μg L^{-1}). [From Lee, G.F., in *Principles and Applications of Water Chemistry*, S.D. Faust and J.V. Hunter (Eds.), John Wiley & Sons, New York, 1967. With permission.]

is activated by substituent groups. Little is known about the kinetics of this reaction and its importance in water supplies.

Three types of electrophilic substitution reactions involving chlorine species are of interest in aquatic systems: (1) substitution into aromatic compounds such as chlorophenols; (2) reaction with nitrogenous compounds (especially amines) to form N-chloro-organic compounds; and (3) reaction with natural dissolved organic matter, including aquatic humus, to form chloroform and other trihalomethanes (THMs). All three categories involve HOCl (at near-neutral pH) acting as the electrophilic agent. The Cl atom in HOCl behaves like Cl$^+$, a strong electrophile, and combines with a pair of electrons in the substrate (which acts as a nucleophile). The substrates, products, and factors affecting these reactions are sufficiently different to make the distinction meaningful.

Formation of Chlorophenols and Other Chlorobenzenes

Phenolic compounds are found commonly in water contaminated by industrial activity. Although phenol itself is an undesirable contaminant (it causes taste and/ or toxicity problems), chlorophenols are much worse. For example, the threshold odor concentration of phenol is 1000 μg L^{-1}, but that of 2-chlorophenol is only 2 μg L^{-1} (Figure 4-23).[121] Phenols react readily with free chlorine, and this can lead to severe taste and odor problems in water supplies. Chlorination of phenols was studied as early as 1926 and recognized as an electrophilic attack of HOCl on phenoxide anions.[122] Our current level of understanding is due primarily to the work of Burtschell et al.,[123] who identified the products of reaction and established a reaction sequence, and Lee and Morris,[121,124] who determined the kinetics of each step. As Figure 4-23 shows, stepwise substitution of Cl at the ortho and para (2, 4, and 6) positions on the ring yields several monochloro- and dichloro-intermediates with low threshold odor concentrations, before forming the relatively low-odor compound 2,4,6-trichlorophenol (TCP). Further reaction of TCP with chlorine results in ring oxidation.

Table 4-8. Rate Constants (k_{obs}) for Chlorination of Phenols[a]

pH	Phenol	2-Chloro-phenol	4-Chloro-phenol	2,4-Dichloro-phenol	2,6-Dichloro-phenol	2,4,6-Tri-chlorophenol
5	2.09E2	4.03E2	9.60E1	2.38E2	6.32E2	1.17E2
6	4.82E2	1.04E3	2.98E2	4.02E2	1.34E3	3.46E2
7	2.23E3	3.16E3	8.93E2	1.76E3	4.95E3	5.16E2
8	6.15E3	8.15E3	1.84E3	2.72E3	2.19E3	1.45E2
9	6.14E3	3.21E3	1.54E3	5.48E2	2.96E2	1.34E1
10	2.84E3	4.30E3	4.15E2	6.32E1	3.09E1	9.05E-1
11	4.73E2	4.60E1	4.70E1	6.37	3.12	5.44E-2
12	4.50E1	4.60	4.54	6.36E-1	3.15E-1	1.81E-3

[a] $T = 25°C$; $I = 0.02$; $[Cl^-] = 10^{-1}$ M. From Lee, G.F., in *Principles and Applications of Water Chemistry*, S.D. Faust and J.V. Hunter (Eds.), John Wiley & Sons, New York, 1967.

The reactions conform to simple second-order rate expressions of general form:

$$-d[Cl]_T = k_{obs}[Cl]_T[Ph]_T \qquad (4\text{-}48)$$

Subscript T refers to the total (analytical) concentration of free chlorine (Cl) or phenolic compound (Ph), irrespective of acid-base form. Rate constants for each step were determined as a function of pH (Table 4-8) by spectrophotometric monitoring of free chlorine disappearance, using its rapid reaction with orthotolidine to yield a yellow product.[124] Because the products of each step also react with chlorine, it was necessary to measure rates over short (initial) time periods to isolate each step. The k_{obs} show substantial pH dependency, in each case having a maximum at some intermediate pH (in the range 7 to 9) and decreasing rapidly on either side of the maximum (Figure 4-24). The pH of the maximum k_{obs} is related to the acidity of the phenol — more acidic compounds have their maximum k_{obs} at lower pH. In the presence of Cl^-, chlorination rates increase rapidly with decreasing pH below 5; Cl_2, formation of which is favored at low pH and high $[Cl^-]$, is a more reactive species than is HOCl. Lee and Morris[124] observed increased ring oxidation by Cl_2 but did not quantify this finding.

The effect of pH on reaction rates can be explained in terms of variations in acid-base forms of chlorine and phenols;[122,124] when rates were expressed in terms of HOCl and phenolate (PhO^-) concentrations, $-d[Cl]_T = k_2[HOCl][PhO^-]$, k_2 was found to be nearly constant with pH over the range 6 to 12. This suggests that the actual reactants are HOCl and PhO^-. The product of the rate constant k_2 times the acid dissociation constants of the chlorophenols was found to be approximately constant and equal to $\sim 10^{-4}$; a similar relationship was found for substituted phenols.[122] Substitutents that make a substituted phenol more acidic (i.e., electron-withdrawing species) thus tend to decrease its rate of reaction with electrophilic chlorine.[121]

At equimolar concentrations of phenol and ammonium, the rate of formation of monochloramine at 25°C and pH 8 is about 10^3 times faster than the formation rate of monochlorophenol. Chloramine reacts with phenol to give the same chlorophenols

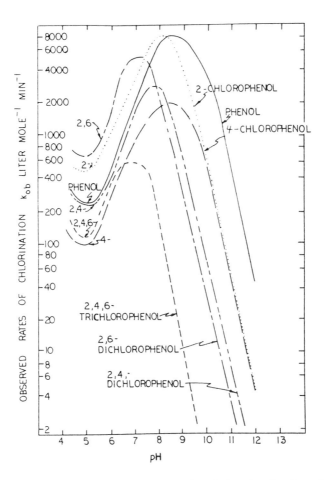

Figure 4-24. Variation in k_{obs} vs. pH for chlorination of phenol and various chlorophenol intermediates. [From Lee, G.F., in *Principles and Applications of Water Chemistry*, S.D. Faust and J.V. Hunter (Eds.), John Wiley & Sons, New York, 1967. With permission.]

as are formed by free chlorine, but the rates with chloramine are *much* slower (days to weeks). Consequently, little chlorophenol will be formed in a water supply when ammonium is present.

Chlorination of benzene, toluene, and similar aromatic compounds requires a stronger electrophile than HOCl and thus does not occur to a significant extent at neutral pH. For example, p-xylene (p-dimethylbenzene) does not react with HOCl at an observable rate, and the order of reactivity of this compound with stronger halogenating agents was found to be $BrCl > Cl_2 \approx Cl_2O \gg Br_2 \gg HOBr$.[125] Because of the superiority of BrCl and H_2OBr^+ as electrophiles, initial rates of bromination of p-xylene are higher than initial rates of chlorination, even when Br^- concentrations are low.[126] Aromatic rings must be activated by substituents like -O- to react with HOCl; the presence of multiple activating groups on the ring accelerates

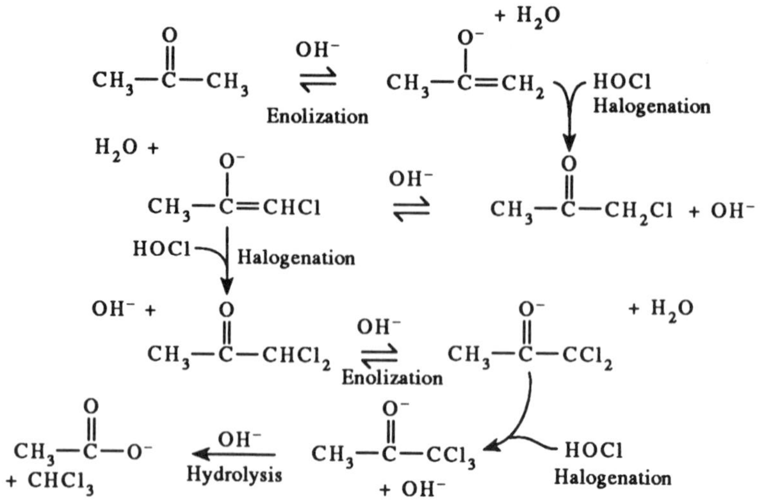

$$CH_3-\overset{O}{\overset{\|}{C}}-CH_3 \underset{Enolization}{\rightleftharpoons}[OH^-] CH_3-\overset{O^-}{\overset{|}{C}}=CH_2 \xrightarrow[Halogenation]{HOCl} +H_2O$$

$$H_2O + CH_3-\overset{O^-}{\overset{|}{C}}=CHCl \underset{}{\rightleftharpoons}[OH^-] CH_3-\overset{O}{\overset{\|}{C}}-CH_2Cl + OH^-$$

$$HOCl \downarrow Halogenation$$

$$OH^- + CH_3-\overset{O}{\overset{\|}{C}}-CHCl_2 \underset{Enolization}{\rightleftharpoons}[OH^-] CH_3-\overset{O^-}{\overset{\|}{C}}-CCl_2 + H_2O$$

$$CH_3-\overset{O}{\overset{\|}{C}}-O^- + CHCl_3 \xleftarrow[Hydrolysis]{OH^-} CH_3-\overset{O^-}{\overset{\|}{C}}-CCl_3 + OH^- \xleftarrow{HOCl \ Halogenation}$$

Figure 4-25. Reaction scheme for production of chloroform from acetone by the classic haloform reaction.

halogenation reactions; for example, resorcinol, *m*-dihydroxybenzene, reacts rapidly and extensively with HOCl, leading to the formation of chloroform.[127]

Formation of Trihalomethanes

The occurrence of trihalomethanes (THMs) at concentrations up to several hundred μg L^{-1} in chlorinated drinking water was first attributed to chlorination of DOM in 1974.[119,120] Much research has been accomplished on factors affecting formation rates of CHCl$_3$ and related THMs and on methods to minimize or prevent their formation or remove them, once formed, from drinking water.[99] Understanding the mechanisms and kinetics of THM formation is made difficult by the fact that the organic substrate is not a single compound or small set of compounds, but a highly complicated mixture of organic structures (including humates, fulvates) that are not precisely definable.

The general mechanism of THM formation from DOM is thought to be described by the classic haloform reaction in which hypohalites react with methyl ketones, successively replacing hydrogens on the carbon alpha to the carbonyl group and forming CHX$_3$ by subsequent hydrolysis (Figure 4-25). This reaction, known since the early 19th century,[128] was invoked by Rook[119] to explain the occurrence of CHCl$_3$ in chlorinated drinking water. The reaction is catalyzed by bases at pH > 5, and it involves an initial step in which proton dissociation from the α-carbon yields an enolate carbanion that is subject to electrophilic attack by HOCl or OCl$^-$. The successive steps of ionization and chlorination are increasingly more rapid, and base-catalyzed hydrolysis of the fully chlorinated methyl group releases CHCl$_3$.[97] The production of mixed THMs, such as bromodichloromethane and dibromochloromethane, and bromoform ($CHBr_3$) is explained by the rapid reaction of free chlorine with traces of bromide ion, which occurs widely in natural water.[97,129] As noted earlier, HOBr is a more effective halogenating agent than is HOCl.

Studies conducted during the past 15 years have shown that simple methyl ketones such as acetone cannot be the primary organic precursors for THMs in drinking water. Chlorination of acetone itself is far too slow to account for much of the THM formation.[130] Trichloroacetone (TCA), the last intermediate before hydrolysis in the classic haloform reaction, has been detected in drinking waters, but its rate of hydrolysis, although faster than the chlorination of acetone, is too slow to account for observed THM concentrations.[131] TCA reacts further with chlorine to give more highly chlorinated intermediates that ultimately yield chloroform, another indication that the simple haloform mechanism does not completely explain the occurrence of THMs in drinking water. Rates of TCA chlorination are comparable to, or faster than, its rate of hydrolysis, but the contribution of both reactions to chloroform formation is still negligible.[131]

As Morris noted,[97] more active structures than those of simple methyl ketones must be involved in THM formation during water chlorination. The organic precursors must have more acidic carbons than those alpha to a single carbonyl group to facilitate the very slow enolization step. Methylene groups located between two carbonyl groups are more acidic, and several model diketones, such as 1,3-cyclohexanedione, have been found to yield chloroform at rapid rates under conditions comparable to those involved in water chlorination.[127] Other organic compounds with acidic carbons that yield carbanions also are important precursors in THM formation. Aquatic humic substances (AH, i.e., fulvic and humic acids) have received the most attention, and 1,3-dihydroxybenzene structures in AH are thought to be the key components.[132,133] Nonetheless, the precise structures of THM precursors remains uncertain, in spite of much research on this subject in the last 15 years.

On the other hand, numerous studies have identified general categories of DOM that give rise to THMs. Aside from AH, these include algal biomass and products of algal metabolism,[134] such as pigments[135] and dissolved organic nitrogen compounds. For example, Scully et al.[136] found that several commercially available proteins (used as models of proteinaceous matter found in natural waters) gave THM yields of about 50 to 1120 µg L^{-1} when solutions containing 20 mg L^{-1} of Cl_2 and 4 to 6 mg L^{-1} of protein were incubated in the dark at pH 7 and 20°C for 5 days. Although these yields seem low, humic acid incubated under the same conditions produced only 135 µg L^{-1} of THM. The structural components of proteins and mechanisms that produce THMs are not well understood.

Enhanced formation of THMs is commonly associated with the occurrence of algal blooms in water supplies, and prevention of algal blooms in raw water supplies is a major strategy for prevention of THM formation in drinking water.[137] Other major control strategies for THM control involve the removal of organic precursors by coagulation or activated carbon prior to chlorination[138] and addition of ammonium during chlorination. Chloramines produce less than 3% of THMs produced by comparable concentrations of free chlorine species,[139] but they do produce signifi-

Table 4-9. Rate Constants for Chlorination of Organic Amines[a]

Compound	k ($M^{-1} s^{-1}$)	A ($M^{-1} s^{-1}$)	E_{act} (kJ mol^{-1})
Methylamine	3.2E8	7.8E9	7.9
Dimethylamine	1.6E8	1.0E11	15.9
Diethylamine	7.2E7	2.4E11	20.1
Sarcosine	1.5E8	1.3E11	16.7
Glycine	8.4E7	3.1E10	14.6
Alanine	6.8E7	2.5E10	14.6
Serine	3.3E7	4.7E10	18.0
Glutamic acid	5.2E7	1.8E9	8.8
Glycylglycine	7.7E6	1.9E8	7.9
Glycine ethyl ester	7.2E6	5.8E8	10.9

From Isaac, R. A. and J. C. Morris, in *Water Chlorination*, Vol. 4, R. L. Jolley et al. (Eds.), Ann Arbor Science, Ann Arbor, MI, 1983, p. 49. With permission.

[a] k at 25°C; A = Arrhenius pre-exponential factor.

cant quantities of nonvolatile, high-weight chlorinated organic material in the presence of AH (fulvic acids).[140]

Reactions with Amines and Amides

Free chlorine reacts with simple amines to form N-chloroamines in much the same way as it does with ammonium. The kinetics and mechanisms of these reactions have been elucidated primarily by Morris and students, and Table 4-9 summarizes kinetic data for common N-organic compounds. For example, formation of CH_3NHCl from HOCl and methylamine is rapid (about 60 times that of NH_2Cl).[106] Formation of CH_3NCl_2 from CH_3NHCl and HOCl is less than 10^{-5} as fast as the formation of CH_3NHCl, but is still slightly faster than formation of $NHCl_2$. In turn, the disproportionation reaction, $2CH_3NHCl \rightarrow CH_3NCl_2 + CH_3NH_2$, is less than 10^{-5} as fast as the reaction of CH_3NHCl with HOCl. As was found for NH_3, formation of the dichloro product from the monochloro species is subject to general- and specific-acid catalysis, but chlorination reactions of the stronger bases, NH_3 and CH_3NH_2, are not.

A good correlation is found between the rate constant for N-chlorination and pK_b, the base dissociation constant for the amine (Figure 4-26). This is a linear free-energy relationship, and it agrees with our general understanding that these are reactions of electrophilic Cl species with nucleophilic nitrogen species and that nucleophilicity toward Cl^+ is correlated with basicity toward H^+.[97] Amides, which are much less basic than amines, react much more slowly with chlorine. For many years it was thought that peptide-N groups, which are very weak bases, are unreactive toward chlorine, as Friend[141] concluded was the case for glycylglycylglycine, and Margerum et al.[98] reported for glycylglycine. However, based on the LFER in Figure 4-26 and an assumed K_b of 0.1 for protonation of the peptide N:

$$RCONHR' + H^+ \rightleftharpoons RCO\overset{+}{N}H_2R'$$

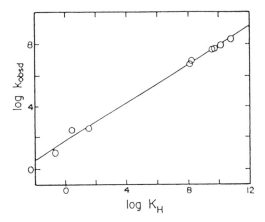

Figure 4-26. Rate constants for HOCl chlorination of organic amines and chloramines are correlated with K_H, the protonation constant (proportional to base strength) of the amines, yielding a Bronsted-type linear free-energy relationship. $I = 0.5$ (NaClO$_4$); T = 25°C. [From Margerum, D.W., et al., in *Organometals and Organometalloids*, F.E. Brickman and J.M. Bellama (Eds.), Symp. Ser. Vol. 82, American Chemical Society, Washington, D.C., p. 278. With permission.]

Ayotte and Gray[142] estimated the second-order rate constant for chlorination of peptide N as 13 M^{-1} s^{-1}, which is relatively small, but not negligible. Using acetylglycine as a model compound for peptides (it has one amide group and no amino N), they found that the amide N combined with free chlorine under acidic conditions to yield a stable compound on a time scale of days. When the chlorinated acidic solutions were made basic, free chlorine was released.

Reactions of chlorine with amino acids typically produce mono- and dichloramines and yield classic breakpoint curves, but rates of decompositon of organic chloramines and bromamines are not well known. Chlorine also reacts with some amino acids to form aldehydes and nitriles; for example, alanine reacts with Cl$_2$ to produce acetaldehyde and acetonitrile.[143] At least one aromatic amine, methylbenzylamine, reacts with chlorine to form THMs.[144]

In contrast to ammonium and simple amines, sulfamic acid (NH$_2$SO$_3$H) reacts more slowly with HOCl at pH > ~4 than does its monochloro product, NHClSO$_3^-$:[106]

$$NH_2SO_3^- + HOCl \rightarrow NHClSO_3^- + H_2O,$$

$$k_1 = 6.5 \times 10^2 \ M^{-1} \ s^{-1}$$

(4-49a)

$$NHClSO_3^- + HOCl \rightarrow NCl_2SO_3^- + H_2O,$$

$$k_2 = 6.5 \times 10^{-2} / \{H^+\} \ M^{-2} \ s^{-1}$$

(4-49b)

The rate constant for the second reaction applies up to about pH 6.3, at which the rate of the second step is about 200 times that of the first step. As is true for other amines, the first step is not acid catalyzed, but the second is. Although the second step is more rapid than the first, $NCl_2SO_3^-$ reacts with excess sulfamic acid to redistribute the chlorine primarily as $NHClSO_3^-$:

$$NCl_2SO_3^- + NH_2SO_3^- \rightarrow 2NHClSO_3^-$$

$$k_{obs} = 2.0 \times 10^{-4}(1 + 6.5 \times 10^3\{H^+\} + 7.7 \times 10^8\{OH^-\})$$

(4-49c)

As the form of k_{obs} implies, the reaction is both acid and base catalyzed, as well as uncatalyzed; a graph of k_{obs} vs. pH is similar to Figure 4-6a.

Reactions of Chlorinated Organic Compounds with Sulfite (Dechlorination)

Many organic chloramines react rapidly with sulfite (time scale of seconds to minutes), and this is a common agent for rapid removal of the chlorine residual from water. However, chlorinated amino-N groups of certain model peptides react with sulfite more slowly (time scales of hours) under environmentally relevant conditions.[145]

Some of the chlorinated byproducts formed when Cl_2 reacts with natural humic substances are of concern because they are mutagens; addition of sulfite to chlorinated water reduces its mutagenic activity and destroys some of these compounds by nucleophilic dehalogenation.[146] Most mutagens are electrophiles; consequently, it is not surprising that they can be destroyed by a nucleophile like sulfite. (Recall that compounds containing sulfur in its lower oxidation states generally are nucleophilic agents.) The mechanism of reaction is attack by sulfite of the electrophilic carbon to which the chlorine is attached, usually resulting in displacement of –Cl by –H (i.e., reductive dehalogenation). Sulfite is lost in the process, ultimately being oxidized to sulfate, and the overall reaction must be more than a simple one-step process. Hydrolysis is an alternative loss mechanism for these compounds.

Croue and Reckhow[146] measured loss rates of some chlorinated compounds produced during drinking water chlorination (Table 4-10). Reactions were first order in the compound and in sulfite ion, and second-order rate constants for the most reactive of the compounds were in the range 10 to 100 M^{-1} s^{-1}. Rates increased with pH over the range 6.1 to 8.5; this was attributed to increased ionization of HSO_3^- to SO_3^{2-} ($pK_2 = 7.18$). Reaction with bisulfite, a much weaker nucleophile than sulfite, was found to be negligible. Based-catalyzed hydrolysis is competitive with sulfite-induced dehalogenation for some of the compounds (e.g., trichloroacetonitrile); others (1,1,1-trichloropropanone and dichloroacetonitrile) undergo hydrolysis, but react negligibly with sulfite. The important mutagen, MX (3-chloro-4-(dichloromethyl)-5-hydroxy-2(5H)-furanone), had unexpectedly low reactivity. This

Table 4-10. Pseudo-First-Order Rate Constants for Decomposition of Some Chlorination Byproducts of Humic Material[a]

Compound	Hydrolysis k_h (10^{-6} s^{-1})		Reductive dehalogenation by sulfite k_s (10^{-6} s^{-1})	
	pH 7.2	pH 8.5	pH 7.2	pH 8.5
CCl_3NO_2 (chloropicrin)	12.8	9.5	1410.0	1950.0
CCl_3CN (trichloroacetonitrile)	17.4	174.0	620.0	950.0
$CHCl_2CN$ (dichloroacetonitrile)	3.1	4.2	0.0	0.5
$CHBr_2CN$ (dibromoacetonitrile)	0[b]	0[b]	1081.0	1310.0
CCl_3COCH_3 (1,1,1-trichloropropanone)	—	73.0	—	0.0
MX (3-chloro-4-(dichloromethyl)-5-hydroxy-2(5H)-furanone)	0.9	—	22.0	—

[a] Summarized from Croue, J.-P. and D. Reckhow, *Environ. Sci. Technol.*, 23, 1412 (1989). Values at 25°C in dark, 5-100 μg L^{-1} of compound, initial sulfite = 25 μM.
[b] Small negative values reported by Croue and Reckhow are not physically meaningful.

compound accounts for a significant fraction (22–44%) of the mutagenic activity of chlorinated drinking water,[147] and sulfite treatment is effective in decreasing the mutagenic activity of such waters by 40 to 80%.[148] Croue and Reckhow concluded that loss of other (unidentified) chlorinated compounds is responsible for the decrease in mutagenicity when water is dechlorinated with sulfite.

4.6 INORGANIC REDOX REACTIONS

4.6.1 Introduction

In this section we describe the kinetics of redox reactions of inorganic sulfur compounds and some heavy metals: primarily iron, manganese, and copper. The emphasis is on abiotic autoxidation (oxygenation) reactions, but oxidation by related oxidants (O_3, H_2O_2) is discussed where information is available. Recent studies on the reduction of Fe[III] and Mn[IV] oxides are reviewed in Section 4.8.4. Microbial oxidation of Fe^{2+} is described in Section 6.4.2, and photochemical aspects of Fe and Mn redox behavior are discussed in Sections 8.4.4 and 8.6.1. For each redox reaction we examine six topics: (1) form of the rate equation, (2) magnitude of k, (3) effect of pH, (4) effects of other solution conditions, (5) evidence for autocatalysis, and (6) mechanistic aspects.

Although the autoxidation kinetics of reduced S, Fe, and Mn differ in many details, there are several general features common to all these reactions. The reactions tend to be mechanistically complicated. Rates are sensitive to a wide variety of solution conditions; homogeneous and heterogeneous catalysis is common; and the oxidation products may vary with reaction conditions. Oxidation rates are particularly sensitive to (i.e., increase with) pH, albeit generally in a nonlinear fashion. Studies conducted within the past decade have demonstrated that the pH dependency in most cases is caused by changing acid-base speciation of the reductant, the various protonated forms of which exhibit different susceptibility to reaction with O_2. The observed rate constants at any pH thus can be expressed as the sum of the products of the intrinsic rate constants for each acid-base species times the fraction, α_i, of reductant in each of its various protonated forms:

$$k_{obs} = \sum_{i=0}^{n} k_i \alpha_i \qquad (4\text{-}50)$$

where i represents the degree of deprotonation; k_0 and α_0 refer to the fully protonated species, and k_n and α_n refer to the fully deprotonated form. The α's can be expressed as functions of acid dissociation constants and $\{H^+\}$ only so that they can be evaluated readily as a function of pH. Derivation of α's for acid-base systems is described in aquatic-chemistry texts[149-151] and is illustrated for sulfide species in the following section.

4.6.2 Oxidation of Sulfide, Pyrite, and Related Compounds

Hydrogen sulfide (H_2S) is formed from the decomposition of S-containing proteins and by microbial reduction of sulfate under anaerobic conditions. H_2S is a noxious, toxic gas and a weak acid ($pK_1 = 7.0$). It frequently occurs in groundwater to which it imparts an unpleasant taste and odor, and it is released to surface waters by diffusion or ebullition from anoxic sediments. Because of the water quality significance of H_2S, many studies have been performed on the kinetics of its oxidation. Chen and Morris[152,153] were the first to apply modern techniques to evaluate H_2S oxidation under natural water conditions. As subsequent workers also reported, they found that oxidation rates decreased rapidly below pK_1; they concluded that HS^- is the form that reacts with O_2 and that H_2S is unreactive (or much less reactive). However, they did not make measurements below pH 6.0. They found a complicated and bizarre pattern above pK_1: a peak near pH 8.0, a sharp minimum near 9.0, another peak near pH 11, and a slower decline up to pH 13. More recent studies have not verified this behavior,[154] and inhibition by heavy metal contaminants in pH buffers has been offered as an explanation.[154,155]

Chen and Morris also reported that the reaction was fractional order in both reactants ($n=0.56$ for O_2 and 1.34 for S^{-II}) and interpreted this as evidence for a chain mechanism. They found other evidence to support this hypothesis: the existence of an induction period before O_2 consumption began; inhibition or catalysis by various heavy metals; and varying stoichiometry and products depending on pH, reactant concentrations, and the presence of impurities in solution. At high ratios of sulfide to O_2, elemental sulfur precipitated, while a low ratio resulted in direct oxidation of sulfide to thiosulfate. Polysulfide species (S_x^{2-}, where x = 2 to 5) formed as intermediates, especially near pH 7. Pyrite, S_2^{2-}, the simplest polysulfide, is formed by reaction of HS^- with elemental sulfur: $HS^- + S \rightarrow S_2^{2-} + H^+$. Higher polysulfides are formed by sequential addition of an S atom to the next-lower polysulfide ion.

Not all of the above results have been verified by more recent studies, but they have documented the complexity of the sulfide-O_2 reaction and the difficulty in obtaining accurate results. The reaction's complexity is exemplified by the variety of sulfur species that have been identified as products in different studies: $S°$, $S_2O_3^{2-}$, SO_3^{2-}, $S_4O_6^{2-}$, and SO_4^{2-}. Some of the problems noted by Chen and Morris (e.g., precipitation of elemental sulfur) were avoided in more recent studies by using lower initial concentrations of sulfide. Chen and Morris used a range of 0.5 to 2.0 $\times 10^{-4}$ M, and under some conditions the rate of $S°$ formation was sufficient to exceed the solubility of elemental S. In contrast, Millero et al.[154] used initial sulfide concentrations of ~25 μM and found no precipitation, but they did find long induction periods at low temperatures.

Recent studies have not verified the fractional order for the sulfide-O_2 reactants. Chen and Morris based their conclusion on experiments in which they varied initial concentrations and measured initial velocities. Other workers typically determined reactant order by plotting temporal data from a given experiment, e.g., by observing a straight-line fit of $log[S^{-II}]$ vs. time under pseudo-first-order conditions ($[O_2]$ "fixed" at a much higher value than the initial $[S^{-II}]$). The initial-velocity method normally is considered more reliable because it avoids complications resulting from build-up of inhibitors and competing pathways over time, but this may not hold for reactions that have an induction period. Some subjectivity may be involved in determining the end of the induction period, and this may cause errors in estimating initial rates.

The most thorough recent study of sulfide autoxidation is that of Millero et al.,[154] who evaluated effects of temperature, pH, and ionic strength on rates of the reaction. The pseudo-first-order rate constant for HS^- oxidation at pH 8 increased by almost three orders of magnitude over the temperature range 5 to 65°C, and E_{act} for oxidation of HS^- was found to be 53.5 kJ mol^{-1} (freshwater) and 63.5 kJ mol^{-1} (seawater). From more limited data at pH 4, E_{act} was estimated to be 43.5 kJ mol^{-1} for H_2S oxidation. The effect of ionic strength (I) was determined at pH 8 by measuring HS^- oxidation in NaCl solutions at I up to 2.5. The data fit a Bronsted plot and Equation 3-68b with a slope of 0.5:

$$\log k = 2.33 + 0.50\sqrt{I} \qquad (4\text{-}51)$$

Recall that the slope of the Bronsted equation is roughly equal to the product of the ionic charges of the reactants. In this case, if HS^- is assumed to be one of the reactants, the other must have a charge of $-1/2$. This, of course, does not make physical sense. Although O_2 has a charge of zero, other intermediates in the reduction of O_2 (such as O_2^-) could be involved in the rate-limiting step; however, it is difficult to imagine a chemical species with a charge of $1/2$! This supports the concern expressed by Moore and Pearson[156] regarding the use of the Bronsted equation to infer reactant charges for complicated, multistep reactions.

Millero et al. measured the oxidation of sulfide over a broad range of pH: 0.95 to 12.0. Although the data are somewhat scattered (Figure 4-27), they suggest that the rate is independent of pH except in the range where S^{-II} speciation is changing from H_2S to HS^- (i.e., in the vicinity of pK_1, pH ~ 6 to 8). The shape of the curve in Figure 4-27 does not support a catalytic role for H^+ or OH^-, but implies that HS^- is

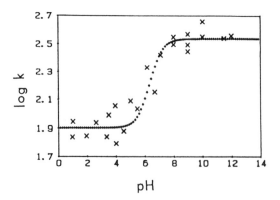

Figure 4-27. Effect of pH on second-order rate constant k (kg H_2O mol^{-1} h^{-1}) for oxygenation of S^{-II} in water at 55°C. [From Millero, F.J., et al., *Environ. Sci. Technol.*, 21, 439 (1987). With permission.]

about four times more reactive toward O_2 than is H_2S. This is a much smaller difference than Chen and Morris observed and also much smaller than that observed for S^{-II} oxidation by H_2O_2.[157] At low pH, $k_{obs} = k_o$ (the rate constant for H_2S); and at high pH, $k_{obs} = k_1$ (the rate constant for HS$^-$); but at intermediate pH values both contribute to k_{obs}. In general,

$$k_{obs} = k_o \alpha_o + k_1 \alpha_1 \qquad (4\text{-}52a)$$

where
$$\alpha_o = [H_2S]/[S^{-II}]_T = [H_2S]/([H_2S]+[HS^-]) \qquad (4\text{-}52b)$$

$$\alpha_1 = [HS^-]/[S^{-II}]_T = [HS^-]/([H_2S]+[HS^-]) \qquad (4\text{-}52c)$$

These α's can be written in terms of {H$^+$} and the first acid-dissociation constant of H_2S, $K_1 = \{H^+\}[HS^-]/[H_2S]$, as follows. Solving K_1 for [HS$^-$] yields [HS$^-$] = $K_1[H_2S]/\{H^+\}$. Substituting this for [HS$^-$] in Equation 4-52b yields

$$\alpha_o = \frac{[H_2S]}{[H_2S]+K_1[H_2S]/\{H^+\}} = \frac{1}{1+K_1/\{H^+\}} \qquad (4\text{-}53a)$$

or
$$\alpha_o = \{H^+\}/(\{H^+\}+K_1) \qquad (4\text{-}53b)$$

Similarly, we can solve K_1 in terms of [H_2S], substitute the result into Equation 4-52c, and simplify to obtain

$$\alpha_1 = 1/(1+\{H^+\}/K_1) \qquad (4\text{-}53c)$$

or
$$\alpha_1 = K_1/(K_1+\{H^+\}) \qquad (4\text{-}53d)$$

Note that the denominators of Equations 4-53b and 4-53d are the same; only the numerators change. This is generally true for α expressions. Equation 4-52a can be arranged into a linear form for plotting purposes by dividing all terms by α_o: $k_{obs}/\alpha_o = k_o + k_1(\alpha_1/\alpha_o)$. From the ratio of Equation 4-53d to 4-53b, it is apparent that $\alpha_1/\alpha_o = K_1/\{H^+\}$. Thus Equation 4-52a becomes

$$k_{obs}/\alpha_o = k_o + k_1 K_1 /\{H^+\} \qquad (4\text{-}54)$$

and a plot of k_{obs}/α_o vs. $1/\{H^+\}$ should be a straight line with slope $= k_1 K_1$ and y-intercept $= k_o$. Millero et al.[154] found $k_o = 80 \pm 17\ M^{-1}\,h^{-1}$ and $k_1 = 344 \pm 7\ M^{-1}\,h^{-1}$ at 55°C (~11 and 48 $M^{-1}\,h^{-1}$ at 25°C) by this approach.

A wide range of rate constants for sulfide autoxidation are found in the literature, and the disparities among published values remain even when results are normalized with respect to temperature and pH. Millero et al. tabulated data from nine studies in terms of half-times for sulfide oxidation at 25°C and pH 8.0 under air-saturated (pseudo-first-order) conditions. They found $t_{1/2}$ ranged from about 18 to 50 h for freshwater studies and ~4 to 65 h (mean = ~30 h) for seawater studies. The mean value is close to that measured by Millero et al.[154] and O'Brien and Birkner,[158] and probably is the best estimate available. Even shorter half-lives were reported in a few studies using electrode methods to measure sulfide loss, but Millero et al. considered them to be experimental artifacts. These wide ranges hint at the difficulties involved in obtaining accurate data for this reaction.

The oxidation of sulfide by O_2 is too slow to be relied on in drinking water treatment, and other oxidants such as H_2O_2 and O_3, are more effective for such purposes. Hoffmann[157] studied the kinetics of sulfide oxidation by H_2O_2 over the pH range 3 to 8. Under alkaline conditions, the product is sulfate (> 99% at pH 8.5), but under neutral and acidic conditions, the principal product is elemental sulfur, which is formed in a colloidal state and consists mainly of S_8, some S_6 (both of these are ring forms), and higher-molecular-weight forms. From 25 to 40% of the product was sulfate in the pH range 4.5 to 6.7. The reaction rate increased with pH over the range 5 to 8 (similar to the autoxidation reaction), and a two-term rate expression fit the data:

$$-d[S^{-II}]_T/dt = k_o[H_2S][H_2O_2] + k_1 K_1 [H_2S][H_2O_2]/\{H^+\} \qquad (4\text{-}55a)$$

or $\quad -d[S^{-II}]_T/dt = k_o[H_2S][H_2O_2] + k_1[HS^-][H_2O_2] \qquad (4\text{-}55b)$

where $k_o = 29\ M^{-1}\,h^{-1}$, and $k_1 = 1740\ M^{-1}\,h^{-1}$. Thus, HS^- is much more reactive toward H_2O_2 (factor of ~60 times) than is H_2S. Not surprisingly, S^{-II} is oxidized by H_2O_2 much more rapidly (~35 times for HS^-) than by O_2.

Spectrophotometric evidence was found for the formation of polysulfide intermediates; solutions turned yellow almost immediately upon addition of H_2O_2, but they eventually cleared under alkaline conditions (sulfate as product) or became cloudy under acidic conditions (colloidal sulfur as product). A proposed mechanism for oxidation under acidic conditions[157] involves nucleophilic attack of H_2O_2 to form a reactive intermediate, HSOH (the rate-limiting step). HSOH reacts further to form a series of polysulfides and ultimately S_8:

$$H_2S + H_2O_2 \rightarrow HSOH + H_2O$$

$$H_2S + HSOH \rightarrow H_2S_2 + H_2O$$

$$H_2S_2 = HS_2^- + H^+; \quad pK_{a1} = 4.7$$

$$HS_2^- + HSOH \rightarrow HS_3^- + H_2O$$

$$HS_3^- + HSOH \rightarrow HS_4^- + H_2O$$

$$\vdots \qquad \vdots \qquad \vdots \qquad \vdots$$

$$HS_x^- + HSOH \rightarrow HS_9^- + H_2O$$

$$HS_9^- \rightarrow S_x + HS^-$$

(4-56)

4.6.3 Oxidation of Sulfur Dioxide in Aqueous Solution

Sulfuric acid is the principal component of acid precipitation. It accounts for about 65 to 70% of the H^+ in rain and is derived from SO_2 emitted during the burning of fossil fuels, especially coal. Although the conversion of SO_2 to sulfuric acid occurs in the atmosphere, the chemical reactions occur primarily in the liquid phase (in liquid aerosols and cloud droplets).[159,160] Indirect evidence for this is available from field studies on SO_2 disappearance from plumes emitted by point sources such as power plant stacks. Loss rates generally have been found to be much higher when the relative humidity is high,[161] suggesting that absorption of SO_2 into atmospheric water droplets is an important loss mechanism.

Depending on pH, S^{IV} exists in three forms in aqueous solution: aquated SO_2 ($H_2O \cdot SO_2$), bisulfite, HSO_3^-, the reactive form of which is thought to be the tautomer $HO\text{-}SO_2^-$,[162] and sulfite, SO_3^{2-}. Aquated SO_2 is a weak diprotic acid: $pK_1 = 1.9$; $pK_2 = 7.2$ (note that sulfurous acid, H_2SO_3, is not considered to be a significant species).

S^{IV} can be oxidized to S^{VI} by several oxidants in liquid phases, including O_2 (autoxidation), O_3, H_2O_2, $OH\cdot$ radicals, and nitrous acid. Although O_3 is present in the atmosphere at much lower concentrations than O_2, O_3 is more soluble in water ($K_H = \sim 10^{-2}\ M$ atm^{-1} for O_3 and $\sim 3 \times 10^{-4}\ M$ atm^{-1} for O_2, at 25°C) and also is a much more reactive species. Ozone is present in the troposphere primarily as the result of photochemical processes (it is a major product of smog reactions), and oxidation of SO_2 by O_3 is the major pathway for H_2SO_4 production in clouds.[160,162-165]

Each form of S^{IV} reacts with O_3 by an independent mechanism, but in general, they can be viewed as nucleophilic attacks by S^{IV} species on electrophilic O_3. The net result is an overall rate expression that is the sum of the second order rate expressions for each form:

Table 4-11. Rate Equation, α Expressions, and Values of Constants for Oxidation on S^{IV} by O_3

1. Equilibrium expressions for S^{IV} species

$$K_1 = \frac{[H^+][HSO_3^-]}{[SO_2 \cdot H_2O]}; \quad K_2 = \frac{[H^+][SO_3^{2-}]}{[HSO_3^-]}$$

$$[S^{IV}]_T = [SO_2 \cdot H_2O] + [HSO_3^-] + [SO_3^{2-}]$$

$$
\begin{aligned}
[SO_2 \cdot H_2O] &= \alpha_0[S^{IV}]_T \\
[HSO_3^-] &= \alpha_1[S^{IV}]_T \\
[SO_3^{2-}] &= \alpha_2[S^{IV}]_T
\end{aligned}
$$

$$\alpha_0 + \alpha_1 + \alpha_2 = 1$$

where

$$\alpha_0 = \frac{[SO_2(aq)]}{[SO_2 \cdot H_2O] + [HSO_3^-] + [SO_3^{2-}]} = \frac{[H^+]^2}{[H^+]^2 + K_1[H^+] + K_1K_2}$$

$$\alpha_1 = \frac{[HSO_3^-]}{[SO_2 \cdot H_2O] + [HSO_3^-] + [SO_3^{2-}]} = \frac{K_1[H^+]}{[H^+]^2 + K_1[H^+] + K_1K_2}$$

$$\alpha_2 = \frac{[SO_3^{2-}]}{[SO_2 \cdot H_2O] + [HSO_3^-] + [SO_3^{2-}]} = \frac{K_1K_2}{[H^+]^2 + K_1[H^+] + K_1K_2}$$

2. Rate equation

$$-d[S^{IV}]/dt = k_0[SO_2 \cdot H_2O][O_3] + k_1[HSO_3^-][O_3] + k_2[SO_3^{2-}][O_3]$$

or $$-d[S^{IV}]/dt = (k_0\alpha_0 + k_1\alpha_1 + k_2\alpha_2)[S^{IV}]_T[O_3]$$

3. Constants[a]

$$
\begin{aligned}
pK_1 &= 1.89;\ pK_2 = 7.22 \\
k_0 &= 2.4(\pm1.1) \times 10^4\ M^{-1}\ s^{-1} \\
k_1 &= 3.7(\pm0.7) \times 10^5\ M^{-1}\ s^{-1} \\
k_2 &= 1.5(\pm0.6) \times 10^9\ M^{-1}\ s^{-1} \\
E_{act}(k_1) &= 46\ kJ\ mol^{-1};\ E_{act}(k_2) = 43.9\ kJ\ mol^{-1}
\end{aligned}
$$

[a] Values recommended by Hoffmann[162] as "best estimates" from five sets of published rate constants; E_{act} values from Erickson, R.E., et al., *Atmos. Environ.*, 11, 813, (1977).

$$-d[S^{IV}]/dt = (k_0\alpha_0 + k_1\alpha_1 + k_2\alpha_2)[S^{IV}]_T[O_3] \qquad (4\text{-}57)$$

where α_0, α_1, and α_2 are the fractions of S^{IV} present as $H_2O \cdot SO_2$, HSO_3^-, and SO_3^{2-}, respectively. The α's are functions of K_1, K_2, and $\{H^+\}$ only (see Table 4-11). The rate coefficients in Equation 4-57 have been estimated by several workers, and a critical review by Hoffmann[162] concluded that the best estimates for the k's and E_{act} are those given in Table 4-11. It is apparent from the k's that reactivity increases greatly with increasing deprotonation of S^{IV}, and consequently, S^{IV} oxidation

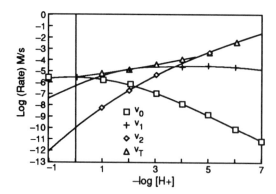

Figure 4-28.   Rate of S^{IV} oxidation by O_3 vs. pH for multiterm rate expression; v_0–v_2 correspond to the three terms of Equation 4-57 (e.g., $v_0 = k_0\alpha_0[S^{IV}]_T[O_3]$), representing contributions from $H_2O \cdot SO_2$, HSO_3^-, and SO_3^{2-}, respectively. v_T represents sum of all three terms. Calculations based on closed system (non-gas-phase interaction considered. [From Hoffmann, M.R., *Atmos. Environ.*, 20, 1145 (1986). With permission.]

increases significantly with pH — almost four orders of magnitude over the pH range 1 to 7 (Figure 4-28). This agrees with the accepted notion that nucleophilic reactivity increases with species basicity. Despite the fact that SO_3^{2-} is the most reactive S^{IV} form, it is not necessarily the most important species in the formation of atmospheric sulfuric acid because it is such a small fraction of S^{IV}_T at the low pH values that may occur in cloud droplets. As Figure 4-28 shows, the third term in Equation 4-57 (representing SO_3^{2-} oxidation) predominates only at pH > 4; the second term (representing bisulfite oxidation) is the predominant term in the pH range 2 to 4. In unbuffered cloud droplets, oxidation of S^{IV} to form sulfuric acid rapidly will depress the pH to the point where the third term becomes negligible, and this implies the reaction exhibits "self-regulating" tendencies (negative feedback).

The situation is much more complicated, however, in the ambient atmosphere, where other oxidants might control the fate of S^{IV}, and basic materials may control the pH of aerosols and cloud droplets. The former include H_2O_2, whose reaction with S^{IV} is pH independent above pH 2,[164] and O_2 catalyzed by Fe^{III} or Mn^{II}.[164,166] The most important pH-controlling atmospheric base is NH_3. Behra et al.[164] recently showed that this important species codetermines the pH of atmospheric water droplets and that absorption of NH_3 from the gas phase counteracts the acidity produced by O_3 oxidation of S^{IV}. In cases where the initial $[NH_3]$ is $\geq 2[SO_2]$, absorption of NH_3 maintains aerosol pH > 5.5, and oxidation by O_3 proceeds rapidly until the SO_2 is exhausted, producing an ammonium sulfate aerosol, $(NH_4)_2SO_4$. When SO_2 is present in excess (relative to NH_3), decreasing aerosol pH during oxidation slows the production of sulfuric acid, but acidic droplets are produced nonetheless. Modeling of this system is complicated by the number of rate and equilibrium equations that must be considered; acid-base reactions and gas-water equilibria involving S^{IV} and NH_3 species at a given pH are instantaneous compared with the oxidation reactions and are treated as equilibrium conditions. Behra et al.[164] used a "reaction progress" numerical approach in which the results from Runge-Kutta solution of the rate equations at each time step were used as input to a chemical equilibrium program to compute pH and

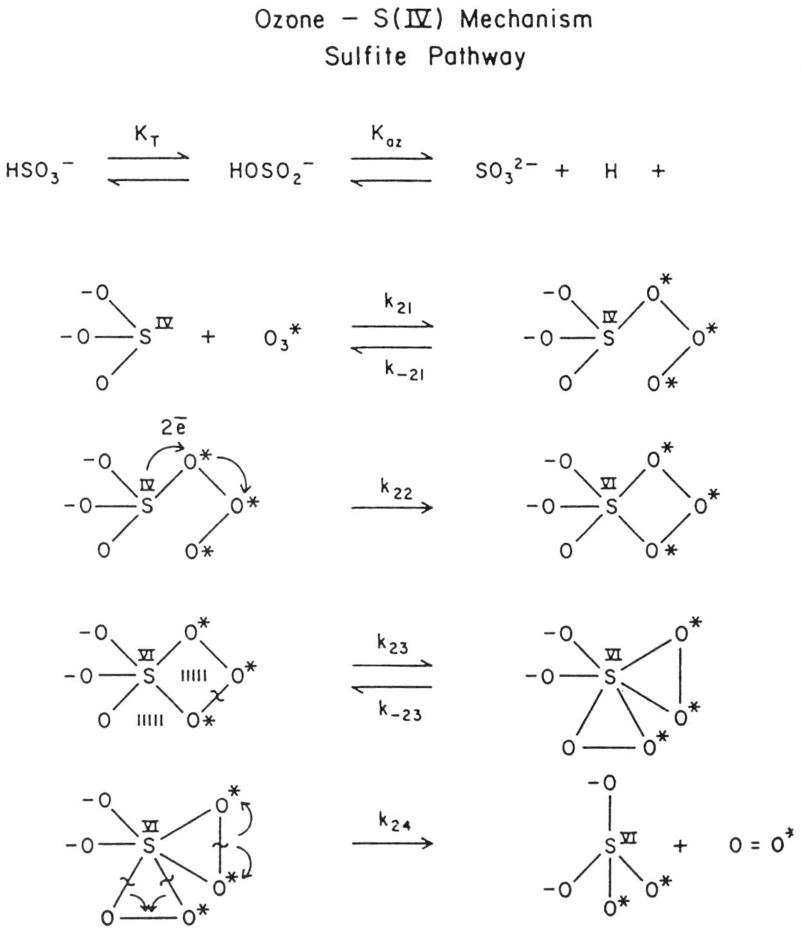

Figure 4-29. Proposed mechanism for oxidation of sulfite ion (SO_3^{2-}) by ozone involves nucleophilic attack by O_3 on S^{IV} center. Asterisks indicate labeled oxygen atoms; the mechanism is consistent with ^{18}O isotope-labeling experiments. [From Hoffmann, M.R., *Atmos. Environ.*, 20, 1145 (1986). With permission.]

acid-base speciation. The latter values then were used as starting conditions for the next iteration of the rate equations.

The net reaction of S^{IV} with ozone can be written as

$$H_nSO_3^{(n-2)} + O_3 \rightarrow nH^+ + SO_4^{2-} + O_2 \tag{4-58}$$

which suggests that one oxygen atom is transferred from O_3 to the reactant sulfur species. However, experiments[167] with ^{18}O-labeled O_3 showed that *two* oxygen atoms are transferred from O_3 to the product sulfate. A mechanism for sulfite oxidation by O_3 consistent with the rate equation and isotope-labeling results is illustrated in Figure 4-29; similar mechanisms for the other two S^{IV} species were described by Hoffmann.[162]

Autoxidation of aquated SO_2 is very slow in the absence of catalysts, but several transition metal ions, including Mn^{2+}, Fe^{3+}, Co^{2+}/Co^{3+}, and Cu^{2+} are effective in catalyzing the reaction. Elegant studies have been reported on the thermodynamic stability of Fe^{III}-S^{IV} complexes and their role in catalyzing the autoxidation.[166,168] Detailed description of this complicated system is beyond our scope, but in brief it appears that concentrations of soluble Fe^{III} in aerosols of the ambient atmosphere are too low to control either the equilibrium speciation of S^{IV} or its oxidation to S^{VI}.[168] Boyce et al.[169] studied the catalytic effect of soluble metal-phthalocyanine complexes on S^{IV} autoxidation and found that Co^{II}-4,4′,4″,4‴-tetrasulfophthalocyanine (Co^{II}-TSP^{2-}) is an effective catalyst at concentrations of 10^{-8} to 10^{-6} M. Other transition metal-TSP complexes were less effective, and differences in catalytic activity were related to the ability of the square-planar metal-TSP complexes to bind O_2 reversibly. Although these reactions are not relevant to the fate of S^{IV} in the environment, they are of potential interest for the removal of SO_2 from stack gases. It is interesting to note that the rate data for the Co-TSP-catalyzed reaction were fit to a bisubstrate enzyme kinetic model. This reaction is light sensitive, but reactions catalyzed by other metal-TSP complexes were not.

4.6.4 Chemical and Geochemical Characteristics of Iron and Manganese

Iron and manganese occupy adjacent positions in the periodic table, and their biogeochemical cycles are qualititatively similar. Table 4-12 summarizes basic information on their chemistry and geology that illustrates their similarities, as well as important differences between them. Both elements have multiple oxidation states and participate in complicated geocycles involving numerous chemical, photochemical, and microbial processes. In both cases the thermodynamic stability of the reduced states increases with decreasing pH (Figure 4-30), and oxidation rates of both elements are strongly pH dependent (decreasing at lower pH). (This correlation between thermodynamic stability and kinetic activity is expressed quantitatively in a linear free-energy relationship.) The reduced forms (Fe^{2+} and Mn^{2+}) are fairly soluble in water, but they are stable only in the absence of O_2. The oxidized forms, Fe^{III} and $Mn^{III,IV}$, have strong tendencies to form oxo and hydroxo species and are highly insoluble at circumneutral pH. As a result, both are only minor constituents in natural waters, even though they are abundant in the Earth's crust.

Iron and manganese are biologically essential, and their low availability in surface waters makes them potentially limiting for planktonic growth. Both elements cause problems (unpleasant tastes and staining of plumbing fixtures) in potable water. Because of their rich and complicated chemistries and inherent water quality significance, both elements have received much scientific study, and a large literature exists on their aquatic behavior.

Qualitative similarities aside, it would be a mistake to think that studies on one of the elements provide much insight into the detailed behavior of the other. From

Table 4-12. Chemical and Geological Properties of Iron and Manganese

Property	Manganese	Iron
Global abundance	10th (0.085%)	4th (5%)
Major minerals	Pyrolusite, MnO_2, manganite, $MnOOH$; hausmanite, Mn_3O_4; rhodochrosite, $MnCO_3$	Hematite, Fe_2O_3; magnetite, Fe_3O_4; limonite, $FeOOH$; siderite, $FeCO_3$; pyrite, FeS_2
Biological importance	Essential; low toxicity; Mn^{II} can be energy source for some bacteria at circumneutral pH	Essential; not toxic at natural water concentrations; Fe^{II} is energy source for some acidophilic bacteria; may limit oceanic primary production
Atomic number	25	26
Outer electron configuration of element	$4s^2 3d^5$	$4s^2 3d^6$
Most common oxidation states in natural waters	II,III,IV	II,III
Redox potentials[b] (V)	(III-II) 1.60 (II-0) −1.18 (IV-II) 1.20	(III-II) 0.771 (II-0) −0.44

For II oxidation state:
  Coordination numbers[c]
  Complexing strength
 	
  Ionic radius (Å)
  $log K_{1a}$ for $M^{II}(H_2O)_6^{2+}$
  Electron configuration
	4,$\underline{6}$,7,8
	Very weak (no ligand field stabilization energy)
	0.80
	−10.6
	$4s^0 3d^5$ ($t_{2g}^3 e_g^2$)
		4,5,$\underline{6}$,8
		Generally weak; strong with N ligands (e.g., CN^-)
		
		0.76
		−9.5
		$4s^0 3d^6$ ($t_{2g}^4 e_g^2$)

For III oxidation state:
  Coordination numbers[c]
  Isoelectric with
  Complexing strength
 	
  $log K_{1a}$ for $M^{III}(H_2O)_6^{3+}$
	5,$\underline{6}$,7
	—
	Forms metastable complexes with $C_2O_4^{2-}$, SO_4^{2-}, EDTA that decompose by Mn oxidation
	−0.4
		3,4,5,$\underline{6}$,7,8
		Mn^{II} (d^5)
		Very low affinity for N ligands Higher affinity for O-containing ligands (PO_4^{3-}) and for halides except F^-
		−3.05

For IV oxidation state:
  Coordination numbers[c]
  Complexing strength
	4,$\underline{6}$
	No soluble complexes in aqueous solution
		—

[a]   Compiled from various sources including Stumm, W. and J.J. Morgan, *Aquatic Chemistry*, 2nd ed., Wiley-Interscience, New York, 1981; and Morgan, J.J., in *Principles and Applications of Water Chemistry*, S.D. Faust and J.V. Hunter (Eds.), John Wiley & Sons, New York, 1967, p. 561.

[b]   Potentials for reduction half reaction indicated in parentheses

[c]   Most common coordination number underlined.

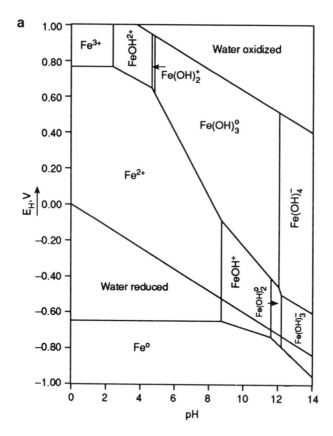

Figure 4-30. Stability-field diagrams for (a) iron and (b) manganese. Lines on the diagrams represent combinations of E_H and pH where two soluble species have equal activities or where a soluble species is at an arbitrarily defined activity in equilibrium with a solid phase. Indicated species are the predominant forms of the element in each region of a diagram. Region between the lines for oxidation and reduction of water is the E_H-pH domain in which water is a stable species. [See reference 150 for details on drawing and interpreting such diagrams.]

quantitative and some qualitative perspectives, the two elements are quite distinct, as some of the parameters in Table 4-12 illustrate. For example, Fe has only one important oxidized state, but Mn has two. Mn is toxic to some organisms at moderate concentrations, but there is no evidence that Fe is toxic at levels encountered in natural waters. Although bacteria mediate the oxidation of both elements, bacterial oxidation of Fe occurs only at low pH, but bacterial oxidation of Mn occurs at circumneutral pH.

From a thermodynamic perspective, Mn^{2+} is much more stable (toward oxidation) than is Fe^{2+}; this is shown by the relative positions of the stability regions for the oxidized and reduced forms of the elements in the pE-pH diagrams of Figure 4-30 and by the higher reduction potentials for Mn^{III} and Mn^{IV} than for Fe^{III} (Table 4-12). The higher stability of reduced Mn is also reflected in the oxidation kinetics of the elements. Although Fe^{II} oxidation decreases rapidly with pH, it nonetheless

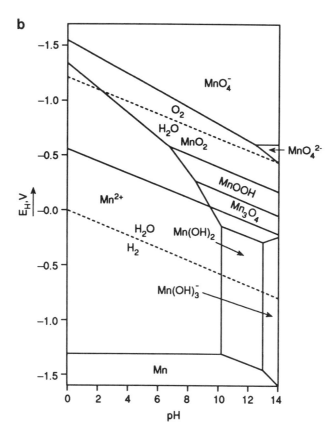

Figure 4-30 (continued).

occurs at measurable rates under acidic conditions. In contrast, Mn^{II} oxidation is exceedingly slow (perhaps nonexistent) at low pH. The mechanisms of electron transfer to O_2 likewise are different for Fe and Mn. The former proceeds by an outer-sphere process (the O_2 molecule is not actually bonded to Fe during the reaction), but electron transfer between Mn^{II} and O_2 most likely occurs by an inner-sphere mechanism, in which O_2 forms a π-σ bond with Mn before the transfer occurs.[170] These thermodynamic and kinetic differences can be understood in terms of the difference in outer orbital electron configurations of the elements. Mn^{II} (d^5; $t_{2g}^3 e_g^2$) has one electron in each of its five d orbitals, a particularly stable situation; Fe^{II} (d^6; $t_{2g}^4 e_g^2$) has four half-filled d orbitals and one filled orbital. The implications of this difference for the reactivity of the metal ions toward O_2 can be explained by frontier-molecular orbital theory[170] (see Figure 4-31).

4.6.5 Autoxidation of Fe^{II}

The kinetics of Fe^{II} oxygenation is one of the most-studied aquatic reactions; this fact reflects both its complexity and its influence on natural water chemistry. Stumm and Lee[171] were among the first chemists to study the reaction under conditions

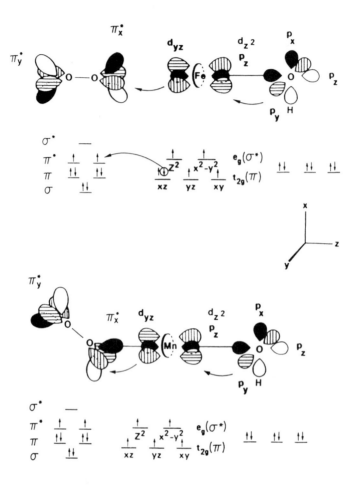

Figure 4-31. Molecular orbital diagrams for autoxidation of Fe^{II} and Mn^{II}. For Fe^{II}, an outer-sphere electron transfer from $\pi(d_{yz})$ to π_y· or $\pi(d_{xz})$ to π_y· is possible, and OH$^-$ bound to Fe^{II} enhances the rate by transfer of electron density through both the σ and π systems, stabilizing the Fe^{III} product. In contrast, an outer-sphere process is not possible when Mn^{II} is in perfect octahedral symmetry (as in Mn(H$_2$O)$_6{}^{2+}$) because: (1) the Mn $e_g(\sigma)$ to O$_2$ π· transfer has unfavorable symmetry (π to π transfers are favored in outer-sphere mechanisms) and (2) the Mn $\pi(t_{2g})$ to O$_2$ π· transfer is energetically unfavorable (the t_{2g} orbital is not the HOMO (highest occupied molecular orbital) of Mn^{II}). Binding of OH$^-$ to Mn^{II} distorts the octahedral symmetry (rearranging the d orbital energies) and donates electron density to the Mn center, thus increasing its basicity. This allows O$_2$ binding by a e_g to π_x·bond and facilitates electron transfer to O$_2$ by the σ or π system. [From Luther, G.W., III, in *Kinetics of Aquatic Chemical Processes*, W. Stumm (Ed.), Wiley-Interscience, New York, 1990, p. 173. With permission.]

applicable to natural waters. In their often-cited study they determined the form of the rate equation, measured the rate constant, and evaluated the effects of solution conditions (including catalysts) on oxidation rates. In brief, the reaction is first order in Fe^{2+} and P$_{O_2}$ and second order in {OH$^-$} at circumneutral pH (Figure 4-32a), leading to the rate expression:

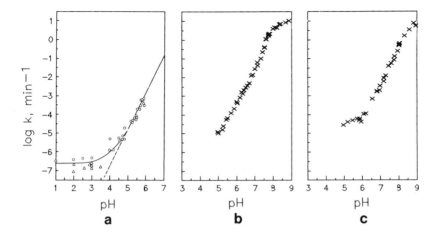

Figure 4-32.   Effect of pH on autoxidation of Fe^{II}: (a) and (b) freshwater conditions; (c) seawater conditions; k is pseudo-first-order rate constant (s^{-1}). [Graph (a) redrawn from Singer, P. and W. Stumm, *Science*, 167, 3921 (1970); graphs (b) and (c) drawn from data in Millero, F.J., et al., *Geochim. Cosmochim. Acta*, 51, 793 (1987). With permission.]

$$-d[Fe^{II}]/dt = k[Fe^{II}]P_{O_2}\{OH^-\}^2 \qquad (4\text{-}59)$$

Ferric iron is highly insoluble at circumneutral pH, and the initial products of Fe^{II} oxidation are lepidocrocite (γ-FeOOH) and amorphous FeOOH.[44] These forms are unstable and age over time to form goethite (α-FeOOH). Although some bacteria oxidize Fe^{2+}, using it as an energy source, several studies have shown that bacterial oxidation is important only at pH <4.5, and chemical oxidation is the only significant mechanism at circumneutral pH.

There is some uncertainty and argument over the lower and upper pH limits for the slope 2 in the plot of *log* k vs. pH. Most rate measurements have been made over the fairly narrow pH range of 6.0 to 7.5 (e.g., references 44,171–173). Roekens and Van Grieken[174] measured the oxidation rate of Fe^{2+} in seawater over the pH range 5.5 to 10 and concluded that second order dependence on $\{OH^-\}$ held only up to pH 7.4. Davison[175] reevaluated their data and concluded that their measured rate constants in the pH range 6.8 to 7.2 were too high (compared with other workers), but their value at pH 8.3 was reasonable if second-order dependency in $\{OH^-\}$ is assumed up to this value. He also concluded that their data were too scattered in the pH range of 7 to 8 to allow an unambiguous interpretation of reaction order, but for the above reason second-order dependency up to pH 8.3 cannot be ruled out. More recently, Millero and co-workers[176] made rate measurements at closely spaced intervals over the pH range 5 to 9 (Figure 4-32b,c) in simulated seawater and freshwater. Regression analyses supported a conclusion of second-order dependency in $\{OH^-\}$ over the pH range 6.0 to 8.0 at 25°C and 7.5 to 9.0 at 5°C. However, inspection of Figure 4-32c indicates that the slope for freshwater at 25°C starts to tail off at pH > 7.5. Millero et al. suggested this could be caused by precipitation of $Fe(OH)_2$ or formation of less reactive complexes such as $FeCO_3^\circ$, but it also could reflect measurement difficulties at high reaction rates. From an environmental viewpoint, the exact pH-dependency under alkaline conditions is not too important, in that the rates are very rapid (time scale of minutes to seconds above pH 7).

The data of Millero et al. show a lower pH dependency at pH < 6 in seawater at 25°C, but not in freshwater (cf. Figures 4-32b and c). In contrast, the data of Singer and Stumm[177] (Figure 4-32a) show second-order dependency down to pH 4.5 and a gradually decreasing slope thereafter until it approaches zero (k independent of pH) at pH ~3. Reasons for these differences are not apparent.

A tremendous range exists in the rate constant for Fe^{II} oxygenation over the pH range of natural waters. For fixed pH and $P_{O_2} = 0.21$ atm, the observed pseudo-first-order rate constant, k', ranges from $10^{-3.5}$ day^{-1} at pH 3 (representative of streams receiving mine drainage) to $10^{1.6}$ day^{-1} ($10^{-1.5}$ min^{-1}) at pH 7.0 and ~$10^{3.8}$ day^{-1} (~4.3 min^{-1}) at pH 8.0. Correspondingly, half-lives range from more than 2000 days at pH 3 to less than 10 s at pH 8.0. Stumm and Lee[171] found $k = 1.5 \times 10^{13}$ M^{-2} atm^{-1} min^{-1} at 20.5°C. Subsequent investigations have produced a wide range of values: ~0.5–21×10^{13} M^{-2} atm^{-1} min^{-1} for synthetic media, and an even wider range for natural waters. According to Davison and Seed,[172] much of the variation in reported k values can be explained by uncertainties in the actual pH of the reacting solution (see Section 2.4.4) and by variations in the ionic composition of the media. A wide variety of ions affects the reaction rate, as described below. Davison and Seed[172] concluded that the best estimate for k in synthetic media without interfering substances is 2.0 (range of 1.5 to 3.0) $\times 10^{13}$ M^{-2} atm^{-1} min^{-1}.

At constant pH and P_{O_2}, the oxygenation rate increases by ten times for a 15°C increase in temperature, leading to an apparent E_{act} of 96 kJ mol^{-1}.[171] However, at constant P_{O_2}, O_2 solubility decreases with increasing temperature, and OH$^-$ activity increases with temperature at a given pH (because K_w increases with temperature). When these changes are taken into account, k is almost constant with temperature,[44,171] implying that $E_{act} \approx 0$. This suggests that the rate-limiting step may involve an ion-radical or molecule-radical reaction, since radicals often react without needing to be activated.

Several inorganic anions affect the rate of ferrous iron oxygenation, and it generally is presumed that they do so by forming complexes that are more or less reactive with O_2 than is free Fe^{II}. Table 4-13 summarizes these effects. Cl$^-$ and SO_4^{2-} form weak complexes with Fe^{2+} and do not affect Fe^{II} speciation and its rate of oxidation at the concentrations found in freshwaters. However, they do have a significant depressing effect on the oxidation rate in seawater. According to calculations by Davison and Seed,[172] the pseudo-first-order half-life of Fe^{II} at pH 8.0 is only 10 to 20 s in freshwater, but is 7 min at this pH in seawater (almost 20 times faster in freshwater). Millero et al.[176] reported a smaller difference between seawater and freshwater rates (~7 times higher in freshwater), and their value is in good agreement with earlier calculations[178] on the extent of Fe^{2+} complexation in seawater by these ions. They concluded that the differences among various investigators may reflect determinate errors in pH measurements among them. The *catalyzing* effects of fluoride and phosphate (at mM concentrations) have been related to the fact that these ions form fairly strong complexes with Fe^{III}, but the precise mechanism whereby they affect the oxidation rate is still uncertain. There

Table 4-13. Ions Affecting the Oxidation of Ferrous Iron[a]

Ion or compound	Effect	Concentration[b]	Number of studies
HCO_3^-	+	0.3–39	3
	0	2.3	1
	−	2.3–50	2
$Si(OH)_4$	+	0.1–1	1
	−	1.2	1
Cl^-	0	0.3	1
	−	100–500	3
SO_4^{2-}	0	0.1	1
	−	30–165	3
NO_3^-, I^-, Br^-	−	100	1
F^-	+	20–100	1
$H_2PO_4^-$	+	0.04–100	2
Cu^{2+}	+	>0.0003	1
Co^{2+}	+	0.04	1
Mn^{2+}	+	0.04	1
MnO_2	+	0.0002	1
Fe^{III}	0	0.1–0.2	2
	+	0.2–1.0	3
Tannic acid	−	0.002–0.1	3
Gallic, glutamic, tartaric acids	−	0.1	1
Citric acid	+	0.1–0.5	2
Glutamine	−	0.1	1
Histidine	0	0.1	1
	−	0.6	1
Humic acid	0	1–3	1
	−	12–145	3
Pyrogallol	−	0.1	1
Syringic, vanillic acids, resorcinol phenol, vanillin	0	0.1	1

[a] Summarized from Davison, W. and G. Seed, *Geochim. Cosmochim. Acta,* 47, 67 (1983); original references listed therein.
[b] Concentrations in m*M* except humic acid (mg L^{-1}).

are discrepancies in the literature regarding the effects of bicarbonate and silica. Of the six studies on bicarbonate, three reported a catalytic effect, two reported inhibition, and one found no effect. Over the range of these species in natural waters, the effects probably are within the range of experimental error.

Several organic ligands, including tannic and humic acids, inhibit the oxidation of Fe^{II} (Table 4-13), but citric acid accelerates the rate.[179,180] The inhibiting effect of organic acid ligands (L_{or}) can be explained by the formation of unreactive (or less reactive) complexes:[179,181]

$$Fe^{II} + L_{or} \underset{k_d}{\overset{k_f}{\rightleftharpoons}} Fe^{II}L_{or} \qquad (4\text{-}60)$$

Pankow and Morgan[181] used computer simulations to illustrate the effects of varying pH, $[L_{or}]_T$, and k_d on the rate of Fe^{II} autoxidation. They assumed that $Fe^{II}L_{or}$ was unreactive toward O_2 and that the rate constant for formation of $Fe^{II}L_{or}$

complexes did not vary with the nature of L_{or}. This is reasonable for ligands with the same charge (see Section 4.4.2). For simple complexation reactions, the stability constant $K = k_f/k_d$, and varying k_d at fixed k_f is tantamount to varying the thermodynamic stability of the $Fe^{II}L_{or}$ complex. All three variables (pH, $[L_{or}]_T$, and k_d) were found to have important effects on Fe^{II} oxidation rates in the range of parameter values examined (Figure 4-33).

The presence of L_{or} lowers the oxidation rate in two possible ways. First, it simply decreases the concentration of free Fe^{II} in solution. Second, release of Fe^{II} from the complex may itself be rate limiting. The first effect is illustrated by the simulation at pH 7.0 in Figure 4-33a. Oxidation is relatively slow at this pH, and complex dissociation is sufficiently rapid that the instantaneous concentration quotient, $Q = [Fe^{II}L_{or}]/[Fe^{II}][L_{or}]$ remains constant and essentially equal to K throughout the time course of reaction. Equilibrium between free and complexed Fe^{II} thus is maintained, and the reaction is slowed by L_{or} to an extent predictable from the equilibrium relationship. In contrast, at pH 8.5, oxidation of free Fe^{II} is so rapid that complex dissociation cannot keep pace with the loss of free Fe^{II}. Q deviates significantly from K (Figure 4-33b), quickly reaching a peak and then settling into a kinetically stable "steady state", Q_{stab}, once most of the Fe^{II} has been oxidized and $[L_{or}] \rightarrow [L_{or}]_T$. Q_{stab} still differs from K, and this represents a constant driving force for the reaction: $\Delta G = -RT \ln Q_{stab}/K$.[181]

A satisfactory explanation for the accelerating effect of citric acid has not been offered. Possibly, Fe^{2+} is held in the citrate complex in a conformation that facilitates attack by O_2, just as phthalocyanine catalysts are thought to facilitate oxygenation of Mn^{II} and Co^{II}.[155,182] It is interesting to note that citric acid (CA) also facilitates the photoreduction of Fe^{III} to Fe^{II}, probably by a ligand-to-metal charge transfer (LMCT) process (Section 8.4.4). A metal-to-ligand charge transfer reaction (Fe^{II}-CA $\rightarrow Fe^{III} + CA$) seems unlikely as an explanation for the effect of CA on Fe^{II} oxidation.

Several heavy metals catalyze Fe^{II} oxygenation, but the concentrations at which catalysis has been reported are above those expected for natural waters, except for Mn. Both Mn^{2+} and MnO_2 act as catalysts at concentrations found in many lakes.[172] The early work of Stumm and Lee[171] indicated that the hydrolyzed Fe^{III} products of Fe^{II} oxygenation catalyze the reaction, and several others have demonstrated this autocatalytic effect.[173,183] Detailed studies by Sung and Morgan[44] showed significant autocatalysis at pH 7.2, $[ClO_4^-] = 0.5 M$ and $[Fe^{II}]_0 = 50 \mu M$ (Figure 4-34). Their data fit a two-term expression (see Section 4.3.2 for details on fitting data to such rate equations):

$$d[Fe^{II}]/dt = (k_1 + k_2[Fe^{III}])[Fe^{II}] \qquad (4-61)$$

where k_1 and k_2 are pseudo-first- and second-order rate constants (applicable at a given pH and P_{O_2}), respectively. They found $k_1 = 0.056$ min^{-1} and $k_2 = 360$ M^{-1} min^{-1} for the above conditions. Their k_1 translates to a value of 0.56×10^{13} M^{-2} atm^{-1} min^{-1}, which is a little lower than the "best-estimate" value given above. Their value for k_2 agrees fairly well with the value of 212 M^{-1} min^{-1} reported by Tamura et al.[173] for similar

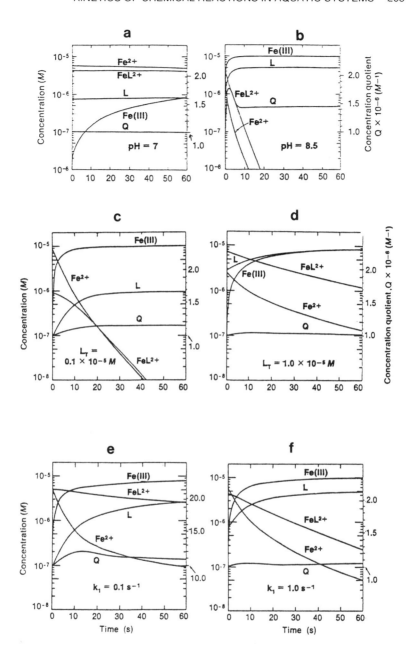

Figure 4-33. Simulations of Fe^{II} autoxidation in presence of a complex-forming ligand, L$_{or}$: (a) and (b) illustrate effect of varying pH on oxidation rate; (c) and (d) show effect of varying [L$_{or}$]$_T$ at constant pH; (e) and (f) show effect of varying k$_d$ (hence, K$_{FeL}$). Q = [Fe^{II}L$_{or}$]/[Fe^{II}][L$_{or}$] is the instantaneous concentration quotient. [From Pankow, J.F. and J.J. Morgan, *Environ. Sci. Technol.*, 15, 1155 (1981). With permission.]

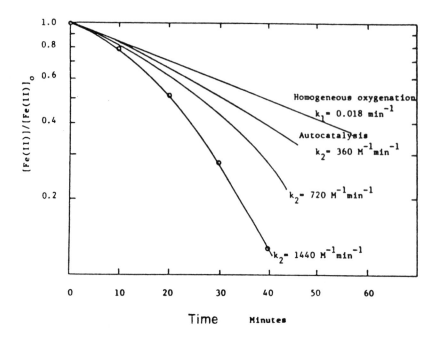

Figure 4-34. Autoxidation of Fe^{II} fits autocatalytic model with k_2 (autocatalytic term) = 1440 M^{-1} min^{-1}. [From Sung, W. and J.J. Morgan, *Environ. Sci. Technol.*, 14, 561 (1980). With permission.]

reaction conditions. Although these results suggest that autocatalysis of Fe^{II} oxygenation may be significant in circumneutral natural waters, Davison and Seed[172] found no evidence for this in Esthwaite Water, a eutrophic English lake. Even though concentrations of Fe^{III} ranged up to several mg L^{-1}, the oxidation rate was enhanced only when synthetic amorphous Fe^{III} oxyhydroxide was added at 10 mg L^{-1} or higher. They suggested that the catalytic effect of natural Fe^{III} oxyhydroxides may be lower than that of synthetic FeOOH, because of sorption of organic matter onto the surface of the natural particles. Autocatalysis thus may not be important in lakes, but it may be important in mine drainage streams where FeOOH may coat the streambed and provide a much greater surface area for catalysis.

The nonlinear relationship between k and pH has been explained[176,178] in terms of differing reactivities of Fe^{II} hydroxy complexes with respect to O_2. In the absence of other ligands and at any given pH

$$[Fe^{II}]_T = [Fe^{2+}] + [FeOH^+] + [Fe(OH)_2] + [Fe(OH)_3^-] \qquad (4\text{-}62)$$

Equation 4-59 can be written in terms of the individual Fe^{II} species:

$$-d[Fe^{II}]_T / dt = (k_0'[Fe^{2+}] + k_1'[FeOH^+] + k_2'[Fe(OH)_2] + k_3'[Fe(OH)_3^-] \qquad (4\text{-}63)$$

where the k'_i are pseudo-first-order rate constants. In terms of α's and $[Fe^{II}]_T$, Equation 4-63 is

$$-d[Fe^{II}]_T / dt = (k'_0\alpha_0 + k'_1\alpha_1 + k'_2\alpha_2 + k'_3\alpha_3)[Fe^{II}]_T \qquad (4\text{-}64)$$

where the α's are the fraction of Fe^{II} present in a given state of deprotonation. Fe^{2+} is a weak Bronsted acid that shows much less tendency to hydrolyze than does Fe^{III}. In fact, Fe^{2+} hydrolyzes only slightly before precipitation of $Fe(OH)_{2(s)}$ or Fe_3O_4 begins,[184] and as a result, hydrolysis constants for Fe^{II} species are not known with much accuracy. Baes and Mesmer[184] supply the following equilibria and estimates of the equilibrium constants:

$$Fe^{2+} + H_2O \rightleftharpoons FeOH^+ + H^+;$$

$$K'_1 = \{H^+\}\{FeOH^+\}/[Fe^{2+}] = 10^{-9.5} \qquad (4\text{-}65a)$$

$$FeOH^+ + H_2O \rightleftharpoons Fe(OH)_2^o + H^+;$$

$$K'_2 = \{H^+\}\{Fe(OH)_2\}/[FeOH^+] = 10^{-20.6} \qquad (4\text{-}65b)$$

$$Fe(OH)_2^o + H_2O \rightleftharpoons Fe(OH)_3^- + H^+;$$

$$K'_3 = \{H^+\}\{Fe(OH)_3^-\}/[Fe(OH)_2] = 10^{-31} \qquad (4\text{-}65c)$$

K'_1–K'_3 are written above as mixed constants (in both concentration and H^+ activity) to be consistent with the fact that Equation 4-63 is expressed in concentrations, but pH is a measure of H^+ activity. The numerical values given above are for zero ionic strength (i.e., they are thermodynamic values — all terms expressed in activity). The distribution of soluble Fe^{II} forms over the pH range 6 to 12 is illustrated in Figure 4-35 for $[Fe^{II}]_T \le 10^{-5}\,M$. It is clear that Fe^{2+} is the predominant species up to pH ~9.8 and that the solubility of $Fe(OH)_{2(s)}$ is exceeded above pH 9.2. $Fe(OH)_2^o$ reaches a maximum concentration of ~$10^{-7.8}\,M$ when solid phase $Fe(OH)_{2(s)}$ is present. The α's for Fe^{II} can be derived readily from the K's in Equation 4-65, by the procedure described for H_2S in Section 4.6.2.

According to Millero,[178] Fe^{2+} is much less reactive toward O_2 than is $FeOH^+$, which is much less reactive than $Fe(OH)_2^o$. Although the latter species is present at only very low concentrations, even at pH 7 to 9, it contributes most to the overall oxidation rate at pH > 5 to 6 (i.e., $k'_2\alpha_2$ becomes the largest term in Equation 4-64). Because $d\log[Fe(OH)_2^o]/d$pH (and $d\log\alpha_2/d$pH) = 2 (see Figure 4-35) up to the pH where $Fe(OH)_{2(s)}$ begins to precipitate (~9.2), the slope of $\log k'_{obs}$ vs. pH likewise is 2.

Similarly, the model indicates that $FeOH^+$ is the most important reactant between pH ~4 and ~5 to 6 (i.e., $k'_1\alpha_1$ > both $k'_0\alpha_0$ and $k'_2\alpha_2$). This causes a slope of 1 for $\log k'_{obs}$ vs. pH (Figure 4-32) because $d\log[FeOH^+]/d$pH (and $d\log\alpha_1/d$pH)=1 in this

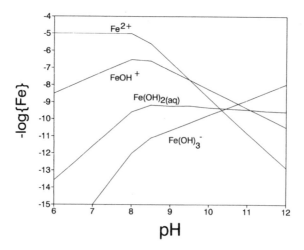

Figure 4-35. Speciation of Fe^{II} from pH 6 to 12 at $[Fe_T] \leq 10^{-5}$ M. [Equilibrium conditions calculated by the computer program MINTEQA2 (Center for Exposure Assessment Modeling, U.S. EPA, Athens, GA, 1993).]

region. Below pH ~3, Fe^{2+} is the most important reactant (i.e., $k'_0\alpha_0 > k'_1\alpha_1$ and $k'_2\alpha_2$), and the rate becomes independent of pH because $\alpha_0 \approx 1$ at pH < 8 (Fe^{2+} is the dominant species over the entire pH range of natural waters). We cannot be very precise about the pH range in which the various $k'_i\alpha_i$ terms are dominant because of large differences in the literature regarding the pH values where the slope of the $\log k'_{obs}$ vs. pH curve changes. According to Millero,[178] the value for k'_0 is very small — ~10^{-6} min^{-1}.[185] He also estimated that $k'_1 = 1.7$ min^{-1} and $k'_0 = 10^{5.63}$ min^{-1} by regression fit of measured rate data to an equation that can be derived from Equation 4-64. A value for k'_3 could not be obtained in this way, and rate measurements in the region of rapid kinetics (pH 8 to 10) will need to be made by stopped-flow methods to determine whether $k'_3\alpha_3$ is a significant term.

The Haber-Weiss mechanism (Section 4.2.2) usually is invoked to describe the autoxidation of Fe^{II}.[35,186,187] According to this mechanism the reduction of O_2 proceeds by a series of four one-electron transfers, yielding several reactive intermediates that also participate in acid-base equilibria:

$$Fe^{II} + O_2 \rightarrow Fe^{III} + O_2^-$$

$$(H^+ + O_2^- \rightleftharpoons HO_2)$$

$$Fe^{II} + O_2^- \rightarrow Fe^{III} + O_2^{2-}$$

$$(2H^+ + O_2^{2-} \rightleftharpoons H_2O_2) \tag{4-66}$$

$$Fe^{II} + H_2O_2 \rightarrow Fe^{III} + \cdot OH + OH^-$$

$$Fe^{II} + \cdot OH \rightarrow Fe^{III} + OH^-$$

The iron forms are intentionally written as generic oxidation states to indicate that several hydrolysis species may be involved as reactants and products, as described earlier. The rate-limiting step most likely is the reaction with O_2 itself, because all the intermediates are unstable and highly reactive. At circumneutral pH, protonation of the superoxide anion radical (O_2^-) is unfavorable ($pK_a = 4.8$ for HO_2), but protonation of the peroxide anion (O_2^{2-}) is highly favored ($pK_1 = 11.7$ for H_2O_2).

The effect of ionic strength (I) on Fe^{II} autoxidation kinetics does not yield information that is useful in enhancing our understanding of the reaction mechanism. Sung and Morgan[44] used perchlorate to examine the effects of I on the kinetics of this reaction.[*] They found that a Bronsted plot ($log\,k$ vs. $\sqrt{I}$) had a slope of -2, suggesting that the product of the charges of ions involved in the rate-limiting step is -2. This implies that uncharged O_2 is not involved in the rate-limiting step. The rate-limiting step could involve a charged intermediate in O_2 reduction, such as O_2^-, but this seems unlikely given the reactivity of this radical anion. Tamura et al.[173] proposed the following rate-limiting step: $FeOH^+ + O_2OH^- \rightarrow Fe(OH)_2^+ + O_2^-$, which should yield a slope of -1 in a Bronsted plot, but this is not supported by the data of Sung and Morgan. Moreover, Millero's model suggests that the principal species involved in Fe^{II} oxygenation above pH 5 to 6 is $Fe(OH)_2$, which would yield a slope of zero in a plot of log k vs. $\sqrt{I}$. We should note that given the complexity of the Fe^{II}-O_2 reaction, different steps could become rate limiting under differing solution conditions, and mechanistic interpretations thus may not be straightforward. As mentioned before, Moore and Pearson[156] cautioned against using Bronsted plots to infer mechanisms of complicated reactions. The significance of the ionic strength data and of Fe^{II} oxygenation thus remains incompletely resolved.

Finally, the kinetics and stoichiometry of the reaction of Fe^{II} with O_3 have been studied by many workers, and its complicated mechanism was unraveled only recently.[188,189] This reaction long has been important analytically as a means of measuring O_3 and could play a role in potable-water treatment. The stoichiometry depends somewhat on pH and the initial $[Fe^{II}]/[O_3]$. The reaction itself is very rapid. O_3 disappears in less than 0.1 s, but oxidation of Fe^{II} by secondary oxidizing species, such as O_3^-, HO_2, HOO_3, and H_2O_2, continues for another ~15 min. The stoichiometric ratio (SR), defined as moles of Fe^{III} produced to moles of O_3 consumed, varies between 1.8 and 2.5, depending on the relative effectiveness of competition for the secondary oxidants by O_3, Fe^{2+}, and H^+.[189]

4.6.6 Autoxidation of Mn^{II}

Much of our current information on this reaction was developed by Morgan and co-workers beginning in the mid-1960s.[190-193] The form of the rate equation for

---

[*] Effects of I must be evaluated by adding salts that do not interact specifically with the reactants. Most anions, including Cl$^-$ and SO_4^{2-} form complexes with Fe^{2+}, but ClO_4^- has almost no tendency to form complexes with metal ions.

Mn^{2+} oxidation by O_2 is similar to that for Fe^{2+}, at least at neutral and slightly alkaline pH; i.e., the rate is first order in Mn^{2+} and O_2 and second order in OH^-. However, there are some important differences. First, whereas Fe^{2+} is oxidized only to Fe^{III}, a mixture of Mn^{III} and Mn^{IV} oxidation states results when Mn^{2+} reacts with O_2. This was demonstrated in the classic study of Morgan,[190,191] who found the reaction was nonstoichiometric, with experimental values for x in the product MnO_x ranging from about 1.3 to 1.9. (x = 1.5 corresponds to Mn^{III}, and x = 2.0 corresponds to Mn^{IV} (MnO_2)). Several more recent studies have agreed that the initial product of Mn^{II} oxygenation is Mn^{III}, specifically (MnOOH, manganite), and dissolved Mn levels in oxic natural waters, which are often supersaturated with respect to Mn^{IV} solid phases,[194] often are modeled to be in equilibrium with this species. However, it is unstable with respect to self-oxidation-reduction:[194]

$$2 MnOOH + 2H^+ \rightarrow MnO_2 + Mn^{2+} + 2H_2O \qquad (4\text{-}67)$$

The above reaction has a ΔG of about -105 kJ mol^{-1} in the pH and dissolved Mn range in seawater.[194] MnOOH thus is an intermediate for the production of the stable oxidized product MnO_2 (birnessite, pyrolusite). Grill[194] suggested that this indirect pathway is followed because E_{act} for the direct conversion of Mn^{II} to MnO_2 is so high that the rate is too low to balance the input of Mn^{2+} to the ocean. Consequently, Mn levels increased until saturation with respect to MnOOH occurred.

Kessick and Morgan[192] proposed that the mechanism for Mn^{II} oxygenation involves the formation of $Mn(OH)_{2(aq)}$ (this would explain the dependence of the rate on $\{OH^-\}^2$) and a subsequent rate-determining step involving a one-electron transfer from O_2:

$$Mn(OH)_{2(aq)} + O_2 \rightarrow MnOOH + HO_2 \qquad (4\text{-}68)$$

with hypothesized transition state: $HO\text{–}Mn_{aq}\text{–}O\text{–}H\text{–}O_2$. Based on published thermodynamic data,[191] they pointed out that $Mn(OH)_{2(aq)}$ should exist only at very low concentrations in free solution, but suggested that similar structures could be formed at the surface of already-precipitated product. This agrees with the model proposed by Diem and Stumm,[195] which is described below.

The second difference between Fe and Mn oxygenation is that the former is only weakly autocatalytic, but the latter is strongly so. Indeed, there is evidence that the homogeneous, uncatalyzed reaction is not important in Mn oxidation in natural waters. Morgan and others described the process by a two-term rate equation consisting of uncatalyzed and catalyzed terms. At constant pH, P_{O_2}, and temperature:

$$-d[Mn^{II}]/dt = k_1'[Mn^{II}] + k_2'[Mn^{II}][MnO_x] \qquad (4\text{-}69)$$

In Morgan's original experiments,[191] k_1' was determined under conditions where the autocatalytic term was thought to be negligible (pH < 9, initial reaction conditions), and a value of 4×10^{12} M^{-2} atm^{-1} day^{-1} was estimated for k_1, where $k_1 = k_1'/[O_2][OH^-]^2$. A large range in k_2 (defined similarly; units of M^{-3} atm^{-1} day^{-1}) is found in the literature: 2.4×10^{14} to 5.0×10^{16} for laboratory studies[191, 196] and 3.2×10^{19} based on *in situ* measurements in a marine environment.[194] This range does not necessarily imply poor accuracy or that one value is better than the others; the differences may simply reflect differences in the catalytic activity of Mn oxides formed under different conditions.[194]

More recent studies suggest that the term for homogeneous oxidation in Equation 4-69 is not correct and that homogeneous oxidation of Mn^{II} either does not occur or is extremely slow (time scale of decades) in solutions that are undersaturated with respect to MnCO$_3$ and Mn(OH)$_2$ and do not have microorganisms or other solids present. The diagram in Figure 4-31 provides an explanation for the stability of Mn^{2+} based on its outer electron orbital configuration.

In an impressive display of patience, Diem and Stumm[195] allowed solutions containing Mn^{2+} to incubate for periods up to 7 years and found no loss of Mn^{2+} in solutions that met the above criteria. According to these authors, the solutions Morgan used[191] to establish the autocatalytic rate law were supersaturated with respect to MnCO$_3$ or Mn(OH)$_2$, but Morgan inferred that they remained homogeneous with respect to Mn^{II}, because of slow nucleation kinetics of Mn^{II} solid phases. In all the experiments of Diem and Stumm where oxidation occurred at measurable rates in the absence of catalysts, the solutions were supersaturated with respect to a solid Mn^{II} phase. They concluded that Mn^{II} oxygenation occurs with a much lower E_{act} than that for aqueous Mn^{2+}, when the Mn^{II} is coordinated in the structure of rhodochrosite (MnCO$_3$), Mn(OH)$_2$, or bound in surface complexes of hydrous oxides:

$$Me - O \diagdown \atop Me - O \diagup Mn^{II}$$

where Me = Fe^{III}, Mn^{III}, or Mn^{IV}. In support of this idea, Davies and Morgan[193] found that Mn^{2+} oxidation rates were enhanced substantially by the presence of several oxyhydroxide surfaces, with the extent of enhancement in the order: γ-FeOOH (lepidocrosite) > α-FeOOH (goethite) > amorphous SiO$_2$ > δ-Al$_2$O$_3$.

Relatively little information is available on the effects of physical factors on Mn^{II} oxidation kinetics. Morgan[191] reported that the rate doubled over the temperature range 11 to 22°C ($E_{act} \approx 50$ kJ mol^{-1}). Higher apparent E_{act} values were found for Mn^{II} oxidation catalyzed by iron oxide surfaces: 120 kJ mol^{-1} (α-FeOOH) and 110 kJ mol^{-1} (γ-FeOOH), but these values include the effect of temperature on the extent of Mn^{II} sorption. When this was factored out, E_{act} for *surface-bound* Mn^{II} was found to be 50 (α-FeOOH) and 35 (γ-FeOOH) kJ mol^{-1}.[193] Although iron oxides are capable of photocatalysis (Section 8.6.1), light did not affect the rate of Mn^{II} oxidation by α- and γ-FeOOH.

Increasing ionic strength decreases Mn oxidation on iron oxides,[193] but this trend also may reflect effects on the extent of Mn^{II} sorption rather than effects on the oxidation reaction per se. A wide variety of ions also affect the rates of Fe-oxide-catalyzed Mn^{II} oxidation. Ions that bind strongly to Fe oxides, such as Ca^{2+}, Mg^{2+}, phosphate, and salicylate, decrease rates by forming surface complexes and decreasing the surface sites available for Mn^{2+}. Other ligands (Cl^-, SO_4^{2-}, HCO_3^-, organic anions) decrease Mn sorption to the oxide surfaces by forming soluble complexes with Mn^{2+}. Adding anions that form complexes with Mn^{2+} to the reaction system causes a decrease in the oxidation rate. In the classic studies of Morgan,[191] sulfate, which forms very weak complexes, had negligible effects, but bicarbonate had a larger effect. The strong complexing agent, $P_2O_7^{4-}$, greatly depresses the oxidation rate, but the extent is greater than can be accounted for by complex formation.[197]

Although most of the quantitative information on Mn oxidation kinetics has been obtained from laboratory studies, careful analysis of field data can yield valuable information. A good example is the elegant analysis by Grill[194] of data on the temporal and vertical distribution of dissolved (reduced) and particulate (oxidized) Mn forms in Saanich Inlet (a fjord on Vancouver Island, British Columbia). From temporal and vertical gradients in particulate Mn (Mn_p), Grill was able to calculate *in situ* rates of Mn precipitation and fit these data to rate equations. He concluded that dissolved Mn (Mn_d) in oceanic waters is controlled by precipitation of manganite (γ-MnOOH).

The fundamental one-dimensional transport equation for nonconservative substances (see Section 5.2.7) applies to this situation (since horizontal transport is negligible compared with vertical processes):

$$\partial[Mn]/\partial t = K_z \partial[Mn]/\partial z - (w + s)\partial[Mn]/\partial z + R \qquad (4\text{-}70)$$

where z is the vertical coordinate (positive downward), w is the downward advective velocity, s is the gravitational sinking velocity of particulate Mn, K_z is the vertical eddy diffusivity, and R is the time rate of change in [Mn] from chemical reaction. For the conditions of Grill's observations, w and K_z were negligible for Mn_p. Hence, Equation 4-70 becomes

$$R_p = \partial[Mn_p]/\partial z + s\partial[Mn_p]/\partial z \qquad (4\text{-}71)$$

where R_p = the rate of Mn precipitation. Grill solved Equation 4-71 by determining values of the time and depth derivatives of $[Mn_p]$ by finite difference, using data from vertical profiles taken during six cruises over 16 months. He found a good correlation (Figure 4-36) between the calculated values of R_p and the average value of $[Mn_p]$ for a given depth-time interval; i.e., $[Mn_p]/R_p$ was roughly constant over about three orders of magnitude variation in both. This ratio is equivalent to the apparent turnover time of Mn_p, the average of which was found to be 13.4 days. This led to the conclusion that $R_p \approx k[Mn_p]$; i.e., the process is autocatalytic. By further analysis of the deviations observed in the relationship between R_p and $[Mn_p]$ at high and low $[Mn_p]$, Grill concluded that a two-term rate expression was needed to explain the data:

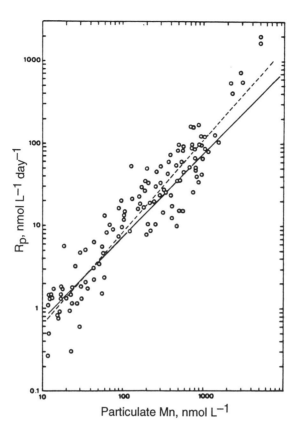

Figure 4-36. Rate of manganese precipitation, R_p vs. concentration of particulate Mn, $[Mn_p]$ in Saanich Inlet, a fjord on Vancouver Island, British Columbia. R_p was calculated from Equation 4-71. Solid line is mean value of the turnover time, $[Mn_p]/Rp$; dashed line is *log* least squares fit of the data. [From Grill, E.V., *Geochim. Cosmochim. Acta,* 46, 2435 (1982). With permission.]

$$R_p = k_2[Mn_p][Mn_d]P_{O_2}[OH^-]^2 + k_3[Mn_p] \qquad (4\text{-}72)$$

The first term is the same as Morgan's autocatalytic expression; the second is independent of $[Mn_d]$, the main reactant, a seeming violation of mass action principles. Grill explained this by hypothesizing a two-step reaction mechanism: a rapid initial step in which Mn^{2+} is adsorbed to the Mn oxide surface, and a slow (rate-determining) step in which adsorbed Mn^{II} is oxidized. The sorption process can be described by a Langmuir isotherm:

$$[Mn_a] = \frac{Nb[Mn_d][Mn_x]}{1 + b[Mn_d]} \qquad (4\text{-}73)$$

where subscripts a and x stand for adsorbed and solid-phase (oxide) Mn, respectively; N = total number of adsorption sites available per mole of Mn_x; and b = the equilibrium binding constant for the adsorption reaction. It is easy to see from Equation 4-73 that at relatively high $[Mn_d]$ (i.e., $b[Mn_d] \gg 1$), $[Mn_a]$ is simply proportional to $[Mn_x]$ (all surface sites are covered) and the rate of oxidation will be independent of $[Mn_d]$. Finally, a similar mechanism and rate expression independent of $[Mn_d]$ could apply to bacterially mediated Mn^{2+} oxidation. Indeed, bacterially mediated processes may be the major mechanism for Mn^{2+} oxidation in natural waters in that observed oxidation rates[194,195] are much higher than those found in Mn^{II}-oversaturated systems without bacteria.

In summary, much has been learned about the kinetics of Mn^{II} oxidation over the past ~25 years. The process is quite complicated and depends on many solution conditions. The effects of many factors have been quantified for specific sets of conditions, but we still lack a comprehensive model to predict oxidation rates over the range of conditions encountered in natural waters.

4.6.7 Oxidation of Cu(I)

Although Cu^{II} is the only stable oxidation state of copper in aqueous solution, Cu^I can be formed by photochemically produced reducing agents like superoxide anion (O_2^-) and H_2O_2:[197]

$$Cu^{2+} + O_2^- \rightarrow Cu^+ + O_2 \qquad (4\text{-}74)$$

There also is evidence that Cu^I is formed intracellularly in some enzymatic processes. As a result of reactions like Equation 4-74, Cu^I exists at low concentrations in surface waters, and its lifetime probably is controlled by oxidation to Cu^{II} by O_2. Overall, the oxidation rate fits the equation:[198]

$$-d[Cu^I]/dt = k[Cu^I][O_2] \qquad (4\text{-}75a)$$

where $\quad \log k = 10.73 + 0.23\,pH - 2373/T - 3.33I^{1/2} + 1.45I \qquad (4\text{-}75b)$

Equation 4-75b was obtained by regression fit of rate data for different temperatures, pH values, and ionic strength. The low coefficient for pH (determined for the range 5.3 to 8.6) suggests that acid or base catalysis is not important and that hydrolysis species of Cu^I (like CuOH) probably are not involved in the oxidation. The apparent E_{act} for the reaction is 45.6 ± 1.7 kJ mol^{-1}. The decrease in k with increasing ionic strength is attributed primarily to the formation of Cu^I chloride complexes that are less reactive than free Cu^+. The observed k in Equation 4-75 can be written in terms of k's and α's for the individual Cu^I species:[178]

$$k = \alpha_{Cu}k_o + \alpha_{CuCl}k_1 + \alpha_{CuCl_2}k_2 + \alpha_{CuCl_3}k_3 \qquad (4\text{-}76)$$

The α_i are molar fractions of the various Cu^I chloro species and can be determined from stability constants for the complexes:

$$\alpha_{Cu} = (1 + \beta_1^*[Cl^-] + \beta_2^*[Cl^-]^2 + \beta_3^*[Cl^-]^3)^{-1}$$

$$\alpha_{CuCl} = \beta_1^*[Cl^-]\alpha_{Cu}$$

$$\alpha_{CuCl_2} = \beta_2^*[Cl^-]^2\alpha_{Cu} \tag{4-77}$$

$$\alpha_{CuCl_3} = \beta_3^*[Cl^-]^3\alpha_{Cu}$$

The β_i^* are concentration-based cumulative stability constants (applicable to a given ionic strength): $\beta_i^* = [CuCl_i^{-i+1}]/[Cu^+][Cl^-]^i$. Millero[178,198] showed that Cu^+ and $CuCl^0$ are the most reactive species. Rates of Cu^+ oxidation are lower in seawater than in NaCl-NaClO$_4$ solutions of comparable I, and further studies[198] showed that Mg^{2+} and Ca^{2+} decrease the oxidation rate, while HCO_3^- increases it, but reasons for these trends are still unknown.

PART III. REACTIONS AT INTERFACES

The processes affecting the composition of aquatic systems involve much more than chemical reactions of substances dissolved in water itself, which were described in previous sections of this chapter. Reactions at air-water and solid-water interfaces also play essential roles in regulating the chemistry of natural waters. In the former category are included the absorption of volatile organic compounds from the atmosphere and their release (volatilization) back to the atmosphere, as well as transfer of unreactive gases (e.g., N_2, O_2) and reactive gases (e.g., CO_2, SO_2, NO_2) across the air-water interface. In this context, reactive gases are those that react with water itself to form species other than the "aquated" gas; unreactive gases may react with other solutes, but do not react with water, and exist in solution as "dissolved gases." Reactions at solid-water interfaces include adsorption and desorption of solutes onto/off of suspended particles and sediments, dissolution of mineral phases, and precipitation and solid-phase growth processes. The following sections describe the kinetics of these processes, focusing on gas transfer and mineral dissolution kinetics.

4.7 GAS TRANSFER

4.7.1 Importance

Gas transfer is important in the biogeochemical cycles of carbon, nitrogen, and sulfur, as well as the transport of volatile organic contaminants through the biosphere. The availability of inorganic carbon for primary production in some lakes may be limited by air-to-water transfer of CO_2. Dry deposition of acidic gases and vapors, such as SO_2, NO_2, and HNO_3, is of concern relative to acidification of aquatic ecosystems. Air-water exchange of NH_3 is of interest as a source or sink for nitrogen in water and also affects the proton balance of water bodies. The primary mechanism for input of organic contaminants, such as PCBs, to pristine water bodies is atmospheric transport and deposition on settling particles or direct gas transfer at the air-water interface. Transfer from water to the atmosphere is an important loss mechanism for volatile organic contaminants from natural waters and is an important treatment technique (air-stripping is used to remove volatile contaminants from wastewaters).

4.7.2 Gas Solubility: Henry's Law

The solubility of gases in water, S_w, is defined by Henry's Law, which states that solubility is directly proportional to the partial pressure of the gas in the atmosphere with which the water is equilibrated:

$$S_w = K_H P \tag{4-78a}$$

For organic compounds at temperatures below their boiling point, P represents the vapor pressure of the compound at the temperature of interest. The simplicity of the relationship in Equation 4-78a is diminished by the fact that the proportionality constant, K_H, depends on temperature, ionic strength, and the nature of the gas. Henry's Law constants for common gases are tabulated in several texts and articles,[200] and corresponding data for classes of organic contaminants have been compiled by various authors,[201] as listed by Mackay et al.[202] K_H commonly is expressed in mol L^{-1} atm^{-1} or sometimes as mg L^{-1} atm^{-1}.

An equally common convention expresses Henry's Law as

$$HS_w = P \tag{4-78b}$$

where H $(= K_H^{-1})$ has units of atm L mol^{-1} or mm Hg L mol^{-1}. The gas-phase content also can be expressed on a mass (or mole) per volume basis in the same units as S_w, in which case H is dimensionless. From the perfect gas law, $PV = nRT$, or $P = RT(n/V)$. The relationship between partial pressure (atm) of a gas-phase component c and its concentration $C_{a(c)}$ (in mg m^{-3}) thus is

$$P = 10^6 C_{a(c)} RT / M_c \tag{4-79}$$

M_c is the molecular weight of c, and R is the gas constant (L atm K^{-1} mol^{-1}).

Values of H for organic compounds are calculated from the measured vapor pressure and measured or calculated aqueous solubility. Difficulties are encountered in measuring S_w accurately for highly hydrophobic compounds, and special measurement techniques have been developed (see reference 202 for a review). Correlation methods also are used to estimate S_w and H from other properties of the compounds[203] (see Chapter 7).

4.7.3 General Gas-Transfer Equation

The kinetics of gas transfer is simple at the macroscopic level; the transfer rate is first order dependent on the difference between the actual concentration and the saturation value:

$$\frac{d[C]}{dt} = \frac{K_1 A}{V}\{[C]_s - [C]\} \tag{4-80}$$

K_1 is the gas-transfer coefficient (units of length time^{-1}, e.g., m day^{-1}, cm s^{-1}), and it varies with the nature of the gas and environmental conditions. For a given body of water, $A/V = 1/\bar{z}$, where $\bar{z}$ is mean depth, and $[C]_s - [C] = D$, the deficit of the dissolved gas with respect to saturation. If $D < 0$, the water is supersaturated, and transfer is from the water to the atmosphere. Thus, we can write,

$$\frac{d[C]}{dt} = -\frac{dD}{dt} = \frac{K_1 D}{\bar{z}} = k_2 D \tag{4-81}$$

where $k_2 = K_1/\bar{z}$ is a first-order rate constant (day^{-1}). The integrated form of Equation 4-81 is that of the reversible first-order rate equation (Equation 2-43b):

$$\ln\frac{[C]_t - [C]_s}{[C]_o - [C]_s} = \ln\frac{D}{D_o} = k_2 t \tag{4-82}$$

k_2 depends on such factors as stream and wind velocity. k_2 can be estimated by semiempirical equations, as described in Section 4.7.8.

4.7.4 Theories for Gas-Transfer Kinetics

Quantification of the factors affecting gas transfer requires an understanding of the mechanism of transfer. The most common mechanistic model for gas transfer is the two-film model developed by Whitman[204] in 1923, which was based on a similar model developed by Nernst in 1904.[205] According to this model, the rate-controlling step occurs in either of two boundary layers (or films) at the air-water interface (Figure 4-37). The two-film model involves four important assumptions:

1. The interface itself is assumed to be at equilibrium so that the gas- and liquid-phase concentrations at the interface are defined by Henry's Law.
2. Transfer through the bulk-gas and liquid phases is assumed to be rapid (non-rate controlling) and controlled by turbulent mixing.
3. Transfer through the surface films is considered to occur by molecular diffusion (the films are considered to be stagnant).
4. The transfer kinetics are assumed to be continuously at steady state.

The physical realism of the two-film model is open to question, and the fourth assumption is particularly problematic. On a microscale, turbulence is *not* a steady phenomenon, as anyone who has tried to battle a gusting wind can testify. The gusts are a large-scale turbulence. A similar unsteadiness occurs in the smaller-scale turbulence associated with the air-water interface, and the thickness of the interfacial films thus varies with time and space. Indeed, it is difficult to imagine an intact liquid film on the surface of a wind- and wave-swept ocean or lake. This realization

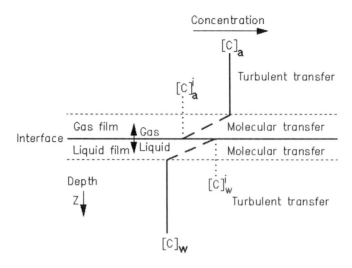

Figure 4-37. Two-film model for gas transfer.

has led to more realistic models involving renewal of the surface film by turbulent mixing that brings small eddies or "packets" of bulk water to the surface.[206]

Several variations of surface-renewal models have been proposed. In Higbie's *penetration* theory,[207] all packets of water reside at the interface for the same amount of time. The *surface-renewal* model of Danckwerts[208] assumes a normal (Gaussian) distribution for the length of time of packets remain at the interface. Both models assume that packets do not remain at the surface long enough for gas solutes to diffuse completely through them. Dobbins[209] combined film theory with the surface-renewal model so that some packets remain at the surface long enough for gas solutes to diffuse across them. More recently, McCready and Hanratty[210] developed a more fundamental relationship based on a numerical solution of the concentration boundary-layer profile. Although they had to make some assumptions that require further testing, their results generally verified the form of the parametric relationships given in the surface-renewal models. These models all lead to the same general form of transfer equation as the two-film model (i.e., Equation 4-80), but they do lead to different functional relationships between gas-transfer rate constants and molecular diffusion coefficients for gases. At least in theory the validity of these models can be tested by examining these relationships. This topic is discussed later, after the two-film model is derived. Gas-transfer kinetics is a mature topic with a very large literature. (For recent reviews see references 211,212.)

4.7.5 Derivation of Two-Film Gas-Transfer Equation

If transport through the films is by molecular diffusion, Fick's first law applies, and the flux of gas across the film is given by

$$F = -D\partial[C]/\partial z \qquad (4-83)$$

F has units of concentration area^{-1} time^{-1}; D is the molecular diffusion coefficient of the gas. In finite form Equation 4-83 is written as

$$F = k_j \Delta[C] \tag{4-84}$$

$\Delta[C]$ is the concentration difference across film j of thickness z, and k_j is a gas exchange coefficient for that film ($= D/z$) with units of velocity (cm s^{-1}). Because D is constant for a given gas and temperature, k_j depends (inversely) on z, the thickness of the film. In turn, z depends on the degree of turbulence in the fluid. The gas-exchange coefficient defines the gas flux across the film for a unit concentration gradient; the reciprocal of k_j is a measure of the *resistance* of a film to gas transfer. The total resistance to transfer is the sum of the resistances across the gas and liquid films.

In that there is no build-up or depletion of gas in either film, the flux across the two films is a steady-state process. Thus, we can equate the fluxes across each film and write them in terms of Equation 4-84:

$$F = k_g ([C]_a - [C]_a^i) = k_1 ([C]_w^i - [C]_w) \tag{4-85}$$

Superscript i refers to concentrations right at the interface; subscripts a and w refer to air and water phases, respectively. The interfacial air and water concentrations, $[C]_a^i$ and $[C]_w^i$, cannot be measured, but they can be eliminated from the flux equation in the following way. Because Henry's Law is assumed to apply at the interface, we can write,

$$H[C]_w^i = [C]_a^i \tag{4-86}$$

By solving the two equalities of F in Equation 4-85 for $[C]_a^i$ and $[C]_w^i$, respectively, and substituting the results into Equation 4-86, we obtain the following equations for F in terms of measurable concentrations only:

$$F = \frac{k_1 k_g}{k_1/H + k_g} \left\{ [C]_a/H - [C]_w \right\} = K_1 \left\{ [C]_a/H - [C]_w \right\} \tag{4-87a}$$

or

$$F = \frac{k_1 k_g}{k_1 + k_g H} \left\{ [C]_a - H[C]_w \right\} = K_g \left\{ [C]_w - H[C]_w \right\} \tag{4-87b}$$

where

$$\frac{1}{K_1} = \frac{1}{k_1} + \frac{1}{Hk_g} \tag{4-88a}$$

and

$$\frac{1}{K_g} = \frac{1}{k_g} + \frac{H}{k_1} \tag{4-88b}$$

Both sets of equations (4-87a/4-88a and 4-87b/4-88b) are equally valid, but most authors express two-film gas-transfer kinetics in terms of Equations 4-87a and 4-88a, and we shall do the same. The units of K_l, k_l, and k_g are identical (e.g., cm h^{-1}). If $[C]_a$ and $[C]_w$ are expressed in identical units (e.g., g m^{-3}), then H is dimensionless. If the air phase of the gas is expressed in pressure, the term RT must be added to the numerator of the second term in Equation 4-88a:

$$\frac{1}{K_l} = \frac{1}{k_l} + \frac{RT}{Hk_g} \tag{4-88c}$$

and H then has units of atm L mol^{-1} or mm Hg L mol^{-1}. To convert Equation 4-87 from flux to change in concentration ($-d[C]_w/dt$), the left side of the equation must be multiplied by (A/V) or $1/\bar{z}$.

The classic two-film gas-transfer model and its surface-renewal relatives implicitly assume that the interface itself poses no barrier to gas transfer; resistance is caused only by diffusion through the gas and/or liquid films. However, when water is covered by films of surface active agents such as C_{14}–C_{20} alcohols, the transfer rates for CO_2 and O_2 are reduced by factors of 1.5 to 4.0.[213,214] Smith et al.[213] suggested that this is evidence that the interface itself provides some resistance to gas transfer. Another possibility is that the surfactant films increase the thickness of the "concentration boundary layer" (i.e., the gas and liquid films), as described by McCready and Hanratty.[210] Conditions right at the interface are particularly important in understanding air-water exchange of hydrophobic organic compounds.[202] Even at equilibrium, concentrations of such compounds may be much higher at the interface than in the bulk water, because this arrangement minimizes disruption of strong water-water hydrogen bonding. The influence of this increased "capacity" for hydrophobic compounds at the interface on water-to-air transport is still uncertain.

For a gas obeying Henry's Law, a third term can be added to Equation 4-88c to account for the resistance of the interface itself:

$$\frac{1}{K_l} = \frac{1}{k_l} + \frac{RT}{Hk_s} + \frac{RT}{Hk_g} \tag{4-89}$$

where k_s is the interface gas-transfer coefficient. From the kinetic theory of gases, Smith et al.[213] proposed that

$$k_s = \alpha(RT / 2\pi m)^{1/2} \tag{4-90}$$

α is the fraction of gas molecules striking the water surface that condense on it. Equation 4-90 suggests that an activation energy is associated with interfacial crossing.

4.7.6 Limits for Liquid- and Gas-Film Control

Equation 4-88a,c shows that the reciprocal of K_l, the overall gas-transfer coefficient, is the sum of terms representing the resistance to gas transfer in the

Table 4-14. Air-Water Transfer Coefficients and Related Data for Some Compounds[a]

Compound	Aqueous solubility mg L^{-1}	Vapor pressure mm Hg	Henry's Law constant atm L mol^{-1}	K$_l$[b] cm h^{-1}
Volatile organic compounds				
Aldrin	2.0E-1	6.0E-6	1.4E-8	3.7E-1
Benzene	1780	95.2	5.5E-6	14.4
Biphenyl	7.5	5.7E-2	1.5E-6	9.2
Cumene	50	4.6	1.5E-5	11.9
DDT	1.2E-3	1.0E-7	3.9E-8	9.3E-1
Dieldrin	2.5E-1	1.0E-7	2.0E-10	5.3E-3
DMS (CH$_3$)$_2$S)	—	—	7.3	19.2
Lindane	7.3	9.4E-6	4.9E-10	1.5E-2
Naphthalene	33	2.3E-1	1.2E-6	9.6
n-octane	6.6E-1	14.1	3.2E-3	12.4
PCBs[c]				
Aroclor 1242	2.4E-1	4.1E-4	5.7E-7	5.7
Aroclor 1248	5.4E-2	4.9E-4	3.5E-6	7.2
Aroclor 1254	1.2E-2	7.7E-5	2.8E-6	6.7
Aroclor 1260	2.7E-3	4.1E-5	7.1E-6	6.7
TCDD (2,3,7,8-tetrachlorodi-benzodioxin)	1.9E-5	7.4E-10	1.6E-2	1.7E-1
Toluene	515	28.4	6.7E-6	13.3
o-xylene	175	6.6	5.3E-6	12.3
Fixed gases				
SO$_2$	—	—	9.3E-1	1.6E3
N$_2$O	—	—	39.1	20
CO	—	—	1223	20
CH$_4$	—	—	1027	20
CCl$_4$	—	—	26.4	10.7
CClF$_3$	—	—	122	11.3

[a] Values at 25°C. Organic compounds extracted or calculated from sources listed in Mackay, D. and P.J. Leinonen, *Environ. Sci. Technol.*, 9, 1178 (1975); and Podoll, R.T., et al., *Environ. Sci. Technol.*, 20, 490 (1986); fixed gases (including DMS) from Liss, P.S. and P.G. Slater, *Nature*, 247, 161 (1974).

[b] K$_l$ varies with environmental conditions (e.g., wind speed); values listed here are relative to K$_l$ for O$_2$ = 20 cm h^{-1}, which is representative of "average" conditions for the open ocean.

[c] Aroclors are mixtures of PCB congeners; the last two digits in the name gives the percent Cl in the mixture; e.g., 1254 contains 54% Cl (by weight).

liquid film and gas films. Although k_g and k_l are not constant among different gases, they do not have wide ranges, and the primary variable that determines whether the controlling resistance is in the liquid or gas film is H, the Henry's Law constant. Table 4-14 summarizes data on H and K$_l$ for some unreactive and reactive gases and some organic compounds with varying volatility. Gas-transfer conditions for compounds that are liquid-film controlled sometimes are expressed in terms of the apparent thickness of the liquid film. As indicated by Equations 4-83 and 4-84, this can be calculated from a measured value of K$_l$ (or k$_l$) and the diffusion coefficient of the substance. Typical values for z_l are in the range of tens of μm for seawater,

Table 4-15. Typical Liquid Film Thickness in Various Water Bodies

Water body	z_l (μm)	Water body	z_l (μm)
Oceans		**Large lakes**	
High latitude	14	Great Salt (UT)	75
Low latitude	63	Pyramid (NV)	150
N. Atlantic	67	Mono (CA)	>200
N. Pacific	31		
S. Pacific	63	**Small lakes**	
Antarctic	24		
		Lake 227 (ELA)	750
		Lake 261 (ELA)	750
		Lake 304 (ELA)	300

Calculated from data tabulated by Emerson[217] from various sources.

a few hundred μm in lakes, and up to 1 mm in small, wind-sheltered water bodies (Table 4-15).

According to Liss and Slater,[216] when H is more than 3500 mm Hg L mol^{-1} (4.4 × 10^{-3} atm m^3 mol^{-1}), flow through the liquid film controls gas transfer in most systems. This applies to unreactive gases and organic compounds with high volatility. When H is less than 10 mm Hg L mol^{-1} (1.2 × 10^{-5} atm m^3 mol^{-1}), transfer in the gas film controls in most flow conditions. This applies to reactive gases and low volatility organic compounds. For example, H ≈ 5.5 × 10^{-5} atm m^3 mol^{-1} for NH_3,[218] and according to the above criteria, transfer of this reactive gas is mostly controlled by the gas film. For compounds with intermediate values of H, both films contribute significantly to gas-transfer resistance.

The basis for the above criteria can be explained as follows. Evaporation and condensation of water is controlled by gas-film resistance only; this statement is true generally for solvents. There is no gradient of water in the "liquid film". Thus, there is no liquid film as far as water molecules are concerned, and the total resistance for water evaporation is equal to k_g. The mean value of K_{H_2O} (hence, $k_{g(H_2O)}$) reported for the open ocean is 3000 cm h^{-1},[216] but ~2000 to 3000 cm h^{-1} is a reasonable range for natural waters. In contrast, gas transfer for the unreactive gas O_2 is controlled by liquid-film resistance only, as shown by the following calculation,[216] which is based on a measured value of H_{O_2} ≈ 30 (dimensionless) and an average value of $K_{l(O_2)}$ ≈ 20 cm h^{-1} for the open ocean (this value applies to moderately windy, wavy conditions). If we assume that $k_{g(O_2)} = k_{g(H_2O)} = 3000$, then

$$\frac{1}{K_l} = 1/20 = 1/k_{l(O_2)} + 1/30(3000)$$

or $k_{l(O_2)} \approx K_{l(O_2)} = 20$ cm h^{-1}. If we assume that $k_{g(H_2O)} = 3000$ cm h^{-1} and $k_{l(O_2)} = 20$ cm/h^{-1} are typical values for k_g and k_l, respectively, then it is easy to show that the above criteria for H (10 and 3500 mm Hg L mol^{-1}) represent the values at which 95% of the resistance to gas transfer occurs in one of the two films.

4.7.7 Predictive Methods for Film and Gas-Transfer Coefficients

Regardless of whether gas transfer is limited by the liquid film, the gas film, or both, transfer rates vary with environmental conditions. The gas-transfer coefficients K_l, k_l, and k_g are *not* constants for a given gas. For example, k_g depends on wind speed, which determines the thickness of the gas film. Similarly, k_l depends on the degree of turbulence in the water, which generally is related to wind speed over natural water bodies; in the two-film model, an increase in water turbulence is assumed to decrease the thickness of the liquid film. It is not practical to measure K_l (or k_l and k_g) for all the substances of interest over the range of conditions needed to predict transfer rates in the ambient environment. Instead, values applicable to field conditions are estimated from relationships based on *reference compounds* whose transfer coefficients have been measured under a wide range of environmental conditions. Water and O_2 are used as reference compounds because the transfer rates of each are controlled exclusively by one of the films (gas for water, liquid for O_2) and because numerous studies have measured these coefficients under field conditions. Two general approaches are used to calculate K_l values for substances from reference compound data: (1) from ratios of diffusion coefficients, based on gas transfer theory; and (2) from ratios of substance and reference transfer rates measured simultaneously under controlled conditions.

Theoretical Approaches

According to two-film theory, k_j is proportional to D, the molecular diffusion coefficient for the compound of interest in the rate-limiting film; i.e., $k_j = D/z$. For any unreactive gas or highly volatile compound c (which will be liquid-film controlled) the theory thus predicts that

$$K_{l(c)} / K_{l(O_2)} = k_{l(c)} / k_{l(O_2)} = D_{l(c)} / D_{l(O_2)} \tag{4-91}$$

In contrast, surface renewal models[207,208] yield equations in which k_l is proportional to $D_l^{1/2}$ (see reference 206 for a derivation). If these models are correct, the reference compound relationship thus should be

$$k_{l(c)} / k_{l(O_2)} = \left(D_{l(c)} / D_{l(O_2)} \right)^{1/2} \tag{4-92}$$

If the mixed two-film, surface-renewal model of Dobbins[209] is correct, the following equation has been shown to apply:

$$k_l = (D_l s)^{1/2} \coth(s L^2 / D_l)^{1/2} \tag{4-93}$$

where s is the fractional rate of packet displacement, and L is packet thickness (similar to film thickness) (see Smith et al.[213] and Holley[219] for further details). Equation 4-93 has two limiting values. When L is small or D_l is large, $k_l \rightarrow D_l/L$, which corresponds to the prediction from two-film theory; and when L is large or D_l is small, $k_l \rightarrow (D_l s)^{1/2}$, which corresponds to the prediction from the surface-renewal theories. Under intermediate conditions, the dependency of k_l on D_l is intermediate. Therefore,

$$k_{l(c)}/k_{l(O_2)} = (D_{l(c)}/D_{l(O_2)})^n, \qquad 0.5 \leq n \leq 1 \qquad (4\text{-}94)$$

Dobbins[209] reported results of lab experiments with O_2 wherein k_l was proportional to $D_l^{1/2}$ at high turbulence and proportional to D_l at low turbulence, thus lending support to the mixed two-film/surface-renewal model. Moreover, Smith et al.[213] found that a plot of $log(k_c/k_{l(O_2)})$ vs. $log(D_{l(c)}/D_{O_2})$ for 12 highly volatile (liquid-film controlled) compounds and gases had a slope (n) of 0.61, which also agrees with predictions from the mixed model.

Several researchers have suggested that K_l for volatile substances is related inversely to the Schmidt number (S_c) according to a power-law expression.[220] S_c is defined as the kinematic viscosity of water (η_w) divided by the molecular diffusion coefficient of a substance. Thus,

$$K_{l(1)}/K_{l(2)} = (S_{c(2)}/S_{c(1)})^n \qquad (4\text{-}95)$$

Because η_w cancels out of the two S_c terms, Equation 4-95 is mathematically equivalent to Equation 4-94. Watson et al.[221] recently evaluated n in Equation 4-95 by conducting field-scale tracer measurements in two lakes at two wind speeds and found $n = 0.51$, which agrees with the surface-renewal model (but does not contradict the mixed model, Equation 4-93). Jähne et al.[222] found that n changed from 0.67 to 0.5 at the onset of waves in wind tunnel experiments, suggesting that wave action induces a fundamental change in hydrodynamic boundary conditions at the surface. They interpreted this change to involve the development of local divergences and convergences at the surface and a transfer of wind energy gained by the waves to near-surface turbulence.

Relationships analogous to Equation 4-91 to 4-94 should apply to compounds of low volatility and reactive gases, which are gas-film controlled, except that gas-phase diffusion coefficients must be used. Tamir and Merchuk[223] found that n in Equation 4-94 varies with gas-phase turbulence for such gases.

If both films affect the transfer rate, which is the case for organic compounds of intermediate volatility, both k_l and k_g must be estimated. This would be easy to do if either two-film or surface-renewal theory were valid over a wide range of turbulence (with n constant at 0.5 or 1). As discussed above, this appears not be be the case, and n varies with the degree of turbulence. It is possible to fit measured data for such compounds to a modified form of Equation 4-87 by nonlinear regression analysis:[224]

$$\bar{z}/k_2 = 1/K_l = 1/k_{l(O_2)}A + RT/Hk_{g(H_2O)}B \qquad (4\text{-}96)$$

where $A = D_{l(c)}/D^n_{l(O_2)}$ and $B = D_{g(c)}/D^{n'}_{g(H_2O)}$. $D_{l(O_2)} = \sim 2.0 \times 10^{-5}$ cm^2 s^{-1} and $D_{g(H_2O)} = 0.24$ cm s^{-1} (both at 20°C). Values of the diffusion coefficients for other compounds can be found in the literature or estimated as described below.

Values of D_l and D_g are available for some small molecules, but in many cases they must be calculated. Diffusion coefficients of organic compounds often are estimated from two well-known formulas: Einstein's diffusion equation and Graham's Law, but neither equation actually is appropriate for calculation of D_l.[213] Einstein's equation for diffusion of particles in a fluid, states that $D_1/D_2 = d_2/d_1$, where d is the molecular diameter of a compound, but this is valid only for large spherical molecules. Liss and Slater[216] estimated D_1/D_2 as $(m_2/m_1)^{1/2}$, where m is the molecular weight of a compound, which is based on Graham's diffusion law, but this is valid only for molecules diffusing in a vacuum. Nonetheless, for most compounds, estimates of D_1/D_2 based on either approximation are within a factor of 2 of the measured values.[213]

A more accurate equation to estimate D is the Othmer-Thakar relationship:[225]

$$D_w^c = 14 \times 10^{-5} \eta_w^{-1.1} V_c^{-0.6} \qquad (4\text{-}97a)$$

where D_c^w (cm^2 s^{-1}) is the diffusion coefficient of c in water; ηv_w (centipoise; 10^{-2} g cm^{-1} s^{-1}) is the viscosity of water; and V_c (cm^3 mol^{-1}) is the molar volume of the solute at its normal boiling point. Values of D_c^w calculated by this equation are within $\pm 11\%$ of measured values.[213] Hayduk and Laudie[226] reevaluated Equation 4-97a with more recent data on D_c^w and modified the coefficients slightly:

$$D_c^w = 13.26 \times 10^{-5} \eta^{-1.4} V_c^{0.589} \qquad (4\text{-}97b)$$

Another common empirical equation for D_c^w is the Wilke-Chang relationship:[227]

$$D_c^w = 7.4 \times 10^{-8} (\chi M_c)^{1/2} T / \eta_w V_c^{0.6} \qquad (4\text{-}98)$$

where χ is an association parameter for the solvent (2.6 for water), T is the absolute temperature, and the other variables are as defined above. The last three equations yield similar values of D_c^w, and both Equation 4-97b and Equation 4-98 are used commonly. Chemical engineers have developed several other equations to estimate aqueous diffusion coefficients. (see references 228,229).

Gulliver et al.[230] used various theoretical relationships described earlier to develop a general indexing equation that relates transfer rates of liquid-film-controlled compounds to the transfer rate of a reference gas, O_2, at 20°C. An indexing factor, f_i, for any gas i, was defined as

$$f_i = \log r_{i(T°C)} / \log r_{O_2(20°C)} = K_{l,i(T°C)} / K_{l,O_2(20°C)} \qquad (4\text{-}99)$$

where $r_{i,T°C}$ is the gas-transfer rate of i at any temperature T. They showed that f_i could be computed from a relationship involving ratios of various physical variables (e.g., diffusion coefficients, temperature, dynamic viscosity) raised to appropriate powers, and no fitted parameters were needed.

Empirical Approaches

The gas-transfer coefficient of a substance for a given set of environmental conditions can be estimated from the corresponding value for a reference compound and the ratio of coefficients for the substance and reference compound measured under controlled laboratory conditions; e.g.,

$$K_{l,env}^{c} = (K_{l,con}^{c} / K_{l,con}^{ref}) K_{l,env}^{ref} \tag{4-100}$$

where c denotes the compound of interest; ref, the reference compound; env, the environmental condition of interest; and con, a condition under which K_l has been measured or calculated for both compounds. It is assumed that the ratio does not vary with environmental conditions. Equation 4-100 applies to compounds whose transfer rates are wholly limited by one of the two films (so that a single reference compound (O_2 or H_2O) applies. For compounds affected by the resistances of both films, a slightly more complicated procedure must be used.

The reference compound approach has been used to estimate film coefficients for ethylene dibromide (EDB), a common soil fumigant and a contaminant in natural waters. The Henry's Law constant for EDB is 0.623 mm Hg L mol^{-1}, which is in the range where gas-film effects dominate, but liquid-film conditions still exert some effects. The gas-film coefficient, k_g, cannot be measured directly for EDB in water because P_{EDB} at the air-water interface cannot be measured. Instead, Rathbun and Tai[231] determined $k_{g(EDB)}$ by measuring the volatilization rate of pure EDB (which is solely gas-film limited) at controlled wind speeds in the laboratory. They compared their values of $k_{g(EDB)}$ to the evaporation rate of water under the same conditions and found that the ratio $\psi = k_{g(EDB)}/k_{g(H_2O)}$ was independent of wind speed over the two liquids and equal to 0.41. In a subsequent study,[232] they determined $K_{l(EDB)}$ in the same laboratory apparatus and calculated $k_{l(EDB)}$ by Equation 4-88c from H_{EDB} and the previous estimate of $k_{g(EDB)}$. They found that $k_{l(EDB)}$ was independent of wind speed, while $k_{g(EDB)}$ was independent of liquid stirring speed in their apparatus. Both findings agree with the two-film theory. By measuring $k_{l(O_2)}$ in their apparatus over a range of wind speeds, they found that the ratio $\Phi = k_{l(EDB)}/k_{l(O_2)}$ was constant with a value of 0.61. Gas-transfer coefficients for EDB under environmental conditions thus can be estimated from the larger body of data and predictive relationships available for O_2 gas transfer under ambient conditions.

A major drawback to O_2 as a reference compound is the difficulty in making reliable measurements of $K_{l(O_2)}$ under field conditions. Many processes besides gas transfer affect O_2 concentrations in surface waters; concentrations in lakes and seawater usually are not far from equilibrium; and the large concentrations of O_2 present in the atmosphere and water make measurement of fluxes a difficult task. Most of the field measurements done to determine the effects of turbulence on $K_{l(O_2)}$

have been performed on streams or rivers affected by discharges of O_2-demanding wastewater effluent, and such studies have not provided satisfactory information on the effects of wind and wave activity on gas-transfer rates for large lakes and the oceans.

Several tracers have been developed for use as reference compounds during the past decade to measure gas-transfer coefficients and modify laboratory-measured or calculated values of $K_{l(c)}$ for field conditions. Such tracers include:

1. Naturally occurring radioisotopes such as radon (^{222}Rn),[233] which is produced in sediments by decay of (^{226}Ra) (see Example 2-1)
2. Naturally occurring trace gases, such as CH_4, which is produced in sediments and anoxic bottom waters[234]
3. Artificial radioisotopes, such as ^{14}C, which was added to a lake to measure rates of CO_2 exchange[235]
4. Stable gases, such as ethylene, sulfur hexafluoride (SF_6), and ^3He,[221,236,237] that are extremely rare or not present in the atmosphere and can be added to water masses and measured over a wide range of concentrations and at very low levels. Other examples of tracers are reviewed in Wilhelms and Gulliver.[211] Although the methods used to obtain $K_L{}^{ref}_{env}$ depend on the nature of the tracer and type of water body, the approach used to relate $K_L{}^{ref}_{env}$ to the compound of interest is generally the same (i.e., Equation 4-99).

4.7.8 Measurement of Gas-Transfer Coefficients for Reference Compounds

In laboratory experiments, $k_{g(H_2O)}$ can be determined by measuring the rate of evaporation of water from a weighed vessel, along with room temperature and relative humidity. The data are applied to the following equation:

$$E_w = k_{g(H_2O)}(P_w^s - P_w)/RT \qquad (4\text{-}101)$$

where P_w^s is the vapor pressure of water at the temperature of interest, and P_w is the actual partial pressure of water (obtained from relative humidity and temperature). It is assumed that P_w^s applies at the air-water interface, and P_w applies at the upper limit of the gas film. The evaporation of water was first described by an equation of this general form by Dalton in 1802.

To estimate $k_{g(H_2O)}$ under field conditions, we must determine the rate of evaporation of water from the surface of the water body. This is one of the fundamental concerns of hydrology and thus is a well-studied process. Many books treat this subject in depth,[238] and our discussion will be brief. For a given water body (such as a lake), the rate of evaporation can be estimated by four methods: (1) water balance, (2) energy balance, (3) pan-lake correlations, and (4) mass-transfer theory. Once E is known, $k_{g(H_2O)}$ can be determined from Equation 4-101 if the relative humidity of the bulk atmosphere is known.

The water balance method is based on the law of mass conservation:

$$\Delta S = P - E + I - O + G_I - G_0 \qquad (4\text{-}102)$$

$$
\begin{array}{l}
\text{change in} \\
\text{storage}
\end{array}
= \text{precipitation} - \text{evaporation} +
\begin{array}{l}
\text{surface} \\
\text{inflows}
\end{array}
-
\begin{array}{l}
\text{surface} \\
\text{outflows}
\end{array}
+
\begin{array}{l}
\text{groundwater} \\
\text{inseepage}
\end{array}
-
\begin{array}{l}
\text{groundwater} \\
\text{recharge}
\end{array}
$$

Provided that all terms but E in Equation 4-102 can be measured accurately, E can be determined by difference. However, measurement errors are associated with each term, and in some cases they may be large. Consequently, the uncertainties in E calculated from water balances may be unacceptable. This is especially true for lakes where groundwater flows (which are difficult to measure accurately) are a large part of the water budget. Because of the large data requirements for accurate water budgets, this is not a common method to estimate E; moreover, the time-scale of such measurements is coarse — generally monthly or longer. Nonetheless, careful water balance measurements do form the basis for the pan-lake correlation method.

The energy balance method is based on the principle of conservation of energy and accounts for the total energy input to a water body by net advection and solar radiation at the water surface. The energy absorbed (and stored) by the water body, and the amounts of energy lost by reflection, longwave exchange, and as sensible heat, are measured or computed. If other energy terms (such as loss by conduction through the lake bottom) are assumed to be negligible, the difference between energy inputs and losses plus storage represents the energy used for evaporation. Because of the difficulties encountered in measuring all the components of the energy balance with sufficient accuracy, this approach has had limited application.

Evaporation pans are common and fairly accurate methods to measure E, and data are available for many sites around the world. However, pans provide a different environment for evaporation than do large water bodies, and pan evaporation is higher than that from lakes and oceans. Pans have limited heat storage capacities and respond quickly to changes in air temperature. Wind affects the surface conditions in a 4-ft. (diameter) pan (the standard size for pans) differently than it affects the surface of a river, lake, or the oceans. Careful studies of pan and lake evaporation at a few sites have produced pan-lake coefficients that are in the general range 0.65 to 0.8. Seasonal variations are exhibited by pan coefficients, and they may be erratic for time intervals less than a month.[238] Consequently, lake evaporation rates estimated by pans are considered accurate as long-term averages, but not for short-term conditions.

Mass-transfer approaches solve Equation 4-101 for E by parameterizing the gas (mass) transfer coefficient directly. Several procedures have been developed, including eddy correlation,[239] which is regarded as the most accurate method. In general, mass-transfer approaches measure physical conditions (e.g., vertical profiles of temperature and wind) in the turbulent boundary layer above the gas film at the water surface to calculate eddy diffusion and transport rates of water vapor. A simple empirical formula that relates E to wind speed was developed from field studies on Lake Hefner, a reservoir in Oklahoma:[240]

$$E = 2.04 \times 10^{-3} U_w (P_w^s - P_w) \qquad (4\text{-}103)$$

Table 4-16. Typical Ranges of k_2 for O_2 in Various Types of Water Bodies[a]

Water body	Range of $k_{2(O_2)}$ at $20°C$[b] day^{-1}
Small ponds and backwaters	0.10–0.23
Sluggish streams; large lakes	0.23–0.35
Large streams (low velocity)	0.35–0.46
Large streams (normal velocity)	0.46–0.69
Swift streams	0.69–1.15
Rapids and waterfalls	>1.15

[a] k_2 is a first-order "reaeration" rate constant (units of t^{-1}); $k_2 = K_l/\bar{x}$, where $\bar{x}$ = ave. stream depth.

[b] For other temperatures, $k_{2(T)} \approx k_{2(20)}1.024^{(T-20)}$. Modified from Tchobanoglous,G. and E.D. Schroeder, *Water Quality*, Addison-Wesley, Reading, MA, 1985.

E is in g (H_2O) cm^{-2} day^{-1}, and U_w is wind velocity in knots (1 knot = 1.15 mph or 0.51 m s^{-1}) measured 9 m above the water surface.

Gas-transfer coefficients for O_2 have been determined under controlled laboratory conditions many times by measuring the rate of increase in the dissolved O_2 concentration of deaerated water. The effects of air and water turbulence are simulated by varying stirring speeds in the water and air velocity by fans (or in wind tunnels), but such techniques cannot reproduce the surface roughness conditions found in natural water bodies exposed to moderate and high winds; the importance of wave action and surface roughness on gas transfer has been noted by many workers. Consequently, there is no substitute for field measurements. Ranges of $K_{l(O_2)}$ for different types of flowing and standing water bodies are listed in Table 4-16. While useful for rough estimates, such tabulations are too coarse to provide accurate rates of gas transfer under ambient environmental conditions, especially in standing water bodies, for which the ranges are large and the database is small.

Both wind tunnel and field experiments have shown that K_l for O_2 and other unreactive gases is a nonlinear function of wind velocity, U (Figure 4-38), and much effort has been devoted to understanding the factors affecting this relationship. A widely cited empirical equation for rivers and streams is[244]

$$k_2 = K_1/\bar{z} = 11.6U/\bar{z}^{1.67} \qquad (4-104)$$

where k_2 is a first-order coefficient (units of t^{-1}) called the reaeration rate constant; U is in ft s^{-1}; and $\bar{z}$ is in ft. Similar expressions have been reported by others (see reference 245 for a review), but none of these relationships has a universal applicability to river and stream gas transfer.

Given that air-water exchange rates for unreactive gases are controlled by the liquid film rather than the gas film, the nonlinearity between U and K_l is understandable. Cohen et al.[246] showed that K_l is a function of the roughness Reynolds number, Re*, for the air above the interface (Figure 4-39a):

$$K_1 = 11.4(\text{Re}^*)^{0.195} - C, \qquad \text{for } 0.11 < \text{Re}^* < 102 \qquad (4-105a)$$

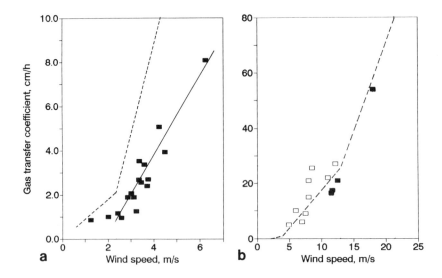

Figure 4-38. Wind tunnel and field experiments show K_l is a nonlinear function of wind speed: (a) closed squares, field data from SF_6 experiment in a small lake; solid line,[236] least-squares fit for data above a wind speed of 2.4 m s⁻¹; short dashed line is from a wind tunnel experiment.[242] (b) CO_2 air-sea transfer rates at 20°C vs. wind speed at height of 10 m. Closed squares from SF_6 experiment in North Sea;[221] open squares, radon flux and other data from various sources; dashed line, theoretical model[243] based on lake and wind tunnel data. [Figure (a) redrawn from Wanninhof, R., et al., *Science*, 227, 1224 (1985); (b) redrawn from Watson, A.J., et al., *Nature*, 349, 145 (1991).]

where
$$Re^* \ (\text{dimensionless}) = Z_o U^* / \eta_a^k \qquad (4\text{-}105b)$$

and Z_o = effective roughness height (cm), U^* = friction velocity (cm s⁻¹), η_a^k is the kinematic viscosity of air (cm² s⁻¹), and C is a constant derived from lab studies (range of 4.1 to 5.0) that depends on stirring speed of the water. Both Z_o and U^* are calculated from a *log-log* plot of measured wind velocity, U, vs. height above the air-water interface, Z (Figure 4-39b). Cohen et al. further proposed that the relationship between K_l and U could be divided into three regions:

1. **U < 3 m s⁻¹.** The water surface is relatively calm and flow is smooth; K_l typically is in the range 1 to 3 cm h⁻¹ and is not strongly related to U. Instead, K_l depends primarily on mixing that originates within the water body (not wind induced) and on the properties (H, D_l) of the gas or volatile compound.
2. **U ≈ 3–10 m s⁻¹.** Shear stress at the interface reaches a critical value and sets the liquid into motion at or near U = 3m s⁻¹, depending on fetch. This produces surface ripples and roughness, which decreases the thickness of the liquid film (or increases the rate of surface renewal). At U > 6 m s⁻¹, flow is completely rough and wave growth is appreciable, causing K_l to increase rapidly. K_l increases from ~3 cm h⁻¹ to ~30 cm h⁻¹ in this region.
3. **U > 10 m s⁻¹.** Wave breaking may occur; bubble entrainment and the presence of spray further enhance gas-transfer rates. These conditions are difficult to simulate in the laboratory. K_l may reach values of 70 cm h⁻¹.

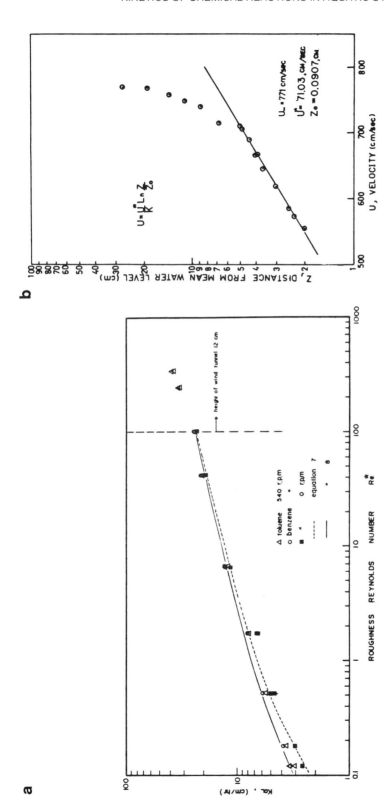

Figure 4-39. (a) Variation in K_l with the air-roughness Reynolds number, Re^*; (b) *log-log* plot of wind speed (U) vs. height above the water surface (Z). Z_0 is obtained from the y-intercept, and U^* from the slope of the linear portion of the curve. [From Cohen, Y., W. Cocchio, and D. Mackay, *Environ. Sci. Technol.*, 12, 553 (1978). With permission.]

Recent field experiments with SF_6 as a tracer for air-water exchange rates (Figure 4-38) have verified the nonlinearity of K_l vs. U relationship and the sharp increase in the dependency of K_l on U at wind speeds above ~3 m s^{-1}. Data at U > 10 m s^{-1} still are so sparse and variable that the nature of the K_l-U relationship is not well defined for these conditions.

4.7.9 Enhancement Factors for Reactive Gases

If a compound reacts rapidly with water, its rate of removal within the liquid film will be enhanced, thus increasing the concentration gradient in the film and enhancing diffusive transport. Compounds in this category include acids (e.g., HNO_3 vapor, SO_2, and CO_2), bases (e.g., NH_3), and substances that undergo redox reactions (e.g., NO_2). The degree of mass-transport enhancement caused by chemical reaction may be quantified by adding a chemical enhancement factor α to the term for liquid-film resistance in Equation 4-88c; i.e.,

$$\frac{1}{K_l} = \frac{1}{\alpha k_l} + \frac{RT}{Hk_g}$$

(4-106)

As α increases above 1.0, the term that defines the amount of resistance caused by the liquid film becomes smaller. The magnitude of α depends on the reactivity of the gas. Strong acids like HNO_3 react with water almost instantaneously (α is large) so that there is virtually no undissociated HNO_3 in the liquid film (hence no concentration gradient in the film), and gas transfer thus is limited by diffusion through the gas film.

At the other extreme, the reaction rate for NO_2 is slow enough[247] that the concentration gradient in the liquid film is not affected, and α is near 1. Because the reaction $2NO_2 + H_2O \rightarrow HNO_3 + HNO_2$ is second order in NO_2, the extent of chemical enhancement varies with P_{NO_2}. Schwarz and Lee[248] concluded that enhancement required $P_{NO_2} >$ ~10^{-8} atm (~20 µg m^{-3}), which is much higher than typical values of NO_2 in nonurban air (1–3 µg m^{-3} for NO_x ($NO + NO_2$) in rural terrestrial areas and much lower in the marine atmosphere). However, NO_2 concentrations can exceed the critical value in polluted urban atmospheres, and chemical enhancement of NO_2 absorption by water could be significant in these cases.

The possibility that chemical enhancement is significant for CO_2 absorption by water has received much attention for two reasons. First, CO_2 is the single-most important greenhouse gas, accounting for about half of the warming predicted by global climate models. The oceans are a major sink for the CO_2 added to the atmosphere by fossil fuel consumption and deforestation, and exchange rates of CO_2 at the air-water interface thus are of great interest. Second, carbon is potentially a limiting element for plant growth in lakes. In the late 1960s and early 1970s some scientists hypothesized that the rate of CO_2 absorption from the atmosphere controlled the severity of algal blooms, especially in softwater lakes. Subsequent studies[249] showed that CO_2 limitation is unlikely, except in extreme circumstances, but the controversy stimulated research on the kinetics of CO_2 transfer across the air-water interface.

Three principal factors determine whether chemical enhancement is significant for CO_2 transfer: pH, alkalinity, and the thickness of the liquid film, z_l. The first two factors affect the mechanism and kinetics of CO_2 hydration (see Section 4.4.1). At pH $\leq \sim 8$, reaction of CO_2 with H_2O is the primary mechanism, but at the high pH values (9 to 10) associated with intense algal blooms, direct reaction of CO_2 with OH^- accelerates the removal of CO_2 from the liquid film. Alkalinity affects the mechanism by buffering pH. The magnitude of z_l determines the steepness of the concentration gradient for CO_2 at the water surface. A thin film implies a steep gradient, thus promoting Fickian diffusion; a thick film allows a more modest gradient, thus slowing diffusion. For typical seawater conditions, the three factors combine to minimize the importance of chemical enhancement for CO_2 transfer. The pH of seawater is well buffered and remains near 8.0; the large fetch of the oceans and normal wind patterns promote rough surface conditions and a thin liquid film (typically less than 100 μm). A mathematical model developed by Quinn and Otto[250] shows that $\alpha < 1.1$ at z_l values typical of seawater, and for practical purposes, CO_2 can be considered an unreactive gas in the context of air-sea exchange rates.

In contrast, several studies show that chemical enhancement occurs in some freshwater situations. Schindler et al.[249] found such evidence in a whole-lake manipulation on Lake 227, in the Experimental Lakes Area of western Ontario. They added inorganic phosphate and nitrate to the epilimnion of this softwater, oligotrophic lake to determine whether N and P additions alone could stimulate algal growth. The lake is small and in a forested watershed, thus promoting protection from wind effects and a large z_l. The amount of CO_2 contributed to Lake 227 from the atmosphere over a 2-week study period was determined by elemental mass balance calculations and found to be 0.19 ± 0.05 g m^{-2} day^{-1} (Example 4-3). Concurrent studies showed that $z_{l(ave)}$ was 295 μm (based on the loss rate of ^{222}Rn from the epilimnion). The rate of CO_2 invasion calculated from $z_{l(ave)}$, assuming no chemical enhancement, was only 0.04 g m^{-2} day^{-1}. These results lead to an enhancement factor $\alpha \approx 5$. The uncertainties involved in basing such estimates on whole-lake mass balance measurements should be recognized, but the results suggest that chemical enhancement should be considered in modeling air-water exchange of CO_2 in small lakes.

Example 4-3. Estimation of $K_{l(CO_2)}$ in a Lake by Mass Balance Method

A.   Epilimnetic masses of carbon and phosphorus were measured in Lake 227 (ELA, Ontario, Canada) on two dates in summer 1970 as follows (values in kg):[249]

Date	Suspended		Dissolved		Total		C:P
	C	P	C	P	C	P	(wt/wt)
4 August	382.3	2.49	23.2	0.62	405.5	3.11	154
18 August	360.8	3.56	50.7	0.83	411.5	4.39	101
Difference	−21.5	1.07	27.5	0.21	6.0	1.29	125 (ave)

B.   Other essential information:
1.   2.28 kg dissolved P was added to the epilimnion as Na_2HPO_4 as part of a nutrient-enrichment experiment over the budget period given in part A.
2.   The nutrient enrichment caused a large bloom of algae (average chlorophyll **a** was 90 µg L^{-1}) that exerted a large demand on the lake's inorganic carbon reservoir; CO_2 depletion in the surface water produced pH values of 9 to 10.
3.   No C was added as fertilizer; the only net sources for algal growth were inorganic carbon in the epilimnion and gas transfer from the atmosphere.
4.   The surface area of Lake 227 was 4.89 ha (4.89×10^4 m²).
5.   The lake had no surface inflows or outflows of water or nutrients during the budget period, and littoral removal of C and P was insignificant.

C.   Mass balance calculations:
1.   In general, $\Delta S = \Sigma I - \Sigma O$, where ΔS = change in storage, I = inputs, O = outflow or loss.
2.   For phosphorus, $1.29 = 2.28 - \Sigma O$; therefore, $\Sigma O = 0.99$ kg.
3.   Because there was no surface outflow, ΣO must represent the loss of P by planktonic uptake and sedimentation.
4.   From the average epilimnetic C:P over the budget period, ~124 kg of C must have sedimented with the 0.99 kg P.
5.   A net increase of 6 kg C was measured in the epilimnion over the 14 days. Therefore, a total of 124 + 6 = 130 kg C must have been added to the lake from the atmosphere during the budget period. This yields an areal mass transfer rate of

$$\left(130 \times 10^3 \, \text{g}\right) / \left(4.89 \times 10^4 \, \text{m}^2 \times 14 \, \text{days}\right) = 0.19 \pm 0.05 \text{g m}^{-2}\text{day}^{-1}$$

A more detailed analysis by Emerson[217] and experiments with in-lake enclosures[251] generally supported the earlier findings on Lake 227; enhancement factors of 5 to 10 were calculated for the range of z_1 (300 to 1000 µm) thought to be appropriate for Lake 227 (Figure 4-40). However, Emerson further indicated that this lake is probably an extreme case, and most freshwaters should have smaller enhancement factors. For large- and medium-size lakes with moderate alkalinity and pH $\leq$ ~8, enhancement factors should be negligible.

4.8 MINERAL DISSOLUTION AND FORMATION

4.8.1 Introduction

Sillen[252] described the inorganic composition of seawater as arising from a "geotitration" of nonvolatile basic rocks by volatile acidic gases. Although this is a somewhat simplified explanation, it is nonetheless true that most of the major ions in both fresh and marine waters originate from chemical attack by acidic aqueous solutions on basic minerals in rocks and soils. This is illustrated by Holland's accounting[253] for the origin of major ions in the world's river water (Table 4-17). The complex set of processes known collectively as mineral weathering produces the major cations from congruent dissolution of carbonate minerals (limestone,

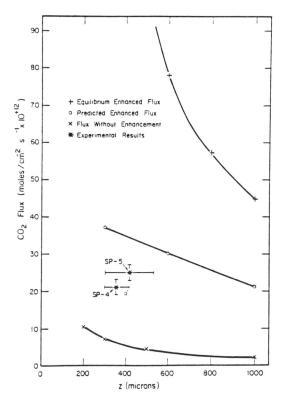

Figure 4-40. CO_2 flux across air-water interface vs. liquid-film thickness for three sets of assumptions and results for two lab experiments; + (equilibrium enhanced flux) assumes CO_2 reacts rapidly enough to achieve equilibrium with HCO_3^- in the liquid film; o (kinetically enhanced flux) assumes reaction and diffusion are of comparable importance (see reference 217 for model equations and parameter values); x (no enhancement) assumes CO_2 hydration is slow compared with diffusion through the liquid film (CO_2 behaves as an unreactive gas); * results of laboratory experiments simulating conditions in ELA Lake 227. [From Emerson, S., *Limnol. Oceanogr.,* 20, 743 (1975). With permission.]

dolomite) and incongruent dissolution of aluminosilicate minerals. (*Congruent* means that no new solid phases are produced; *incongruent* means that new solid phases result from the weathering reaction.) The compounds that produce acidity in the attacking solutions include CO_2, sulfuric acid (from natural sources as well as anthropogenic acid rain), and organic acids that are products of microbial metabolism; these acids contribute a significant portion of the anions found in natural waters. Over geologic time, the chloride in freshwaters that now is attributed to cyclic sea salt, the dissolution of marine evaporites, and the use of salt in human diets and industry originated as volcanic HCl.

In general, heterogeneous reactions involving solid and aqueous phases are slow to reach equilibrium, especially at normal temperatures for the surface environment. Such reactions usually involve a five-step sequence:

Table 4-17. Sources of Major Solutes in Average River Water

| Source | Anions (meq kg⁻¹) | | | Cations (meq kg⁻¹) | | | Neutral species | |
	HCO_3^-	SO_4^{2-}	Cl^-	Ca^{2+}	Mg^{2+}	Na^+	K^+	SiO_2[a]
Atmosphere	0.58	0.09	0.06	0.01	<0.01	0.05	<0.01	<0.01
Weathering or dissolution of								
Silicates	0	0	0	0.14	0.20	0.10	0.05	0.21
Carbonates	0.31	0	0	0.50	0.13	0	0	0
Sulfates	0	0.07	0	0.07	0	0	0	0
Sulfides	0	0.07	0	0	0	0	0	0
Chlorides	0	0	0.16	0.03	<0.01	0.11	0.01	0
Organic matter	0.07	0	0	0	0	0	0	0
Total	0.96	0.23	0.22	0.75	0.35	0.26	0.07	0.22

From Holland, H.E., *The Chemistry of the Oceans and Atmosphere,* Wiley-Interscience, New York, 1978.

[a] Exists as $Si(OH)_4$, $pK_1 = 9.3$.

1. Diffusion of reactants through the boundary layer to the solid surface
2. Adsorption of reactants onto reactive sites on the surface
3. Chemical reaction on the surface (e.g., bond formation or breakage involving crystal lattice)
4. Desorption of products from the surface
5. Diffusion of products through the boundary layer to the bulk solution

In the case of dissolution reactions, the inward-diffusing reactants may be acids or complexing agents that promote breakdown of the crystal lattice; the outward-diffusing products are constituents of the dissolving mineral. For mineral formation, the inward-diffusing reactants are constituents of the growing mineral, and the outward-diffusing products may be residual ions (e.g., H^+) or H_2O left from the reaction of solutes to form the solid phase. In theory, any of the five steps could be rate limiting. In practice, the adsorption/desorption steps are rapid compared with the others. Except when the driving force for chemical reaction is very large (i.e., when the system is far from equilibrium), diffusion is rapid compared with the surface reaction step. For example, inward diffusion of the attacking agent H^+ controls the dissolution rate of $CaCO_3$ only in highly undersaturated, acidic solutions (pH < ~4).

Numerous reactions of solid phases influence the composition of natural waters. We will consider the kinetics of four types of reactions that span the range of complexity for this class of reactions:

1. Dissolution and precipitation of silica, a simple, congruent hydration/dehydration reaction that is largely unaffected by solution composition
2. Dissolution of calcium carbonate, a congruent reaction that varies with pH and concentration of complex-forming ions such as phosphate

3.   Redox-related dissolution reactions of iron and manganese oxides and sulfides, which are influenced by such solution conditions as pH, redox status, the presence of oxidizable organic ions, and microbial activity
4.   Dissolution of aluminosilicate minerals such as feldspars, which are complicated, incongruent reactions that are affected by numerous solution conditions and involve many steps, reactants, and products

4.8.2 Silica-Water Reactions

Silica is one of the most abundant minerals in the Earth's crust, and the equilibrium and kinetic properties of silica-water reactions are important in many geochemical processes — from high-temperature water-rock reactions in deep crustal zones to low-temperature weathering processes in near-surface regions. Silica has four important solid phases, and all are only sparingly soluble in water. The most stable (thermodynamically) and least soluble is quartz; the least stable and most soluble is amorphous silica (Table 4-18).

The dissolution and formation of solid silica phases are perhaps the simplest example of mineral dissolution/precipitation reactions that are relevant to aquatic chemistry. The reactions themselves are simple hydration/dehydrations:

$$SiO_{2(s)} + 2H_2O \rightleftharpoons Si(OH)_{4(aq)} \tag{4-107}$$

Over the pH range of most natural waters, $Si(OH)_4$ is essentially undissociated[150] ($pK_1 = 9.46$ at 25°C and $I = 0.5$), and complexation and hydrolysis reactions do not affect the rate of either the forward or reverse reaction.[254] The reactions are pH independent at circumneutral pH, but low-pH fluids are much more aggressive in driving silica into solution. Silica phases are pure (there are no solid solutions of silica), and SiO_2 dissolves congruently to yield $Si(OH)_{4(aq)}$, i.e., no solid products are formed.

Reaction 4-107 is a reversible reaction, and the equilibrium constant is

$$K = \{Si(OH)_{4(aq)}\}/\{SiO_{2(s)}\}\{H_2O\}^2 \tag{4-108}$$

In dilute aqueous solutions at low temperatures and pressures, the activity of solid silica and water and the activity coefficient of undissociated $Si(OH)_4$ essentially equal unity, and K is equal to the molar solubility of silica. The reversible first-order rate equations described in Section 2.3.1 can be used to develop a rate expression for the silica-water reactions. Rimstidt and Barnes[254] expressed the change in $[Si(OH)_4]$ with time as the net of the opposing dissolution and precipitation reactions:

$$R = -d\{Si(OH)_4\}/dt$$

$$= (A/M)\left[\gamma_{Si(OH)_4} k_d \{SiO_2\}\{H_2O\}^2 - k_p\{Si(OH)_4\}\right] \tag{4-109}$$

Table 4-18. Equilibrium and Kinetic Constants for Silica-Water Reactions[a]

A. Equilibrium relationships

	$G_f°$ (kJ mol^{-1})[b]	$logK$[c] (25°C)	$logK(T)$[c]
Quartz	−855.86	−3.958	$1.881 - 2.028 \times 10^{-3}T - 1560/T$
α-Cristobalite	−855.06	−3.348	$-0.0321 - 988.2/T$
β-Cristobalite	—	−2.919	$-0.2560 - 793.6/T$
Amorphous silica	−849.92	−2.716	$0.338 - 7.889 \times 10^{-4}T - 840.1/T$

B. Kinetic relationships[d]

	$logk_d$ (25°C) (s^{-1})	$logk_d(T)$ (s^{-1})	E_{act} (kJ mol^{-1})
Quartz	−13.38	$1.174 - 2.028 \times 10^{-3}T - 4158/T$	71.9
α-Cristobalite	−12.77	$-0.739 - 3586/T$	68.6
β-Cristobalite	−12.35	$-0.963 - 3586/T$	64.8
Amorphous silica	−11.99	$-0.369 - 2.89 \times 10^{-4}T - 3438/T$	62.7

	$logk_p$ (25°C)	$logk_d(T)$	E_{act}
All phases	−9.43	$-0.707 - 2598/T$	49.7

[a] $G_f°$ from Stumm, W. and J.J. Morgan, *Aquatic Chemistry,* 2nd ed., Wiley-Interscience, New York, 1981; other data from Rimstidt, J.D. and H.L. Barnes, *Geochim. Cosmochim. Acta,* 44, 1683 (1980).
[b] $G_f°$ = standard free energy of formation of solid phase from the elements; $G_f°$ = −1315.45 kJ mol^{-1} for $Si(OH)_{4(aq)}$ and −236.96 kJ mol^{-1} for $H_2O_{(l)}$.
[c] For dissolution reaction, $SiO_2 + 2H_2O \rightarrow Si(OH)_{4(aq)}$; $K \approx$ molar solubility of $Si(OH)_{4(aq)}$.
[d] k_d = rate constant for dissolution of solid phase; k_p = rate constant for precipitation of $Si(OH)_4$ (independent of nature of solid phase formed).

where A = surface area of silica in the system; M = mass of water ($\approx V_{H_2O}$); and subscripts d and p stand for dissolution and precipitation, respectively. A/M is called the "extent of the system". Equation 4-109 is valid at constant T, P, and M, and it implies that both the precipitation and dissolution reactions occur simultaneously. Dissolution occurs even when the solution is supersaturated but is less rapid than precipitation. In dilute systems at low T and P, $\gamma_{Si(OH)_4}$, $\{SiO_2\}$, and $\{H_2O\} \approx 1$, and Equation 4-109 simplifies to

$$R = (A/M)\left[k_d - k_p\{Si(OH)_4\}\right] \qquad (4\text{-}110)$$

Because $K_{eq} = k_d/k_p$ for reaction 4-107, we can rewrite the above equation as

$$R = (A/M)k_d(1 - \{Si(OH)_4\}/K) = (A/M)k_d(1 - S) \qquad (4\text{-}111)$$

where S = $\{Si(OH)_4\}/K$ is the degree of saturation. Equation 4-111 cannot be integrated directly, but by dividing both sides by K, we obtain a form that can be: $-dS/dt = (A/M)k_p(1 - S)$, which integrates to

$$\ln\left[\frac{1-S}{1-S_o}\right] = -k_p t \tag{4-112}$$

If $S_o = 0$ (i.e., the initial $[Si(OH)_4] = 0$), then Equation 4-112 simplifies to

$$\ln(1-S) = -k_p t \tag{4-113}$$

Thus a plot of $ln[(1-S)/(1-S_o)]$ or $ln(1-S)$ vs. t should yield a straight line with slope $= k_p$. It is interesting to note that this relationship holds whether the starting solution is supersaturated (precipitation dominates) or undersaturated (dissolution dominates) (Figure 4-41a,b). Under the latter conditions, however, an initial steep slope always was observed by Rimstidt and Barnes[254] (Figure 4-41b). They interpreted this as evidence for a higher solubility surface layer that dissolves more rapidly than bulk quartz, and they showed that in terms of its solubility and dissolution kinetics, this layer behaves like amorphous silica.

If the solubility of amorphous SiO_2 is used to calculate the degree of saturation (S) for the steep region, similar values of k_p are obtained as those for quartz and amorphous silica. Thus, the rate of precipitation of dissolved silica is the same, regardless of the nature of the precipitating phase. From the law of microscopic reversibility and $K = k_d/k_p$, it then is apparent that the dissolution rate constant is directly proportional to K (i.e., to the solubility of the solid SiO_2 phase). Moreover, the activation energies for dissolution will increase with the stability of the solid phase. The basis for these relationships can be understood by examining the diagram of energy vs. reaction coordinate in Figure 4-42. The values of k_d, K, and E_{act} in Table 4-18 agree with the above conclusions.

Over the broad range of temperature (0 to 300°C) in which silica-water reactions have been studied, rates and equilibria (solubility) vary widely (Table 4-18). In brief, the solubility of the solid phases increases with temperature, and high concentrations can be achieved in geothermal conditions. For example, quartz solubility is $10^{-2.16}$ M (440 mg L^{-1} as SiO_2) at 250°C, but only $10^{-3.96}$ M (7 mg L^{-1}) at 25°C. Corresponding values for amorphous silica are $10^{-1.68}$ M (1334 mg L^{-1}) at 250°C and $10^{-2.72}$ M (123 mg L^{-1}) at 25°C. Similarly, rates of dissolution and precipitation are very slow below 100°C (10^{-13} to 10^{-12} s^{-1} at 25°C), but comparatively rapid (up to 10^3 higher) under geothermal conditions (Figure 4-43). For example, at A/M = 100 and T = 300°C, a solution effectively will reach equilibrium in less than 15 h, but at 25°C it would take nearly 6 years. These numbers are based on characteristic times ($\tau_c = 1/k$). At $t = 3\tau_c$, the solution has reached 95% of its equilibrium value. As a result, high, metastable concentrations of $Si(OH)_{4(aq)}$ can be achieved when geothermal fluids are cooled suddenly.[254]

It also is clear from Equations 4-109 to 4-111 that the rate of silica dissolution, measured in terms of the change in silica concentration in solution, is directly proportional to A/M, the extent of the system. This ratio can vary widely in both

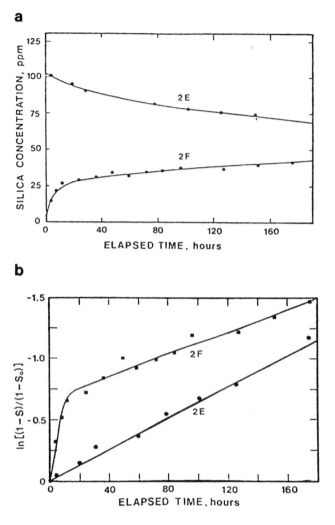

Figure 4-41. (a) Dissolved SiO_2 vs. time in distilled water-quartz sand system at 105°C approaches the equilibrium solubility of 58 mg L^{-1} from both super- and undersaturated conditions. (b) Plot of $ln[(1-S)/(1-S_o)]$ vs. time yields straight-line relationships for both super- and undersaturated conditions, but an initial steep slope occurs in the latter case (run 2F), apparently because of rapid dissolution of a higher-solubility surface layer of amorphous silica. [From Rimstidt, J.D. and H.L. Barnes, *Geochim. Cosmochim. Acta,* 44, 1683 (1980). With permission.]

laboratory and natural systems, depending on the size and porosity of solid silica particles.

The activation energies calculated for silica dissolution and precipitation (209 to 297 kJ mol^{-1}) indicate that the rate-limiting step is a surface chemical reaction (probably involving the breakage of a covalent Si-O bond) rather than diffusion of reactants to or from the silica surface. Lasaga and Gibb[255] have characterized the

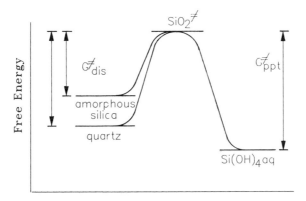

Reaction Coordinate

Figure 4-42.   Energy profiles for forward (dissolution) and reverse (precipitation) directions of silica-water reaction. [Redrawn from Rimstidt, J.D. and H.L. Barnes, *Geochim. Cosmochim. Acta*, 44, 1683 (1980).]

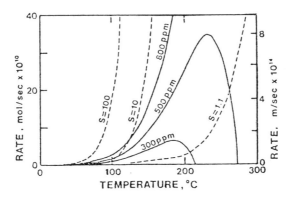

Figure 4-43.   Rate of quartz precipitation vs. temperature in system where activities of water and silica, $\gamma_{Si(OH)_4}$, and A/M are one. Solid lines are for constant values of Q, the activity product; dashed lines are for constant values of S, the saturation ratio. [From Rimstidt, J.D. and H.L. Barnes, *Geochim. Cosmochim. Acta*, 44, 1683 (1980). With permission.]

nature of the transition state for this reaction by *ab initio* statistical mechanical methods (see Figure 3-10).

4.8.3 Dissolution of Calcium Carbonate

By far the most important sedimentary mineral is calcium carbonate. It is the primary inorganic constituent of sediments formed in hardwater lakes and the most common mineral formed in the oceans. Extensive limestone deposits formed in earlier geological times serve as the main groundwater aquifers for much of the U.S. Deposits of $CaCO_3$ are a common problem in water distribution systems of areas with hardwater. According to Table 4-17, about two thirds of the Ca^{2+} in average river water arises from dissolution of carbonate minerals, primarily limestone (impure $CaCO_3$) and secondarily dolomite $(CaMg(CO_3)_2)$. Factors affecting $CaCO_3$ dissolution kinetics are of interest from the perspective of fundamental geochemistry and for practical reasons — many buildings and statues are of limestone and are subject to weathering/dissolution processes initiated by natural agents (CO_2 and water), as well as acidic air pollutants (SO_2, acid rain). Limestone is used to treat lakes and watersheds that have been acidified by atmospheric acid deposition, and the kinetics of $CaCO_3$ dissolution is of interest in designing effective treatment programs.[256] (For reviews of the extensive literature on $CaCO_3$ dissolution, see references 257–260.)

Virtually all areas of the world's oceans are supersaturated with respect to calcium carbonate in their near-surface regions. This suggests that precipitation of $CaCO_3$ is widespread in the oceans. Indeed, sediments containing more than 30% $CaCO_3$ (by weight) do cover more than a third of the ocean bottom (~25% of the Earth's surface).[257] Most of this $CaCO_3$ is biogenic, however, and is formed as protective shells in surface waters by foraminifera (small marine animals) and coccolithophores (microscopic plants), both of which produce calcite, or pteropods, animals that produce aragonite. Only a small portion of the biogenic $CaCO_3$ produced in marine surface waters is preserved in the sediments; the majority (75 to 95%) dissolves before it reaches the bottom or while residing at the sediment-water interface before burial by later-arriving sediment particles.[257,258] The $CaCO_3$ content of marine sediments varies markedly with depth, representing up to 90% of the sediment at the start of the pelagic zone and gradually decreasing with increasing water depth until it virtually disappears at depths of more than ~6000 m. The depth where $CaCO_3$ becomes a minor component (<1%) in marine sediment, is called the *calcium carbonate compensation depth* (CCD). At this depth its input rate to the sediment approximately matches its rate of dissolution. The *aragonite compensation depth* (ACD) is ~3 km shallower than the CCD because aragonite is more soluble than calcite.

The above facts suggest that bottom waters of the oceans must be undersaturated with respect to $CaCO_3$; otherwise the entire ocean floor would be covered by biogenic $CaCO_3$. Indeed, careful measurements and equilibrium calculations show that the oceans are undersaturated with respect to $CaCO_3$ at depths greater than a few hundred meters.[150,257] This provokes a question that occupied aqueous geochemists for many years: if the bulk of the ocean is undersaturated with respect to $CaCO_3$, why is it so common in marine sediments except at depths greater than ~6 km?

Answers to the above question were obtained by laboratory studies on factors affecting the kinetics of $CaCO_3$ dissolution.[261-263] However, the impetus for these studies was an ingenious *in situ* experiment by Peterson,[264] who measured dissolution rates of polished calcite spheres suspended at various depths in the central Pacific Ocean for 4 months. Results of this and several later experiments (Figure 4-44) showed that dissolution began at ~0.5 km, near the depth where unsaturated conditions began, but rates were low until a depth of ~3.8 km and then increased rapidly with depth. The depth of rapid increase in dissolution differs for various locations in the oceans and is correlated with a depth called the foraminiferal lysocline (FL).

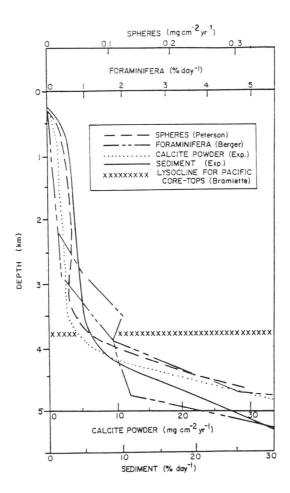

Figure 4-44. Dissolution rate of various calcite sources as function of depth in the Pacific Ocean. Data for calcite spheres and foraminifera from *in situ* experiments. Data for calcite powder and CaCO₃-rich sediment from laboratory experiments (see Figure 4-45a), plotted as function of equivalent depth based on the ΔpH-depth relationship for the site of the calcite-sphere experiment. [From Morse, J.W. and R.C. Berner, *Am. J. Sci.,* 272, 840 (1972). With permission.]

The FL is based on work by Berger[265] showing that different species of foraminifera dissolve at dissimilar rates. He divided these species into two classes — resistant and easily dissolved species — and found that the ratio of soluble-to-resistant species in sediments declined as water depth increased. Definite maxima in the rate of change in this ratio were found at particular depths (the FL) in plots of the ratio vs. water depth for different regions of the oceans. The existence of the FL suggests that dissolution of easily dissolved species increases rapidly at some water depth. The depth at which dissolution increased rapidly for Peterson's calcite spheres was similar to the FL at that location (Figure 4-44). Finding an explanation for the FL and for the depth of rapid increase in $CaCO_3$ dissolution was the stimulus for laboratory experiments that unraveled the kinetics of $CaCO_3$ dissolution.

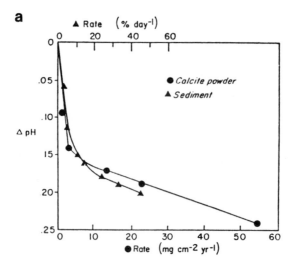

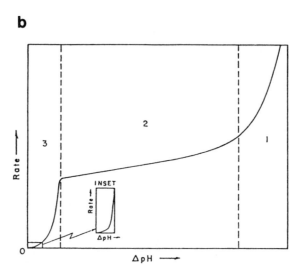

Figure 4-45.   (a) Dissolution of $CaCO_3$-rich sediment and calcite powder increased rapidly after ΔpH reached a critical value of ~0.15 in laboratory experiments. (b) Schematic of R_{dis} vs. pH developed by Berner and Morse from laboratory experiments over a broad range of pH shows that $CaCO_3$ dissolution increases with decreasing pH in a complicated fashion. Dissolution is diffusion-controlled in region 1 and chemically controlled in regions 2 and 3; the discontinuity on the inset represents the chemical lysocline. Only region 3 is relevant to seawater conditions. (c) The concentration of dissolved inorganic phosphate (μM) has a major effect of the value of ΔpH where $CaCO_3$ dissolution begins to increase. [Graph (a) from Morse, J.W. and R.C. Berner, *Am. J. Sci.*, 272, 840 (1972); graphs (b) and (c) from Berner, R.C. and J.W. Morse *Am. J. Sci.*, 274, 108 (1974). With permission.]

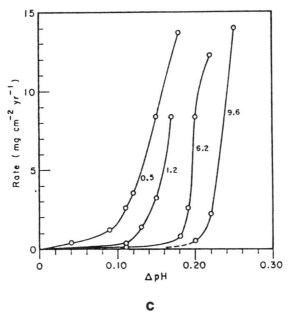

C

Figure 4-45 (continued).

The development of the pH-stat technique[266] facilitated these studies. This technique allows the accurate measurement of low dissolution rates under a constant (and known) state of disequilibrium. At any given temperature and total pressure, the solubility of $CaCO_3$ is fixed by three variables: $[Ca^{2+}]$, P_{CO_2}, and pH. The Ca^{2+} content of seawater is high and can be held effectively constant even in a vessel where $CaCO_3$ is dissolving by limiting the extent of reaction. P_{CO_2} can be fixed by bubbling a gas mixture with fixed P_{CO_2} through the solution; and pH can be monitored accurately with a glass electrode and meter that activates syringes to add acid or base to the solution, as necessary, to maintain a preselected pH. Morse[266] described a number of precautions that must be taken in the construction and use of such devices to obtain accurate measurements of $CaCO_3$ dissolution rates.

An early step in unraveling the kinetics of $CaCO_3$ dissolution was the realization that the rate depends in a nonlinear way on the degree of $CaCO_3$ unsaturation (symbolized Ω), such that the rate is low when Ω is small and increases rapidly once a critical value, Ω_c, is exceeded. In general, Ω is expressed as[261,267]

$$\Omega = \frac{\text{measured ion product}}{\text{saturation ion product}} = \frac{[Ca^{2+}]_m[CO_3^{2-}]_m}{[Ca^{2+}]_s[CO_3^{2-}]_s} = \frac{IP}{{}^cK_{s0}/\gamma_{CaCO_3}} \quad (4\text{-}114)$$

Concentrations are in molality; subscripts s and m denote saturation and measured values, respectively; and ${}^cK_{s0}/\gamma_{CaCO_3}$ is the concentration-based solubility product of $CaCO_3$, corrected for the activity of the solid phase. If $\Omega > 1$, the solution is

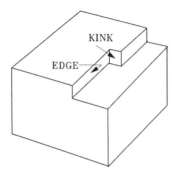

Figure 4-46.   Sketch of crystal cube, showing edges and kinks where surface reactions such as dissolution are most likely to occur.

supersaturated; if $\Omega < 1$, it is undersaturated. It can be shown that Ω is related to a term called ΔpH:

$$\Delta pH = pH_s - pH_m = -0.5 \log \Omega \qquad (4\text{-}115)$$

where pH_s is the pH at $CaCO_3$ saturation for the given concentration of Ca^{2+} and total carbonate in the water, and pH_m is the measured pH. Positive values of ΔpH thus express the degree of unsaturation. In laboratory studies Morse and Berner[261] found that dissolution rates (R_{dis}) of calcite powder and $CaCO_3$-rich sediment were very slow at $\Delta pH < 0.15$ and increased rapidly beyond that value (Figure 4-45a). Similarly, they showed that ΔpH remained just below this critical value throughout the depths where R_{dis} was slow in Peterson's sphere experiment (Figure 4-44) and increased linearly with depth starting at the FL, where *in situ* dissolution began to increase. The depth at which ΔpH reaches the critical value for rapid increase in R_{dis} is called the *chemical lysocline*.[268]

Berner and Morse[262] found that R_{dis} increased in a complicated manner with increasing ΔpH (Figure 4-45b). R_{dis} was much lower (by factors of 0.01 to 0.1) than predictions based on diffusion-limited dissolution, except at pH $< \sim4$, and thus turbulence is not a limiting factor for $CaCO_3$ dissolution in the ocean. The sharp increase in R_{dis} at $\Delta pH \approx 0.15$ suggests that a change in mechanism (or in the rate-controlling step) occurs at this point. The additional finding that R_{dis} is highly sensitive to the presence of impurities that adsorb to reactive sites on the surface[262] led to a mechanistic interpretation of the reaction in terms of "kink" theory.[269] According to this theory, both dissolution and crystal growth occur at kinks in the crystal lattice (Figure 4-46); migration of kinks gradually leads to step retreat and dissolution of the crystal.

Kinks are site of excess surface energy, because bonding of ions to the crystal lattice is less complete there than elsewhere on the surface. However, kinks are preferred sites not only for reaction, but also for chemisorption of impurities. For example, phosphate ions (P_i) bind strongly at kinks, and this inhibits dissolution dramatically. As Figure 4-45c shows, a larger ΔpH (higher unsaturation) is required to induce dissolution as $[P_i]$ increases in solution. Berner and Morse[262] explained these findings by a model in which the binding of P_i at kinks prevents detachment of the crystal ion at that kink, thus immobilizing the kink and inhibiting dissolution. If all the kinks are immobilized by bound P_i, dissolution can proceed only by formation of new kinks — a slow process compared with kink migration.

In contrast to Berner and Morse, who found no difference in R_{dis} over a wide range of P_{CO_2}, Plummer et al.[263] found that R_{dis} increased with P_{CO_2} under moderately unsaturated conditions (pH 5 to 6). They performed experiments in pH-stats at values of Ω where the reaction can be considered unidirectional (dissolution only) and found evidence for three separate mechanisms. Their results (Figure 4-47a) suggest that dissolution occurs by H^+ attack at pH < ~3; rates are proportional to $\{H^+\}$ and independent of P_{CO_2}. Above pH 3, dissolution becomes less dependent on $\{H^+\}$, but does depend on P_{CO_2}. However, at P_{CO_2} less than 0.01 to 0.03 atm and pH > ~7, R_{dis} becomes independent of P_{CO_2}. Mechanisms corresponding to these findings are

$$CaCO_{3(s)} + H^+ \rightarrow Ca^{2+} + HCO_3^- \qquad (4\text{-}116a)$$

$$CaCO_{3(s)} + H_2CO_3^* \rightarrow Ca^{2+} + 2HCO_3^- \qquad (4\text{-}116b)$$

$$CaCO_{3(s)} + H_2O \rightarrow Ca^{2+} + HCO_3^- + OH^- \qquad (4\text{-}116c)$$

The overall rate of dissolution is the sum of the three mechanisms:

$$R_{dis} = k_{H^+}\{H^+\} + k_{H_2CO_3^*}\{H_2CO_3^*\} + k_{H_2O}\{H_2O\} \qquad (4\text{-}117)$$

where R_{dis} is in mmol cm^{-2} s^{-1}. The rate constant for the H^+-catalyzed mechanism depends on the stirring rate and has an activation energy of only 8 kJ mol^{-1}, indicative of a diffusion-limited process. In contrast, rate constants for the mechanisms involving $H_2CO_3^*$ and H_2O do not depend on stirring rates. For reaction 4-116b, $E_{act} = 42$ kJ mol^{-1}, and E_{act} for reaction 4-116c was reported to be only 6.3 kJ mol^{-1} below 25°C and 33 kJ mol^{-1} above 25°C.[259] The change in E_{act} with temperature suggests a change in the reaction mechanism, and the low E_{act} below 25°C at first may suggest the change is from diffusion to chemical control. However, R_{dis} for the H_2O-driven reaction is much lower than that which would occur if diffusion were limiting. The low E_{act}, if real, may signify a multistep mechanism with a pre-equilibrium condition (see Section 3.2.4).

Figure 4-47b shows the pH-P_{CO_2} regions where one of the three reactions in Equation 4-116 is the dominant dissolution reaction, and the boundaries where each process balances the other two. The reaction with H_2O clearly is the dominant process for conditions found in most natural waters. Unfortunately, R_{dis} is less well defined in this region than the above analysis may imply. As solutions approach saturation, precipitation becomes increasingly important, and net reaction must be viewed as the difference between forward (dissolution) and reverse (precipitation) reactions:

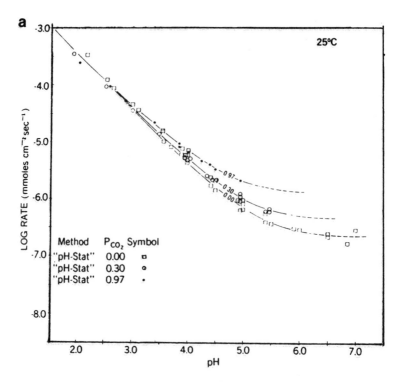

Figure 4-47. (a) Variation in R_{dis} vs. pH under conditions far from saturation shows that dissolution is not affected by P_{CO_2} at low pH, where the reaction is diffusion-controlled, but is affected by P_{CO_2} at pH > ~3. R_{dis} becomes pH-independent above pH ~5.5. (b) pH-P_{CO_2} regions where each of the three mechanisms of $CaCO_3$ dissolution are dominant and boundaries where contributions from a given mechanism balance the sum of the rates of the other two. More than one mechanism contributes significantly in the gray region. (c) $(R_{net} - k_{H+}\{H^+\})$ vs. $\{Ca^{2+}\}\{HCO_3^-\}$ ion product from "free-drift" dissolution experiments at moderate-to-slight undersaturation. The slope of the line for the data yields k_p (see Equation 4-118). [From Plummer, L.N., et al., in *Chemical Models*, E.A. Jenne (Ed.), ACS Symp. Ser. 93, American Chemical Society, Washington, D.C., 1979, p. 537. With permission.]

$$R_{net} = k_{H^+}\{H^+\} + k_{H_2CO_3^*}\{H_2CO_3^*\} + k_{H_2O}\{H_2O\} - k_p\{Ca^{2+}\}\{HCO_3^-\} \qquad (4\text{-}118)$$

where k_p is the rate constant for $CaCO_3$ formation, expressed in terms of the bulk reactants, Ca^{2+} and HCO_3^-. Although k_p can be calculated from empirical data, as described below, the way it varies with temperature and P_{CO_2} is complicated. Expressions for k_p have been derived from mechanistic models of $CaCO_3$ dissolution-precipitation,[259] but these models require information on $\{H_2CO_3^*\}$ and $\{H^+\}$ at the $CaCO_3$-H_2O interface, which is not readily obtained.

Experiments under near-equilibrium conditions are often conducted by the "free-drift" method; i.e., P_{CO_2} and temperature are fixed, but pH, $\{Ca^{2+}\}$, $\{HCO_3^-\}$, and $\{CO_3^{2-}\}$ are allowed to vary as the reaction proceeds toward equilibrium. Data

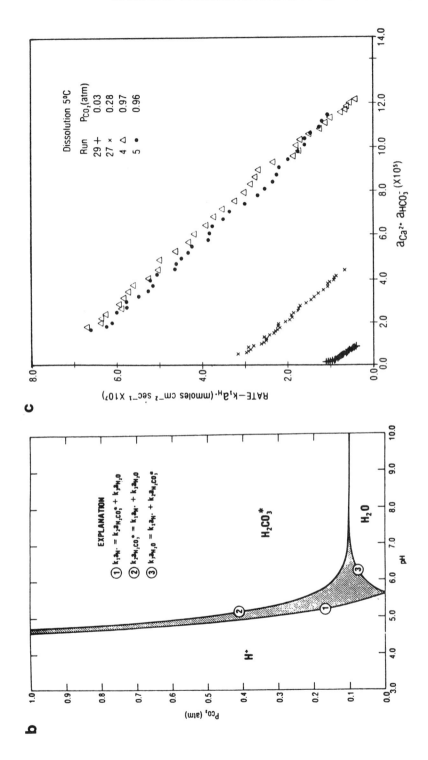

Figure 4-47 (continued).

from such experiments show a linear relationship between $\mathbf{R}_{net} - k_{H^+}\{H^+\}$ and the ion activity product $\{Ca^{2+}\}\{HCO_3^-\}$ (Figure 4-47c). The y-intercept of the figure represents the contribution to dissolution from the second and third terms of Equation 4-118 (i.e., attack by $H_2CO_3^*$ and H_2O) in solutions far from equilibrium. The slope yields k_p for a given temperature and P_{CO_2}. Values of k_p thus calculated are nonlinear functions of temperature and P_{CO_2}.[259]

Experiments with unsaturated solutions that are fairly close to equilibrium thus can yield estimates of the *precipitation* rate constant. Calcite and aragonite (also quartz, see Section 4.8.2) are considered well behaved in this regard; i.e., they usually are assumed to reach a dissolution equilibrium rapidly in a closed system.[258] Each forward reaction then is balanced by a corresponding back reaction, as required by the principle of microscopic reversibility, and the kinetic-thermodynamic relationship, $K_{s0} = k_{dis}/k_p$, applies. Nonetheless, measured values of k_{dis} and k_p do not always yield the correct values of K_{s0} for calcite and aragonite. Wollast[258] suggested this may indicate that the solids precipitated under near-equilibrium conditions are disordered and not completely dehydrated, but an alternate explanation, that the concentration of H^+ at the surface is not the same as that in the bulk solution, is equally plausible (see the model described in the next paragraph).

Dolomite $[CaMg(CO_3)_2]$ is not well-behaved with respect to dissolution equilibria. It does not form from supersaturated solutions in the laboratory, and its dissolution is fractional order in H^+, as also occurs with oxides and silicates (see following sections). This suggests that complicated reactions occur at the mineral surface.

Current mechanistic models of $CaCO_3$ dissolution describe the process in terms of three solution layers: an adsorption layer right at the $CaCO_3$ surface, the hydrodynamic boundary layer, and the bulk solution.[259] The adsorption layer is assumed to be only a few molecules thick, and ions in this layer are loosely bound to reactive sites occurring at kinks on the $CaCO_3$ surface. Consequently, ions in the adsorption layer have restricted mobility compared with ions in the boundary layer. At low pH and relatively high P_{CO_2}, dissolution is fast and dominated by H^+ attack. $\{H^+\}$ in the adsorption layer will be smaller than that in the bulk solution, but $\{H_2CO_3^*\}$ is roughly the same in all three layers. The surface activities of Ca^{2+}, HCO_3^-, H^+, and related species are assumed to be in equilibrium with $CaCO_3$ at the activity of $H_2CO_3^*$ in the surface layer, which in turn is assumed to be the same as the *measurable* activity of $H_2CO_3^*$ in the bulk solution. These assumptions lead to a model (not derived here) that allows one to predict k_p:[259]

$$k_p = (K_2/K_{s0})(k'_{H^+} + [k_{H_2CO_3^*}\{H_2CO_3^*\}_s + k_{H_2O}\{H_2O\}_s]/\{H^+\}_s) \qquad (4\text{-}119)$$

K_2 is the second dissociation constant of $H_2CO_3^*$; K_{s0} is the solubility product of calcite or aragonite (whichever is the solid phase); subscript s denotes activities in the adsorption layer; and k'_{H^+} is the intrinsic rate constant for H^+-driven $CaCO_3$ dissolution in that layer (not the same as the experimentally defined constant k_{H^+}). As noted above, $\{H_2CO_3^*\}_s$ and $\{H_2O\}_s$ are assumed to be the same as the bulk

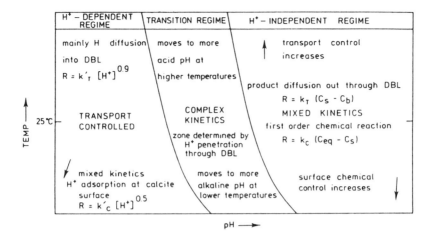

Figure 4-48. Summary diagram of calcite dissolution processes as function of pH and temperature. [From Sjöberg, E.L. and D.T. Rickard, *Geochim. Cosmochim. Acta*, 48, 485 (1984). With permission.]

solution values, and $\{H^+\}_s$ is computed from $CaCO_3$ equilibrium in the adsorption layer. Plummer et al. found similar temperature and P_{CO_2} trends in measured values of k_p and values of k_p computed from Equation 4-119. Because k'_{H^+} cannot be measured directly, they replaced it with the experimentally determined value, k'_{H^+}. Values of k_p computed in this way were consistently lower than measured values, and k'_{H^+} was estimated to be 10 to 20 × k'_{H^+}.

Unfortunately, $CaCO_3$ precipitation (hence, k_p) is **not** important at the conditions where the above model applies, i.e., for the assumptions described in the previous paragraph. In contrast, for the near-equilibrium conditions of natural waters, precipitation is a significant component of $\mathbf{R}_{net}$, but the assumption that $\{H_2CO_3^*\}$ is the same in the bulk solution and adsorption layer is not valid. As a result, the model cannot be used to predict $\mathbf{R}_{net}$ for these important conditions.

Sjöberg and Rickard[260,270] used a rotating-disk technique to distinguish between mass-transfer and chemical processes in $CaCO_3$ dissolution and were able to compute separate E_{act} values for the two processes. Their findings with respect to pH (Figure 4-48) agree generally with those described previously. At pH < ~4, $\mathbf{R}_{dis}$ was limited by H^+ transport to the surface, and E_{act} was 16 kJ mol⁻¹. However, the rate was proportional to $\{H^+\}^{0.9}$ rather than strictly first order in $\{H^+\}$, and this was interpreted to mean that product diffusion from the surface also influenced $\mathbf{R}_{dis}$. At pH > ~5.5, dissolution was pH independent and characterized by mixed kinetics, i.e., control by both product diffusion through the boundary layer and chemical reaction at the surface. E_{act} for chemical reaction was 54 kJ mol⁻¹ (Carrara marble) and 46 kJ mol⁻¹ (Iceland spar), and E_{act} for transport was 27 kJ mol⁻¹. This relatively high value suggests that transport limitation involves product diffusion. Dissolution is generally more transport dependent at high temperatures and chemical-reaction dependent at low temperatures, over the whole pH range studied (Figure 4-48). This leads to complicated temperature dependencies for the overall dissolution process.

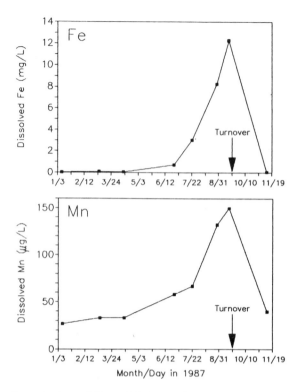

Figure 4-49. Concentrations of Fe and Mn in near-bottom water of Little Rock Lake (WI) increase rapidly following onset of anoxia. (C.E. Mach and P.L. Brezonik, unpublished data.)

4.8.4 Reductive Dissolution of Metal Oxides

The oxidized states of several environmentally important metals are highly insoluble compared with the reduced forms: Fe^{III}/Fe^{II} and $Mn^{III,IV}/Mn^{II}$ are the best known, but Co^{III}/Co^{II} and Ni^{III}/Ni^{II} also fit this situation. Compared with the wealth of information available on the oxidation reactions of metal ions (especially Fe and Mn), little is known about the reduction kinetics of the higher-oxidation-state oxides, even though these processes are equally important in the geocycles of these elements. The paucity of studies reflects the difficulty of the subject; because the reactants are (hydr)oxide solids, kinetic analysis cannot be achieved solely through conventional solution-phase methods.

This section focuses on recent studies to unravel the kinetics of reductive dissolution of Mn and Fe (hydr)oxides. Based on observed rates of increases in soluble (reduced) Fe and Mn in lake hypolimnia (Figure 4-49) after oxygen is depleted,[271,272] we can conclude that the reduction of Fe and Mn (hydr)oxides is fairly rapid under anoxic conditions; dissolved Fe^{2+} and Mn^{2+} concentrations may reach several hundreds (or even thousands) of $\mu g\ L^{-1}$ on a time scale of weeks after the onset of anoxic conditions. Such observations tell us nothing about the mechanisms of reduction/solubilization, however, or the factors affecting the rates.

The major reductant in the reductive dissolution of Fe and Mn (hydr)oxides in aquatic systems is natural organic matter. It may play a direct role, via abiotic chemical reactions, or an indirect role, as a carbon and energy source for bacteria that use Fe or Mn (hydr)oxides as terminal electron acceptors. In addition, photoreduction of Fe and Mn (hydr)oxides also occurs in surface waters (see Section 8.6.1). In a broad geochemical context, the redox cycles of Fe and Mn may be viewed as playing a catalytic role in the oxidation of organic matter by O_2. The metal (hydr)oxides oxidize certain organic compounds, and the soluble, reduced metal-ion products are reoxidized relatively rapidly by O_2. Direct abiotic oxidation of organic compounds by O_2 generally is very slow.

Abiotic Mechanisms

Organic compounds that are effective as metal-oxide reductants share two properties: (1) they have functional groups that form surface complexes, and (2) they have an electron-rich group that can transfer an electron to the metal center and become a radical or radical ion in the process. The functional groups are not necessarily ionic, and the complexes may be inner or outer sphere in nature.[273] The former involve direct bonding of the organic ligand to the metal center, and the latter have coordinated -OH or H_2O between the two. For example, Mn^{III} (hydr)oxide forms surface complexes with phenolate ions:

Outer sphere: $\quad >Mn^{III}OH + {}^-OPhen \rightarrow >Mn^{III}OH - {}^-OPhen$ $\qquad$ (4-120a)

Inner sphere: $\quad >Mn^{III}OH + HOPhen \rightarrow >Mn^{III}OPhen + H_2O$ $\qquad$ (4-120b)

Two lines of evidence point to the importance of surface complex formation in reductive dissolution: (1) adsorbed calcium and phosphate ions strongly inhibit dissolution,[274] and (2) small changes in the structure of the reductant may cause large changes in the reaction rate.[275]

The extent to which the oxide surface sites are positively charged, neutral, or anionic depends on pH; protonation equilibria can be written to describe this:

$$>Mn^{III}OH_2^+ \rightarrow >Mn^{III}OH + H^+, \quad K_{a1}^s; \qquad (4\text{-}121a)$$

$$>Mn^{III}OH \rightarrow >Mn^{III}O^- + H^+, \quad K_{a2}^s \qquad (4\text{-}121b)$$

If adsorption of charged ligands is not involved, oxide surfaces have a net positive charge when protonated sites ($>Mn^{III}OH_2^+$) are more numerous than deprotonated sites ($>Mn^{III}O^-$) and a net negative charge when the converse is true. The pH where the number of positively and negatively charged sites is equal is called the pH_{zpc} (zero point of charge).[150] Reactions analogous to Equations 4-120a,b can be written for all protonation states of the surface sites.

Classes of organic compounds that fit the above-mentioned requirements for reductants of Fe and Mn (hydr)oxides include phenols, thiols, and oxygen-rich

Table 4-19. Inner- and Outer-Sphere Mechanisms for the Reductive Dissolution of Trivalent Metal Oxide (>Me^{III}OH) Surfaces by Phenols (HA)

Inner-sphere mechanism	Outer-sphere mechanism

(1) Precursor complex formation

$$>Me^{III}OH + HA \underset{k_{-1}}{\overset{k_1}{\rightleftharpoons}} >Me^{III}A + H_2O \qquad\qquad >Me^{III}OH + HA \underset{k_{-1}}{\overset{k_1}{\rightleftharpoons}} >Me^{III}OH,HA$$

(2) Electron transfer

$$>Me^{III}A \underset{k_{-2}}{\overset{k_2}{\rightleftharpoons}} >Me^{II\cdot}A \qquad\qquad >Me^{III}OH,HA \underset{k_{-2}}{\overset{k_2}{\rightleftharpoons}} >Me^{II}OH^-,HA^{\cdot+}$$

(3) Release of oxidized organic product

$$>Me^{II\cdot}A + H_2O \underset{k_{-3}}{\overset{k_3}{\rightleftharpoons}} >Me^{II}OH_2 + A^\cdot \qquad\qquad >Me^{II}OH^-,HA^{\cdot+} \underset{k_{-3}}{\overset{k_3}{\rightleftharpoons}} >Me^{II}OH_2 + A^\cdot$$

(4) Release of reduced metal ion

$$>Me^{II}OH_2 + 2H^+ \underset{k_{-4}}{\overset{k_4}{\rightleftharpoons}} >Me^{III}OH + Me^{2+} \qquad\qquad >Me^{II}OH_2 + 2H^+ \underset{k_{-4}}{\overset{k_4}{\rightleftharpoons}} >Me^{III}OH + Me^{2+}$$

From Stone, A.T. and J.J. Morgan, in *Aquatic Surface Chemistry*, W. Stumm (Ed.), Wiley-Interscience, New York, 1988, p. 221.

organic acids, such as pyruvate, oxalate, and ascorbate. Inorganic sulfide also is an effective reductant of Mn oxides.[276] Some of these — especially sulfide and the organic acids — occur in sedimentary organic matter, as a consequence of microbial metabolism,[277] thus providing evidence for indirect involvement of microorganisms in reductive dissolution. Phenols are found in humic matter, and thiols have been identified in marine and freshwater sediments.[90,278]

Under limited sets of circumstances, rates of reductive dissolution of Mn (hydr)oxides are correlated with the reduction potential (E°) for the organic reductant, and within sets of related compounds, E° varies according to the electron-withdrawing or -donating characteristics of the substituents that differentiate individual compounds within the set. This results in linear free-energy relationships (LFER) that can be used for predictive purposes. For example, rates of reductive dissolution of Mn$^{III/IV}$ oxides decrease as the Hammett σ constants of ring substituents (see Section 7.3.1) become more positive, reflecting trends in reductant basicity, nucleophilicity, and reduction potentials.[279] However, as the model described below demonstrates, such correlations occur only when the rate-limiting step in the dissolution process is electron transfer from the organic reductant to the metal center. That is not a universal characteristic of these reactions.

Table 4-19 describes general mechanisms for reductive dissolution of metal (hydr)oxides via the formation of outer- and inner-sphere complexes. Stone and Morgan[273] analyzed several limiting cases for these mechanisms by computer simulations (Example 4-4) and determined the effects of such variables as pH and reductant concentration on rates of metal solubilization. Experimental verification of the results from such simulations is hindered by the fact that analytical methods are not readily available to measure hypothesized intermediates on the (hydr)oxide surfaces (i.e., determine "surface speciation"). This prevents direct measurement of rate constants for the elementary steps, and values used in simulations should be regarded as illustrative estimates only.

Example 4-4. Reaction Mechanisms for Reductive Dissolution of Metal Oxides[273]

Case I. Consider the following simplification of the inner-sphere mechanism in Table 4-19. Assume that all steps except complex formation are irreversible and that release of reduced metal and oxidized organic products is rapid compared with complex formation and electron transfer. Because reduced metal is released as soon as it forms, $[>Me^{II \cdot}A]$ and $[Me^{II}OH_2]$ are negligible, and the general mass balance for surface sites simplifies to

$$(1) \qquad S_T = [>Me^{III}OH] + [>Me^{III}A]$$

Rates of metal release are proportional to the number of surface sites complexed by A^-:

$$(2) \qquad d[Me^{2+}]/dt = k_2[>Me^{III}A]$$

In turn, $[>Me^{III}A]$ depends on relative rates of adsorption, desorption, and electron transfer:

$$(3) \qquad d[>Me^{III}A]/dt = k_1[HA][>Me^{III}OH] - k_{-1}[Me^{III}A] - k_2[>Me^{III}A]$$

If [HA] is assumed constant, an integratable form of Equation 3 is achieved by solving Equation 1 for $[>Me^{III}OH]$ and substituting the result into Equation 3. This yields

$$(4) \qquad [>Me^{III}A] = S_T\left[\frac{k_1[HA]}{k_1[HA]+k_{-1}+k_2}\right](1-\exp\{-(k_1[HA]+k_{-1}+k_2)t\})$$

Substituting Equation 4 into Equation 2 yields a rate equation for metal solubilization in terms of rate constants and solution parameters ([HA]) only.

According to Equation 4, $[>Me^{III}A]$ reaches a steady-state value of

$$(5) \qquad [>Me^{III}A]_{ss} = S_T k_1[HA]/(k_1[HA]+k_{-1}+k_2)$$

when t is large compared with the characteristic time (defined as the reciprocal of the rate coefficient term, $k_1[HA]+k_{-1}+k_2$). Under steady-state conditions, the rate of reduced metal release is given by

$$(6) \qquad d[Me^{2+}]/dt = k_2[>Me^{III}A]_{ss} = k_1k_2S_T[HA]/(k_1[HA]+k_{-1}+k_2)$$

Inspection of Equation 6 shows that electron transfer is the rate-limiting step when $k_2 \ll k_1[HA]+k_{-1}$, and adsorption is rate-limiting when $k_1[HA] \ll k_{-1}+k_2$.

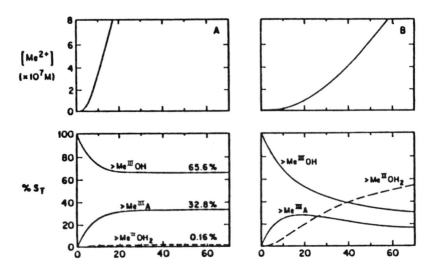

Figure 4-50.   Results of simulation for Mn^{II} release from Mn (hydr)oxide for cases I and II in Example 4-4. Reductant adsorption and electron transfer are rate limiting in (A); metal ion release is rate limiting in (B). Values used for simulations: $S_T = 5 \times 10^{-6}$ M; $k_1 = 500$ M^{-1} min^{-1}; $k_{-1} = k_2 = 0.05$ min^{-1}; $k_4 = 1.0$ min^{-1} in (A) and 0.01 min^{-1} in (B). [From Stone, A.T. and J.J. Morgan, in *Aquatic Surface Chemistry,* W. Stumm (Ed.), Wiley-Interscience, New York, 1988, p. 221. With permission.]

Case II. Alternatively, consider the situation where release of reduced-metal ions from the lattice is rate limiting and the other conditions of **Case I** are the same. Reaction 4 (Table 4-19) now must be included in the kinetic analysis, and the mass balance for surface sites becomes

(7) $$S_T = [>Me^{III}OH] + [>Me^{III}A] + [>Me^{II}OH_2]$$

Four rate constants are required for this case (k_1, k_{-1}, k_2, and k_4); release of the oxidized organic (step 3, Table 4-19) is assumed to be rapid and thus can be ignored. It is simpler to solve this case numerically than analytically. *Acuchem* (Section 2.3.2) or other differential equation solvers can be used if numerical values of the rate coefficients can be supplied. Such simulations (Figure 4-50) show that [>Me^{II}OH$_2$] reaches a steady state that accounts for only a small portion of S_T when k_4 is large and metal release is not rate limiting. Conversely, when k_4 is small and metal release is rate-limiting, [>Me^{II}OH$_2$] increases throughout the simulation and eventually accounts for most of S_T.

Microbial Reduction

The existence of microorganisms capable of reducing Fe and Mn (hydr)oxides has been known for many decades (see reference 280 for a review on the microbiology and biochemistry of these reactions). Isolates that reduce Fe and/or Mn oxides have been reported from at least 18 common genera of bacteria, including aerobes, facultative anaerobes, and strict anaerobes, as well as a few genera of fungi. In some cases activity is inhibited by O_2 and/or NO_3^-, but in others, metal-oxide reduction is not affected by these alternative (and thermodynamically preferable) electron acceptors. Available evidence suggests that several processes are involved in microbial reduction of these metal oxides, including (1) direct reactions with bacterial respiratory enzymes in the electron transport system (ETS) or with other reductases not associated with the ETS, and (2) nonenzymatic reactions with metabolic products such as pyruvate.

Precisely how enzymatic reduction occurs is something of a mystery. Although direct contact of bacterial cells with oxide surfaces can be seen in photomicrographs,[280] the amorphous solid oxides cannot be transported directly into the cells. Therefore, the enzymes that are involved must be extracellular or membrane bound. It should be noted that cellular binding to (hydr)oxide surfaces will enhance both enzymatic mechanisms and those involving nonenzymatic reduction by metabolic products. In the case of Fe oxides, many bacteria apparently use nitrate reductase enzymes, but this is less certain for bacterial reduction of Mn oxides.[280] Based on an analysis of laboratory experiments by enzyme kinetic models, Arnold et al.[281] concluded that reduction of ferric oxide by a *Pseudomonas* isolate involves both soluble and solid Fe^{III} species as substrates. A two-site model was proposed in which both enzymatic sites are functionally equivalent, but most of the sites are sterically limited to interact with only soluble Fe^{III} (e.g., $Fe(OH)_2^+$, $Fe(NTA)(OH)_2^{2-}$). This could happen, for example, if some sites are located on an outer membrane or cell wall (thus being accessible to the (hydr)oxide surface), but most of the sites are inside a membrane that is permeable only to soluble species.

Whether cells are able to obtain energy by coupling the reduction to the formation of ATP in the ETS has not been demonstrated conclusively. Cells could grow using Fe or Mn oxides as electron acceptors and still not obtain energy directly from the reduction reaction. In such cases, energy would be obtained by substrate-level phosphorylation — as occurs in fermentation — and the reaction would simply be a means to eliminate excess reductant without producing ATP. Arnold et al.[281] argued against coupled ATP formation on energetic grounds. Based on their calculations the energy yield from ETS-driven reduction of ferric (hydr)oxide is only barely sufficient to drive the reaction $ADP + P_i \rightarrow ATP$, and an implausibly high process efficiency (~85%) would be needed for the coupled reaction.

The relative importance of microorganisms in the reductive dissolution of metal (hydr)oxides has been evaluated in several studies in which microbial and enzymatic inhibitors were added to anaerobic sediments.[282,283] These studies indicate that microorganisms play a major role in reducing Fe oxides in such environments, but insufficient information is available to make similar statements about the reduction of Mn oxides.[280]

4.8.5 Weathering of Aluminosilicate Minerals

As a class, the aluminosilicate minerals are much more complicated structurally and chemically than the minerals described in previous sections, and the kinetics of their dissolution (weathering) is correspondingly complex. This section provides an introduction to the topic; more detailed treatments are given elsewhere.[284-288] As a class, the aluminosilicates are very diverse and include primary minerals such as feldspars, which are major components of crystalline granitic rocks, and secondary minerals and clays that are formed from weathering reactions of the primary minerals. Stumm and Morgan[150] provide a general introduction to the nature of the aluminosilicate minerals and to the stoichiometry and equilibria of reactions involved in weathering.

The stabilities of aluminosilicates vary widely, even within subcategories of the entire class. Jackson et al.[289] ranked common minerals in order of their reactivity and proposed a weathering sequence:

gypsum > calcite > hornblende > Ca-feldspar > K-feldspar >

Na-feldspar > montmorillinite > mica > kaolinite

The first two are not aluminosilicates, and this is only a partial list. In general, more-stable (less-soluble) minerals weather more slowly,[290] a case where thermodynamic and kinetic properties are in accord.

Weathering is often described as a heterogeneous acid-base reaction in which basic mineral phases and dissolved acids are the reactants. The acids may be strong mineral acids (e.g., from acid rain), $H_2CO_3^*$, or natural organic acids. The products are dissolved base cations, bicarbonate, dissolved silica, and secondary aluminosilicates (clays). A typical weathering reaction is

$$\underset{albite}{2\,NaAlSi_3O_8} + 11H_2O + CO_2$$

$$\to 2\,Na^+ + HCO_3^- + 4Si(OH)_4 + \underset{kaolinite}{Al_2Si_2O_5(OH)_4} \qquad (4\text{-}122)$$

Weathering is a major mechanism for alkalinity generation in soils and contributes to the neutralization of acid rain in terrestrial environments. Aluminosilicate weathering also is a major source of alkalinity and cations in natural waters. According to the calculations in Table 4-17, aluminosilicate weathering accounts for about 70% of the K^+ and 57% of the Mg^{2+} in global-average river water, and smaller, but still important, fractions of Na^+ (40%) and Ca^{2+} (19%).

Among the difficulties encountered in studying the kinetics of aluminosilicate dissolution are the following:

1. Rates are low to very low at neutral pH and ambient temperatures.
2. Rates depend on many solution conditions, including pH and concentrations of organic ligands that attack reactive sites on the mineral surface.
3. Rates depend on many mineral conditions, including particle size and the presence of organic or oxide surface coatings.
4. Rates and products may change over time (i.e., the reactions are not stoichiometric).

In addition, weathering rates in the natural environment depend on the degree to which mineral surfaces are exposed to aggressive water (i.e., water with low pH or

Table 4-20. Comparison of Laboratory Methods to Measure Weathering Rates

Method	Advantages	Disadvantages
Batch flasks	Simple, inexpensive	pH and solutes uncontrolled; difficult to obtain rate constants that can be transferred to other systems
Flow-through column	Provide flow and water/solid ratio representative of natural environment	pH and solute concentrations may vary over length of column; difficult to obtain kinetic formulations from measured data
pH stat	Facilitates determination of reaction order with respect to H^+	Solute concentrations increase over time and may result in secondary mineral formation
Fluidized bed reactor with fast water recycle and continuous (slow) sample bleed-off	Allows maintenance of steady-state concentrations of pH and solutes at levels low enough to avoid secondary mineral formation	Operating conditions limited by need to keep mineral grains suspended (size of mineral grains must be balanced to desired flow rate)
Recirculating column with recycle	Same as fluidized bed and flow-through column	Constant pH and solute concentrations throughout column can be obtained by using small length:diameter column and rapid recycle

Summarized from Schnoor, J.L., in *Aquatic Chemical Kinetics,* W. Stumm (Ed.), Wiley-Interscience, New York, 1990, p. 475, which provides diagrams and references.

attacking ligands). Consequently, water-flow paths through soils and rocks are important factors in ambient weathering rates and greatly affect the export rates of weathered cations from watersheds.

Several devices have been developed over the past two decades for use as reactors to study weathering kinetics; Table 4-20 summarizes their characteristics. Each device was developed to avoid certain experimental difficulties, but none solves all the problems. In addition, various procedures have been developed to overcome the difficulties listed above and obtain reproducible dissolution data in the laboratory. Experiments occasionally are conducted at elevated temperatures to accelerate the process. According to Nesbitt and Young,[290] there is no evidence that weathering mechanisms change with temperature, and the qualitative aspects of weathering (including stability fields of the reactants and products) do not change greatly in the range 0 to 50°C. However, greatly elevated reactor temperatures are not desirable (nor necessary). Uncertainties in measuring E_{act} limit the temperature range over which rates can be extrapolated accurately. The use of low pH to accelerate dissolution is common, but unless enough measurements are done to define the relationship between rate and pH (which is not linear over a broad range of pH), measurements at low pH are of little use in predicting rates at circumneutral pH.

The nonstoichiometry of weathering reactions is particularly a problem for aluminosilicates because they have more than one structure-forming metal center (e.g., Al, Si, Fe, Mg). Many aluminosilicates consist of several layers, and the metal centers exist in both octahedral or tetrahedral bonding arrangements. The ease of removing metal ions from the lattice varies considerably, depending on their structural arrangements. Important insights into the dissolution of these complicated structures have been obtained through model studies on simpler metal oxides, such as aluminum oxide[291,292] and ferric oxide.[293]

As the weathering reaction for albite (Equation 4-122) indicates, the overall stoichiometry of aluminosilicate weathering involves formation of a second aluminosilicate phase, a clay mineral — in this case kaolinite. In reality, reaction 4-122 can be viewed as the sum of two separate reactions: dissolution of albite to form soluble products (and perhaps a disordered solid residue on the mineral surface), and formation of a new mineral phase from some of the soluble products (and perhaps the solid residue). The second step is slow relative to the time scale of measurement, and for practical purposes, weathering reactions studied in the laboratory do not involve the formation of clay minerals. Nonetheless, formation of secondary solid phases may occur in laboratory kinetic studies by at least two mechanisms:

1. Precipitation of amorphous Al (or Fe) hydroxides, if release of soluble Al (or Fe) from the mineral matrix causes the (pH dependent) solubility of these hydroxides to be exceeded;
2. Formation of residual solid phases on the mineral surface as partial dissolution of the mineral proceeds.

These processes complicate the interpretation of kinetic data, as described in the next section.

Mechanisms

Many early studies (see references 284 and 294 for reviews) found that mineral dissolution rates were linear in $t^{1/2}$; i.e., arithmetic plots of amount of dissolved vs. time are parabolic. Such plots generally were assumed to mean that the overall process is diffusion limited. Weathering reactions are far too slow for diffusion in the aqueous boundary layer to be rate limiting, but models were developed in which diffusion of soluble products through a surface layer of the mineral was the initial rate-limiting process. According to these models,[288] dissolution proceeds in two stages. In the initial stage, diffusion of cations and/or solubilized matrix metals (Al, Fe, or silica) from the surface results in incongruent dissolution, with the development of a "leached surface layer". As this layer thickens, diffusion of solubilized products through it eventually becomes as slow as the surface-controlled detachment of the central metal ion, and a pseudo-steady state results. Dissolution then proceeds congruently (no more solid products). An alternative explanation for parabolic kinetics involves formation of an "armored precipitate layer",[285] i.e., a secondary metal hydroxide precipitate such as amorphous $Al(OH)_3$ formed on the surface of the weathering mineral when the metal's solubility is exceeded as a result of release from the dissolving mineral.

Berner and co-workers[294,295] concluded in the late 1970s that the parabolic kinetics are merely an artifact of mineral grinding. Mineral samples usually are ground in ball mill to obtain a uniform particle size. This produces disrupted grain surfaces and ultrafine particles that dissolve much more quickly than larger particles with smooth surfaces. The net effect is a decreasing rate of dissolution with time (and a parabolic plot). When particles were cleaned extensively to remove fines and treated to remove disrupted surfaces, they observed linear (zero-order) dissolution rates. Evidence that supports this interpretation was obtained using scanning electron microscopy and X-ray photoelectron spectroscopy. The former measurements[295] showed that dissolution occurs preferentially at crystal defects, screw dislocations, and etch pits rather than uniformly on the mineral surface. In the latter measurements,[296,297] the investigators were not able to detect the presence of a cation-depleted layer thick enough to account for diffusion limitation.

Nonetheless, evidence for an initial cation-leaching stage has been developed recently in experiments designed to avoid experimental artifacts from grinding and secondary precipitation. Chou and Wollast[298] measured rates of albite dissolution in a fluidized-bed reactor in which dissolved product concentrations were kept well below saturation with respect to possible precipitates. They found that parabolic kinetics could be induced repetitively in a given experiment simply by changing the pH. This finding cannot be explained by the grinding hypothesis, but could be explained by a mechanism in which incongruent weathering solubilizes base cations and some Si and Al, leaving a residual layer on the weathered surface. These authors used a mass-balance approach to calculate the thickness of the feldspar layer from which various components were released after a given experimental period (Figure 4-51). The calculations were based on the amounts of Al, Si, and Na released to solution, the surface area available for dissolution, and the molar volume of albite. Different thicknesses of the dissolved layer were obtained for each component, that for Na being the greatest and that for Si being the least.

Similarly, Schnoor[288] used a pH-stat reactor to study weathering rates of a size-fractionated B-horizon soil. By simultaneously monitoring the release of Ca, Mg, Al, and Si, he found evidence that ion exchange was the predominant reaction for the first 100 h, and chemical weathering of aluminosilicates predominated thereafter.

The surface-controlled dissolution of aluminosilicates does not just happen by spontaneous breakdown of lattice constituents to dissolved species, however. Instead, surface acid-base and coordination reactions, like those described previously for the reductive dissolution of Fe and Mn (hydr)oxides, play a key role. These reactions occur preferentially at surface sites with excess free energy — crystal defects, kinks, etch pits, etc. The mechanism of H^+-promoted dissolution can be viewed as a two-step process:

$$\equiv|-OH + H^+ \xrightarrow{\text{fast}} \equiv|-OH_2^+ \qquad (4\text{-}123a)$$

$$\equiv|-OH_2^+ \xrightarrow{\text{slow}} Me(H_2O)_x^{z+} + \equiv|- \qquad (4\text{-}123b)$$

where $\equiv|-OH$ is a reactive site on a hydrous metal (Al, Si, Fe, Mn) oxide or aluminosilicate with a functional OH group, and Me is a central metal ion of valence z. In step 1, hydrogen ions from solution "attack" and protonate surface –OH groups. This weakens the Me-O bonds holding central metal ions in the mineral

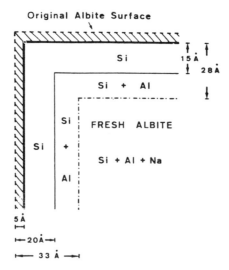

Figure 4-51. Schematic representation of a grain of albite (Na feldspar) after 500 h of dissolution at pH 5.1 (255 h) and 3.1 (245 h), showing differential loss rates of Al, Si, and Na. Shaded area represents layer in which all three components have dissolved, and heavy line represents new surface of the grain. Enough additional Na and Al have dissolved to yield a 15 Å residual layer of Si, and additional Na has been leached to produce a cation-depleted layer that is 13 Å thick. [From Chou, L. and R. Wollast, *Geochim. Cosmochim. Acta*, 48, 2205 (1984). With permission.]

lattice and promotes detachment of an "activated" coordination complex. The process of detachment renews the surface for further attack and dissolution. Figure 4-52 is a two-dimensional representation of reaction sequence 4-123 for an aluminum oxide surface.

Addition of more than one H^+ to the site is necessary to weaken lattice bonds sufficiently for detachment to occur. Indeed, the dissolution rate is thought to be proportional to the degree of surface protonation to the z power: $R_{dis} \propto [\equiv\!l\!-\!OH_2^+]^z$. In turn, the degree of surface protonation depends in nonlinear fashion on bulk-solution pH. Surface protonation can be viewed as an adsorption reaction. At low bulk-solution pH (high $\{H^+\}$), the surface becomes saturated with H^+, and the rate is independent of pH. At intermediate pH values, protonation can be described by the Freundlich isotherm:

$$\log[\equiv\!|\!-\!OH_2^+] = n \log\{H^+\} + \log K_{ads} \qquad (4\text{-}124)$$

where n is the slope of the isotherm, and K_{ads}, the y-intercept, is the adsorption equilibrium constant. Over the pH range where Equation 4-124 is valid, the dissolution rate thus depends on the bulk solution $\{H^+\}$ as follows: $R_{dis} \propto \{H^+\}^m$, where m = nz. Stumm et al.[299] showed that m = 0.4, n = 0.13, and z = 3.1 for dissolution of aluminum oxide; Schnoor[288] reported m = 0.5 for a B-horizon soil

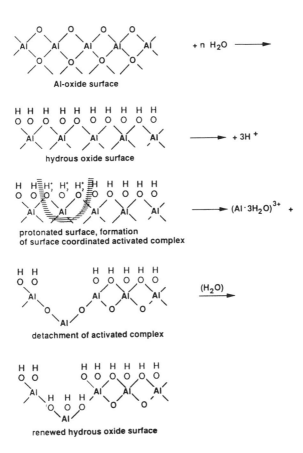

Figure 4-52. Schematic representation of hydrolysis and dissolution mechanism for an aluminum oxide surface. Dissolution occurs when an Al center becomes sufficiently protonated to form $Al(H_2O)_3^{3+}$, which detaches from the lattice and renews the surface. [From Schnoor, J.L., in *Aquatic Chemical Kinetics*, W. Stumm (Ed.), Wiley-Interscience, New York, 1990, p. 475. With permission.]

containing 25% albite, 5% orthoclase and 10% mica. Surface-controlled dissolution by H^+ attack thus leads to fractional order dependence on bulk-solution $\{H^+\}$.

Similarly, ligand attack occurs by the following mechanism:

$$\equiv\!\!|\!-\!OH + HA^- \xrightarrow{\ fast\ } \equiv\!\!|\!-\!A^- + H_2O \qquad (4\text{-}125a)$$

$$\equiv\!\!|\!-\!A^- \xrightarrow{\ slow\ } \equiv\!\!|\!-\! + M(A)^{(z-2)+} \qquad (4\text{-}125b)$$

This leads to dissolution rates proportional to bulk-solution concentrations of the ligand. The strength of complex formation by HA^- affects the rate of dissolution. For example, bidentate ligands that form five- and six-membered rings with the metal center (e.g., oxalate, malonate, salicylate) are more effective in enhancing dissolution of d-Al_2O_3 than are ligands that form seven-membered rings (e.g., succinate,

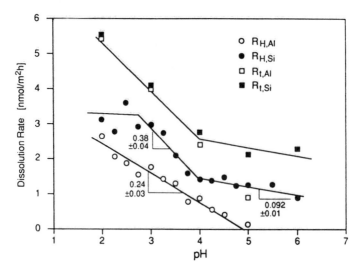

Figure 4-53.   The dissolution of kaolinite is nonstoichiometric over experimental periods of 10 to 15 days, and Si is released at higher rates than Al is. Release rates of both decrease with increasing pH, but R_{Al} is linear with pH, while R_{Si} exhibits a complicated dependency on pH. R_{Si} exceeds R_{Al} at all pH values. Addition of oxalate increases the dissolution of both Al and Si beyond that induced by H^+. [From Stumm, W. and E. Wieland, in *Aquatic Chemical Kinetics*, W. Stumm (Ed.), Wiley-Interscience, New York, 1990, p. 367. With permission.]

phthalate); and monodentate ligands such as benzoate do not enhance dissolution.[291] If surface binding by the ligand follows an adsorption isotherm, the dissolution rate may be fractional order in [HA$^-$]. Furthermore, if the pK_a of HA$^-$ is in the pH range of interest for dissolution, [HA$^-$] will be pH dependent at a fixed total ligand concentration.

Overall, the two mechanisms lead to "mixed" kinetics:

$$R_{dis} = k_{H^+}\left\{H^+\right\}^m + k_{HA}\left[HA^-\right]^{m'}$$ (4-126)

At low pH, H^+ attack predominates, and at high pH, ligand attack predominates; i.e., dissolution becomes pH independent if [HA$^-$] is not pH dependent. The pH at which this occurs depends on the nature of the mineral and ligand(s), as well as the concentration of the ligand(s). Information is available to quantify this for some aluminosilicate minerals. Dissolution of potassium feldspar (orthoclase) is H^+ dependent below pH 5 and pH independent in the range 5 to 10;[284] dissolution of a basalt glass was found to be pH dependent below pH 4, pH independent in the range 4 to 8, and increased with pH above pH 8.[300] Plots of R_{dis} vs. pH can assume more complicated shapes, depending on the mineral constituent being measured. Figure 4-53 illustrates this for H^+ and oxalate-promoted dissolution of kaolinite; the difference between the H^+-promoted release rates for Si and Al indicates that dissolution is nonstoichiometric, at least over reaction times of 10 to 15 days.

Table 4-21. Comparison of Field and Laboratory Weathering Rates

Mineral/ watershed	Weathering rate		Cation measured	Comments	Reference
	mol Si m^{-2} s^{-1}	mol M^{n+} m^{-2} s^{-1}			
Laboratory studies					
Albite	5E-12	2.5E-11	Na^+	25°C,	303
Oligoclase	5E-12	1.7E-12	Na^+	pH 4,	303
Anorthite		3.2E-13	Na^+	P_{CO_2} =1 atm	303
Orthoclase		5.5E-13	Na^+		303
Muscovite	2.4E-13	6.5E-15	Na^+		304
Olivine	7E-12		Mg^{2+}	pH 4.5	305
Gibbsite		2.6E-12	Al^{3+}	pH 3.2, 25°C	306
Kaolinite	3.6E-13	3.6E-14	Al^{3+}	pH 5 ｛see	287
	3.9E-13	2.5E-13	Al^{3+}	pH 4 ｛Fig. 4-53	287
Field (watershed) studies					
Bear Brook, ME	9E-15			Plagioclase, bio- tite, hornblende	288
Coweeta, NC	7E-13			Plagioclase	307
Filson Creek, MN	5E-15			Plagioclase	308
Cristallina, Switzerland	6E-14			Plagioclase biotite	309
Trnavka River, Czechoslovakia					
Forested catchment		8.9E-15	Na^+	Oligoclase	302
Agricultural catchment		4.2E-14	Na^+	Oligoclase	302
Elbe River Basin					
in 1892		5.2E-15	Na^+	Oligoclase	302
in 1976		1.6E-14	Na^+	Oligoclase	302

However, for reaction times up to 40 days, kaolinite dissolution was found to be only "slightly incongruent".[301]

Comparison of Field and Laboratory Weathering Rates

The methodological advances described earlier have allowed geochemists to obtain reliable and reproducible dissolution rates in laboratory experiments (Table 4-21). Interpretation of these data has substantially furthered our understanding of the mechanisms of aluminosilicate weathering over the past decade (although this understanding is still incomplete). However, extrapolation of laboratory measurements to the natural environment is problematic. In large part, the laboratory experiments succeeded because complicating factors such as organic or oxide coatings were eliminated and water was supplied at nonlimiting rates. These factors may limit mineral weathering in terrestrial watersheds, and consequently, laboratory data may be a poor predictor of field rates.

Catchment mass-balance studies have been used to determine weathering rates in the field. This approach, which was pioneered by Pačes,[302] involves measurement of cation or silica fluxes from catchments (small drainage basins) of uniform geological composition. Mass inputs of major cations from atmospheric deposition (wet and dryfall) are subtracted from mass export rates that are determined as the product of hydrologic outflow and cation concentrations in the outflow. Cation contributions from fertilizer or other anthropogenic sources and losses attributable to biomass harvesting also must be taken into account in some watersheds. Measurements are made over annual periods, and sampling must be sufficiently frequent to capture short-term events such as snowmelt and storm runoff that may dominate the water and ion budgets, even though they occur over a small fraction of the year.

Weathering rates determined in this way for catchments in the Bohemian Massif of central Europe were $\sim 5 \times 10^{-15}$ to 7×10^{-13} (mean of $\sim 10^{-14}$) mol m^{-2} s^{-1};[302] oligoclase (sodium feldspar) is the major weatherable mineral in this region. These values are about two orders of magnitude lower than laboratory-derived weathering rates for oligoclase (1.7×10^{-12} mol m^{-2} s^{-1}).[303] Similar discrepancies are found in other comparisons of field and laboratory weathering rates based on cation and silica export (Table 4-21), and laboratory studies thus yield much higher rates than observed in the field. Lower temperature and P_{CO_2} in the field do not account for the large differences.

The most likely explanations are

1. Uncertainties in the actual surface area of weatherable minerals exposed to water under field conditions
2. The preponderance of old minerals with smooth surfaces and few defects in the field compared with fresh surfaces having more kinks and edges that dissolve faster in the laboratory[302]
3. The presence of Al (hydr)oxide coatings on mineral surfaces and high dissolved Al levels in soil-water, which inhibits aluminosilicate dissolution[288]

Unsaturated flow conditions in soils and the presence of soil macropores, which afford a rapid conduit for percolating water, also tend to limit the exposure of weatherable minerals to water.

The notion that either hydrology or surface reaction kinetics may control mineral weathering rates is suggested by the plot of areal export rates for dissolved silica vs. the ratio of flow rate to mass of soil in Figure 4-54. At flow-rate:mass ratios less than $\sim 10^{-3.4}$ L day^{-1} g^{-1}, areal silica export depends strongly on hydrologic flow rate, but at ratios greater than this value, silica export approaches an asymptotic value of 10^{-12} to 10^{-11} mol m^{-2} s^{-1}. The critical ratio corresponds to a flushing rate of about one volume of water per volume of soil per day, and it may represent a limit above which surface chemical reactions control weathering rates and below which hydrologic control is dominant. Laboratory studies on pure minerals generally are conducted at flow rates above the critical value and yield weathering rates corresponding to the asymptotic range.

It is important to note that even though the uncertainty in weathering rates measured in the laboratory is large (about one order of magnitude), hydrologic

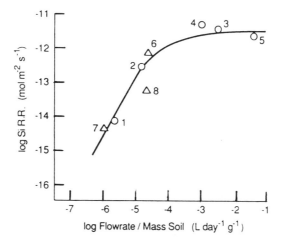

Figure 4-54. Weathering rates of aluminosilicate minerals expressed in terms of dissolved silica release vs. flow-rate/mass ratio. Circles are for soils from Bear Brook watershed (ME) studied in laboratory reactors; triangles are field measurements of Si export from watersheds. [From Schnoor, J.L., in *Aquatic Chemical Kinetics*, W. Stumm (Ed.), Wiley-Interscience, New York, 1990, p. 475. With permission.]

conditions and corresponding uncertainties related to these conditions may cause field rates to be up to two orders of magnitude lower than those measured in the laboratory. Additional studies thus are necessary if we wish to develop models that can accurately predict mineral dissolution rates under environmentally relevant conditions.

PROBLEMS

1. Explain why the pH of maximum k_{obs} decreases with increasing acidity of phenols in the chlorination of phenols.

2. Early studies on the hydrolysis reaction of Cl_2 indicated that k decreased as the reaction proceeded. On this basis, Morris [*J. Am. Chem. Soc.*, 68, 1692 (1946)] proposed that the actual reaction involved hydroxide ions: $Cl_2 + OH^- \rightarrow HCl + Cl^-$. He computed a second-order rate constant for this reaction of 5×10^{14} M^{-1} s^{-1} and found that it had a zero activation energy. Comment on the reasonableness of these values. More recent work [Spalding, C.W., *Am. Inst. Chem. Eng. J.*, 8, 685 (1962)] indicates that $k_2 = 1 \times 10^6$ M^{-1} s^{-1} for this reaction. Estimate the contribution of the OH^--induced mechanism to Cl_2 hydrolysis at pH 7.

3. Compute the rate constant for the reverse reaction of Cl_2 hydrolysis, e.e., for HOCl + $H^+ + Cl^- \rightarrow Cl_2 + H_2O$. Compare the rate of this reaction with the rate of Br^- oxidation by HOCl in seawater at pH 7 and 8, $[Cl^-]$, = 0.55 M, $[Br^-] = 8.1 \times 10^{-4}$ M.

4. From the ratio of rate constants for the formation and hydrolysis of NH_2Cl, compute the equilibrium constant for the reaction and determine the equilibrium composition of a solution containing equimolar concentrations of ammonium and free chlorine (10^{-5} M) at pH 7.0.

5.   The disproportionation of monochloramine is subject to both specific and general (Brønsted) acid catalysis. Given the following data, estimate $k_{H_3PO_4}$:

k_{obs} (M^{-1} s^{-1})	12.5	15	17	21	25.5
$(H_3PO_4)_T$	45	70	90	135	180

All the data were obtained at pH 3.00; $pK_1 = 2.1$ for phosphoric acid. Both H_3PO_4 and $H_2PO_4^-$ may catalyze the reaction, but assume that only the completely protonated form acts as a Brønsted catalyst. [Data from R. Valentine and C. Jafvert, *Env. Sci. Technol.*, 22, 691 (1988)].

6.   Some authors suggest that one can obtain $1/k_g$ and $1/k_1$ as the intercept and slope of a plot of $1/K_L$ vs. RT/H (Equation 4-88c). Is this approach valid? If so, what are the limitations to the accuracy of estimates for k_g and k_1 obtained in this manner?

7.   Derive a rate equation for the Ag^+-catalyzed oxidation of metal ion M^+ by persulfate and show the conditions under which the rate is independent of $[M^+]$.

8.   Given the following reactions and rate constants (all at 25°C),

$$CO_2 + OH^- \rightarrow HCO_3^-; \quad k_f = 7.1 \times 10^3 \ M^{-1} \ s^{-1}$$

$$HCO_3^- \rightarrow CO_2 + OH^-; \quad k_r = 1.8 \times 10^{-4} \ s^{-1}$$

$$CO_2 + H_2O \rightarrow H^+ + HCO_3^-; \quad k_h = 0.037 \ s^{-1}$$

(a) compute the pH at which the two hydration mechanisms for CO_2 are of equal importance; (b) estimate K_1, the apparent first acid dissociation constant for $CO_{2(aq)}$. Note: $K_w = 1.0 \times 10^{-14}$ for $H_2O \rightleftharpoons H^+ + OH^-$ at 25°C.

9.   Derive the general rate equation for anation reactions (Equation 4-22) by both the equilibrium constant and steady-state approaches.

10. Estimate K_{os} for the following outer sphere complexes, Al^{3+}-F^-, $AlOH^{2+}$-F^-, Al^{3+}-HF, and $AlOH^{2+}$-HF, using Equation 4-23 and compare your results to the values in Table 4-3.

11. Calculate the dissociation rate constant for AlF^{2+} and $FeCl^{2+}$ complexes and compare the rate for AlF^{2+} to that reported by Plankey et al. [*Env. Sci. Technol.*, 20, 160 (1986)].

12. The overall rate of Cu-EDTA formation from Ca-EDTA (by ligand exchange) is given by Equation 4-26. Using values if the rate and equilibrium constants provided in the text, sketch a graph to show how the observed rate constant for disappearance of Cu^{2+} from seawater varies with pH and $[Ca^{2+}]$.

13. Estimate the half-life of Fe-desferal complex toward complete dissociation at pH 8.

14. According to the data of Millero et al. [*Environ. Sci. Technol.*, 21, 439 (1987)] autoxidation of HS^- is about four times that of H_2S at 55°C, and $E_{act} = 10.4$ and 12.8 kcal/mol, respectively, for autoxidation of H_2S and HS^- in water. If $log k = 2.50$ for pseudo-first-order rate constant of HS^- at 55°C, estimate k_{H_2S} and k_{HS^-} at 25°C. How do these results compare with the relative reactivities of H_2S and HS^- toward H_2O_2 at 25°C and the results of Chen and Morris [*Env. Sci. Technol.*, 6, 579 (1972)] for H_2S autoxidation at 25°C?

15. Given the values of k_1 and k_2 for SO_2 listed in Table 4-11, compute α_0, α_1, and α_2 at unit pH values over the range 1–6. Determine the overall (observed) rate constant for ozonation of SO_2 and compute the fractional contributions of the three S^{IV} species to the overall rate at the same pH values.

16. Use *Acuchem* or some other kinetics package to simulate the effects of complex formation by ligand L on the rate of Fe^{II} autoxidation. Assume that the Fe^{II}-L complex does not undergo autoxidation and that pseudo-first-order conditions apply for free Fe^{II} (fixed pH and P_{O_2}), and compute rate constants at pH 7.0, 7.5, and 8.0 from information given in the text. Assume that the rate constant for Fe^{II}-L formation, $k_f = 10^6 M^{-1} s^{-1}$, and the rate constant for complex dissociation, $k_d = 1.0 s^{-1}$. Also, evaluate the effects of varying complex stability on the kinetics of Fe^{II} autoxidation at pH 7.0. Assume that k_f is independent of the nature of L and k_d is inversely proportional to the stability constant, K_f, of the complex. Vary k_d over the range 10^{-1} to $10 s^{-1}$. *Note:* this problem was studied by Pankow and Morgan [*Env. Sci. Technol.*, 15, 1155 (1981)], who provide further details and simulations. Compare your results to theirs; see Figure 4-33.

REFERENCES

1. Bell, R.P., *Acid-Base Catalysis*, Oxford University Press, Oxford, 1941.
2. Adamson, A.W., *A Textbook of Physical Chemistry*, 2nd ed., Academic Press, New York, 1979.
3. Frost, A.A. and R.G. Pearson, *Kinetics and Mechanism*, 2nd ed., John Wiley & Sons, New York, 1961.
4. Laidler, K.J., *Chemical Kinetics*, McGraw-Hill, New York, 1965.
5. Purcell, K.F. and J.C. Kotz, *Inorganic Chemistry*, Saunders, Philadelphia, 1977.
6. Valentine, R.L. and J.T. Jafvert, *Environ. Sci. Technol.*, 22, 691 (1988).
7. Hand, V.C. and D.W. Margerum, *Inorg. Chem.*, 22, 1449 (1983).
8. Buckingham, D.A., D.M. Foster, and A.M. Sargeson, *J. Am. Chem. Soc.*, 90, 6032 (1968).
9. Buckingham, D.A., F.R. Keene, and A.M. Sargeson, *J. Am. Chem. Soc.*, 96, 4981 (1974).
10. Hoffmann, M.R., *Environ. Sci. Technol.*, 14, 1061 (1980).
11. Wells, M.A., G.A. Rogers, and T.C. Bruice, *J. Am. Chem. Soc.*, 98, 4336 (1976).
12. Nakamura, A. and M. Tsutsui, *Principles and Applications of Homogeneous Catalysis*, Wiley-Interscience, New York, 1980.
13. Fife, T.H., T.J. Przystas, and V.L. Squillacote, *J. Am. Chem. Soc.*, 101, 3017 (1979).
14. Fife, T.H. and T.J. Przystas, *J. Am. Chem. Soc.*, 102, 7297 (1980).
15. Fife, T.H. and V.L. Squillacote, *J. Am. Chem. Soc.*, 100, 4787 (1978).
16. Green, C.L., R.P. Houghton, and D.A. Phipps, *J. Chem. Soc. Perkin Trans.*, I, 2623 (1974).
17. Ketelaar, J.A.A., H.R. Gersman, and M.M. Beck, *Nature*, 177, 392 (1956).
18. Prue, J.E., *J. Chem. Soc.*, II, 2331 (1952).
19. Gelles, E. and A. Salama, *J. Chem. Soc.*, II, 3684 (1958).
20. Thilo, E., *Adv. Inorg. Chem. Radiochem.*, 4, 31 (1962).
21. Healy, R.M. and M.L. Kilpatrick, *J. Am. Chem. Soc.*, 77, 5258 (1955).
22. Bender, M.L., *Mechanisms of Homogeneous Catalysis from Protons to Proteins*, John Wiley & Sons, New York, 1971; Bender, M.L. and L.J. Brubacher, *Catalysts and Enzyme Action*, McGraw-Hill, New York, 1973.
23. Schrauzer, G.N., *Transition Metals in Homogeneous Catalysis*, Marcel Dekker, New York, 1971.
24. Busch, D.H. (Symp. chair.), *Reactions of Coordinated Ligands and Homogeneous Catalysis*, Adv. Chem. Ser. 37, Am. Chem. Soc., Washington, D.C., 1963.

25. King, E.L., Catalysis in homogeneous reactions in a liquid phase, in *Catalysis, Vol. II, Fundamental Principles* (Part 2), P.H. Emmett (Ed.), Reinhold, New York, 1955.

26. Chandra Singh, U. and K. Venkatarao, *Int. J. Chem. Kinetics*, 13, 555 (1981).

27. Yost, D.M., *J. Am. Chem. Soc.*, 48, 152 (1926).

28. Willard, H.H. and L.L. Merritt, *Ind. Eng. Chem. Anal. Ed.*, 14, 486 (1942).

29. Rodriguez, P.A. and H.L. Pardue, *Anal. Chem.*, 41, 1369,1376 (1969).

30. American Public Health Association, *Standard Methods for the Examination of Water and Wastewater,* 15th ed., New York, 1981.

31. Kolthoff, I.M. and J. Lingane, *Polarography*, Vol. II, 2nd Ed., John Wiley & Sons, New York, 1951.

32. Wilson, S.A. and J.H. Weber, *Chem. Geol.*, 26, 345 (1979).

33. Schwarzenbach, R.P. et al. *Environ. Sci. Technol.*, 24, 1566 (1990).

34. Taube, H., *J. Am. Chem. Soc.*, 68, 611 (1946).

35. Weiss, J., *Naturwissenschaften*, 25, 64 (1935).

36. Wilshire, J. and D. T. Sawyer, *Acct. Chem. Res.*, 12, 105 (1979).

37. Duke, F.R. and R.C. Pinkerton, *J. Am. Chem. Soc.*, 73, 3045 (1951).

38. Libby, W.F., *J. Phys. Chem.,* 56, 863 (1952); Gryder, J.W. and R.W. Dodson, *J. Am. Chem. Soc.*, 71, 1845 (1949).

39. Taube, H., *Science*, 226 1028 (1984).

40. Taube, H., *Electron Transfer Reactions of Complex Ions in Solution*, Academic Press, New York, 1970.

41. Taube, H., H. Myers, and R.L. Rich, *J. Am. Chem. Soc.*, 75, 4118 (1953); Taube, H. and H. Myers, *J. Am. Chem. Soc.*, 75, 2103 (1954).

42. Hill, C. G., Jr., *An Introduction to Chemical Engineering Kinetics and Reactor Design*, John Wiley & Sons, New York, 1977.

43. Mahler, H.R. and E.H. Cordes, *Biological Chemistry*, 2nd ed., Harper & Row, New York, 1971.

44. Sung, W. and J.J. Morgan, *Environ. Sci. Technol.*, 14, 561 (1980).

45. Mabey, W. and T. Mill, *J. Phys. Chem. Ref. Data*, 7, 383 (1978).

46. Johnson, K.S., *Limnol. Oceanogr.*, 27, 849 (1982).

47. Jones, P., M.L. Haggett, and J.L. Longridge, *J. Chem. Ed.*, 41, 610 (1964); Kern, D.M., *J. Chem. Ed.*, 37, 14 (1960).

48. Goldman, J.C. and M.R. Dennett, *Science*, 220, 199 (1983).

49. Morgan, J.J. and W. Stumm, in *Metals and their Compounds in the Environment*, E. Merian (Ed.), VCH Verlag., Weinheim, New York, 1991, pp. 67–103.

50. Brezonik, P.L., S.O. King, and C.E. Mach in *Metal Ecotoxicology: Concepts and Applications,* M.C. Newman and A.W. McIntosh (Eds.), Lewis Publishers, Chelsea, MI, 1991, pp. 1–29.

51. Margerum, D.W., G.R. Cayley, D.C. Weatherburn, and G.K. Pagenkopf, in *Coordination Chemistry*, Vol. 2, A.E. Martell (Ed.), ACS Monograph 174, Am. Chem. Soc., Washington, D.C., 1978, pp. 1–220.

52. Langford, C.H. and H.B. Gray, *Ligand Substitution Processes*, Benjamin, New York, 1965; Stengle, T.R. and C.H. Langford, *Coord. Chem. Rev.*, 2, 349 (1967); Kustin, K. and J. Swinehart, *Prog. Inorg. Chem.*, 13, (1970); Langford, C.H. and M. Parris, in *Comprehensive Chemical Kinetics*, Vol. 7, C.H. Bamford and C.F. Tipper (Eds.), Elsevier, New York, 1972, pp. 1–55; Burgess, J. (Ed.), *Inorganic Reaction Mechanisms*, Vols. 1–4, Specialist Periodical Reports, Chemical Society, London, 1971–1974.

53. Hunt, J.P., *Coord. Chem. Rev.*, 7, 1 (1971).
54. Neely, J.W., Ph.D. thesis, University of California, Berkeley, 1971.
55. Fuoss, R.M., *J. Am. Chem. Soc.*, 80, 5059 (1958).
56. Eigen, M., *Z. Phys. Chem.*, NF1, 176 (1954).
57. Haim, A., R.J. Grassi, and W.K. Wilmarth, *Adv. Chem. Ser.*, 49, 31 (1965); also see p. 717 in ref. 5.
58. Plankey, B.J., H.H. Patterson, and C.S. Cronan, *Environ. Sci. Technol.*, 20, 160 (1986).
59. Margerum, D.W. and G.R. Dukes, in *Metal Ions in Biological Systems*, Vol. 1, H. Sigel (Ed.), Marcel Dekker, New York, 1973.
60. Biruš, M., Z. Bradić, G. Krznarić, N. Kujundšić, M. Pribanić, P.C. Wilkins, and R.G. Wilkins, *Inorg. Chem.*, 26, 1000 (1987).
61. Hering, J.G. and F.M.M. Morel, *Environ. Sci. Technol.*, 22, 1469 (1988).
62. Margerum, D.W., D.L. Janes, and H.M. Rosen, *J. Am. Chem. Soc.*, 87, 4463 (1965).
63. Margerum, D.W., *Record Chem. Progr.*, 24, 237 (1963).
64. Brúcher, E. and G. Laurenczy, *Inorg. Chem.*, 22, 338 (1983).
65. Maljković, D. and M. Branica, *Limnol. Oceanogr.*, 16, 779 (1971).
66. Anderson, D.M. and F.M.M. Morel, *Limnol. Oceanogr.*, 23, 283 (1978).
67. Hering, J.G. and F.M.M. Morel, *Geochim. Cosmochim. Acta*, 53, 611 (1989).
68. Turner, D.R., M.S. Varney, M. Whitfield, R.F.C. Mantoura, and J.P. Riley, *Geochim. Cosmochim. Acta*, 50, 289 (1986); Perdue, E.M. and C.R. Lytle, *Environ. Sci. Technol.*, 17, 654 (1983); Ephraim, J. and J.A. Marinsky, *Environ. Sci. Technol.*, 20, 367 (1986); Dzombak, D.A., W. Fish, and F.M.M. Morel, *Environ. Sci. Technol.*, 20, 669 (1986); Fish, W, D.A. Dzombak, and F.M.M. Morel, *Environ. Sci. Technol.*, 20, 676 (1986).
69. Choppin, G.R. and K.L. Nash, *J. Inorg. Nucl. Chem.*, 43, 357 (1981).
70. Mak, M.K.S. and C.H. Langford, *Can. J. Chem.*, 60, 2023 (1982).
71. Plankey, B.J. and H.H. Patterson, *Environ. Sci. Technol.*, 21, 595 (1987).
72. Olson, D.L. and M.S. Shuman, *Geochim. Cosmochim. Acta*, 49, 1371 (1985).
73. Cabaniss, S.E. and M.S. Shuman, *Geochim. Cosmochim. Acta*, 52, 185 (1988); Hering, J. and F.M.M. Morel, *Environ. Sci. Technol.*, 23, 1234 (1988).
74. Fish, W., Ph.D. thesis, Massachusetts Institute Technology, Cambridge, 1984.
75. Clesciri, N. and G.F. Lee, *Int. J. Air Water Poll.*, 9, 743 (1965).
76. Heinke, G.W., Proc. 12th Conf. Great Lakes Res., Int. Assoc. Great Lakes Research, 1969, Ann Arbor, MI, pp. 766–773.
77. Van Wazer, J.R., E.J. Griffith, and J.F. McCullough, *J. Am. Chem. Soc.*, 77, 287 (1955).
78. Schwarzenbach, R. and P.A. Gschwend, in *Kinetics of Aquatic Chemical Processes*, W. Stumm (Ed.) Wiley-Interscience, New York, 1990, pp. 199–233.
79. Burlinson, N.E., L.A. Lee, and D.H. Rosenblatt, *Environ. Sci. Technol.*, 16, 627 (1982).
80. Haag, W.R. and T. Mill, *Environ. Sci. Technol.*, 22, 658 (1988).
81. Vogel, T.M. and M. Reinhard, *Environ. Sci. Technol.*, 20, 992 (1986).
82. Vogel, T.M., C.S. Criddle, and P.L. McCarty, *Environ. Sci. Technol.*, 21, 722 (1987).
83. Dilling, W.L., N.B. Tefertiller, and G.J. Kalos, *Environ. Sci. Technol.*, 9, 833 (1975).
84. Wolfe, N.L., R.G. Zepp, J.A. Gordon, G.L. Baughman, and D.M. Cline, *Environ. Sci. Technol.*, 11, 88 (1977).
85. Wolfe, N.L., R.G. Zepp, D.F. Paris, G.L. Baughman, and R.C. Hollis, *Environ. Sci. Technol.*, 11, 1077 (1977).

86. Zepp, R.G., N.L. Wolfe, J.A. Gordon, and G.L. Baughman, *Environ. Sci. Technol.*, 9, 1144 (1975).

87. Chapmen, R.A. and C.M. Cole, *J. Environ. Sci. Health*, Part B, B17, 487 (1982); Lemley, A.T. and W.Z. Zhong, *J. Environ. Sci. Health*, B18, 189 (1983).

88. Miles, C.J. and J.J. Delfino, *J. Agric. Food Chem.*, 33, 455 (1985).

89. Schwarzenbach, R.P., W. Giger, C. Schaffner, and O. Wanner, *Environ. Sci. Technol.*, 19, 322 (1985).

90. Barbash, J.E. and M. Reinhard, in *Biogenic Sulfur in the Environment*, E.S. Saltzman and W.J. Cooper (Eds.), ACS Symp. Ser. 393, Am. Chem. Soc., Washington, D.C., 1989, pp. 101–138.

91. Haag, W.R. and T. Mill, *Environ. Toxicol. Chem.*, 7, 917 (1988).

92. Perdue, E.M. and N.L. Wolfe, *Environ. Sci. Technol.*, 17, 635 (1983).

93. Khan, S.U., *Pestic. Sci.*, 9, 39 (1978).

94. Perdue, E.M. and N.L. Wolfe, *Environ. Sci. Technol.*, 16, 847 (1982).

95. Wolfe, N. L., in *Dynamics, Exposure and Hazard Assessment of Toxic Chemicals*, R. Haque (Ed.), Ann Arbor Science, Ann Arbor, MI, 1980, pp. 163–178.

96. Epstein, J., *Science*, 170, 1396 (1970).

97. Morris, J.C., in *Water Chlorination*, R.L. Jolley (Ed.), Ann Arbor Science, Ann Arbor, MI, 1978, pp. 21–33.

98. Margerum, D.W., E.T. Gray, and R.P. Huffman, in *Organometals and Organometalloids. Occurrence and Fate in the Environment*, F.E. Brinckman and J.M. Bellama (Eds.), Symp. Ser. Vol. 82, Amer. Chem. Soc., Washington, D.C., 1978, pp. 278–291,

99. Jolley, R.L. (Ed.), *Water Chlorination: Environmental Impact and Health Effects*, Vol. 1, Ann Arbor Science, Ann Arbor, MI, 1978 (proceedings of 1976 conference); Jolley, R.L., H. Gorchev, and D.H. Hamilton, Jr. (Eds.), *Ibid.*, Vol. 2, Ann Arbor Science, Ann Arbor, MI, 1978 (proceedings of 1977 conference); Jolley, R.L., W.A. Brungs, and R.B. Cummings (Eds.), *Ibid.*, Vol. 3, Ann Arbor Science, Ann Arbor, MI, 1980, (proceedings of 1979 conference); Jolley, R.L., W.A. Brungs, J.A. Cotruvo, R.B. Cummings, J.S. Mattice, and V.A. Jacobs (Eds.), *Ibid.*, Vol. 4, Ann Arbor Science, Ann Arbor, MI, 1983 (proceedings of 1981 conference); Jolley, R.L., R.J. Bull, W.P. Davis, S. Katz, M.H. Roberts, Jr., and V.A. Jacobs (Eds.), *Water Chlorination: Chemistry, Environmental Impact and Health Effects*, Vol. 5, Lewis Publishers, Chelsea, MI, 1985 (proceedings of 1984 conference).

100. Aieta, E.M. and P.V. Roberts, in *Water Chlorination*, Vol. 5, R.L. Jolley et al. (Eds.), Lewis Publishers, Chelsea, MI, 1985, pp. 783–794.

101. Kumar, K. and D.W. Margerum, *Inorg. Chem.*, 26, 2706 (1987).

102. Farkas, L., M. Lewin, and R. Bloch, *J. Am. Chem. Soc.*, 71, 1988 (1949).

103. Aieta, E.M. and P.V. Roberts, *Environ. Sci. Technol.*, 20, 50 (1986).

104. Taube, H. and H. Dodgen, *J. Am. Chem. Soc.*, 71, 3330 (1949).

105. Weil, I. and J.C. Morris, *J. Am. Chem. Soc.*, 71, 1664, 3123 (1949).

106. Morris, J.C., in *Principles and Applications of Water Chemistry*, S.D. Faust and J.V Hunter (Eds.), John Wiley & Sons, New York, 1967.

107. Morris, J.C. and R.A. Isaac, in *Water Chlorination*, Vol. 4, R. L. Jolley et al. (Eds.), Ann Arbor Science, Ann Arbor, MI, 1983, pp. 49–62.

108. Jolley, R.L. and J.H. Carpenter, in *Water Chlorination*, Vol. 4, R. L. Jolley et al. (Eds.), Ann Arbor Science, Ann Arbor, MI, 1983, pp. 3–47.

109. Isaac, R.A. and J.C. Morris, in *Water Chlorination*, Vol. 4, R. L. Jolley et al. (Eds.), Ann Arbor Science, Ann Arbor, MI, 1983, pp. 63–75.

110. Kumar, K., R.W. Shinness, and D.W. Margerum, *Inorg. Chem.*, 26, 3430 (1987).

111. Granstrom, M.L., Ph.D. thesis, Harvard University, Cambridge, 1954.

112. Sanguinsin, J.L.S. and J.C. Morris, in *Disinfection of Water and Waste Water,* J.D. Johnson (Ed.), Ann Arbor Science, Ann Arbor, MI, 1975, pp. 277–299.

113. Dotson, D. and G.R. Helz, in *Water Chlorination,* Vol. 5, R.L. Jolley et al. (Eds.), Lewis Publishers, Chelsea, MI, 1985, pp. 713–722.

114. Shaw, M.P. and W.J. Snodgrass, in *Water Chlorination,* Vol. 4, R.L. Jolley et al. (Eds.), Ann Arbor Science, Ann Arbor, MI, 1983, pp. 113–123.

115. Yiin, B.S., D.M. Walker, and D.W. Margerum, *Inorg. Chem.,* 26, 3435 (1987).

116. Lister, M.W. and P. Rosenblum, *Can. J. Chem.,* 39, 1645 (1961).

117. Valentine, R.L., in *Water Chlorination,* Vol. 5, R.L. Jolley et al. (Eds.), Lewis Publishers, Chelsea, MI, 1985, pp. 975–984.

118. Symons, J.M., T.A. Bellar, J.K. Carswell, J.DeMarco, K.L. Kropp, G.G. Robeck, D.R. Seeger, C.L. Slocum, B.L. Smith, and A.A. Stevens, *J. Am. Water Works Assoc.,* 67, 634 (1975).

119. Rook, J.J., *Water Treatment Exam.,* 23, 234 (1974).

120. Bellar, T.A., J.J. Lichtenberg, and R.C. Kroner, *J. Am. Water Works Assoc.,* 66, 703 (1974).

121. Lee, G.F., in *Principles and Applications of Water Chemistry,* S.D. Faust and J.V Hunter (Eds.), J. Wiley & Sons, New York, 1967.

122. Soper, F.G. and G.F. Smith, *J. Chem. Soc.,* p. 1582, (1926).

123. Burtschell, R.H. et al., *J. Am. Water Works Assoc.,* 51, 205 (1959).

124. Lee, G.F. and J.C. Morris, *Int. J. Air Water Poll.,* 6, 419 (1962).

125. Voudrias, E.A. and M. Reinhard, *Environ. Sci. Technol.,* 22, 1049 (1988).

126. Voudrias, E.A. and M. Reinhard, *Environ. Sci. Technol.,* 22, 1056 (1988).

127. Rook, J.J., *J. Am. Water Works Assoc.,* 68, 168 (1976).

128. Fuson, R.C. and B.A. Bull, *Chem. Rev.,* 15, 278 (1934).

129. Rook, J.J. et al., *J. Environ. Sci. Health,* 13, 91 (1978); Amy, G.L., P.A. Chadik, Z.K. Chowdhury, P.H. King, and W.J. Cooper, in *Water Chlorination,* Vol. 5, R.L. Jolley et al. (Eds.), Lewis Publishers, Chelsea, MI, 1985, pp. 907–922.

130. Stevens, A.A., C.J. Slocum, D.R. Seeger, and G.G. Robeck, in *Water Chlorination,* Vol. 1, R.L. Jolley (Ed.), Ann Arbor Science, Ann Arbor, MI, 1978, pp. 77–101.

131. Gurol, M.D., A. Wowk, S. Meyers, and I.H. Suffet, in *Water Chlorination,* Vol. 4, R.L. Jolley (Eds.), Ann Arbor Science, Ann Arbor, MI, 1983, pp. 269–284.

132. de Leer, E.W.B., J.S. Sinninghe Damste, and L. de Galan, in *Water Chlorination,* Vol. 5, R.L. Jolley et al. (Eds.), Lewis Publishers, Chelsea, MI, 1985, pp. 843–857.

133. Christman, R.F., J.D. Johnson, J.R. Hoss, F.K. Pfaender, W.T. Liao, D.L. Norwood, and H.J. Alexander, in *Water Chlorination,* Vol. 1, R.L. Jolley (Ed.), Ann Arbor Science, Ann Arbor, MI, 1978, pp. 15–28 ; Norwood, D.L., J.D. Johnson, R.F. Christman, J.R. Hoss, and M.J. Bobenreith, *Environ. Sci. Technol.,* 14, 187 (1980).

134. Briley, K.F., R.F. Williams, K.E. Longley, and C.A. Sorber, in *Water Chlorination,* Vol. 3, R.L. Jolley et al. (Eds.), Ann Arbor Science, Ann Arbor, MI, 1980, pp. 117–129; Oliver, B.G. and D.B. Schindler, *Environ. Sci. Technol.,* 14, 1502 (1980); Hoehn, R.C., D.B. Barnes, B.C. Thompson, C.W. Randall, T.J. Grizzard, and T.B. Shaffer, *J. Am. Water Works Assoc.,* 72, 344 (1980).

135. Wachter, J.K., Sc.D. dissertation, University of Pittsburgh, Pittsburgh, PA, 1982.

136. Scully, F.E., Jr., R. Kravitz, G.D. Howell, M.A. Speed, and R.P. Arber, in *Water Chlorination,* Vol. 5, R.L. Jolley (Eds.), Lewis Publishers, Chelsea, MI, 1985, pp. 807–820 .

137. Briley, K.F., R.F. Williams, and C.A. Sorber, Alternative Water Disinfection Schemes for Reduces Trihalomethane Formation, Vol. II, Algae as Precursors for Trihalomethanes in Chlorinated Drinking Water, EPA-600/S2-84-005, U.S. EPA, Cincinnati, OH, 1984.

138. Amy, G.L. and P.A. Chadik, *J. Am. Water Works Assoc.*, 75, 527 (1983); Chadik, P.A. and G.L. Amy, *Ibid.*, 75, 532 (1983); Glaze, W.H. and J.L. Wallace, *Ibid.*, 76, 68 (1984).

139. Fleischacker, S.J. and S.J. Randtke, *J. Am. Water Works Assoc.*, 75, 132 (1983).

140. Jensen, J.N., J.J. St. Aubin, R.F. Christman, and J.D. Johnson, in *Water Chlorination*, Vol. 5, R.L. Jolley et al. (Eds.), Lewis Publishers, Chelsea, MI, 1985, pp. 939–949.

141. Friend, A.G., Ph.D. thesis, Harvard University, Cambridge, 1954.

142. Ayotte, R.C. and E.T. Gray, Jr., in *Water Chlorination*, Vol. 5, R.L. Jolley et al. (Eds.), Lewis Publishers, Chelsea, MI, 1985, pp. 797–806.

143. Le Cloirec, C. and G. Martin, in *Water Chlorination*, Vol. 5, R.L. Jolley et al. (Eds.), Lewis Publishers, Chelsea, MI, 1985, pp. 821–834.

144. Cooper, W.J. and D.M. Kaganowicz, in *Water Chlorination*, Vol. 5, R.L. Jolley et al. (Eds.), Lewis Publishers, Chelsea, MI, 1985, pp. 895–906.

145. Stanbro, W.D. and M.J. Lenkevich, *Science*, 215, 967 (1982).

146. Croue, J.-P. and D. Reckhow, *Environ. Sci. Technol.*, 23: 1412 (1989).

147. Holmbom, B., L. Kronberg, P. Backlund, V.-A. Langvik, J. Hemming, M. Reunanen, A. Smeds, and L. Tikkanen, Proc. 6th Conf. Water Chlorination, Oak Ridge, TN, 1987.

148. Cheh, A.M., J. Skochdopole, J. Koski, and L. Cole, *Science*, 207, 90, (1980).

149. Pankow, J.F., *Aquatic Chemistry Concepts*, Lewis Publishers, Chelsea, MI, 1991.

150. Stumm, W. and J.J. Morgan, *Aquatic Chemistry*, 2nd ed., Wiley-Interscience, New York, 1981.

151. Snoeyink, V.L. and D. Jenkins, *Water Chemistry*, John Wiley & Sons, New York, 1980.

152. Chen, K.Y. and J.C. Morris, *Environ. Sci. Technol.*, 6, 579 (1972).

153. Chen, K.Y. and J.C. Morris, *J. Sanit. Eng. Div. Amer Soc. Civil Eng.*, 98, 215 (1972).

154. Millero, F.J., S. Hubinger, M. Fernandez, and S. Garnett, *Environ. Sci. Technol.*, 21, 439 (1987).

155. Hoffmann, M.R. and B.C. Lim, *Environ. Sci. Technol.*, 13, 1406 (1979).

156. Moore, J.W. and R.G. Pearson, *Kinetics and Mechanism*, 3rd ed., John Wiley & Sons, New York, 1981.

157. Hoffmann, M.R., *Environ. Sci. Technol.*, 11, 61 (1977).

158. O'Brien, D.J. and F.G. Birkner, *Environ. Sci. Technol.*, 11, 1114 (1977).

159. Seinfeld, J.H., *Atmospheric Chemistry and Physics of Air Pollution*, John Wiley & Sons, New York, 1986.

160. Lamb, D., D.F. Miller, N.F. Robinson, and A.W. Gertler, *Atmos. Environ.*, 21, 2133 (1987).

161. Kadowski, S., *Environ. Sci. Technol.*, 20, 1249 (1986).

162. Hoffmann, M.R., *Atmos. Environ.*, 20, 1145 (1986).

163. Waldman, J.M. and M.R. Hoffmann, in *Sources and Fate of Aquatic Pollutants*, R.A. Hites and S.J. Eisenreich (Eds.), Adv. Chem. Ser. 218, Am. Chem. Soc., Washington, D.C., 1987, pp. 70–129.

164. Behra, P., L. Sigg, and W. Stumm, *Atmos. Environ.*, 23, 2691 (1989).

165. Erickson, R.E., L.M. Yates, R.L. Clark, and D. McEwen, *Atmos. Environ.*, 11, 813 (1977).

166. Conklin, M.R. and M.R. Hoffmann, *Environ. Sci. Technol.*, 22, 899 (1988).

167. Espenson, J.H. and H. Taube, *Inorg. Chem.*, 4, 704 (1965).

168. Jacob, D.J. and M.R. Hoffmann, *J. Geophys. Res. C: Oceans Atmos.*, 88, 6611 (1983).

169. Boyce, S.D., M.R. Hoffmann, P.A. Hong, and L.M. Moberly, *Environ. Sci. Technol.*, 17, 602 (1983).

170. Luther, G.W., III, in *Kinetics of Aquatic Chemical Processes*, W. Stumm (Ed.), Wiley-Interscience, New York, 1990, pp. 173–198.

171. Stumm, W. and G.F. Lee, *Ind. Eng. Chem.*, 53, 143 (1961).

172. Davison, W. and G. Seed, *Geochim. Cosmochim. Acta*, 47, 67 (1983).

173. Tamura, H., K. Goto, and M. Nagayama, *J. Inorg. Nucl. Chem.*, 38, 113 (1976).

174. Roekens, E.J. and R.E. Van Grieken, *Mar. Chem.*, 13, 195 (1983).

175. Davison, W., *Mar. Chem.*, 15, 279 (1984).

176. Millero, F.J., S. Sotolongo, and M.Izaguirre, *Geochim. Cosmochim. Acta*, 51, 793 (1987).

177. Singer, P. and W. Stumm, *Science*, 167, 3921 (1970).

178. Millero, F.J., *Geochim. Cosmochim. Acta*, 49, 547 (1985).

179. Theis, T.L. and P.C. Singer, *Environ. Sci. Technol.*, 8, 569 (1974).

180. Miles, C.J. and P.L. Brezonik, *Environ. Sci. Technol.*, 15, 1089 (1981).

181. Pankow, J.F. and J.J. Morgan, *Environ. Sci. Technol.*, 15, 1155 (1981).

182. Lever, A.B., J.P. Wilshire, and S.K. Quan, *Inorg. Chem.*, 20, 761 (1981).

183. Takai, T., *J. Jpn. Water Works Assoc.*, 466, 22 (1973); Tamura, H., K. Goto, and M. Nagayama, *Corros. Sci.*, 16, 197 (1976).

184. Baes, C.F., Jr. and R.E. Mesmer, *The Hydrolysis of Cations*, Wiley-Interscience, New York, 1976.

185. Lowson, R.T., *Chem. Rev.*, 82, 461 (1982).

186. Haber, F. and J. Weiss, *Proc. R. Soc. London*, A147, 332 (1934).

187. Kester, D.A., R.H. Byrne, and Y.J. Liang, in *Marine Chemistry in the Coastal Environment*, T.M. Church (Ed.), ACS Symp. Ser. No. 18, Am. Chem. Soc., Washington, D.C., 1975, pp. 56–79.

188. Hart, E.J., K. Sehested, and J. Holeman, *Anal. Chem.*, 55, 46 (1983).

189. Yang, T.C. and W.C. Neely, *Anal. Chem.*, 58, 1551 (1986).

190. Morgan, J.J., Ph.D. thesis, Harvard University, Cambridge, MA, 1964.

191. Morgan, J.J., in *Principles and Applications of Water Chemstry*, S.D. Faust and J.V. Hunter (Eds.), John Wiley & Sons, New York, 1967, pp. 561–622.

192. Kessick, M.A. and J.J. Morgan, *Environ. Sci. Technol.*, 9, 157 (1975).

193. Davies, S.H.R. and J.J. Morgan, *J. Coll. Interfac. Sci.*, 129, 63 (1989).

194. Grill, E.V., *Geochim. Cosmochim. Acta*, 46, 2435 (1982).

195. Diem, D. and W. Stumm, *Geochim. Cosmochim. Acta*, 48, 1571 (1984).

196. Matsui, I., Ph.D. thesis, Lehigh University, Bethlehem, PA, 1973; cited by ref. 194.

197. Stumm, W. and J.J. Morgan, *Aquatic Chemistry*, 1st ed., John Wiley & Sons, New York, 1970, p. 294.

198. Moffett, J.W. and R.G. Zika, *Environ. Sci. Technol.*, 21, 804 (1987).

199. Sharma, V.K. and F.J. Millero, *Environ. Sci. Technol.*, 22, 768 (1988).

200. Gerrard, W., *Gas Solubilities: Widespread Applications*, Pergamon Press, Elmsford, New York, 1980 (various gases); Weiss, R.F., *Deep Sea Res.*, 17, 721 (1970) (N_2, O_2, Ar); Am. Publ. Health Assoc., *Standard Methods for the Examination of Water and Wastewater*, 16th ed., APHA, AWWA, WPCF, Washington, D.C. (O_2).

201. Suntio, L.R., W.Y. Shiu, D. Mackay, and J.N. Seiber, *Rev. Environ. Contam. Toxicol.*, 103, 1 (1988) (pesticides); Burkhard, L.P., D.E. Armstrong, and A.W. Andren, *Environ. Sci. Technol.*, 19, 590 (1985) (PCBs); Dunnivant, F.M. and A.W. Elzermann, *Chemosphere*, 17, 525 (1988) (PCBs); Shiu, W.Y. and D. Mackay, *J. Phys. Chem. Ref. Data*, 15, 911 (1986) (PCBs); Leighton, D.T., Jr. and J.M. Calo, *J. Chem. Eng. Data*, 26, 382 (1981) (volatile chlorinated organic compounds); Jonson, J.A., J. Vejrosta, and J. Novak, *Fluid Phase Equil.*, 9, 279 (1982) (*n*-alkanes); Snider, J.R. and G.A. Dawson, *J. Geophys. Res.*, 90, 3797 (1985) (alcohols); Leuenberg, C., M.P. Ligocki, and J.F. Pankow, *Environ. Sci. Technol.*, 19, 1053 (1985) (phenols); Mackay, D. and W.Y. Shiu, *J. Phys. Chem. Ref. Data*, 10, 1175 (1981) (various organic contaminants).

202. Mackay, D., W.-Y. Shiu, K.T. Valsaraj, and L.J. Thibodeaux, in *Gas Transfer at Water Interfaces*, W. Brutsaert and G. H. Girka (Eds.), D. Reidel, Dordrecht, 1984, pp. 34–56.

203. Nirmalakhandan, N.N. and R.E. Speece, *Environ. Sci. Technol.*, 22, 1349 (1988).

204. Whitman, W.G., *Chem. Metall. Eng.*, 29, 146 (1923); Lewis, W.K. and W.G. Whitman, *Ind. Eng. Chem.*, 16, 1215 (1924).

205. Nernst, W., *Z. Phys. Chem.*, 47, 52 (1904).

206. Gulliver, J.S., in *Gas Transfer at Water Interfaces*, W. Brutsaert and G. H. Girka (Eds.), D. Reidel, Dordrecht, 1984, pp. 1–7.

207. Higbie, R., *Trans. Am. Inst. Chem. Eng.*, 31, 365 (1935).

208. Danckwerts, P.V., *Ind. Eng. Chem.*, 43, 1460 (1951); *AIChE J*, 1, 456 (1955).

209. Dobbins, W.E., *J. Sanit. Eng. Div. ASCE*, 90, (SA3) 53 (1964).

210. McCready, M.J. and T.J. Hanratty, in *Gas Transfer at Water Interfaces*, W. Brutsaert and G.H. Girka (Eds.), D. Reidel, Dordrecht, New York, 1984; also see Hanratty, T.J., in *Air-Water Mass Transfer*, S. C. Wilhelms and J. G. Gulliver (Eds.), Proc. 2nd Int. Symp., American Society of Civil Engineers, New York, 1991, pp. 10–31.

211. Wilhelms, S.C. and J.S. Gulliver (Eds.), *Air-Water Mass Transfer*, Proc. 2nd Int. Symp., American Society Civil Engineers, New York, 1991.

212. Brutsaert, W. and G.H. Girka (Eds.), *Gas Transfer at Water Surfaces*, D. Reidel, Dordrecht, NY, 1984.

213. Smith, J.H., D.C. Bomberger, Jr., and D.L. Haynes, *Environ. Sci. Technol.*, 14, 1332 (1980).

214. Davies, J.T. and E.K. Rideal, *Interfacial Phenomena*, Academic Press, New York, 1961; Asher, W.E. and J.F. Pankow, pp. 68–80 in ref. 210.

215. Mackay, D. and P.J. Leinonen, *Environ. Sci. Technol.*, 9, 1178 (1975); Podoll, R.T., H.M. Jaber, and T. Mill, *Environ. Sci. Technol.*, 20, 490 (1986).

216. Liss, P.S. and P.G. Slater, *Nature*, 247, 161 (1974).

217. Emerson, S., *Limnol. Oceanogr.*, 20, 743 (1975).

218. Jayaweera, G.R. and D.S. Mikkelsen, *Soil Sci. Soc. Am. J.*, 54, 1447 (1990); Jayaweera, G.R., D.S. Mikkelsen, and K.T. Paw U, *Soil Sci. Soc. Am. J.*, 54, 1462 (1990).

219. Holley, E.R., *Water Res.*, 7, 559 (1973).

220. Holmen, K. and P.S. Liss, *Tellus*, B36, 92 (1984).

221. Watson, A.J., R.C. Upstill-Goddard, and P.S. Liss, *Nature*, 349, 145 (1991).

222. Jähne, B., K.O. Münnich, R. Bösinger, A. Dutzli, W. Huber, and P. Libner, *J. Geophys. Res.*, 92, 1937 (1987).

223. Tamir, A. and J.C. Merchuk, *Chem. Eng. Sci.*, 33, 1371 (1978).

224. Smith, J.H., D.C. Bomberger, Jr., and D.L. Haynes, *Chemosphere*, 10, 281 (1981).

225. Othmer, D.F. and M.S. Thakar, *Ind. Eng. Chem.*, 45, 589 (1953).

226. Hayduk, W. and H. Laudie, *AIChE J.*, 20, 611 (1974)

227. Wilke, C.R. and P. Chang, *AIChE J.*, 1, 264 (1955).

228. Reid, R.C., J.M. Prausnitz, and T.K. Sherwood, *The Properties of Gases and Liquids,* 2nd ed., McGraw-Hill, New York, 1977.

229. Cornwell, D.A., in *Water Quality and Treatment. A Handbook of Community Water Supplies,* 4th ed., F.W. Pontius (Ed.), American Water Works Association, McGraw-Hill, New York, 1990, pp. 229–268.

230. Gulliver, J.S., J.R. Thene, and A.J. Rindels, *J. Environ. Eng.,* 116, 503 (1990).

231. Rathbun, R.E. and D.Y. Tai, *Environ. Sci. Technol.,* 20, 949 (1986).

232. Rathbun, R.E. and D.Y. Tai, *Environ. Sci. Technol.,* 21, 248 (1987).

233. Broecker, W.S., T. Takahashi, H.J. Simpson, and T.-H. Peng, *Science,* 206, 409 (1979).

234. McDonald, J.P. and J.S. Gulliver, in *Gas Transfer at Water Interfaces,* W. Brutsaert and G. H. Girka (Eds.), D. Reidel, Dordrecht, 1984, pp. 267–277.

235. Peng, T.-H., W.S. Broecker, G. Mathieu, and Y.-H. Li, *J. Geophys. Res.,* 84, 2471 (1979).

236. Wanninhhof, R., J.R. Ledwell, and W.S. Broecker, *Science,* 227, 1224 (1985).

237. Wanninhkof, R., J. Ledwell, and J. Crusius, pp. 441–458 in ref. 210.

238. Gray, D. M. (Ed.), *Handbook on the Principles of Hydrology,* Water Information Center, Port Washington, NY, 1970.

239. Chow, V. T., D. R. Maidment, and L. W. Mays, *Applied Hydrology,* McGraw-Hill, New York, 1988.

240. Marciano, J.J. and G.E. Harbeck, in *Water-Loss Investigations: Lake Hefner Studies,* USGS Prof. Paper No. 269, Washington, D.C., 1954, pp. 46–70 .

241. Tchobanoglous, G. and E.D. Schroeder, *Water Quality,* Addison-Wesley, Reading, MA, 1985.

242. Broecker, W.S., J. Peterman, and W. Siems, *J. Mar. Res.,* 36, 595 (1978).

243. Liss, P.S. and L. Merlivat, in *The Role of Air-Sea Exchange in Geochemical Cycling,* P. Buat-Menard (Ed.), Reidel, Dordrecht, 1986, pp. 113-127 .

244. Churchill, M.A., H.L. Elmore, and R.A. Buckingham, *J. Sanit. Eng. Div., Proc. ASCE,* SA4, Paper 3199 (1962).

245. Thomann, R.V. and J.A. Mueller, *Principles of Surface Water Quality Modeling and Control,* Harper & Row, New York, 1987.

246. Cohen, Y., W. Cocchio, and D. Mackay, *Environ. Sci. Technol.,* 12, 553 (1978).

247. Lee, Y.-N. and S.E. Schwarz, *J. Phys. Chem.,* 85, 840 (1981).

248. Lee, Y.-N. and S.E. Schwarz, *J. Geophys. Res.,* 86, 11971 (1981).

249. Schindler, D.W., G.J. Brunskill, S. Emerson, W.S. Broecker, and T.H. Peng, *Science,* 177, 1192 (1972).

250. Quinn, J.A. and N.C. Otto, *J. Geophys. Res.,* 76, 1539 (1971).

251. Emerson, S., *J. Geophys. Res.,* 20, 754 (1975).

252. Sillen, L.G., in *Oceanography,* M. Sears (Ed.), Publ. No. 67, Am. Assoc. Adv. Science, Washington, D.C., 1961, pp. 549–581.

253. Holland, H.E., *The Chemistry of the Oceans and Atmosphere,* Wiley-Interscience, New York, 1978.

254. Rimstidt, J.D. and H.L. Barnes, *Geochim. Cosmochim. Acta,* 44, 1683 (1980).

255. Lasaga, A.C. and G.V. Gibbs, in *Aquatic Chemical Kinetics,* W. Stumm (Ed.), Wiley-Interscience, New York, 1990, pp. 259–289.

256. Sverdrup, H., *Chemica Scripta,* 22, 12 (1983); Sverdrup, H., R. Rasmussen, and I. Bjerle, *Chemical Scripta,* 24, 53 (1984).

257. Morse, J.W. and R.C. Berner, in *Chemical Models,* E.A. Jenne (Ed.), ACS Symp. Ser. 93, Am. Chem. Soc., Washington, D.C., 1979, pp. 499–535.

258. Wollast, R.W., in *Aquatic Chemical Kinetics*, W. Stumm (Ed.) Wiley-Interscience, New York, 1990, pp. 431–445.

259. Plummer, L.N., D.L. Parkhurst, and T.M.L. Wigley, in *Chemical Models,* E.A. Jenne (Ed.), ACS Symp. Ser. 93, Am. Chem. Soc., Washington, D.C., 1979, pp. 537–573.

260. Sjöberg, E.L. and D.T. Rickard, *Geochim. Cosmochim. Acta*, 48, 485 (1984).

261. Morse, J.W. and R.C. Berner, *Am. J. Sci.*, 272, 840 (1972).

262. Berner, R.C. and J.W. Morse, *Am. J. Sci.*, 274, 108 (1974).

263. Plummer, L.N., T.M.L. Wigley, and D.L. Parkhurst, *Am. J. Sci.*, 278, 179 (1978).

264. Peterson, M.N.A., *Science*, 154, 1542 (1966).

265. Berger, W.H., *Mar. Geol.*, 8, 111 (1970); *Science*, 156, 383 (1967).

266. Morse, J.W., *Am. J. Sci.*, 274, 97 (1974).

267. Berner, R.C. and P. Wilde, *Am. J. Sci.*, 272, 826 (1972).

268. Morse, J.W., *Am. J. Sci.*, 274, 638 (1974).

269. Burton, W.K. and N. Cabrera, *Faraday Soc. Disc.*, V, 33 (1949); Burton, W. K., N. Cabrera, and F.C. Frank, *Trans. R. Soc. London,* A243, 299 (1951).

270. Rickard, D.T. and E.L. Sjöberg, *Am. J. Sci.*, 283, 815 (1983).

271. Mach, C.E., Ph.D. thesis, University of Minnesota, Minneapolis, 1992.

272. Hoffmann, M.R. and S.J. Eisenreich, *Environ. Sci. Technol.*, 15, 339 (1981).

273. Stone, A.T. and J.J. Morgan, in *Aquatic Surface Chemistry*, W. Stumm (Ed.), Wiley-Interscience, New York, 1988, pp. 221–254.

274. Stone, A.T. and J.J. Morgan, *Environ. Sci. Technol.*, 18, 450 (1984).

275. Stone, A.T. and J.J. Morgan, *Environ. Sci. Technol.*, 18, 617 (1984).

276. Burdige, D.J. and K.H. Nealson, *Geomicrobiology*, 4, 361 (1986).

277. Stone, A.T., *Geochim. Cosmochim. Acta*, 51, 919 (1987).

278. Mopper, K. and B.F. Taylor, in *Organic Marine Geochemistry*, M.L. Sohn (Ed.), Am. Chem. Soc. Symp. Ser. 305, Washington, D.C., 1986, pp. 324–339.

279. Stone, A.T., *Environ. Sci. Technol.*, 21, 979 (1987).

280. Ghiorse, W., in *Environmental Microbiology of Anaerobes*, A.J.B. Zehnder (Ed.), Wiley-Interscience, New York, 1986, pp. 305–331.

281. Arnold, R.G., T.M. Olson, and M.R. Hoffmann, *Biotech. Bioeng.*, 28, 1657 (1986).

282. Sørensen, J., *Appl. Environ. Microbiol.*, 43, 419 (1982).

283. Jones, J.G., S. Gardener, and B.M. Simon, *J. Gen. Microbiol.*, 129, 131 (1983); 130, 45 (1984).

284. Aagaard, P. and H.C. Helgeson, *Am. J. Sci.*, 282, 237 (1982).

285. Helgeson, H.C., W.M. Murphy, and P. Aagaard, *Geochim. Cosmochim. Acta*, 48, 2405 (1984).

286. Wieland, E., B. Wehrli, and W. Stumm, *Geochim. Cosmochim. Acta*, 52, 1969 (1988).

287. Stumm, W. and E. Wieland, in *Aquatic Chemical Kinetics*, W. Stumm (Ed.), Wiley-Interscience, New York, 1990, pp. 367–400.

288. Schnoor, J.L., in *Aquatic Chemical Kinetics*, W. Stumm (Ed.), Wiley-Interscience, New York, 1990, pp. 475–504.

289. Jackson, M.L., S.A. Tyler, A.C. Willis, G.A. Burbeau, and R.P. Pennington, *J. Phys. Coll. Chem.,* 52, 1237 (1948).

290. Nesbitt, H.W. and G.M. Young, *Geochim. Cosmochim. Acta*, 48, 1523 (1984).

291. Furrer, G. and W. Stumm, *Geochim. Cosmochim. Acta*, 50, 1847 (1986).

292. Žutić, and W. Stumm, *Geochim. Cosmochim. Acta*, 48, 1493 (1984).

293. Zinder, B., G. Furrer and W. Stumm, *Geochim. Cosmochim. Acta*, 50, 1861 (1986).

294. Berner, R.A. and G.R. Holdren, Jr., *Geology*, 5, 369 (1977).

295. Berner, R.A., E.L. Sjoberg, M.A. Velbel, and M.D. Krom, *Science*, 207, 1205 (1980).
296. Petrović, R., R.A. Berner, and M.B. Goldhaber, *Geochim. Cosmochim. Acta*, 40, 537 (1976).
297. Holdren, G.R., Jr. and R.A. Berner, *Geochim. Cosmochim. Acta*, 43, 1161 (1979).
298. Chou, L. and R. Wollast, *Geochim. Cosmochim. Acta*, 48, 2205 (1984).
299. Stumm, W., G. Furrer, and G. Kunz, *Croat. Chim. Acta*, 56, 593 (1983).
300. Schott, J., in *Aquatic Chemical Kinetics*, W. Stumm (Ed.), Wiley-Interscience, New York, 1990, pp. 337–365.
301. Carroll-Webb, S.A. and J.V. Walther, *Geochim. Cosmochim. Acta*, 52, 2609 (1988).
302. Pačes, T., *Geochim. Cosmochim. Acta*, 47, 1855 (1983).
303. Busenberg, E. and C.V. Clemency, *Geochim. Cosmochim. Acta*, 40, 41 (1976).
304. Lin, F.-C. and C.V. Clemency, *Geochim. Cosmochim. Acta*, 45, 571 (1981).
305. Granstaff, D.E., *Geochim. Cosmochim. Acta*, 41, 1097 (1977).
306. Bloom, P.R., *Soil Sci. Soc. Am. J.*, 47, 164 (1983).
307. Velbel, M.A., *Am. J. Sci.*, 285, 904 (1985).
308. Siegel, D.I. and H.O. Pfannkuch, *Geol. Soc. Am. Bull.*, 95, 1446 (1984).
309. Giovanoli, R., J.L. Schnoor, L. Sigg, W.Stumm, and J. Zobrist, *Clays Clay Min.*, 36, 521 (1989).

"All is flux, nothing stays still."

Heraclitus (540–480 BC)

CHAPTER 5

Reactors, Mass Transport, and Process Models

PART I. REACTOR THEORY

5.1 FUNDAMENTALS OF CHEMICAL REACTORS

5.1.1 Introduction

Rate equations developed in preceding chapters were for reactions in closed systems that do not exchange material with their surroundings. In such systems, reactant concentrations are highest at the beginning of an experiment; reactions rates also are maximum at the beginning, except in autocatalyzed and consecutive reactions. Laboratory experiments normally are conducted in such systems for simplicity and ease of analysis. Material inputs and outflows can be ignored, thus simplifying the rate equations. Environmental systems such as lakes sometimes can be treated as closed systems, but most often they are open, and material inflows and outflows must be considered. Engineered systems like wastewater treatment plants usually are operated as continuous-flow open systems. Continuous-flow systems are used in some kinetic studies; e.g., to measure microbial growth (Chapter 6). Flow-through reactors are not always macroscopic; microbial cells also can be treated as continuous-flow reactors.

This chapter describes the kinetics of open systems, with emphasis on the two main kinds of open reactors: plug-flow and completely mixed systems. Applications of reactor principles to biogeochemical processes in aquatic systems are described next, and simple mass-balance models are presented for nutrients, organic contaminants, and alkalinity in lakes. Finally, the concepts of flow-through reactors are extended to multicompartment systems, and the use of matrix methods to model their behavior is illustrated.

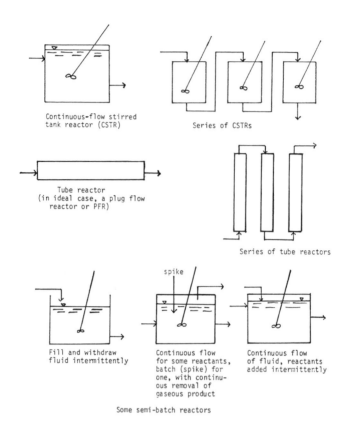

Figure 5-1. Classification of chemical reactors based on physical shape and flow characteristics.

5.1.2 Types of Reactors

Chemical reactors can be divided into three categories based on flow characteristics: (a) no-flow (closed or batch) reactors; (b) continuous-flow reactors; and (c) intermittent or semibatch systems. Continuous-flow reactors normally are assumed to have constant flow rates. Based on physical shape, reactors can be classified as tubes or tanks. Figure 5-1 illustrates the common types of reactors based on combinations of flow and shape. Tube reactors behave ideally as plug-flow devices such that a volume of fluid entering the inlet remains a discrete entity as it flows through the system. These reactors are known as plug-flow reactors (PFRs). Tank reactors ideally are completely mixed at all times such that reactant concentrations are the same throughout their volume. These systems are called completely-mixed reactors or **continuous-flow stirred tank reactors** (CFSTRs). The concentrations of reactants and products in effluent of a CFSTR are the same as those in the reactor. If the reactor is effective in producing change, reactant concentrations will be much lower in the reactor than in the inflow; influent concentrations are diluted to effluent values essentially instantly.

Industrial reactors may be joined in series for flexibility in process control (Figure 5-1). Staged reactors also occur in nature. A series of pools in a river may behave like a series of PFRs, and a chain of lakes may behave like a series of CFSTRs. In terms of product formed, a series of tube reactors is equivalent to a single PFR with a volume equal to the total volume of the PFRs in series, but this is not true for a series of CFSTRs (Section 5.3.1). As the number of CFSTRs in series increases, the behavior of the system approaches that of a PFR with volume equal to the total for the CFSTRs.

Some of the semibatch systems illustrated in Figure 5-1 are used in industrial applications. For example, rates of highly exothermic reactions can be controlled and temperature maintained at a desirable level by batch loading some reactants and feeding others gradually.[1] Semibatch fill-and-draw reactors are used to treat toxic wastes, for lab-scale waste treatment studies, and for growing microorganisms in laboratory and industrial applications. Semibatch systems are more difficult to analyze mathematically and rarely are used in kinetic studies. One exception, the continuously fed batch reactor, is used to measure microbial growth parameters (see Figure 6-14).

5.1.3 Hydraulic Characteristics of Reactors

An important characteristic of continuous-flow reactors is the hydraulic residence time, τ_w, the ratio of its volume V to the flow rate (Q) into it:

$$\tau_w = V/Q_{in} \tag{5-1}$$

For constant-volume systems, Q applies to the entire system, and τ_w is equivalent to the average time an element of fluid spends in the system. At steady state, $Q_{in} = Q_{out}$, and flow can be measured at either end of the system. Lakes in which evaporation is an important component of the water balance are exceptions; in these cases $Q_{in} > Q_{out}$ (based on surface and subsurface outflows). Because solutes are not lost by evaporation, τ_w is based on Q_{out} when we are interested in relating it to retention of solutes; this is equivalent to a conservative ion residence time, τ_C. The reciprocal of residence time is

$$1/\tau_w = Q/V = D \text{ or } \rho_w \tag{5-2}$$

This is called the **flushing coefficient** (ρ_w) or **dilution rate** (D) (time^{-1}); ρ_w and D are equivalent to first-order rate constants for fluid flow.

Another useful parameter for tank reactors is the **areal hydraulic loading rate** (also known as the overflow rate). Defined as the fluid inflow rate divided by reactor surface area, $q_a = Q/A$, this is an important design parameter for sedimentation tanks in water and wastewater treatment. It also is used in models of substances that are lost from lakes by settling of suspended matter.

5.2 PRINCIPLES OF MASS BALANCE ANALYSIS

5.2.1 Concepts and Importance of Mass Balances

The primary tools chemical engineers use to analyze reactor behavior are mass and energy balances. Temperature is fixed externally in natural waters, and reactant concentrations are too low for reactions to have significant effects on the energy balance of most natural systems. Consequently, energy balances are not very useful in analyzing the behavior of natural reactors. The main tool thus is the mass balance, which derives from the concept of mass conservation. For any reacting system the general mass-balance equation is

$$\Delta S \quad = \quad F_{in} \quad - \quad F_{out} \quad + \quad R$$

| Rate of accumulation within system (a) | = | Rate of reactant inflow (b) | − | Rate of reactant outflow (c) | + | Rate of change by reaction within system (d) | (5-3) |

Equation 5-3 is simple and self-evident; indeed the concept of mass balance is so basic that the verbal expression may seem trivial. To be useful for calculations, each term in the equation must be expressed in quantitative, functional form, attention paid to boundary conditions, and consistent units used for all terms. In some cases, mass-balance analysis leads to simple algebraic expressions, but in other cases (e.g., nonsteady-state conditions) the mathematics become complicated. Flow terms may include both advective and diffusive components. The former are given by the product of fluid flow (Q, m³ s⁻¹) and reactant concentration ($g\ m^{-3}$ or mol m⁻³) at the system boundary. The latter are given by the product of the Fickian flux (F_d, $g\ m^{-2}\ s^{-1}$) and the area (A, m²) of the system boundary across which diffusion occurs. F_d is given by Fick's first law as $-D_S d[S]/dx$, where D_S is the diffusion coefficient of S, and the concentration gradient is measured at the system boundary. In engineered systems the reaction term may be a simple first- or second-order expression, but in natural systems where reactants may disappear by a number of physical, chemical, and biological processes, this term may include several functional expressions.

The general mass-balance equation can be simplified for some conditions and types of reactors. For example, terms (b) and (c) are not needed for batch reactors. For continuous-flow reactors at steady state we can ignore term (a). CFSTRs are homogeneous, and one mass-balance equation applies to the entire reactor. In PFRs composition depends on position in the reactor, and the mass balance is written for a differential element of volume, in which concentrations are considered homogeneous, and integrated over the entire reactor.

Reactor analysis is more complicated if temperature or volume changes during the reaction. Natural aquatic systems have reasonably uniform temperatures, and the dilute nature of aquatic reactants means that the heat absorbed or evolved during reaction will not affect the system's temperature measurably. Temperature variations caused by exo- or endothermic reactions may have large effects on the performance of industrial reactors. Volume changes are unimportant in liquid systems and in atmospheric reactions, where the reactants and products are present at low (ppb to ppm) mixing ratios. Volume changes are important for gas-phase reactions in industrial reactors when the number of moles of reactants and products is not the same. These topics are discussed in chemical engineering texts.[1,2]

5.2.2 Mass-Balance Derivation of Kinetic Equations for PFRs

The ideal plug-flow reactor is a long tube through which the fluid flows with no longitudinal dispersion and perfect mixing from the wall to center of the tube. For plug-flow conditions (advective movement only) with constant flow of fluid and reactant S, the mass balance for S on an element of volume ΔV is

$$\begin{array}{ccccc} \text{Accumulation} \\ \text{in } \Delta V \end{array} = \begin{array}{c} \text{Solute flow} \\ \text{rate into } \Delta V \end{array} - \begin{array}{c} \text{Solute flow} \\ \text{rate out of } \Delta V \end{array} + \begin{array}{c} \text{Reaction} \\ \text{rate in } \Delta V \end{array} \qquad (5\text{-}4a)$$

$$\Delta V \partial [S] / \partial t = F_s - (F_s + \Delta F_s) + R_s \Delta V \qquad (5\text{-}4b)$$

R_S is positive for products and negative for reactants. Mass flow rate (F_S) is the product of volumetric flow rate times concentration, Q[S], where Q is constant; ΔF_S = $Q\Delta[S]$, where $\Delta[S]$ is the change in [S] from the inlet to the outlet side of the volume element; and $\Delta V = A\Delta x$, where A is cross-sectional area, and Δx is the length of the volume element. Substituting these relationships into Equation 5-4b and taking the limit as $\Delta x \to 0$, we obtain

$$\frac{\partial [S]}{\partial t} = -\frac{Q \partial [S]}{A\, \partial x} + R_s \qquad (5\text{-}4c)$$

At steady state, the accumulation term $\partial[S]/\partial t = 0$, and we can write Equation 5-4c as

$$A dx / Q = d[S] / R_s \qquad (5\text{-}5)$$

Integrating over the length of the reactor we obtain

$$\frac{Ax}{Q} = \frac{V}{Q} = \tau_w = \int_{[S]_{in}}^{[S]_{out}} \frac{d[S]}{R_s} \qquad (5\text{-}6a)$$

This equation defines the residence time (combination of volume and flow rate) needed for a given outflow concentration of S, i.e., a given conversion of reactant to product. To use the equation we must express loss rate R_S in a way that the integral can be solved. If the reaction is first order, $R_S = -k[S]$, and Equation 5-6a integrates to

$$k\tau_w = -\ln\frac{[S]_{out}}{[S]_{in}} \qquad (5\text{-}6b)$$

For the general case of $R_S = -k[S]^n$, the integrated solution to Equation 5-6a is

$$\left[\frac{1}{[S]_{out}^{n-1}}\right] - \left[\frac{1}{[S]_{in}^{n-1}}\right] = k(n-1)\tau_w \qquad (5\text{-}6c)$$

In analyzing PFRs, Equation 5-6a often is reformulated in terms of the fraction of S converted to product, f_S:[1]

$$\tau_w = V/Q = [S]_{in}\int_{f_{S_{in}}}^{f_{A_{out}}}\frac{df_s}{R_s} \qquad (5\text{-}7a)$$

Normally, $f_{S_{in}} = 0$; $f_{S_{out}}$ is set to the desired conversion factor, and the equation is solved for the residence time needed for that conversion. R_S must be expressed in terms of the fraction of S converted to product. For a first-order reaction this is $R_S = -k[S]_{in}(1 - f_S)$, and the integrated expression is

$$\tau_w = -(1/k_1)\ln(1 - f_s) \qquad (5\text{-}7b)$$

or $\qquad [S]_{out}/[S]_{in} = 1 - f_s = \exp(-k\tau_w) = \exp(-k_1 V/Q) \qquad (5\text{-}7c)$

5.2.3 Mass-Balance Derivation of Kinetic Equations for CFSTRs

CFSTRs are tanks (cylindrical to facilitate mixing) whose contents are stirred continuously to achieve uniform composition. Fresh fluid with reactants is added at one side of the tank, and spent fluid is withdrawn from an outlet located opposite the inlet (Figure 5-1). Perfect mixing has two important consequences. First, inflowing fluid has an equal probability of being anywhere in the reactor shortly after entry, and thus all fluid elements have an equal probability of leaving the reactor at any time. This leads to a broad distribution of substance residence times in CFSTRs. The *average* residence time of a conservative substance is the same as that of the fluid, but only a small fraction of molecules are in the reactor for this time. Second, the effluent has the same composition as fluid in the reactor. This has important implications regarding reaction rates in CFSTRs vs. rates in PFRs (see Section 5.3.1).

The nonsteady-state mass balance for a reactant in a CFSTR is similar to that for a PFR, but because the CFSTR composition is uniform, the mass balance is written for the entire reactor rather than a small element of volume:

$$\text{Accumulation } = \text{ inflow} - \text{outflow} + \text{reaction} \tag{5-8a}$$

$$Vd[S]/dt = Q[S]_{in} - Q[S] + R_s V \tag{5-8b}$$

The following derivation assumes that input is constant over time, but solutions also can be derived where inputs are simple functions of time[3,4] see next section). For a first-order reaction term, $R_S = -k[S]$, we can write

$$d[S]/dt + \{k + Q/V\}[S] = (Q/V)[S]_{in} \tag{5-8c}$$

or
$$d[S]/dt + \phi[S] = \rho_w[S]_{in} \tag{5-8d}$$

where $\phi = k + \rho_w$ is a "total-elimination" coefficient (sum of reaction plus washout). This equation is mathematically equivalent to the rate equation for a reversible first-order reaction (cf. Equation 2-39), which has a standard integral form. Equation 5-8d is in the general form of the first-order ordinary differential equation, $dy/dt + P(t)y = Q(t)$, which is useful to illustrate the application of integrating factors to solve differential equations.[3-5] Integrating factors have the form $\exp(\int P dt)$. In the above case $P = \phi$, and the integrating factor is $e^{\phi t}$. Multiplying both sides of Equation 5-8d by $e^{\phi t}$ yields

$$e^{\phi t}\{[S]' + \phi[S]\} = \rho_w[S]_{in} e^{\phi t} \tag{5-8e}$$

where $[S]'$ is $d[S]/dt$. Note that the left side of Equation 5-8e can be written as $(e^{\phi t}[S])'$ since $(e^{\phi t}[S])' = e^{\phi t}[S]' + \phi[S]e^{\phi t}$. Equation 5-8e thus becomes

$$(e^{\phi t}[S])' = \rho_w[S]_{in} e^{\phi t} \tag{5-8f}$$

which can be integrated to

$$e^{\phi t}[S] = \rho_w[S]_{in}\int e^{\phi t} dt = \frac{\rho_w}{\phi}[S]_{in} e^{\phi t} + C \tag{5-8g}$$

where C is a constant of integration. Solving Equation 5-8g at $t = 0$, where $[S] = [S]_o$ yields: $C = [S]_o - (\rho_w/\phi)[S]_{in}$. Substituting this into Equation 5-8g yields the nonsteady-state solution for reactant S in a CFSTR:

$$[S] = [S]_{out} = \frac{\rho_w}{\phi}[S]_{in}\{1 - \exp(-\phi t)\} + [S]_o \exp(-\phi t) \tag{5-8h}$$

The steady-state solution for CFSTRs is simple and can be obtained from Equation 5-8h by setting $t = \infty$, which causes all exponential terms to approach zero:

$$[S]_{out} = [S]_{in} \rho_w / \phi = [S]_{in} / \tau_w (k + \rho_w) = [S]_{in} / (1 + k\tau_w) \qquad (5\text{-}9)$$

Alternatively, we can write the mass balance:

$$\text{accumulation} = 0 = \text{inflow} - \text{outflow} + \text{reaction} \qquad (5\text{-}10a)$$

or
$$0 = F_{S_{in}} - F_{S_{out}} + R_S V \qquad (5\text{-}10b)$$

Substituting the product of concentration times flow for reactant mass flow and rearranging Equation 5-10b yields:

$$V/Q = \tau_w = \{[S]_{out} - [S]_{in}\} / R_s \qquad (5\text{-}10c)$$

For first order reactions, $R_S = -k[S]_{out}$, and Equation 5-10c can be rearranged to Equation 5-9. Reaction rates in CFSTRs thus can be determined directly from influent and effluent concentrations, reactor volume, and flow rate. This makes CFSTRs attractive for studying the kinetics of biological processes.

5.2.4 Solutions of CFSTR Equation for Simple Loading Functions

Analytical solutions for the CFSTR model can be derived for a variety of simple input functions of substances whose behavior can be modeled as a first-order process or the sum of several first-order processes.[3,4,6]

Constant Mass Loading. In this case $L(t) = Q[S_{in}]$ where both Q and $[S]_{in}$ are constant. The time-varying solution is Equation 5-8h, which we can rewrite as

$$[S_t] = [S_0] \exp(-\phi t) + \frac{[S_{in}]}{1 + k\tau_w} \{1 - \exp(-\phi t)\} \qquad (5\text{-}11)$$

Step Change. If the input concentration changes from a constant value $[S_{1in}]$ to a new value $[S_{2in}]$ at some time t_{12} and hydraulic inflow is unchanged, the equation for $[S_t]$ has a form similar to Equation 5-11:[3,4]

$$[S_t] = [S_1] \exp[-\phi(t - t_{12})] + \frac{[S_{2in}]}{1 + k\tau_w} \{1 - \exp[-\phi(t - t_{12})]\} \qquad (5\text{-}12)$$

$[S_1]$ is the steady-state concentration in the CFSTR for input concentration $[S_{1in}]$. Alternatively, if the CFSTR has not achieved steady state for this input concentration, $[S_1]$ is the value of $[S_t]$ computed from Equation 5-11 for $[S_{in}] = [S'_{in}]$ at $t = t_{12}$.

Pulse Input. This function applies to contaminant spills that occur over very short periods of time (relative to τ_w and substance reaction times). For a pulse input of S into a CFSTR at $t = t_o$, concentrations over time can be obtained from Equation 5-11 by assuming that $[S_{in}] = 0$ at $t > t_o$. The pulse is mixed instantaneously through the entire reactor, and $[S_o]$ is given by M/V where M is the total mass of the pulse discharge. This assumes the reactor is not influenced by previous spills or inputs. The initial concentration decays exponentially with total elimination coefficient ϕ:

$$[S_t] = [S_o]\exp\{-\phi(t - t_o)\} = (M/V)\exp\{-\phi(t - t_o)\} \; for \; t > t_o \qquad (5\text{-}13)$$

A more rigorous definition of pulse inputs is the Dirac delta function:[4]

$$\delta(t - t_o) = 0, t \neq t_o; \qquad (5\text{-}14a)$$

and

$$\int_{-\infty}^{+\infty} d(t - t_o)dt = 1 \qquad (5\text{-}14b)$$

It is apparent from Equation 5-14b that the function $\delta(t - t_o)$ has units of reciprocal time. Together, Equations 5-14a,b define an input that is zero for all t except $t = t_o$ and appproaches infinity at t_o, with the integrated area of the pulse equal to unity. The Dirac delta function must be multiplied by the total mass discharge to give the pulse input to a reactor:[4]

$$L(t) = M\delta(t - t_o) \qquad (5\text{-}15)$$

General Solution. For other integratable functions of L(t), a general solution can be written based on the integrating factor method:[3]

$$[S_t] = [S_o]\exp(-\phi t) + \frac{1}{V}\int_o^t \exp(-\phi(t - \tau)L(\tau)d\tau) \qquad (5\text{-}16)$$

where τ is a "dummy" variable. This form has been used to derive several useful equations for $[S_t]$.

Linearly Increasing or Decreasing Input. This case has the form: $L(t) = L_o \pm \omega_m t$, where ω_m is the rate of change in mass loading, and integration of Equation 5-16 with this function yields[7]

$$[S_t] = [S_o]\exp(-\phi t) + \frac{[S_{in}^o]}{1 + \phi\tau_w}\{1 - \exp(-\phi t)\} \mp \frac{\omega_c}{\delta^2\tau_w}\{1 - \exp(-\phi t) - \phi t\} \quad (5\text{-}17a)$$

where $\omega_c = \omega_m\tau_w/V$ is the rate of change in input concentration:

$$[S_{in}](t) = [S_{in}^o] \pm \omega_c t \qquad (5\text{-}17b)$$

Exponential Decay. This is a useful loading function for organic contaminants such as pesticides. For example, if a pesticide had been applied to a land area for some time and then was removed from use (by legislative ban), its export rate to receiving waters may decline exponentially because loss processes such as degradation by soil microorganisms, hydrolysis, and volatilization usually fit first-order expressions.[6] In this case, $L(t) = L_o \exp(-\omega_1 t)$, where ω_1 is the rate coefficient for declining export of the contaminant due to the sum of all first-order loss processes in the watershed. Under these conditions, Equation 5-16 becomes

$$d[S_t]/dt = (L_o/V)\exp(-\omega t) - \phi[S_t] \qquad (5\text{-}18)$$

This equation can be solved by change of variable (cf. derivation of Equation 2-46) or by substituting $L(t)$ into general solution Equation 5-16 and integrating:

$$[S_t] = [S_o]\exp(-\phi t) + \frac{S_{in}^o}{\varepsilon}\{\exp(-\omega t) - \exp(-\phi t)\} \qquad (5\text{-}19)$$

where $\varepsilon = 1 + k\tau_w - \omega_1\tau_w$. Section 5.6 describes the application of this equation to dieldrin concentrations in a reservoir after the use of that organochlorine pesticide was banned. Exponential-decay loading functions also apply to the downstream lakes in a chain of lakes if the loading of a contaminant to the first lake is stopped and the concentration of the contaminant in this lake decays by first-order processes (see references 3 and 4 for examples of time-variable responses of lakes in series to such loading changes).

5.2.5 Mass-Balance Derivation of the Diffusion-Reaction Equation

Fluid and reactant exchange occurs by diffusion rather than advection in the movement of substances between sediment pore water and the overlying water column. This situation also applies to transport of nutrients into cells. The mass-balance principle can be used with Fick's first law to derive equations for these conditions. Consider a tube reactor at steady state (accumulation is zero). If there is no advection, the mass balance for a volume element with cross sectional area πr^2 and thickness Δx (Figure 5-2) is

Input	=	Output	−	Reaction
$-\pi^2 D_s d[S]/dx\vert_x$	=	$-\pi^2 D_s d[S]/dt\vert_{x+\Delta x}$	−	$R_s \Delta x \pi r^2$
$cm^2 \times cm^2 s^{-1} \times g / cm^3 \div cm$		$cm^2 \times cm^2 s^{-1} \times g / cm^3 \div cm$		$g(cm^{-3}s^{-1}) \times cm \times cm^2 = g/s$

$$(5\text{-}20a)$$

Dividing both sides by $-\pi r^2$ and Δx yields

$$D_s\frac{\left\{d[S]/dx\vert_x - d[S]/dx\vert_{x+\Delta x}\right\}}{\Delta x} = R_s \qquad (5\text{-}20b)$$

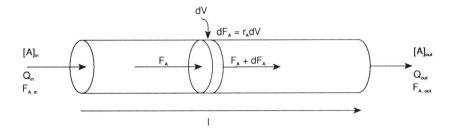

Figure 5-2. Volume element for a plug flow reactor.

Taking the limit as $\Delta x \to 0$, we obtain:

$$D_s d^2 [S]/dx^2 = -R_s \qquad (5\text{-}21)$$

Equation 5-21, a second-order ordinary differential equation, is the fundamental one-dimensional diffusion-reaction equation. The ease with which Equation 5-21 can be integrated depends on the functional form of R_S and the boundary conditions. For first-order reactions, Equation 5-21 becomes

$$D_s d^2 [S]dx^2 = k_1 [S] \qquad (5\text{-}22)$$

This equation is used for substrate diffusion into biofilms (Section 6.6.6).

Example 5-1. Derivation of Diffusion-Reaction Equation in Spherical Coordinates

This equation is important in describing substrate uptake within biological floc. Assume the floc can be approximated as a sphere with radius r_f as shown at the right, and bacteria are distributed uniformly throughout. Also, assume the permeability of the floc is uniform and that D_{sf}, the diffusion coefficient for substrate in the floc, is constant.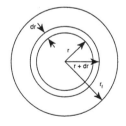

The steady-state mass balance for substrate S on the shell between r and dr is

$$(1) \quad \left(-D_{sf} \cdot \frac{d[S]}{dr} \cdot 4\pi r^2 \right)_r - \left(-D_{sf} \cdot \frac{d[S]}{dr} \cdot 4\pi r^2 \right)_{r+dr} = R \cdot 4\pi r^2 dr$$

Areal flux at inner boundary layer of shell times surface area of inner layer	Areal flux at outer boundary layer of shell times surface area of outer layer	Volumetric reaction rate times volume of shell
$(cm^2 s^{-1})(gcm^{-3} \cdot cm^{-1})(cm^2)$	$(cm^2 s^{-1})(gcm^{-3} \cdot cm^{-1})(cm^2)$	$(gcm^{-3}s^{-1})(cm^2)cm$
$g\,s^{-1}$	$g\,s^{-1}$	$g\,s^{-1}$

R is the volumetric reaction rate. Dividing through by $4\pi dr$ yields

(2)
$$D_{sf} \frac{\left(r^2 d[S]/dr_{|r+dr} - r^2[S]/dr_{|r} \right)}{dr} = r^2 R$$

Taking the limit as $dr \to 0$ yields

(3)
$$D_{sf} \frac{d}{dr}\left(r^2 \frac{d[S]}{dr} \right) = r^2 R$$

Differentiating by the product formula yields

(4)
$$D_{sf}\left(r^2 \frac{d^2[S]}{dr^2} + 2r \frac{d[S]}{dr} \right) = r^2 R$$

or (5)
$$D_{sf}\left(\frac{d^2[S]}{dr^2} + \frac{2d[S]}{rdr} \right) = R$$

To integrate this equation we must express R in terms of [S]. Analytical solutions are possible when R is zero- order or first-order in [S]; numerical methods must be used for other functions (see Section 6.6.5).

5.2.6 Steady-State One-Dimensional Mass-Transport Equation

Mass transport in some flowing systems occurs by both dispersion and advection. For the one-dimensional case the steady-state mass balance then becomes

| **Input** | = | **Output** | − | **Reaction** |

$$\left[-D_L \frac{d[S]}{dx} + u[S]\pi r^2 \right]_x = \left[-D_L \frac{d[S]}{dx} + u[S]\pi r^2 \right]_{x+\Delta x} - \left(R_s\right)\pi r^2 \Delta x$$

(5-23)

D_L is a dispersion coefficient that reflects the combined influence of molecular diffusion and turbulent dispersion (the latter predominates in flowing systems), and $u =$ flow velocity (m s^{-1}; $u\pi r^2 = Q$, m^3 s^{-1}). Dividing by πr^2 and Δx, and taking the limit as $\Delta x \to 0$, we obtain

$$D_L \{d^2[S]/dx^2 - u\{d[S]/dx\} + R_s = 0$$ (5-24)

This equation is a model for reaction in a flowing tubular reactor with longitudinal dispersion. Analytical solutions are available when $R_S = 0$ (S is conservative) and when R_S is first order.[1,8] Numerical solutions are available for other orders.

5.2.7 General One-Dimensional Mass-Transport Equation

An additional level of complexity in a tubular reactor occurs when it is not at steady state, i.e., when accumulation is nonzero. The mass balance on a volume element gives rise to the fundamental one-dimensional mass-transport equation:

$$D_L \partial^2[S]/\partial x^2 - u\partial[S]/\partial x - \partial[S]/\partial t + R_S = 0 \tag{5-25}$$

This is a partial differential equation because there are two independent variables (x and t). A function must be substituted for R_S to solve the equation. The four terms in Equation 5-25 reflect changes due to diffusion, advection, accumulation, and reaction, respectively.

Analytical solutions for Equation 5-25 are difficult to obtain. Detailed analysis is beyond our scope, but a few comments can be made. First, solutions are available[9] for various initial and boundary conditions and zero- and first-order expressions for R_S. Second, the equation forms the basis for many water quality models that simulate the behavior of pollutants in aquatic systems. These models usually are solved numerically because the simulations are done for varying conditions of temperature, light, and flow, and the coefficients (u, D_L, k's) are not constant (perhaps not even simple functions of time). Third, analytical solutions of the equation can be used to describe flow of reacting fluids through porous media (activated carbon, groundwater aquifers) if the coefficients and flow rates are assumed constant.[10,11]

Integration of equations like Equation 5-25 can be simplified by reformulating them into a dimensionless form. This is done by defining a series of "normalized" (dimensionless) variables and collecting the coefficients in a way to get dimensionless group parameters. Consider Equation 5-25 with R_S as a first-order expression. First, we divide all terms by $[S]_{in}$ and divide the denominator of each term by l^2, where l is a "characteristic" length, e.g., reactor length. Next, define $[S^*] = [S]/[S]_{in}$ and $\xi = x/l$ as dimensionless concentration and dimensionless length, respectively. Dividing the result by D_L, we obtain

$$\frac{\partial^2[S^*]}{\partial \xi^2} - \frac{ul\partial[S^*]}{D_L \partial \xi} - \frac{kl^2[S^*]}{D_L} - \frac{l^2\partial[S^*]}{D_L \partial t} = 0 \tag{5-26a}$$

We can define the following dimensionless groups and substitute them into Equation 5-26a to obtain a dimensionless equation:

$$P = 2\alpha = ul/D_L; \quad \Phi = (k/D_L)^{-1/2}; \quad \tau' = D_L t/l^2;$$

$$\frac{\partial^2[S^*]}{\partial \xi^2} - 2\alpha\frac{\partial[S^*]}{\partial \xi} - \Phi^2[S^*] - \frac{\partial[S^*]}{\partial \tau'} = 0 \tag{5-26b}$$

P is called the **Peclet number**. It describes the relative importance of advection to diffusion in a reactor. Φ is the **Thiele modulus**, and it expresses the importance of chemical reaction compared with diffusion (see Section 6.6.5). Equation 5-26b can be solved by Laplace transform for various initial and boundary conditions. Application of the mass-transport equation to analysis of pollutant behavior in groundwater aquifers is described in many books.[5,10]

Example 5-2. Solution of One-Dimensional Mass-Transport Equation

The one-dimensional mass transport equation (Equation 5-26b) can be solved by Laplace transform under the following conditions: at $\xi > 0$ and $\tau' = 0$, $[S^*] = 0$; for $\xi = 0$ and $\tau' \geq 0$, $[S^*] = 1$; and as $\xi \to \infty$ at $\tau' \geq 0$, $[S^*] \to 0$. Applying the Laplace transform yields:[11]

(1) $d^2[S^*]/d\xi^2 - 2\alpha d[S^*]/d\xi - (\Phi^2 + w)[S^*] = 0$

with boundary conditions: at $\xi = 0$, $[S^*] = 1/w$; as $\xi \to \infty$, $[S^*] \to 0$, and where

(2) $[S^*] = \displaystyle\int_0^\infty \exp(wr')d\tau'$

is the Laplace transform and w is a complex parameter. The solution of Equation 2 is

(3) $[S^*] = \exp(\alpha\xi)\{\exp(-\xi\sqrt{w + \alpha^2 + \Phi^2})\}/w$

The inverse transform of Equation 3 is obtained from standard tables[9] and provides the solution for $[S^*]$:

(4) $[S^*] = \dfrac{1}{2}\exp[-\xi(m - \alpha)]\left\{\exp(2\xi m) \times erfc(\dfrac{\xi}{2\sqrt{\tau}} + m\sqrt{\tau}) + erfc(\dfrac{\xi}{2\sqrt{\tau}} - m\sqrt{\tau})\right\}$

where $m = (\alpha^2 + \Phi^2)^{1/2}$ and $erfc(\xi)$ is the complementary error function:

(5) $erfc(\xi) = 1 - erf(\xi) = 1 - \dfrac{2}{\sqrt{\pi}}\displaystyle\int_0^\xi \exp(-t^2)dt = \dfrac{2}{\sqrt{\pi}}\displaystyle\int_\xi^\infty \exp(-t^2)dt$

Tabulations of erf(x) and erfc(x) for different values of x are available.[4,12] Note: erf(∞) = 1, and erf($-\infty$) = -1. Therefore, erfc(∞) = 0, and erfc($-\infty$) = 2. Thus when $\tau \to \infty$ (i.e., at steady-state), Equation 4 becomes:

(6) $[S^*] = \exp[-\xi(m - \alpha)]$

5.3 COMPARATIVE BEHAVIOR OF CFSTRs AND PFRs

5.3.1 Extent of Reaction and Volume Requirements

If the extent of reaction is considerable, the reactant concentration in a CFSTR and its effluent is small compared with that in the incoming fluid. As mentioned

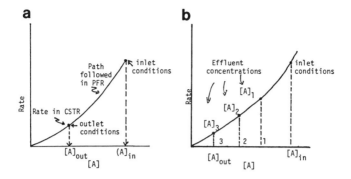

Figure 5-3. (a) Hypothetical plot of reaction rate vs. reactant concentration for a PFR and a CFSTR achieving the same degree of reaction. (b) Reaction rate vs. reactant concentration for three PFRs in series; [S]₁–[S]₃ are outflow concentrations for the reactors; closed circles are rates occurring in each reactor if operated as a CFSTR. (Modified from Hill, C.G., Jr., *Chemical Engineering Kinetics and Reactor Design*, John Wiley & Sons, New York, 1977.)

earlier, reaction rates decrease over distance in a PFR as the reactant is used up, but rates are constant in a CFSTR and equal to the rate at the outlet (Figure 5-3a). For equal extents of reaction, the rate throughout a CFSTR is equal to the lowest rate (that at the outlet) in a comparably operated PFR. To achieve the same extent of reaction, a CFSTR thus must be larger than a PFR. Exceptions may occur in autocatalytic reactions where the product catalyzes the reaction. The added volume requirements of CFSTRs can be minimized by placing several CFSTRs in series so that small step changes in reactant concentration are obtained between reactors (Figure 5-3b; Example 5-3). If many CFSTRs are in series, the system approaches plug-flow behavior in which each CFSTR is equivalent to an element of volume in a PFR.

Example 5-3. Extent of Reaction in a Series of CFSTRs

According to Equation 5-9 the outflow concentration for a substance that reacts by first-order kinetics in a CFSTR depends on its inflow concentration, the hydraulic residence time of the reactor and the first-order rate constant of the reaction: $[S]_{out} = [S]_{in}/(1 + k\tau_w)$. Consider a series of three CFSTRs linked so that the outflow from one is the inflow to the next. The CFSTRs could represent sequential pools in a river or reservoir or a chain of several well-mixed lakes. Outflow concentrations for each reactor are as follows:

(1)
$$[S]_{out,1} = \frac{[S]_{in}}{(1 + k\tau_{w,1})}$$

(2)
$$[S]_{out,2} = \frac{[S]_{out,1}}{(1 + k\tau_{w,2})} = \frac{[S]_{in}}{(1 + k\tau_{w,1})(1 + k\tau_{w,2})}$$

(3)
$$[S]_{out,3} = \frac{[S]_{out,2}}{(1 + k\tau_{w,3})} = \frac{[S]_{in}}{(1 + k\tau_{w,1})(1 + k\tau_{w,2})(1 + k\tau_{w,3})}$$

In general, for n reactors in series,

$$(4) \qquad [S]_{out,n} = \frac{[S]_{in}}{\prod_{i=1}^{n}(1+k\tau_{w,i})}$$

or for equally sized reactors,

$$(5) \qquad [S]_{out,n} = \frac{[S]_{in}}{(1+k\tau_w)^n}$$

In spite of the larger volume requirements of CFSTRs, they have many advantages over PFRs. They are simple to construct and operate; material costs usually are a minor consideration in laboratory and small-scale systems. Because some fluid elements have short residence times in CFSTRs, the outlet composition responds rapidly to changes in inlet conditions, and it is easier to develop control systems for CFSTRs. Finally, the rate equations for CFSTRs are simpler than those of PFRs, and it is easier to analyze kinetic data from CFSTRs.

The relative volume requirements of CFSTRs and PFRs to consume the same amount of reactant are given by the ratio of Equation 5-10c to Equation 5-6a:[1,2]

$$\frac{V_{CFSTR}}{V_{PFR}} = \frac{\{[S]_{out}-[S]_{in}\}/R_S}{\int_{[S]_{in}}^{[S]_{out}} \frac{d[S]}{R_S}} \tag{5-27}$$

Numerical values for the volume ratios can be obtained by substituting an appropriate rate equation for R_S and values for $[S]_{in}$ and $[S]_{out}$. The solution is made easier by expressing $[S]_{out}$ and the rate equation in terms of the fraction (f_S) of $[S]_{in}$ converted to product: $[S]_{out} = [S]_{in}(1-f_S)$. The denominator on the right side of Equation 5-27 then becomes Equation 5-7a. If we assume that $f_S = 0$ at the inlet, that $[S]_{in}$ is the same for both reactors and that the reaction is nth order in $[S]$, Equation 5-27 becomes

$$\frac{V_{CFSTR}}{V_{PFR}} = \frac{\dfrac{[S]_{in}(1-f_S)-[S]_{in}}{-k\{[S]_{in}(1-f_S)\}^n}}{\dfrac{[S]_{in}}{k}\displaystyle\int_0^{f_S}\dfrac{df_S}{-\{[S]_{in}(1-f_S)\}^n}} = \frac{\dfrac{f_S}{(1-f_S)^n}}{\displaystyle\int_0^{f_S}\dfrac{df_S}{(1-f_S)^n}} \tag{5-28}$$

Levenspiel[2] evaluated Equation 5-28 as a function of f_S for zero- to third-order reactions (Figure 5-4). For $n > 0$, the CFSTR must be larger than the PFR for all f_S. The size disparity increases with increasing n and increasing f_S. For zero-order reactions, reactor size is independent of reactor type.

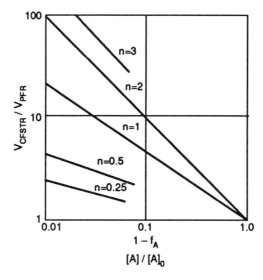

Figure 5-4. Ratio of volumes required by CFSTR and PFR reactors to achieve the same degree of reactant removal for nth-order reactions: $S \rightarrow P$; rate $= -k[S]^n$. Input concentrations and volumes assumed identical; $f_S =$ fraction of reactant consumed. (Redrawn from Levenspiel, O., *Chemical Reaction Engineering*, 2nd ed., John Wiley & Sons, New York, 1972.)

Table 5-1. $F(t)^{\cdot}$ **for Various Reactor Types**

PFR	$F(t) = 0$ for $0 < t < \tau_w$
	$F(t) = 1$ for $t > \tau_w$
CFSTR	$F(t) = 1 - \exp(-t/\tau_w)$
Laminar flow	$F(t) = 1 - (\tau_w/2t)^2$

$\cdot$ $F(t) =$ volume fraction of fluid leaving reactor that has been in it for a time $< t$.

5.3.2 Distribution of Residence Times in Different Reactor Types

The time that reactants spend in an open system is important in analyzing the behavior of the system. The average residence time for reactors is the same regardless of flow configuration and is given by Equation 5-1 ($\tau = V/Q$), but the distribution of residence times for discrete molecules or fluid elements depends on the type of reactor. Residence time distributions can be determined by applying a step change or pulse input of a conservative substance to a reactor and measuring the response in outlet concentration.

Distribution time of fluids in reactors is expressed as $F(t)$, the volume or weight fraction of fluid in the outflow that has been in the reactor for a time less than t.[13] The differential of $F(t)$, $dF(t)$, is called the residence time distribution function.[1] $F(t)$ functions for common reactor types are listed in Table 5-1. The limits on $F(t)$ for any reactor are 0 at $t = 0$ and 1.0 at $t = \infty$. From probability theory it can be shown that τ is related to $F(t)$:

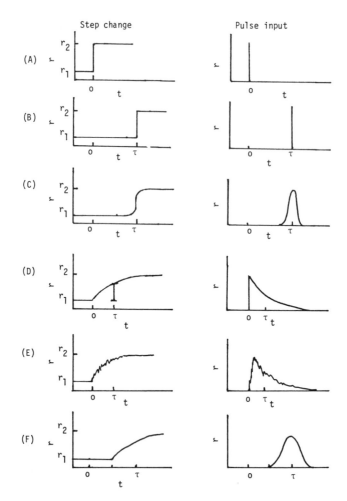

Figure 5-5. Response curves of various reactor types to (A: left) step change in input rate of a conservative substance from r_1 to r_2 at t = 0 and (A: right) a pulse input at t = 0: (B) ideal PFR; (C) real tube reactor with dispersion; (D) CFSTR; (E) real tank reactor with imperfect mixing; (F) laminar flow reactor. (Adapted from Hill, C.G., Jr., *Chemical Engineering Kinetics and Reactor Design*, John Wiley & Sons, New York, 1977.)

$$\tau = \int_{F(t)=0}^{F(t)=1} t\,dF(t) = \int_{t=0}^{t=\infty} \left(\frac{t\,dF(t)}{dt} \right) dt \tag{5-29}$$

Responses ($F(t)$ curves) of ideal and nonideal tank and tube reactors to step and pulse inputs of a conservative tracer are shown in Figure 5-5. A step change in input concentration to a PFR does not affect the outlet concentration until $t = \tau_w$, when a step change equal to the input change occurs. Similarly, a pulse input results in an outflow pulse at $t = \tau_w$ (Figure 5-5b). $F(t)$ is simple for an ideal PFR: τ_w is the same for all fluid elements, and $F(t) = 1.0$ at $t > \tau_w$. In real systems, mixing (longitudinal dispersion) occurs, and the fluid elements entering the reactor at a given time do not all leave at exactly the same time. This causes rounding of the response curve (Figure 5-5c).

The response of an ideal CFSTR to step or pulse change in input is essentially instantaneous (Figure 5-5d), but some time is required to achieve a new steady state after a step change. The exponential nature of the response curve (Table 5-1) implies that 63% $(1 - e^{-1})$ of a pulse input to a CFSTR is lost through the outlet in one τ_w, 86.5% is lost in two τ_w and 95% in three τ_w. In a similar manner, 63% of the change in the outlet response to a step change in the input occurs in one τ_w. Ideal mixing is not obtained in real reactors, and a time lag may occur between an input change and outlet response. Short-circuiting also may allow some inflowing fluid to reach the outlet without being completely mixed. These conditions lead to irregular response patterns (Figure 5-5e).

If the flow in a tubular reactor is laminar rather than plug, a gradient exists in radial velocity. The velocity, u, at any distance, r, from the center of the tube is related to the centerline velocity, u_c, by[1]

$$u = u_c \left[1 - (r/R)^2 \right] \qquad (5\text{-}30)$$

R is the radius of the tube. The mean fluid velocity, $\bar{u}$, is equal to $0.5\, u_c$. Because of the variation in fluid velocity with tube radius, response curves for laminar flow tubular reactors (Figure 5-5f) are qualitatively similar to those of tube reactors with dispersion.

5.3.3 Effects of Dispersion on Behavior of Tubular Reactors

The most important type of nonideal behavior in flow-through tubular reactors is caused by longitudinal dispersion. This is characterized by a longitudinal dispersion coefficient (D_L) that accounts for both a molecular diffusion and turbulent eddies; i.e., $D_L = D_m + D_e$, where D_m is the molecular diffusion coefficient and D_e is an eddy dispersion coefficient. The latter is far more important except at low flows. D_L has units of length^2/time (cm^2 s^{-1}). Values of D_m are in the range 10^{-6} to 10^{-5} cm^2 s^{-1} (10^{-10} to 10^{-9} m^2 s^{-1}) for small solutes in aqueous solution. D_e depends on flow velocity and stream dimension (see next section), but numerically it can be many orders of magnitude larger than D_m. The extent to which dispersion affects reactor behavior is described by the Peclet number, $P = u l/D_L$ (a dimensionless parameter), or its reciprocal, P_r. A value of 0 for P_r implies plug flow; a value of infinity implies perfect mixing.

The steady-state behavior of a reactant in a tubular reactor with longitudinal dispersion is defined by Equation 5-23, and its analytical solution for first-order reactions is[8]

$$\frac{[S]}{[S]_{in}} = \frac{4\beta \exp(P/2)}{(1+\beta)^2 \exp(\beta P/2) - (1-\beta)^2 \exp(-\beta P/2)} \qquad (5\text{-}31)$$

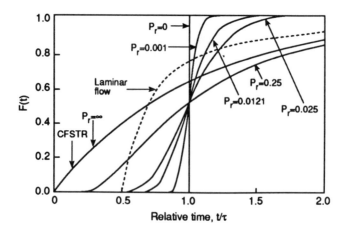

Figure 5-6. Residence time distribution curves for CFSTR; laminar flow, and tube reactors with varying dispersion, as defined by P_r. (Adapted from Levenspiel, O., *Chemical Reaction Engineering*, 2nd ed., John Wiley & Sons, New York, 1972. With permission.)

$\beta = \{1 + 4k(P_r)l/u\}^{1/2}$. Equation 5-23 is solved numerically for higher-order reactions, and approximate solutions are available[1] for small values of P_r.

Figure 5-6 shows residence time distributions for a CFSTR and a tubular reactor at various values of P_r. Nonideal behavior (deviation from plug flow) occurs at quite low values of P_r. Significant rounding of the $F(t)$ curves occurs at P_r as low as 0.001, and the curves approach CFSTR behavior at $P_r = \sim 1$. Moreover, while response curves for pulse inputs are symmetric at $P_r < \sim 0.01$, tailing increases at higher values. The $F(t)$ curve for $P_r = 0.25$ in Figure 5-6 indicates that only ~80% of a pulse tracer would be flushed from a reactor with this P_r when $t = 2\tau_w$. In general, $P_r < 0.01$ is indicative of small dispersion; $P_r = 0.05$ to 0.5 indicates large dispersion; and $P_r > 0.5$, very large dispersion.[14]

Because dispersion allows some fluid to leave a tubular reactor in less than one hydraulic residence time, the degree of conversion of reactants to products is lower in a reactor with dispersion than in an ideal PFR with the same τ_w. In order to achieve the same conversion, a reactor with dispersion must have a larger volume than a PFR. Levenspiel and Bischoff[15] compared results from Equation 5-31 with those for the plug flow case (Equation 5-6b) and computed the ratios of volumes, V_{disp}/V_{PFR}, required to obtain the same conversion as a function of f_S at several values of P_r. As Figure 5-7 shows, the volume ratio increases with increasing P_r and increasing f_S.

5.3.4 Estimation and Measurement of D_L

Dispersion is a spreading phenomenon caused by variations in the velocity field. Although it is often represented by a Fickian relationship, the magnitude of D_L (or D_c) depends on the scale being considered in transport. Dispersion coefficients are directly related (approximately) to flow velocity: $D_L = a\bar{u}$, where $\bar{u}$ is the cross

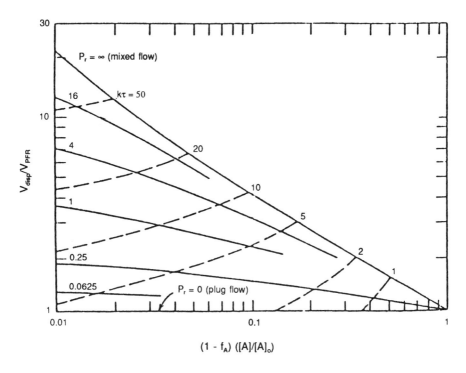

Figure 5-7. Ratios of volumes required by dispersed flow vs. ideal plug-flow reactors to achieve the same degree of reactant removal for first-order reactions; f_S = fraction of S converted to product. Dispersion increases with P_r. Dashed lines show effect of P_r on degree of reaction for constant residence times (or reactor volumes). (Redrawn from Levenspiel, O., *Chemical Reaction Engineering*, 2nd ed., John Wiley & Sons, New York, 1972.)

sectional mean velocity (ft s^{-1} or m s^{-1}), and a is a length scale (ft or m) called the dispersivity.

D_L for natural streams can be estimated from the formula[16]

$$D_L \ (ft^2 \ s^{-1}) = 0.011\bar{u}^2 W^2 / zu^* \tag{5-32a}$$

$$D_L \ (m^2 \ s^{-1}) = 1 \times 10^{-3} \bar{u}^2 W^2 / zu^* \tag{5-32b}$$

u^* is the bottom shear velocity (ft s^{-1} or m s^{-1}), W is the channel width (ft or m), and z is the channel depth (ft or m). In turn, u^* can be estimated from the open channel slope (or hydraulic gradient) S: $u^* = (gr_h S)^{1/2}$, where g = acceleration of gravity (32.2 ft s^{-2} or 9.81 m s^{-2}), and r_h is the hydraulic radius of the stream (ft or m). Fischer et al.[16] provide a more detailed treatment of mixing in natural streams.

D_L cannot be predicted accurately from the physical shape of constructed reactors and must be determined empirically from the outflow response curve for a step or pulse input of a conservative tracer.[1,2,15] (Dye tracer studies are widely used to measure dispersion in engineered reactors as well as natural streams.) Because response curves are similar to the normal probability curve at low P_r values, the known characteristics and equation of this curve can be used to calculate D_L. The standard deviation σ of the normal probability curve is a quantitative measure of spread or dispersion; 68% of the total area under the normal curve lies within $\pm 1\sigma$ of the mean. Levenspiel and Bischoff[15] derived an equation relating σ_t^2, the variance of the concentration-time curve, and D_L:

$$\sigma_t^2 = \tau_w^2\{(2D_L/u\ell) + 8(D_L/u\ell)^2\} = \tau_w^2(2P_r + 8P_r^2) \tag{5-33}$$

For small values of P_r ($< \sim 0.01$), the second term is small compared with the first, and Equation 5-33 simplifies to $\sigma_t^2 \approx 2\tau_w^2 P_r$.

The variance of a concentration-time response curve can be estimated graphically or calculated from standard equations.[1] For example, if the outlet concentration is measured at n equally spaced times over a certain response period, the variance is given by

$$\sigma_t^2 = \sum_{i=1}^{n} t_i^2 [A]_i / \sum_{i=1}^{n} [A]_i - \tau_w^2 \tag{5-34}$$

Equation 5-34 is based on the assumption that a perfect pulse input is added at the reactor inlet. Of course, in real situations that is not possible. In fact, to avoid hydraulic perturbations, the tracer should be added slowly. One way to avoid this dilemma is to use an imperfect pulse and measure the change in σ^2 between two downstream points in the reactor:[17] $\Delta\sigma^2 = \sigma_2^2 - \sigma_1^2$. P_r is related to the change in variance by

$$P_r = \Delta\sigma^2 \tag{5-35}$$

This technique is useful in studies on natural streams, where perfect pulse additions are precluded by turbulence and stream geometry.[16]

PART II. APPLICATIONS OF REACTOR MODELS TO NATURAL AQUATIC SYSTEMS

Reactor models for CFSTRs and PFRs are widely used to describe biological waste treatment systems. This subject is covered in many texts[14] (also see Chapter 6). PFR principles have been used to model water quality in streams for many decades; the Streeter-Phelps oxygen sag model[4,18] is the earliest application of this approach and is the forerunner of the complicated, computer-based (numerical) models of water quality used widely today. The first major application of CFSTR models in natural aquatic systems was to describe the behavior of nutrients in lakes. More recently, they have been used to describe the behavior of organic contaminants, radionuclides, and alkalinity-related constituents in lakes. The following sections describe reactor models for substances in aquatic systems and discuss their usefulness and limitations as water quality management tools. The primary focus is on lakes and on modeling biogeochemical processes based on reactor principles rather than on details of the processes being modeled. The texts by Chapra and Reckhow[3] and Thomann and Mueller[4] treat this subject in greater detail and are excellent sources of further information on water quality modeling.

5.4 EQUILIBRIUM VS. KINETIC MODELS FOR OPEN SYSTEMS

Aquatic chemists often use equilibrium models for aquatic environments, but natural waters are open systems that exchange matter and energy with their external environment. The time-invariant state for closed systems is the equilibrium state, which is quantified by thermodynamic principles, but the time-invariant state for open systems is the **steady state**. The most fundamental characteristic of equilibrium states — zero entropy production — is not a characteristic of steady states. Prigogine[19] described steady-state chemical systems as having minimum entropy production consistent with other system constraints, but this hypothesis may not apply to biological or geochemical systems, or even to all chemical systems.[20] In any case it has not been used to generate models to quantify steady-state conditions. Equilibrium models obviously are inadequate to describe systems that change over time. But what of time-invariant open systems? The applicability of equilibrium models to them depends on the extent to which they can be approximated as closed systems, and this depends on the rates of pertinent reactions relative to rates of fluid flow through the system.

The degree to which simple reactions approach equilibrium in CFSTRs was treated by Morgan[21] and Hoffmann.[22] Consider a reaction, $A \rightleftharpoons B$, that follows first-order kinetics in a CFSTR with volume V and inflow Q. Inflow concentrations are $[A]_o$ and $[B]_o$, and outflow concentrations are $[A]$ and $[B]$. τ_w is the water residence time, and $t_{1/2} = 0.69/k_f$ is the reactant half-life. The equilibrium constant of the reaction is the ratio of the forward and reverse rate constants: $K = k_f/k_r$. Mass-balance rate expressions for A and B are

$$d[A]/dt = Q[A]_o/V - Q[A]/V - k_f[A] + k_r[B] \qquad (5\text{-}36a)$$

$$d[B]/dt = Q[B]_o/V - Q[B]/V + k_f[A] - k_r[B] \qquad (5\text{-}36b)$$

At steady state, $d[A]/dt = d[B]/dt = 0$, and Equations 5-36a,b become algebraic expressions. If $[B]_o = 0$, then $[A]_o - [A] = [B]$, and the following equation can be derived for $[A]/[B]$ by equating the two expressions and rearranging:

$$k_f[A] = k_r[B] + [B]Q/V$$

or

$$\frac{[A]}{[B]} = \frac{(k_r + 1/\tau_w)}{k_f} = \frac{1}{K} + \frac{t_{1/2}}{0.69\tau_w} \qquad (5\text{-}37)$$

This equation can be used to predict whether an open, reacting first-order system will approach equilibrium. From it we see that the steady-state ratio $[A]/[B]$ approaches the equilibrium ratio when $\tau_w \gg t_{1/2}$. Under these circumstances the system behaves like a closed, equilibrium system. Reactions that are not first order can be made "pseudo-first order" if one reactant is much lower in concentration than the other reactants.

For second-order reactions of type $A + B \rightleftharpoons C$, the mass balance equations for a CFSTR are

$$d[A]/dt = (Q/V)[A]_o - (Q/V)[A] - k_f[A][B] + k_r[C] \tag{5-38a}$$

$$d[B]/dt = (Q/V)[B]_o - (Q/V)[B] - k_f[A][B] + k_r[C] \tag{5-38b}$$

$$d[C]/dt = (Q/V)[C]_o - (Q/V)[C] - k_f[A][B] - k_r[C] \tag{5-38c}$$

Conditions are the same as in the first-order case. At steady state, the equations are equal to zero, and conditions are defined by the resulting algebraic equations. Setting Equation 5-38c equal to zero and rearranging yields[22]

$$[A][B] = (k_r/k_f)[C] + (Q/Vk_f)\{[C] - [C]_o\} \tag{5-39a}$$

or $\qquad\qquad [A][B]/[C] = k_r/k_f + Q\{1 - [C]_o/[C]\}/Vk_f \tag{5-39b}$

From Equation 5-39b, we see that when $[C]_o = 0$,

$$[A][B]/[C] = 1/K + 1/\tau_w k_f \tag{5-39c}$$

If $[A]_o = [B]_o$, the concept of half life applies, and $t_{1/2} = 1/k_f[A]_o$. Thus, $k_f = 1/t_{1/2}[A]_o$, and the steady-state reaction quotient Q_r becomes

$$Q_r = \frac{[C]}{[A][B]} = \frac{\tau_w K}{\tau_w + Kt_{1/2}[A]_o} \tag{5-40}$$

Equation 5-40 shows that Q_r approaches K (chemical equilibrium) if τ_w is large compared with the product: $t_{1/2}$ times the reaction equilibrium constant times the initial reactant concentration. The validity of equilibrium approximations for open second-order systems can be assessed by inserting values of τ_w, and $[A]_o$ into Equation 5-40, along with values of K and k for a reaction. Because many reactions in natural waters are affected by metal or acid/base catalysis or by microbial or photochemical processes, Equations 5-37 and 5-40 should be used with caution. Rate constants for some common reactions are listed in Tables 3-3 and 3-4, and water residence times for typical aquatic systems are shown in Figure 1-2 (also see reference 23). Comparison of these figures shows that some simple inorganic reactions can be described by equilibrium models in aquatic systems, but many redox and organic reactions are slow and do not approach equilibrium.

5.5 NUTRIENT MODELS

5.5.1 Basic Mass-Balance Model for Phosphorus

Most nutrient loading models developed for lakes are for phosphorus[3,24-28] because it commonly is assumed that phosphorus controls lake eutrophication, but the modeling concepts are applicable to other nutrients such as nitrogen.[29] These models have been used to develop nutrient loading criteria for lakes, and the criteria have been used widely in water quality management. The basic reactor model for phosphorus assumes that the lake behaves as a CFSTR. The rate of change in the mass of phosphorus (P) in the lake is described as the difference between the sums of the inputs and losses:

$$dP/dt = L_p - S - O \tag{5-41}$$

L_p is the total mass flow of P into the lake from all sources (point and nonpoint surface flows, atmospheric deposition, groundwater inflow); S is the internal loss rate of P from the water (normally assumed to be sedimentation); and O is the rate of P loss from the lake by surface water outflows.

In order to use Equation 5-41 for predictions, several assumptions must be made and the terms must be parameterized. In the basic model it is assumed that all external sources can be measured and summed up in a single term, L_p. The time scale of interest usually is 1 year, and L_p has units of g year^{-1}. The only significant internal sink for P is assumed to be sedimentation. For simplicity, the earliest phosphorus model[24] assumed that S is proportional to the amount of P in the lake at any time:

$$S = \sigma P \tag{5-42a}$$

S is in g year^{-1}, P is in g, and σ is a first-order loss (sedimentation) coefficient (year^{-1}). If the lake behaves as a CFSTR, outflow loss can be represented as the product of the in-lake P concentration, [P], times the hydraulic outflow rate, Q_{out} (m^3 year^{-1}):

$$O = [P]Q_{out} = Q_{out}P/V = \rho_w P \tag{5-42b}$$

V (m^3) is the lake volume, and $Q/V = \rho_w$ is the flushing coefficient (year^{-1}). If $Q_{in} = Q_{out}$, we can drop the subscripts on Q. This condition is met for many lakes and requires approximate balances between (1) precipitation on the lake surface and evaporation and (2) groundwater inflows and outflows. These four terms are small components of the water balances of many drainage lakes.

Substituting the terms for S and O from Equations 5-42a,b into Equation 5-41 yields

$$dP/dt = L_p - \sigma P - \rho_w P = J_p - \phi P \tag{5-43}$$

where $\phi = \sigma + \rho_w$ is a first-order total elimination coefficient (cf. Equation 5-8d). Equation 5-43 has a standard integral form (cf. Equations 2-55,56 and 5-8):

$$P(t) = P_o \exp\{-(\sigma + \rho_w)t\} + [L_p / (\sigma + \rho_w)][1 - \exp\{-(\sigma + \rho_w)t\}] \tag{5-44a}$$

or $\quad P(t) = P_o \exp(-\phi t) + (L_p/\phi)(1 - \exp\{-\phi t\}) \tag{5-44b}$

Equation 5-44 states that the mass of P in a lake at any time is a function of the initial mass, the input rate (assumed constant over time), and values for the loss coefficients, ρ_w and σ. If the input is zero, the second term on the right side of the equation disappears, and P(t) decreases exponentially from its initial value P_o. At large values of t (relative to ϕ^{-1}), the exponential terms approach 0, and P(t) approaches a steady state:

$$P_{ss} = L_p/(\sigma + \rho_w) \tag{5-45}$$

This equation can be obtained directly from Equation 5-43 by setting dP/dt = 0 and solving the resulting algebraic equation for P. Steady-state conditions often are assumed in evaluating the trophic status of a lake (because algebraic expressions are easier to use and require much less information). Equation 5-44 can be used to compute the temporal response to a step change in P loading that results from diversion of sewage effluent out of a drainage basin, or development of P removal systems in wastewater treatment plants.

We are more interested in phosphorus concentration, [P], than in the total mass of phosphorus in a lake. The equations can be expressed in terms of [P] by dividing all terms by lake volume V. Thus, Equation 5-43 becomes

$$(1/V)dP/dt = d[P]/dt = L_V - \sigma[P] - \rho_W[P] \tag{5-46}$$

$L_V = L_p/V$ is the lake's volumetric phosphorus loading rate (g m^{-3} year^{-1}). The steady-state solution is

$$[P]_{ss} = L_V(\sigma + \rho_w) \tag{5-47}$$

Nutrient loading rates usually are expressed on an areal rather than a volumetric basis. Because $V = A\bar{z}$ (A = area, $\bar{z}$ = mean depth), we can rewrite Equation 5-47 in terms of the areal loading, L_A (g m^{-2} year^{-1}). Moreover, the product $\bar{z}\rho_w$ (units of m year^{-1}) is equivalent to the areal water loading to a lake, q_a (= Q/A). (Recall that $\rho_w = Q/V = Q/A\bar{z}$.) Thus,

$$[P]_{ss} = L_A/\bar{z}(\sigma + \rho_w) = L_A/(\bar{z}\sigma + q_a) \tag{5-48}$$

This equation was first derived by Vollenweider[24,25] and is a common form of the steady-state P loading/concentration relationship. It forms the basis for models to predict lake nutrient concentrations and trophic state.

5.5.2 Residence Times for Elements in Lake CFSTR Models

The residence time of a conservative substance in a lake is the same as the water residence time τ_w (if evaporation is a negligible component of the water budget), but for nonconservative substances like phosphorus, both internal loss mechanisms and hydraulic losses must be taken into account.[30] In general,

$$\tau_i = \frac{\text{average mass in system}}{\sum \text{internal and hydraulic loss rates of } i} \tag{5-49}$$

For phosphorus in the CFSTR model, we can write

$$\tau_P = \frac{[P]V}{[P]Q + V[P]\sigma} = \frac{V}{Q + V\sigma}$$

or
$$\tau_P = 1 / (\rho_w + \sigma) \tag{5-50}$$

Equation 5-50 can be used to develop an equation for $[P]_t$ in terms of $[P]_o$, the initial in-lake concentration, and $[P]_{in}$, the average inflow P concentration.[30] To do this we divide all terms in Equation 5-44a by V to convert mass to concentration, substitute τ_P^{-1} for $\rho_w + \sigma$, and substitute $[P]_{in}Q/V = [P]_{in}/\tau_w$ for J_P/V. After rearranging terms, we obtain

$$[P]_t = \frac{\tau_P}{\tau_w}[P]_{in} + \left([P]_o - \frac{\tau_P}{\tau_w}[P]_{in}\right)\exp(-t/\tau_P) \tag{5-51}$$

When $t \gg \tau_P$, the exponential term approaches zero, and the steady-state P concentration is given by

$$[P]_{ss} = (\tau_P/\tau_w)[P]_{in} \tag{5-52}$$

Equation 5-51 then may be expressed in terms of $[P]_{ss}$:

$$[P]_t = [P]_{ss} + \{[P]_o - [P]_{ss}\}\exp(-t/\tau_P) \tag{5-53a}$$

or
$$\frac{[P]_t - [P]_{ss}}{[P]_o - [P]_{ss}} = \exp(-t/\tau_P) \tag{5-53b}$$

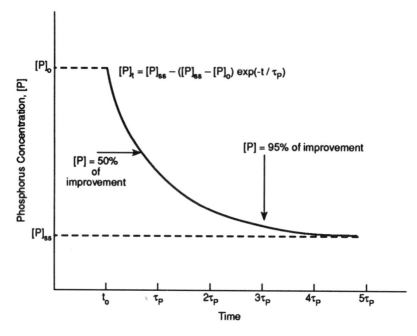

Figure 5-8. Approach to new steady state [P] in lake following a step reduction in P loading described by CFSTR model. [Redrawn from Sonzogni, W.C., P. Uttormark, and G.F. Lee, *Water Res.*, 10, 429 (1976). With permission.]

The approach to steady-state conditions thus is a simple exponential function of the phosphorus residence time, τ_P, which is inversely related to the sum of the loss coefficients, ρ_w and σ. Equation 5-53b is mathematically equivalent to the kinetic expression for a reversible first-order reaction, Equation 2-43b. From Equation 5-52 it is apparent that $[P]_{ss}$ is directly proportional to the average input concentration. Therefore, a given percentage change in the latter will induce the same percentage change in the former. Equation 5-53a can be used to describe a lake's approach to a new steady state following a step reduction in P loading (Figure 5-8; cf. Figure 5-5d). From the exponential nature of the response curve, we know that 63% of the in-lake change in [P] will occur in one τ_P, and 95% will occur within $3\tau_P$.

5.5.3 Problems in Measuring σ and Alternative Expressions

In principle, all the terms in on the right side of Equation 5-48 except σ are measurable, although extensive sampling may be required to accurately determine water and nutrient budgets of lakes with large and complex watersheds. Estimation of σ, however, presents both practical and conceptual difficulties. Use of sediment traps to measure settling rates of nutrient elements in lakes is difficult (because of spatial and temporal variations in settling rates), and they are especially unreliable in shallow lakes. Moreover, other mechanisms besides sedimentation remove phosphorus from lake water: uptake through the food web into fish that are subsequently harvested, uptake of dissolved P by macrophytes and periphyton, and sorption of dissolved P by sediments. Loss coefficient σ is a composite measure of all these processes.

If all the terms in Equation 5-48 are known except σ, it can be determined by fitting data to the equation. (This approach assumes that the in-lake [P] is in steady state with the P loading.) We usually are interested in predicting [P] rather than σ, however, and in some cases (e.g., proposed reservoirs), data on [P] do not exist. Vollenweider[24] used Equation 5-48 and data on [P], L_A, $\bar{z}$, and ρ_w to estimate σ for several lakes and found an inverse correlation with mean depth: $\sigma \approx 10/\bar{z}$. He proposed this as a general predictive relationship for σ.

Phosphorus Retention Coefficient, R_P

An alternative approach to quantifying σ was taken by Dillon and Rigler,[27] who recognized that the term $\sigma[P]$ in Equation 5-46 accounts for all internal sinks of phosphorus and thus represents the amount of P retained in a lake. They defined R_P as the fraction of P input that is retained in a lake and called it the phosphorus retention coefficient:

$$R_P = \sigma P / L_P = \left(L_P - \rho_w P\right)/ L_P \tag{5-54}$$

R_P can be calculated readily from phosphorus mass balance. Alternatively, $R_P = \sigma[P]/L_V$. Solving this relationship for σ, substituting into Equation 5-47 and rearranging leads to

$$[P] = L_V (1 - R_P)/\rho_w = L_A (1 - R_P)/\bar{z}\rho_w \tag{5-55a}$$

or $\qquad [P] = L_A (1 - R_P)/q_a \tag{5-55b}$

The terms on the right side of Equation 5-55 can be measured directly from water and nutrient budgets. However, use of the equation to predict a new steady-state [P] resulting from a change in P loading requires the assumption that R_P is independent of loading rates. Problems also arise when mass-balance data are not available or if predictions are desired for a reservoir not yet in existence. In either case R_P cannot be determined directly. Several studies have shown that R_P can be estimated from simple morphometric and hydrologic variables, and many correlations of R_P vs. functions of q_a, ρ_w, and $\bar{z}$ have been reported (Table 5-2). By equating Equation 5-48 to Equation 5-55b and rearranging we see that:

$$R_P = \frac{\sigma}{\rho_w + \sigma} \tag{5-56}$$

Table 5-2. Predictive Equations for the Phosphorus Retention Coefficient, R_P

Equation	Reference
$R_P = 0.426\exp(-0.271q_a) +$ $0.584\exp(-0.00949q_a)$	31
$R_P = 0.854 - 0.142q_a$	32
$R_P = 0.482 - 0.112 ln\tau_w$	32
$R_P = \dfrac{1}{1 + \tau_w}$	32
$R_P = \dfrac{10}{10 + q_a}$	32
$R_P = \dfrac{16}{16 + q_a}$	33

Unfortunately, many of the predictive equations have limited accuracy when applied outside the database from which they were developed,[47] and estimated R_P values should be used cautiously in predicting [P]. Some discrepancies between predicted and measured R_P reflect inaccurate phosphorus budget data (measured R_P is wrong) or lack of steady-state conditions. Others reflect the fact that P retention in lakes is a function of many mechanisms, and the simple relationships in Table 5-2 do not take these factors into account.

Phosphorus Settling Velocity, v_P

Chapra[33] argued that although σ, the (apparent) first-order loss coefficient for P, varies among lakes, the phosphorus settling velocity is constant. If $\sigma \approx \text{constant}/\bar{z}$ (as Vollenweider showed), it is obvious that $\sigma\bar{z}$ is a constant. This product has units of velocity (m year^{-1}) and can be interpreted as the velocity of phosphorus settling (v_P). Use of v_P allows us to reformulate the sedimentation term of Equation 5-46 more realistically as $v_P A[P]$, i.e., an output to the lake bottom, which should be proportional to the bottom area (assumed equal to the surface area). In contrast, $\sigma[P]$ implies that phosphorus removal is uniform throughout the lake volume. Equation 5-48 thus can be rewritten,

$$[P] = L_A / (v_P + q_a) \tag{5-57}$$

If the numerator and denominator of Equation 5-56 are both multiplied by $\bar{z}$, we find

$$R_P = \frac{v_P}{q_a + v_P} \tag{5-58}$$

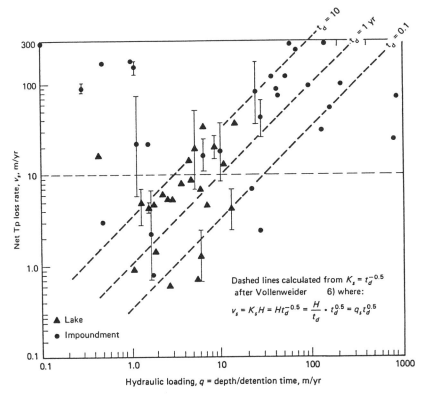

Figure 5-9. Apparent settling velocity for phosphorus, v_P, vs. areal water loading, q_a, in lakes and reservoirs; dashed lines: best fit relationships for lakes with stated water residence times. (From Thomann, R.V. and J.A. Mueller, *Principles of Surface Water Quality Modeling and Control*, Harper & Row, New York, 1987. With permission.)

If we assume that v_P is constant among lakes (which is true only on an order of magnitude basis), it is apparent that R_P is primarily a function of q_a. Chapra[33] fit nutrient budget data for 15 lakes in North America to Equation 5-58 and concluded that $v_P \approx 16$ m year^{-1}; this compares with a value of 10 m year^{-1} based on Vollenweider's correlation. Using a larger database (n = 29), Dillon and Kirchner[34] found $v_P \approx 13.2$ m year^{-1} by least-squares fit to Equation 5-58, and Chapra and Tarapchak[35] used an estimate of 12.4 m year^{-1} for v_P. Subsequent studies[4,36-38] have shown that v_P varies widely among lakes and even more widely among reservoirs (Figure 5-9), and site-specific information is needed for accurate prediction of in-lake [P]. A very crude correlation exists between v_P and q_a, but the scatter is too large to develop a meaningful predictive equation. According to the data plotted in Figure 5-9, the range in v_P for lakes is ~1 to 30 m year^{-1}, and values as high as 300 m year^{-1} are found in impoundments. In general, v_P is higher in impoundments than in lakes; this probably reflects higher loadings of particulate-P associated with high-suspended solids (from soil erosion) in the tributaries to impoundments.[4]

5.5.4 Phosphorus Loading Criteria

Equation 5-57 can be rearranged to predict allowable phosphorus loading rates for target concentrations of phosphorus in lakes:

$$L_{A.crit} = [P]_{target}(v_P + q_a) \qquad (5\text{-}59a)$$

For example, annual average total P (TP) concentrations of 10 and 20 mg m^{-3} are commonly accepted as the upper limit for oligotrophic conditions and lower limit for eutrophic conditions, respectively, in temperate lakes. Substituting these values for TP and $v_P \approx 13$ m year^{-1} into Equation 5-59a yields

$$L_{A.oligo} = 0.01(13 + q_a) \qquad (5\text{-}59b)$$

$$L_{A.cu} = 0.02(13 + q_a) \qquad (5\text{-}59c)$$

Figure 5-10 plots these critical loading rates as a function of q_a; the curvilinear nature of the plots indicates there are two factors affecting L_{crit}. At low q_a, L_{crit} is constant (for a given target value of [P]), and the internal loss rate of P (v_p) controls trophic state. At high q_a, L_{crit} is proportional to q_a, and the driving factor is hydraulic flushing of P from the lake. Both factors are important at an intermediate range of q_a.

5.5.5 Relationships Between [P] and Other Trophic Conditions

The usefulness of simple phosphorus loading models to predict lake trophic state relies on two major assumptions: (1) that algal production in most lakes is limited by P and (2) that in-lake [P] is correlated with other indicators of trophic state. Various lines of evidence indicate that most lakes are P limited; important exceptions include highly oligotrophic lakes (Lake Tahoe is N limited) and eutrophic lakes that are heavily loaded with municipal wastewater. The N/P ratio in wastewater (~3 to 5, by weight) is well below the optimum N/P ratio for plant growth (~8 to 10, by weight). Various short-term nutrient bioassays can be used to determine which of the major or minor nutrients is limiting algal growth in lake water at a given point in time.[39] Extrapolation of this information to say whether controlling the external loading rate of a given nutrient will affect lake productivity is much more difficult. For management purposes, however, selection of the limiting nutrient usually is restricted to the two major nutrient elements (N and P), and the choice between them usually is based simply on either concentration ratios ([N]/[P]) in the lake (averaged over the year) or loading ratios J_N/J_P. Concentration ratios may be based on either TN and TP or inorganic N and P; answers may differ depending on which ratio is used, and both have their advantages and disadvantages. Commonly used criteria are as follows:[40]

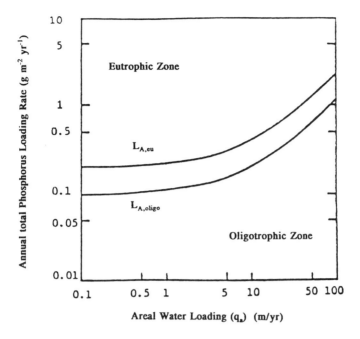

Figure 5-10. Critical loading rates for phosphorus as function of areal water loading, q_A. [Redrawn from Chapra, S.C. and S.J. Tarapchak, *Water Resour. Res.*, 12, 1260 (1976).]

$$[TN]/[TP] \geq 20 \qquad \text{phosphorus limited}$$
$$10 < [TN]/[TP] < 20 \qquad \text{mixed limitation}$$
$$[TN]/[TP] \leq 10 \qquad \text{nitrogen limited}$$

If phosphorus is the limiting nutrient, we should expect a reasonable relationship between [TP] in lake water and algal biomass. In turn, the latter is proportional to the concentration of chlorophyll **a** (chl **a**). Numerous [TP]-chl **a** correlations, spanning several orders of magnitude, have been reported in the literature (see Table 5-3). Much of the variation in coefficients among the correlations and much of the unexplained variance in a given correlation probably can be explained by differences in methods or types of lakes used to obtain the data. These include spring maximum TP vs. averages over the ice-free season or summertime averages, mean epilimnetic concentrations vs. surface values, stratified vs. unstratified lakes, natural lakes vs. impoundments (which often have short water residence times and may be light limited because of high levels of inorganic turbidity), and inclusion of lakes that are not strictly P limited in the database. However, sufficient information is not always provided on the nature of the database to delineate the cause(s) of differences in the correlations. Furthermore, recent studies[48] show that a large portion of the unexplained variance in TP-chl **a** relationships is caused by food web variations among lakes, in particular the presence or absence of large-bodied

Table 5-3. Trophic Indicator Relationships in Lakes

Relationship	r^2	n	Reference
$log(\text{chl a})_{su} = -1.134 + 1.583 log[\text{TP}]_{sp}$	0.95	30	(41)
$log(\text{chl a})_{su} = -1.136 + 1.449 log[\text{TP}]_{sp}$	0.90	46	(42)
$log[\text{chl a}]_{su} = -1.09 + 1.46 log[\text{TP}]_{an}$	0.98	143	(28)
$log[\text{chl a}] = -0.194 + 0.807 log[\text{TP}]_{me}$		55	(43)
$log[\text{chl a}] = -0.259 + 0.76 log[\text{TP}]$			(38)
$log[\text{chl a}] = -1.55 log\left[\dfrac{6.404}{0.334 + 0.0204(\text{TN}/\text{TP})}\right] + 1.55 log[\text{TP}]$			(44)
$log[\text{chl a}]_{su} = -0.676 + 1.119 log[\text{TP}]_{sp}$			(45)
$log[\text{chl a}]_{su} = -0.661 + 1.146 log[\text{TP}]_{su}$			(45)
$log[\text{chl a}]_{su} = -0.751 + 1.299 log[\text{TP}]_{sp}$ (unstratified lakes)	0.59	22	(46)
$log[\text{chl a}]_{su} = -0.528 + 0.996 log[\text{TP}]_{sp}$ (stratified lakes)	0.56	27	(46)
$log[\text{chl a}]_{su} = -0.16 + 0.71 log[\text{TP}]_{sp}$	0.57	63	(36)
$log[\text{chl a}]_{an} = -0.41 + 0.79 log[\text{TP}]_{an}$	0.72	100	(36)
$log[\text{chl a}]_{an} = 1.03 + 1.46 log[\text{TN}]_{an}$ (subtropical lakes)	0.77	100	(36)
$logSD_{su} = 0.803 - 0.473 log[\text{chl a}]_{su}$			(38)
$logSD_{su} = 0.887 - 0.68 log[\text{chl a}]_{su}$	0.86	147	(47)
$logSD_{an} = 0.55 - 0.47 log[\text{chl a}]_{an}$ (subtropical lakes)	0.70	100	(36)
$[\text{chl a}]_{su} = 367 \left[\dfrac{L_{P,A}/q_a}{1 + \sqrt{\tau_w}}\right]^{0.91}$			(26)
$[\text{chl a}]_{su} = 1866 \left[\dfrac{L_{P,A}}{q_a + 12.4}\right]^{1.449}$			(35)

Subscripts: su = summer, sp = spring; an = annual average; me = annual median; no subscript = information not available. Units: [chl a] and [TP] in mg/m^3; [TN] in mg/L; SD in m; $L_{P,A}$ in g m^{-2} year^{-1}.

zooplankton that prey on phytoplankton and lower the chl a/TP ratio. The abundance of large-bodied zooplankton in a lake depends on the abundance of herbivorous fish, which in turn depends on the presence or absence of top carnivores in the food web.[49]

Finally, chlorophyll levels are correlated with the common physical measure of trophic state, Secchi disk (SD) transparency (Table 5-3). The latter parameter is important because it relates directly to user perception of water quality. Transparency is affected by the presence of dissolved organic color (aquatic humus) and by inorganic turbidity, as well as by the size distribution of the algae in which the chlorophyll is "packaged". Nonetheless, the trophic indicator correlations in Table 5-3 and phosphorus models described previously provide a reasonable basis for predicting trophic conditions in a lake if its basic morphometry, hydrology and nutrient loading rates are known. Several authors have combined these relationships to obtain equations that predict chlorophyll levels directly from phosphorus loading rates (see Table 5-3).

5.5.6 A Simple PFR Model for Phosphorus in Rivers and Reservoirs

Many lakes do not resemble CFSTRs in shape and hydraulic behavior. Riverine reservoirs often are long and narrow and not well mixed. In these cases a plug-flow model is more appropriate. Consider the mass balance for phosphorus in a unit volume of such a system:

$$A_{yz} \Delta x d[P]/dt = [P]Q - ([P] + \Delta[P])Q - v_P[P]A_{s.el} \qquad (5\text{-}60)$$

A_{yz} is the cross-sectional area, $A_{s.el}$ is the surface area, and Δx is the longitudinal thickness of the volume element. If we divide the equation by Δx, take the limit $\Delta[P]/\Delta x = d[P]/dx$, and assume steady state, we obtain

$$Q d[P]/dx = -v_P[P]A_{s.el} \qquad (5\text{-}61)$$

If the reservoir width is constant ($A_{s.el}$ is independent of x), the integrated form of Equation 5-61 is simply

$$\ln[P]_x/[P]_{in} = -v_P x A_{s.el}/Q = -v_P A_{s.x}/Q \qquad (5\text{-}62a)$$

$x A_{s.el} = A_{s.x}$ is the reservoir surface area from the upstream end to downstream distance x. Over the entire length of the reservoir Equation 5-62a becomes

$$\ln[P]_{out}/[P]_{in} = -v_P A_s/Q = -v_P/q_a \qquad (5\text{-}62b)$$

or $$[P]_{out} = [P]_{in} \exp(-v_P/q_a) \qquad (5\text{-}62c)$$

According to Equation 5-62c, when $v_P \gg q_a$, $[P]_{out} \ll [P]_{in}$, and when $q_a \gg v_P$, $[P]_{out}$ approaches $[P]_{in}$.

Reservoirs with complicated geometry and multiple phosphorus inputs along the reservoir cannot be treated either as CFSTRs or as simple PFRs. These cases require spatially resolved models in which the system is divided into coupled homogeneous compartments. The equations for such systems generally are too complicated for analytical solution and are solved numerically (Example 5-3 is a simple exception). Many computer simulation models of river and reservoir water quality are available.[3,4,50] Information on water and mass transfers between compartments may be difficult to obtain, and although such models can be written with relative ease, they are difficult to parameterize and solve.

5.5.7 Phosphorus Models for Stratified Lakes

The assumption of homogeneity in lakes is violated most commonly because of thermally induced vertical stratification. Stratification effects can be accounted for empirically in CFSTR models by multiplying outflow term $\rho_w[P]$ in Equation 5-46 by the ratio of the annual average [P] in the upper (mixed) layer to the annual average [P] in the whole lake.[30] Use of this approach to make predictions requires the assumption that the ratio is constant from year to year and over a range of loading rates, and this may not be so. To model the phosphorus dynamics of stratified lakes in greater detail (e.g., to simulate seasonal variations in [P]), it is necessary to consider the major in-lake processes in the phosphorus cycle and recognize that phosphorus exists in several chemical forms that behave differently. It also is necessary to recognize that the phosphorus cycle is coupled closely to other elemental cycles (e.g., N, C) that involve not just phytoplankton, but the entire aquatic food web.

As a minimum, we need to consider two forms of phosphorus: soluble P (P_S) and particulate P (P_P), because they are functionally different in terms of biological activity and they are subject to different loss mechanisms. Only P_S or some fraction of it is directly available for plant growth, but only P_P is lost from the water column by gravitational settling. P_S accumulates in the hypolimnion during summer as settling seston decompose and sediments release P_S, especially under anoxic conditions. The higher [P_S] in the hypolimnion induces a diffusional flux of P_S into the epilimnion; erosion of the thermocline as summer progresses also transfers hypolimnetic P_S into the epilimnion.

Example 5-4. Two-Species, Two-Box Phosphorus Model for Stratified Lakes

The next level of complexity beyond simple "one-box" CFSTR models in simulating the phosphorus dynamics of lakes thus has two physical compartments and two forms of phosphorus.[51-53] This model consists of four coupled differential equations (separate equations for P_S and P_P in each compartment) for the period of summer stratification. Two equations (for the two P forms) are adequate for the time of year when the lake is mixed. The equations are much more complicated than that of the one-form, one-box model (Equation 5-46) because they must account for transfers between compartments and transformations between P forms.

Epilimnion
Soluble phosphate, P_S

(1) $\dfrac{d[P_s]_e}{dt} = \dfrac{1}{V_e} \sum_i J_i(P_s) - \rho_e[P_s]_e - p_e[P_s]_e + \dfrac{D_t A_t}{z_t V_e}\{[P_s]_h - [P_s]_e\}$

Particulate phosphate, P_P

(2) $\dfrac{d[P_p]_e}{dt} = \dfrac{1}{V_e} \sum_i J_i(P_p) - \rho_e[P_p]_e + p_e[P_s]_e - v_{p.e}\dfrac{A_t}{V_e}[P_p]_e + \dfrac{D_t A_t}{z_t V_e}\{[P_p]_h - [P_p]_e\}$

Hypolimnion
Soluble phosphate

(3) $\dfrac{d[P_s]_h}{dt} = r_h[P_p]_h + \dfrac{D_t A_t}{z_t V_h}\{[P_s]_e - [P_s]_h\}$

Particulate phosphate

(4) $\dfrac{d[P_p]_h}{dt} = v_{p.e}\dfrac{A_t}{V_h}[P_p]_e - v_{p.h}\dfrac{A_s}{V_h}[P_p]_h - r_h[P_p]_h + \dfrac{D_t A_t}{z_t V_h}\{[P_p]_e - [P_p]_h\}$

Subscripts: e = epilimnion; t = thermocline; h = hypolimnion; s = sediment-water interface; r_c = flushing coefficient for epilimnion; z_t = thickness of the thermocline; D_t = vertical exchange coefficient in the thermocline region (m^2 day^{-1}), which includes molecular diffusion, eddy dispersion, and thermocline erosion; p_c = net primary production coefficient (day^{-1}); r_h = net respiration coefficient (day^{-1}); v_P = effective settling velocity for P_P (m day^{-1}).

For simplicity, the model assumes that net production and net respiration can be described by first-order coefficients and that net production occurs only in the epilimnion and net respiration in the hypolimnion.

Counterintuitive results were obtained in the original formulations of two-box models for stratified lakes.[51,52] For given TP and water loadings, deeper lakes were predicted to have higher epilimnetic P_S concentrations than shallow lakes. These problems were resolved[52] by making D_t and $v_{p.h}$ functions of mean depth ($\bar{z}$). Phosphorus exchange across the thermocline is proportional to D_t, which was found to be correlated with lake mean depth, $\bar{z}$:[52]

(5) $\qquad\qquad \ln D_t = 1.118 \ln \bar{z} - 7.13 \quad (r^2 = 0.85; n = 14)$

A further approximation based on Equation 5 and an estimate of the dependence of the thickness of the thermocline (z_t) on mean depth ($\bar{z}$) was used to run the model:

(6) $\qquad\qquad\qquad D_t^{'} \,(m/day) = D_t/z_t = 0.005\bar{z}$

The net effect is to move P_p from the epilimnion to the hypolimnion more quickly in deeper lakes than in shallow ones.

A simple term also was developed to model the effects of natural coagulation on P_p sedimentation in the hypolimnion.[52] Coagulation causes particles to aggregate and settle more rapidly the longer they remain in the water column (i.e., the greater $\bar{z}_h$ is):

$$(7) \qquad\qquad\qquad v_{P,h} = v_P\left(1 + f\,\bar{z}_h\right)$$

f is a coagulation coefficient (m^{-1}). The model describes phosphorus data for Lake Ontario reasonably well[52] for the following coefficient values:

$$v_{P,e} = 0.1 \text{ m/day} \qquad\qquad f = 0.05 \text{ m}^{-1}$$

$$v_{P,h} = 0.05\left(1 + 0.5\bar{z}_h\right) \qquad\qquad p_e = 2.0 \text{ day}^{-1}$$

$$r_h = 0.03 \text{ day}^{-1}$$

As Figure 5-11 shows, the predicted values of spring [TP] based on the model agree closely with observed values for a variety of temperate lakes.

A similar reactor-based model[54] divides TP into three forms (soluble, planktonic, and nonliving organic P). Analytical solutions are possible for these "second-generation" models, at least for simple functional forms of the loading term (constant loading, step functions, linear changes, exponential changes)[54] (Section 5.2.4), but the complexity of the equations makes computer solutions more expedient (cf. references 54 and 55). In addition, phosphorus inputs to real systems often cannot be expressed as simple mathematical functions.

The coagulation term in the two-box model is simplistic, but more realistic expressions are much more complicated. For example, O'Melia and co-workers[56] divided particles into 25 size classes in a three-box (epilimnion, thermocline, hypolimnion) model of particle dynamics in lakes. Because of its complexity, the model is solved numerically. The model treats the kinetics of coagulation by the basic Smoluchowski expression (Equation 1-39) and accounts for the removal of particles from small-size classes and their incorporation into larger classes. Larger particles are removed from the water column by gravitational settling, which is described by Stokes Law (Equation 1-40). To run the model, information is needed on particle-size distributions and particle stability (i.e., their tendency to aggregate, a function of solution conditions such as concentrations of calcium and dissolved natural organic matter).

Overall, the two-box model in Example 5-4 is still highly simplified compared with the complexity of nutrient cycling dynamics in lakes. As a result, it is not very successful in portraying short-term variations in concentrations. In particular, it

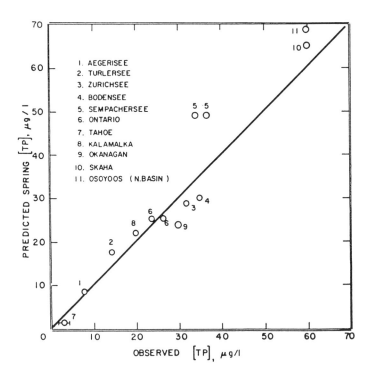

Figure 5-11. Observed spring concentrations of phosphorus in several lakes vs. values predicted from two-compartment phosphorus model. [From Snodgrass, W.J. and C.R. O'Melia, *Environ. Sci. Technol.*, 9, 937 (1975). With permission.]

ignores biotic detail and hydrodynamic aspects. Regarding the former, nutrient uptake kinetics, growth characteristics, settling rates, and suitability as food for herbivores vary greatly among planktonic species, especially between colonial or filamentous blue-green algae and free-living microscopic green algae and diatoms. Some of these issues are not of direct concern if one is interested only in modeling nutrient concentrations, but often one is interested in relating particulate P to plankton levels and other water quality conditions. In addition, the model ignores the effects of predation by zooplankton and the possibility that control of production dynamics may be "top down" rather than resource limited. The hydrodynamics of thermal stratification is important not only in vertical nutrient transport, but also in controlling the depth of the mixed layer relative to the photic zone. By definition, horizontal variability is not addressed in CFSTR-based models. Biotic and hydrodynamic aspects are complicated and beyond the scope of this book. Further treatment of food-web dynamics is given in books on ecological models.[3,50,57] Hydrodynamic aspects of models for nutrient-plankton kinetics are described in some simulation models[58] and several texts.[3,4,59]

5.6 MODELS OF ORGANIC CONTAMINANT TRANSPORT AND FATE

5.6.1 A CFSTR Model for Organic Contaminants

The fate of organic contaminants such as PCBs, pesticides, and various types of industrial chemicals in aquatic systems has received much attention in the past two decades because of the potentially toxic effects of such compounds, both to aquatic organisms and to human consumers of fish and water from contaminated systems. CFSTR models can be used to obtain first-level approximations of the fate of organic contaminants in lakes. Aside from input and output rates, such models include terms for major physical, chemical, and biological transformation processes: chemical hydrolysis, photolysis, volatilization, sorption onto solids (and subsequent sedimentation), biodegradation, and bioaccumulation in the food chain (especially fish). Functional expressions for these processes are given elsewhere in this book: photolysis kinetics are treated in Section 8.2, volatilization in Section 4.7, hydrolysis in Section 4.4.4, sorption and partitioning in Section 7.7.5, and biodegradation in Chapter 6. In addition, Chapter 7 describes ways to predict the behavior of organic compound (e.g., estimate rate constants for the above processes) based on readily measured physical or chemical properties. These estimation methods are particularly important in modeling the fate of organic compounds in aquatic systems, because measured values are not available for thousands of organic compounds of potential interest as aquatic contaminants.

If we momentarily ignore sorption/desorption processes, we can write a reactor model for organic contaminants as follows:

$$d[CT]/dt = W/V - \rho_w[C_T] - \sum k_i[C_d] - k_s[C_p] - k_1[C_d] \qquad (5\text{-}63)$$

Subscripts T, d, and p stand for total, dissolved, and particulate forms of contaminant C; W is the mass loading rate of C into the lake; Σk_i is the sum of rate constants for loss of C due to hydrolysis, photolysis, biodegradation, and volatilization; k_s is a sedimentation rate constant; and k_1 is a bioaccumulation rate constant for fish (assumed to be the only significant biotic reservoir for the contaminant in the lake). Equation 5-63 and the following discussion is adapted from Schnoor.[6] Each rate constant in the summation of Equation 5-63 can be written as a function of physical and/or chemical conditions and thus is a function of time, but initially we will assume that the processes can be represented as time-averaged first-order constants.

Compared with other loss and transport processes, rates of sorption and desorption to and from suspended matter are often thought to be rapid. This is not always so, but if we make this assumption, the relationship between dissolved and particulate concentrations of contaminants can be described in equilibrium rather than kinetic terms. Although mechanistic models of sorption are complicated and nonlinear, the behavior of organic contaminants at low concentrations can be described approximately as a *phase partitioning* process characterized by partition coefficients, which are simple linear relationships:[6,60]

$$K_p = m_a / [C_d] \tag{5-64a}$$

i.e.,

$$m_a = K_p[C_d]; \quad or \quad [C_p] = K_p[C_d]S_{TS} \tag{5-64b}$$

m_a is the mass of contaminant adsorbed per unit mass of particulate matter, K_p is an adsorption equilibrium constant or partition coefficient, and S_{TS} is the total concentration of suspended particles in the water (assumed constant). Because hydrophobic organic contaminants sorb much more strongly to organic solids than to mineral surfaces, the partitioning often is described in terms of the mass of organic particulate matter ($f_{oc} S_{TS}$) rather than to S_{TS} itself; f_{oc} is the fraction of organic carbon in the suspended matter (cf. Section 7.7.5). Partition coefficients for hydrophobic organic compounds are not constant, but vary inversely with suspended particle concentrations, as one would expect for a phase partitioning process.[4,61,62] From Equation 5-64b and the mass balance, $[C_T] = [C_d] + [C_p]$, we can write expressions for $[C_d]$ and $[C_p]$ in terms of $[C_T]$:

$$[C_d] = \frac{[C_T]}{1 + K_p S_{TS}} \qquad [C_p] = \frac{K_p S_{TS}[C_T]}{1 + K_p S_{TS}} \tag{5-65}$$

and define the following fractions:

$$f_1 = \frac{[C_d]}{[C_T]} = \frac{1}{1 + K_p S_{TS}} \qquad f_2 = \frac{[C_p]}{[C_T]} = \frac{K_p S_{TS}}{1 + K_p S_{TS}} \tag{5-66}$$

Equation 5-63 thus can be rewritten as

$$\frac{d[C_T]}{dt} = \frac{W}{V} - \rho_w[C_T] - \frac{\sum k_i}{1 + K_p S_{TS}}[C_T] - \frac{k_s K_p S_{TS}}{1 + K_p S_{TS}}[C_T]$$
$$- \frac{k_1}{1 + K_p S_{TS}}[C_T] \tag{5-67a}$$

i.e.,

$$d[C_T]/dt = L_V - \left\{ \rho_w + f_1 \left(k_1 + \sum k_i \right) + f_2 k_s \right\}[C_T] \tag{5-67b}$$

$$d[C_T]/dt = L_V - (\rho_w + \alpha)[C_T] \tag{5-67c}$$

or

$$d[C_T]/dt = L_V - \delta[C_T] \tag{5-67d}$$

L_V is the volumetric loading rate; α is the group parameter, $f_1(k_1 + \Sigma k_i) + f_2 k_s$; and $\delta = \rho_w + \alpha$. The similarity of this model to the CFSTR model for phosphorus is obvious when the coefficients are grouped together, as in Equations 5-67c,d; i.e., the concentration of a contaminant in a lake depends on its loading rate (L_V), the hydraulic outflow (ρ_w), and a group parameter (α, a pseudo-first-order rate constant) that expresses the sum of internal loss processes. Analytical solutions are straightforward if the coefficients are constant. The steady-state solution is

$$\left[C_T \right]_{ss} = L_v / \delta = L_v / \left(\rho_w + \alpha \right) = \left[C_{in} \right]_{ss} / \left(1 + \alpha \tau_w \right) \tag{5-68}$$

where $[C_{in}]$ is the inflow concentration. The last equality in the above expression is derived by multiplying the numerator and denominator of the previous equality by τ_w (cf. derivation of Equation 5-9). A more complicated CFSTR model that treats the dissolved and particulate components of a contaminant in separate equations is described by Thomann and Mueller.[4]

It should be noted that Equation 5-63 includes terms for contaminant uptake from water by fish and contaminant loss by gravitational settling of particles, but it does not include terms for return of the contaminant to the water by depuration or sediment resuspension. Thus, the equation does not completely represent the dynamic mass balance of contaminants in lakes. In many cases depuration probably is insignificant; bioaccumulation factors for hydrophobic organic contaminants are high, and uptake tends to be irreversible. However, depuration could be important in lakes where water column concentrations are declining because of curtailed inputs.

The time-varying solutions of the basic CFSTR model for simple loading functions (Section 5.2.4) apply directly to the above contaminant model. For example, the solution for exponentially declining input (Equation 5-19) yielded a good fit to mean concentrations of the chlorinated pesticide dieldrin in an Iowa reservoir[6] after its use was banned in the watershed (Figure 5-12). Mean values of $[C_d]$ and $[C_p]$ can be obtained from $[C_T]$ and Equation 5-65. However, it is apparent from the wide range of concentrations in the reservoir at any given time that mean values do not tell the whole story. It also should be noted that the good fit for mean values does not mean the model is physically correct. It is useful to recognize that (for the assumptions of the basic model) four loss-process terms in Equation 5-63 collapse to one term with a single coefficient in Equation 5-67d. Nature is not really that simple! Nonetheless, reactor models are useful in first-level analyses in water quality management even if they do not portray all details of a phenomenon. In fact, it can be argued that they are especially useful in decision making because they are not cluttered with detail.

5.6.2 Relative Importance of Contaminant Loss Mechanisms

The simplicity of reactor models makes it easy to conduct sensitivity analyses for the coefficients. In addition, the relative importance of various loss mechanisms for a contaminant are readily compared by examining the magnitudes of the pseudo-

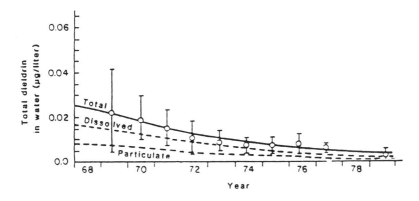

Figure 5-12. Field data and model results for dieldrin in an Iowa reservoir. Model treats the reservoir as a CFSTR with an exponentially declining inflow concentration of dieldrin: $\omega = 0.164$/year; $k_s = 0.18$/day; $\tau_w = 14$ days; $[C_T]_{in,0} = 0.05$ µg L^{-1}; $K_p S_{TS} = 0.50$. [From Schnoor, J.L., *Science,* 211, 840 (1981). With permission.]

first-order rate constants (the k_i in Equation 5-63) used in the model. The main challenge in using such models to predict contaminant behavior in aquatic systems is in deriving reliable estimates of the k's. In many cases they are not measured directly, but computed from some simple functions of measured physical and/or chemical properties of the contaminant. Example 5-5 describes such calculations for a few contaminants and illustrates their use in evaluating the dominant loss mechanism for a contaminant in a given system.

The fact that reactor models for organic contaminants treat losses as pseudo-first-order processes rather than more complicated time-varying functions limits the reliability of this approach. The relative importance of loss mechanisms should be evaluated under realistic conditions by numerical simulation models, such as EXAMS.[64] Although these models are based on the same principles as the reactor model described here, they allow more parameters to be specified and varied, and they represent loss processes as time-varying functions rather than simple coefficients.

Example 5-5. Estimation of Loss Coefficients of Phthalate Esters for a Contaminant Fate Model And Evaluation Of Their Loss Mechanisms

Phthalate esters are manufactured in large quantities and used widely as plasticizers in consumer goods. As a result, they are widespread contaminants in the environment. As a class, the phthalates exhibit a wide range of chemical reactivities and physical properties, and it would not be appropriate to make generalizations about their fate in aquatic systems from studies on one or two phthalate compounds. Wolfe et al.[63] used the fate and transport model EXAMS[64] to evaluate the relative reactivities of five phthalate esters in several model aquatic systems and to estimate the relative importance of different loss mechanisms. The compounds they studied were dimethylphthalate (DMP), diethylphthalate (DEP), di-*n*-butylphthalate (DNBP), di-*n*-octylphthalate (DNOP), and di-(2-ethylhexyl)phthalate (DEHP). EXAMS is capable of simulating complicated aquatic systems under varying environmental conditions (e.g., light, temperature, pH, turbulence), but constant conditions representative of midday summertime values for the southeastern U.S. were selected for the comparative study. (Under constant conditions, EXAMS behaves much like a CFSTR model with constant coefficients.) The steady-state behavior of the five chemicals was calculated at input concentrations of 0.1 mg L^{-1} each to several water bodies with the following physical characteristics:

	Pond	Eutrophic lake	Oligotrophic lake
Volume (m^3)	2×10^4	5×10^6	5×10^6
Water input ($m^3 h^{-1}$)	28	1×10^3	1×10^3
τ_w (days)	30	200	200
Suspended solids (mg L^{-1})	30	50	10
pH	8.0	8.0	6.0
Bacterial population			
water (cells/mL)	1×10^3	1×10^5	1×10^2
sediment (cells/100 g [dry wt])	1×10^8	1×10^{10}	1×10^7
Sediment organic carbon (%)	4	4	1

The evaluation exercise compared four internal loss processes: chemical hydrolysis, photolysis, biodegradation, and volatilization. Rate constants for each process and other physico-chemical data needed to run the model are listed in Table 5-4. Not all the constants were measured; as described in the footnote to the table, some were estimated by linear free-energy or property-activity relationships described in Chapter 7.

The results summarized in Table 5-5 show large differences in behavior among the five compounds in the three surface waters. DMP was the most reactive (least conservative) phthalate in all three water bodies, but the relative reactivities of the other four compounds depended on environmental conditions in the three water bodies. The accumulation factors in Table 5-5 represent the number of days of mass loadings (e.g., in kg day^{-1}) that would yield the total mass in the system at steady state[63] (see Section 5.10.3). The low accumulation factor for DMP reflects its high reactivity; conversely, the high factors for DEHP reflect low chemical and biological reactivities and a high tendency to accumulate in the sediments (high K_{oc}, low fraction in water column at steady state). The relative importance of the loss mechanisms in most cases reflects the relative magnitudes of the rate constants in Table 5-4. Biodegradation was the dominant loss process except for DEHP, which has the lowest k_B. Hydrolysis generally was slower than biodegradation for all the compounds in all three systems, and volatilization was important only for DNOP and for DNBP in systems where biodegradation was slow. Photolysis was more important than biodegradation in the relatively clear oligotrophic lake (except for DMP).

5.6.3 Other Applications of Compartment Models for Organic Contaminants

Transfer of organic contaminants from water into sediment and fish reservoirs can be described by simple compartment models. For example, the exchange dynamics of contaminants between fish and water are described by a two-compartment model with first-order processes of uptake and release (depuration):

$$dF/dt = k_1 f_1 \left[C_T \right] / B - k_2 F \qquad (5\text{-}69)$$

Table 5-4. Rate Constants and Physico-Chemical Characteristics of Phthalate Esters Used for Modeling Exercise in Example 5-5

Constant	DMP	DEP	DNBP	DNOP	DEHP
k_{H+} (M^{-1} h^{-1})	4.0E-2	4.0E-2	4.0E-2	4.0E-2	4.0E-2
k_{OH-} (M^{-1} h^{-1})	2.5E2	7.9E1	3.8E1	5.9E1	4.0E1
k_d (h^{-1})	2.0E-4	2.0E-4	2.0E-4	2.0E-4	2E-4
k_b (mL org^{-1} h^{-1})	5.2E-6	3.2E-9	2.9E-8	3.1E-10	4.2E-12
H (atm m^3 mole^{-1})	1.1E-6	2.0E-8	1.3E-6	5.5E-6	4.4E-7
K_B	2.6E1	7.3E1	8.0E3	2.9E3	8.9E3
K_{oc}	1.6E2	4.5E2	6.4E3	1.9E4	5.7E4
S_w (mg/L)	4.3E3	8.9E2	1.3E1	3.0E0	4.0E-1

Summarized from reference 63.

k_{H+} = acid-catalyzed hydrolysis constant.

k_{OH-} = base-catalyzed hydrolysis constant; value for DNOP estimated from relationship 7 in Table 7-5.

k_d = direct photolysis constant, all assumed to be the same as measured value for DMP.

k_b = microbial biodegradation constant; value for DMP calculated from relationship 5 in Table 7-11.

H = Henry's law constant (measure of volatility; see Section 4.7), all calculated from measured water solubilites (S_w) and vapor pressures extrapolated to 25°C.

K_B = biosorption constant, all estimated from a reported correlation between K_B and K_{ow} (octanol-water partition coefficient): $logK_B = -0.907logK_{ow} - 0.361$.

K_{ow} estimated from the following correlation: $logK_{ow} = -0.65logS_w + 0.889$.

K_{oc} = sediment-water partition coefficient (normalized to organic content of sediment); estimated from reported correlations between K_{oc} and S_w and correlations between K_{oc} and K_{ow} (see Section 7.7.5 and Equation 7-25).

F = whole body fish residue (mg kg^{-1}); $f_1 = [C_d]/[C_T]$ (defined in Equation 5-66); and B = fish biomass concentration (kg L^{-1} lakewater). The time-varying solution of this two-compartment model is derived by matrix methods in Example 5-7. The steady-state solution of this model is simply

$$F = k_1 f_1 [C_T] / k_2 B \qquad (5-70)$$

Analytical solutions are also possible for some nonsteady-state conditions, e.g., when contaminant loading can be represented by simple functions of time, and $[C_T]$ varies with time.[6] However, solutions even for this simple model are mathematically complicated, and it usually is easier to solve the equations numerically. This also provides the advantage of modeling more complicated variations in inputs and having time-dependent values of coefficients. The fraction of contaminant present in dissolved form (f_1) may vary, for example, because suspended solids concentrations change over time. In addition, rate coefficients are functions of temperature.

5.7 CFSTR MODELS FOR SULFATE AND ALKALINITY IN LAKES

Over the past decade much research has been directed toward defining the factors affecting the alkalinity of lakes sensitive to acidification by atmospheric deposition. The alkalinity (or acid-neutralizing capacity, ANC) in such lakes is naturally low (<100 μeq L^{-1}) because of geochemical and hydrologic factors. In this section we

Table 5-5. Simulation Results for Fate of Phthalate Esters in Surface Waters

	DMP	DEP	DNBP	DNOP	DEHP
Load reduction (%)					
Pond	80.5	9.9	42.7	27.4	4.8
Eutrophic lake	100	62.4	92.3	51.0	11.4
Oligotrophic lake	73.2	14.8	24.7	33.3	16.0
Accumulation factor (days)					
Pond	5.8	39	130	521	1910
Eutrophic lake	0.08	65	12	316	1564
Oligotrophic lake	52	174	186	235	544
Percent in water column[a]					
Pond	99.98	69.7	13.3	4.2	1.5
Eutrophic lake	100.0	99.8	96.4	28.3	11.2
Oligotrophic lake	100.0	98.3	81.0	57.2	31.0
Recovery time (months)					
Pond	0.67	7	19	67	228
Eutrophic lake	6.7 h	8	1.3	47	228
Oligotrophic lake	6	20	23	32	72
Hydrolysis loss (%)					
Pond	3.5	2.8	3.3	1.4	0.0
Eutrophic lake	0.1	6.7	2.1	5.6	0.2
Oligotrophic lake	0.3	0.2	0.3	0.1	0.0
Photolysis loss (%)					
Pond	0.4	1.8	1.2	1.4	1.8
Eutrophic lake	0.0	0.7	0.2	0.8	1.4
Oligotrophic lake	4.6	13.9	12.3	11.0	13.7
Biodegradation loss (%)					
Pond	74.5	5.1	31.8	0.5	0.1
Eutrophic lake	99.9	55.0	89.1	28.6	0.7
Oligotrophic lake	65.6	0.6	4.9	0.0	0.0
Volatilization loss (%)					
Pond	2.2	0.2	6.2	24.0	2.8
Eutrophic lake	0.0	0.1	0.9	16.0	2.2
Oligotrophic lake	2.7	0.1	7.2	22.2	2.3

All results for steady-state conditions; see Example 5-5 for details; summarized from reference 63.

[a] Difference between these values and 100% represents % of compound in sediment compartment.

describe applications of CFSTR models to predict concentrations of alkalinity and related substances in acid-sensitive lakes.

5.7.1 Alkalinity Production and Consumption Processes in Lakes

From a simple mass-balance perspective, the alkalinity in a lake at any time is the net sum of the inflows of alkalinity and acidity from atmospheric deposition, watershed runoff, and groundwater inflow, minus the loss of alkalinity by hydraulic outflow (groundwater recharge and surface outflow), plus the net production of alkalinity by chemical and biological processes within the lake. The two main acids in atmospheric precipitation are nitric and sulfuric acids, which are produced from NO_x and SO_2 by chemical reacions in the atmosphere. A variety of microbial processes in terrestrial watersheds and aquatic systems consume these acids (and produce alkalinity). In general, alkalinity is produced by reduction of the oxidized forms of various elements and consumed by oxidation of the reduced forms of the elements. Alkalinity also is generated by various mineral-weathering reactions in soils and sediments. Table 5-6 lists the important alkalinity-generating and -consuming reactions that occur in lake-watershed systems.

Nitrate is highly nonconservative in most ecosystems, and much of the annual deposition is assimilated by plants or microorganisms. Nitrate reduction and assimilation consumes a proton (Equations 1,4; Table 5-6). Consequently, the HNO_3 component of acidic deposition does not contribute significantly to the long-term acidification of lakes, at least at moderate deposition rates. However, at high loading rates (such as occur in southern Scandinavia and parts of the northeastern U.S.) aquatic primary producers are unable to assimilate all of the nitrate (acid-sensitive lakes tend to be oligotrophic — hence, low in phosphorus), and the accumulating nitrate contributes to long-term acidification. NO_x emissions and deposition of HNO_3 also are of concern for other reasons, including short-term pH depressions associated with spring snow melt, direct phytotoxicity of NO_x, and the role of NO_x in the formation of other air pollutants, such as ozone.

Both assimilation and oxidation of ammonium ion produce acidity (Equations 2 and 3; Table 5-6). Most of the ammonium oxidized to nitrate in lakes is assimilated by plants and reduced back to ammonium, thus completing a null cycle. The net effect of atmospheric inputs of inorganic N on the alkalinity balance of a lake can be estimated from the ratio of ammonium to nitrate inputs if it is assumed that both are nearly completely consumed, which is the case if τ_w is greater than 0.5 to 1.0 year and nitrate loading rates are relatively low. In many cases deposition of ammonium and nitrate are roughly in balance, and retention of N inputs has relatively little net effect on a lake's alkalinity.[65]

Until recently, SO_4^{2-} was regarded as essentially conservative in lakes, and deposition of acid sulfate was thought to contribute stoichiometrically to the long-term acidification of low-ANC waters. Sulfate is not conservative, however, and is reduced biologically to sulfide and other reduced sulfur forms in anoxic sediments and bottom waters. The effect of SO_4^{2-} reduction on acid-base chemistry is analogous to that for NO_3^- reduction (Equations 5,6; Table 5-6). H_2S and FeS are not the only products (or even the main products) of SO_4^{2-} consumption in anoxic systems; organic-S forms, metal pyrites, and ester sulfates also are produced (Equations 8–11, Table 5-6). The relative importance of each still is uncertain,[66] but the exact nature of the products is not important here. For the present it is sufficient to note that net proton consumption occurs as long as the reduced S is not reoxidized to SO_4^{2-}. Metal sulfides and organic-S compounds are chemically and biologically stable to oxidation only when O_2 is absent, and their net retention in the sediments depends on the following conditions: they are formed in the anoxic zone, they are

Table 5-6. Reactions that Produce or Consume Alkalinity in Natural Systems

Nitrate assimilation:
$$106CO_2 + 122H_2O + 16H^+ + 16NO_3^- \rightarrow C_{106}H_{260}O_{106}N_{16} + 138O_2 \tag{1}$$

Ammonium assimilation:
$$106CO_2 + 106H_2O + 16NH_4^+ \rightarrow C_{106}H_{260}O_{106}N_{16} + 106O_2 + 16H^+ \tag{2}$$

Nitrification:
$$NH_4^+ + 2O_2 \rightarrow NO_3^- + H_2O + 2H^+ \tag{3}$$

Denitrification:
$$5CH_2O + 4NO_3^- + 4H^+ \rightarrow 5CO_2 + 2N_2 + 7H_2O \tag{4}$$

Assimilatory sulfate reduction:
$$106CO_2 + 16NO_3^- + HPO_4^{2-} + 122H_2O + 19H^+ + 0.5SO_4^{2-} \rightarrow$$
$$C_{106}H_{264}O_{110}N_{16}P_1S_{0.5} + 139O_2 \tag{5}$$

Sulfate reduction (H_2S production):
$$2H^+ + SO_4^{2-} + CH_2O \rightarrow H_2S + 2H_2O + CO_2 \tag{6}$$

Formation of FeS:
$$4Fe(OH)_3 + 9CH_2O + 8H^+ + 4SO_4^{2-} \rightarrow 4FeS + 9CO_2 + 19H_2O \tag{7}$$

Formation of elemental S:
$$SO_4^{2-} + 2H^+ + 3H_2S \rightarrow 4S_{(s)} + H_2O \tag{8}$$

Formation of pyrite:
$$FeS + S_{(s)} \rightarrow FeS_2 \tag{9}$$

Formation of carbon-bonded sulfur:
$$H_2S + RC-OH \rightarrow RC-SH \text{ (a thiol)} + H_2O \tag{10}$$

Formation of ester sulfate:
$$R-OH + SO_4^{2-} + H^+ \rightarrow ROSO_3^- + H_2O \tag{11}$$

Oxidation of H_2S:
$$H_2S + 2O_2 \rightarrow 2H^+ + SO_4^{2-} \tag{12}$$

Congruent dissolution of minerals:
$$CaCO_3 + H^+ \rightarrow Ca^{2+} + HCO_3^- \tag{13}$$
$$Me(OH)_3 + 3H^+ \rightarrow Me^{3+} + 3H_2O \text{ (Me = Al, Fe)} \tag{14}$$

Incongruent dissolution of minerals (aluminosilicate weathering):
$$2KAlSi_3O_8 + 2H^+ + 9H_2O \rightarrow 2K^+ + Al_2Si_2O_5(OH)_4 + 4Si(OH)_4 \tag{15}$$
$$\text{or } 2KAlSi_3O_8 + 2H_2CO_3 + 9H_2O \rightarrow 2K^+ + 2HCO_3^- + Al_2Si_2O_5(OH)_4 + 4Si(OH)_4 \tag{16}$$

Cation exchange:
$$H^+aq + Me^+-X^- \rightarrow Me^+ + H^+-X^- \tag{17}$$
(X = negatively charged mineral or organic particle)

Precipitation of organic acids (humic acid):
$$A^- + H^+ \rightarrow HA_{(s)} \text{ (A}^- = \text{organic anion)} \tag{18}$$

Sulfate adsorption:
$$\begin{array}{c} Al-OH \\ \quad\quad + 2H^+ + SO_4^{2-} \rightarrow \\ Al-OH \end{array} \begin{array}{c} Al-O \\ \quad\quad S \\ Al-O \end{array} \begin{array}{c} O \\ \quO \end{array} + 2H_2O \tag{19}$$

not mobile, and penetration of O_2 into the anoxic zone is prevented (or minimized) by continuous decomposition of sedimenting organic matter at the sediment-water interface. Oxidation of H_2S at pH < 6 is sufficiently slow (Section 4.6.1) that H_2S diffusing from sediments could escape to the atmosphere before being oxidized in shallow acidic lakes. The importance of H_2S and other volatile S compounds as products of sulfate reduction and their fate in low-ANC lakes are not well known, but there is some evidence to suggest that they are of minor significance.

Base cations (mainly Ca^{2+} and Mg^{2+}) are released from sediments primarily by three processes: mineral weathering, cation exchange, and decomposition of sedimenting biomass. The quantitative significance of these mechanisms is uncertain. Although the cation production reactions are diverse and complicated, they all have the same net effect on alkalinity generation (Equations 13–19; Table 5-6). Factors affecting the kinetics of mineral weathering are discussed in Chapter 4. Net cation release from sediments by ion exchange is likely to be important in only lakes where water column concentrations of cations, including H^+, change significantly over time, thus disrupting the equilibrium condition for the reaction ($M–Sed + H^+_{aq} \rightleftharpoons H^+–Sed + M^+_{aq}$). This situation applies to lakes that are *becoming* more acidic and essentially represents the transfer of acidity from the water column to the sediments. In lakes that are becoming less acidic (because of decreasing acid deposition), the reverse exchange reaction will occur, and the sediments will be a sink for cations and a source of acidity. Cation release from settling biomass is likely to be balanced on an annual basis by planktonic uptake in the water column.

5.7.2 CFSTR Model for Sulfate Loss in Lake Enclosures

Whereas NO_3^- reduction in lakes is complete on a time scale of weeks to months, SO_4^{2-} reduction is much slower, and the extent of loss is much less. Quantifying SO_4^{2-} loss within lakes is important in predicting the extent of acidification for a given acid loading. Several studies have evaluated the factors affecting SO_4^{2-} reduction rates in lakes,[65,67-70] and CFSTR models have been developed to predict SO_4^{2-} losses by microbial reduction and the associated production of alkalinity in low-ANC lakes.[71-73]

The behavior of SO_4^{2-} has been studied in large experimental enclosures (4 to 5-m diameter tubes made with polyethylene sheeting) that are left open to the sediment, but otherwise sealed from the rest of the lake.[74,75] Repeated additions of H_2SO_4 are required to maintain water column pH at levels below the ambient lake pH in such enclosures. As Figure 5-13 shows, $[SO_4^{2-}]$ reaches steady-state values in such enclosures in spite of repeated H_2SO_4 additions. The SO_4^{2-} apparently was consumed by diffusion into the sediment and subsequent biological reduction. Data on $[SO_4^{2-}]$ in enclosures were fit to a simple reactor model consisting of three terms:[74]

$$\begin{array}{ccccccc} \text{change in} \\ \text{concentration} \end{array} = \begin{array}{c} \text{loading} \\ \text{rate} \end{array} - \begin{array}{c} \text{hydraulic} \\ \text{loss rate} \end{array} - \begin{array}{c} \text{loss rate by} \\ \text{internal reaction} \end{array}$$

$$d\left[SO_4^{2-}\right]/dt = L_{SO_4} - \rho_w\left[SO_4^{2-}\right] - k_{SO_4}A\left[SO_4^{2-}\right] \qquad (5\text{-}71)$$

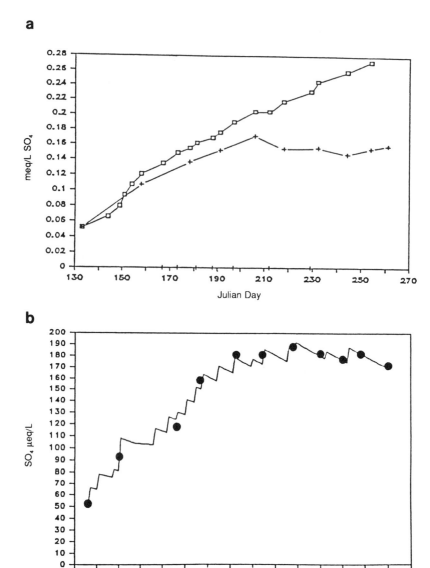

Figure 5-13. (a) Sulfate concentration in the water column of a 4-m diameter littoral enclosure in Little Rock Lake, WI, approached a steady-state value in spite of repeated additions of sulfuric acid to maintain the pH at a target value of 4.5; (b) observed sulfate levels in the enclosure fit a reactor model (Equation 5-71) with $k_{SO_4} \approx 0.008$ m month^{-1} and $\rho_w \approx 0$. (Redrawn from Perry, T.E., M.S. thesis, University of Minnesota, Minneapolis, 1987.)

L_{SO_4} is the input rate (meq L^{-1} day^{-1}) to an enclosure; ρ_w was determined by adding a conservative tracer (NaCl) to each enclosure; k_{SO_4} is a rate coefficient for SO$_4^{2-}$ loss by internal mechanisms (analogous to v_P in Equation 5-57), and A is the enclosure surface area. Because each enclosure was oxygenated from top to bottom, no sulfate reduction occurred in the water column. Loss of SO_4^{2-} thus is portrayed as an areal flux to the sediment and is proportional to sediment surface area. Good fit of the sulfate data to Equation 5-71 was obtained with $k_{SO_4} \approx 0.03$ to 0.04 m month^{-1} (Figure 5-13b). (Data on [Cl$^-$] indicated that leakage was negligible and ρ_w was near zero.) Values of k_{SO_4} were obtained by trial and error. Because H_2SO_4 inputs to the enclosures were pulses at 1- to 2-week intervals rather than continuous inputs, it was impractical to solve Equation 5-71 analytically, and the results in Figure 5-13b were obtained by a computer spread-sheet program.

5.7.3 CFSTR-Based Model for Internal Alkalinity Generation in Lakes

Extension of CFSTR models to other constituents affecting alkalinity in lakes is accomplished in a manner similar to that described above.[72] Table 5-7 lists input-output equations for nitrate, ammonium, base cations, and sulfate — the major inorganic species affecting the alkalinity budgets of low-ANC lakes. In each case the mass balance is composed of a source term, an outflow term, and an internal reaction (production or loss) term. In turn, the equations for sulfate, nitrate, ammonium, and base cations are linked to a fifth equation to define the alkalinity balance. The latter equation likewise is composed of loading and outflow terms plus the reaction terms from Equations 1–4.

The loading terms (L_i) in Equations 1–5 are total annual loads from all sources (wet plus dry atmospheric deposition plus stream inputs plus groundwater inflow) and must be evaluated by field measurements or from a separate watershed model. As in Equation 5-71, the model assumes that sulfate loss occurs by diffusion of sulfate from the water column into the sediments (and subsequent microbial reduction and retention of reduced S in the anoxic zone of the sediments). The loss rate thus is proportional to lake area. If sulfate reduction also occurs in the hypolimnion, a separate term should be added to Equation 1 to account for this loss.

Equations 2 and 3 assume that losses of NH_4^+ and NO_3^- occur throughout the lake water (by planktonic assimilation) and thus are proportional to lake volume. This approach is appropriate for most lakes, in which inorganic N exhibits a pronounced seasonal cycle with maximum concentrations in spring and depletion during summer. In lakes where the nitrate loading exceeds the assimilation capacity of the plankton, seasonal variations in [NO_3^-] are less pronounced. In such lakes, nitrate loss may occur primarily by diffusion across the sediment/water interface and subsequent denitrification at the top of the anoxic zone of the sediment. This mechanism should be modeled like sulfate loss — on an areal, rather than a volumetric, basis. Kelly et al.[73] modeled nitrate loss in this way. Losses of sulfate and nitrogen forms to the sediments are assumed to be irreversible, but it is known that significant recycling (e.g., oxidation of NH_4^+ to NO_3^-) occurs in the region of the sediment-water (oxic-anoxic) interface. The loss coefficients thus represent *net* terms. The functional form of the rate equation for cation production is not well understood, and for simplicity, the sum of the cation production processes is modeled as a zero-order term; W must be estimated for each lake or sediment type.

The loading term in Equation 5 is the sum of the alkalinity inputs from atmospheric deposition (negative if rainfall is acidic), watershed runoff, and

Table 5-7. Equations for CFSTR-Based Lake Alkalinity Generation Model

(1) $d[SO_4^{2-}]/dt$ $=$ $(1/V)\{L_{SO_4} - [SO_4^{2-}](S_{out} + k_{SO_4}A)\}$

(2) $d[NO_3^-]/dt$ $=$ $(1/V)\{L_{NO_3} - [NO_3^-](S_{out} + k_{NO_3}V)\}$

(3) $d[NH_4^+]/dt$ $=$ $(1/V)\{L_{NH_4} - [NH_4^+](S_{out} + k_{NH_4}V)\}$

(4) $d[\Sigma BC]/dt$ $=$ $(1/V)\{L_{\Sigma BC} - [\Sigma BC]S_{out} + WA\}$

(5) $dANC/dt$ $=$ $(1/V)\{L_{ANC} - [ANC]S_{out} + k_{SO_4}A[SO_4^{2-}]$
                  $+ k_{NO_3}V[NO_3^-] + WA - k_{NH_4}V[NH_4^+]\}$

V = lake volume (m^3)
A = lake surface area (m^2)
S_{out} = rate of water outflow from lake (m^3/year), here assumed to be due to
       groundwater recharge (outseepage)
L_i = loading of species i to the lake (meq/year)
k_i = first order loss (reaction) constant for species i (year^{-1} for NO$_3^-$ and NH$_4^+$,
       m/year for SO$_4^{2-}$)
W = zero order weathering rate for base cations in lake sediments (meq m^{-2} year^{-1}) $\Sigma[BC]$ =
total concentration of weatherable base cations (meq/m^3).

Steady-state conditions are derived readily from Equations 1-5:

(6) $[SO_4^{2-}]_{ss}$ $=$ $L_{SO_4}/\{S_{out} + k_{SO_4}A\}$

(7) $[NO_3^-]_{ss}$ $=$ $L_{NO_3}/\{S_{out} + k_{NO_3}V\}$

(8) $[NH_4^+]_{ss}$ $=$ $L_{NH_4}/\{S_{out} + k_{NH_4}V\}$

(9) $[\Sigma BC]_{ss}$ $=$ $L_{\Sigma BC}/\{S_{out} + WA\}$

(10) $[ANC]_{ss} = -\dfrac{1}{S_{out}}\left\{L_{ANC} + k_{SO_4}A[SO_4^{2-}]_{ss} + WA + k_{NO_3}V[NO_3^-]_{ss} - k_{NH_4}V[NH_4^+]_{ss}\right\}$

The following estimates of the k_i were obtained by Baker and Brezonik[72] by calibrating the
model equations with ion budget data from some low-ANC lakes in North America and
Scandinavia:

k_{SO_4} = 0.52 (±0.34) m/year; k_{NO_3} = 1.33 (±0.94) year^{-1}; k_{NH_4} = 1.5 (±1.2) year^{-1}

---

groundwater inseepage (generally positive because weathering reactions produce
alkalinity in the soil). Equations 1-5 are solved for steady-state concentrations of
each substance under conditions of constant loading in Equations 6-10 of Table 5-
7. These equations are similar to the steady-state equations in the phosphorus model
(cf. Equations 5-58, 5-64). The steady-state alkalinity is given by the ratio of the
alkalinity input to the sum of the coefficients for hydraulic loss plus internal losses
or gains from chemical or biological reactions.

Although the model of Table 5-7 is simple, it does account for the major
processes affecting alkalinity in low-ANC lakes. The coefficients can be determined from mass-balance measurements, and if they are constant (or fall in a narrow
range), the model may be useful as a predictive tool. Estimates obtained by Baker

and Brezonik[72] are listed in Table 5-7. It is interesting to note that the coefficient for sulfate loss ($k_{SO_4} \approx 0.5$ m year^{-1}) is much lower than the analogous term for phosphorus ($v_P \approx 13$ m year^{-1}). This is not surprising in that phosphorus losses occur by gravitational settling of seston, and sulfate loss occurs primarily by diffusion from lake water into anoxic sediments.

In situations where watershed inputs of water and alkalinity are unimportant (i.e., seepage lakes), the model can be used to predict the effects of changes in acid deposition on a lake's chemistry. For example, Figure 5-14a shows the rate of reduction of in-lake alkalinity with increasing acid loading to a hypothetical lake. It is obvious that the first order sulfate loss model provides more buffering than a zero-order (constant rate) model does. Moreover, if no internal sulfate sink exists, the lake is much more sensitive to acidification from acid deposition. Similarly, we can use the steady-state equation for sulfate and the estimated k_{SO_4} (0.5 m year^{-1}) to show that in-lake reduction has little effect on sulfate retention (hence, little effect on $[SO_4^{2-}]_{ss}$) at $q_a > \sim 2$ m year^{-1}. This is equivalent to $\tau_w < \sim 2$ year for lakes with $\bar{z} \leq 4$ m (Figure 5-14b). In rapidly flushed lakes, $[SO_4^{2-}]_{ss}$ is controlled primarily by the sulfate loading rate and hydraulic flushing, but internal sinks have a significant effect on $[SO_4^{2-}]_{ss}$ in lakes with longer water residence times.

Input-output and CFSTR concepts have been used to derive models for alkalinity in lakes and watersheds.[76-79] Because the mechanisms of alkalinity production by mineral weathering in watersheds are very complicated and poorly quantified, these models necessarily are simplified representations of reality. Nonetheless, they are more realistic than empirical models based on simple titration concepts[80] and have much smaller data requirements than complicated simulation models for alkalinity, such as the ILWAS model.[81] Mechanistic and kinetic aspects of mineral weathering important in understanding the functional form and limitations of alkalinity models are discussed in Chapter 4.

5.8 OTHER APPLICATIONS OF REACTOR MODELS IN LAKES

Several radionuclides found in lakes result from natural processes of formation and transport in the atmosphere (e.g., ^{14}C, ^{40}Ar, ^{210}Pb); others are present as a result of nuclear weapons testing in the atmosphere from 1945 to 1962 (^{90}Sr, ^{137}Cs). The dynamic behavior of these isotopes in lakes is of interest for several reasons. The nature of their formation and long-range transport in the atmosphere makes them widespread at fairly uniform levels throughout the biosphere. Because of their "self-disclosing properties," they are detectable at ultralow concentrations. Most important, they serve as tracers of biogeochemical cycles and as chronometers of various processes; i.e., their disappearance from the water column is helpful in evaluating mixing and sedimentation dynamics, and their profiles in sediments can be used to infer rates of sediment accumulation. Well-mixed conditions usually apply to these elements in lakes so that CFSTR modeling approaches can be used. The development of such models to simulate temporal trends of radionuclides in the Great Lakes

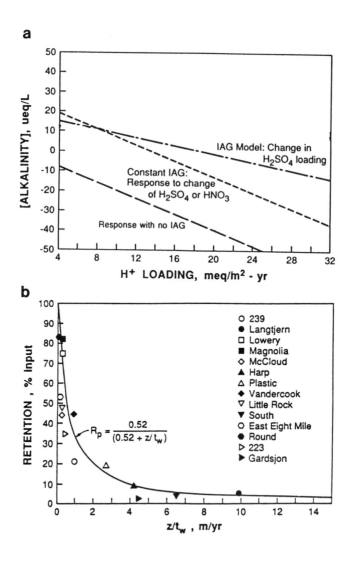

Figure 5-14. (a) Predicted alkalinity in a hypothetical lake with $\bar{z} = 5$ m and $\tau_w = 10$ years to change in sulfuric acid loading. Upper line: response based on internal alkalinity generation (IAG) model in Table 5-7 with $k_{SO_4} = 0.5$ m year^{-1}; middle line: response from holding IAG constant at predicted initial value; lower line: response if !AG = 0. (b) Effect of τ_w on sulfate retention in lakes. [From Baker, L.A. and P.L. Brezonik, *Water Resour. Res.*, 24, 65 (1988). With permission.]

has been a popular exercise,[82-85] partly because long-term monitoring data are available on these lakes.

The complexity of the model needed to describe the occurrence of radioisotopes in lakes and sediments depends on the chemical nature of the element (which defines the nature of the removal processes). For example, ^{90}Sr can be modeled with just a hydraulic outflow term and a radioisotope decay term.[82] The fraction of input ^{90}Sr lost to the sediments (by chemical precipitation, uptake by organisms, or sorption to settling mineral particles) is small (at least in the Great Lakes); Sr thus behaves like Ca. In contrast, ^{137}Cs interacts significantly with suspended particles, and models of its fate require both a sorption term[83,84] and a term to describe its recycling from the sediments. Description of the fate of plutonium in the Great Lakes is still more involved[84,85] because this radionuclide exhibits complicated redox chemistry, in addition to sorption and biotic interactions. Numerical models generally are more practical for such complicated simulations.

Finally, Engstrom[86] developed a simple mass-balance model analogous to the one-box phosphorus model to evaluate effects of hydrologic and geographic factors on concentrations of aquatic humus (AH) in Labrador lakes. Humic matter was measured in the field in terms of apparent water color. The major trend in water color of lakes in this region — and increase from north to south — was found to reflect differences in external loadings of AH from watershed soils and vegetation. In particular, color increased in hyperbolic fashion with increasing values of A_w/A_l, the ratio of the watershed area to lake surface area, but the nature of the dominant vegetation in the watershed affected the shape of the relationship. (Peatland watersheds produced much more AH per unit area than woodland watersheds, which in turn produced more than tundra watersheds.) Two hydrologic factors were found to exert important controls on water color. Direct precipitation on the lake surface dilutes the color (AH) exported from the terrestrial catchment in lakes with low values ($< \sim$10) of A_w/A_l. Similarly, a long residence time in lakes with small catchments promotes internal loss of AH by photolysis, biodegradation, and sedimentation processes. The magnitude of the internal loss coefficient (k_{-AH}) for AH is not known with much accuracy. Loss rates of AH from lakewater depend on light and temperature (and perhaps other factors such as pH). Thus k_{-AH}, is not likely to be constant from region to region (nor even among lakes in a region); Engstrom used a range of 0.05 to 0.4 year^{-1} for k_{-AH} in his modeling exercise.

PART III. KINETICS OF MULTICOMPARTMENT SYSTEMS

5.9 NATURE OF MULTICOMPARTMENT SYSTEMS

5.9.1 Types and Uses of Compartment Models

Although many aquatic systems can be treated as well-mixed reactors (CFSTRs) or tubular plug-flow reactors, others are more complex. For example, we may be interested in analyzing the movement of a substance among several compartments, each of which may be treated as a separate CFSTR. The compartments may represent a series of lakes within a drainage basin or discrete parts of a single lake with complex morphometry. Alternatively, the compartments may represent the amounts of an element in different functional groups of organisms within a food web or more physically heterogeneous components of the biosphere.

As a minimum, we are interested in two features of such compartmentalized systems: the amount of a substance in each compartment, and its rates of transfer between compartments. At one extreme we may assume steady-state conditions and use simple algebraic mass balances; at the other extreme we may wish to simulate the dynamics of substance transfers under varying conditions of temperature, light, and water flows. In some cases we may wish to quantify concentrations in each compartment and intercompartmental flows of water with varying concentrations of a substance. Transfer relationships in these cases are complicated by hydraulic factors such as turbulent mixing; substance transfer rates in many multicompartment environmental systems are controlled by the physical movement of water. In other cases we may ignore flow volumes and concentrations and focus on mass transfers and masses in compartments. Transfer relationships in such models involve simple constant-value coefficients analogous to rate constants. Finally, systems may be closed, with a fixed total quantity of substance, or open, with one or more compartments receiving inputs from outside the system and/or exporting the substance to external reservoirs.

As implied above, compartment models are diverse. Their mathematical form and complexity depend on the type of system being modeled. Such models are common in several areas of aquatic science:

1. **Water quality/eutrophication models** relate loadings of contaminants to their concentrations and fate in lakes, rivers, and estuaries. Nutrient models relate N and P loadings to levels in lakes and to planktonic response. Some of these models have strong hydrodynamic components; as a minimum, they require data on water budgets and basin morphometry. A large literature, including many texts, exists on this subject.[3,4,50]

2. **Food web models** in aquatic ecology simulate flows of energy and cycling of nutrient elements through trophic levels and functional compartments of food chains and webs. Effects of predation and species competition for nutrients may be examined; more complicated models may simulate food web dynamics for varying physical conditions (temperature, light). Chapter 6 describes transfer relationships between abiotic and biotic compartments of such models and develops several compartment models for microbial growth. Likewise, a large literature exists on this subject.[57]

3. **Geochemical cycling models** divide the world into such reservoirs as the atmosphere, shallow and deep ocean layers, terrestrial and marine biomass, and dead organic matter. Cycling of elements among the compartments is described by linear or simple nonlinear transfer relationships. Effects of human activity on reservoir sizes, residence times of elements in reservoirs, and transfer rates are examined. Most of these models consider one element at a time. Coupled models that describe the behavior of interrelated elements are more complicated to analyze, but more interesting and realistic. Although the actual behavior of elements in geological systems is complicated, geocycling models tend to be simple because functional relationships are not well understood. Our current level of knowledge permits only crude descriptions of element cycles involving a few compartments and simple transfer relationships. The time scales of geocycling models justify some simplification of transfer relationships. For example, because turnover times of most geoscale compartments are several-to-many years long, we can assume that temperature is constant and ignore its effects on transfer coefficients.

4. **Tracer kinetic models** are used in aquatic biology to calculate transfer rates (e.g., biomass production rates) and sizes of autotrophic, heterotrophic, and abiotic compartments from data obtained by adding a radiotracer to a compartment and sequentially sampling for tracer activity in the compartments. Tracer kinetic models are

important in medical fields such as pharmacology, and much of the developmental work on these models is in the biomedical literature.[87-91] Some applications to aquatic systems are given in Sections 5.6.4–5.

Two general approaches are used to solve the systems of differential equations used in compartment models: numerical methods and analytical solutions. Numerical methods are the most common and have the widest range of applicability. Complicated water quality and ecosystem models that have time-varying inputs and coefficients that depend on physical conditions are solved numerically by computers. Time-varying physical data are read into the programs as raw data or simulated by simple subroutines (e.g., sinusoidal or normal distribution functions). Numerical methods range from simple one-step Euler integration to more accurate Runge-Kutta and predictor-corrector methods. Many programs are available to solve sets of ordinary differential equations,[92,93] and many texts describe simulation models for water quality and ecosystem dynamics.

Some multicompartment models are sufficiently simple that closed-form (analytical or quasianalytical) solutions are possible. Such solutions are attractive because they allow analysis of system behavior in exact mathematical terms, and they provide a certain intellectual satisfaction compared with numerical methods, which are inherently approximate. (In practical terms, there may be no difference in accuracy between the two types of solutions.) The feasibility of analytical solutions for coupled differential equations is enhanced by the use of matrix methods. This section focuses on such simple compartment models, which are useful in analyzing geochemical cycles and in tracer kinetics. These models are similar to the systems of multireaction kinetics described in Chapter 2, except that the latter involve sparser matrices and larger temporal ranges of compartment sizes (concentrations of reactants, intermediates, and products). Some reactants go to zero in chemical sequences; concentrations of some intermediates start at zero and end at or near zero. In contrast, geochemical compartments usually do not go to zero.

5.9.2 Transfer Relationships

The most common way to describe the flow of a substance between compartments is by first-order kinetics. The rate of substance transfer from compartment i to compartment j is assumed to be proportional to the amount of substance in i:

$$dA_{i \to j}/dt = k_{ij}A_i \qquad (5\text{-}72)$$

A_i is the amount (or concentration) of A in the donor compartment, A_j is the amount in the acceptor compartment, and k_{ij} is a transfer coefficient (t^{-1}) analogous to a first-order rate constant. The first subscript in a transfer coefficient refers to the donor compartment, and the second to the acceptor.

The assumption of first-order (donor) transfer rates has a strong basis in chemical kinetics and often is justified as a first approximation in geochemical and ecosystem models, provided that the time scale is such that effects of varying physical

Table 5-8. Transfer Relationships for Compartment Models

1. Linear donor	$F_{ij}{}^a$	$=$	$k_{ij}A_i$
2. Linear acceptor	F_{ij}	$=$	$k_{ij}A_j$
3. Second order	F_{ij}	$=$	$k_{ij}A_iA_j$
4. Lotka-Volterra (γ is an efficiency coefficient for capture of A_i by A_j)	F_{ij}	$=$	$k_{ij}A_i - \gamma A_iA_j$
5. Monod (Michaelis-Menten) (K_A is a half-saturation constant with the same units as A_i)	F_{ij}	$=$	$\dfrac{k_{ij}A_iA_j}{K_A + A_i}$

$^aF_{ij} = dA_{i \rightarrow j}/dt$

conditions on transfer rates can be ignored. In geochemical cycling models it also is necessary to assume that the elemental composition of each reservoir either does not change over time or that individual elemental cycles are decoupled. For example, in a linear model of the carbon cycle, we must assume that the rate of primary production is proportional to the concentration of atmospheric CO_2 and that nutrient limitation is unimportant (or at least constant). These assumptions limit the realism of simple linear models; where interactions among elements are important, more complicated coupled and nonlinear models must be used.

For some transfers, other functions listed in Table 5-8 are more appropriate. Predation rates may depend more on the density of predators than the density of prey, yielding a first-order (acceptor) relationship. A more realistic way to describe predator-prey transfers is as second-order processes — dependent on the size of both compartments. Such transfers are common in biological systems, and they are mathematically equivalent to autocatalytic chemical reactions[94] because one of the reactants is also a product. For example, in the reaction, $A_1 + A_2 \rightarrow 2A_2$, the product catalyzes its own formation. A simple example is A_1 = phytoplankton and A_2 = zooplankton. Such reactions provide strong positive feedback. The lack of negative feedback in first- and second-order transfer functions can cause unstable behavior in model simulations, and the negative feedback term in the Lotka-Volterra relationship improves the stability of predator-prey models. Microbial uptake of organic substrates or nutrients may be first order in both donor and acceptor compartments (second order overall) when the donor compartment is small. When the latter is large, the transfer rate depends only on the acceptor compartment size. The Monod relationship displays this behavior and is used in models of substrate or nutrient uptake.

5.10 ANALYSIS OF COMPARTMENT MODELS BY MATRIX METHODS

5.10.1 Matrix Treatment of Linear Models

The total flow out of compartment i is the sum of the individual flows out of i and into all other compartments j, and total flow into compartment i is the sum of the flows from all other compartments j into i. If the transfers between compartments are described by the relationship in Equation 5-31, the net change of substance in compartment i is obtained by mass balance as

$$dA_i / dt = \text{sum of inputs} - \text{sum of outflows}$$

$$dA_i / dt = \sum_{j=1}^{n} k_{ji} A_j - A_i \sum_{j=1}^{n} k_{ij}, \qquad i = 1,\ldots,n \qquad (5\text{-}73)$$

n is the number of compartments in the system. To describe the whole system, n equations must be written (one for each compartment). This can be cumbersome for systems with many compartments, especially if they have numerous interconnections. Writing and solving the equations is simplified by expressing them in matrix form. Unfortunately, matrix notation is so compact that it tends to be cryptic. We will try to remedy this by explaining the notation and by simple examples. Readers with no background in matrix methods are referred to texts on this topic.[95-97]

The coefficients for all n rate equations form an $n \times n$ matrix $\mathbf{K}$, where K_{ij} is the element of the ith row and jth column of the matrix. k_{ij} and k_{ji} are zero if compartments i and j are not connected directly; in models with many compartments there may be many such 0's; i.e the matrix may be sparse. The ith row of the matrix represents the set of coefficients for the ith rate equation (one equation for each compartment), and the jth column represents the coefficients associated with transfer from the jth compartment to the ith compartment or (when j = i) the sum of the coefficients for transfer out of the ith compartment. The K_{ij} thus are defined as follows:

$$K_{ij} = k_{ji} \text{ when } i \neq j; \quad K_{ij} = -\sum_{i \neq l} k_{il} \text{ when } i = j \qquad (5\text{-}74)$$

Equation 5-73 then can be written as

$$dA_i / dt = \sum_{j} K_{ij} A_j \qquad (5\text{-}75a)$$

or $$dA/dt = \mathbf{KA} \qquad (5\text{-}75b)$$

$\mathbf{A}$ is the $n \times 1$ column matrix (vector) $(A_1,\ldots,A_n)$. Use of these concepts is illustrated for a compartment model of the global carbon cycle in Example 5-6.

Example 5-6. Five-Compartment Model of Global Carbon Cycle

Holland[98] described the global carbon cycle as a five-compartment system, as shown in Figure 5-15, and Lasaga[94] expressed the cycle as a linear model in the following matrix form:

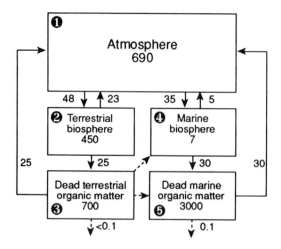

Figure 5-15. Simple five-compartment model of the global carbon cycle. [From Lasaga, A.C., *Geochim. Cosmochim. Acta*, 44, 815 (1980). With permission.]

$$dA_1/dt = -\left(k_{12} + k_{14}\right) A_1 + k_{21} A_2 + k_{31} A_3 + k_{41} A_4 + k_{51} A_5$$

$$dA_2/dt = k_{22} A_1 - \left(k_{21} + k_{23}\right) A_2$$

$$dA_3/dt = k_{23} A_2 - k_{31} A_3$$

$$dA_4/dt = k_{14} A_1 - \left(k_{41} + k_{45}\right) A_4$$

$$dA_5/dt = k_{45} A_4 - k_{51} A_5$$

with matrix of rate coefficients:

$$K = \begin{vmatrix} -\left(k_{12} + k_{14}\right) & k_{21} & k_{31} & k_{41} & k_{51} \\ k_{12} & -\left(k_{21} + k_{23}\right) & 0 & 0 & 0 \\ 0 & k_{23} & -k_{31} & 0 & 0 \\ k_{14} & 0 & 0 & -\left(k_{41} + k_{45}\right) & 0 \\ 0 & 0 & 0 & k_{45} & -k_{51} \end{vmatrix}$$

Rate constant k_{ij} (year^{-1}) for transfer from compartment i to compartment j is obtained from F_{ij}/A_i, where A_i is the size of the donor compartment (in 10^{15} g), assumed to be at steady state, and F_{ij} is the (steady-state) flux from i to j (in 10^{15} g year^{-1}). From the values in Figure 5-15, $k_{12} = 48/690 = 0.06956$ year^{-1}, and

$k_{14} = 0.05072$ year^{-1} $k_{23} = 0.05555$ year^{-1} $k_{41} = 0.71428$ year^{-1} $k_{51} = 0.01000$ year^{-1}

$k_{21} = 0.05111$ year^{-1} $k_{31} = 0.03571$ year^{-1} $k_{45} = 4.28571$ year^{-1}

and

$$
K = \begin{vmatrix}
-0.12028 & 0.05111 & 0.03571 & 0.71428 & 0.01000 \\
0.06956 & -0.10666 & 0 & 0 & 0 \\
0 & 0.05555 & -0.03571 & 0 & 0 \\
0.05072 & 0 & 0 & -5.0000 & 0 \\
0 & 0 & 0 & 4.28571 & -0.01000
\end{vmatrix}
$$

Because the system is closed and at steady state, the sum of each column of coefficients in **K** is (and must be) zero.

Solution of matrix differential Equation 5-75b is an "eigenvalue" problem.[94,96,97] In general, for constant k_{ij}, the solutions are sums of exponentials in time, and the coefficients of the exponents are given by the eigenvalues, λ_i, of matrix **K**:

$$A(t) = \psi \exp(\lambda t) \psi^{-1} A^0 \tag{5-76a}$$

$$A(t) = a\psi \exp(\lambda t) \tag{5-76b}$$

$A(t)$ is the $n \times 1$ column matrix of solution values (compartment sizes) at time t and $\exp \lambda t$ is the $n \times n$ diagonal matrix:

$$
\begin{vmatrix}
\exp(\lambda_1 t) & 0 & \cdots & 0 \\
0 & \exp(\lambda_2 t) & \cdots & 0 \\
\vdots & \vdots & \ddots & \vdots \\
0 & 0 & \cdots & \exp(\lambda_n t)
\end{vmatrix}
$$

The λ_i are eigenvalues (see explanation below) of matrix **K**; ψ is the matrix of associated eigenvectors of **K**; ψ^{-1} is the inverse of the matrix of eigenvectors; and **a** is a vector of coefficient values equal to $\psi^{-1} \times A^0$. An eigenvector of n dimensions is associated with each of the n eigenvalues; matrix ψ is formed by lining up the n eigenvectors in columns. Finally, A^0 is the $1 \times n$ row matrix of initial compartment sizes. Equation 5-76 is derived in texts on differential equations.[99] The time-varying solution for any given compartment i is

$$A_i(t) = \psi_i \exp(\lambda t) \psi^{-1} A^0 \tag{5-77}$$

ψ_i is the $1 \times n$ row matrix of elements from eigenvector matrix ψ.

5.10.2 Fundamentals of Eigenvalues and Eigenvectors

Eigenvalues (also known as characteristic values or latent roots) and their corresponding eigenvectors are important properties of matrices. The theoretical aspects of eigenvalues are complicated, and they are difficult to calculate, especially for large matrices. Once calculated, however, they greatly simplify a variety of problems involving matrices. Some basic properties of eigenvalues and eigenvectors are described below. In general, if A is an $n \times n$ matrix and ψ is a vector, we wish to find a value λ to satisfy the equation

$$A\psi = \lambda\psi \tag{5-78}$$

If vector $\psi = 0$, any value of λ is a solution of Equation 5-78. Values of λ for which Equation 5-78 has nonzero solutions of ψ are called eigenvalues of matrix A. Thus, we wish to find values for the elements of ψ such that the product matrix $A\psi$ is a vector proportional to itself; the proportionality constant is λ. The diagonal matrix of λ contains all the values that satisfy Equation 5-78.

It is relatively simple to show that an eigensystem equation of the form

$$A(t) = e^{\lambda t}\,\psi \tag{5-79}$$

is a general solution of matrix differential equation $dA/dt = KA$.[90] Differentiation of Equation 5-79 yields $dA/dt = \lambda e^{\lambda t}\psi$. Substituting this result into Equation 5-75b for dA/dt and Equation 5-79 for $A(t)$, we obtain $\lambda e^{\lambda t}\psi = Ke^{\lambda t}\psi$. Elimination of the common scalar $e^{\lambda t}$ yields: $\lambda\psi = K\psi$. According to the definition in Equation 5-78, Equation 5-79 is a solution of matrix differential Equation 5-75b if λ is an eigenvalue of matrix K and ψ is a corresponding eigenvector.

The eigenvalues of any matrix A are the roots of its corresponding characteristic polynomial (defined below). Consider the $n \times n$ expansion of Equation 5-78:

$$
\begin{aligned}
a_{11}\psi_1 &+ \quad\cdots\quad + \quad a_{1n}\psi_n &= \lambda\,\psi_1 \\
a_{21}\psi_1 &+ \quad\cdots\quad + \quad a_{2n}\psi_n &= \lambda\,\psi_2 \\
&\cdots\cdots\cdots\cdots\cdots\cdots\cdots\cdots\cdots\cdots\cdots\cdots \\
a_{n1}\psi_1 &+ \quad\cdots\quad + \quad a_{nn}\psi_n &= \lambda\,\psi_n
\end{aligned}
\tag{5-80a}
$$

Transferring terms to the left side of each equation yields

$$
\begin{aligned}
\left(a_{11} - \lambda\right)\psi_1 &+ \cdots\cdots\cdots\cdots + a_{1n}\psi_n = 0 \\
a_{21}\psi_1 &+ \left(a_{22} - \lambda\right)\psi_2 + \cdots\cdots + a_{2n}\psi_n = 0 \\
&\cdots\cdots\cdots\cdots\cdots\cdots\cdots\cdots\cdots\cdots\cdots\cdots \\
a_{n1}\psi_1 &+ \cdots\cdots\cdots\cdots + \left(a_{nn} - \lambda\right)\psi_n = 0
\end{aligned}
\tag{5-80b}
$$

which in matrix notation becomes

$$(A - \lambda I)\psi = 0 \tag{5-80c}$$

I is the $n \times n$ identity matrix (all diagonal elements of $I = 1$; all off-diagonal elements $= 0$). From Cramer's theorem for solution of linear equations by determinants, Equation 5-80c has a nonzero solution only if the corresponding determinant of the coefficients is zero:[96]

$$\det(A - \lambda I) = \begin{vmatrix} a_{11} - \lambda & a_{12} & \cdots & a_{1n} \\ a_{21} & a_{22} - \lambda & \cdots & a_{2n} \\ \cdot & \cdot & \cdots & \cdot \\ a_{n1} & a_{n2} & \cdots & a_{nn} - \lambda \end{vmatrix} = 0 \tag{5-81}$$

Development of the determinant by cofactors ultimately yields a polynomial of nth degree in λ. The roots of this polynomial are the values of λ that satisfy Equation 5-78, i.e., the eigenvalues of matrix A. It follows that A has at least one and at most n distinct eigenvalues.

Example 5-7. Matrix Solution of Simple Two-Box Model[*]

Consider a simple two-box system:

(1)
$$\boxed{A_1} \underset{k_{21}}{\overset{k_{12}}{\rightleftharpoons}} \boxed{A_2}$$

with rate equations

(2)
$$dA_1/dt = -k_{12} A_1 + k_{21} A_2$$
$$dA_2/dt = k_{12} A_1 - k_{21} A_2$$

or (3)
$$\frac{d}{dt}\begin{bmatrix} A_1 \\ A_2 \end{bmatrix} = \begin{bmatrix} -k_{12} & k_{21} \\ k_{12} & -k_{21} \end{bmatrix}\begin{bmatrix} A_1 \\ A_2 \end{bmatrix} = K\begin{bmatrix} A_1 \\ A_2 \end{bmatrix}$$

The general solution to Equation 3 has the form

(4)
$$A(t) = a_1 \exp(\lambda_1 t)\psi_1 + a_2 \exp(\lambda_2 t)\psi_2$$

λ_1 and λ_2 are the eigenvalues, and ψ_1 and ψ_2 the eigenvectors, of the matrix K defined in Equation 3; coefficients a_1 and a_2 are determined from initial conditions.

---

[*] Summarized from Lasaga.[94]

The determinant of $\mathbf{K}$ is

$$\begin{vmatrix} -k_{12} - \lambda & k_{21} \\ k_{12} & -k_{21} - \lambda \end{vmatrix} = 0$$

or

$$\left(k_{12} + \lambda\right)\left(k_{21} + \lambda\right) - k_{12}\,k_{21} = 0$$

This quadratic equation in λ has the solutions $\lambda = 0$ and $\lambda = -(k_{12} + k_{21})$. The eigenvectors are obtained by solving the equation $\mathbf{K}\psi_i = \lambda_i\psi_i$, or

(5)
$$\begin{bmatrix} -k_{12} & k_{21} \\ k_{12} & -k_{21} \end{bmatrix}\begin{bmatrix} a \\ b \end{bmatrix} = \lambda\begin{bmatrix} a \\ b \end{bmatrix}$$

For $\lambda = 0$, the right side of the above equation is zero, and we obtain

$$-k_{12}a + k_{21}b = 0$$

$$k_{12}a - k_{21}b = 0$$

or (6)
$$b = (k_{12}/k_{21})a$$

Since a can have any value, we can set it equal to 1, and $\psi_1 = \begin{bmatrix} 1 \\ k_{12} \\ k_{21} \end{bmatrix}$
Similarly, to find ψ_2, we must solve $\mathbf{K}\psi_2 = \lambda_2\psi_2$:

(7)
$$\begin{bmatrix} -k_{12} & k_{21} \\ k_{12} & -k_{21} \end{bmatrix}\begin{bmatrix} a \\ b \end{bmatrix} = -(k_{12} + k_{21})\begin{bmatrix} a \\ b \end{bmatrix}$$

or $-k_{12}a + k_{21}b = -(k_{12} + k_{21})a$, and $k_{12}a - k_{21}b = -(k_{12} + k_{21})b$, both of which reduce to $a = -b$. Setting $a = 1$, we obtain

$$\psi_2 = \begin{bmatrix} 1 \\ -1 \end{bmatrix}$$

The solution to Equation 4 thus is

(8)
$$A(t) = \begin{bmatrix} A_1(t) \\ A_2(t) \end{bmatrix} = a_1\,\exp(0 \cdot t)\begin{bmatrix} 1 \\ k_{12} \\ k_{21} \end{bmatrix} + a_2\,\exp\left\{-\left(k_{12} + k_{21}\right)t\right\}\begin{bmatrix} 1 \\ -1 \end{bmatrix}$$

Finally, coefficients a_1 and a_2 are determined from initial conditions in the system. Solving Equation 8 for $t = 0$, we obtain

(9a) $$A_1^o = a_1 + a_2$$

and (9b) $$A_2^o = (k_{12}/k_{21})a_1 - a_2$$

These equations can be solved algebraically (or by simple matrix manipulation) to yield

(10) $$a_1 = \frac{k_{21}(A_1^o + A_2^o)}{k_{12} + k_{21}}; \quad a_2 = \frac{k_{12}A_1^o - k_{21}A_2^o}{k_{12} + k_{21}}$$

The final equations are obtained by combining Equations 8 and 10

(11a) $$A_1(t) = \frac{k_{21}(A_1^o + A_2^o)}{k_{12} + k_{21}} + \frac{k_{12}A_1^o - k_{21}A_2^o}{k_{12} + k_{21}} \exp\{-(k_{12} + k_{21})t\}$$

(11b) $$A_2(t) = \frac{k_{12}(A_1^o + A_2^o)}{k_{12} + k_{21}} + \frac{k_{12}A_1^o - k_{21}A_2^o}{k_{12} + k_{21}} \exp\{-(k_{12} + k_{21})t\}$$

Example 5-7 shows that we can obtain eigenvalues for 2×2 matrices analytically by solving a quadratic equation. It is increasingly difficult to obtain the roots as n becomes larger, and analytical solutions are not possible for matrices of the size needed in most environmental models ($n > 3$). Except for simple two- and three-compartment systems, eigenvalues are *always* estimated by numerical approximations and not by determining the roots of the characteristic equation. Thus, Equation 5-76 is a true analytical solution for a set of rate equations only if we can obtain the λ analytically.

Many methods have been developed to solve eigensystem problems, but efficient and numerically stable computer methods were not developed until the mid-1960s. Numerical codes now are readily available[92,100] to compute the eigenvalues of a matrix and calculate its eigenvectors. It is important to realize that if ψ is an eigenvector of $\mathbf{K}$, then so is $\alpha\psi$ for any number α; eigenvectors define directions but not length.[101] Eigenvectors produced by computer programs often are normalized so that the sums of squares of the vector elements are unity. Inversion of matrix ψ, which is needed to obtain the time-varying compartment sizes by Equation 5-76, also is done numerically. Inversion of small matrices can be done manually, but this is very tedious for large matrices.

For matrices of rate constants in compartment models, the eigenvalues all are negative or zero. This insures the stability of the system; otherwise compartment values would become unbounded at large values of t. In closed compartment systems with only linear transfer relationships (as in Equations 5-73 and 5-75), one (and only one) eigenvalue always has a value of 0.[94] As can be seen by inspecting the matrix solution of a set of linear differential equations (Equation 5-76), this λ

determines the steady-state compartment values, because at large t, all exponential terms with negative λ approach zero. Thus, for $\lambda = 0$

$$\lim A(t) = (\psi)^{-1} A_o \psi_o \qquad (5\text{-}82)$$

ψ_o is the eigenvector corresponding to $\lambda = 0$. Acceptance of the statement that there is only one zero eigenvalue for linear systems is equivalent to saying that the system has only one steady state (which is uniquely defined by Equation 5-82). (For proof of this, see reference 94.) This is not the case for nonlinear and open systems, which may not have steady states.

The total quantity of material in a closed system is constant. Thus,

$$A_T = \sum_{i=1}^{n} A_i \qquad (5\text{-}83\text{a})$$

Therefore, one of the A_i can be eliminated:

$$A_n = A_T - \sum_{i=1}^{n-1} A_i \qquad (5\text{-}83\text{b})$$

This reduces the system to n–1 linear nonhomogeneous differential equations,[90] the solution of which is the sum of n–1 exponential terms (one for each nonzero λ) plus a constant that represents the zero λ term.

The matrix concepts described above can be understood more easily by applying them to environmental examples. Although most aquatic systems contain more than two compartments, the two-compartment system described in Example 5-7 has some important applications, such as bioaccumulation of contaminants by fish.

Example 5-8. Eigenvalue Solution of Five-Compartment Global Carbon Cycle

The carbon cycle described in Figure 5-15 and quantified in matrix form in Example 5-6 can be solved as an eigenvalue problem.[94] Eigenvalues and eigenvectors for the 5×5 symmetric matrix $\mathbf{K}$ of transfer coefficients were calculated here by the program EIGRS from the IMSL MATH/PC LIBRARY[92] as follows:

	λ_0	λ_1	λ_2	λ_3	λ_4
$\lambda =$	-5.00731	-0.16096	-0.08480	0.00000	-0.01957

	ψ_0	ψ_1	ψ_2	ψ_3	ψ_4
$\psi =$	0.657073	−1.265663	−0.256036	0.580487	−0.009026
	0.428498	−1.010887	−0.814360	−0.743716	0.000128
	0.666575	−3.480243	0.921595	0.329850	−0.000001
	0.006665	−0.012889	−0.002642	0.006084	0.062497
	2.856738	5.769373	0.151385	−0.172733	−0.053598

The inverse of the eigenvector matrix is easily obtained:[102]

$$
\psi^{-1} = \begin{vmatrix}
0.216652 & 0.216642 & 0.216655 & 0.216662 & 0.216664 \\
-0.058809 & -0.117530 & -0.130154 & 0.044421 & 0.061421 \\
-0.743409 & -0.363641 & 0.540826 & -0.021373 & 0.099391 \\
1.018757 & -0.661845 & -0.290460 & 0.090606 & -0.067486 \\
-0.165835 & 0.001715 & 0.001191 & 15.977099 & 0.000332
\end{vmatrix}
$$

From the vector $\mathbf{A}^\circ$ of initial comparment sizes and ψ^{-1} we can compute the vector of coefficients $\mathbf{a}^\circ$ from Equation 5-76. For example, a_0°, the coefficient associated with ψ_0, is obtained by multiplying the first row of ψ^{-1} times $\mathbf{A}^\circ$ and summing the product, which is 1050.145. The reader can verify that coefficients a_1° through a_4° all approach zero in this case (because the initial state of the system is at steady state). Thus, $\mathbf{A}(t)$ simply becomes $\mathbf{A}(t) = a_0^\circ \psi_0$, or

$$
\begin{vmatrix}
690 \\
450 \\
700 \\
7 \\
3000
\end{vmatrix} = 1050.145 \times \begin{vmatrix}
0.657073 \\
0.428498 \\
0.666575 \\
0.006665 \\
2.856738
\end{vmatrix}
$$

More interesting situations occur if we perturb the system to an initial nonsteady-state condition. For example, suppose we increase the atmospheric compartment by 50% (a simple "greenhouse" warming scenario). The initial vector of compartment sizes then becomes

$$
\mathbf{A}^{per} = \begin{vmatrix}
345 \\
0 \\
0 \\
0 \\
0
\end{vmatrix} + \begin{vmatrix}
690 \\
450 \\
700 \\
7 \\
3000
\end{vmatrix} = \begin{vmatrix}
1035 \\
450 \\
700 \\
7 \\
3000
\end{vmatrix}
$$

Because the transfer coefficients are unchanged, $\mathbf{K}$ is the same as before and so are the eigenvalues and eigenvectors of $\mathbf{K}$. To obtain the time evolution of the perturbation, all we need are values of $\mathbf{a}^{per}$, obtained as the product of $\mathbf{A}^{per}$ times ψ^{-1}. $\mathbf{A}(t)$ thus becomes

$$
\mathbf{A}(t) = 1124.89\psi_0 - 20.2897\psi_1 \exp(-0.01957t) - 246.82\psi_2 \exp(-0.08479t)
$$
$$
+ 351.44\psi_3 \exp(-0.16096t) - 57.198\psi_4 \exp(-5.00731t)
$$

The new steady state occurs at $t \to \infty$, (i.e., all exponential terms $\to 0$). Thus,

$$
\mathbf{A}_{ss} = 1124.89\psi_0 = \begin{vmatrix}
739.1 \\
482.0 \\
749.8 \\
7.5 \\
3213.5
\end{vmatrix}
$$

Furthermore, Equation 1 can be used to compute the compartment sizes at any time, as the new steady state is approached.

The rate at which the carbon cycle attains steady state is given by the smallest nonzero eigenvalue, because this is the last term to decay in the integrated equation for $A(t)$:

$$\tau_{sys} = 1/\left|\lambda_i\right|_{min} \tag{5-84}$$

τ_{sys} is the response time of the system, and it is analogous to characteristic times of reactants, i.e., it is the time required to reduce a perturbation to $1/e$ of its original size. In the five-compartment carbon cycle, $\tau_{sys} = 1/|\lambda_2| = 1/0.01957$ year^{-1} or 51.1 years. This is substantially smaller than the characteristic time obtained from the smallest individual transfer rate ($k_{51} = 0.0100$ year^{-1}; $\tau_5 = 100$ years) and leads to an important generalization about multicompartment systems: the collective behavior of a cycle is faster than that predicted by the slowest individual transfer rate. However, the collective nature of the response causes some reservoirs to have longer response times than predicted by individual rate constants. For example, the atmospheric CO_2 compartment has a residence time of 8.3 years $(690/(35 + 43))$ based on its reservoir size and transfer (outflow) rates.

5.10.3 Matrix Solution of Steady-State Mass Balance Equations

Solution of mass-balance equations for multicompartment systems is simplified if steady-state conditions can be assumed. In these cases the differential equations become simple algebraic equations, and solution of simultaneous algebraic equations by matrix methods simply involves inversion of a matrix of coefficients. Consider the following CFSTR representation of the Great Lakes:

$$
\begin{array}{l}
1 \quad \lambda_{13} \\
\qquad 3 \xrightarrow{\lambda_{34}} 4 \xrightarrow{\lambda_{45}} 5 \xrightarrow{\lambda_{5SLR}} SLR \\
2 \quad \lambda_{23}
\end{array}
\tag{5-85}
$$

where 1 = Lake Superior, 2 = Lake Michigan, 3 = Lake Huron, 4 = Lake Erie, 5 = Lake Ontario, and SLR = St. Lawrence River. Mass-balance differential equations for the five-compartment lake system at steady state are

$$0 = dS_1/dt = L_1 - \lambda_{13}S_1$$

$$0 = dS_2/dt = L_2 - \lambda_{23}S_2$$

$$0 = dS_3/dt = L_3 + \lambda_{13}S_1 + \lambda_{23}S_2 - \lambda_{34}S_3$$

$$0 = dS_4/dt = L_4 + \lambda_{34}S_4 - \lambda_{45}S_4$$

$$0 = dS_5/dt = L_5 + \lambda_{45}S_4 - \lambda_{5SLR}S_5$$

$$\tag{5-86}$$

Solving for L_i and rewriting in matrix form yields

$$
\begin{vmatrix} L_1 \\ L_2 \\ L_3 \\ L_4 \\ L_5 \end{vmatrix} = \begin{vmatrix} \lambda_{13} & 0 & 0 & 0 & 0 \\ 0 & \lambda_{23} & 0 & 0 & 0 \\ -\lambda_{13} & -\lambda_{23} & \lambda_{34} & 0 & 0 \\ 0 & 0 & -\lambda_{34} & \lambda_{45} & 0 \\ 0 & 0 & 0 & -\lambda_{45} & \lambda_{5SLR} \end{vmatrix} \times \begin{vmatrix} S_1 \\ S_2 \\ S_3 \\ S_4 \\ S_5 \end{vmatrix} \quad (5\text{-}87a)
$$

or

$$ L = \lambda S \quad (5\text{-}87b) $$

where L and S are the column matrices (vectors) of loading rates and steady-state mass quantities in the lakes, and λ is the matrix of transfer coefficients. The solution of Equation 5-87b is

$$ S = \lambda^{-1} L \quad (5\text{-}88) $$

Inversion of the coefficient matrix λ can be done by many computer programs; even some PC-based spreadsheets[102] have matrix inversion commands.

If the elements of S and L have units of mass, then λ has units of mass·time^{-1}/mass, and λ^{-1} has units of mass/mass·time (or just time if S and L are expressed in the same units of mass). The columns of λ^{-1} are equivalent to unit response coefficients; i.e., the first column represents the steady-state responses of all the S_i to a unit load to compartment 1; the second column represents the responses of all S_i to a unit load to compartment 2, and so on.[4]

5.10.4 Matrix Descriptions of Open and Nonlinear Compartment Systems

The preceding concepts can be expanded to include systems with inputs and/or outflows or nonlinear intercompartment transfers. If an n-compartment system has outflows but no inputs, the outflows can be treated as if they are going into another compartment (the environment), and the whole system can be treated as a closed n + 1 compartment system.[90] Such systems also have a zero eigenvalue. If all n compartments in such a system are linked reversibly, the only stable steady state is the trivial one: complete washout from all compartments (except the external environment). Examples of this occur in kinetic studies when a radiotracer is added as a pulse to one compartment of a "leaky" system.

Analytical solutions have been derived for small open systems (two and three compartments) with inputs.[90] These are of interest primarily in studying the distributional kinetics of tracer-labeled substances in biological systems. Matrix methods were extended to open linear geochemical compartment models by Lasagna,[94] and his approach is summarized briefly below. All the inputs and outflows for a given compartment are combined into a net time function, $f_i(t)$ (+ for net input; – for net outflow). The input function vector (outflows are regarded as negative inputs) thus is

$$F(t) = \begin{matrix} f_1(t) \\ : \\ f_n(t) \end{matrix}$$

(5-89)

The general rate equation for the ith compartment thus is

$$dA_i / dt = \sum_j k_{ji} A_j - \sum_j k_{ij} A_i + f_i(t)$$

(5-90)

In matrix notation, the set of rate equations for all n compartments is

$$dA(t)/ dt = KA(t) + F(t)$$

(5-91)

Solution of Equation 5-91, which we will not derive, has the general form:

$$A(t) = \sum_r a_r(t) \exp(\lambda_r t) \psi_r$$

(5-92)

The fixed coefficients ($\psi^{-1} A_o = a_o$) of Equation 5-76 thus are replaced by functions of time, $a_r(t)$:

$$a_r(t) = a_{ro} + \int_o^t \exp(-\lambda_r t) g_r(t) dt$$

(5-93)

The fixed coefficients, a_{ro}, still are determined by initial compartment sizes

$$a_{ro} = \psi^{-1} A_o$$

(5-94)

while

$$g(t) = \psi^{-1} F(t)$$

The geochemical cycles of the major biogenic elements (C,N,P,O,S) are highly coupled. For example, the rates of N and P storage in lake and marine sediments are linked to rates of primary production (i.e., to the carbon cycle) and also depend on O_2 concentrations in bottom waters and surficial sediments. Anaerobic decomposition is slower and less efficient than aerobic decomposition, but release of ammonium and inorganic phosphate from sediments is much higher when the sediment-water interface is anoxic. Modeling of coupled cycles is not yet well understood, but coupling generally leads to nonlinear transfer relationships. This leads to the possibility of multiple steady states, of which only some may be stable. (Stability in this case refers to the tendency of a system to return to its original steady state after a small perturbation and is analogous to the ecological concept of resiliency.) Unstable steady states are associated with positive eigenvalues. Based on an analysis of the thermodynamics of irreversible chemical processes,[103] Lasaga[94,101] suggested that a geochemical cycle will *not* exhibit unstable behavior near a steady state unless it has an autocatalytic transfer relationship.

Coupling of cycles also increases their overall stability by causing faster response times compared with uncoupled linear cycles. The existence of multiple steady states in coupled cycles has been demonstrated for cycles that follow simple rate equations like those for elementary chemical reactions. These equations describe natural systems only approximately, however, and the extent to which the dynamics of natural systems can be predicted from the constrained mathematical behavior of simple compartment models is not certain.

5.11 THE INVERSE PROBLEM

In modeling geochemical cycles, we normally know the number and sizes of the reservoirs and values of the transfer coefficients before the modeling exercise begins. (How well we know them is another matter.) If data are lacking for one or a few rate constants, we may be able to assume steady-state conditions and estimate the missing value(s) by mass-balance analysis on a compartment. Modeling of the system then can proceed in straight-forward fashion by numerical simulation or by the matrix methods described in the previous section. The situation is different in compartment analyses based on the distribution kinetics of radiotracers. In these cases the modeler may have only limited information about the structure of the system and magnitude of individual rate constants, and the model must be developed from the data. This is the so-called inverse problem. Conceptually, it is more difficult than the analytical problem (i.e., solution of a model with specified structure and rate constants).

There are two extremes to the inverse problem. On the one hand, one may have independent knowledge of system structure and need only to estimate model parameters (initial compartment sizes and transfer coefficients). For example, one may know that the system can be treated as four compartments linked in a particular manner. The problem then is one of *parameter estimation*. On the other hand, one may not know the structure of the system (number of compartments and their linkages), and then the problem is a more formidable one of *system identification*. Intermediate situations often exist where the structure is known in part, but not in detail. Obviously, the more that one knows about the system, the more accurate the model will be. Poorly defined systems may require several iterations of experimental design, data gathering, and model development before the system is understood. Results from an initial model, even though crude, can help design a second set of experiments to characterize the system more accurately.

Data from radiotracer experiments on multicompartment systems usually involve a change over time in isotope activity in one or several compartments after a pulse addition of tracer to one compartment. When a tracer is added to a one-compartment system, the loss in activity usually follows first-order kinetics, and a plot of ln(activity) vs. time is a straight line. The transfer rate constant (export coefficient) is obtained from the slope and the y-intercept gives the initial activity. In contrast, plots of ln(activity) vs. time for multicompartment systems usually are curvilinear, and the rate constants for intercompartment transfers are imbedded in the curve. In simple cases one can deconvolve the curve graphically as the sum of several straight lines, as described in Figure 2-13. Criteria for successful application of the "curve peeling" method include (1) the system should consist of few compartments (two or three at most); (2) the decay constants should differ from each other by at least a factor of two; and (3) the reciprocals of the constants should not be very small or very large compared with the duration of the experiment.[89]

The curve-peeling method depends on the assumption that decay curves are the sum of n exponential terms, where n is the number of compartments in the system. Thus,

$$A(t) = a_1 \exp(-\lambda_1 t) + a_2 \exp(-\lambda_2 t) + a_3 \exp(-\lambda_3 t) + \ldots + a_n \exp(-\lambda_n t)$$

(5-95)

$$\lambda_1 > \lambda_2 > \lambda_3 > \lambda_n > 0$$

If the λ are sufficiently different in size, the first n−1 terms of Equation 5-95 become negligible compared with the last as t becomes large. Thus, at large t, a plot of $ln[A(t)]$ vs. t approaches a straight line with slope $-\lambda_n$ and y-axis intercept a_n. Subtracting $a_n \exp(-\lambda_n t)$ from A(t) for all data points not on the straight line yields a new equation with n−1 terms. If the λ are sufficiently different in size, the first n−2 terms now are negligible compared with the last (n−1) term of the new equation at large t. Again, the curve approaches a straight line with slope λ_{n-1} and y-intercept a_{n-1}. The term $a_{n-1} \exp(\lambda_{n-1} t)$ can be subtracted from data points on the curved part of the second line, and the process is repeated until all the data points fit on a straight line. If the decay curve for a compartment reaches a steady-state value, (i.e., it approaches a horizontal line), then $\lambda_n = 0$. This means the system is either closed, or it has a "trap" subsystem with no excretion and no outward transfers to other compartments.

In practice, experimental errors and random noise limit the number of coefficients that can be fitted and the accuracy of the fitted values. The most accurate values are a_n and λ_n (the first fitted term and smallest rate coefficient), and the accuracy decreases for each successively estimated pair. Even the use of least-squares methods to fit the lines does not solve the major deficiencies of graphical curve peeling: (1) increasing errors in parameter estimation for each pair of extracted coefficients and (2) the small number of compartments that can be resolved. Parameter estimation from tracer data now can be done by computerized nonlinear least-squares fitting routines;[88,104] graphical methods are used primarily to obtain initial estimates for these programs, which determine all parameters simultaneously and thus provide equal accuracy for all the parameters.

It is interesting to note that the λ determined by curve peeling are eigenvalues of the compartment system, and the a_j are elements of the corresponding eigenvectors.

Each compartment i measured in the above manner yields a row vector of eigenvector elements a_i. If all n compartments are accessible to measurement of tracer activity as a function of time, one can obtain all n^2 elements of the eigenvector matrix ψ. In a closed system, only $n(n-1)$ eigenvector elements are independent, but we also have n values for λ, so that the total number of independent values is n^2.[89] Consequently, one can determine all n^2 rate constants (the k_{ij}) comprising matrix $\mathbf{K}$ in a single experiment, if all compartments are accessible. This is essentially the inverse of the original matrix problem of Section 5.10.2. The matrix solution for $\mathbf{K}$ is[89]

$$K = A\,\psi\,\lambda\,\psi^{-1}A^{-1} \tag{5-96}$$

$\mathbf{A}$ is the diagonal matrix of compartment sizes (assumed constant in steady-state tracer analyses), ψ is the matrix of eigenvectors (the a_{ij} determined by curve fitting from the experimental data), and λ is the diagonal matrix of eigenvalues. The solution for an element of $\mathbf{K}$ is

$$k_{ij} = \frac{A_i}{A_j} \sum A_{il}\lambda\frac{D_{jl}}{D} \tag{5-97}$$

where D is the determinant of the eigenvector matrix ψ, and D_{jl} is the cofactor of the jth row and lth column of D.

The analysis of tracer data from nutrient cycling experiments[97,105] is a common example of the inverse problem in aquatic biology. In these experiments, the soluble inorganic pool of some element (e.g., C or P) is spiked with a radiotracer at $t = 0$, and the appearance of label is followed over time in various operationally defined compartments such as total particulate matter (seston), size fractions of seston, and the dissolved organic and inorganic pools. A related inverse problem arises in estimating contaminant input histories (rates over time) to lakes from sedimentary records (concentration profiles with depth in a sediment). This problem has been addressed for heavy metals and PCBs in Great Lakes sediments by Christensen et al.[106]

The above treatment of tracer kinetics for compartment systems is only a brief introduction to a complicated subject. Interested readers should consult the monographs by Atkins,[89] Jacquez,[90] and Rubinow,[91] who describe mathematical solutions for simple compartment systems and applications in biological sciences. Smith and Horner[105] provide a useful example of analyzing real radiotracer data in planktonic carbon-cycling kinetics.

PROBLEMS

1. Derive the temporal solutions for a contaminant in a well-mixed lake under conditions of constant input (Equation 5-11) and exponentially decreasing input (Equation 5-18) from the general solution (Equation 5-16) for a reactor model with time-varying input functions.

2. Everett Jordan Lake, a reservoir on the New Hope and Haw Rivers in North Carolina, may be described approximately by four CFSTRs, denoted as Pools I-IV. Pools II-IV

are in series in the New Hope River drainage basin. Pool IV is the uppermost and receives river input directly; Pools III and II receive inputs from the next higher pool. Pool I receives water from the Haw River plus water from Pool II of the New Hope drainage basin. Causeways divide the reservoir into these pools, which had the following morphometric, hydrologic, and trophic conditions in 1983:

	Surface area (10^6 m^2)	Mean depth (m)	Volume 10^6 m^3	$[TP]_{ave}$ µg L^{-1}	$[Chla]$ µg L^{-1}	TN/TP wt/wt
IV. (NHR Upper pool)	13.76	2.6	36.09	76	57	9
III. (NHR Middle pool)	15.19	4.5	68.78	42	41	19
II. (NHR Lower pool)	19.69	5.5	107.92	45	33	19
I. (HR Pool)	7.78	6.8	52.57	105	39	10
Total Reservoir	56.42	4.7	256.36	—	—	—

Data from C.M. Weiss, D.E. Francisco, and P.H. Campbell, Proc. Triangle Conf. Environ. Technol., Duke University, Durham, NC, 1984.

The total drainage area for the resevoir is 1,580 mi^2, of which 1,300 mi^2 is in the Haw River drainage basin and the balance in the New Hope. The total water inflow to the resevoir in 1983 was 1.69×109 m^3. Assume that (1) $Q_{in} = Q_{out}$ for each pool (i.e., precipitation on the lake surface balances evaporation from the lake surface) and (2) inflows from the rivers are proportional to their drainage areas. Compute the water residence time (τ_w), flushing coefficient (ρ_w), and areal water loading rate (q_a) for each segment.

Given that the average concentration of TP in the New Hope River was 0.60 mg L^{-1} in 1983 and the corresponding value for the Haw River was 0.411 mg L^{-1}, estimate the annual average TP concentration in each pool and the corresponding average chlorophyll values, using the average value of $v_p = 13$ m year^{-1} discussed in Section 5.5.3. Compare your predicted values with the values reported above. Discuss reasons for the differences between predicted and measured values. Assuming that steady state existed in the resevoir when the above data were collected, estimate the value v_p required to give accurate predictions of $[TP]_1$ in the pools.

3. According to the constant-rate-of-supply model [Appleby and Oldfield, *Catena*, 5:1 (1978)] used to date sediments by ^{210}Pb, the total quantity of unsupported ^{210}Pb in a sediment (expressed areally, pCi cm^{-2}), $M_{Pb,u}$, times the first-order rate constant for ^{210}Pb decay, k_1, is equal to the average rate of supply (pCi cm^{-2} year^{-1}) of unsupported ^{210}Pb to the sediment, $R_{Pb,u}$:

$$R_{Pb,u} = k_1 M_{Pb,u} \qquad (1)$$

$M_{Pb,u}$ is obtained by summing (integrating) the activity of unsupported ^{210}Pb over depth in a sediment core, as determined by α-spectroscopy (see Example 2-1). Unsupported ^{210}Pb is defined as that derived from atmospheric deposition; in contrast, supported ^{210}Pb is derived from radioactive decay of radium isotopes present naturally in sediment materials. (In practical terms, a sediment core usually is segmented into

1- or 2-cm intervals, and 15 to 20 segments over the top 30 to 50 cm of the sediment core are analyzed for ^{210}Pb activity. This depth usually is sufficient to reach sediments so old that essentially all the ^{210}Pb derived from atmospheric deposition (the unsupported ^{210}Pb) has decayed; the constant, low level of ^{210}Pb activity approached asymptotically at depth in the sediment represents the supported ^{210}Pb).

(a) Show why Equation 1 is valid by writing the dynamic mass balance equation of mass of ^{210}Pb in a sediment core and solving for the steady-state condition.

(b) Baker et al. [*Limnol. Oceanogr.*, 37:689 (1992)] found that $m_{Pb,u}$ ranged from 12.5 to 15.9 pCi cm^{-2} in three epilimnetic sediment cores from Little Rock Lake in northern Wisconsin. What are the average ^{210}Pb fluxes $R_{Pb,u}$ for these cores? The half-life of ^{210}Pb is 22 years. How well do these values agree with the estimated mean annual deposition rate (0.2 to 1.0 pCi cm^{-2} year^{-1}) for atmospheric fallout of ^{210}Pb?

REFERENCES

1. Hill, C.G., Jr., *Chemical Engineering Kinetics and Reactor Design*, John Wiley & Sons, New York, 1977.

2. Levenspiel, O., *Chemical Reaction Engineering*, 2nd ed., John Wiley & Sons, New York, 1972.

3. Chapra, S.C. and K.H. Reckhow, *Engineering Approaches for Lake Management Vol. 2: Mechanistic Modeling*, Butterworths, Boston, 1983.

4. Thomann, R.V. and J.A. Mueller, *Principles of Surface Water Quality Modeling and Control*, Harper & Row, New York, 1987.

5. Tchobanoglous, G. and E.D. Schroeder, *Water Quality*, Addison-Wesley, Reading MA, 1985.

6. Schnoor, J.L., *Science,* 211, 840 (1981).

7. O'Connor, D.J. and J.A. Mueller, *J. San. Eng. Div. ASCE*, 96, 955 (1970).

8. Wehner, J.F. and R.H. Wilhelm, *Chem. Eng. Sci.*, 6, 89 (1959).

9. Smith, J.M., *Chemical Engineering Kinetics*, 2nd ed., McGraw-Hill, New York, 1970.

10. Freeze, R.A. and J.A. Cherry, *Groundwater*, Prentice-Hall, Englewood Cliffs, N.J., 1979.

11. Overman, A.R., R.-L. Chu, and W.G. Leseman, *J. Water Pollut. Contr. Fed.*, 48, 880 (1976).

12. Carslaw, H.S. and J.C. Jaeger, *Conduction of Heat in Solids*, 2nd ed., Oxford University Press, New York, 1959.

13. Dankwerts, P.V., *Chem. Eng. Sci.*, 2, 1 (1953).

14. Reynolds, T.D., *Unit Operations and Processes in Environmental Engineering*, Brooks/Cole Engineering Division, Monterey, CA, 1982; Schroeder, E.D., *Water and Wastewater Treatment,* McGraw-Hill, New York, 1977.

15. Levenspiel, O. and K. B. Bischoff, *Adv. Chem Eng.*, 4, 95 (1963).

16. Fischer, H.B., E.J. List, R.C.Y. Koh, J. Imberger, and N.H. Brooks, *Mixing in Inland and Coastal Waters*, Academic Press, New York, 1979.

17. Bischoff, K.B. and O. Levenspiel, *Chem. Eng. Sci,*. 17, 245 (1962).

18. Streeter, H.W. and E.B. Phelps, A Study of the Pollution and Natural Purification of the Ohio River. III. Factors Concerned in the Phenomena of Oxidation and Reaeration, U.S. Pub. Health Serv., Pub. Health Bull. 146, 1925.

19. Prigogine, I., *Thermodynamics of Irreversible Processes*, 2nd ed., Wiley-Interscience, New York 1961.
20. Denbigh, K.G., *The Thermodynamics of the Steady State*, John Wiley & Sons, New York, 1951.
21. Morgan, J.J., in *Equilibrium Concepts in Natural Water Systems*, W. Stumm (ed.), Adv. Chem. Ser. 67, Am. Chem. Soc., Washington, D.C., 1967, pp. 1–29.
22. Hoffmann, M.R., *Environ. Sci. Technol.*, 15, 345 (1981).
23. Morgan, J.J. and A.T. Stone, in *Chemical Processes in Lakes*, W. Stumm (ed.), Wiley-Interscience, New York, 1985, pp. 389–426; Imboden, D.M. and R.P. Schwarzenbach, *Ibid*, pp. 1–30.
24. Vollenweider, R.A., *Arch. Hydrobiol.*, 66, 1 (1969).
25. Vollenweider, R.A., *Schweiz. Z. Hydrol.*, 37, 53 (1975).
26. Vollenweider, R.A., *Mem. Ist. Ital. Idrobiol.*, 33, 53 (1976).
27. Dillon, P.J. and R.H. Rigler, *J. Fish. Res. Bd. Can.*, 31, 1771 (1974); 32, 1519 (1975).
28. Jones, J.R. and R. Bachman, *J. Water. Pollut. Contr. Fed.*, 48, 2176 (1976).
29. Baker, L.A., P.L. Brezonik, and C.R. Kratzer, in *Lake and Reservoir Management*, Proc. 4th annual conference, N. Am. Lake Management Society, Washington, D.C., 1984, pp. 253–257.
30. Sonzogni, W.C., P. Uttormark, and G.F. Lee, *Water Res.*, 10, 429 (1976).
31. Kirchner, W.B. and P.J. Dillon, *Water Resour. Res.*, 11, 182 (1975).
32. Larson, D.P. and H.T. Mercier, Nat. Eutroph. Surv., Working Paper 174, U.S. EPA, Corvallis, OR, 1975; *J. Fish. Res. Bd. Can.*, 33, 1742 (1976).
33. Chapra, S.C., *Water Resour. Res.*, 11, 1033 (1975).
34. Dillon, P.J. and W.B. Kirchner, *Water Resour. Res.*, 11, 1035 (1975).
35. Chapra, S.C. and S.J. Tarapchak, *Water Resour. Res.*, 12, 1260 (1976).
36. Baker, L.A., P.L. Brezonik, and C.R. Kratzer, Nutrient Loading-Trophic State Relationships in Florida Lakes, Publ. 56, Wat. Resour. Res. Center, University of Florida, Gainesville, 1981.
37. Higgins, J.M. and B.R. Kim, *Water Resour. Res.*, 17, 571 (1981).
38. Rast, W. and G.F. Lee, Summary Analysis of the North American (U.S. Portion) OECD Eutrophication Project: Nutrient Loading-Lake Response Relationships and Trophic State Indices, Rept. EPA-600/3-78-008, U.S. EPA, Corvallis, OR, 1978.
39. Lean, D.R.S. and F.R. Pick, *Limnol. Oceanogr.*, 26, 1001 (1981).
40. Chiaudini, G. and M. Vighi, *Water Res.*, 8, 1063 (1974); Porcella, D.B. and A.B. Bishop, *Comprehensive Management of Phosphorus Water Pollution*, Ann Arbor Science, Ann Arbor, MI, 1975.
41. Sakamoto, M., *Arch. Hydrobiol.*, 62, 1 (1966).
42. Dillon, P.J. and F.H. Rigler, *Limnol. Oceanogr.*, 19, 767 (1974).
43. Bartsch, A.F. and J.H. Gakstatter, in Symposium on Use of Mathematical Models to Optimize Water Quality Management, EPA-600/9-78-024, U.S. EPA, Gulf Breeze, FL, 1978, pp. 371–393.
44. Smith, V.H. and J. Shapiro, *Environ. Sci. Technol.*, 15, 444 (1981).
45. Prepas, E.E. and J. Vickery, *Can. J. Fish. Aquat. Sci.*, 41, 351 (1984).
46. Riley, E.T. and E.E. Prepas, *Can. J. Fish. Aquat. Sci.*, 42, 831 (1985).
47. Carlson, R.E., *Limnol. Oceanogr.*, 22, 361 (1977).
48. Quiros, R., *Can. J. Fish. Aquat. Sci.*, 47, 928 (1990).
49. Shapiro, J., *Hydrobiologia*, 200/201, 13 (1990); Gulati, R.D., E.H.R.R. Lammens, M.-L. Meijer and E. van Donk. 1990. *Biomanipulation: Tool for Water Management*, Kluwer, Boston.

50. Thomann, R.V., *Systems Analysis and Water Quality Management*, McGraw-Hill, New York, 1972; O'Connor, D.J., Dynamic Water Quality Forecasting and Management, EPA-600/3-73-009, U.S. EPA, Washington, D.C., 1973; Middlebrooks, E.J., et al. (Eds.), *Modeling the Eutrophication Process*, Ann Arbor Science, Ann Arbor, MI, 1974; Thomann, R.V., D.M. Di Toro, R.P. Winfield, and D.J. O'Connor, Mathematical Modeling of Phytoplankton in Lake Ontario, EPA-660/3-75-005, U.S. EPA, Washington, D.C., 1975; Di Toro, D.M. and W.F. Matystik, Jr., Mathematical Models of Water Quality in Large Lakes, Part 1: Lake Huron and Saginaw Bay, EPA-600/3-80-056, U.S. EPA, Duluth, 1980; Di Toro, D.M. and J.P. Connolly, Ibid., Part 2: Lake Erie, EPA-600/3-80-065, U.S. EPA, Duluth, 1980; Walker, W.W., Jr., Empirical Methods for Predicting Eutrophication in Impoundments, Rept 4, Phase III, Applications Manual, Tech. Rept. E-812-9, U.S. Army Corps of Engineers, Waterways Experiment Station, Vicksburg, MS, 1987. (This report describes BATHTUB and FLUX, PC-based models for nutrient loadings to lakes.)

51. Imboden, D.M., *Schweiz. Z. Hydrol.*, 35, 29 (1973); *Limnol. Oceanogr.*, 19, 297 (1974).

52. Snodgrass, W.J. and C.R. O'Melia, *Environ. Sci. Technol.*, 9, 937 (1975).

53. Imboden, D.M., in: Internat. Congr. Lakes Pollution and Recovery, European Water Pollut. Control Association, Rome, 1985, pp. 29–40.

54. Schnoor, J.L. and D.J. O'Connor, *Water Res.*, 15, 1651 (1981).

55. DePinto, J.V., T.C. Young, and D.K. Salisbury, *Hydrobiol. Bull.*, 20, 225 (1986).

56. O'Melia, C.R. and K.S. Bowman, *Schweiz. Z. Hydrol.*, 46, 64 (1984); O'Melia, C.R., in *Chemical Processes in Lakes*, W. Stumm (Ed.), Wiley-Interscience, New York, 1985, pp. 207–224; Weilenmann, U., C.R. O'Melia, and W. Stumm, *Limnol. Oceanogr.*, 33, (1988).

57. Odum, H.T., *Systems Ecology: An Introduction*, Wiley-Interscience, New York, 1980; Hall, C.A.S. and J.W. Day, Jr. (Eds.), *Ecosystem Modeling in Theory and Practice*, Wiley-Interscience, New York, 1977; Patten, B.C. (Ed.), *Systems Analysis and Simulation in Ecology*, four vols., Academic Press, New York, 1971, 1972.

58. Gulliver, J.S. and H. Stefan, *J. Environ. Eng. Div. ASCE*, 108, 864 (1982).

59. Henderson-Sellers, *Engineering Limnology*, Pitman Advanced Publ., Boston, 1984.

60. Karickoff, S.W., D.S. Brown, and T.S. Scott, *Water Res.*, 13, 421 (1979).

61. O'Connor, D.J. and J.P. Connolly, *Water Res.*, 14, 1517 (1980).

62. Di Toro, D.M., *Chemosphere*, 14, 1503 (1985).

63. Wolfe, N.L., L.A. Burns, and W.C. Steen, *Chemosphere*, 9, 393 (1980).

64. Burns, L.A., D.M. Cline, and R.R. Lassiter, Exposure Analysis Modeling System (EXAMS): User Manual and System Documentation, EPA-600/3-82-023, U.S. EPA, Athens, GA, 1982; Burns, L.A., in *Validation and Predictability of Laboratory Methods for Assessing the Fate and Effects of Contaminants in Aquatic Systems*, T.P. Boyle (Ed.), ASTM STP 865, American Society of Testing and Materials, Philadelphia, PA, 1985, pp. 176–190.

65. Brezonik, P.L., L.A. Baker, and T.E. Perry, in *Sources and Fates of Aquatic Pollutants*, R.A. Hites and S.J. Eisenreich (Eds.), Adv. Chem. Ser. 216, Am. Chem. Soc., Washington, D.C., 1987, pp. 229–260.

66. Urban, N.R. and P.C. Brezonik, *Canad. J. Fish Aquat. Sci.*, 50, in press (1993); Baker, C.A., D.R. Engstrom, and P.L. Brezonik, *Limnol. Oceanogr.*, 37, 689 (1992).

67. Schindler, D.W., M.A. Turner, M.P. Stainton, and G.A. Linsey, *Science*, 232, 844 (1986).

68. Cook, R.B., C.A. Kelly, D.W. Schindler, and M.A. Turner, *Limnol. Oceanogr.*, 31, 134 (1986).

69. Perry, T.E., L.A. Baker, and P.L. Brezonik, in *Lake and Reservoir Management*, Vol. 2, J. Taggart and L. Moore (Eds.), North American Lake Management Society, Washington, D.C., 1986, pp. 309–312.

70. Perry, T.E., C.D. Pollman, and P.L. Brezonik, in *Impact of Acid Deposition on Aquatic Biological Systems*, Rep. STP 928, B.G. Isom, S.D. Dennis, and J.M. Baktes (Eds.), American Society of Testing and Materials, Philadelphia, PA, 1986, pp. 67–83.

71. Baker, L.A., P.L. Brezonik, and C.D. Pollman, *Water Air Soil Pollut.*, 25, 215 (1985).

72. Baker, L.A. and P.L. Brezonik, *Water Resour. Res.*, 24, 65 (1988).

73. Kelly, C.A., J.W.M. Rudd, R.H. Hesslein, D.W. Schindler, P.J. Dillon, C.T. Driscoll, S.A. Gherini, and R.E. Hecky, *Biogeochemistry*, 3, 129 (1987).

74. Perry, T.E., M.S. thesis, University of Minnesota, Minneapolis, 1987.

75. Schiff, S.L. and R.F. Anderson, *Can. J. Fish. Aquat. Sci.*, 44, (1987).

76. Schnoor, J.L., W.D. Palmer, Jr., and G.E. Glass, in *Modeling of Total Acid Precipitation Impacts,* J.L. Schnoor (Ed.), Butterworths Boston, MA, 1984.

77. Schnoor, J.L. and W. Stumm, in *Chemical Processes in Lakes*, W. Stumm (Ed.), Wiley-Interscience, New York, 1985, pp. 311–338.

78. Cosby, B.J. et al., *Water Resour. Res.,* 21, 51, 1591 (1985), 22, 1283 (1986); *Environ. Sci. Technol.,* 19, 1144 (1985); Hornberger et al., *Water Resour. Res.,* 21, 1841 (1985), 22, 1293 (1986).

79. Christopherson, N., and R.F. Wright, *Water Resour. Res.,* 17, 377 (1981); Christopherson, N., H.M. Seip, and R.F. Wright, *Ibid.,* 18, 977 (1982).

80. Henriksen, A., in *Ecological Impacts of Acidic Precipitation*, D. Drablos and A. Tollan (Eds.), SNSF Project, Oslo, Norway, 1980, pp. 68-74.

81. Gherini, S.A., L. Mok, R.J.M. Hudson, and G.F. Davis, *Water Air Soil Pollut.*, 26, 425 (1985).

82. Eadie, B.J. and J.A. Robbins, in *Sources and Fates of Aquatic Pollutants*, R.A. Hites and S.J. Eisenreich (Eds.), Adv. Chem. Ser. 216, Am. Chem. Soc., Washington, D.C., 1987, pp. 319–364.

83. Lerman, A., *J. Geophys. Res.*, 77, 3256 (1972).

84. Tracy, B.L. and F. Prantl, *Water Air Soil Pollut.*, 19, 15 (1983).

85. Thomann, R.V. and D. DiToro, *J. Great Lakes Res.*, 9, 474 (1983).

86. Engstrom, D., *Can. J. Fish. Aquat. Sci.*, 44, 1306 (1987).

87. Sheppard, C.W. and A.S. Householder, *J. Appl. Phys.*, 22, 510 (1951); Berman, M. and R. Schoenfeld, *J. Appl. Phys.*, 27, 1361 (1956); Berman, M., M.F. Weiss, and E. Shahn, *Biophys. J.*, 2, 275 (1962).

88. Berman, M. and M.F. Weiss, *SAAM User's Manual*, National Institutes of Health, Bethesda, 1974. (SAAM is a series of computer programs to fit tracer kinetic data to compartment models; it can be obtained (gratis) from the Mathematical Reseach Branch, NIAMD, NIH, Bethesda, MD 20014.

89. Atkins, G.L., *Multicompartment Models for Biological Systems*, Methuen, London, 1969.

90. Jacquez, J.A., *Compartmental Analysis in Biology and Medicine*, Elsevier, New York, 1972.

91. Rubinow, S.I., *Introduction to Mathematical Biology*, Wiley-Interscience, New York, 1975.

92. Rice, J.R., *Numerical Methods, Software, and Analysis: IMSL Reference Edition*, McGraw-Hill, New York, 1983; IMSL, *User's Manual Math/PC Library*, Houston, TX, 1985.

93. Gear, C.W., *Numerical Initial Value Problems in Ordinary Differential Equations*, Prentice-Hall, Englewood Cliffs, NJ, 1971.

94. Lasaga, A.C., *Geochim. Cosmochim. Acta*, 44, 815 (1980).

95. Jennings, A., *Matrix Computation for Engineers and Scientists*, Wiley-Interscience, New York, 1977; Searle, S.R., *Matrix Algebra for the Biological Sciences (Including Applications in Statistics)*, John Wiley & Sons, New York, 1966; Gere, J.M. and W. Weaver, *Matrix Algebra for Engineers*, Van Nostrand Reinhold, New York, 1965.

96. Kreyszig, E., *Advanced Engineering Mathematics*, 5th ed., John Wiley & Sons, New York, 1983.

97. Machin, D., *Biomathematics: An Introduction*, Macmillan, London, 1976; Wilkinson, J.H., *The Algebraic Eigenvalue Problem*, Clarendon Press, Oxford, 1965.

98. Holland, H., *The Chemistry of the Atmosphere and Oceans*, Wiley-Interscience, New York, 1978.

99. Bellman, R., *Stability Theory of Differential Equations*, Dover, New York, 1969.

100. Smith, B.T., et al., *Matrix Eigensystem Routines — EISPACK Guide*, Lecture Notes in Computer Science No. 51, Springer-Verlag, New York, 1976. EISPACK is a collection of 52 FORTRAN programs for solving eigensystem problems available from National Energy Software Center, Argonne Nat. Lab., Argonne, IL 60439 for a nominal charge.

101. Lasaga, A.C., in *Kinetics of Geochemical Processes*, A.C. Lasaga and R.J. Kirkpatrick (Eds.), Rev. Mineralogy, vol. 8, Mineralogical Society of America, Washington, D.C., 1982, pp. 69–109.

102. Such common spreadsheet programs as Lotus 1-2-3 and Quatro-Pro have matrix inversion functions in them.

103. Glansdorff, P. and I. Prigogine, *Thermodynamic Theory of Structure, Stability, and Fluctuations*, Wiley-Interscience, New York, 1971.

104. Gomeni, R. and C. Gomeni, *Comput. Biol. Med.*, 9, 39 (1979); Estreicher, J., C. Revillard, and J.-R. Scherrer, *Comput. Biol. Med.*, 9, 49, 66 (1979).

105. Smith, D.F. and S.M.J. Horner, Physiological Basis of Phytoplankton Ecology, T. Platt (Ed.), *Can. Bull. Fish. Aquat. Sci.* 210, 113 (1982).

106. Christensen, E.R. and R.H. Goetz, *Environ. Sci. Technol.*, 21, 1088 (1987); Christensen, E.R., P.K. Bhunia, and M.H. Hermanson, in *Heavy Metals in the Environment*, Vol. 1, T.D. Lekkas (Ed.), Proc. 5th Int. Conf., CEP Consultants, Edinburgh, U.K., Athens, 1985, pp. 116–118.

CHAPTER 6

Kinetics of Biochemical Reactions and Microbial Processes in Natural Waters

OVERVIEW

Microbial activities directly control the concentrations of inorganic nutrients and many organic compounds in aquatic systems. Most other substances are affected indirectly by microbial effects on pH and redox status. Consequently, an understanding of microbial processes is essential to an understanding of natural water chemistry. This chapter treats the kinetics of biological processes (primarily microbial) in natural waters. Because nearly all metabolic processes are catalyzed by enzymes, enzyme kinetics is treated first. Microbial growth kinetics is described next; growth equations are developed for both batch and continuous cultures, and the basic concepts of growth kinetics are used to develop more complicated growth models. Effects of mass transport and diffusion limitation on rates of microbial processes are treated subsequently. Factors affecting the kinetics of biological processes in aquatic environments are described throughout the chapter.

PART I. BIOCHEMICAL KINETICS: ENZYME-CATALYZED REACTIONS

6.1 REACTIONS INVOLVING A SINGLE SUBSTRATE

6.1.1 Historical Development

Our understanding of the mechanisms and kinetics of enzyme catalysis dates back to the late 19th century. In 1892 and 1902 Brown found that enzymatic reactions did not follow simple second-order kinetics and that rates of sucrose hydrolysis were independent of sucrose concentrations.[1] He concluded that the reaction rate was limited by an ephemeral enzyme-reactant complex. In 1903 Henri[2] theorized that enzymes catalyze reactions by forming a complex with the reactant (called the substrate). He assumed that the complex was in equilibrium with free enzyme and free substrate and derived an equation having the same form as the single-substrate enzyme-catalyzed rate equation used today. Although Brown and Henri reached essentially correct conclusions, their experiments were open to criticism.[3] A decade later, Michaelis and Menten[4] made the study of enzyme kinetics more reliable by using a buffer to maintain constant pH and by measuring initial reaction rates. They derived an equation similar to Henri's and used the equilibrium constant approach to eliminate intermediates in the equation. In 1925 Briggs and Haldane[5] used the steady-state approach and derived the same equation.

6.1.2 The Michaelis-Menten Equation

Regardless of the approach used for derivation and the contributions of early workers, the rate expression for single-substrate enzyme-catalyzed reactions is known as the Michaelis-Menten equation. For such reactions the simplest mechanism is

$$E + S \rightleftharpoons ES \rightleftharpoons E + P \qquad (6\text{-}1)$$

E represents the enzyme, S the substrate, and P the product. Equation 6-1 is identical to the general mechanism for reactions of recyclable catalysts (see Section 4.3.1 for derivation of the rate equation). In brief, the rate of product formation can be obtained by using the steady-state assumption ($d[ES]/dt = 0$) and a mass balance for enzyme ($E_t = [E] + [ES]$), and by assuming that the reverse reaction ($E + P \rightarrow ES$) can be ignored initially (since $[P]_0$ can be set to zero). With these assumptions we obtain

$$-d[S]/dt = d[P]/dt = \frac{k_1 k_3 E_t [S]}{k_2 + k_3 + k_1 [S]} \qquad (6\text{-}2)$$

It is not practical to measure the microscopic rate constants separately, and a group constant known as the Michaelis or half-saturation constant is defined as $K_S = (k_2 + k_3)/k_1$. For a given E_t, $k_3 E_t$ defines the maximum reaction rate and is symbolized as V or V_{max}. The rate of product formation is $k_3[ES]$, and $k_3 E_t$ is the rate of forward reaction when all the enzyme is present as the "active complex." Substituting K_S and V into Equation 6-2 and defining $-d[S]/dt = v$ yields the usual form of the Michaelis-Menten equation:

$$v = \frac{V[S]}{K_S + [S]} \qquad (6\text{-}3)$$

Five conditions must hold for this equation to apply:

1. The reaction must involve a single substrate, or all other substrate concentrations must be constant.
2. Because the reverse reaction $(P + S \rightarrow ES)$ was ignored in deriving the equation, initial velocities (v_o) must be measured at varying $[S]_o$; v_o is the extrapolated value of $(-d[S]/dt)_{t=0}$.
3. The concentration of enzyme must be constant (V is proportional to E_t).
4. E_t must be much smaller than $[S]_o$; otherwise it may not be possible to invoke the steady-state assumption.
5. Other conditions (temperature, pH, ionic strength) must be constant.

The Michaelis constant is symbolized in several ways: K, K_S, K_s, K_M, K_m. The terms are not all equivalent, however, and their precise meaning depends on the mechanistic model from which they were derived. In the derivation by Michaelis and Menten (by the equilibrium constant approach), K $(= k_2/k_1)$ represents the dissociation constant for enzyme-substrate complexes. This is valid only when k_3 $\ll k_2$, in which case K is an inverse measure of enzyme affinity for substrate. Interpretation of K as a dissociation constant is not generally valid, however. Some texts use K_M for dissociation constants and K_S for constants based on the Briggs-Haldane steady-state assumption $(K_S = (k_2 + k_3)/k_1)$, but this convention is not universal.

The Michaelis-Menten equation is a mixed zero- and first-order expression. When $[S] \gg K_S$, $v \approx V$, and zero-order kinetics apply. When $K_S \gg [S]$, first-order kinetics apply; i.e., $v \approx (V/K_S)[S] = k'[S]$. A plot of v vs. $[S]$ is hyperbolic for reactions fitting Equation 6-3 (Figure 6-1). K_S has units of concentration; this follows from its definition as the ratio of first-order constants, k_2 and k_3, to a second-order constant, k_1. It also is apparent that when $[S] = K_S$, $v = 0.5V$. K_S thus is often called a half-saturation constant. The plot of v vs. $[S]$ in Figure 6-1a shows that v approaches V rather slowly. Expressing $[S]$ in multiples of K_S, we find that $v = 0.5V$ at $[S] = K_S$, $0.67V$ at $[S] = 2K_S$, $0.9V$ at $[S] = 10K_S$, and $0.99V$ at $[S] = 100K_S$. Thus, the largest changes in v occur within a range of $[S] = 0$ to $\sim 3K_S$. The Michaelis-Menten equation is consistent with the mechanisms of enzyme-catalyzed reactions. At high $[S]$, enzyme active sites are saturated, and rates are independent of $[S]$. At low $[S]$, the concentration of "active complex," $[ES]$, is proportional to $[S]$, and the reaction is first order in $[S]$.

6.1.3 Parameter Estimation

The Michaelis-Menten equation has only two fitted coefficients, but it is a good example of the difficulties involved in accurate parameter estimation. A large literature has been developed on this topic (see reference 6). Three general approaches have been applied to the problem:

1. Graphical analysis of linear transformations of the equation
2. "Robust" methods that require few assumptions about the statistical structure of the data
3. Direct fit to the equation by nonlinear regression analysis (NLRA)

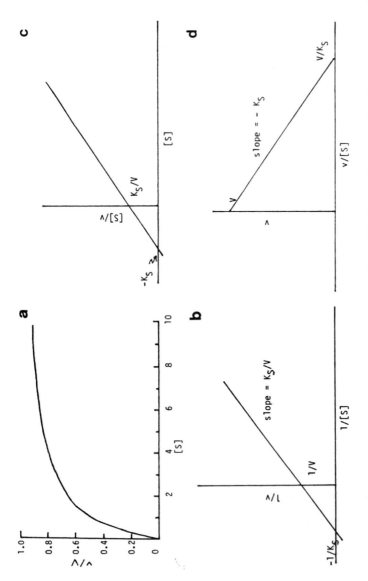

Figure 6-1. (a) v vs. [S] for Michaelis-Menten equation. v is expressed relative to V, and [S] is in units of K_S. Linear plots of enzyme rate data: (b) Lineweaver-Burk plot; (c) Hanes or Dixon plot; (d) Eadie-Hofstee plot.

Linearized Transformations

As noted above, the Michaelis-Menten equation yields a nonlinear plot of v vs. [S]. This is not useful in estimating K_S and V, because accurate fitting of an asymptotic curve is difficult, especially for data with experimental uncertainty. The historical solution to this problem has been to recast Equation 6-3 to obtain linear transformations (Figure 6-1b–d). The best known is the "double reciprocal" form of Lineweaver and Burk:[7]

$$\frac{1}{v} = \frac{1}{V} + \frac{K_S}{V}\frac{1}{[S]}$$ (6-4)

which is obtained by inverting both sides of Equation 6-3. A plot of 1/v vs. 1/[S] is a straight line with slope K_S/V, y-intercept 1/V, and x-intercept $-1/K_S$. Multiplying Equation 6-4 by [S] yields the Hanes[8] equation:

$$\frac{[S]}{v} = \frac{K_S}{V} + \frac{1}{V}[S]$$ (6-5)

A plot of [S]/v vs. [S] is linear with slope 1/V, y-intercept K_S/V, and x-intercept $-K_S$. This plot has some advantages over the Lineweaver-Burk plot and often is used in applying enzyme kinetics to aquatic systems. In some cases, enzymes are inhibited by high [S], while in others, excess S activates the enzyme. Such anomalies lie to the right on Hanes plots rather than near the y-intercept, as on Lineweaver-Burk plots. Extrapolating linear portions of the data to obtain K_S and V thus is done more easily with Hanes plots.

If Equation 6-4 is multiplied by Vv and rearranged, a third linear transformation, the Eadie-Hofstee plot,[9] is obtained:

$$v = -K_S(v/[S]) + V$$ (6-6)

A plot of v vs. v/[S] is linear with slope $-K_S$, y-intercept V, and x-intercept V/K_S. This plot has the advantage of magnifying departures from linearity that might be overlooked in the other two plots.

The reliability of the three linear transformations in estimating K_S and V has been the subject of several analyses. Dowd and Riggs[10] concluded that Lineweaver-Burk plots give deceptively good fits to unreliable data points (outliers are less noticeable than in the other plots); nonetheless, these plots give the poorest estimates of V and K_S. Based on a set of hypothetical data in which computer-generated normally distributed error was applied to values of v, they found that Hanes plots were slightly superior to Eadie-Hofstee plots, when the error in v is small (Figure 6-2). For larger errors in v, the reverse is true.

The deficiencies of linear transformations are related to problems of properly weighting experimental data. In enzyme rate studies [S] is assumed to be known

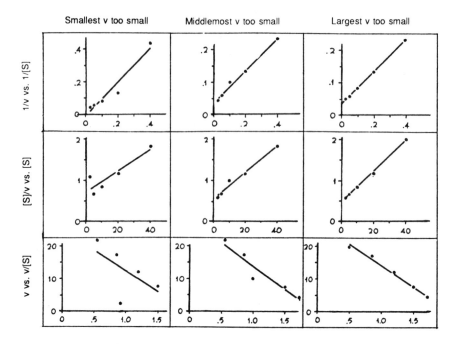

Comparison of three linear transformations in left column

Method	Slope	Intercept	Estimated K_s	V_{max}	Error of estimate (% of true value) K_s	V_{max}
1/v vs. 1/[S]	1.036	-0.0091	-114	-110	860	467
[S]/v vs. [S]	0.0262	0.712	27.2	38.2	81	27
v vs. v/[S]	-12.65	24.85	12.65	24.85	16	17

Figure 6-2.   Errors in linear transformations of enzyme data when one of five values of v is reduced by 2 units in a synthetic data set based on $K_s = 15$ and $V = 30$ units, with [S] ranging from 2.5 to 40 units, and v ranging from 4.29 to 21.82 units (unaltered "true" values). Three graphs on left: error affects only smallest v; middle graphs: only middle v affected; graphs on right: only largest v affected. Other values of v are theoretical values for the given K_S and V. Table below graphs gives analysis of K_S and V for graphs at left. [Redrawn from Dowd, J.E. and D.S. Riggs, *J. Biol. Chem.*, 210, 863 (1965). With permission of the Am. Soc. of Biochemistry and Molecular Biology.]

exactly, but v is subject to measurement error. The weight assigned (in linear regression) to a measurement, v_i, should be inversely proportional to s_i^2, the variance of v_i.[11] If measurement errors are constant (s^2 independent of v), all data points should be weighted equally. However, if equal weighting of v is correct, then 1/v values should be weighted proportional to v^4.[11] Lineweaver and Burk showed this in an early analysis of weighting problems associated with their double-reciprocal equation.[12] On the other hand, if the coefficient of variation s/v is constant (error is relative; s^2 is proportional to v^2), then the values of v should be weighted proportional to $1/v^2$, and $1/v$ should be weighted proportional to v^2. Unfortunately, we seldom know the true error distribution associated with v, although constant error often is assumed.

Li[13] reported that Hanes plots gave unreliable results for data on heterotrophic activity by aquatic microbes. He concluded that direct-fitting NLRA procedures should be used to estimate kinetic parameters for the Michaelis-Menten equation. According to Li, use of a Hanes plot is equivalent to assuming constant uncertainty in substrate turnover time ([S]/v) over all [S], but direct fit of unweighted data assumes the uncertainty in turnover rate (v) is independent of [S]. Considering the variables that can be controlled and those subject to errors, Li concluded that the latter is true in most experiments.

The Hanes and Eadie-Hofstee equations involve plots of a variable vs. a function of the same variable. Such plots can lead to spurious self-correlations and should be avoided. Under certain conditions plots of this type yield high correlation coefficients even when the original variables are completely uncorrelated.[14,15] The Eadie-Hofstee plot has the particular disadvantage that the dependent variable, v, which presumably has larger measurement errors than [S], appears on both axes. The Lineweaver-Burk procedure does not suffer from spurious self-correlations, but for other reasons it is still the poorest of the three methods for linearizing enzyme rate data. Overall, the Hanes plot probably is the best of the three methods, but none of the transformations is recommended for accurate estimates.

Robust Methods

Robust parameter estimation methods require few assumptions about the distribution of errors or statistical structure of the data. One example is the direct-linear plot method,[16] in which the Michaelis-Menten equation is rearranged to express coefficients V and K_s as the variables:

$$V = v + (v/[S])K_s \qquad (6\text{-}7)$$

For any values of v and [S] one can plot V vs. K_s as a straight line with slope v/[S] and intercepts v on the V axis and −[S] on the K_s axis (Figure 6-3). In practice the method involves drawing a line for each experimental observation to connect the measured v (plotted on the ordinate) for a given [S] (plotted as a negative value on the abscissa). This line gives all values of V and K_s that satisfy the observed v and given [S]. If there is no error associated with v and [S], similar lines drawn for all measurements will intersect at a common point, the coordinates of which give the only values of V and K_s that satisfy all observations. In practice the point of intersection is less well defined than shown in Figure 6-3 because of experimental errors, but if most of the data are good, a bad observation will be obvious as one that does not intersect near the majority. Arithmetic averaging of intersection points for all pairs of lines could be done to obtain a "best estimate" of K_s and V, but this is

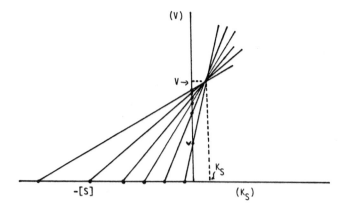

Figure 6-3.   Plot of unisubstrate enzyme rate data according to direct linear plot method.[16] Implicit axes, V and K_S, are in parentheses. Lines connect hypothetical values of v and [S] (closed circles) plotted on axes. Dashed lines give values of V and K_S.

not recommended[11] because the statistical distribution of the points is unknown. Instead, median values of intersection points for all pairs of lines should be used because they are not sensitive to outliers. The direct-linear method has the advantages of simplicity and robustness to assumptions about distribution of measurement errors and requires only simple calculations to determine V and K_S. This method also could be used for other equations with two coefficients.

Another robust method is the replicate design.[17] If one assumes the data obey Equation 6-3, one can readily calculate K_S and V by solving the simultaneous equations: $v_1 = V[S_1]/(K_S + [S_1])$ and $v_2 = V[S_2]/(K_S + [S_2])$. In practice, replicate measurements of v are made at $[S_1]$ and $[S_2]$ to obtain more reliable estimates of v_1 and v_2. If the error distribution is unknown, the median values of v should be used. If the variance in v is assumed constant, it can be shown that the optimal design for estimating K_S and V is achieved by making $[S_1] \approx K_S$ and $[S_2] \gg K_S$.[17] If for practical reasons, $[S_2]$ cannot be made much greater than K_S, the optimal design (for constant variance in v) is obtaining by making $[S_2]$ as large as possible and setting $[S_1] = [S_2]K_S/([S_2] + 2K_S)$. If error is proportional to v, the optimal design occurs by maximizing $[S_2]$ and making $[S_1]$ as low as possible, consistent with analytical sensitivity in measuring v. It thus is useful to have a preliminary estimate of one of the parameter values (K_S) one is attempting to determine.

Nonlinear Regression Analysis

All linear transformation and robust methods of parameter estimation have some drawbacks when applied to the Michaelis-Menten equation, and the best approach is to fit data directly to the equation. Direct fitting cannot be done accurately by graphical procedures (which is the reason transformed equations were developed). In contrast to the situation for linear equations, there are no analytical solutions for least-squares estimates of parameters in nonlinear equations (i.e., equations that are nonlinear in the parameters). However, numerical methods are available that fit data directly to nonlinear equations (Section 2.4.3), and these methods should be used in research using Michaelis-Menten kinetics.[18,19]

6.1.4 Reversible Reaction Kinetics for One Substrate

Because the back reaction (E + P $\rightarrow$ ES) was ignored in deriving the Michaelis-Menten equation, we restricted its validity to initial conditions where [P]=0. Thus, v, V, and [S] in previous sections all should be interpreted as initial values (v_o, V_o, and [S]$_o$). Equation 6-3 may be valid at later reaction times for irreversible reactions if P does not act as a deadend competitive inhibitor (Section 6.3.3). In the more general case, a reaction may proceed in both directions. The rate equation for this case is derived in Example 6-1.

Example 6-1. Derivation of Rate Equation for Reversible Single-Substrate Reaction.

The simplest mechanism we can consider is Equation 6-1 with both steps reversible:

(1)
$$S + E \underset{k_2}{\overset{k_1}{\rightleftharpoons}} ES \underset{k_4}{\overset{k_3}{\rightleftharpoons}} E + P$$

We can write rate equations for the three species, S, P, and ES:

(2) $-d[S]/dt = k_1[S][E] - k_2[ES]$

(3) $d[P]/dt = k_3[ES] - k_4[E][P]$

(4) $d[ES]/dt = k_1[S][E] + k_4[E][P] - k_2[ES] - k_3[ES]$

Invoking the steady-state assumption for [ES], we obtain

(5)
$$[ES] = \frac{\{k_1[S] + k_4[P]\}[E]}{k_2 + k_3}$$

From the mass balance, $E_t = [E] + [ES]$, and Equation 5 we obtain

(6)
$$[E] = \frac{(k_2 + k_3)E_t}{k_2 + k_3 + k_1[S] + k_4[P]}$$

and (7)
$$[ES] = \frac{(k_1[S] + k_4[P])E_t}{k_2 + k_3 + k_1[S] + k_4[P]}$$

Substituting Equations 6 and 7 into Equation 2 or 3, we obtain

(8)
$$v = -d[S]/dt = d[P]/dt = \frac{(k_1 k_3[S] - k_2 k_4[P])E_t}{k_2 + k_3 + k_1[S] + k_4[P]}$$

This equation can be put into a more useful form by defining $V_1 = k_3 E_t$ (as in the Michaelis-Menten equation), $V_2 = k_2 E_t$, and dividing numerator and denominator by $(k_2 + k_3)$. Defining $(k_2 + k_3)/k_1 = K_S$, as before, and $(k_2 + k_3)/k_4 = K_P$ leads to:

(9)
$$v = \frac{K_P V_1[S] - K_S V_2[P]}{K_S K_P + K_P[S] + K_S[P]}$$

Equation 9 consists of a forward component (v_1) and reverse component (v_2) such that $v = v_1 - v_2$, where

(10)
$$v_1 = \frac{V_1[S]}{K_S + [S] + [P]K_S / K_P}$$

(11)
$$v_2 = \frac{V_2[P]}{K_P + [P] + [S]K_P / K_S}$$

The single-substrate reversible rate equation (Equation 9) reduces to the Michaelis-Menten equation with initial velocity v_{10} when $[P] = 0$ (Equation 10) and to an analogous expression (Equation 11) for the reverse reaction with initial velocity v_{20} when $[S] = 0$. As the nature of the mechanism and rate Equations 9–11 indicate, the system is symmetrical. The term $[P]K_S/K_P$ in Equation 10 reflects a slowdown in forward rate caused by increasing immobilization of E_t in an enzyme-product complex as $[P]$ increases. $[S]K_P/K_S$ is interpreted similarly for the reverse reaction. The effect of $[P]$ on the forward reaction rate is clarified by writing the Lineweaver-Burk form of Equation 10:

$$\frac{1}{v_1} = \frac{1}{V_1} + \frac{K_S}{V_1}\left[1 + \frac{[P]}{K_P}\right]\frac{1}{[S]} \tag{6-8}$$

The y-intercept of a Lineweaver-Burk plot is unaffected by $[P]$, but the slope increases by a factor of $(1 + [P]/K_P)$. Plots of $1/v_1$ vs. $1/[S]$ at different $[P]$ yield a pattern called competitive inhibition (see Section 6.3.3 and Figure 6-8). This occurs whenever an inhibitor I (in this case I = P) combines with the same enzyme form as the substrate being studied.

Complete description of single-substrate reversible reactions requires four parameters: V_1, V_2, K_S, and K_P. These can be determined from initial-velocity measurements on the forward and reverse reactions, with $[P] = 0$ and $[S] = 0$, respectively. If $[E_t]$ (in mol [active site]/L) also is known, values for microscopic rate constants k_1 to k_4 can be determined. From the definitions of K_S, K_P, V_1, and V_2, it follows that $k_2 = V_2/E_t$, $k_3 = V_1/E_t$, $k_1 = (V_1 + V_2)/K_S E_t$, and $k_4 = (V_1 + V_2)/K_P E_t$. By definition, $[P]_{eq}/[S]_{eq} = K_{eq}$. Because $v = 0$ at equilibrium, it follows that

$$K_{eq} = k_1 k_3 / k_2 k_4 \qquad (6\text{-}9)$$

and from equation 9 (Example 6-1) that

$$K_{eq} = K_P V_1 / K_S V_2 \qquad (6\text{-}10)$$

Equation 6-10 is called the Haldane relationship. It reinforces the principle that the kinetic parameters of reversible enzymatic reactions are not all independent, but are constrained by the equilibrium constants for the reactions.

6.1.5 A More Realistic Single-Substrate Mechanism

The mechanism in Equation 6-1 is too simple because it has only one intermediate (ES) that breaks down to form free enzyme and free product. It is not reasonable to suppose that the complex formed on initial combination of E and P is the same as that formed by initial combination of E and S. A more realistic mechanism adds an additional complex (EP) formed by the reverse reaction

$$S + E \underset{k_2}{\overset{k_1}{\rightleftharpoons}} ES \underset{k_4}{\overset{k_3}{\rightleftharpoons}} EP \underset{k_6}{\overset{k_5}{\rightleftharpoons}} E + P \qquad (6\text{-}11)$$

Derivation of the rate equation for this mechanism is briefly summarized here. If the steady-state assumption is invoked ($d[ES]/dt = d[EP]/dt = 0$), three algebraic equations can be written, including the mass balance for E_t. These can be solved simultaneously for [ES], [EP], and [E], which then can be substituted into the overall rate equation. Solution of simultaneous equations is tedious when many terms are involved, and some short-cut methods are described in Section 6.2 and Example 6-2. Expressed in terms of microscopic rate constants, the rate of mechanism 6-11 is

$$v = \frac{(k_1 k_3 k_5 [S] - k_2 k_4 k_6 [P]) E_t}{k_2 k_5 + k_2 k_4 + k_3 k_5 + k_1 (k_3 + k_4 + k_5)[S] + k_6 (k_2 + k_3 + k_4)[P]} \qquad (6\text{-}12)$$

This equation and Equation 8 (Example 6-1) for the simpler reversible mechanism have the general form

$$v = \frac{(\varepsilon_1 [S] - \varepsilon_2 [P]) E_t}{\delta_0 + \delta_1 [S] + \delta_2 [P]} \qquad (6\text{-}13)$$

where ε_i and δ_i are coefficients of the numerator and denominator, respectively, and both lead to the same rate equation (9; Example 6-1)) when the intrinsic constants are expressed in terms of measurable kinetic parameters. In general, the introduction of any number of unimolecular (isomerization) steps between different enzyme forms does not alter the form of the rate equation. Thus, it is not possible to differentiate among isomerization mechanisms on the basis of fit of data to a given rate equation.

The coefficients in Equation 6-13 are composed of different intrinsic rate constants in the two mechanisms, and the kinetic parameters (K's and V's) thus involve different sets of microscopic constants. An interesting distinction exists between the two mechanisms. As noted above, it is possible to measure four kinetic parameters (K_S, K_P, V_1, and V_2) for a single-substrate single-product reaction. The simple mechanism has only four intrinsic rate constants, and explicit solutions for the k's are possible. The more realistic mechanism has six intrinsic contstants (more than the number of experimental parameters). Evaluation of the intrinsic rate constants thus is not possible from kinetic measurements on the net reaction.

Example 6-2. Derivation of Rate Equations for Enzymatic Reactions

Consider the single-substrate reaction with two enzyme complex intermediates:

(1)
$$S + E \underset{k_2}{\overset{k_1}{\rightleftharpoons}} ES \underset{k_4}{\overset{k_3}{\rightleftharpoons}} EP \underset{k_6}{\overset{k_5}{\rightleftharpoons}} E + P$$

The rate equation for net product formation is

(2)
$$v = d[P]/dt = k_5[EP] - k_6[E][P]$$

The **King-Altman** approach[20] rewrites the mechanism as a geometric figure (1) and takes all subset graphs of the figure that link all possible enzyme forms, using one fewer line than the number of forms. Closed loops are not allowed. For the above mechanism, three graphs are obtained, (2)–(4):

(1)   (2)   (3)   (4)

In general, a mechanism will have $m!/(n-1)!(m-n+1)!$ graphs, where m = number of interconversion steps (pairs of arrows) in the mechanism, and n = number of enzyme forms. For each enzyme form, we write an expression of form E_i/E_t = numerator/Σ, where the numerator is the sum over all graphs of the products of rate constants and concentration terms (if any) on the paths leading to that enzyme form. Σ is the sum of all numerator terms. For the above case we have

(3)
$$E/E_t = (k_2 k_5 + k_2 k_4 + k_3 k_5)/\Sigma$$

(4)
$$ES/E_t = (k_1 k_5[S] + k_1 k_4[S] + k_4 k_6[P])/\Sigma$$

(5)
$$EP/E_t = (k_2 k_6[P] + k_3 k_6[P] + k_1 k_3[S])/\Sigma$$

Equations 3–5 define the proportion of enzyme in any one form in terms of individual rate constants. Substituting into Equation 2 yields

(6)
$$v = \frac{(k_1 k_3 k_5 [S] - k_2 k_4 k_6 [P]) E_t}{k_2 k_5 + k_2 k_4 + k_3 k_5 + k_1 (k_3 + k_4 + k_5)[S] + k_6 (k_2 + k_3 + k_4)[P]}$$

The equation can be simplified by defining macroscopic constants (K's and V's), as described earlier in the text.

The rate equation for mechanism 1 also can be derived by **matrix algebra and determinants**. The mass conservation equation for E_t is

(7)
$$E_t = [E] + [ES] + [EP]$$

Rate equations for each enzyme form at steady state are written with the forms on the right sides in order of Equation 7 and common factors of enzyme forms collected together:

(8)
$$d[ES]/dt = k_1 [S][E] - (k_2 + k_3)[ES] + k_4 [EP] = 0$$

(9)
$$d[EP]/dt = k_6 [P][E] + k_3 [ES] - (k_4 + k_5)[EP] = 0$$

(10)
$$d[E]/dt = -(k_1 [S] + k_6 [P])[E] + k_2 [ES] + k_5 [EP] = 0$$

Together, Equations 7–10 "overdetermine" the system, which has only three unknowns: [E], [ES], and [EP]. Although [S] and [P] are variables, it is assumed that they can be measured and thus are not unknowns. Equations 8–10 are a system of homogeneous algebraic equations. They cannot be solved by determinants, but by substituting Equation 7, which has a nonzero solution, for one of Equations 8–10, we obtain a nonhomogeneous system that can be solved by determinants. For example, substituting Equation 7 for Equation 10 and writing the system in matrix form yields

(11)
$$\begin{bmatrix} E_t \\ 0 \\ 0 \end{bmatrix} = \begin{bmatrix} 1 & 1 & 1 \\ k_1 [S] & -(k_2 + k_3) & k_4 \\ k_6 [P] & k_3 & -(k_4 + k_5) \end{bmatrix} ([E] + [ES] + [EP])$$

Equation 11 can be solved by applying Cramer's rule to the determinant D of the 3×3 coefficient matrix. In general, for 3×3 matrices,

(12)
$$D = \begin{vmatrix} a_{11} & a_{12} & a_{13} \\ a_{21} & a_{22} & a_{23} \\ a_{31} & a_{32} & a_{33} \end{vmatrix} = a_{11} M_{11} - a_{21} M_{21} + a_{31} M_{31}$$

$$D = a_{11} a_{22} a_{33} - a_{11} a_{32} a_{23} + a_{21} a_{32} a_{13} - a_{21} a_{12} a_{33} + a_{31} a_{12} a_{23} - a_{31} a_{22} a_{23}$$

M_{ij} is the minor (matrix) associated with element a_{ij}. The determinant of the matrix in Equation 11 thus is

(13) $D = k_2 k_4 + k_2 k_5 + k_3 k_5 + k_1 (k_3 + k_4 + k_5)[S] + k_6 (k_2 + k_3 + k_4)[P]$

By Cramer's rule we can solve for any enzyme form by replacing the column of coefficients for that form by the column vector of equation values, $\begin{bmatrix} E_t \\ 0 \\ 0 \end{bmatrix}$, e.g.,

$$[E] = D_1 / D \qquad\qquad\qquad [EP] = D_3 / D$$

$$D_1 = \begin{vmatrix} E_t & 1 & 1 \\ 0 & -(k_2 + k_3) & k_4 \\ 0 & k_3 & -(k_4 + k_5) \end{vmatrix} \qquad D_3 = \begin{vmatrix} 1 & 1 & E_t \\ k_1[S] & -(k_2 + k_3) & 0 \\ k_6[P] & k_3 & 0 \end{vmatrix}$$

$$D_1 = (k_2 k_4 + k_2 k_5 + k_3 k_5) E_t \qquad D_3 = (k_1 k_3 [S] + k_2 k_6 [P] + k_3 k_6 [P]) E_t$$

The overall rate equation (Equation 2) can now be written as

(14) $$d[P]/dt = k_5[EP] - k_6[E][P] = \frac{k_5 D_3}{D} - \frac{k_6[P]D_1}{D}$$

Substitution of the above results for D, D_1, and D_3 into Equation 14 yields the same rate equation given by the King-Altman method (Equation 6).

6.1.6 The Time-Integrated Michaelis-Menten Equation

Biochemists traditionally have used differential forms of rate equations for enzyme kinetics. This is based on the recommendation of Michaelis and Menten to measure initial velocities because of confounding effects of product inhibition at longer reaction times. However, in some situations product inhibition is not important. For example, microbial uptake of nutrients and organic substrates probably is irreversible under normal conditions. The Michaelis-Menten equation can be integrated by separation of variables:

$$v = \frac{-d[S]}{dt} = \frac{V[S]}{K_s + [S]} \quad \text{or} \quad \frac{K_s}{[S]} d[S] + d[S] = -V dt$$

Integration between t = 0 and t yields

$$K_s \ln[S] / [S]_o + [S] - [S]_o = -Vt \qquad\qquad (6\text{-}14)$$

This expression can be rearranged in several ways to plot [S] as a function of t from a single experimental run. For example,[3]

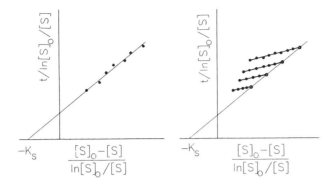

Figure 6-4.   Time-integrated plots of enzyme rate data. (a) Simple (uninhibited) model, Equation 6-15. Open circle = initial value, which is indeterminate but extrapolated from experimental points back to 0% of reaction. (b) Competitive product inhibition model, Equation 6-16. Sets of data are plotted as in (a) for various $[S]_o$. Extrapolating lines from each experiment to the abscissa value $([S]_o - [S])/ln([S]_o/[S]) = [S]_o$ generates the point $[S]_o$, $[S]_o/v_o$. The extrapolated initial values lie on straight line with slope $1/V$ and y-intercept K_S/V. (Modified and redrawn from Cornish-Bowden, A., *Principles of Enzyme Kinetics*, Butterworths, London, 1976.)

$$\frac{t}{ln[S]_o / [S]} = \frac{1}{V}\left[\frac{[S]_o - [S]}{ln[S]_o / [S]}\right] + \frac{K_S}{V} \qquad (6\text{-}15)$$

A plot of $t/ln([S]_o/[S])$ vs. $([S]_o - [S])/ln([S]_o/[S])$ is a straight line with slope $1/V$ and intercept K_S/V (Figure 6-4a). Thus, Equation 6-15 is analogous to Equation 6-5, and Figure 6-4a is analogous to a Hanes plot. For accurate estimates of K_S and V, $[S]_o$ should be more than K_S, and measurements should be made until $[S] < K_S$.

The above expression should be used cautiously because of the possibility of product inhibition, the effects of which on v cannot be distinguished from decreases caused by decreasing $[S]$ with time. A time-integrated expression for reactions exhibiting competitive product inhibition is (3):

$$\qquad\qquad\qquad\qquad\qquad\qquad\qquad\qquad\qquad\qquad\qquad (6\text{-}16)$$

$$\frac{t}{ln[S]_o/[S]} = \frac{1}{V}(1 - K_S/K_P)\left[\frac{[S]_o - [S]}{ln[S]_o/[S]}\right] + \frac{K_S}{V}(1 + [S]_o/K_P)$$

A plot of $t/ln\{[S]_o/[S]\}$ vs. $([S]_o - [S])/ln[S]_o/[S]$ yields a straight line (as with Equation 6-15), but the slope and intercept are functions of three parameters: K_S, K_P, and V. Three constants cannot be determined from a single straight line, but they can be determined from a series of experiments at several values of $[S]_o$, as shown in Figure 6-4b. Although this method is more complicated than a conventional series of initial velocity experiments, it provides useful information on product inhibition and yields more accurate estimates of $[S]_o/v_o$ than one can obtain from initial velocity studies.

Alternatively, Equation 6-16 can be cast into a linear form and solved by multiple linear regression,[21,22] and the parameter estimates obtained in this way can be used as initial values for more accurate analysis of Equation 6-16 by nonlinear regression methods. Barrio-Lage et al.[22] used this approach to analyze data on microbial degradation of trichloroethene (TCE) in microcosms containing organic-rich sediments. First-order rate constants, k_1 (h^{-1}), calculated as V/K_S, varied only over a factor of 2.5 for four samples ($3.4-8.7 \times 10^{-4} h^{-1}$), but K_S varied over a factor of five ($10-50 \ \mu M$), and K_P varied more than three orders of magnitude on the same samples ($1.1 \times 10^{-2} \ \mu M$ to $41 \ \mu M$). In addition, these workers found that TCE degradation data for microcosms containing crushed calcareous rock and water (low organic content) fit the simpler time-integrated expression (Equation 6-15) with no product inhibition. It is difficult to explain these results, especially the large variations in K_P, in terms of simple enzyme kinetics, and the approach may simply represent sophisticated "curve-fitting."

6.2 MULTISUBSTRATE REACTIONS

6.2.1 Rate Equations

Enzymatic reactions with two or three substrates and products are common, and a few even involve four substrates and/or products. The mechanisms for such reactions can be quite complicated, but if the concentrations of all but one substrate are held constant, it still will fit the Michaelis-Menten equation. The kinetic parameters in the equation then are functions of more fundamental constants and concentrations of the nonvaried substrates. Although each substrate in such reactions can be fit to the simple Michaelis-Menten equation, complete rate expressions are useful for making inferences about reaction mechanisms. For example, the type of product inhibition patterns obtained when $[S_i]$ and $[P_i]$ are varied and v_o is measured depends on the reaction mechanism. Complete rate equations have been derived for many multisubstrate reactions.[3,23-26]

Derivation of rate equations for such reactions follows the principles outlined for the single-substrate case, but they are complicated by addition of more intermediates (each representing a different enzyme form), and more terms occur in the rate equations. In addition, several mechanisms (and rate equations) may be written for a stoichiometric expression. The general approach, however, is straightforward: we write a rate equation for each intermediate (each enzyme form), invoke the steady-state assumption, and solve the resulting algebraic equations. Solution of the equations is facilitated by writing them in matrix form; the coefficients multiplying concentrations of each enzyme form are elements of the matrix. The determinant of the matrix can be obtained by conventional methods (Example 6-2), but this approach is laborious for mechanisms that involve many enzyme forms and many simultaneous equations, each with a large number of terms. A simpler graphical technique[20] facilitates solutions for more complicated mechanisms (Example 6-2). The complete rate equations are long and complicated for most multisubstrate mechanisms, especially when both forward and reverse reactions are considered. A general approach for deriving such equations was developed by Cleland.[26,27]

Our examination of multisubstrate kinetics is limited to bisubstrate, biproduct reactions. Rate equations are not derived here, but derivations are available elsewhere. For a bisubstrate reaction,

$$A + B = P + Q \qquad (6\text{-}17)$$

initial rates in the forward and reverse directions are[23]

$$v_1^0 = \frac{V_1[A][B]}{K_{iA}K_B + K_B[A] + K_A[B] + [A][B]} \qquad (6\text{-}18a)$$

$$v_2^0 = \frac{V_2[P][Q]}{K_{iQ}K_P + K_Q[P] + K_P[Q] + [P][Q]} \qquad (6\text{-}18b)$$

The singly subscripted K's are Michaelis constants for subscripted species, and K_{iA} and K_{iQ}, are dissociation constants for EA and EQ, respectively. This terminology assumes that A is the first substrate to form a complex with E, and $K_{iA} = k_1/k_2$. The full rate equation (including forward and reverse components) for this reaction is more complicated.[26] A complete kinetic description of bisubstrate reactions requires four parameters for each direction (V_1, K_A, K_{iA}, and K_B for the forward reaction). K_{iA} and K_{iQ} are called inhibition constants, and they reflect product inhibition by the subscripted species. Inhibition constants can be written for each substrate and product in a reaction sequence; the composition of the constants (in terms of intrinsic rate constants) depends on the mechanism and varies for different substrates in a reaction.

Biochemists traditionally have analyzed bisubstrate rate data by Lineweaver-Burk plots. The Lineweaver-Burk form of Equation 6-18a is

$$\frac{1}{v_i} = \frac{1}{V_1}\left[1 + \frac{K_A}{[A]} + \frac{K_B}{[B]} + \frac{K_{iA}K_B}{[A][B]}\right] \qquad (6\text{-}19)$$

This equation can be rearranged to factor out one substrate to be considered a variable while the other one is held constant:

$$\frac{1}{v_1} = \frac{K_A}{V_1[A]}\left[1 + \frac{K_{iA}K_B}{K_A[B]}\right] + \frac{1}{V_1}\left[1 + \frac{K_B}{[B]}\right]; \quad \frac{1}{v_1} = \frac{K_B}{V_2[B]}\left[1 + \frac{K_{iA}}{[A]}\right] + \frac{1}{V_1}\left[\frac{K_B}{1+[A]}\right] (6\text{-}20)$$

Hanes plots are preferred over Lineweaver-Burk plots for statistical reasons. For variable substrate A, the corresponding equation is

$$\frac{[A]}{v_1} = \frac{K_A}{V_1}\left[1 + \frac{K_{iA}K_B}{K_A[B]}\right] + \frac{[A]}{V_1}\left[1 + \frac{K_B}{[B]}\right] \qquad (6\text{-}21)$$

The slope term in a Lineweaver-Burk plot thus is the intercept term of a Hanes plot, and the intercept term of a Lineweaver-Burk plot is the slope of a Hanes plot. For consistency with the biochemical literature, Lineweaver-Burk plots are used in the following discussion.

Equation 6-20 defines the four kinetic parameters for the forward reaction in terms of the slopes and y-intercepts of plots of $1/v_1$ vs. $1/[A]$. Kinetic evaluation of multisubstrate reactions thus uses the isolation approach described in Chapter 2 for multireactant reactions; i.e., one reactant is varied while others are held constant. The values of K_S and V computed from such experiments are apparent constants that are functions of the reaction's intrinsic kinetic parameters and concentrations of nonvaried substrates. In Equation 6-20, K_S and V are functions of K_A, K_{iA}, K_B, and V_1 and [B]. Values of the kinetic parameters can be evaluated by a series of measurements of v_{10} vs. [A] at several levels of constant [B], followed by similar measurements of v_{10} vs. [B] at several levels of constant [A]. Appropriate design reduces this to a factorial experiment, but optimal parameter estimation may require a sophisticated experimental design.[28] The slopes and intercepts of the two "primary" plots of the rate data (Figure 6-5a,b) are replotted vs. $1/[A]$ and $1/[B]$, as in Figure 6-5c–f. These secondary plots provide eight measurements (four slopes and four intercepts) from which the kinetic parameters K_A, K_{iA}, K_B, and V_1 can be obtained with some redundancy.[23]

6.2.2 Mechanisms for Bisubstrate Reactions

Bisubstrate reactions can proceed by several pathways (Figure 6-6), some of which follow the same general rate equation (Equation 6-18a), or simplifications of it. The major types of mechanisms are described below, along with diagnostic procedures based on kinetic expressions to distinguish among the mechanisms.

In **ping-pong** mechanisms (Equation 3; Figure 6-6), at least one product is released before all substrates have combined with the enzyme. For bisubstrate ping-pong mechanisms, this means the enzyme or a co-enzyme is converted into a different chemical form E'. As a result, denominator terms $K_{iA}K_B$ and $K_{iQ}K_P$ are lost from the general bisubstrate rate equations (Equation 6-18a,b), and interaction terms involving [A][B] are eliminated from the reciprocal equations, yielding

$$\frac{1}{v_1} = \frac{1}{V_1}\left[1 + \frac{K_A}{[A]} + \frac{K_B}{[B]}\right] = \frac{K_A}{V_1}\frac{1}{[A]} + \frac{1}{V_1}\left[1 + \frac{K_B}{[B]}\right] \qquad (6\text{-}22a)$$

$$\frac{1}{v_2} = \frac{1}{V_2}\left[1 + \frac{K_P}{[P]} + \frac{K_O}{[Q]}\right] \qquad (6\text{-}22b)$$

Double-reciprocal plots of $1/v_1$ vs. $1/[A]$ or vs. $1/[B]$ are parallel lines with identical slopes for all values of fixed substrate in bisubstrate ping-pong mechanisms (Figure 6-5g), and Hanes plots of $[A]/v_1$ vs. [A] or $[B]/v_2$ vs. [B] have a common y-intercept

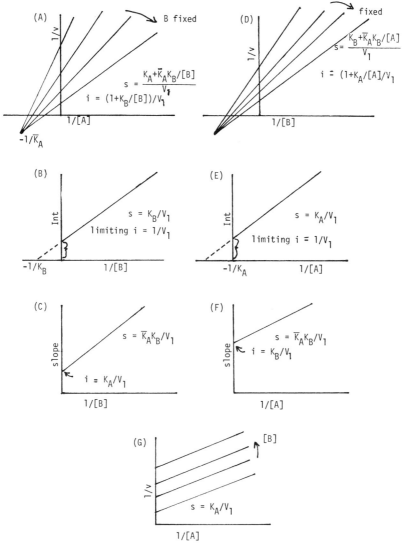

Figure 6-5.   Reciprocal plots for bisubstrate reactions. (A) Primary Lineweaver-Burk plot for varying [A] with [B] fixed at several levels; (B) intercept (i) of (a) vs. 1/[B]; (C) slope (s) of (A) vs. 1/[B]; (D)–(F) analogous plots for varying [B], fixed [A]; (G) primary plot for ping-pong mechanism, Equation 6-22a. Similar graph occurs for 1/v vs. 1/[B], with $s = \bar{K}_B/V_1$. (After Mahler, H.R. and E.H. Cordes, *Biological Chemistry*, 2nd ed., Harper & Row, New York, 1971.)

and varying slope. Ping-pong mechanisms also can be identified by inhibition patterns of products P and Q (Section 6.3.3).

In **sequential** mechanisms all substrates are added to the enzyme before any product is released. The two main types of sequential mechanisms are **ordered** and **random**. By convention, substrate "A" combines with the free enzyme before "B", A is called the leading substrate. Similarly, the first product is designated P. The

Ping-pong mechanisms are of the general type:

$$
\text{(1)} \qquad A + E \underset{k_2}{\overset{k_1}{\rightleftharpoons}} (EA \rightleftharpoons E'P) \underset{k_4}{\overset{k_3}{\rightleftharpoons}} P + E'
$$

$$
\text{(2)} \qquad E' + B \underset{k_6}{\overset{k_5}{\rightleftharpoons}} (E'B \rightleftharpoons EQ) \underset{k_8}{\overset{k_7}{\rightleftharpoons}} Q + E
$$

Such sequences also can be portrayed by a line graphic notation.[23] Reactants and products are shown with vertical arrows to or from a line representing the enzyme forms and complexes:

$$
\text{(3)} \qquad
\begin{array}{ccccc}
A & P & B & & Q \\
\downarrow & \uparrow & \downarrow & & \uparrow \\
\hline
E & (EA \rightleftharpoons E'P) & E' & (E'B \rightleftharpoons EQ) & E
\end{array}
$$

Ordered mechanisms are of the general type:

$$
\text{(4)} \qquad A + E \underset{k_2}{\overset{k_1}{\rightleftharpoons}} AE,
$$

$$
\text{(5)} \qquad AE + B \underset{k_4}{\overset{k_3}{\rightleftharpoons}} (AEB \rightleftharpoons QEP) \underset{k_6}{\overset{k_5}{\rightleftharpoons}} QE + P,
$$

$$
\text{(6)} \qquad QE \underset{k_8}{\overset{k_7}{\rightleftharpoons}} Q + E
$$

or

$$
\text{(7)} \qquad
\begin{array}{ccccc}
A & B & & P & Q \\
\downarrow & \downarrow & & \uparrow & \uparrow \\
\hline
E & AE & (AEB \rightleftharpoons QEP) & QE & E
\end{array}
$$

The rapid-equilibrium random mechanism involves rapid, reversible formation of binary complexes with either A or B. The rate-limiting step is interconversion of the ternary complex:

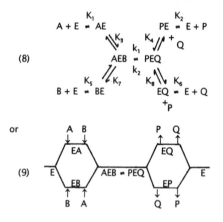

(8)

or

(9)

Figure 6-6.   Mechanisms for bisubstrate reactions.

second product (Q) must be the product of A, since both species combine with enzyme form E. Ordered bisubstrate, biproduct (bi-bi) reactions (Equation 7, Figure 6-6) follow Equation 6-18. Many enzyme-catalyzed redox reactions follow ordered bi-bi mechanisms. In reductions of organic substrates, the leading reactant is the electron-transporting intermediate NADH or NADPH, nicotinamide adenine dinucleotide (phosphate), and the second product is NAD^+ or $NADP^+$; the reverse is true for oxidations.

To evaluate the kinetic parameters for an ordered bi-bi reaction, it is necessary to know which substrate is the leading one. This can be determined from equilibrium binding studies or analysis of product inhibition patterns. Only the leading substrate is competitively inhibited by its product (Q) (see Section 6.3.3). The slopes and y-intercepts of primary reciprocal plots of $1/v_{10}$ vs. $1/[A]$ and $1/[B]$ (Figure 6-5a, b) have the same interpretations relative to the kinetic parameters, and one can evaluate V_1, K_A, and K_B from secondary plots of y-intercepts vs. $1/[A]$ and $1/[B]$ (Figure 6-5c,e) without knowing whether A or B is the leading substrate. However, the meaning of the horizontal coordinate of the intersection point for lines in primary plots depends on whether the leading or second substrate is varied (Figure 6-5a,b). Consequently, the identity of A must be known to evaluate K_{iA} from a plot of $1/v_{10}$ vs. $1/[A]$.

If AEB and QEP in ordered mechanisms are rapidly interconvertible and interconversion rates can be ignored, the mechanism has only eight microscopic rate constants. Because eight parameters can be determined from the plots in Figure 6-5, the microscopic rate constants can determined from measured parameters (if E_t is known). The **Theorell-Chance** mechanism is a special case of sequential ordered mechanisms in which the concentration of ternary complex (AEB = QEP) is very low and the first product appears to be formed directly from collision of B with EA. This can occur if $k_5 >> k_7, k_4 >> k_2, (k_5 + k_7) > k_6$, and $(k_2 + k_4) >> k_3$. Reactions having this mechanism yield primary plots like other sequential mechanisms, but have product inhibition patterns like ping-pong mechanisms, i.e., competitive inhibition of A by Q (not P) and of B by P (not Q) (see Table 6-1). Such a mechanism was proposed for the nonenzymatic oxidation of HS^- by O_2 in the presence of a catalytic Co^{II} phthalocyanine chelate.[29]

Sequential mechanisms with alternative pathways of substrate addition or product release are called random; the simplest is the rapid-equilibrium random mechanism. It involves rapid, reversible formation of binary E-S complexes with either reactant, and breakdown of the ternary complex is the rate-limiting step. The K's in Equation 8 (Figure 6-6) are dissociation constants for each step (ratios of k_{rev} to k_{for}). The rate equation becomes complicated if ternary complex interconversion is not rate limiting. If it is rate limiting, the overall rate equations for forward and reverse reactions are given by Equation 6-18, and $V_1 = k_1E_t$ and $V_2 = k_2E_t$. Furthermore, each kinetic parameter is equal to an equilibrium constant for one of the individual steps.[23] The ratio K_{iA}/K_A determines the intersection point of lines on double reciprocal plots (Figure 6-5a,b). The usual case is $K_{iA} > K_A$, and the lines intersect above the x-axis on reciprocal and Hanes plots. The slopes in plots of v_{10} vs. $1/[A]$ increase more rapidly than the y-intercepts for increasing $[B]$. This implies that addition of one substrate facilitates addition of the second. From the relationships between the dissociation constants and the kinetic parameters, if $K_{iA} > K_A$, then $K_1 > K_7$. Thus, EA dissociates to E + A more readily than AEB dissociates to EB + A.

6.3 ENZYME INHIBITION

6.3.1 Classes of Inhibitors

Three major classes of enzyme inhibitors exist: alternate substrates, products, and dead-end inhibitors. Conventional inhibitors affect enzyme activity by direct

Table 6-1. Product Inhibition Patterns for Bisubstrate-Biproduct Mechanisms (A + B → P + Q)[a]

Mechanism	Inhibitory product	Fixed substrate A		Variable substrate B	
		U[b]	S[b]	U	S
Ordered bi-bi	P	NC	UC	NC	NC
(Equation 6-27)	Q	C	C	NC	—
Ping-pong	P	NC	—	C	C
(Equation 6-25)	Q	C	C	NC	—
Random					
Rapid equilibrium	P or Q	NC[c]	NC	NC[c]	NC
Rapid equilibrium					
random (Equation 6-28)	P or Q	C	—	C	—
Theorell-Chance	P	NC	—	NC	—
	Q	C	C	C	C

After Mahler, H.R. and E.H. Cordes, *Biological Chemistry,* 2nd ed., Harper & Row, New York, 1971.

[a] NC = noncompetitive; UC = uncompetititve; C = competitive; — = no inhibition.
[b] U = unsaturated with fixed substrate; S = saturated with fixed substrate.
[c] Reciprocal plots are nonlinear.

interaction with the active site; allosteric inhibitors modify enzyme activity by binding at sites remote from the catalytic site, thus causing conformational changes (see Section 6.3.4). Alternate substrates are chemically similar to the normal substrate and form alternate products that also can act as inhibitors. Products inhibit forward reactions by recombining with the enzyme form from which they were released, thus increasing the steady-state concentration of that form. Dead-end inhibitors combine with one or more forms of an enzyme, but do not undergo further reaction. The net effect of inhibitors is to redistribute E_t among its forms and complexes, decreasing the amount of enzyme in the form of the rate-limiting step. Enzymatic reactions also may be inhibited by direct reaction of S with another soluble species (forming an unreactive complex). Enzyme inhibition is important physiologically as a means for cells to control reaction rates and maintain homeostasis; e.g., product inhibition allows cells to regulate product pools. Enzyme inhibition patterns are of analytical importance in that they provide the chief means of establishing the order of reactant addition and product release.

6.3.2 Substrates as Self-Inhibitors: The Haldane Equation

High concentrations of the normal substrate inhibit some reactions. In such cases, reaction rates reach a maximum at some intermediate [S] and then decline as [S] increases. For single-substrate reactions this behavior is described by the Haldane mechanism:[30]

$$E + S = \overset{\overset{\textstyle S}{\textstyle +}}{\underset{\underset{\textstyle ES_2}{\Updownarrow}}{ES}} \rightarrow E + P \qquad (6\text{-}23)$$

Substrate inhibition is important in describing the kinetics of microbial growth on compounds like phenols, which are toxic at moderate to high concentrations (this subject is treated in Section 6.6.9).

Example 6-3. Rate Equation for Photosynthesis Based on the Haldane Equation

Photosynthesis is a highly complicated process, but it can be described by relatively straightforward models based on enzyme kinetics. The two main products are O_2 and reduced (organic) carbon, and both are used in measuring rates of photosynthesis in aquatic systems. The carbon-fixation step can proceed independently (in the dark) if a source of reducing power (biochemical intermediates like $NADPH_2$) is built up in the light. As a result, it is difficult to relate carbon fixation directly to light intensity. O_2 production is intrinsically connected to the production of reducing power in the light steps that occur in a section of photosynthetic apparatus called Photosystem II.

The rate of conversion of light energy into chemical energy by plants, as measured by O_2 production, is proportional to light intensity (I) at low I, but reaches a peak or plateau at I near or below full sunlight, and then declines with increasing I. This pattern is similar to that for substrate inhibition of enzymatic reactions. The Haldane mechanism and equation have been used to describe this mechanistically.[31] The process is modeled by the scheme

(1)
$$T \underset{k_{-1}}{\overset{I.k_1}{\rightleftharpoons}} T^* \xrightarrow{H_2O.A.k_2} T + O_2 + AH_2$$

with T^* reversibly (k_{-3}, $I.k_3$) connected to T^{**}.

T represents the trap for light energy (a structure including the main photosynthetic pigment chlorophyll a), I is light irradiance (Ei m^{-2} h^{-1}), and A is a biochemical acceptor of reduced H from water (e.g., NADP). According to the scheme, T absorbs a photon of light and is converted to an excited state T^* that can accept electrons from water, thus producing O_2 and AH_2 in a second, irreversible step. Alternatively, T^* can return to the ground state at rate k_{-1} by fluorescence and other processes, without splitting water, or it can accept another photon and become an unreactive excited state T^{**} at rate k_3. The competing reaction to produce T^{**} also is a reversible step, but it prevents a part of the pigment trap from participating in O_2 production.

The specific rate of oxygenic photosynthesis (mmol O_2 [mg chl a]$^{-1}$ h^{-1}) is $P = k_2T^*$. Differential equations can be written for T, T^*, and T^{**}, and a mass balance for T_T is $T_T = T + T^* + T^{**}$. The rate equation for this mechanism is easily derived by the methods described in Example 6-2:

(2)
$$P(I) = \frac{vI}{K_1 + I + I^2/K_2}$$

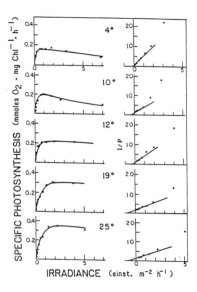

IRRADIANCE (einst. m⁻² h⁻¹)

Figure 6-7.   Effects of light intensity on oxygenic photosynthesis at five temperatures. Curves at left were fit to Equation 2 of Example 6-3, from parameter estimates obtained from I/P vs. I (Hanes-type) plots on right. [From Megard, R.O., D.W. Tonkyn, and W.H. Senft, Jr., *J. Plankton Res.*, 6, 325 (1984). With permission.]

where $v = k_2 T_T, K_1 = (k_{-1} + k_2)/k_1$, and $K_2 = k_{-3}/k_3$. Multiplication of Equation 2 by the chlorophyll concentration yields the volumetric rate of photosynthesis. Adjustment of inhibition parameter K_2 can change the shape of photosynthesis-light curves to match those measured for various algae (Figure 6-7). Whether the mechanism of light inhibition of oxygenic photosynthesis is as simple as implied by Equation 1 is unknown, and other mechanisms have been postulated based on flash photolysis experiments (see references in 31). However, these mechanisms lead to steady-state equations of the same form as Equation 2, which thus can be accepted as reasonable on both mechanistic and predictive grounds.

6.3.3 Kinetic Patterns of Inhibition

Inhibition patterns usually are evaluated by Lineweaver-Burk plots (in spite of the statistical problems mentioned earlier). Three major classes of effects are observed (Figure 6-8). An inhibitor may affect the slope, y-intercept, or both parameters. These patterns are called competitive, uncompetitive, and noncompetitive (or mixed) inhibition, respectively. From the nature of the Lineweaver-Burk plot, it is clear that competitive inhibition does not affect V, since the y-intercept defines 1/V. Saturating levels of substrate thus can overcome competitive inhibitors. Conversely, no amount of substrate is able to overcome effects of uncompetitive inhibitors, and V decreases as [I] increases. Noncompetitive inhibition is simply the sum of the first two types. Product inhibition adds terms containing [P] to the denominator of the rate equation, and dead-end inhibitors add terms in [I] to the denominator. Rate equations can be modified to account for dead-end inhibitors

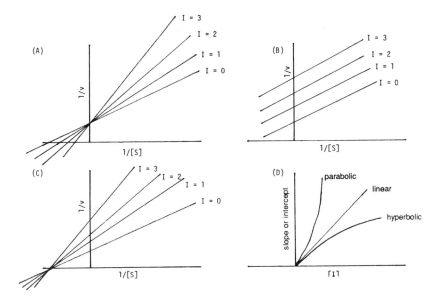

Figure 6-8.   Enzyme inhibition patterns: (a) competitive; (b) uncompetitive; (c) simple linear noncompetitive (I = 1–3 represent increasing inhibitor concentrations); (d) patterns of effects on slope and intercept vs. [I].

(provided the enzyme form(s) that combine with I are known) simply by multiplying the denominator terms corresponding to those forms by $(1 + [I]/K_{i1})$. K_{i1} is the dissociation constant for the complex of that form with I. Alternate substrate inhibitors have more complicated effects because they induce alternative reaction sequences and a random mechanism. In such cases terms containing [I] can appear in both the numerator and denominator.

The slopes and y-intercepts from primary Lineweaver-Burk plots (Figure 6-8a–c) for the inhibition patterns can be plotted vs. [I], and the resulting secondary plots (Figure 6-8d) may be linear, parabolic, or hyperbolic. To distinguish among these types, the nature of secondary plot is prefixed to the inhibition pattern (e.g., linear competitive). Linear patterns are the easiest to interpret. Physical explanations for the patterns are described elsewhere.[23,27]

For single-substrate and pseudo-single-substrate reactions (where all but one reactant are fixed), the three types of inhibition are described by the following modified Michaelis-Menten and Lineweaver-Burk equations:

Competitive:
$$v = \frac{VK_{il}[A]}{K_A K_{il} + K_A[I] + K_{il}[A]} \qquad (6\text{-}24a)$$

$$\frac{1}{v} = \frac{K_A}{V}\left[1 + \frac{[I]}{K_{il}}\right]\frac{1}{[A]} + \frac{1}{V} \qquad (6\text{-}24b)$$

Uncompetitive:
$$v = \frac{VK_{i2}[A]}{K_A K_{i2} + [A]K_{i2} + [A][I]}$$
(6-25a)

$$\frac{1}{v} = \frac{K_A}{V[A]} + \frac{1}{V}\left[1 + \frac{[I]}{K_{i2}}\right]$$
(6-25b)

Noncompetitive
$$v = \frac{VK_{i1}K_{i2}[A]}{K_A K_{i2}(K_{i1} + [I]) + K_{i1}[A](K_{i2} + [I])}$$
(6-26a)

$$\frac{1}{v} = \frac{K_A}{V}\left[1 + \frac{[I]}{K_i}\right]\frac{1}{[A]} + \frac{1}{V}\left[1 + \frac{[I]}{K_{i2}}\right]$$
(6-26b)

K_{i1} and K_{i2} are dissociation constants for different complexes of I with E. The Lineweaver-Burk forms clearly show the nature of the inhibitory effect (on the slope in Equation 6-24b, the intercept in Equation 6-25b, and both in Equation 6-26b). In noncompetitive inhibition, the lines on Lineweaver-Burk plots always intersect to the left of the ordinate. If the lines intersect on the x-axis (simple linear noncompetitive inhibition), $K_{i1} = K_{i2}$, and Equation 6-26b simplifies to

$$\frac{1}{v} = \left[1 + \frac{[I]}{K_i}\right]\left[\frac{K_A}{V[A]} + \frac{1}{V}\right]$$
(6-26c)

For single-substrate reactions, the inhibition patterns have the following interpretation. A dead-end inhibitor that combines with the free enzyme (with which S combines) yields competitive inhibition. Saturation by S overcomes the effects of I, and V is not affected. Uncompetitive inhibition is rare for single-substrate reactions, but could occur if I combined only with ES, leading to an unreactive ternary complex, EIS. Noncompetitive inhibition may result if I combines with E at a site different from the active site, so that both EI and EIS can be formed. If EI has a lower affinity for S and a lower reaction rate, both the slope and intercept of a Lineweaver-Burk plot are affected.

Interpretation of inhibition patterns for multisubstrate reactions is more complicated, but two simple rules[23,24,27] aid in relating inhibition patterns to reaction mechanisms:

1. I affects the slopes of reciprocal plots when it and the variable substrate (S_v) combine with the same enzyme form or are separated in the reaction sequence only by reversible steps so that each can affect the amount of E available to the other by displacing intervening equilibria. Addition of I thus lowers the rate of the step involving S_v, but this can be overcome by raising [S_v].

2. I affects the intercept of a reciprocal plot when it and S_v combine with different forms of E. The presence of I lowers the amount of E available for distribution among the enzyme forms in a way that saturation with S_v cannot overcome.

Release of product at $[P] = 0$ in initial-velocity studies and the addition of saturating substrate are irreversible steps. These effects can occur separately, leading to competitive and uncompetitive inhibition, or jointly, leading to noncompetitive patterns. Application of these rules in interpreting product inhibition patterns for bisubstrate reactions is illustrated in Table 6-1. Equations 6-23 to 6-25 have been used widely to analyze mechanisms of enzymatic reactions, as well as to analyze inhibition patterns of metabolic activity in microbial cultures (Section 6.6.9).

6.3.4 Allosteric Modifiers

Allosteric modifiers affect the catalytic activity of enzymes either positively (activators) or negatively (inhibitors), even though they bind to enzymes at positions remote from the active site. Such compounds have been found to regulate many key metabolic enzymes,[23] and they cause anomalous kinetic behavior; in particular, plots of v vs. [S] yield sigmoid, rather than hyperbolic, curves (Figure 6-9). The degree of sigmoid curvature depends on the concentration and nature of the allosteric agent. In general, a sigmoid v vs. [S] curve implies cooperative behavior; i.e., binding of each molecule of the substrate increases the affinity of the enzyme for next. Such behavior was first recognized in the binding of O_2 by hemoglobin, but many other examples have been found in the past 25 years. Enzymes susceptible to this behavior typically consist of several protomeric units (polypeptide chains) aggregated together into an oligomeric structure.[*] Binding of allosteric agents at specific sites on each protomeric unit is presumed to cause conformational changes that enhance or inhibit enzyme activity. The concerted transition (all-or-none) equilibrum binding model of Monod et al.[32] is widely used to analyze the kinetics of allosteric interactions.

The cooperativity feature of allosteric interactions usually is expressed in terms of the Hill equation, which was first used to explain cooperativity of O_2 binding by hemoglobin:[33]

$$y = Kp^n / (1 + Kp^n) \qquad (6\text{-}27a)$$

$$\log y/(1 - y) = \log(-Kp^n) \qquad (6\text{-}27b)$$

y is the fractional saturation of hemoglobin with O_2, p is the partial pressure of O_2, n is the number of O_2 molecules bound per hemoglobin molecule, and K is a binding constant. In enzyme kinetics an equivalent Hill equation is[23]

$$\log v / (V - v) = n \log[S] - \log K \qquad (6\text{-}27c)$$

---

[*]Protein structure is described in terms of four levels of organization. Primary structure is the sequence of covalently bonded amino acids in polypeptide chain(s). Secondary structure orders peptide chains into helices or sheets by hydrogen bonding between carbonyl or amide groups on amino acids. Tertiary structure is the three-dimensional position of peptide chains, which is determined by hydrogen bonding and other weak interactions between peptide segments and covalent links between sulfhydryl groups on nonadjacent amino acids. Quaternary structure aggregates peptide units into complex enzymes. Small changes in temperature affect weak structural interactions most readily.

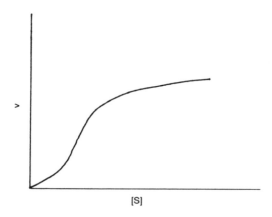

Figure 6-9. Pattern of allosteric inhibition: v vs. S. (After Mahler, H.R. and E.H. Cordes, *Biological Chemistry*, 2nd ed., Harper & Row, New York, 1971.)

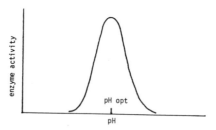

Figure 6-10. Generalized effects of pH on enzyme activity.

For ideal cases, n is the slope of a *log-log* plot of v/(V − v) vs. [S] and is equal to the number of binding sites (Figure 6-9), but actual plots often deviate substantially from linearity. When n is greater than 1, binding is cooperative; when n = 1, there is no cooperativity, and normal hyperbolic plots of v vs. [S] are obtained. In the rare case of n less than 1, there is "negative cooperativity".

6.3.5 Effect of pH on Enzymatic Reactions

In general, enzymatic reactions exhibit a narrow pH range of optimum activity within the range of physiological conditions (Figure 6-10). Three phenomena contribute to the pH optimum and the bell-shaped curve of activity vs. pH. First, the substrate may have varying degrees of protonation with different affinities for the enzyme. Second, pH affects basic protein structure and may decrease the binding of coenzymes or alter the physical configuration of the active site. Third, pH may affect the substrate-binding abilities of the enzyme and the reactivity of the ES complex. For example, the active site may exist in several protonated forms, only one of which combines with substrate. Mathematical treatment of the latter effects has been developed for unisubstrate reactions under conditions where the first two classses of effects are ignored.[23,25] Quantitative models of pH effects are complicated for multisubstrate reactions, and it is sufficient to note here that because enzymatic reactions are pH sensitive, the reaction medium must be well buffered at an appropriate pH in kinetic studies.

6.4 THEORETICAL ASPECTS OF ENZYME CATALYSIS

6.4.1 Effect of Temperature on Enzyme-Catalyzed Reactions

Within fairly narrow temperature limits, enzymatic reactions obey the Arrhenius equation, but above and below these limits activity drops rapidly because of changes in enzyme stability. The fact that enzymes are proteins makes them subject to temperature-dependent changes in structure. Enzymes exist in one of two possible states: a native state with catalytic activity, and a denatured state of low or no catalytic activity, in which the active site is no longer intact. The lower temperature limit may reflect conformational changes induced by changes in the structure of water, loss of enzyme or reactant solubility at low temperature, or loss of reactant mobility as water freezes. The upper limit reflects a loss of enzyme structural integrity (denaturation) over a narrow temperature range. These trends give rise to a typical relationship between activity and temperature (Figure 6-11a).

The concepts of activation energy and transition states apply to enzymatic reactions, but even single-substrate reactions proceed through several elementary steps. Consequently, treatment of reaction energetics rapidly becomes complicated, and interpretion of phenomenological variables, like the overall E_{act} in terms of elementary processes, is not possible. Figure 6-11b illustrates the energy diagram for a single-substrate reaction. Even the simplest case has three steps and associated energy barriers: (1) formation of ES from free E and S; (2) conversion of ES to EP; and (3) dissociation of EP to free E and P. The second step is the actual catalyzed chemical reaction. Complete description of the reaction requires seven values of each thermodynamic function (ΔG, ΔH, ΔS) in each direction.[23] Each step is characterized by ΔH's of reaction and activation for forward and reverse directions, and there is a ΔH_r for both directions of the overall reaction.

The 14 ΔH functions are related as sums or differences, and only six independent measures (six elementary rate constants) are needed to determine the rest (see Figure 6-11b). Enthalpies are determined by plotting the *log* of rate or equilibrium constants vs. 1/T (Section 3.2). We stated earlier that it is not possible to distinguish kinetically between the simplest two-step unisubstrate reaction sequence (Equation 6-1) and sequences involving unimolecular isomerization steps (Equation 6-10), and we further noted that only four rate constants can be determined from steady-state kinetic studies on unisubstrate reactions. Consequently, two further measurements are required to define the system: equilibrium constants for binding of S and P, or rates of complex formation or dissociation. In the energy diagram of Figure 6-11b, the second step has the largest E_{act} and is rate limiting. This is not always the

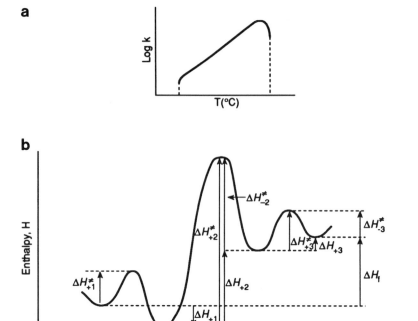

Figure 6-11.   (a) Generalized effects of temperature on rates of enzyme reactions. Upper limit caused by denaturation, lower limit by solubility, denaturation, or freezing of solution. Middle range obeys Arrhenius equation. (b) Enthalpy changes for single substrate enzymatic reaction. (Redrawn from Mahler, H.R. and E.H. Cordes, *Biological Chemistry*, 2nd ed., Harper & Row, New York, 1971. With permission.)

case; binding of S or dissociation of P also can be rate limiting. When E_{act} of one step is much greater than the E_{act} for all other steps, it controls the overall reaction rate, and the observed E_{act} tends to be about equal to the E_{act} for that step.

6.4.2 Enhanced Transition-State Binding: The Basis for Enzyme Catalysis

Enzymes are highly potent catalysts. Their effectiveness and selectivity are in many ways truly astounding, and biochemists long have sought to understand how they work. This section develops an understanding of enzyme catalysis in terms of transition-state theory, from which it can be shown that the catalytic activity of enzymes results from the fact that they bind transition states more tightly than they bind substrate(s).[34-37]

Consider the following unisubstrate enzyme-catalyzed reaction:

$$E + S \xrightleftharpoons{K_N^{\ddagger}} E + S^{\ddagger'} \rightleftharpoons E + P$$

$$\Updownarrow K_{Seq} \qquad \Updownarrow K_{Teq} \qquad \Updownarrow$$

$$ES \xrightleftharpoons{K_E^{\ddagger}} ES^{\ddagger} \rightleftharpoons EP \tag{6-28}$$

K_{Seq} and K_{Teq} are equilibrium constants for association of substrate S and transition-state $S^{\ddagger'}$ with E to form ES and $ES^{\ddagger}$, respectively, and $K_N^{\ddagger}$ and $K_E^{\ddagger}$ are equilibrium constants for transition-state formation in the nonenzymatic and enzymatic reactions, respectively. The equilibrium constants are algebraically related in the following way:

$$K_{Teq} / K_{Seq} = K_E^{\ddagger} / K_N^{\ddagger} \tag{6-29}$$

The equilibrium constant between a reactant and its transition state is proportional to the rate constant of the reaction ($k_i = \kappa_i \nu_i K_i^{\ddagger}$), where κ is the transmission coefficient, and ν is the frequency of the "normal mode" oscillation of the transition state along the reaction coordinate (Section 3.5.2). If $\kappa_E \nu_E \approx \kappa_N \nu_N$ (which is reasonable but unproven[37]), Equation 6-29 implies that the effectiveness of enzyme catalysis, expressed as the ratio k_E/k_N, is equivalent to a tighter binding of the transition state compared with the substrate:

$$k_E / k_n \approx K_{Teq} / K_{Seq} \tag{6-30}$$

Typical values for k_E/k_N are in the range 10^8 to 10^{14} or greater;[35-37] and K_{Seq} commonly is in the range 10^3 to 10^5 M^{-1}. Values of K_{Teq} thus must be very large, on the order of 10^{11} to 10^{19} M^{-1}. The central idea of this analysis then is that enzymes are designed (by evolution) to be "precisely complementary to the reactants in their transition-state geometry, as distinct from their ground-state geometry";[37] i.e., E binds $S^{\ddagger}$ more strongly than S by a factor roughly equal to the catalytic factor of the enzyme. In doing so, it greatly increases the concentration of the transition state (this is called transition-state stabilization) and thus accelerates the reaction. This concept was first expressed by Pauling in the 1940s,[34] but it has become the prevailing viewpoint only in recent years. It leads to the very useful idea that stable compounds that structurally resemble transition states — so-called transition-state analogues — should bind much more strongly to enzymes than substrates themselves. Such compounds provide important information on the mechanisms of enzyme catalysis, and this has led to the design of improved enzyme inhibitors.

According to Lienhard,[35] five factors could affect the relative magnitudes of the binding constants K_{Seq} and K_{Teq}: (1) changes in the basic structure of the transition state, (2) entropy changes involving substrate molecules, (3) interactions with solvent water, (4) interactions with the enzyme, and (5) conformational changes of the enzyme. Lienhard concluded that factor (1) usually is not important and that $S^{\neq\prime}$ and $S^{\neq}$ (transition states of the uncatalyzed and catalyzed reactions, respectively) are similar in structure and energy. Enzymatic reactions that proceed by fundamentally different mechanisms from their uncatalyzed counterparts are exceptions to this rule.

Both S and $S^{\neq}$ lose translational and rotational entropy on binding to E, and this lowers the binding strength for each species. The enhanced binding of $S^{\neq}$ implies that the entropy loss from binding is less for $S^{\neq}$ than for S. In some cases internal rotational freedom is already restricted in converting S to $S^{\neq}$ nonenzymatically, and $S^{\neq}$ has less entropy to lose on binding with E than does S. The difference in loss of rotational entropy allows tighter binding of $S^{\neq}$ and contributes to catalysis. Interactions with solvent molecules can cause additional entropy changes that affect K_{Tcq}/K_{Scq}. If more water molecules are released on binding $S^{\neq}$ to E than are released on binding S, there is a net increase in entropy associated with formation of $ES^{\neq}$ compared with formation of ES. Again, this leads to tighter binding of the transition state and to catalysis. Entropy changes resulting from solvent interactions are complicated, but generally reflect the relative binding strengths of S and $S^{\neq}$. If $S^{\neq}$ binds water more strongly than S, entropy increases favor tighter binding of $S^{\neq}$.

The fourth category of factors deals with relative strengths of noncovalent interactions between E and S or $S^{\neq}$. These include hydrogen bonding, van der Waals forces, and electrostatic interactions. If the structure of the active site is optimum for interaction with $S^{\neq}$, it cannot be optimum for interaction with S, which has a slightly different size, shape, and/or electronic configuration than $S^{\neq}$. This explains much of the catalytic activity of lysozyme, an enzyme that catalyzes the hydrolysis of glycoside linkages in bacterial cell walls.[23,35,38] Repulsive steric interactions destabilize the ES complex relative to $ES^{\neq}$. The reactant-state sugar residue being hydrolyzed, which is in the chair conformation, does not fit into the active site as well as the transition state, which is in the half-chair conformation.[37]

The final factor causing tighter binding of $S^{\neq}$ compared with S involves conformational changes in structure of E when S or $S^{\neq}$ bind to E. These must be energetically unfavorable; otherwise E would occur in the altered conformation in its free state. If the conformational state of the protein in the $ES^{\neq}$ state has a lower energy than that of the ES state, formation of $ES^{\neq}$ is favored, and that contributes to the catalytic activity.

The molecular configurations of enzyme active sites and mechanisms of enzymatic reactions have been subjects of intensive investigation for decades. Structural details of active sites are known for several enzymes,[24,38] including the identity of amino acids and/or coenzymes at the active site and steric information such as distances between functional groups involved in reactions at the active site. Detailed mechanisms, including intermediates and transition states, have been described for many reactions, and biochemists are able to use such information to alter some enzymes to catalyze both synthesis and decomposition reactions of organic compounds that are not normal metabolites.[39]

PART II. KINETICS OF MICROBIAL PROCESSES

6.5 KINETICS OF NUTRIENT-LIMITED MICROBIAL GROWTH

6.5.1 Introduction

The biochemistry of microbial growth is very complicated, but microbial growth often follows simple rate equations. Growth models that do not consider variations in microbial composition in response to changing conditions are called unstructured models. These are the most common and simplest growth models, and this chapter focuses on them. Structured models consider changes in the composition of biomass caused, for example, by intracellular storage of substrate or excretion of metabolites. These models yield more detailed pictures of microbial growth under varying conditions, but they do so at the expense of added complexity (Section 6.6.7). Structured models may be thought of as a "second generation" of microbial growth models.

This section presents equations for microbial growth under batch-culture conditions. First we consider growth under optimal conditions, but the main focus is on growth under conditions of substrate or nutrient limitation. Microbial processes are mediated by enzymes, and the equations presented in Section 6.1 provide the basis for microbial growth kinetics. Batch cultures are the simplest way to grow microorganisms in the laboratory. Because microbial populations and substrate levels change temporally in complex ways in batch reactors, it is easier to evaluate microbial growth coefficients by continuous culture methods, as described in Section 6.6. Numerous reviews are available on these subjects and the related topic of nutrient transport.[40-42]

6.5.2 Growth of Microorganisms in Batch Culture

When no nutrients are limiting and physical conditions are constant, microbial growth follows simple first-order kinetics:

$$dM/dt = \mu M \qquad (6\text{-}31)$$

or
$$\ln M/M_o = \mu t \qquad (6\text{-}32)$$

μ is the specific growth rate or growth rate constant (units of t^{-1}), analogous to k in chemical kinetics. Under these conditions the population increases exponentially, and the doubling time G of the population is constant:

$$G = 0.69/\mu \qquad (6\text{-}33)$$

Optimal growth conditions cannot be maintained indefinitely in batch cultures. The population eventually depletes the nutrients needed for growth, and the growth rate becomes dependent on the concentration of some nutrient. In the following development we consider growth to be limited by one nutrient; other nutrients are assumed to be at nonlimiting levels. The possibility that several nutrients may limit growth simultaneously is discussed in Section 6.9.3.

The general relationship between change in microbial biomass and change in concentration of the growth-limiting substance is

$$d[M]/dt = -yd[S]/dt \qquad (6\text{-}34)$$

[M] is the microbial concentration, [S] is the limiting substrate concentration, and y (dimensionless) is the yield coefficient (gram biomass produced per gram of substrate consumed). If both sides of Equation 6-34 are divided by [M], we obtain the specific growth rate relationship:[41]

$$\frac{1}{[M]}\frac{d[M]}{dt} = -y\frac{1}{[M]}\frac{d[S]}{dt} \qquad (6\text{-}35)$$

or
$$\mu = yk_T \qquad (6\text{-}36)$$

where $k_T = -(1/[M])d[S]/dt$ is the specific transfer rate of limiting substrate (t^{-1}), i.e., the rate of change in [S] per unit of [M], and $(1/[M])d[M]/dt = \mu$ is the specific growth rate, as defined by Equation 6-31. The specific transfer rate is a function of external and intracellular substrate concentrations and many other factors. Equation 6-36 is too general to be useful in calculations, and parameterization of k_T and y is described below.

6.5.3 The Monod Equation

The most common substrate transfer relationship is the Michaelis-Menten equation, which assumes irreversible transfer of substrate into cells by a recyclable catalyst. In active transport, nutrients and organic substrates are transported across cell membranes by energy-requiring enzymes, and the mechanism *is* that of a recyclable catalyst. In such cases we can write

$$-\frac{d[S]}{dt} = \frac{k_{Tmax}[M][S]}{K_{Su} + [S]} \qquad (6\text{-}37)$$

k_{Tmax} {mg day^{-1} (mg organisms)$^{-1}$} is the maximum specific substrate uptake constant, and K_{Su} is the half-saturation constant for *uptake* of limiting nutrient. At constant [M], substrate uptake follows saturation kinetics with respect to [S] (see Figure 6-12a). Although reactions "downstream" from the presumed rate-limiting transport step may influence the transport flux, this influence is small, and transport kinetics still follows the Michaelis-Menten equation.[42] Combining Equations 6-34 and 6-37 gives

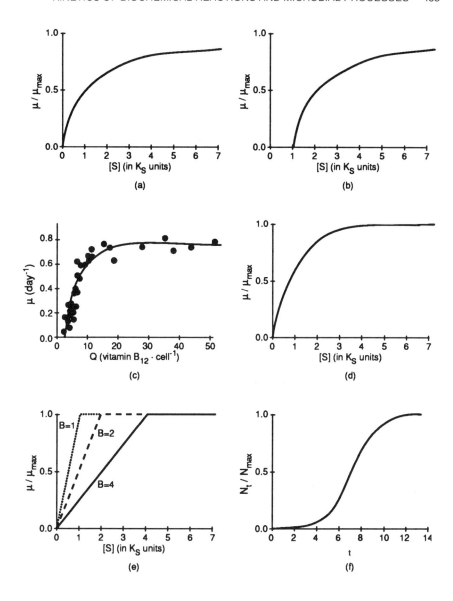

Figure 6-12. Graphical portrayal of microbial growth rates by various models: (a) simple Monod expression (Equation 6-40); (b) Monod expression with threshold requirement, $T = 1.0K_S$ (Equation 6-43); (c) cell quota model (Equation 6-46) with actual data for growth rate of *Pavlova lutheri* as function of cell quota for vitamin B_{12} [Droop, M.R., *J. Mar. Biol. Assoc. U.K.*, 48, 689 (1968)]; (d) Tessier model (Equation 6-47); (e) Blackman model (Equation 6-48) with empirical B = 1,2, or 4; and (f) logistic equation (6-49b) with $r = 1.0t^{-1}$ and A (dimensionless) = 7. A defines the position of the curve on the x-axis; at $rt = A$, $N_t/N_{max} = 0.5$.

$$\frac{d[M]}{dt} = \frac{yk_{T\max}[M][S]}{K_{Su} + [S]}$$

(6-38)

or in terms of specific growth rate,

$$\mu = \frac{yk_{T\max}[S]}{K_{Su} + [S]}$$

(6-39)

where $yk_{T\max}$ has units of t^{-1}, y is dimensionless, and $k_{T\max}$ is in $mass(mass)^{-1}t^{-1}$. The term is equivalent to the maximum specific growth rate, $\mu_{\max}$, which occurs when $[S]$ is at saturating levels. Thus, we can write,

$$\mu = \frac{\mu_{\max}[S]}{K_{Su} + [S]} \approx \frac{\mu_{\max}[S]}{K_{Sg} + [S]}$$

(6-40)

K_{Sg} is the half-saturation constant for nutrient-limited *growth*. It is equal to K_{Su} only when nutrient transport (uptake) limits $\mu_{\max}$, but the distinction between K_{Sg} and K_{Su} is not always made in the literature. For simplicity, we will ignore the distinction except in discussion of parameter values. Equations 6-38 to 6-40 are variations of the Monod equation,[43] but strictly speaking, the last form of Equation 6-40 (with K_{Sg}) is the Monod equation. It expresses the specific rate of microbial growth as a function of limiting substrate concentration (Figure 6-12a) and is the basis for most microbial growth models. This type of growth is called transport limited. Although Monod developed it on empirical grounds (it fit his data), a mechanistic basis for the equation can be provided by assuming that one enzymatic reaction in a series of steps involved in microbial growth is slower than the rest (and thus growth limiting) and that y and $k_{T\max}$ do not change over the growth conditions. Zero- and first-order growth models are limiting cases of the Monod equation for high and low concentrations of limiting substrate, respectively.

Equation 6-38 can be integrated to yield an expression for the change in $[S]$ and $[M]$ with time in a batch reactor.[44] The integration is somewhat complicated because both $[M]$ and $[S]$ are variables, but changes in the two terms are related by the yield coefficient (Equation 6-34). Consequently, Equation 6-38 can be converted into a one-variable equation, and the resulting equation is

$$K_{T\max}t = \left[\frac{K_S}{[S]_o + [M]_o/y} + 1 \right] \ln\frac{[M]}{[M]_o} - \left[\frac{K_S}{[S]_o + [M]_o/y} \right] \ln\frac{[S]}{[S]_o}$$

(6-41)

Equation 6-41 yields an S-shaped curved for $[M]$ vs. t, similar to Figure 6-12f, but the integrated expression is not useful in evaluating K_S and $k_{T\max}$. The Monod equation also has some disadvantages in describing biomass changes within batch cultures. Three fittable parameters (μ_{max}, K_S, and y) and two equations (6-34 and 6-40) are needed to describe changes in substrate and biomass. In batch cultures where initial [S] is high and K_S is low, the growth rate is independent of [S] during much of the culture period. [S] decreases rapidly only after the population has become large, and approaches K_S only at the end of the growth period, as [S] becomes exhausted. Consequently, it may be difficult to determine the parameters accurately from batch-culture data. For this reason continuous-culture methods are preferred for parameter estimation.

From the assumption made in deriving Equation 6-40, it follows that Monod kinetics should be observed only when substrate transfer follows unidirectional Michaelis-Menten kinetics. If substrate can diffuse out of cells (and if the concentration gradient allows this), μ follows a reversible rate expression:[44]

$$\mu = yk_{T\,max} \left[\frac{[S]_e}{K_{S,in} + [S]_e} - \frac{[S]_i}{K_{S,o} + [S]_i} \right] \qquad (6\text{-}42)$$

Subscripts on [S] indicate external and intracellular concentrations, and those on K_S denote uptake and release of S. Measurement of $K_{S,o}$ is difficult, and little is known about this parameter. Although Equation 6-42 adds a measure of realism to substrate uptake, it is difficult to use. It appears to be an unnecessary refinement in most cases.

6.5.4 Substrate Affinity

The specific rate of substrate uptake, v_S, normalized to substrate concentration is a measure of microbial affinity, a_S, for the substrate. In general, a_S has units of L (g cells)$^{-1}$ h^{-1}: $v_S/[S]$ = (g substrate)(g cells)$^{-1}$ h^{-1} ÷ (g substrate) L^{-1}. Button[42] tabulated 14 different definitions and equations for this concept, but many involve only subtle differences. Healey[45] recommended using $a_S = V_{max}/K_S$, which ties the concept to Michaelis-Menten kinetics. For systems where V_{max} is difficult to measure (e.g., when uptake enzymes are inducible), this definition is problematic. Button[42] recommended that affinity be defined as the initial slope of a $v_S/[S]$ curve: $a_S^o = v_S/[S]$ as [S] → 0. This definition specifies no particular form of kinetics and has the advantage of comparing the abilities of organisms to accumulate substrate at the lowest (most limiting) conditions. On the other hand, there may be significant measurement problems with a_S^o; the estimated value may depend on how low one can measure both v_S and [S]. In this respect, use of V_{max}/K_S, which is less sensitive to inaccuracies in low v_S and [S], is a preferable measure.

6.5.5 Addition of a Threshold Requirement to the Monod Model

The threshold concentration (T) is defined as the minimum concentration of substrate or nutrient at which growth is observed. Endogenous metabolism — the need for cells to spend energy for maintenance — provides the basis for a carbon threshold for heterotrophs. Endogenous metabolism includes energy expended for osmoregulation (maintenance of energized membranes), motility, and turnover and repair of cellular materials — synthesis of new RNA, protein, maintenance of other metabolic pools. Endogenous metabolism is assumed constant with time and depends only on biomass. It often is expressed as a rate constant, k_e (Section 6.6.6). The net effect of adding a threshold to the Monod equation

$$\mu = \frac{\mu_{max}([S]-T)}{K_S + [S] - T} \tag{6-43}$$

is to move the growth rate curve to the right by T units on a plot of μ vs. [S] (Figure 6-12b). The size of T can be estimated from endogenous requirements and the specific affinity:[42]

$$T \approx k_e / a_S^\circ \tag{6-44}$$

Although the theoretical basis for threshold levels of nutrients like N and P is arguable, microbial growth data sometimes fit better to models that include a threshold concentration term.[46]

6.5.6 The Cell Quota, Q: A Micronutrient-Limited Growth Rate Model

Rates of microbial growth in cultures limited by trace nutrients such as iron often are described by an empirical model based on a concept called the cell quota Q, i.e., the amount of the limiting nutrient per cell. For steady-state conditions in a chemostat (Section 6.6.3), Q is given by

$$Q = ([S]_{in} - [S]_{out}) / [M] \tag{6-45}$$

In most cases $[S]_{out} << [S]_{in}$ for the limiting nutrient, and Q can be determined from the steady-state cell density and influent concentration of limiting nutrient. Growth rates are related to Q in hyperbolic fashion (Figure 6-12c):

$$\mu = \mu'_{max}(Q - Q_o)/Q \tag{6-46}$$

Q_o is the subsistence cell quota, i.e., the cellular content of limiting nutrient when $\mu = 0$. This model was proposed by Droop[47] for vitamin B_{12}-limited growth of a marine flagellate, *Pavlova lutheri*, but it is based on the earlier finding[48] that the amount of phosphorus per algal cell decreased as it became more limiting. The model has been used widely in phytoplankton ecology,[49] and many examples have been reported where algal growth kinetics follow Equation 6-45.[50] μ'_{max} is a theoretical (fitted) maximum rate; μ_{max}, the maximum observable growth rate, is always less than μ'_{max}. If $Q_{max} \gg Q_o$, the difference between μ'_{max} and μ_{max} is small. For example, Rhee[51] estimated the difference was less than 3% for iron-limited algal growth. On the other hand, the range between Q_o and Q_{max} is smaller for cellular constituents like nitrogen and silica, and much larger differences occur between μ_{max} and μ'_{max}. Carbon has an invariant Q over a wide range of growth rates ($Q_o \approx Q_{max}$), and in this case μ'_{max} approaches infinity.[49] Similar expressions have been used for phosphorus uptake and P-limited growth by blue-green algae.[52]

6.5.7 Other Growth Models

Several other empirical growth expressions have been proposed (Figure 6-12d,e); and some workers prefer them to the Monod expression for some applications. The Tessier model[53] relates μ to an exponential function of $[S]$:

$$\mu - \mu_{max}\left\{1 - \exp(-[S]/K)\right\} \qquad (6\text{-}47)$$

The much older Blackman model[54] is a linear approximation of the hyperbolic Monod relationship:

$$\mu = \mu_{max} \text{ for } [S] > \mu_{max}B;$$
$$\mu = [S]/B \text{ for } [S] < \mu_{max}B \qquad (6\text{-}48)$$

K and B are empirical constants. These models and the Monod equation describe experimental observations from batch and continuous cultures in similar ways,[55] but the Monod equation is by far the most widely used expression and is the basis for more complicated and realistic models of microbial growth.

A common growth equation in population models for higher organisms is the logistic equation, first described by Verhulst in 1844 and applied to population dynamics by Lotka:[56]

$$dN/dt = rN(N_{max} - N)/N_{max} \qquad (6\text{-}49a)$$

N is the population size (cell numbers per unit volume); r is the intrinsic growth rate (equivalent to μ in Equation 6-31); and N_{max} is the maximum population (approached asymptotically, as in Figure 6-12f). In population ecology N_{max} is called the carrying capacity. The logistic equation defines a sigmoid growth curve in which growth of the population declines as it approaches its maximum. The equation is fundamentally different from the other models in this section, in that it treats changes in population rather than changes in intrinsic growth rates. The integrated form of the equation (assuming constant r) is

$$N_t = N_{max}/\{1 + \exp(A - rt)\} \qquad (6\text{-}49b)$$

or
$$\ln\{(N_{max} - N)/N\} = A - rt \tag{6-49c}$$

A is a constant of integration that defines the position of the curve relative to the origin. The logistic equation has advantages of simplicity and realism (population curves usually are sigmoid). The differential form has two fitted parameters, but the integrated equation has three. The equation is flexible and well suited for batch-population data, but statistical problems are encountered in estimating the constants. Moreover, it is essentially empirical, and as we showed earlier, r cannot be assumed constant in microbial growth.

6.6 CONTINUOUS CULTURE OF MICROORGANISMS

6.6.1 Introduction

Although batch culturing of microorganisms is a common practice, this method has several disadvantages for kinetic studies. The physiological state of organisms changes continually in batch culture, making it difficult to repeat experiments; organism response to external stimuli depends on physiological state and cell age. Kinetic parameters (e.g., K_S for nutrient uptake) obtained from one batch culture thus may not apply to another batch system with a different cell age or physiological state. If nutrient concentrations can be measured accurately over a range of dilution rates, the (steady-state) kinetics of nutrient uptake and microbial growth can be measured without disturbing the system. Because conditions in natural systems often change fairly slowly (relative to microbial lifetimes), such steady-state parameters are useful in describing processes in many natural aquatic environments.[42] In industrial applications where microorganisms are grown to yield a product or metabolize a substrate, it is more convenient and economical to operate continuously. These comments apply equally well to biological waste treatment processes.

The microbial population in a steady-state continuous culture is in a constant physiological state that is controlled by varying reactor residence time, substrate input, and the degree of cell recycle. Of course, not all the individuals in a continuous culture are in the same state; a continuum exists from recently divided cells to those about to divide and those in the process of dividing. Because of the large number of cells in the reaction vessel at any time, the distribution among these phases can be maintained constant. In certain cases all the individuals can be forced into the same growth phase, so that cell division occurs synchronously. Such cultures of phytoplankton can be obtained by manipulating light-dark cycles (Section 6.6.8).

The simplest type of continuous-culture device is the continuous-flow stirred tank reactor (CFSTR) (see Chapter 5). CFSTRs are the most common reactors used for continuous culture of microorganisms in the laboratory; such systems usually are called chemostats. Plug-flow biological reactors are rarely used in lab or industrial applications, but are common in wastewater treatment. A novel compromise between batch and continuous-flow systems is the continuously fed batch reactor (CFBR), which has some useful properties for determining microbial growth constants.[57] Figure 6-13 compares some characteristics of CFSTRs, PFRs, and CFBRs. In practice, continuous-flow reactors often are neither perfectly mixed nor perfect plug flow, but exhibit dispersed flow behavior.

6.6.2 Microbial Growth in CFSTRs

The most important parameters characterizing growth in CFSTRs are dilution rate D and specific growth rate μ. Both have dimensions of time^{-1}. D is the ratio of inflow rate to reactor volume and is equal to the flushing coefficient (ρ) defined in Chapter 5: $D = \rho = Q/V$. The reciprocal of D is the average residence time of fluid in the CFSTR, $\tau_w = 1/D = V/Q$. Because of the well-mixed conditions in a CFSTR, the fate of individual molecules of the culture medium is random. Some are swept out as soon as they enter the reactor; by chance, others remain for a long time. Solvent molecules have a continuum of residence times from near zero to infinity, but mean residence time is defined by V/Q.

The behavior of conservative substances in a CFSTR is described by first-order kinetics where D replaces rate constant k. Consider a CFSTR with a conservative substance at concentration [C] and an input solution that contains none of the substance. For time period dt, flow into and out of the vessel is Qdt, and the substance leaves the vessel at a rate of [C]Q. The loss rate thus is equal to the mass in the reactor times the dilution rate:

$$-dm/dt = [C]Q = (m/V)Q = mD \qquad (6\text{-}50a)$$

or
$$m = m_o \exp(-Dt) = m_o \exp(-t/\tau_w) \qquad (6\text{-}50b)$$

m is the mass of conservative substance. According to Equation 6-50b, when $t = \tau_w$, $m/m_o = e^{-1} = 0.37$. In a biological context, this means that if the cells in a CFSTR are not growing, only 37% of the initial population remains after passage of one reactor volume. For the population to remain constant, it must grow at precisely the rate required to balance the outflow loss. Microbial growth is described by Equation 6-31 as $dM/dt = \mu M$, and dilutional loss by Equation 6-50a as $-dM/dt = DM$ (M is in mg/L). It is apparent that when growth = loss, $\mu = D$.

6.6.3 Maintenance of Steady-State Conditions in CFSTR Cultures

Steady-state populations can be achieved in CFSTRs operated at constant D, with little or no external manipulation. It is apparent that the specific growth rate μ must adjust automatically to a value equal to D. How does this happen? We know from Section 6.5.3 that μ is not constant, but depends on the concentration of the growth-limiting substance according to the Monod equation. The relationships among [S], μ, D, and rate of substrate uptake by the microbes cause convergence to a steady state as long as physical conditions and input of fresh medium are maintained constant.[58]

Consider a CFSTR inoculated with a small population of microorganisms and

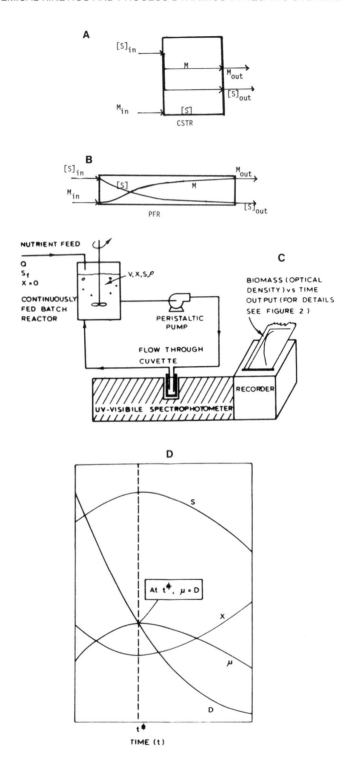

Figure 6-13.

operated at dilution rate D and input concentration [S] for the growth-limiting nutrient. After an initial lag, the population will increase at a rate equal to its instantaneous concentration (in number or biomass) times the specific growth rate; i.e., $dM/dt = \mu M$. Since cells are lost to the effluent at a rate of DM, the net rate of change in population is

$$dM/dt = M(\mu - D) \qquad (6\text{-}51)$$

For simplicity, consider that the initial concentration of limiting substrate is sufficiently high that $\mu \approx \mu_{max}$. Then initially, $dM/dt = M(\mu_{max} - D)$. If $D > \mu_{max}$, washout must occur, the steady-state population will approach zero, and steady-state culture will not be possible.

If $D < \mu_{max}$, net growth will occur, and the biomass will increase. This cannot persist indefinitely, since eventually the cells will consume enough incoming substrate to reduce its concentration in the CFSTR, thus causing μ to decrease. As long as $\mu > D$, however, M will increase, and [S] will decrease, until $\mu = D$, and $dM/dt = 0$. From the Monod equation we can write

$$\mu = \frac{\mu_{max}[\hat{S}]}{K_S + [\hat{S}]} = D \qquad (6\text{-}52a)$$

or
$$[\hat{S}] = K_S D / \{\mu_{max} - D\} \qquad (6\text{-}52b)$$

$[\hat{S}]$ is the steady-state concentration of S. Tilman et al.[46] used a similar approach to define $\hat{R}$, the resource requirement, in a resource-based species competition model for phytoplankton (Section 6.9.4). $\hat{R}$ is equivalent to $[\hat{S}]$ in Equation 6-52b and represents the nutrient concentration at which the growth rate of a species equals its loss rate by dilution in a CFSTR. According to resource competition theory, the species with the lowest $\hat{R}$ will displace its competitors because it can have net growth at nutrient concentrations where other species cannot maintain stable populations. From the relationship between yield and substrate use ($M = M_o + y([S]_o - [S])$), we can write

$$\hat{M} = y([S]_{in} - [\hat{S}]) \qquad (6\text{-}53)$$

Figure 6-13. Schematic diagrams of (A) CFSTR; (B) PFR; and (C) CFBR. Concentration profiles of limiting substrate, [S], and microbial biomass are shown as function of location in CFSTR and PFR. CFBR with initial volume V_o first is run in batch mode to build up biomass to measurable level $[M]_o$. Then substrate is fed into reactor at flow rate Q such that $Q/V_o > \mu_{max}$. Transients in [S], M, μ, and D are shown in (D). [M] decreases and [S] increases when substrate is added because $D > \mu$. [S] has a maximum and [M] a minimum at some intermediate time t', at which $\mu = D$. Equations to calculate $\mu([S])$ from this point are given by Webster. [(C) and (D) from Webster, I.A., *Biotech. Bioeng.*, 25, 2981 (1983). With permission.]

$\hat{M}$ is the steady-state biomass concentration. Solving Equation 6-53 for $[\hat{S}]$ we have

$$[\hat{S}] = [S]_{in} - \hat{M} / y \qquad (6\text{-}54)$$

The steady-state concentration of limiting substrate thus is equal to the input concentration less that used to produce the steady-state biomass. Substituting Equation 6-52b into 6-53 we obtain an alternative expression for $\hat{M}$:

$$\hat{M} = y \left[[S]_{in} - \frac{K_S D}{\mu_{max} - D} \right] \qquad (6\text{-}55)$$

Noting that μ_{max}/D (or $\tau_w \mu_{max}$) is a dimensionless relative residence time, we can rearrange Equation 6-55 in terms of this quantity:[41,44]

$$\frac{\mu_{max}}{D} = \tau_w \mu_{max} = \frac{K_S + [S]_{in} - \hat{M} / y}{[S]_{in} - \hat{M} / y} \qquad (6\text{-}56)$$

The dependence of $[\hat{S}]$ and $\hat{M}$ on D and on $\tau_w \mu_{max}$ is shown in Figure 6-14. At low D ($<< \mu_{max}$), $[\hat{S}]$ is very small (Equation 6-52), and $\hat{M}$ approaches a maximum of $y[S]_{in}$. As D increases, $\hat{M}$ decreases and $[\hat{S}]$ increases. The upper limit for $[\hat{S}]$ is $[S]_{in}$; the lower limit for $\hat{M}$ is 0. The upper limit for D, above which cell washout is complete, is found by equating Equation 6-55 to 0 and solving for D:

$$D_{crit} = \frac{\mu_{max} [S]_{in}}{K_S + [S]_{in}} \qquad (6\text{-}57)$$

Since both K_S and $[S]_{in}$ are finite and $[S]_{in}$ can be made much greater than K_S, the maximum value for D_{crit} is slightly lower than μ_{max}. As $[S]_{in}$ decreases and approaches K_S, D_{crit} decreases accordingly.

According to Equation 6-52b, $[\hat{S}]$ is independent of $[S]_{in}$ for all $D < D_{crit}$. Furthermore, it is apparent from Equation 6-53 that $\hat{M}$ increases as $[S]_{in}$ increases, as long as other conditions are nonlimiting. This is not always simple to achieve. Oxygen demand is great in dense cultures of heterotrophs, and extreme efforts may be required to transfer enough O_2 to maintain aerobic growth. Practical aspects of O_2 transfer and measurement of transfer rates are described in texts on biochemical engineering.[59]

6.6.4 Efficiency of Substrate Utilization

Recall that the substrate concentration in the effluent of a CFSTR is equal to the concentration in the reactor. The supply rate of S is $[S]_{in}Q$, and the loss rate is $[\hat{S}]Q$.

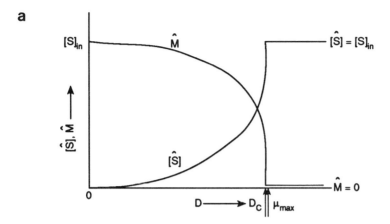

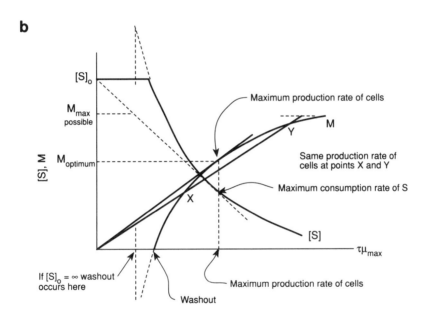

Figure 6-14. (a) Relationships between $[\hat{S}]$ vs. D and $[\hat{M}]$ vs. D in CFSTR growth of microorganisms for Monod model. (b) [S] and [M] vs. mean relative residence time $(\tau\mu_{max})$ in CFSTR showing maximum production rates and optimum concentrations. [From Levenspiel, O., *Biotech. Bioeng.*, 22, 1671 (1980). With permission.]

The rate of substrate use for growth thus is $([S]_{in} - [\hat{S}])Q$, and the efficiency of substrate use, E, is

$$E = \frac{[S]_{in} - [\hat{S}]}{[S]_{in}} \qquad (6\text{-}58)$$

E tends to be high at high $[S]_{in}$ and low D, where $[\hat{S}]$ is small. For a given $[S]_{in}$, there is an optimum value of D at which biomass production and E are maximum. From the definition of $[\hat{S}]$ in Equation 6-52b and the fact that $Q = VD$, the amount of substrate used for cell growth can be rewritten as

$$([S]_{in} - [\hat{S}])Q = V\left[D[S]_{in} - \frac{K_s D^2}{\mu_{max} - D} \right] \tag{6-59}$$

The value of D at which this expression is maximum is obtained by differentiating with respect to D and setting the result equal to zero:[58]

$$0 = \frac{d}{dD}\{VD([S]_{in} - [\hat{S}])\} = V\left[[S]_{in} - \frac{2K_s D}{\mu_{max} - D} - \frac{KD^2}{(\mu_{max} - D)^2} \right]$$

Solving for $D = D_{max}$, subject to the constraint $D_{max} < \mu_{max}$, we obtain

$$D_{max} = \mu_{max}\left[1 - \left(\frac{K_S}{K_S + [S]_{in}} \right)^{0.5} \right] \tag{6-60}$$

Provided that $[S]_{in} \gg K_S$, D_{max} and D_{crit} approach μ_{max} as an upper limit, as shown by the following calculations:

Parameter	Predictive equation	$[S]_{in}$	
		$10K_S$	$100K_S$
D_{max}	6-60	$0.70\mu_{max}$	$0.90\mu_{max}$
D_{crit}	6-57	$0.91\mu_{max}$	$0.99\mu_{max}$
$[\hat{S}]$	6-52b	$2.3K_S$	$9K_S$
E	6-58	0.77	0.91

6.6.5 Operation of CFSTRs at Low $[S]_{in}$: The Value of Cell Recycle

According to the Monod equation, as $[S]_{in}$ decreases, μ decreases in hyperbolic fashion. Equally important, D_{crit} also decreases, as shown by Equation 6-57, which is valid whether we consider endogenous metabolism or not (see Section 6.6.6).

Microbial growth kinetics thus requires that D be small at low $[S]_{in}$. CFSTRs often must be operated at low $[S]_{in}$ because of solubility or toxicity limitations or a desire to simulate natural conditions. Such operations are the norm in biological waste treatment. Low D leads to uneconomical operating conditions (large reactor volumes) at high flow rates, because $Q = VD$. A simple way to avoid the large volume requirement is to harvest cells from the effluent and recycle them into the reactor. Suppose we harvest and waste $1/n$th of the effluent cells and return the rest to the reactor. No medium is returned, and fresh medium is added as before. For each harvested cell, $n - 1$ cells are returned to the reactor, and the concentration change is only $1/n$th of that without cell recycle: $-dM/dt = (1/n)DM$. The average residence time of cells, symbolized Θ_c, thus is n times the hydraulic residence time ($\Theta_c = n\tau_w$).

To maintain the same net harvest as before, we must withdraw medium n times as fast as the rate without recycle. If the harvest (and production) rate is the same with and without recycling, the amount of substrate used also is the same. Because n times as much medium passes through the reactor per unit time with cell recycle, only $1/n$th as much nutrient is used per unit volume of medium, and $[S]_{in}$ can be much lower. Without recycle, the substrate used for growth is $([S]_{in}^{nr} - [\hat{S}])DV$. With harvest of $1/n$th of the cells and a dilution rate n times as great, the amount used is $([\hat{S}]_{in}^r - [\hat{S}])nDV$. For equal substrate use with and without recycle we have

$$([S]_{in}^r - [\hat{S}])nDV = ([S]_{in}^{nr} - [\hat{S}])DV$$

For equal growth rates, $[\hat{S}]$ and $\hat{M}$ must be the same in both reactors. Solving for $[S]_{in}^r$, we find

$$[S]_{in}^r = (1/n)\{[S]_{in}^{nr} - (n-1)[\hat{S}]\}$$ (6-61)

If $[\hat{S}] << [S]_{in}$, the input concentration with recycling is approximately $1/n$th of that without recycling for the same biomass yield. Therein is the advantage of cell recycle: dilute media can be used with high cell concentrations and high hydraulic throughput (high D). This is the basis for the activated-sludge method of wastewater treatment.

6.6.6 Effect of Variable Yield Factors on Monod Models

A major limitation of the Monod model is its assumption that y is constant for all $[S]$. Two exceptions can be considered. If the limiting substrate is a nutrient and the energy source is present in excess, heterotrophs may change their metabolism and continue to grow. For example, if nitrogen is limiting, they may synthesize polysaccharides, accumulation of which may be interpreted as cell growth if nonspecific parameters like turbidity or suspended solids are used to estimate biomass. The yield coefficient (mg biomass produced per mg N consumed) thus increases as $[S]$ decreases, as shown in Figure 6-15a. A simple mathematical relationship between y and $[S]$ is not available in these cases, and the figure is only qualitative. Structured growth models that consider variations in cellular composition can describe this quantitatively.

In contrast, if the limiting substrate is the energy source, y decreases with decreasing $[S]$ because a greater proportion of S is consumed for cellular maintenance (endogenous metabolism) as the growth rate decreases (Figure 6-15b). Simple formulations have been developed to account for the change in y in this

a

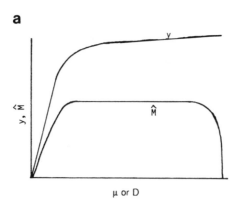

b

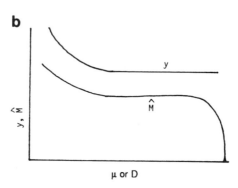

Figure 6-15. Deviations from Monod equation for microbial growth. (a) y decreases at low
growth or dilution rate in CFSTR when S is an energy substrate that is used
increasingly for cell maintenance at low μ (or D), and steady-state biomass $\hat{M}$
decreases to 0 at low D. (b) y increases at low D (hence, low [S] in CFSTR) when
nitrogen is limiting, because synthesis of polysaccharide continues, increasing
apparent biomass $\hat{M}$. [After Powell, E.O., *Lab. Pract.*, 1145 (1965).]

situation.[60,61] For example, k_T can be considered the sum of specific rates for growth
(k_g) and maintenance (k_e). The specific growth rate $\mu = yk_T$ thus becomes[61]

$$\mu = y(k_g + k_e) \tag{6-62}$$

The specific growth rate still should be proportional to the specific transfer rate
of the limiting substrate, $\mu \propto k_g$, or $\mu = Yk_g$, where Y is a proportionality factor. It
is apparent that y (now a variable) and Y (a constant) are related to each other.
Substituting Yk_g for μ in Equation 6-62 and rearranging yields

$$y = Y\left[\frac{k_g}{k_g + k_e}\right] \tag{6-63}$$

Y, the maximum value of y, occurs when $k_e \ll k_g$ and is the maximum yield factor.

Because $k_g = \mu/Y$, we can substitute this into Equation 6-63 and obtain

$$y = Y\left[\frac{\mu}{k_{Tm}Y + \mu}\right] = Y\left[\frac{\mu}{\mu_e + \mu}\right] \tag{6-64}$$

$\mu_e = Yk_e$ is the endogenous metabolism rate constant or specific maintenance rate (t^{-1}); k_e, the maintenance coefficient (units of mass mass^{-1} t^{-1}),is considered independent of growth rate and constant for a given organism and temperature. A Monod-like expression with a term for maintenance is[60]

$$\mu = \frac{Yk_{Tg}[S]}{K_S + [S]} - \mu_e = \frac{\mu_{g,max}[S]}{K_S + [S]} - \mu_e \tag{6-65}$$

The maximum observed specific growth rate, μ_{max}, occurs when [S] is saturating (>> K_S); thus, $\mu_{max} = \mu_{g,max} - \mu_e$. Equation 6-65 implies that endogenous metabolism can be treated as a negative growth rate.

Equation 6-65 and the concept of μ_e were derived by Herbert[60] prior to the derivation by Pirt[61] of Equations 6-62 to 6-64 and the concept of k_e. If we define $\mu_e = Yk_e$, as above, the two relationships are mathematically equivalent, but the mechanisms implied by the two relationships are different.[55] The Herbert model implies that maintenance energy is supplied by consumption of biomass, but the Pirt model assumes that it is provided by consumption of external substrate. The latter assumption is reasonable as long as external substrate is available, but cells still must maintain themselves when external substrate is exhausted, and the Herbert model allows this. Although some maintenance functions probably are influenced by growth rate, k_e normally is considered to be constant. Experimental data support this assumption over a wide range of growth rates, but k_e apparently declines at the very low growth rates occurring in some wastewater treatment systems. Energy expenditure for maintenance is well established from both experimental data and from a fundamental thermodynamic concept that states that energy is required to keep an open system (such as a living cell) in an ordered (low-entropy) state. Esener et al.[55] discussed this topic in greater detail.

The concept of variable cell yield when the limiting substrate is the energy source can be applied to CFSTRs to improve the range of substrate loading over which microbial growth can be predicted. Because k_e is constant, y decreases as [S]$_{in}$ (and μ) decreases (Equation 6-64). Because Equation 6-52a is still valid at steady state (i.e., $\mu = D$), we can substitute D for μ in Equation 6-64 and obtain

$$y = Y\left[\frac{D}{\mu_e + D}\right] \tag{6-66a}$$

Thus, y decreases as D decreases (assuming μ_e is constant). The reciprocal of Equation 6-66a yields a Lineweaver-Burk type of equation:

$$1/y = 1/Y + (\mu_e/Y)(1/D) \tag{6-66b}$$

Plots of y^{-1} vs. D^{-1} are linear when Y and μ_c are constant and independent of D.[59] Y and μ_c can be obtained from the slopes and intercepts of such plots. A similar equation can be derived for k_c:[42]

$$1/D = (1/k_e)(1/y) - 1/y_g k_e \tag{6-66c}$$

y_g, the yield of cells from substrate used for growth, is equivalent to Y.

The effect of varying y on the steady-state biomass $\hat{M}$ in a CFSTR can be found by substituting Equation 6-66a for y into Equation 6-55:

$$\hat{M} = Y\left[\frac{D}{\mu_e + D}\right]\left[[S]_{in} - \frac{K_s D}{\mu_{max} - D}\right] \tag{6-67a}$$

or

$$\hat{M}(\mu_e + D) = YD[S]_{in} - \frac{YK_s D^2}{\mu_{max} - D} \tag{6-67b}$$

As D approaches 0, the right side of Equation 6-67a,b approaches 0; therefore $\hat{M}$ approaches 0 (Figure 6-15a). Because D_{crit} occurs where $\hat{M} \rightarrow 0$ whether or not we consider endogenous metabolism, the solution of Equation 6-67a at $\hat{M} = 0$ yields the same expression for D_{crit} as the one derived without considering endogenous metabolism (Equation 6-57).

Example 6-4. Growth Parameters for Fe²⁺ Oxidizers from Continuous Culture Data

Microbial oxidation of ferrous iron to the ferric state is the cause of acidity in acid mine drainage and is an important process in microbial leaching of ore bodies. In contrast to the situation at circumneutral pH where abiotic oxidation of Fe^{II} by O_2 is rapid and microbial oxidation is unimportant, chemical oxidation is much slower than microbial oxidation at pH < 3 to 4. The organism responsible for this reaction, *Thiobacillus ferrooxidans*, is unusual in several respects. Most notably, it is amazingly tolerant of acidic conditions (good growth rates[62] as low as pH 1.0) and high levels (>1 g L^{-1}) of several heavy metals, including copper, zinc, and uranium. A true acidophile, *T. ferrooxidans*, does not grow on Fe^{II} at pH > 4, and it is an obligate aerobe and mesophile (temperature range of 20 to 40°C). It also is unusual in being a chemolithotroph; i.e., it obtains its energy by oxidizing an inorganic ion (Fe²⁺) and its carbon by reducing inorganic CO_2. It also is capable of growth on H_2S and other reduced inorganic sulfur compounds. Similar to other chemolithotrophs (such as nitrifying bacteria), it is slow growing and has low yield coefficients. The stoichiometry of its growth is approximately

(1)
$$Fe^{2+} + 0.2240_2 + 0.0045CO_2 + 0.011HCO_3^- + H^+ + 0.0011NH_4^+ \rightarrow$$
$$Fe^{3+} + 0.0011C_5H_7O_2N + 0.4989H_2O$$

$C_5H_7O_2N$ is the approximate stoichiometry of bacterial cells. Equation 1 assumes an electron transfer efficiency of 30% at pH 2.[62] Although Equation 1 shows the consumption of one H^+ per Fe^{2+} oxidized, it must be remembered that Fe^{3+} is itself a fairly strong acid and tends to react with water to form $Fe(OH)_3 + 3H^+$.

Smith et al.[62] studied the growth kinetics of *T. ferrooxidans* growing on Fe^{2+} in batch and continuous cultures over the pH range 1.0 to 2.4. The upper pH limit was selected to prevent precipitation of $Fe(OH)_3$, which would have complicated calculation of cell yields and growth rates from volatile suspended solids (VSS) data. In batch cultures at 30°C and $[Fe^{2+}]_o$ = 2400 mg l⁻¹, the depletion time for Fe^{2+} was roughly the same (40 h) from pH 1.9 to 2.4. A small lag in Fe^{2+} oxidation was observed at pH 1.0, and 50 h was needed for complete oxidation. Steady-state results in CFSTRs showed little difference in growth at 25 and 30°C: cell washout occurred at $\tau_w < {\sim}0.3$ days, and Fe^{2+} use was 90% complete at $\tau_w = 0.6$ days (Figure 6-16a). Growth was slower at 20°C: washout occurred at $\tau_w \approx {\sim}0.7$ days, and complete Fe^{2+} uptake at $\tau_w = 1.0$ day. Yield coefficients were obtained by applying CFSTR data to the following equation, which is based on Equation 6-53:

$$(2) \qquad D = 1/\tau_w = V/Q = \frac{y}{\hat{M}V}\left\{[Fe^{2+}]_{in}Q_{in} - [Fe^{2+}]_{out}Q_{out}\right\}$$

where y is the yield coefficient (mg cells per mg Fe^{2+} oxidized), which was assumed not to vary with D (i.e., y = Y). $\hat{M}$, the steady-state biomass (mg L⁻¹), was measured as VSS or total organic carbon (TOC) in the reactor effluent; Q is in L day⁻¹, V in liters, and $[Fe^{2+}]$ in mg L⁻¹. Q_{in} and Q_{out} were measured separately to account for evaporative losses in the reactors. A plot of D vs. the term $\{[Fe^{2+}]_{in}Q_{in} - [Fe^{2+}]_{out}Q_{out}\}/\hat{M}V$ generally fits a straight line (Figure 6-16b) with slope = y. Yield coefficients were small, as typical of chemolithotrophs, and agreed well with results of thermodynamic calculations:

T (°C)	pH	y (TOC basis)	y (VSS basis)
20	2.0–2.1	0.0013	0.0031
25	2.0–2.1	0.0014	0.0023
30	1.8–2.1	0.0016	0.0023
Thermodynamic model (30% efficiency)		0.0013	0.0024

These calculations were made under the assumption that endogenous metabolism was negligible ($k_c \approx 0$ day⁻¹). If typical values of k_c (0.05 day⁻¹) for bacteria apply to *T. ferrooxidans*, much longer reactor residence times would be needed to measure k_c reliably.[62]

The half saturation constant (K_{Sg}, mg L⁻¹) and maximum specific rate of substrate use (k_{Tmax}, mg Fe^{2+}/mg cell·day) were determined by applying CFSTR data to the equation

$$(3) \qquad [Fe^{2+}]_{out}\,y/D = \frac{[Fe^{2+}]_{out}}{k_{T\max}} + \frac{K_{Sg}}{k_{T\max}}$$

which can be obtained by rearranging Equation 6-52a. A plot of the left side of Equation 3 vs. $[Fe^{2+}]_{out}$ (Figure 6-16c) has a slope of k_{Tmax} and y-intercept of K_{Sg}/k_{Tmax}. Results for the three temperatures are

a

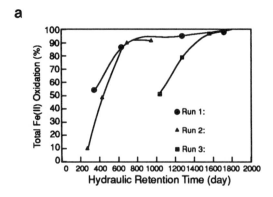

b

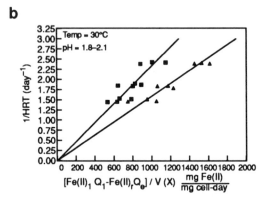

c

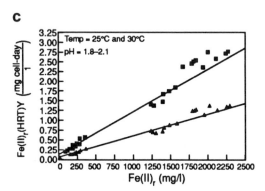

Figure 6-16. (a) Disappearance of Fe^{II} in batch cultures of *Thiobacillus ferrooxidans*: circles, 30°C, pH 1.0–2.1; triangles, 25°C, pH 2.0–2.1; squares, 20°C, pH 2.0–2.1; (b) plot of Equation 2 (Example 6-4) to obtain yield coefficient for *T. ferrooxidans*: squares, volatile suspended solids (VSS) basis; triangles, TOC basis; (c) plot of Equation 3 (Example 6-4) to obtain K_{sg} and k_{Tmax} for same organism: squares, VSS basis; triangles, TOC basis. [From Smith, J.R., R.G. Luthy, and A.C. Middleton, *J. Water Pollut. Contr. Fed.*, 60, 518 (1988). With permission of Water Environment Federation.]

T (°C)	TOC basis		VSS basis	
	K_{Sg}	k_{Tmax}	K_{Sg}	k_{Tmax}
20	60	1030	51	500
25	90	1960	90	980
30	190	2200	190	1100

These results confirm the earlier statement that growth was essentially the same at 25 and 30°C (i.e., k_{Tmax} values are similar at these temperatures). In contrast, k_{Tmax} is about half as large at 20°C. K_{Sg} varied by a factor of three over the temperature range, but it should be noted that this parameter was difficult to determine accurately from the data. Estimates of K_{Sg} depend on k_{Tmax} and on the y-intercept of the graph, both of which are subject to uncertainty.

6.6.7 Structured Growth Models

The concepts of endogenous metabolism and variable yield coefficients have been helpful in expanding the useful range of the Monod model. Nonetheless, this unstructured model does not adequately describe the transient behavior of mixed cultures, even when endogenous metabolism is considered. For example, intracellular storage and excretion of metabolites produce changes in cellular composition over time in response to changing input conditions in mixed culture systems. Unstructured growth models assume that growth is an instantaneous function of [S] (i.e., uptake of S and growth are simultaneous), but in reality these are separate processes. Growth may follow uptake of S only after some lag; this implies that growth depends on the *history* of substrate concentration rather than instantaneous values of [S]. Structured models consider these factors explicitly and describe the behavior of such systems more accurately than unstructured models.

Multistage models that separate nutrient uptake and cell growth have been developed for phytoplankton,[63] and many structured models have been described for growth of heterotrophic bacteria.[64-66] Most include terms for synthesis of storage products, and some include terms for substrate inhibition and enzyme repression when several substrates are present.[66] Others are concerned with growth energetics and vary substrate yield coefficients according to the redox status of the substrate.[67] In-depth treatment of structured models is beyond our scope, but Example 6-5 depicts a relatively simple structured model of heterotrophic bacterial growth and substrate use. The detail provided by such models is obtained at considerable expense: the number of coefficients needed to run them is large, and parameter evaluation is a problem. Consequently, they are of more theoretical interest than practical importance.

Example 6-5. Structured Growth Model for Biological Waste Treatment

Unstructured models are unreliable for modeling microbial growth and substrate removal in wastewater treatment systems because they directly link the distinct processes of substrate removal and microbial growth. In waste treatment systems that receive time-varying inputs

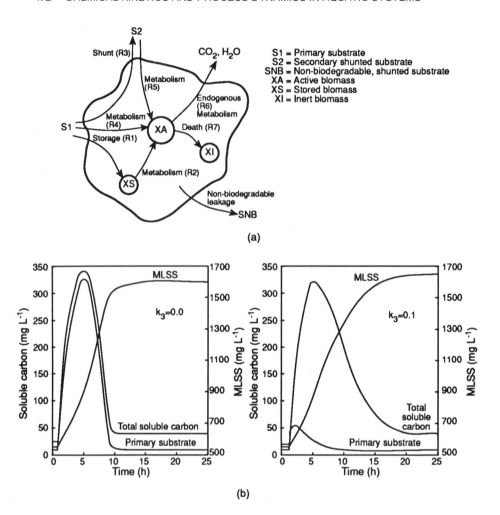

Figure 6-17. (a) Conceptual diagram of structured-growth model for activated sludge. (b) Model simulations: (left) no shunt product ($k_3 = 0$); (right) shunt product formed with $k_3 = 0.1$. [From Dennis, R.W. and R.L. Irvine, *Water Res.*, 15, 1363 (1981). With permission.]

of substrate, microbes remove substrate from solution rapidly and form storage products as insoluble polymers within their cells or the floc in which they reside. The storage products are available for metabolism later when external supplies are low. Extracellular release of simple organic compounds such as acetate and lactate ("shunt products") also may occur, and these may serve as secondary substrates for the same or other organisms in the system.

A conceptual "stoichiometric model" of bacterial growth in activated sludge that accounts for storage and shunt products[64] is described in Figure 6-17a. The model consists of six components and seven metabolic reactions: formation of insoluble storage and soluble shunt products; oxidation of primary substrate, shunt products, and storage products; endogenous maintenance; and dieoff. Changes in amounts of primary and secondary substrate, nonbiodegradable substrate, active and inert biomass, and storage products are modeled as sums of the appropriate rate processes. Reactions defining the system stoichiometry and rate equations for the reactions are given in Table 6-2. Note that formation of active biomass, M_A, is imbedded in the three oxidation reactions, r_2, r_4, and r_5; and loss of M_A is imbedded in two reactions, endogenous maintenance, r_6, and dieoff, r_7. Formation of inert biomass, M_I, is the result of dieoff. Yield coefficients and typical values of rate coefficients used in simulations are given in the table. The model is too complicated for analytical solution and was solved numerically by Runge-Kutta integration.[64] Model simulations with and without shunt (secondary) product formation (Figure 6-17b,c) show how the existence of a shunt pathway can slow the formation of biomass and extend the period of elevated total soluble carbon levels in a reactor.

6.6.8 Synchronous Growth Kinetics: Cyclostats

Chemostats are widely used to obtain kinetic parameters for microbial growth and nutrient uptake, and they have many advantages over batch growth techniques. Nonetheless, the steady-state conditions of chemostats are not representative of growth conditions microorganisms experience in nature. This is especially true for algae, which experience light/dark cycles in nature rather than continuously lit conditions. The natural 24-h light/dark (l/d) cycle tends to induce similar timing in reproduction cycles of algae. Effects of nutrient limitation on events occurring during the cell cycle can be studied most conveniently in the laboratory with a "cyclostat," which is simply a chemostat operated on a light/dark cycle. Algal growth in cyclostats tends to become synchronous with strongly rhythmic cell cycles and cell division in phase (Figure 6-18). Mathematical description of nutrient-limited growth in algal cyclostats is similar to that for chemostats, but the time-dependency of growth must be considered.[68] For example, the instantaneous population growth rate, $\mu(t)$, is not instantaneously equal to the dilution rate D, as in steady-state chemostats,[69] but the period-averaged growth rate $<\mu>_T$ is equal to D:

$$< \mu >_T = D = (1/T)\int_0^T \mu(t)dt \qquad (6\text{-}68)$$

$<\mu>_T$ denotes the period-averaged μ and T is the period length (e.g., 24 h for daily cell-division). DT, the integrated specific growth over period T, is numerically equivalent to the dilution rate (D) of the cyclostat. (DT is dimensionless, but D has units of t^{-1}). Because dilution occurs continuously but cell division occurs periodically (e.g., at the start of a light period), the population density varies during the l/d period and is constant (in steady state) only on a period-averaged basis. As a result, mathematical description of instantaneous cell numbers in cyclostats is complicated (see references 68,69 for further details).

6.6.9 Application of Enzyme Inhibition Kinetics to Microbial Growth

Inhibition by Nonutilizable Substrates and Toxicants

The equations describing enzyme inhibition (Section 6.3.3) can be used to analyze inhibition of metabolic processes in whole microbes. Three examples will

Table 6-2. Equations for Structured Growth Model of Activated Sludge

A. Stoichiometric reactions[a,c]	B. Kinetic expressions[b,d]
R1 Storage product manufacture: $-S1 + Y_{1MS}M_S + Y_{1SNB}SNB = 0$	$r_1 = k_1[M_A][S1]$
R2 Storage product oxidation: $-M_S - Y_{2O_2}O_2 + Y_{2MA}M_A + Y_{2CO_2}CO_2 = 0$	$r_2 = k_2[M_A]$
R3 Shunt product formation: $-S1 + Y_{3S_2}S2 + Y_{3SNB}SNB = 0$	$r_3 = k_3[M_A][S1]$
R4 Primary substrate oxidation: $-S1 - Y_{4O_2}O_2 + Y_{4MA}M_A + Y_{4SNB}SNB + Y_{4CO_2}CO_2 = 0$	$r_4 = k_4[M_A][S1]/(K_{S4} + [S1])$
R5 Secondary substrate oxidation: $-S2 - Y_{5O_2}O_2 + Y_{5MA}M_A + Y_{5SNB}SNB + Y_{5CO_2}CO_2 = 0$	$r_5 = k_5[M_A][S2]/(K_{S5} + [S2] + K_{iS1}[S1])$
R6 Endogenous maintenance: $-M_A - Y_{6O_2} + Y_{6CO_2} = 0$	$r_6 = k_6[M_A]$
R7 Natural death: $-M_A + Y_{7MI}M_I = 0$	$r_7 = k_7[M_A]$

a   Model components: S1 = primary substrate; S2 = secondary (shunted) substrate; SNB = nonbiodegradable shunted substrate; M_A = active biomass; M_S = storage products; M_I = inert biomass; Y_{kC} = yield coefficient for component C in reaction k.

b   Rates (r_k) in mg L^{-1} h^{-1}; concentrations of all components in mg L^{-1}. Constraints are placed on several reactions to avoid instabilities and contradictions to the model's logic: $r_1 = 0$ when [S1] < a limiting value or storage capacity of system is exceeded; $r_2 = 0$ when either S1 or S2 is present or when $M_S = 0$; $r_3 = 0$ when [S1] < some limiting value; $r_6 = 0$ when $r_2 = 0$.

c   Values of the yield coefficients used by Dennis and Irvine[64] are: $Y_{1MS} = 1.0$, $Y_{1SNB} = 0$ (no release of nonbiodegradable carbon during storage); $Y_{2O_2} = 2.67$, $Y_{2MA} = 0.0$, $Y_{2CO_2} = 1.0$ (no growth associated with storage product use; storage products are oxidized at respiratory quotient of 1.0 to supply maintenance energy during starvation; $Y_{3S_2} = 1.0$, $Y_{3SNB} = 0.0$ (no release of nonbiodegradable carbon during shunt); $Y_{4O_2} = 0.92$, $Y_{4MA} = 0.64$, $Y_{4SNB} = 0.02$, $Y_{4CO_2} = 0.34$, $Y_{5O_2} = 0.34$, $Y_{5MA} = 0.92$, $Y_{5SNB} = 0.02$, $Y_{5CO_2} = 0.34$ (these values are based on production of 4 mol ATP per mol O_2 consumed, 10.5 g dry cells produced per mol ATP, and respiratory quotient = 1); $Y_{6O_2} = 3.7$, $Y_{6CO_2} = 1.0$ (the respiratory coefficient is 0.72; $Y_{7MI} = 1.0$ (no shunt during death).

d   Values of the rate coefficients used for the simulations in Figure 6-17 are k_1 (L mg^{-1} h^{-1}) 0.0; k_2 (h^{-1}) 0.018; k_3 (L mg^{-1} h^{-1}) 0.0; k_4 (h^{-1}) 0.66; K_{S4} (mg L^{-1}) 5.0; k_5 (h^{-1}) 0.15; K_{S5} (mg L^{-1}) 200; k_6 (h^{-1}) 0.013; k_7 (h^{-1}) 0.005; K_{iS1} (dimensionless) 0.0.

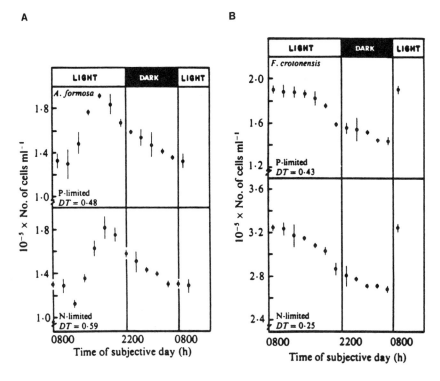

Figure 6-18. Cell number oscillations averaged over three photocycle for P- and N-limited cultures of (A) *Asterionella formosa* and (B) *Fragilaria crotonensis* grown in cyclostats at 19°C on 14/10 h l/d cycle. DT is the integrated growth rate; bars are standard errors of means. [From Gotham, I.J. and G.-Y. Rhee, *J. Gen. Microbiol.*, 128, 199 (1982). With permission.]

be mentioned. First, the equation for competitive enzyme inhibition (Equation 6-24) describes glucose uptake by anaerobic yeast in the presence of L-sorbose;[70] K_s and K_i are $5.6 \times 10^{-4}\,M$ and $0.18\,M$, respectively. Uptake of glucose under aerobic conditions was inhibited by L-sorbose, but did not follow unidirectional Monod kinetics. Second, effects of heavy metals on the metabolic activity of activated sludge organisms have been described by enzyme inhibition kinetics.[71] However, the mechanisms of heavy metal toxicity are complicated and involve various enzymes and cellular processes. Consequently, caution must be observed in ascribing molecular processes to inhibition patterns. Third, Rhee[72] found that phosphate uptake by *Scenedesmus* in a P-limited chemostat could be described only when a term was included for internal phosphate. The result is equivalent to the equation for noncompetitive enzyme inhibition.

Growth Inhibition at High Substrate Concentration

The Monod model implies that μ asymptotically approaches a value μ_{max} that remains constant indefinitely as [S] increases. However, all substrates that are sufficiently soluble depress μ at high concentration (Figure 6-19) if for no other

a

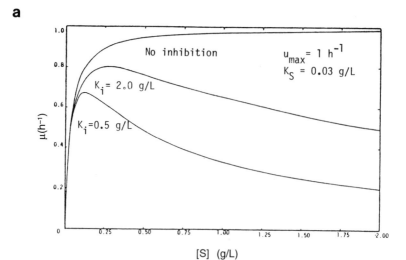

b

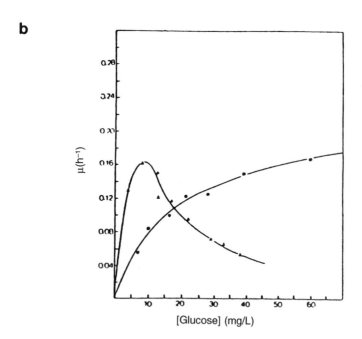

Figure 6-19. Growth curves for substrate inhibition. (a) Hypothetical plots of Equation 6-69 for several values of K_i. [From Andrews, J.F., *Biotech. Bioeng.*, 10, 707 (1968). With permission.] (b) Growth of mixed microbial batch culture on glucose alone (circles) and glucose plus 2,4-D (triangles), and (c) 2,4-D alone (circles) and 2,4-D in presence of glucose. Initial [2,4-D] in (b) ≈ 55 mg L⁻¹; initial [glucose] in (c) ≈ 45 mg L⁻¹. [Both (b) and (c) from Papanastasiou, A.C. and W.J. Maier, *Biotech. Bioeng.*, 24, 2001 (1982). With permission.] (d) Measured rates of phenol uptake (μ) vs. phenol concentration for *Pseudomonas* cells taken from chemostats operated at various residence times: □ 5.25 h, ▽ 3.85 h, + 3.2 h, △ 3.0 h, ○ 2.7 h. Curves are best fits according to simplified Haldane equation (6-72). [From Sokol, W. and J.A. Howell, *Biotech. Bioeng.*, 23, 2039 (1981). With permission.]

c

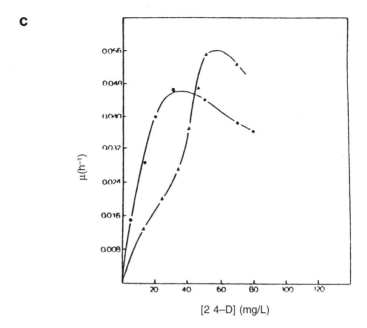

[2 4–D] (mg/L)

d

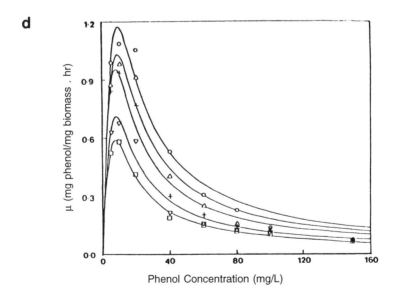

Phenol Concentration (mg/L)

Figure 6-19 (c,d).

reason than osmotic stress. In some cases the plateau of optimum growth is broad; concentrations of simple organic substrates can be varied over two or three orders of magnitude without depressing μ. In contrast, some trace metals are toxic at concentrations only slightly higher than their optimum for growth. Cu^{2+} inhibits some phytoplankton at concentrations as low as 10^{-9} M, and optimum concentrations are in the range 10^{-10} to 10^{-9} M.[73] As Figure 6-19 shows, there are two possible values of [S] for every value of μ — a suboptimal value corresponding to Monod kinetics and a supraoptimal value corresponding to inhibition. Inhibition can be caused by numerous physical-chemical stresses, and a single equation to predict μ is not likely to fit the entire range of concentrations for all substrates and nutrients.

Haldane's equation[30] for enzyme inhibition at high [S] is the most common model for inhibitory effects of high concentrations of organic substrates on microbial growth:[64,74-78]

$$\mu = \frac{\mu_{max}[S]}{[S]+K_S+[S]^2/K_i} = \frac{\mu_{max}}{1+K_S/[S]+[S]/K_i} \tag{6-69}$$

μ_{max} is the maximum growth rate in the absence of inhibition, K_S is the lowest concentration of S at which μ is half the maximum rate in the absence of inhibition, and K_i is an inhibition constant equal to the highest [S] at which μ is half the maximum (uninhibited) rate. At low [S] (the usual condition for steady-state operation of a CFSTR), [S]/K_i is small compared to K_S/[S], and Equation 6-69 reduces to the Monod expression. However, [S] may be high in batch or plug-flow cultures during the early stages of growth and in CFSTRs under transient or start-up conditions. Figure 6-19a shows plots of Equation 6-69 for several values of K_i. Even when $K_i \gg K_S$, the maximum observed growth rate is lower than that possible with no substrate inhibition. Operation of a CFSTR in the supraoptimal range promotes process instability, and operation of a batch culture in this range leads to long lag times before growth commences.

The maximum growth rate under conditions of substrate inhibition (μ_{max}^{in}) can be found by differentiating Equation 6-69 with respect to [S] and setting the result equal to zero,[76] from which we find [S] = $(K_iK_S)^{0.5}$. This is the value of [S] at μ_{max}^{in}. Substituting this value into Equation 6-69, we obtain

$$\mu_{max}^{in} = \mu_{max}/(1+2(K_S/K_i)^{0.5}) \tag{6-70}$$

Other equations have been proposed to describe inhibitory effects of high [S] on microbial growth. Edwards and Wilke[79] compared several more complicated models with the Haldane equation for goodness of fit and concluded that models with four parameters did little better in fitting growth data than the three-parameter Haldane model. The difficulty of fitting data to the Haldane equation also has been described,[77,80] and simpler equations that represent limiting conditions of Equation 6-69 have been found to provide a better fit than the complete equation in some cases. Difficulties in fitting data to Equation 6-69 can be illustrated by writing the equation in its inverse form:

$$\frac{1}{\mu} = \frac{1}{\mu_{max}} + \frac{K_S}{\mu_{max}}\frac{1}{[S]} + \frac{[S]}{K_i\mu_{max}} \tag{6-71}$$

One or more of the terms on the right side of Equation 6-71 may contribute insignificantly to the value of $1/\mu$ at some values of [S]. For example, the first term contributes $< x\%$ to $1/\mu$ at all [S] when $(K_S/K_i)^{0.5} > 50/x$.[80] If the errors that affect the experimental values of $1/\mu$ are $> x\%$, the first term is not statistically significant, and a simpler version of the Haldane equation is more appropriate. Note that when $[S] >> K_S$, a plot of $1/\mu$ vs. [S] (Equation 6-71) yields a straight line with y-intercept $= 1/\mu_{max}$. If one can obtain data under these conditions, one can evaluate the intercept and its standard deviation, thus determining whether $1/\mu_{max}$ is significant or not (within experimental error). Unfortunately, practical difficulties in operating at high [S] may preclude obtaining the asymptotic straight line for some substrates.

We noted earlier (Section 6.1.3) that fitting experimental data to nonlinear equations is more appropriately done with nonlinear regression fitting routines than by transforming equations to forms that can be plotted linearly. Unfortunately, nonlinear numerical methods are sensitive to starting conditions. Sokol and Howell[77] found this to be the case in using the Marquardt algorithm to fit data on phenol oxidation rates by continuous cultures of *Pseudomonas* to the Haldane equation; more than one local solution was found for the parameter values. The data were found to fit better to a two-parameter version of the Haldane equation (standard errors of predicted μ were lower):

$$\mu = \frac{\mu_{max} K_i [S]}{K_S K_i + [S]^2} = \frac{k_1 [S]}{k_2 + [S]^2} \tag{6-72}$$

This corresponds to the situation in enzyme inhibition (Equation 6-23) where $[ESS] >> [ES]$ and E_t is divided mainly between the free form and the inhibited state, with ES present only in small amounts. This implies strong inhibition by the substrate and also corresponds to the case in Equation 6-71 where $1/\mu_{max}$ is negligible compared with the other two terms. It is obvious that fitted parameters k_1 and k_2 in Equation 6-72 are not intrinsic constants of the Haldane equation, but are implicit functions of the three constants in that equation.

The Haldane model has been used to describe effects of undissociated volatile acids on microbial growth in anaerobic waste treatment systems.[75] Volatile fatty acids are important substrates in methane formation, but their undissociated forms are toxic to the bacteria that metabolize them. Several workers[66,78,81] have used the Haldane equation for continuous-culture models of microbial growth in the presence of two (or more) substrates, each of which inhibits the uptake of the other (see Section 6.8.2). Inhibition of microbial growth by the buildup of toxic products was treated by Levenspiel.[44]

6.6.10 Kinetics of Cell Growth in Plug-Flow Reactors

Microbial growth in plug-flow reactors follows the expressions for growth in batch cultures, except that the term dV/Q (or $d\tau_w$) replaces dt. Equations 6-31 to 6-41 apply to PFRs, with dV/Q substituted for dt. Equations for cell growth and substrate use in PFRs are

$$\mu = \frac{Q}{M}\frac{dM}{dV} = yk_{T\max}\frac{[S]}{K_S + [S]} \tag{6-73a}$$

and
$$-\frac{Q}{M}\frac{d[S]}{dV} = k_{T\max}\frac{[S]}{K_S + [S]} \tag{6-73b}$$

$$k_{T\max}V/Q = \left[\frac{K_S}{[S]_o + [M]_o/y} + 1\right]\ln\frac{[M]}{[M]_o} - \left[\frac{K_S}{[S]_o + [M]_o/y}\right]\ln\frac{[S]}{[S]_o} \tag{6-73c}$$

Equations 6-73a–c often may be simplified to the limiting cases: $[S] \ll K_S$ (first order) and $[S] \gg K_S$ (zero order). In biological waste treatment (the principal application of PFRs), $[S]$ is high compared with K_S, and zero-order kinetics apply: $dM/M = yk_{T\max}dV/Q$. Integrating over reactor volume V yields

$$\ln M_{out}/M_{in} = \mu_{\max}V/Q = \mu_{\max}\tau_w \tag{6-74}$$

Also
$$-d[S]/M = k_{T\max}dV/Q$$

$$-d[S] = k_{T\max}M_{in}\exp(\mu_{\max}V/Q)dV/Q$$

or
$$[S]_{out} = [S]_{in} + (1/y)M_{in}\{1 - \exp(\mu_{\max}\tau_w)\} \tag{6-75}$$

Because zero-order substrate uptake was assumed to obtain Equation 6-75, it applies only where $[S]_{in} \gg K_S$, and it is useful primarily when $M_{in} \ll [S]_{in}$. This condition is necessary to attain significant cell growth (several times the inoculum concentration).

Derivation of $[S]_{out}$ is simpler for PFRs used in wastewater treatment, because a high value of M is maintained by cell recycle. If $M_{in} \gg [S]_{in}$, then M will not change much even if all the substrate is consumed in the reactor. Consequently, M can be treated as a constant ($= \hat{M}$), and

$$-d[S] = k_{T\max}\hat{M}dV/Q$$

or
$$[S]_{out} = [S]_{in} - k_{T\max}\hat{M}V/Q = [S]_o - k_{T\max}\hat{M}\tau_w \tag{6-76a}$$

Since $V = AL$, where A is the cross-sectional area and L is the axial distance, we can compute $[S]$ at any distance along the PFR:

$$[S]_L = [S]_{in} - k_{T\max}\hat{M}(A/Q)L \tag{6-76b}$$

The growth equations for PFRs can be modified to account for longitudinal dispersion (Section 5.3.3), as characterized by the dimensionless Peclet number, P $= \mu L/D_L$, or its reciprocal, P_r. u is axial velocity, and D_L, the effective dispersion coefficient, combines the effects of molecular diffusion and turbulent dispersion. D_L cannot be predicted accurately from reactor geometry and flow rates, but must

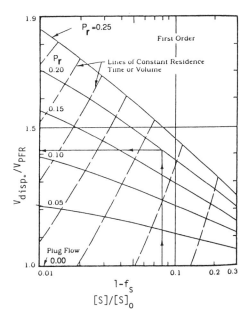

Figure 6-20. Comparison of dispersed-flow and plug-flow reactors for first-order reaction under dispersion conditions found in activated-sludge tanks. P_r is the reciprocal Peclet number. Ordinate is the ratio of volumes of dispersed-flow reactor to volume of PFR needed to achieve a certain substrate use (expressed as the fraction of substrate remaining). [From Reynolds, T.D., *Unit Operations and Processes in Environmental Engineering*, Brooks/Cole Engineering Division, Monterey, CA, 1982. Based on Levenspiel, O. and K.B. Bischoff, *Ind. Eng. Chem.*, 51, 1431 (1959). With permission.]

be evaluated experimentally. At $P_r = 0$, we have perfect plug flow. For reactors used in wastewater treatment, $P_r = \sim\!0.05$ to 0.20.[82]

For the same degree of substrate removal, the detention time must be higher in a dispersed flow reactor than in a PFR of the same dimensions. If flows are the same, the ratio of detention times for the two reactors is the same as the ratio of their volumes. For first-order reactions, Figure 6-20 can be used to determine the volume or detention time ratio (V_d/V_p or τ_d/τ_p) required to obtain the same substrate removals in a PFR and a dispersed-flow reactor of given P_r.[84] For example, assume $P_r = 0.10$ and $[S]_{out}$ is 10% of $[S]_{in}$. According to Figure 6-20, V_d/V_p (or τ_d/τ_p) is 1.22, and τ_d must be $1.22\tau_p$ for equal removal of S. If reactor volumes are the same, flow into the dispersed flow reactor must be lower than that to the PFR by the same factor.

6.7 EFFECTS OF TEMPERATURE ON MICROBIAL GROWTH

6.7.1 Temperature Model for μ_{max} Based on Enzyme Inactivation

Temperature is the most important physical variable affecting microbial growth. It affects the growth parameters μ_{max}, K_S, Y, and k_c in different ways, and temperature sensitivity varies widely among organisms and microbial processes.

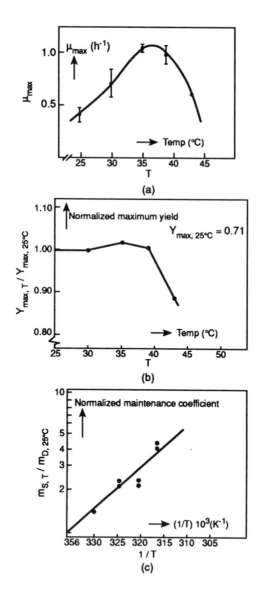

Figure 6-21. Effects of temperature on microbial growth constants: (a) μ_{max} vs. T; (b) normalized Y vs. T; (c) *log*-normalized m_s vs. 1/T (Arrhenius-type plot). [From Esener, A.A., J.A. Roels, and N.W.F. Kossen, *Biotech. Bioeng.*, 23, 2093 (1983). With permission.]

Most workers have used the Arrhenius equation to relate μ_{max} and T, but it is applicable only over a limited range. Above some optimum temperature, T_{opt}, which varies among organisms, μ_{max} decreases rapidly with increasing temperature. This trend is similar to that for effects of temperature on enzyme activity (cf. Figures 6-11a and 6-21a). In fact, inactivation (denaturation) of some key enzyme has been invoked as the explanation of the rapid decrease in μ_{max} above T_{opt},[85] but the decrease may reflect effects on many enzymes. Inactivation of enzymes is not the only possible explanation; changes in membrane permeability also could cause μ_{max} to decrease with increasing T.

The hypothesis that enzyme inactivation causes the decrease in μ_{max} above T_{opt} can be expressed as a quasitheoretical model.[86,87] Assume that the key enzyme or set of enzymes exists in active (A) and inactive (I) forms that are in equilibrium with each other. The effect of temperature on this equilibrium is

$$K_{eq} = f_I / f_A = C \exp(-\Delta H_I / RT)$$ (6-77)

(van't Hoff's equation); f_I and f_A are the fractions of enzyme in the two configurations; C is a constant; and ΔH_I is the enthalpy change for inactivation. From the fact that $f_I + f_A = 1$, it follows that $f_A^{-1} = 1 + C\exp(-\Delta H_I/RT)$. Similarly, μ_{max} can be described by the Arrhenius equation:

$$\mu_{max} = A \exp(-\Delta H_{act} / RT)F'$$ (6-78)

ΔH_{act} is the enthalpy of activation ($\approx E_{act}$), and F' is the fraction of biomass represented by the active form of the key enzyme(s). Finally, $F' = Ff_A$, where F is the fraction of biomass represented by the amount of key enzyme(s) in both active and inactive forms. These relationships lead to

$$\mu_{max} = \frac{A' \exp(-\Delta H_{act} / RT)}{1 + C \exp(-\Delta H_i / RT)}$$ (6-79)

The model fits observed trends in μ_{max} vs. T (e.g., Figure 6-21a) for reasonable parameter values.[87] However, other mechanisms (e.g., changes in membrane permeability) could lead to the same model, and other relationships could be derived to fit the observations.

6.7.2 Effects of Temperature on Kinetic Parameter Values

The effect of temperature on μ_{max} is expressed quantitatively (for the range in which the Arrhenius equation applies) by E_{act}, an *apparent* activation energy for growth, or the related parameters, θ_T and Q_{10} (Section 3.2.6). Values of E_{act} tabulated by Button[42] range from 38 to 280 kJ mol^{-1} (median = 64.8) for six heterotrophs, and 45 to 253 kJ mol^{-1} (median = 89) for five algal species. The values do not appear to be related to μ_{max}, the temperature of μ_{max}, or cell size. Temperature affects growth rates of microorganisms to widely different extents. Microbial processes also have a wide range of temperature sensitivity (Table 6-3), which seems to depend on environmental conditions. The situation is complicated in nature by the presence of mixed communities in which various species may have different values of Q_{10} and T_{opt}. Consequently, it is risky to generalize about temperature effects on growth and metabolic process rates.

Few data are available regarding the effects of temperature on K_S. In some cases

Table 6-3. Temperature Coefficients for Some Microbial Processes and Species

A. Processes

	θ_T	Q_{10}	Comments/conditions
μ_{max} for algae		1.88–2.19	
Filtration rate			
of *Mytilus edulis*		2.15	(277–287 K)
		1.17	(283–293 K)
		1.001	(288–298 K)
		0.059	(293–303 K)
Nitrification			
NH_4^+ oxidation	1.083		range = 1.055–1.10 (n = 5)
NO_2^- oxidation	1.06		
Ammonification			
org N → NH_4^+	1.08		
Benthic O_2 demand	1.075		range = 1.04–1.09, n = 9
BOD exertion	1.055		range = 1.047–1.075, n = 4
Sulfate reduction		3.4–3.9	marine sediments
Detritus decomposition	1.04		
Coliform dieoff	1.04		
Denitrification	1.05		
Denitrification	1.13		

B. K_S Values

Organism/community	Substrate	Q_{10}
E. coli	glucose	1.33
A. aerogenes	glucose	0.51
Activated sludge	glucose	1.35
Activated sludge	wastewater	2.29
Activated sludge	syn. waste	2.62
Activated sludge	linoleic acid	3.10
Anaerobic sludge	acetic acid	0.17
Anaerobic sludge	complex waste	0.50

C. Species

Organism	Type	μ_{max} h^{-1}	Temp. °C	E_{act} kcal mol^{-1}
E. coli	enteric bacterium	1.97	42	14.2
Pseudomonas sp.	psychrophilic bacterium	1.2	32	9.1
Enterobacter cloacae	freshwater bacterium	0.56	25	67
Cytophaga sp.	freshwater bacterium	0.31	25	17
Rhodotorula sp.	marine yeast	0.17	18	26
Pseudomonas fluorescens	freshwater bacterium	0.13	10	61
Thalassiosira weisflogii	marine diatom	0.075	20	24.5
Phaeodactylum tricornutum	marine diatom	0.066	20	10.8
Chaetoceros simplex	marine diatom	0.058	20	11.6

Parts A and B summarized from Jorgenson, S.E., *Handbook of Environmental Data and Ecological Parameters*, Pergamon Press, New York, 1979; part C summarized from Button, D.K., *Microb. Rev.*, 49, 270 (1985).

K_S increases with temperature; in others it decreases. This is not surprising in that K_S is a *ratio* of rate coefficients. Temperature effects on K_S sometimes are reported as an apparent E_{act}, but because K_S is not a rate constant, this is not proper. Q_{10} values (ratio of K_S at 30°C to K_S at 20°C) range from 0.17 to 3.1 for various heterotrophic processes[55,88,89] (Table 6-3). K_S for bacterial ammonium oxidation ranges from 35 to 150 μM over the range 10 to 35°C,[90] but a consistent trend is not apparent. A few data have been reported for the effects of temperature on K_S for nutrient-limited algal growth. K_S for SiO_2-limited growth of the diatom *Synedra ulna* was 5.0 μM and independent of T in the range 8 to 24°C, and K_S (SiO_2) for *Asterionella formosa* was constant between 4 and 13°C, but increased fivefold at 24°C.[46] It has been proposed that K_S for growth is independent of T below T_{opt}, but increases sharply above T_{opt}. Minimum cell quotas (Q_o) for nutrients (Equation 6-46) also vary with temperature. For example, Q_o increased for both limiting and nonlimiting nutrients with decreasing T in N- and P-limited cultures of two diatoms[91] and reached minimum values for C and N in five marine phytoplankton at T_{opt} for cell division.[92]

Effects of temperature on yields of microbial biomass are exerted through effects on Y, the maximum yield coefficient, and k_c, the maintenance coefficient. Limited observations suggest that Y is a discontinuous function of T — nearly constant over some range and then dropping rapidly as the organism's temperature limit is approached (Figure 6-21b). This pattern is expected if there are no significant changes in metabolism over the range in which the organism is adapted, but large metabolic changes as it reaches its temperature limit.

In contrast, data for k_c (a first-order rate coefficient) fit the Arrhenius relationship (Figure 6-21c).[88] Temperature sensitivity of k_c is expressed in terms of E_{act}. Such values have no intrinsic meaning, but are a convenient way to quantify the temperature dependence of k_c. A wide range of $E_{act}(k_c)$ values has been reported for heterotrophic microorganisms, and the use of a mean value in simulations could lead to large errors. Esener et al.[88] reported ten values of $E_{act}(k_c)$ ranging from 38 to 284 kJ mol^{-1}. Excluding one high value, the mean and 95%-confidence range was 76.1 ± 18.4 kJ mol^{-1}. Heijnen and Roels[93] reported a mean and 95%-confidence range of 38.9 ± 18.0 kJ mol^{-1}.

6.8 KINETIC MODELS OF HETEROTROPHIC GROWTH IN AQUATIC SYSTEMS

Prediction of microbial growth rates and rates of microbially mediated processes in natural and managed aquatic systems is important in many branches of science, including ecology, biogeochemistry, and environmental engineering. In this section we apply the relationships described in Sections 6.5 and 6.6 to this task. Our emphasis is on relatively uncomplicated single-species growth models rather than complex simulations of populations in ecosystems.

6.8.1 Kinetic Parameters for Heterotrophs

Y and k_c have been determined for heterotrophic bacteria in batch and continuous-culture experiments,[42,55,88] and K_{Sg} (or K_{Su}) and μ_{max} have been determined for many organisms and substrates in pure cultures. Tabulations of these parameters are given by Jorgensen[94] and Button,[42] and Table 6-4 lists values for some bacteria and substrates of interest in aquatic systems. Rates predicted with these parameters should be considered rough estimates for several reasons. Natural populations are

Table 6-4. Kinetic Constants for Growth and Substrate Uptake by Bacteria

A. Organic substrates

Organism/ community	Substrate	$K_{s,u}$ μM	$K_{s,g}$ μM	μ_{max} h⁻¹	Y	k_e h⁻¹	Ref.[a]
Ancalomicrobium sp.	acetate	13–19					94
Arthrobacter sp.	glucose					0.009	42
Azotobacter sp.	glucose			0.3–0.4			94
Candida sp.	glucose					0.0038	42
Corynebacterium	glucose		2.2–2.6		0.42		42
Denitrifiers (mixed)				0.075			94
E. coli	glucose	19–550					94
E. coli (sw)[b]	glucose	44					94
E. coli	glucose					0.015–0.025	42
E. coli	lactose	111					94
E. coli	lactose	50					94
E. coli (sw)[b]	lactose	6–24					94
Mixed community	toluene						94
Pseudomonas sp.	glucose		0.47		0.28	0.002	42
Pseudomonas putida	phenol	10				0.012–0.013	42
Pseudomonas putida	p-chlorophenol	10					42
Spirillum sp.	glucose					0.006	42
Sulfate reducer	acetate		95	0.014			94
Sulfolobus				0.11–0.14			94

B. Nitrogen cycle reactions[c]

	Temp. °C	K_S N reaction μM	K_S O₂ uptake μM	μ_{max} h⁻¹
Nitrosomonas sp. (NH₄⁺ oxidation)	293	86	9	0.33–0.70
	298	250		0.33
	303	715	16	1.5–2.2
Nitrobacter sp. (NO₂⁻ oxidation)	291–3	360–570	8	0.14 (act. sludge)
	298			0.65 (act. sludge)
	303–5		16–31	1.39 (act. sludge)
Denitrification (NO₃⁻ reduction)	308	220		(lake sediment)

[a] See references 42 and 94 for references to original literature; reference 42 provides a longer list of K_S values.

[b] sw = seawater.

[c] Summarized from Jorgensen, S.E., *Handbook of Environmental Data and Ecological Parameters*, Pergamon Press, New York, 1979.

Table 6-5. Threshold Concentrations for Growth and Substrate Uptake

Substrate	Organism(s)	T
Arginine	*Corynebacterium* sp.	0.53 μM
Glucose (+ amino acids)	*Corynebacterium* sp.	<0.16 nM
Glucose (+ arginine)	*Corynebacterium* sp.	0.83 μM
Glucose (alone)	*Corynebacterium* sp.	1.3 μM
Glucose	*Rhodotorula rubra*	0.55 μM
Lactose	*Spirillum serpens*	50 μM
Phosphate	*Rhodotorula rubra*	3 nM
Phosphate	*Nitzschia actinastroides*	10 nM
Phosphate	*Selenastrum capricornutum*	15 nM
Silicate	six species of diatoms	0.3–1.3 μM ave. = 0.5 μM

Summarized from Button, D.K., *Microb. Rev.*, 49, 270 (1985); original references cited therein.

mixtures of many species; ambient temperatures likely are different from those used to determine the values; and other factors not considered in the growth equations may influence ambient rates.

For most parameters, a wide range of values exists in the literature, and considerable uncertainty arises in selecting a value for a mixed culture or previously unmeasured species. For example, the range of k_e in Table 6-4 is 0.002 to 0.025 h^{-1}, and the mean and standard deviation of the eight values is 0.0107 ± 0.007 h^{-1}. Only a few values of threshold concentrations for heterotrophic growth are available (Table 6-5), and the wide range of values does not encourage extrapolation to other species or substrates. The adaptive ability of microorganisms leads to problems in applying literature values of μ_{max} and K_S outside the original measurement conditions. For example, slowly increasing the dilution rate over a period of weeks can double the value of μ_{max} for both heterotrophs[95] and nutrient-limited algae.[96]

The applicability of Monod kinetics to mixed microbial populations and mixtures of natural substrates (e.g., algal excretory products) has been the subject of much discussion.[13,97-102] Kinetic diversity within and among different natural microbial populations may cause variations in μ_{max} and K_S if substrate concentrations vary widely,[97,99] and these parameters should not be used for conditions greatly different from those used to obtain them. Nonetheless, Bell[100] reported that uptake of ^{14}C-labeled algal extracellular products by aquatic bacteria fit the single-substrate uptake model. This suggests that either the excretory products were dominated by a single compound or that the individual compounds were equally available to the bacteria; i.e., V and K_S were constant over the range of compounds and bacterial species. In contrast, phosphate uptake by lake plankton has been found to fit a compound Michaelis-Menten model with multiple sets of V and K_{S_u}.[101,102]

6.8.2 Heterotrophic Growth in the Presence of Several Substrates

The growth models described heretofore assume only one substrate is available. In natural ecosystems and reactors used in waste treatment, organisms are exposed to many substrates simultaneously, and microbial growth in the presence of multiple substrates thus is an important topic. The subject is especially relevant to degradation of toxic substances, which usually occur in water at much lower concentrations than normal substrates and are degraded (cometabolized) by the same microorganisms that metabolize the latter substrates.

A general equation for microbial growth on multiple substrates is[81]

$$\mu = \sum_{i=1}^{n} [\mu_{\max i} [S]_i / (K_i + \sum_{j=1}^{n} a_{ij} [S]_j)] \tag{6-80}$$

Coefficients a_{ij} represent the inhibitory effect of the jth substrate on uptake of the ith substrate; $a_{ii} = 1$. If $a_{ij} = 1$, the jth substrate has the same effect on uptake of the ith substrate as the ith substrate itself. If $a_{ij} > 1$, the jth substrate inhibits uptake of the ith substrate, and if $a_{ij} < 1$, it promotes uptake of the ith substrate.

Example 6-7. Two-Substrate Continuous-Culture Model

Yoon et al.[81] developed a CFSTR model for microorganisms that can grow on two energy substrates, S_1 and S_2, supplied at a constant rate. Mass-balance equations for $[S_1]$, $[S_2]$, and biomass (M), are

(1)
$$dM / dt = -DM + \mu M$$

(2)
$$d[S_1] / dt = D\{[S_1]_{in} - [S_1]\} - (1/y_1)\mu_1 M$$

(3)
$$d[S_2] / dt = D\{[S_2]_{in} - [S_2]\} - (1/y_2)\mu_2 M$$

The specific growth rate for the organism is

(4)
$$\mu = \mu_1 + \mu_2$$

where (5)
$$\mu_1 = \frac{\mu_{\max 1}[S_1]}{K_{S1} + [S_1] + a_{12}[S_2]}$$

(6)
$$\mu_2 = \frac{\mu_{\max 2}[S_2]}{K_{S2} + [S_2] + a_{21}[S_1]}$$

It is interesting to note that although $\mu = \mu_1 + \mu_2$, $\mu_{\max}$ is less than $\mu_{\max 1} + \mu_{\max 2}$ for all combinations of $[S_1]$ and $[S_2]$. Both μ_1 and μ_2 are functions of $[S_1]$ and $[S_2]$ because of inhibiting effects each substrate has on uptake of the other. If S_1 is the preferred substrate, it has a strong inhibitory effect on uptake of S_2, and $a_{21} > 1$. The enzyme mechanism on which Equations 4 and 5 are based assumes $a_{12} = 1/a_{21}$, which means that evaluation of one coefficient specifies the other. This may not be true for microbial uptake of substrates, however. For example, concurrent use of glucose and benzoic acid yielded greater growth rates than occurred with either substrate alone.[103] This implies that both a_{12} and a_{21} are less than 1 and that they behave independently. Further work is needed to evaluate relationships between the substrate inhibition coefficients. This is important in modeling the biodegradation of refractory organic compounds in contaminated systems such as groundwaters. Such compounds are not primary substrates, but are cometabolized by bacteria growing on conventional substrates.

At steady state, $\mu = D$, and Equations 2–4 yield

(7)
$$\hat{M} = y_1 \left\{ [S_1]_{in} - [\hat{S}_1] \right\} + y_2 \left\{ [S_2]_{in} - [\hat{S}_2] \right\}$$

î is the steady-state value of i. $[\hat{S}_1]$ and $[\hat{S}_2]$ cannot be evaluated from the steady-state solutions of Equations 2 and 3 because μ_1 and μ_2 are not measured directly, but the total specific growth rate (μ) is. Algebraic substitution of Equations 5 and 7 into Equation 2 and Equations 6 and 7 into Equation 3 at steady state yields

(8) $$0 = y_1 D \left\{ [S_1]_{in} - [\hat{S}_1] \right\} - \frac{\mu_{max\,1}[\hat{S}_1] y_1 \left\{ [S_1]_{in} - [\hat{S}_1] \right\} + y_2 \left\{ [S_2]_{in} - [\hat{S}_2] \right\}}{K_{S1} + [\hat{S}_1] + a_{12}[\hat{S}_2]}$$

(9) $$0 = y_2 D \left\{ [S_2]_{in} - [\hat{S}_2] \right\} - \frac{\mu_{max\,2}[\hat{S}_2] y_1 \left\{ [S_1]_{in} - [\hat{S}_1] \right\} + y_2 \left\{ [S_2]_{in} - [\hat{S}_2] \right\}}{K_{S2} + [\hat{S}_2] + a_{21}[\hat{S}_1]}$$

Equations 8 and 9 define the steady-state concentrations of S_1 and S_2 as functions of D; all other terms are constants or input parameters. a_{12} and a_{21} can be calculated from these equations if $[\hat{S}_1]$ and $[\hat{S}_2]$ are measured for a given D. The equations can be solved for $[\hat{S}_1]$ and $[\hat{S}_2]$ at various $D < \mu_{max}$ by trial and error or by numerical methods.

Biodegradation of the herbicide 2,4-dichlorophenoxyacetic acid (2,4-D) by mixed bacterial cultures in the presence of glucose (Figure 6-19b,c) has been described by a model similar to that in Example 6-7. This system may serve as a simple model for the behavior of moderately refractory organic chemicals in wastewater treatment systems. The Haldane equation (6-69) was used to describe growth on 2,4-D, and the equation was modified to describe mutual uptake inhibition by the two substrates. The inhibition term for glucose in the 2,4-D equation is nonlinear, however, in contrast to the linear terms in Example 6-7. In spite of the glucose inhibition factor in the 2,4-D growth equation, the presence of glucose led to more rapid use of 2,4-D in batch cultures because more biomass was produced. Application of these results to a CFSTR model[104] led to the conclusion that "mixed" growth (concurrent uptake of glucose and 2,4-D) allows removal of 2,4-D at higher dilution rates and lower mean cell residence times than possible in the absence of glucose. In addition, the model suggested that process stability could be maintained over a greater range of conditions when glucose was present. These findings have interesting implications for the biodegradation of similar compounds in engineered systems.

6.8.3 Species Competition for Substrates

Just as there normally is more than one substrate in a real world system, there normally are several (often many) species apparently coexisting in a natural aquatic ecosystem. Coexistence is not a problem if the populations do not interact, but if such interactions occur, ecological theory suggests that they should lead (at equilibrium) to extinction of all but one species. The question arises: "Why does the theory not prevail?" In aquatic ecology, this is known as the "paradox of the plankton".[105] Microbial interaction patterns can be grouped into two types: (1) competition for resources (food, light), which reduces resource availability and thus reduces growth rates (this is sometimes called indirect competition); and (2) interference by release of toxic chemicals into the growth environment (sometimes called direct competition). The latter is called antagonism when two species affect mutual releases, and amensalism when only one species releases a toxin. Many possibilities exist within these major categories (see reference 106 for further details).

Mathematical treatment of species competition has been developed only for rather simple cases, like that between two species for a single substrate and for two substrates in chemostats. Aris and Humphrey[107] showed that steady-state coexistence of two species growing on a common substrate can occur in a well-mixed environment only when the growth curves (μ vs. $[S]$) intersect, as can happen when one species is inhibited by high $[S]$ while the other is not (Figure 6-19b) or when both μ_{max} and K_S are higher for one species than another (Figure 6-22a). The conditions for coexistence in a chemostat are $\mu_1 = \mu_2$, $D = D_{ss}$, and $\mu_1 - D_{ss} = \mu_2 - D_{ss} = 0$. In theory, this represents a stable situation, but in practice it is not. In any real chemostat, D is subject to random fluctuations that may be biased ($D_{av} \neq D_{ss}$). Model analysis[108] has shown that exclusion of one or the other population always occurs under such circumstances.

Example 6-8. Two-Species/Two-Substrate Microbial Growth Model

Yoon et al.[81] also derived a model for competition of two species growing on two substrates. Four steady states are possible: coexistence of the species, washout of both species, and washout of one species and survival of the other. For coexistence, the steady-state growth rates, population densities, and substrate concentrations must satisfy the equations in part B.

A. Rate Equations

Let M_1 and M_2 be mass concentrations of microbial species 1 and 2 growing on substrates S_1 and S_2, both of which can be used by both species. Then

(1) $$dM_1/dt = -DM_1 + \mu_1 M_1$$

(2) $$dM_2/dt = -DM_2 + \mu_2 M_2$$

(3) $$d[S_1]/dt = D([S_1]_{in} - [S_1]) - (1/y_{11})\mu_{11} M_1 - (1/y_{21})\mu_{21} M_2$$

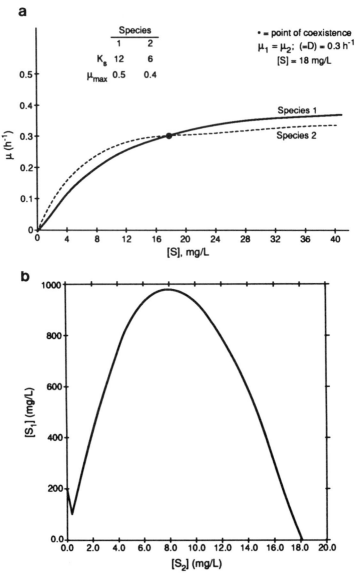

Figure 6-22. (a) Two species competing for a single substrate can coexist at one value of [S] if both μ_{max} and K_S are larger for one species than the other (see Example 6-8). (b) Phase diagram of substrate concentrations $[S_1]$ and $[S_2]$ that satisfy $\mu_1 = \mu_2$ for two-species, two-substrate model (Example 6-8) for parameter values given below. (c) Steady-state concentrations of substrates and species vs. dilution rate for the model. Species 2 dominates in region I; the two coexist in region II; and species 1 dominates at the high dilution rates in region III. $X = X_1 + X_2$. [(b) and (c) from Yoon, H., G. Klinzing, and H.W. Blanch, *Biotech. Bioeng.*, 19, 1193 (1977). With permission.]

	Species 1	Species 2
μ_{max} (h^{-1})	$\mu_{max11} = \mu_{max12} = 0.5$	$\mu_{max21} = \mu_{max22} = 0.4$
K_S (mg/L)	$K_{11} = K_{12} = 12$	$K_{21} = K_{22} = 6$
y	$y_{11} = y_{12} = 0.15$	$y_{21} = y_{22} = 0.15$
a_{ijk}	$a_{211} = 50$	$a_{212} = 50$
	$a_{121} = 1/a_{211}$	$a_{122} = 1/a_{212}$

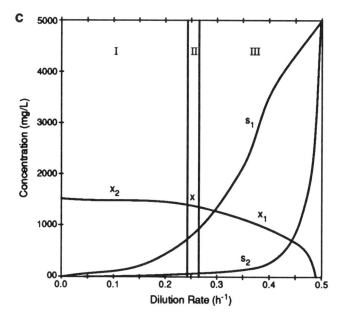

Figure 6-22(c).

(4) $d[S_2]/dt = D([S_2]_{in} - [S_2]) - (1/y_{12})\mu_{12}M_1 - (1/y_{22})\mu_{22}M_2$

(5) $\mu_1 = \mu_{11} + \mu_{12}$

(6) $\mu_2 = \mu_{21} + \mu_{22}$

μ_1 and μ_2 are the aggregate specific growth rates of species 1 and 2. All the component growth rates are functions of $[S_1]$ and $[S_2]$, as follows:

(7) $\mu_{11} = \mu_{max11}[S_1]/(K_{S11} + [S_1] + a_{121}[S_2])$

(8) $\mu_{12} = \mu_{max12}[S_2]/(K_{S12} + [S_2] + a_{211}[S_1])$

(9) $\mu_{21} = \mu_{max21}[S_1]/(K_{S21} + [S_1] + a_{122}[S_2])$

(10) $\mu_{22} = \mu_{max22}[S_2]/(K_{S22} + [S_2] + a_{212}[S_1])$

The a_{ijk} are inhibition coefficients representing the effect of the jth substrate on use of the ith substrate by species k.

B. Steady-State Solution for Coefficients of Species 1 and 2

If the two species co-exist at steady state, the following equations apply:

(11) $$\mu_1 - D = 0; \qquad \mu_2 - D = 0; \qquad \text{i.e.,} \quad \mu_1 = \mu_2$$

(12) $$D\left(\left[S_1\right]_{in} - \left[S_1\right]\right) - \left(1/y_{11}\right)\mu_{11}M_1 - \left(1/y_{21}\right)\mu_{21}M_2 = 0$$

(13) $$D\left(\left[S_2\right]_{in} - \left[S_2\right]\right) - \left(1/y_{12}\right)\mu_{12}M_1 - \left(1/y_{22}\right)\mu_{22}M_2 = 0$$

Equations 7-10, 12, and 13 can be solved numerically for the values of $[S_1]$ and $[S_2]$ at which $\mu_1 = \mu_2$, and the results can be plotted as a "phase diagram" of $[S_2]$ vs. $[S_1]$, as illustrated in Figure 6-22b. The $[S_1]$–$[S_2]$ plane is called a phase plane, a term derived from the use of such plots in mechanics.[109] The fact that coexistence occurs over a range of $[S_1]$ and $[S_2]$ implies that there is a range of dilution rates at which the two species coexist (Figure 6-22c). When only one substrate is fed into a reactor, coexistence occurs only at a single value of D or μ (0.3 h^{-1} in Figure 6-22a). This can be verified by substituting values for the kinetic parameters into the equations for μ_1 and μ_2 at $[S_1]$ or $[S_2] = 0$ and solving for the nonzero $[S]$ when $\mu_1 = \mu_2$. Substitution of this value into one of the growth equations yields the desired answer. When both S_1 and S_2 are present, coexistence is possible over a range of D, but the maximum D at which coexistence occurs is shifted downward (to 0.25 h^{-1} in the figure).

It is sometimes stated that for steady-state coexistence of m species in a homogeneous environment, m nutrients must exert controlling effects. However, several other mechanisms may lead to stable coexistence if the number of controlling nutrients is less than m. Periodic disturbance of inputs to a spatially homogeneous system may help prevent competitive exclusion, but mathematical analysis[106,110] suggests that by itself this cannot explain the coexistence of large numbers of species on a few limiting resources. Selective predation or parasitism may help eliminate competitive exclusion, and the ability of organisms to form dormant (resting) cells may allow them to persist during unfavorable conditions and become active when conditions again become favorable. Production by competing populations of metabolic byproducts with stimulatory effects on growth rates can prevent competitive exclusion; several examples of this type involve two species of heterotrophic bacteria.[106]

Heterogeneity of environmental conditions has been invoked frequently to explain species coexistence; in recent years this argument has been extended to microscale and shortlived perturbations in nutrient levels.[111] In unmixed environments, spatial gradients may exist in $[S]$, and organism motility may be a factor in competition. The case of two randomly motile organisms growing in an unmixed system with diffusion of S into the growth region from an external source has been modeled[112] and could apply to bacteria growing in aqueous films around soil or sediment particles. Coexistence of the species can occur even though one species has a smaller intrinsic μ at all nutrient concentrations if it also has a smaller random motility coefficient. (Random motility decreases population size in unmixed regions, because it tends to disperse bacteria away from the nutrient source.) Positive chemotaxis would help slower growing organisms outcompete faster growing ones under nonsteady-state conditions in a reactor with a spatial gradient in [S]. Random motility also could help slower growing organisms in a homogeneous but unstirred environment by preventing substrate depletion in the boundary layer around cells.

6.8.4 Predator-Prey Relationships

Species interactions may occur by predation rather than competition for common resources. Predator-prey interactions are common among higher trophic levels in aquatic and terrestrial ecosystems and also are important for some microbial heterotrophs (e.g., protozoa graze on bacteria in activated sludge). Selective predation by zooplankton is an important factor in phytoplankton community composition. Predation may lead to oscillating populations of the predator and prey populations rather than steady-state conditions. Such populations dynamics have been observed for larger organisms (e.g., the well-known lynx and hare dynamics[113]), but oscillating populations of microbial predator-prey pairs also can be observed in laboratory systems[114] (Figure 6-23a).

Predator-prey population dynamics often are described by the Lotka-Volterra model,[115] which dates from the 1920s and consists of two rate equations:

$$dn_1/dt = k_1 n_1 - \upsilon n_1 n_2 \qquad (6\text{-}81a)$$

$$dn_2/dt = k_2 n_2 - y\upsilon n_1 n_2 \qquad (6\text{-}81b)$$

n_1 and n_2 are the population densities of prey and predator, respectively; k_1 is the growth rate coefficient of prey, and k_2 the death rate coefficient of predator. The second term of Equation 6-81a is a second-order loss term for prey; υ is a measure of the efficiency of prey capture on a given encounter. The second term in Equation 6-81b describes growth of predator by predation on n_1; y is a yield coefficient (increase in predator population per capture of prey).

The model, which is described in many ecology texts,[116] usually is solved numerically but also can be simulated on analog computers.[117] Steady-state solutions for n_1 and n_2 can be obtained by setting the rate equations equal to zero and solving:

$$n_{1ss} = k_2/y\upsilon; \qquad n_{2ss} = k_1/\upsilon \qquad (6\text{-}81c)$$

The model does not converge to these steady-state values, however, when initial values of n_1 and n_2 and assumed values of the coefficients are used to run the model. Instead, the two populations oscillate out of phase with each other, as in Figure 6-23a. This behavior can be understood by examining a phase plane diagram for n_1 and n_2, which can be derived as follows. The two equations of the Lotka-Volterra model

a

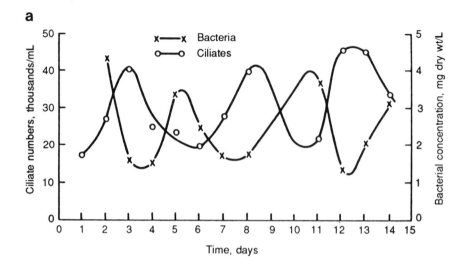

b

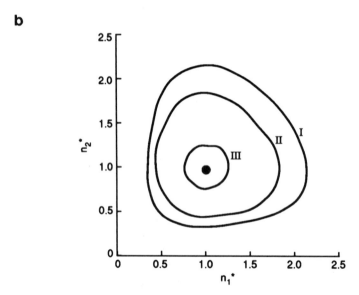

Figure 6-23. (a) Predator-prey oscillations of *Aerobacter aerogenes* (bacteria) and *Tetrahymena pyriformis* (protozoa) grown in a chemostat. [From Meers, J.L., in *Microbial Ecology*, A.I. Laskin and H. Lechvalier (Eds.), CRC Press, Boca Raton, FL, 1974. With permission.] (b) Phase-plane diagram for Lotka-Volterra model for parameter values: $k_1 = k_3 = 0.01$, $k_2 = 0.001$, $y = 0.5$, which yields $n_{1ss} = 20$ and $n_{2ss} = 10$. Curve I is for initial populations $n_{1,0} = 15$, $n_{2,0} = 5$; curve II is for $n_{1,0} = 10$, $n_{2,0} = 5$; curve III is for $n_{1,0} = 18$, $n_{2,0} = 8$. Dot represents steady-state condition. (c) Experimental values and model simulation for bacteria-amoeba predator-prey system by Lotka-Volterra model modified for Monod kinetics. [From Tsuchiya, H.M., et al., *J. Bacteriol.*, 110, 1147 (1972). With permission of the American Society of Microbiology.]

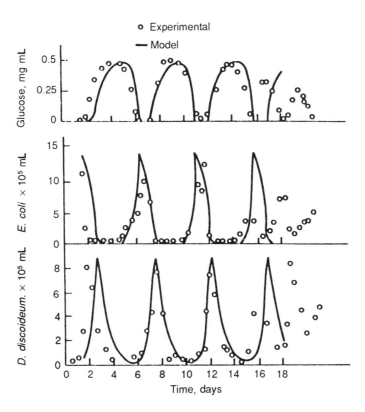

Figure 6-23(c).

(Equations 6-81a,b) first are simplified by normalizing n_1 and n_2 to the steady-state values defined by Equation 6-81c: $n_1^* = n_1/n_1^{ss}$; $n_2^* = n_2/n_2^{ss}$. This results in the following equations:

$$dn_1^* / dt = k_1\left(1 - n_2^*\right)n_1^* \qquad (6\text{-}82a)$$

$$dn_2^* / dt = -k_3\left(1 - n_1^*\right)n_2^* \qquad (6\text{-}82b)$$

Dividing Equation 6-82b by Equation 6-82a eliminates time (dt) and yields a phase plane differential equation that relates changes in n_2^* to the change in n_1^*:

$$\frac{dn_2^*}{dn_1^*} = \frac{-k_3 n_2^*\left(1 - n_1^*\right)}{k_1 n_1^*\left(1 - n_2^*\right)} \qquad (6\text{-}83a)$$

Algebraic rearrangement yields

$$\frac{1}{k_3}\left[\frac{1}{n_2^*}-1\right]dn_2^* = \frac{1}{k_1}\left[1-\frac{1}{n_1^*}\right]dn_1^*$$

(6-83b)

Integration of Equation 6-83b yields

$$k_1 \ln n_2^* - k_1 n_2^* = k_3 n_1^* - k_3 \ln n_1^* + C$$

(6-84a)

or

$$\left[\frac{n_1^*}{e^{n_1^*}}\right]^{k_3}\left[\frac{n_2^*}{e^{n_2^*}}\right]^{k_1} = e^C$$

(6-84b)

where C is a constant of integration that depends on initial population sizes. As Figure 6-23b shows, Equation 6-84 describes an ellipsoidal pathway for movement of n_1^* and n_2^* in the phase plane.

Although the Lotka-Volterra model successfully portrays the oscillating behavior of predator-prey systems, it has several deficiencies. First, the periodicity and amplitude of the oscillations depend on initial conditions, and a small disturbance in the system produces a permanent change in these parameters.[59] This does not seem reasonable in that some populations exhibit fairly stable oscillations in nature in spite of small, frequent perturbations. Second, in the absence of predation, the prey population is uncontrolled exponential growth. Both of these deficiencies are eliminated in a model (Example 6-9) that incorporates Monod relationships to limit the growth of prey by substrate and the growth of the predator feeding on prey. Stability properties of predator-prey models have been analyzed in further detail elsewhere[118] (for a general discussion of stability in multi-species ecosystem models, see reference 59).

Example 6-9. Predator-Prey Model with Monod Kinetics

Tsuchiya and co-workers[119] developed a more realistic model for predator-prey relationships between microbes in chemostats. It incorporates Monod kinetics for predator growth on prey and prey growth on a limiting substrate:

(1)
$$d[S]/dt = D\{[S]_o - [S]\} - \frac{1\mu_{S\max}[S]n_1}{y_S(K_S + [S])}$$

(2)
$$dn_1/dt = -Dn_1 + \frac{\mu_{S\max}[S]n_1}{K_S + [S]} - \frac{1\mu_{p\max}n_1 n_2}{y_p(K_p + n_1)}$$

(3)
$$dn_2/dt = -Dn_2 + \frac{\mu_{p\max}n_1 n_2}{K_p + n_1}$$

D = dilution rate (Q/V); y_S and y_p are yield factors (assumed constant) for prey growth on S and predator growth on prey, respectively; n_1 and n_2 are populations (#/mL) of prey and predator, respectively.

The model was fit to data obtained on a two-species system in the laboratory: *E. coli* (prey) and the amoeba *Dictyostelium discoideum* (predator). Good results were found (Figure 6-23c) for the first several cycles (up to about day 16) with the following kinetic constants:

$$E.\ coli: \quad \mu_{Smax} = 0.25\ h^{-1}; K_s = 0.5\ mg\ L^{-1} (glucose)$$

$$y_s = 3.3 \times 10^{-10}\ mg\ glucose/bacterium$$

$$D.\ discoideum: \quad \mu_{pmax} = 0.24\ h^{-1}; K_p = 4 \times 10^8\ E.\ coli/mL$$

$$y_p = 1.4 \times 10^3\ E.\ coli$$

6.8.5 Modeling Chaos in Biological Systems

The erratic nature of some predator-prey populations (e.g., Figure 6-23a,c) is reminiscent of the phenomenon called chaos, and population biologists have been trying for the past 15 years to determine whether chaos truly exists in natural ecosystems. Somewhat difficult to describe precisely, chaos is a mathematical term for behavior that is seemingly random (i.e., apparently unpredictable), but actually is deterministic, i.e., defined by straightforward mathematical equations. Chaotic behavior is highly sensitive to the choice of initial conditions, just as the Lotka-Volterra model is. Perhaps the simplest population model that yields chaotic behavior is a single equation for one species with generations that do not overlap each other.[120] It applies to insect populations that complete their life cycle over the course of 1 year (including some with aquatic life stages). According to this model, the population in a given generation (N_t) determines the population of the succeeding generation (N_{t+1}) as follows:

$$N_{t+1} = aN_t - bN_t^2 \tag{6-85a}$$

where a corresponds to a birth rate, and b is an overcrowding factor that causes the population to decrease when it becomes too large for its resources. By proper selection of units and normalizing N_t, Equation 6-85a can be simplified to

$$N_{t+1}^* = a'\left(N_t^* - N_t^{*2}\right) \tag{6-85b}$$

where $0 < N_t^* < 1$. Depending on the values of a', this equation can have simple to very complicated behavior (see Figure 6-24). For $0 < a' < 1$, the population dies faster than it is replaced, and N_t^* goes to zero. For $1 < a' < 3$, N_t^* approaches a fixed steady-state value independent of the initial N, but dependent on a'. For $3 < a' < 3.4$, the population oscillates between two values, and as a' increases above 3.4, N_t^*

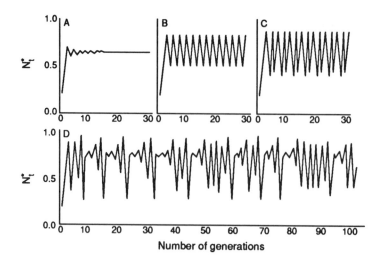

Figure 6-24. Stability of simple population model (Equation 6-85) depends on magnitude of coefficient a'. (A) For $a' = 2.8$, model achieves steady state; (B) for $a' = 3.3$, model oscillates between two values; (C) for $a' = 3.53$, a four-cycle oscillation results; (D) for $a' = 3.7$, chaos results. [From Pool, R., *Science*, 243, 310 (1989). With permission of the American Association for the Advancement of Science.]

oscillates among an increasing number of fixed values — 4, 8, 16, and so on — so-called period doubling. Finally, at $a' > 3.57$, there are no fixed values of N_t^* and the population behaves chaotically.

The key to whether a population model will exhibit chaotic behavior lies in the values chosen for its parameters (birth, survival rates, etc.). In simple models the parameter values must be unrealistically high for chaos to appear, and this led many population biologists to dismiss chaos as a natural phenomenon. The erratic behavior in ambient populations is explained as a direct response to environmental "noise" rather than an intrinsic feature of the populations. However, in more complex, multispecies models, it is easier to get chaos with more reasonable parameter values.[121] Although it still is questionable whether natural ecosystems exhibit chaos, the positive feedback processes inherent in them are said to possess the seeds of chaos,[122] and it is possible to force them into chaotic-like regimes by increasing these feedback parameters. Human activities, like the addition of pesticides to aquatic ecosystems, possibly could induce chaos, but further work is needed on this subject.

6.9 KINETIC MODELS FOR AUTOTROPHIC GROWTH AND NUTRIENT UPTAKE

6.9.1 Kinetic Parameters for Autotrophs

Uptake of inorganic nutrients by phytoplankton usually is described by Michaelis-Menten kinetics. Many simulation models for aquatic ecosystems use Monod kinetics for algal growth, and an extensive literature exists on this subject. K_{Sg}, K_{Su}, and μ_{max} values for NH_4^+, NO_3^-, orthophosphate, and silica for some marine and freshwater species are summarized in Table 6-6, and threshold values for phosphate- and silica-limited algal growth are given in Table 6-5 (longer compilations of kinetic parameters are given in references 42 and 94). The activities of enzymes like phosphatases are highly pH sensitive, and it is likely that transport rate parameters for ionic nutrients like phosphate and nitrate are also pH dependent (because of coupling with H^+ transport).

Maximum specific uptake rates of nutrients by algae (V_{max}) are expressed in h^{-1}. Volumetric rates depend on biomass and are obtained by multiplying V_{max} by some measure of biomass such as dry weight, protein, particulate carbon, or chlorophyll a. Biomass is preferable to cell numbers, since algal size varies greatly both within and among species.

Maximum uptake rates for nutrients are inversely related to cell quotas (Figure 6-25a). Gotham and Rhee[123] found that the apparent V_{max} of NO_3^- uptake at a given μ was inversely related to $Q - Q_o$, where Q_o is the minimum cell quota of N (Equation 6-46). At low Q, large differences occur between V_{max} and V_Q, the steady-state N uptake rate for a given cell quota (Figure 6-25b). Uptake capacity and uptake rate approach each other at high Q, where $\mu \to \mu_{max}$, $Q \to Q_{max}$, and nutrient-saturated growth occurs. Similar results were found for uptake of phosphate vs. Q.[124] These results imply that cells synthesize more of the enzyme(s) used to take up a nutrient in response to their needs for it. Synthesis of more transporting enzymes increases the specific affinity (a_S) and V_{max} and possibly decreases K_{Sg} (but not K_{Su}).

Dugdale[125] was the first to apply enzyme kinetics to algal nutrient uptake. His work led to efforts to evaluate kinetic parameters on natural assemblages of algae and to explore their usefulness in explaining the occurrence of species or assemblages in particular habitats (i.e., explaining competitive abilities). For example, MacIsaac and Dugdale[126] reported that K_S values for NO_3^- uptake by marine phytoplankton from nutrient-poor waters were lower than K_S values for populations in nutrient-rich waters. They hypothesized that the organisms in nutrient-poor waters adapted to become efficient at nutrient uptake under ambient conditions or that selection of strains efficient at low nutrient concentrations had occurred. Other studies[127] have verified that K_S varies with the nutrient status of the water containing the organisms, and there appear to be two general types of uptake systems in algae.[42] Those with high affinity (low K_S) also tend to have low V_{max}, and those with low affinity (high K_S) usually have high V_{max}. Button[42] explained this inverse relationship in terms of a mobil carrier hypothesis analogous to the simple enzyme mechanism in Equation 6-1: as the substrate residence time on the carrier increases (as the lifetime of [ES] increases), k_3 decreases. This decreases V_{max} ($= k_3 E_T$) and also decreases K_S ($= (k_2 + k_3)/k_1$). Thus, a small K_S implies a slow transport system.

As mentioned previously, K_S for nutrient-limited growth is not necessarily the same as K_S for nutrient uptake by a species; separate enzyme systems are involved in nutrient transport into cells and assimilation within cells. Evidence for this exists for N and P, but not for silica.[47] The existence of separate enzymes for uptake and assimilation does not imply that the processes are wholly independent of each other,

Table 6-6. Growth and Nutrient Uptake Parameters for Species and Communities of Algae

	K_S (μM)			V (uptake) (10^{-14} mol cell^{-1} h^{-1})			$\mu_{max}(T)$ [day^{-1}(°C)]
	NH_4^+	NO_3^-	PO_4	NH_4^+	NO_3^-	PO_4	
A.			**Growth**				
Chaetoceros gracilis	—	—	0.8				
Flagellates	—	0.6–9.2	—				
Phytoplankton (L. Erie)	—	—	0.3				
Phytoplankton (L. Ontario)	—	—	0.06				
Selenastrum capricornutum	5.2–5.5	—	—				
B.			**Uptake**				
Anabaena cylindrica	—	70	—	—	—	—	—
Asterionella formosa	—	—	—	—	—	0.8	—
Asterionella japonica	1.0	1.0	0.12	—	—	—	—
Chaetoceros debilis	0.5–2.2	—	—	0.17	—	—	—
Chaetoceros gracilis	0.4	0.2	0.12	—	—	—	2.6 (298)
Chlorella pyrenoidosa	—	—	4.0–5.0	—	—	—	2.1 (298)
Coccolithus huxlei	0.1	0.1	—	—	—	—	1.75
Coscinodiscus lineatus	2.0	2.6	—	—	—	—	—
Coscinodiscus gigas	—	—	0.06–0.12	—	—	—	0.51–0.73
Cyclotella nana	0.4	0.4–1.9	0.03–0.58	—	0.3–1.6	—	2.1 (298)
Dinobryon cylindricum	—	—	0.7–0.8	—	—	—	—
Ditylum brightwellii	1.1	0.6	0.11	—	120	0.20	1.2–2.3
Dunaliella sp.	—	0.3–1.0	—	—	—	—	1.93 (288)
Dunaliella tertiolecta	0.2–0.6	0.2–1.4	—	2.6–10.6	2.2–3.7	—	1.83 (298)
Euglena gracilis	—	—	0.3–16	—	—	24	—
Fragilaria pinnata	—	0.6–1.6	—	—	—	—	—
Freshwater phytoplankton	—	—	0.2–0.8	—	—	—	0.1–2.1
Gonyaulax polyedra	5.5	9.5	—	—	—	—	—
Gymnodinium splendens	1.1	1.0–6.5	—	—	—	—	0.81 (298)

	K_s (μM)	V (uptake) (10^{-14} mol cell^{-1} h^{-1})					μ_{max} (T) [day^{-1} (°C)]
Microcystis aeruginosa	0.3	–	~0.3	–	–	–	–
Monochrysis lutheri	–	0.5–0.6	–	2.1–3.8	1.4	–	0.84 (292)
Neritic bacillariophyceae	–	6.3–28	–	–	–	–	–
Nitzschia closterium	–	–	–	–	–	–	1.75 (300)
Oceanic species	–	1.4–7	–	–	–	–	–
Oscillatoria agardhii	–	–	–	–	–	–	0.86
Oscillatoria theibautii	6.7	–	–	4.2	–	–	–
Pediastrum duplex	–	–	–	–	–	–	–
Phaeodactylum tricornutum	–	2.6	0.3	–	–	4.0	–
Rhizosolenia robusta	7.5	3.0	–	–	–	0.07–0.14	–
Rhizosolenia stolterfothii	0.5	1.7	–	–	–	–	–
Scenedesmus spp.	–	–	0.03–0.6	–	–	–	1.5–2.0
Selanastrum capricornutum	–	–	0.16	–	–	1.2	1.54 (300)
Skeletonema costatum	0.5–1.3	0.5	0.008–0.04	0.1–0.16	150–830	–	1.27 (292)
Thalassiosira gravida	0.3–0.5	–	–	0.03–0.16	–	–	1.5
Thalassiosira pseudonana	–	0.4–1.9	0.6–0.7	0.03–0.16	–	–	1.1–3.6

C. Silica	K_s (μM)	V (uptake) (10^{-14} mol cell^{-1} h^{-1})	μ_{max} (T) [day^{-1} (°C)]
Ditylum brightwellii	2.96	95	1.2–2.3
Nitzschia actinastroides	3.5	–	–
Nitzschia alba	–	0.25–0.57	–
Skeletonema costatum	0.8	0.34	1.27 (292)
Thalassiosira pseudonana	0.2–2.9	0.26	1.1–3.6

Summarized from Jorgensen, S.E., *Handbook of Environmental Data and Ecological Parameters*, Pergamon Press, New York, 1979, who compiled the data from the original literature; for additional values, also see Button, D.K., *Microb. Rev.*, 49, 270 (1985); original references cited therein.

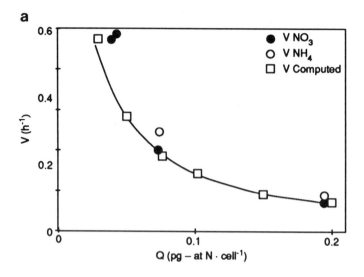

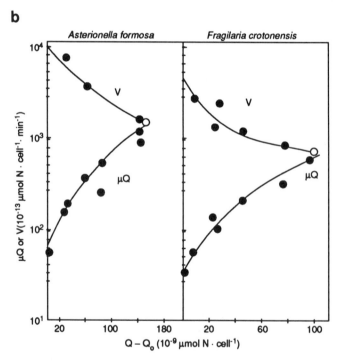

Figure 6-25.   (A) Maximum uptake velocity V (h⁻¹) for inorganic nitrogen vs. cell quota Q for *T. pseudonana*. [From Dugdale, R.C., B.H. Jones, Jr., J.J. MacIsaac, and J.J. Goering, in *Physiological Bases of Phytoplankton Ecology*, T. Platt (Ed.), Can. Bull. Fish. Aquat. Sci., 210, Ottawa.] (B) Semilog plots of maximum uptake velocity for nitrate (V) and steady-state nitrogen uptake rates μQ (growth rate times cell quota) vs. cell quota for nitrogen for two algal species grown in turbidostats. [From Gotham, I.J. and G.-Y. Rhee, *J. Phycol.*, 17, 309 (1981).] Note: μQ vs. Q is a plot of a variable vs. a function of that variable and could lead to spurious self-correlation.[15]

but the steps may not be simultaneous. Dugdale et al.[128] found that algal uptake of nutrients can be controlled either by external concentrations or by internal (cell) conditions. When large spikes of nutrients were added to algae in chemostats, nonlimiting nutrients were taken up at the dilution rate (which is the same as the steady-state specific growth rate), but the limiting nutrient was taken up at higher rates. K_{S_g} usually is determined from steady-state growth measurements in chemostats. K_{S_u} may be calculated from (1) steady-state chemostat measurements or (2) may be measured by adding the nutrient to subsamples from the chemostat and measuring the change in concentration over short periods of time. In case (1), K_{S_g} and K_{S_u} are likely to be similar in value if other nutrients are nonlimiting over the range of measurements and if concentrations of the limiting nutrient in the chemostat can be measured accurately. In case (2), there is no reason to expect concordance between the two constants.

6.9.2 Limitations of Monod/Michaelis-Menten Models for Algal Growth and Nutrient Uptake

Nutrient uptake and algal growth data obtained from chemostats do not always fit the Michaelis-Menten/Monod model, and many modelers have used the cell quota model instead. At steady state the two models are equivalent,[129] but practical limitations in some cases favor analysis of data by the cell quota model. Several problems arise in measuring kinetic constants for nutrient uptake or growth in chemostats. Steady-state nutrient concentrations in chemostats are very low in many cases and thus difficult to determine accurately. K_S may be below analytical detection limits for some nutrients, making it difficult to define the μ vs. [S] curve.[49] In such cases μ_{max} may occur at concentrations only slightly above the detection limit, and large relative changes in [S] at these low values may not affect μ significantly (Figure 6-26). Such results invite the conclusion that growth does not follow Monod kinetics, when in reality the problem is analytical.

The simplicity of the Monod and cell quota models places limits on their ability to describe nutrient-limited algal growth under transient conditions. It would be better to use more complicated structured models that separate nutrient uptake and growth processes for such conditions than to contrive explanations for the seemingly anomalous behavior of the simple models under conditions for which they were not intended. The cell quota model is a simple structured model in that uptake depends on nutrient content, but it does not relate uptake to external conditions or model lags between uptake and growth.

Short-term batch experiments on nutrient-spiked samples are used to avoid the analytical problems encountered in estimating nutrient uptake constants from CFSTR data. However, exposure of organisms acclimated to one nutrient concentration to different concentrations may lead to altered physiological responses, and values of kinetic parameters obtained in this way may not be representative of steady-state growth. The nutrient-perturbation experiments of Dugdale et al.[128] described above are evidence of this behavior.

Similarly, Tarapchak and Herche[101] found many examples where phosphate (P_i) uptake by Lake Michigan plankton deviated substantially from simple Michaelis-

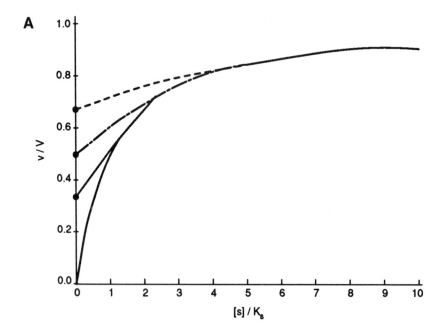

Figure 6-26. (A) Hypothetical shape of uptake rate vs. substrate concentration for Monod kinetics at increasing detection limits for [S]; — det. lim. << K_S; --- det. lim. > $0.5K_S$; -·- det. lim. > $1.0K_S$; __ det. lim. > $2K_S$. In the last three cases it is assumed that the uptake rate can be measured, but that [S] = 0 at [S] < det. limit. (B) Growth rate (μ) of diatom *T. pseudonana* as function of measured NH_4^+ concentration in chemostat. [From Goldman, J.C. and J.J. McCarthy, *Limnol. Oceanogr.*, 23, 695 (1978).]

Menten kinetics. The instantaneous P_i uptake at low added $[P_i]$ typically was higher (by as much as 300 times) than values extrapolated from Hanes plots of uptake at high $[P_i]$ (Figure 6-27a). This was explained by the presence of taxa with a wide range of K_{Su} for P_i. The authors suggested that plankton adapt to P-limited environments by synthesizing uptake systems with K_{Su} values at least an order of magnitude lower than those measured in laboratory culture studies. More recently, Tarapchak and Herche[102] proposed a "compound" Michaelis-Menten model, reflecting the presence of species with differing K_{Su} values, to analyze P_i uptake by natural mcirobial assemblages. According to this model (Figure 6-27b), uptake constants at low $[P_i]$ (region A in the figure) differ from those associated with high $[P_i]$ (region C). Constants determined from data at high $[P_i]$ should not be used to model P_i uptake at low concentrations, and the converse is also true. The simple Michaelis-Menten model is likely to hold over a wide range of $[P_i]$ only when P is not limiting or when one species dominates the plankton community (e.g., extreme bloom conditions).[102]

Several practical problems may be encountered in attempting to measure nutrient uptake constants of plankton. Substantial nutrient depletion may occur even during short incubations at low nutrient concentrations. McCarthy[49] reported

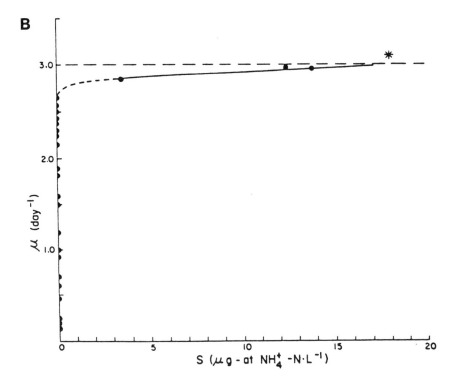

Figure 6-26(B).

more than 50% depletions in 5-min exposures of the marine alga *Thalassiosira pseudonana* to $^{15}NH_4^+$ at concentrations below 4 µ*M* when the culture was growing at 1.55 d^{-1}. In addition, nutrient concentrations may be overestimated because of contamination or lack of specificity in the analytical method. (The lack of specificity in methods for orthophosphate is well known.) The net effect of overestimating the available nutrient concentration is to overestimate uptake rates at the lowest concentrations. This artifact can be detected in data sets obtained at several initial nutrient concentrations. Finally, failure to appreciate the nonequivalence of K_S values for uptake and growth leads to poor fit of data to Monod models when constants obtained for one process are used for the other. Other considerations in measuring kinetic parameters of growth and nutrient uptake are discussed by Button[42] and Platt.[130]

6.9.3 Multiple Nutrient Limitation in Algae

Both the Monod and cell quota models assume that only a single nutrient limits growth at any given time, and both models thus are quantitative expressions of the principle first proposed by Liebig in 1835: plant growth is limited by the nutrient present in the least amount relative to plant needs. It is common, however, for several nutrients to be present at suboptimal concentrations (near or below K_S) in aquatic systems, especially in oligotrophic waters. This suggests that several

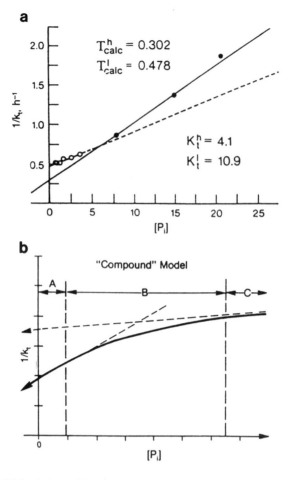

Figure 6-27. (a) Deviations of P_i uptake by Lake Michigan plankton from simple Michaelis-Menten kinetics (Hanes plot) at low $[P_i]$ can be explained by the presence of multiple species with different K_{Su} values. [From Tarapchak, S.J. and L.R. Herche, *Can. J. Fish. Aquat. Sci.*, 43, 319 (1986). With permission.] (b) Hanes-type plot of a compound Michaelis-Menten model: data at low $[P_i]$ and high $[P_i]$ yield different values of kinetic parameters that apply only within those regions; nonlinear region occurs at intermediate $[P_i]$. [From Tarapchak, S.J. and L.R. Herche, *J. Environ. Qual.*, 18, 17 (1989). With permission.]

nutrients could limit growth rates simultaneously, in spite of the Liebigian principle. Some computer models[131] simulate algal growth as multiplicative functions of Monod terms for several nutrients; e.g.,

$$\mu = \mu_{max}\left[\frac{[P]}{K_P+[P]}\right]\left[\frac{[N]}{K_N+[N]}\right]\left[\frac{[CO_2]}{K_C+[CO_2]}\right]\cdots\cdots \tag{6-86}$$

Experiments to determine the validity of multiplicative models have been performed on aquatic organisms only recently. Some support for the single limiting-nutrient model has been obtained in chemostats with combinations of potentially limiting nutrients. Droop[50] reported no evidence for multiplicative effects of phosphate and vitamin B_{12} on growth of *Pavlova lutheri*; growth was controlled by the nutrient in shorter supply. In contrast, Rhee[132,133] found only partial support for the single limiting-nutrient concept in studies with *Scenedesmus*. This organism had an optimum cellular N:P ratio* of 30:1 (atomic basis). At higher ratios the alga was limited by P; at lower values growth was limited by N. Ratios of N:P input to chemostats with *Scenedesmus* were varied between 5:1 and 80:1 by holding [P] constant and increasing $[NO_3^-]$. No additive or multiplicative effects were observed on growth measured as cell numbers (Figure 6-28a). Below the optimum ratio, growth was proportional to the N:P ratio. At N:P > 30:1, cell numbers were constant because the limiting nutrient remained constant, and the transition from N to P limitation was sharp. However, cell volume continued to increase (at a slower rate) in the P-limited region (Figure 6-28b), and the surplus N was stored primarily as protein. Thus, there is some interaction between the two nutrients in terms of biomass, but not cell numbers. Cell N was constant under N limitation (up to N:P = 30:1) and increased linearly at higher ratios (Figure 6-28c). Cell P had an analogous pattern: constant under P limitation, and a rapid increase under N limitation.

Interactive effects of nutrients and light on algal growth have been described,[135-138] but the functional nature of the interaction is still uncertain. Light has a saturation effect on photosynthesis and can be modeled by Monod-like expressions (Example 6-4). Nyholm[136] modeled light and nutrient-limited algal growth as a multiplicative relationship like Equation 6-86, but Falkowski[137] combined the effects of light and nutrients into a bisubstrate enzymatic equation. Rhee and Gotham[138] found that the effects of light and nitrogen were neither additive nor multiplicative. Nitrogen cell quotas for constant μ and subsistence quotas increased with decreasing light under nitrate limitation, and within certain limits, light and changing cell quotas compensated each other in maintaining growth rates.

6.9.4 Competition Among Autotrophic Species

The data in Tables 6-4 and 6-6 explain why nitrification does not occur in lake surface waters during summer. K_S for ammonium oxidation by nitrifying bacteria has a range of about 0.5 to $4 \times 10^{-4} M$, whereas K_S values for ammonium uptake by phytoplankton are at least a factor of ten lower (1 to $5 \times 10^{-6} M$). Nitrifying bacteria also have intrinsically low growth rates ($\mu_{max} = 0.5–2.2$ day^{-1}) and are unable to compete successfully with phytoplankton at low $[NH_4^+]$. Once the spring maximum in ammonium is depleted by early blooms of phytoplankton, epilimnetic $[NH_4^+]$ is maintained at low levels by growing algae, which efficiently remove NH_4^+ added to the water from internal recycling and external inputs.

---

*Optimum N:P ratios vary widely among freshwater algae. Values ranging from 7:1 to 30:1 (atomic basis; mean = 17:1) were reported for seven species.[134] In general, low ratios were found for diatoms and high values for green algae.

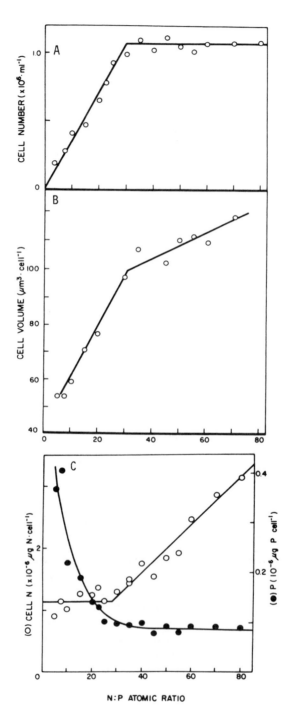

Figure 6-28.   Steady-state cell numbers (A), cell volume (B), and cell N and P contents (C) in a chemostat culture of *Scenedesmus* growing at $\mu = 0.59\ d^{-1}$ as function of N:P ratio in inflow medium. N:P input ratios varied by holding [P] constant and increasing N (nitrate) concentration. [From Rhee, G.-Y., *Limnol. Oceanogr.*, 23, 10 (1978). With permission.]

Aquatic ecologists have used both the Monod and cell quota models to explain competitive elimination among species of phytoplankton under various conditions of light, temperature, and limiting nutrients. The idea that plankton community composition is controlled by differing requirements among species for available resources has received much attention in recent years. An alternative view is that community composition is controlled primarily by selective predation ("top-down" vs. resource-based control). (For a review of these opposing views, see reference 139.) The following discussion focuses on the latter view; predator control theory is discussed in ecology texts.

Epply et al.[127] combined light response curves of four marine phytoplankton and kinetic parameters for NO_3^- and NH_4^+ uptake into a model to predict which species would predominate at various light conditions and inorganic N concentrations (assumed to be the growth-limiting nutrient). The light curves (Figure 6-29a) suggest that Monod kinetics can be used to describe μ as a function of light intensity (I). The kinetic parameters for N uptake are given in Table 6-6, and predicted growth rates for varying $[NO_3^-]$ at low and high I are shown in Figure 6-29b,c. *Coccolithus huxlei* has a high μ_{max} at low I, and the lowest K_S for NO_3^- of the four algae; thus, at low I it has the fastest growth rate at all $[NO_3^-]$. At high I the pattern is more complicated. *Coccolithus* still grows the fastest at low $[NO_3^-]$, because of its low K_S, but at higher $[NO_3^-]$ the diatom *Skeletonema costatum*, which has the highest μ_{max} at high I, grows the fastest. Extrapolation of results like these to predict species predominance in mixed natural populations should be done cautiously. The model does not consider possible control of net growth by selective predation, physical factors like sinking rates, or other nutrients. In addition, temporal variations in physical and chemical conditions may enable several species to compete successfully; conditions favoring predominance by one species may not exist for a sufficient duration to allow a steady state to develop.

The ability of Monod models to predict species dominance in mixed cultures has been shown in some laboratory experiments. Holm and Armstrong[140] determined nutrient uptake and growth parameters for the diatom *Asterionella formosa* and the blue-green alga *Microcystis aeruginosa* in batch cultures under P and SiO_2 limitation. The diatom had a higher μ_{max} and lower K_S for growth under P limitation than did the blue-green, and it switched from Si to P limitation at Si:P > 100. Semicontinuous cultures with the two species grown together at various Si:P ratios yielded results that agreed with predictions based on the pure culture data. *Asterionella* dominated when both species were P limited, but *Microcystis* dominated when the diatom was limited by Si.

Tilman and co-workers[46] extended the use of nutrient-growth kinetics into a model for algal competition, called resource competition theory.[141,142] The model is based on the Monod equation modified to include a threshold (Equation 6-43). This fit silica-limited growth by diatoms more accurately than a simple Monod equation. According to this theory the species with the lowest resource requirement $\hat{R}$ becomes dominant because it can lower the limiting nutrient concentration below that required for growth by all other species. $\hat{R}$ is defined by steady-state CFSTR theory and the Monod equation (Equation 6-52b):

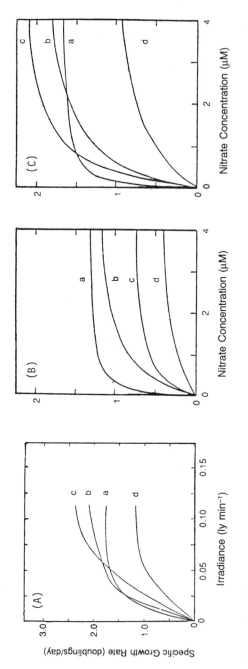

Figure 6-29. Specific growth rate (A) measured as function of light intensity, and calculated as function of nitrate concentration at light intensity of (B) 1/20 and (C) 1/5 of surface sunlight for four marine algal species: a = *Coccolithus huxlei*; b = *Ditylum brightwellii*; c = *Skeletonema costatum*; d = *Duneliella tertiolecta*. [From Epply, R.W., J.N. Rogers, and J.J. McCarthy, *Limnol. Oceanogr.*, 14, 912 (1969). With permission.]

$$\hat{R} = \frac{K_S D}{\mu_{max} - D} \qquad (6\text{-}87)$$

D may include losses from other mechanisms besides dilution in a chemostat (death, sinking out of the photic zone). Continuous-culture experiments verified a major prediction of competition theory: independent of initial population densities, the species with the lowest $\hat{R}$ became dominant because of its ability to grow at lower concentrations of the limiting nutrient than other species. Resource competition theory is described in detail elsewhere[141] and has been used to describe competitive exclusion and coexistence of algae along gradients of various nutrient ratios (Figure 6-30). The extent to which predictions based on this steady-state theory can be applied to a nonsteady-state world is still an open question.

Species-specific optimum nutrient ratios have been used as an alternative explanation for exclusion or coexistence among algal species.[134] For example, if species A and B have optimum N:P ratios of 10 and 30, respectively, and both have similar μ_{max}, they could coexist at N:P ratios between 10 and 30 because they would be limited by different nutrients. Both would be P limited at N:P > 30, but B likely would outcompete A, which would be more P limited.

The cell quota model also has been used with some success to predict algal dominance in natural systems.[143] In general, for a given nutrient the species with the smallest subsistence cell quota (Q_o) would be expected to outcompete other species. Of course, the cell quota and Monod models are not completely independent, and a low value of Q_o is likely to be associated with a low value of K_S and a low value of $\hat{R}$. The magnitude of the difference $\Delta Q = Q_{max} - Q_o$ is a measure of the growth capacity of an organism in time of nutrient depletion. Other factors being equal, we might expect species with the largest ΔQ to become predominant during such periods. The cell quota model is less useful for nutrients that do not vary greatly in cellular content (e.g., Si, N) than for those that do (P, vitamins), and it is useless for elements (like carbon) that have nearly constant cellular concentrations.

6.10 DIFFUSION-LIMITED TRANSPORT AND GROWTH

Substrate transport to cells frequently is limited by physical diffusion. The rate-limiting step may be transport through a boundary layer surrounding free-living cells or passive transport across cell membranes. For microorganisms growing in films or flocculent suspensions, the limiting step may be transport through the organic matrix. The equations describing transport and growth differ for each of these situations.

For free-living cells the problem is one of sequential transport and reaction. Diffusion occurs through a boundary layer in which no reaction is occurring, and reaction occurs in the cell interior where concentrations are considered homogeneous. Models that treat transport and reaction separately are called *lumped models*.

When microorganisms are distributed within flocs and films, transport and reaction occur throughout the matrix, and the two steps cannot be separated. A similar situation prevails in sediments, where substrates and/or oxidants diffuse into

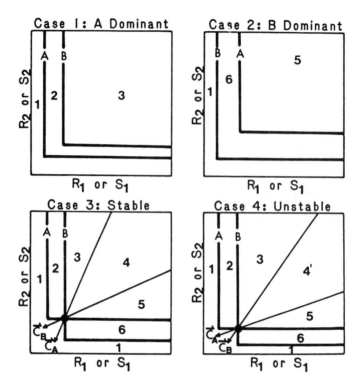

Figure 6-30. Competitive exclusion and coexistence of species along gradients of nutrient ratios can be explained by resource competition theory: a simple case of two species (A, B) competing for two resources (R_1, R_2). Thick lines labeled A, B in each graph represent the minimum resource concentrations ($\hat{R}$) for each species and resource, i.e., the values of R at which zero net growth occurs; these lines are "zero net growth isoclines", ZNGIs. **Case 1**: ZNGI of species A is inside that of species B, and species A can reduce resources to levels below those at which B can survive. Habitats with initial resource concentrations (called resource supply points) in region 1 do not have sufficient R_1 and R_2 to support either species. Habitats with resource supply points in regions 2 and 3 will be dominated by A. **Case 2**: Similar to case 1 except that the ZNGI for species B is inside that of A, and B will dominate at all resource concentrations and ratios where net growth can occur. **Case 3**: The ZNGIs cross at a two-species equilibrium point (black dot). The point is locally stable because each species consumes more of the resource that limits its growth at the equilibrium point. A will dominate habitats with resource supply points in regions 2 and 3; the species can coexist in habitats with resource supply points in region 4; and B will dominate habitats with resource supply points in regions 5 and 6. **Case 4**: Similar to case 3 except that the equilibrium point is locally unstable because each species consumes more of the resource that limits the other species. Results of competition are similar to case 3 except that either species may dominate in region 4', depending on initial conditions. (From Tilman, D., *Resource Competition and Community Structure*, Princeton University Press, Princeton, NJ, 1982. With permission.)

the sediment from the overlying water and reactions occur over some thickness of sediment. These cases are treated by *distributed models* in which reaction conditions vary with location in the floc or film. These models are more complicated than lumped models, and their form depends on boundary conditions. This section describes diffusion-controlled growth and discusses its importance in controlling microbial processes in aquatic systems. Diffusion limitation is especially important in sedimentary environments and in dense microbial cultures characteristic of engineered systems (industrial reactors, waste treatment systems). This subject is treated in many texts on biochemical engineering and wastewater treatment.[59,82,83,144,145]

6.10.1 Fick's Laws

As described in Chapter 3, diffusion follows Fick's first law, which in one dimension is $F_i = -D_i d[C_i]/dx$. F_i is the flux of i (typically mol cm^{-2} s^{-1}); D_i is its diffusion coefficient (typically cm^2 s^{-1}); and $d[C_i]/dx$ is the concentration gradient in the direction of diffusion (mol cm^{-3} cm^{-1}). If the gradient is spherically symmetric, as when coccoidal microorganisms consume a substrate, concentration is a function of r, the distance from the center of the cell, and Fick's law becomes

$$F_i = -D_i (d[C_i]/dr) \tag{6-88}$$

If $[C_i]$ changes temporally, the flux varies with time, and Fick's second law, a time-dependent diffusion equation, is required: $d[C_i]/dt = D_i d^2[C_i]/dr^2$. In spherical coordinates this becomes

$$\frac{d[C_i]}{dt} = \frac{D_i dr^2}{r^2 dr} \left[\frac{d[C_i]}{dr} \right] \tag{6-89}$$

These equations can be used to evaluate concentration gradients of diffusing substances by integrating over limits with $d[C_i]/dt = 0$.

6.10.2 Diffusion-Limited Specific Uptake Rates of Substrates

When the substrate is transferred to cells by passive diffusion, the specific rate of substrate uptake (k_T) is proportional to the concentration difference between the bulk solution and a receptor site in the cell:

$$k_T = D_{sp}([S]_o - [S]_i) \tag{6-90}$$

D_{sp}, the specific diffusion coefficient, is defined as the mass of nutrient transferred into a unit of biomass per unit time under unit molar concentration difference between the solution and cellular receptor site. If transport into cells is active (mediated by enzymes), diffusion through a stagnant boundary layer around the cells still could limit transfer and thus limit growth. In this case, $[S]_i$ would be the concentration at surface permeases responsible for active transport. The boundary layer may include a gelatinous sheath and (in passive transport) the cell wall and cytoplasmic membrane.

D_{sp} has units of $M^{-1} t^{-1}$ and is a function of the molecular diffusion coefficient of substrate in the boundary layer (D_S^b), the cell surface area per unit biomass, and the gradient G through which substrate diffuses. D_S^b is related to D_S^w, the aqueous diffusion coefficient of a substance, which for small solutes is $\sim 1 \times 10^{-5}$ cm^2 s^{-1}. In general, D_b is smaller than D_w by a factor related to the fraction of solid matter in the boundary layer. The surface area per unit biomass of spherical cells is $4N\pi r_c^2$, where N is the number of cells per gram (dry wt.) and r_c is the cell radius (cm). G has units of length^{-1} and is inversely proportional to δ, the width of the boundary layer. For spherical cells with receptor sites near the surface, the gradient is modeled as $G = r_c/(r' - r_c)r'$ or $r_c/\delta(r_c + \delta)$, where $r' = r_c + \delta$ is the outer radius of the boundary layer.[146,147] This leads to the formula,

$$D_{sp} = 4(10^{-3})M_w N\pi r_c r'D_S/(r_c - r') \qquad (6\text{-}91)$$

M_w is the substrate molecular weight. The factor $10^{-3}M_w$ converts the right side of Equation 6-91 from cm^3 g^{-1} s^{-1} to M^{-1} s^{-1}.

6.10.3 A Lumped Substrate Diffusion-Reaction Model for Free-Living Organisms

Consider a culture of spherical cells with radius r_c and stagnant layer thickness δ. The steady-state concentration gradient of S diffusing into cells is obtained by integrating Equation 6-89 with $dC_i/dt = 0$:[59,144]

$$\frac{D_S}{r^2} \int_{dr}^{d} \left[r^2 \frac{d[S]}{dr} \right] = 0 \qquad (6\text{-}92a)$$

or

$$r^2 d[S]/dr = C_1 \qquad (6\text{-}92b)$$

C_1 is a constant of integration. Subsequent integration of Equation 6-92b yields

$$[S] = -C_1/r + C_2 \qquad (6\text{-}92c)$$

C_1 and C_2 are evaluated from the boundary conditions $[S] = [S]_{rc}$ at $r = r_c$, and $[S] = [S]_o$ at $r = r_c + \delta$. Substitution into Equation 6-92c yields

$$C_1 = ([S]_o - [S]_{rc})(r_c + \delta)r_c/\delta$$

and insertion of this into Equation 6-92b yields

$$d[S]/dr\big|_{r=r_c} = ([S]_o - [S]_{rc})(r_c + \delta)/r_c\delta = ([S]_o - [S]_{rc})\theta_g \qquad (6\text{-}92d)$$

where $\theta_g = (r_c + \delta)/r_c\delta$. At steady state, the rate of substrate use by cells equals the rate of substrate transport into the cells. The former is defined by a Monod expression and the latter by Fick's first law. A mass balance of substrate for the cell thus can be written,

$$\begin{bmatrix} \text{volume} \\ \text{of} \\ \text{cell} \end{bmatrix} \times \begin{bmatrix} \text{density} \\ \text{of} \\ \text{cell} \end{bmatrix} \times \begin{bmatrix} \text{specific rate} \\ \text{of substrate} \\ \text{uptake by cells} \end{bmatrix} = \begin{bmatrix} \text{substrate flux} \\ \text{across surface} \\ \text{area of cell} \end{bmatrix} \qquad (6\text{-}93a)$$

or
$$\left(\frac{4}{3}\pi r_c^3\right)\rho k_T = 4\pi r_c^2 D_S d[S]/dr\big|_{r=r_c} \qquad (6\text{-}93b)$$

For simplicity, one can assume $\rho = 1$. Because $k_T = \mu/y$ and μ is a Monod function of $[S]_{rc}$, we can write,

$$\frac{4\pi r_c^3}{3y}\left[\frac{\mu_{max}[S]_{rc}}{K_S + [S]_{rc}}\right] = 4\pi r_c^2 D_S d[S]/dr\big|_{r=r_c} \qquad (6\text{-}93c)$$

Substitution of Equation 6-92d for the concentration gradient at the cell surface results in the following relationship between $[S]_i$ and $[S]_0$:

$$(r_c\rho/3y)\left[\frac{\mu_{max}[S]_{rc}}{K_S + [S]_{rc}}\right] = D_S([S]_o - [S]_{rc})/\theta_g \qquad (6\text{-}93d)$$

Because $[S]_{rc}$ cannot be measured, we need an expression for substrate use that contains only $[S]_o$. Equation 6-93d can be solved for $[S]_{rc}$ as a somewhat complicated function of $[S]_o$:[144,148]

$$[S]_{rc} = 0.5(K_S + a\theta_g / D_S - [S]_o)\{-1 + [1 + 4K_S[S]_o(K_S + a\theta_g / D_S$$
$$-[S]_o)^{-2}]^{\frac{1}{2}} \qquad (6\text{-}93e)$$

where, for convenience, $a = \mu_{max}\rho r_c/3y$. Returning to Equation 6-93d, we see that if we multiply both sides by $3/r_c$, the left side becomes equal to the rate of substrate use per unit volume of microorganisms, R_S, which is experimentally measurable. Substituting Equation 6-93e for $[S]_i$ on the left side of the equation then leads to the following expression for R_S as a function of bulk solution substrate concentration:[59,144,148]

$$R_S = \frac{\mu_{max}}{2yd}\left[1 + \sigma + \frac{[S]_o}{K_S}\right] - \left[\left[1 + \sigma - \frac{[S]_o}{K_S}\right]^2 + \frac{4[S]_o}{K_S}\right] \qquad (6\text{-}94)$$

$\sigma = \mu_{max}\rho r_c\theta_g/3yK_SD_S$ is a dimensionless parameter that simplifies the equation.

Although Equation 6-94 is long, most of its terms are measurable when diffusion is not limiting. Cell radius r_c can be observed directly by microscope; μ_{max} and y can be determined at high [S], where diffusion should not be important; R_S is measured; and $[S]_o$ is controlled or measured by the experimenter. This leaves three parameters (D_S, K_S, δ) to be fit to the equation. D_S can be estimated from data in the literature or determined independently. Powell[148] derived Equation 6-94 and compared it with other growth equations. In most cases he found it gave superior fit compared with the simple Monod model.

Example 6.10. Diffusion-Limited Nutrient Uptake by Phytoplankton

Nutrient uptake at the low concentrations in surface waters of lakes or oceans may be sufficient to deplete a region around the cells. Uptake then will be diminished to reflect the lower concentration at the cell surface compared with the bulk solution. Pasciak and Gavis treated this subject in a lumped model for spherical[149] and nonspherical[150] cells and showed that it is formally analogous to a well-known electrostatic problem: the potential of a conducting spheroid of given total charge in a dielectric medium.[151] They developed criteria to evaluate the importance of transport limitation on nutrient uptake by plankton and concluded that organism motility lessens the importance of diffusion limitation, but does not always eliminate it.

I. Nonmotile spherical cells. Development of this case is similar to that of Section 6.10.3. Consider a spherical cell suspended in a dilute solution of nutrient that is absorbed uniformly over the cell surface. Assume the system is in steady state, the cells are motionless, and they reduce the nutrient concentration from $[S]_o$ in the bulk solution to $[S]_{rc}$ at the cell surface. Quiescent diffusion controls the transport of nutrient to the cells. The steady-state nutrient concentration in the boundary layer is obtained by setting Fick's second law to zero:

(1)
$$d[S]/dt = (D_S/r)\frac{d}{dr}\left[\frac{r^2 d[S]}{dr}\right] = 0$$

Integrating from $r = r_c$, where $[S] = [S]_{rc}$, to $r = \infty$, where $[S] = [S]_o$, yields

(2)
$$\frac{[S]-[S]_o}{[S]_{rc}-[S]_o} = \frac{r_c}{r}$$

The flux of nutrient to the cell surface, F_{rc}, is defined by Fick's first law. From Equations 2 and 6-88 we obtain

(3)
$$F_{rc} = 3.6 D_S([S]_o - [S]_{rc})/r_c$$

F_{rc} is in $\mu mol\ cm^{-2}\ h^{-1}$, D_S in $cm^2\ s^{-1}$, [S] in μM, and 3.6 is a conversion factor for time and volume (3600 s h^{-1}/1000 $cm^3\ L^{-1}$). D_S/r_c is a mass transfer coefficient (cm s^{-1}).

In general, the rate of nutrient transport to cells of any geometry (Q_R, $\mu mol\ h^{-1}$) is given by

(4)
$$Q_R = F_{rc}A = kA([S]_o - [S]_{rc})$$

k is a mass-transfer coefficient and A is the area over which transport occurs. For spherical cells, $A = 4\pi r_c^2$, and $Q_R = 14.4\pi r_c D_S([S]_o - [S]_{rc})$. Nutrient uptake also is given by the Michaelis-Menten equation, $v = V[S]_{rc}/(K_S + [S]_{rc})$. We can equate v and Q_R and obtain an equation that can be solved for $[S]_{rc}$:

(5) $$[S]_{rc}^2 + (V/14.4\pi RD_S + K_S - [S]_o)[S]_{rc} - K_S[S]_o = 0$$

This equation can be simplified by expressing [S] in units of K_S, $[S] = [S]/K_S$, and by defining a dimensionless parameter $P = 14.4\pi r_c D_S K_S/V$, which yields

(6) $$\left[\tilde{S}\right]_{rc}^2 + \left(1/P + 1 - \left[\tilde{S}\right]_o\right)\left[\tilde{S}\right]_{rc} - \left[\tilde{S}\right]_o = 0$$

Equation 6 can be solved by the quadratic formula to give the normalized substrate concentration at the cell surface, $[\tilde{S}]_{rc}$, as a function of the corresponding bulk solution value, $[S]_o$. These values can be inserted into the Monod equation to obtain the diffusion-controlled uptake rate in terms of measurable parameters (K_S, V, r_c, D_S). Figure 6-31 illustrates the normalized nutrient uptake rate v/V as a function of $[S]_o$ for several values of P. When P is large, V tends to be small, and uptake is not affected by diffusion ($[S]_{rc} \rightarrow [S]_o$). V increases with decreasing P for constant values of the other parameters, and $[S]_{rc}$ becomes smaller than $[S]_o$. Diffusion becomes a significant factor when $P < \sim2$.

The importance of enzymatic vs. diffusion limitation may be evaluated by examining apparent activation energies for substrate uptake over a range of concentrations. Mierle[152] found $E_{act} = 49.7$ kJ mol^{-1} for P_i uptake by P-starved *Synechococcus leopoliensis* (a blue-green alga) at substrate saturation, but E_{act} was only 23.4 kJ mol^{-1} at zero-added P_i (measured with carrier-free ^{32}P). These values are consistent with chemical and diffusion control, respectively.

II. Nonspherical cells. The above treatment was extended to nonspherical cells,[150] in particular disks and cylinders, which can be approximated as oblate and prolate spheroids, respectively. $[\tilde{S}]_{rc}$ for such cells is still given by Equation 6, but P is modified to $P' = a(14.4\pi D_S r_c K_S/V)$, where a is a shape factor. Equations defining a for prolate and oblate spheroids are given in Figure 6-32, which shows a as a function of the eccentricity (e) of spheroids. When e = 0, the spheroid is a sphere, and a = 1.

Values of P, P' and related parameters for some spherical and nonspherical phytoplankton are listed in Table 6-7. It is apparent that diffusion limitation is potentially significant for some phytoplankton (those with P or $P' < \sim2$), but not for others. Whether diffusion is *actually* important depends on the nutrient concentration, as well as the relative velocity between cells and surrounding medium (see Part III). Boundary-layer diffusion control becomes negligible at high concentrations, and uptake approaches simple Monod kinetics. This is illustrated for uptake of NO_3^- by a diatom in Figure 6-33. At $[NO_3^-] > \sim4$ μM, uptake follows Monod kinetics, but at lower concentrations, uptake is lower than predicted by the Monod equation.

At very small $[S]_o$, Equation 6 can be simplified to

(7) $$\frac{[S]_{rc}}{[S]_o} = \frac{P'}{P' + 1}$$

Substitution of Equation 7 into the Monod equation and simplifying for the condition $[S]_o \rightarrow 0$ yields

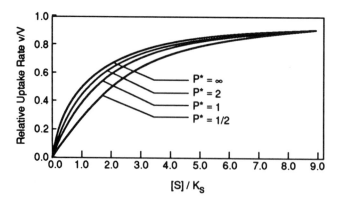

Figure 6-31. Relative substrate uptake (v/V) vs. relative substrate concentration ([S]/K$_S$) for various values of P, the diffusion control parameter defined in Example 6-9. [From Pasciak, W.J. and J. Gavis, *Limnol. Oceanogr.*, 19, 881 (1974). With permission.]

(8)
$$v = \left[\frac{P'}{P'+1} \right] \frac{[S]_o V}{K_S} \quad \text{or} \quad \left[\frac{[S]_o}{v} \right] = \left[\frac{P'+1}{P'} \right] \frac{K_S}{V}$$

Diffusion thus reduces v by a limiting factor of $P'/(P' + 1)$ and increases the y-intercept of a Hanes plot by the factor $(P' + 1)/P'$ compared with the predicted Monod intercept.[150] The term $1/(1 + P')$ represents the proportion of the total resistance to nutrient uptake caused by diffusion resistance.[152]

III. Effects of turbulence and cell motility on diffusion limitation. The diffusion parameter P can be modified to consider relative motion of cells in the medium.[149] The effect of motion is to increase the rate of nutrient transport and decrease the importance of diffusion limitation. Relative motion may occur because of cell motility, buoyancy, or turbulence. For moving spheres the mass transfer coefficient becomes[153]

(9)
$$k = (D_S/r_c)\{1 + 0.5 r_c u/D_S\}$$

where u (cm s^{-1}) is the relative velocity. This equation is valid only for small values of u and r_c, but applicable values encompass the range of environmental interest. Combining Equation 9 and the definition of P' yields

(10)
$$P'' = 14.4\pi a(r_c/V)D_S K_S\{1 + 0.5 r_c u/D_S\}$$

P" thus increases with u (other factors remaining constant), and diffusion control becomes less important. The minimum relative velocity u_{min} required to prevent diffusion limitation of nutrient uptake can be estimated by dividing P' by P" and by assuming that diffusion independence occurs at P" = 2.0:

$$P''/P' = 2/P' = 1 + u_{min} r_c/2D_S$$

or (11)
$$u_{min} = (2D_S/r_c)\{(2/P')-1\}$$

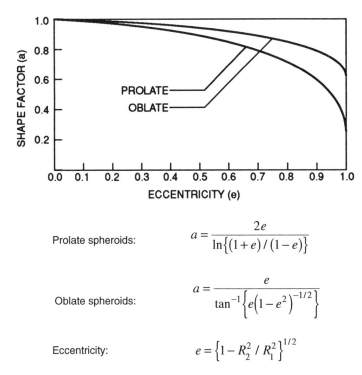

Prolate spheroids:

$$a = \frac{2e}{\ln\left\{(1+e)/(1-e)\right\}}$$

Oblate spheroids:

$$a = \frac{e}{\tan^{-1}\left\{e\left(1-e^2\right)^{-1/2}\right\}}$$

Eccentricity:

$$e = \left\{1 - R_2^2 / R_1^2\right\}^{1/2}$$

Figure 6-32. Shape factor a vs eccentricity e for prolate and oblate spheroids. R_1 = major axis; R_2 = minor axis [From Pasciak, W.J. and J. Gavis, *Limnol. Oceanogr.*, 20, 211 (1975). With permission.]

By definition, $u_{min} = 0$ at $P' = 2$, and u_{min} increases as P' decreases. Values of P' as low as ~0.3 are given for a few algae in Table 6-7. At $P' = 0.2$, u_{min} ranges from ~0.02 to ~2 cm s^{-1} over the size range of unicellular algae (100 μm > r_c > 1 μm). The lowest u_{min}, which applies to the largest sizes, is in the range of swimming speeds reported for motile algae (~50 to 300 μm s^{-1}). Pasciak and Gavis[149] concluded that motility lessened the effect of transport limitation on nutrient uptake by the dinoflagellate *Gymnodinium splendens*.

Turbulence increases nutrient uptake rates at low concentrations where diffusion limitation can occur, but quiescent and turbulent uptake rates approach each other at higher concentrations (Figure 6-34a). Moreover, uptake increases with increasing turbulence (as measured by the shear rate $\underline{s}$) only at low $\underline{s}$ (Figure 6-34b). The shear rate for cells in a fluid can be estimated from

(12) $$\underline{s} = u/r_c$$

u and r_c have consistent units of distance; $\underline{s}$ has units of t^{-1}. At high turbulence, uptake is not diffusion-limited, and increasing $\underline{s}$ has no effect on v. Values of $\underline{s}$ necessary to achieve diffusion independence for $P' = 0.2$ can be calculated by substituting Equation 11 for u in Equation 12. Values range from ~2 × 10^4 s^{-1} to 2 s^{-1} for spherical algae of radius 1 to 100 μm. Values of $\underline{s}$ in relatively calm surface waters are near the low end of this range.

Finally, the time required to achieve steady-state diffusion in a quiescent suspension of microorganisms can be estimated as follows:

Table 6-7.   Transport Limitation Parameters for Nutrient Uptake by Phytoplankton[a]

Organism	Nutrient	Shape	e	a	R	V	K_s	P or P'
Coccolithus huxleyi	NO_3^-	S	—	—	2.5	0.046	0.1	3.6
Coscinodiscus lineatus	NO_3^-	O	1	0.64	25	90	2.8	0.34
Cyclotella nana 3H	NO_3^-	S	—	—	2	0.074	1.87	34.4
Cyclotella nana 13–1	NO_3^-	S	—	—	4	0.084	0.38	12.3
Cyclotella nana 7–15	NO_3^-	S	—	—	4	0.042	1.19	77.2
Cyclotella nana 3H	SiO_2	S	—	—	2	0.05	0.98	27
Cyclotella nana 13–1	SiO_2	S	—	—	4	0.3	0.19	18
Ditylum brightwellii	NO_3^-	P	0.94	0.54	75	12.5	0.6	1.3
Ditylum brightwellii	NO_2^-	P	0.94	0.54	75	17	4.0	6.5
Duneliella tertiolecta	NO_3^-	S	—	—	4	0.107	1.4	35
Gymnodinium splendens	NO_3^-	O	1	0.64	20	400	15	0.33
Nannochloris sp.	NO_3^-	S	—	—	2	0.015	7.5	680
Nitzschia closterium	NO_3^-	P	0.97	0.46	15	0.13	2.8	101
Prorocentrum minimum	NO_3^-	O	1.0	0.64	10	0.45	8.5	32
Rhizosolenia robusta	NO_3^-	P	0.97	0.46	42	227	9.3	0.55
Skeletonema costatum	NO_3^-	P	0.87	0.65	4	0.285	0.4	2.5

[a]   Symbols and units: S = spherical; O = oblate (disk); P = prolate; e = eccentricity (Figure 6-32; Equation 3); a = shape factor (Figure 6-32, Equations 1,2); R = cell radius (μm); V = maximum uptake velocity (μmol/cell·h); K_s = half-saturation constant (μM); P and P' = transport parameter for sperical and nonsperical cells, respectively. Tabul⁻:.ɘd from Pasciak, W.J. and J. Gavis, Limnol. Oceanogr., 19, 881 (1974); Limnol. Oceanogr., 20, 211 (1975).

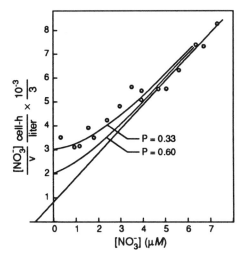

Figure 6-33. Hanes plot of nitrate uptake by *Ditylum brightwellii* under quiescent conditions; $K_S = 0.7 \, \mu M$; $V = 3.2 \times 10^{-6}$ µmol/cell-h. Monod kinetics followed at $[NO_3^-] > 4$ µM; P ≈ 0.33 provides best fit of data. [From Pasciak, W.J. and J. Gavis, *Limnol. Oceanogr.*, 20, 211 (1975). With permission.]

$$(13) \qquad\qquad X = (D_S t_{ss})^{1/2}$$

X is a characteristic dimension of the system, and t_{ss} is the time to achieve steady state.[149] In this case the characteristic dimension is cell radius r_c (~1 to 100 µm). D_S for nutrient ions typically is ~0.5 to 1.0×10^{-5} cm² s⁻¹, and t_{ss} thus ranges from ~1×10^{-3} s to ~0.1 s. A nutrient-depleted zone thus is established rapidly if it occurs at all, and in such cases continuous mixing is required to prevent its occurrence.

6.10.4 Distributed Model for Diffusion and Reaction in Microbial Systems

When diffusion and chemical/biological reactions cannot be separated into spatially distinct processes, distributed models must be used, and a reaction term must be included in the differential equation. These equations are more difficult to integrate and yield more complicated expressions than do lumped models in which the reaction term is a boundary condition of the diffusion process. The starting point for distributed models is the general steady-state **diffusion-reaction equation**. For biological films on flat surfaces, diffusion occurs in only one dimension, and the following equation applies:

$$D_S d^2[S]/dx^2 = R \qquad\qquad (6\text{-}95)$$

The analogous equation for spherical flocs is

$$D_S \left[\frac{d^2[S]}{dr^2} + \frac{2d}{r}\frac{[S]}{dr} \right] = R \qquad\qquad (6\text{-}96)$$

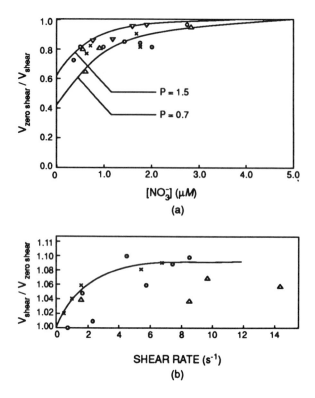

Figure 6-34.   Ratios of (a) uptake under quiescent and controlled-shear conditions (s = 13.5 s⁻¹) vs. [NO₃⁻] for culture of *D. brightwellii*, and (b) NO₃⁻ uptake under shear and quiescent conditions vs. shear rate for same culture. [From Pasciak, W.J. and J. Gavis, *Limnol. Oceanogr.*, 20, 211 (1975). With permission.]

This equation can be derived from a mass balance on a thin shell of thickness dr in the sphere of diffusion-reaction[59] (see Example 5-1). Integration of Equations 6-95 and 6-96 requires replacement of **R** with a functional form of substrate uptake. In general, we can say that **R** is a function of [S], which, in turn, is a function of time and distance (x or r) on the axis along which diffusion is occurring. At steady state, however, t is no longer a variable. Analytical solutions for zero- and first-order substrate uptake are presented in subsequent sections. When **R** is nonlinear (e.g., the Monod equation), the equation must be solved numerically or approximated by its limiting cases: first order at low [S], and zero order at high [S].

The concept of catalyst effectiveness is useful in describing substrate uptake by biological flocs and films. The effectiveness factor is defined as

$$E_f = \frac{R_{obs}}{R_{[S]_b}} = \frac{\text{observed rate of substrate uptake}}{\text{rate if S had no concentration gradient}} \quad (6\text{-}97)$$

When diffusion effects are not important, $E_f = 1$. Functional expressions of E_f for various-order reactions are given in the following section.

6.10.5 Diffusion of Substrate into Suspensions of Microbial Aggregates

Microorganisms often grow in suspended aggregates (flocs). Blue-green algae may grow in colonies that clump together into macroscopic particles (mm size) containing millions of cells. A complex community of microbes (bacteria, yeast, fungi, protozoa) and metazoa inhabits the flocculent particles of activated sludge. Metabolic activity and microbial growth within the floc depend on diffusion of organic substrates, nutrients, and O_2 from the bulk solution.

The factors affecting substrate and O_2 consumption by activated sludge floc are of great interest to wastewater engineers, and many diffusion-reaction models have been developed (e.g., references 59,145,148,154–156). Although the highest organism concentrations likely are near the surface,[155] uniform distribution of cells throughout the floc usually is assumed for mathematical ease. The models also usually assume that floc particles are spherical, but this is far from true, as microscopic examination of floc shows. Departures from sphericity increase the surface area to volume ratio and thus decrease the potential for diffusion limitation.

Diffusivities reported for O_2 and organic substrates in biological floc range from about 8 to 90% of the corresponding values for pure water (Table 6-8). The literature contains conflicting reports on the effects of C/N ratio and sludge age on diffusivities. Matson and Characklis[155] reported that O_2 diffusivity decreased with increasing sludge age and C/N feed ratios and suggested that the decrease was caused by higher production of extracellular polysaccharides. However, these variables had little effect on glucose diffusivity. In contrast, Pipes[157] found lower glucose diffusivity at high C/N ratios. Differences in microbial species composition might explain some of the variability among studies.[155] D_S increases relatively slowly with increasing temperature. For example, D_S for glucose in a biological floc increased from 0.42×10^{-6} to 0.69×10^{-6} cm^2 s^{-1} over the range 20 to 30°C ($Q_{10} = 1.6$).[154]

Zero-Order Model for Substrate Uptake by Floc

A diffusion-reaction model for substrate uptake by microbial floc can be developed as follows.[155] Consider a spherical floc containing a homogeneous concentration of microorganisms at steady state. Equation 6-96 applies to this case. Analytical solution is simple for R as a zero-order reaction and more complicated for first-order reactions. For high [S], zero-order kinetics is sufficient, and Equation 6-96 becomes

$$D_{Sf}\left[\frac{d^2[S]}{dr^2} + \frac{2d[S]}{rdr}\right] - \rho_f k_o = 0 \qquad (6\text{-}98)$$

D_{Sf} is the diffusion coefficient of S in the floc, ρ_f is the floc dry density (mg L^{-1}), and k_o is a zero-order uptake rate constant. Boundary conditions are: (1) [S] finite for all r, and (2) [S] = [S]$_o$ (the bulk solution concentration) at r = r$_f$, the floc radius. Integration yields

Table 6-8. Diffusion Coefficients for Substrates in Microbial Floc and Film

Substrate	Microbial source	Temp. (°C)	D_{Sf} ($10^{-5} \times cm^2\ s^{-1}$)	D_{H_2O} ($10^{-5} \times cm^2\ s^{-1}$)	$\dfrac{100\ D_{Sf}}{D_{H_2O}}$	Ref.
Glucose	Mixed culture[a]	20				155
	(C/N = 5)		0.11–0.17	0.7	16–24	
	(C/N = 50)		0.13–0.21	0.7	19–30	
Mixed	var		0.06–0.6	var	10–100	155[b]
culture	var		0.06–0.21	var	10–30	155[b]
	Biofilm	22	0.13–0.49	0.7	19–70	166
	Domestic sewage plant	20	0.25	0.7	36	155
	Mixed domestic and industrial waste plant	20	0.35	0.7	50	
	Petrochem waste treatment plant	20	0.18	0.7	26	
	Zooglea ramigera	20	0.048	0.7	7	154
Methanol	Mixed culture					
	C/N = 5	20	0.36	1.5	24	154
	C/N = 50		0.36	1.5	24	
	C/N = 5	30	0.48	1.9	25	
	C/N = 50		0.46	1.9	24	
NH_4^+	Nitrifier culture		1.3		80	155[b]
NO_3^-	Nitrifier culture		1.4		90	155[b]
O_2	Nitrifier culture		2.2		90	155[b]
	Bacterial slime		1.5		70	155[b]
	Zooglea ramigera	20	0.21		8	155[b]
	Mixed culture	var	0.4–2.0	var	10–30	155[b]

[a]   Sludge age ranging from 1 to 20 days.
[b]   Original reference cited therein.

$$[S]_o - [S] = \rho_f k_o \{r_f^2 - r^2\}/6D_{Sf} \tag{6-99}$$

The depth in the floc at which S disappears can be obtained from Equation 6-99 by setting [S] = 0 and solving for r:

$$r_{|S|=0} = \{r_f^2 - 6[S]_o D_{Sf}/\rho_f k_o\}^{1/2} \tag{6-100}$$

The depth of substrate penetration is $r_f - r_{|S|=0}$. Equation 6-100 can be further rearranged to compute the combination of r_f and $[S]_o$ required to prevent transport limitation of substrate uptake. For zero-order reactions the limiting condition is that S just disappears at the center of the floc. If S disappears at r > 0, diffusion clearly is limiting the overall reaction rate. However, if [S] > 0 at the floc center, diffusion

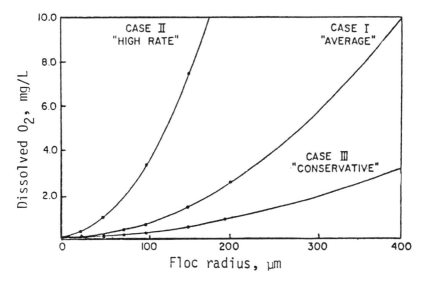

Figure 6-35. Dissolved oxygen concentration at which diffusion becomes limiting vs. floc size for three cases:

	ρ_f	D_{Sf}	k_o
Case I (average)	7.5	1.0	1.8
Case II (high rate)	10.0	0.5	3.6
Case III (conservative)	5.0	2.0	1.8

where ρ_f is in mg/L; D_{Sf} is in 10^{-5} cm²/s; and k_o is in 10^{-3} mg O_2/L per mg MLVSS/L per h. [From Matson, J.V. and W.G. Characklis, *Water Res.*, 10, 877 (1976). With permission.]

is more than sufficient because R is independent of [S] for all [S] > 0. The limiting condition thus is defined by setting $r_{[S]=0} = 0$ (i.e., the center of the floc) in Equation 6-100:

$$[S]_o = \rho_f k_o r_f^2 / 6 D_{Sf} \qquad (6\text{-}101)$$

Calculations with Equation 6-101 to compute transport-limiting concentrations of dissolved O_2 vs. floc size are shown in Figure 6-35 for several values of ρ_f, k_o, and D_{Sf}.[155] It is apparent that $[O_2] > 2$ to 3 mg L^{-1} is needed to maintain aerobic conditions throughout the floc even for small floc sizes (r < 100 μm). Because of the assumptions used to make the calculations, results in Figure 6-35 should not be taken too literally, but they illustrate the potential importance of diffusion control for metabolic processes in microbial aggregates.

The effectiveness factor for a zero-order reaction in spherical floc is equal to the ratio of floc volume, with [S] > 0 to the total floc volume because the reaction rate is uniform everywhere that [S] > 0. Thus,

$$E_f = \{r_f^3 - r_{[S]=0}^3\} / r_f^3 \qquad (6\text{-}102)$$

First-Order Model for Substrate Uptake by Floc

If R is first order, the diffusion-reaction equation for spherical floc is

$$D_{Sf}\left[\frac{d^2[S]}{dr^2} + \frac{2d[S]}{rdr}\right] - k_1[S] = 0 \tag{6-103}$$

The analytical solution for this equation is difficult to derive,[83,145] but the following expression has been obtained for [S]:

$$[S] = \frac{[S]_o r_f \sinh\{r(k_1/D_{Sf})^{1/2}\}}{r \sinh\{r_f(k_1/D_{Sf})^{1/2}\}} \tag{6-104}$$

The effectiveness factor can be obtained from the rate at which substrate enters the floc, F_{obs}, divided by the reaction rate that would occur if $[S] = [S]_o$ throughout the floc. The latter is equal to the volume of the floc times the reaction rate at the surface: $F = (4/3)\pi r_f^3 k_1 [S]_o$. F_{obs} is given by

$$F_{obs} = D_{Sf}\, 4\pi r_f^2 d[S]/dr\Big|_{r=r_f} \tag{6-105a}$$

Differentiation of Equation 6-104 with respect to r yields

$$\frac{d[S]}{dr} = -\frac{r_f[S]_o \sinh\left\{r(k_1/D_{Sf})^{1/2}\right\}}{r^2 \sinh\left\{r_f\left(k_1/D_{Sf}\right)^{1/2}\right\}} + \frac{r_f[S]_o (k_1/D_{Sf})^{1/2}}{r \tanh\left\{r_f(k_1/D_{Sf})^{1/2}\right\}} \tag{6-105b}$$

Recall that $d(sinhx) = coshx$, and $tanhx = sinhx/coshx$. Evaluating the derivative at $r = r_f$, we obtain[83,145]

$$\frac{d[S]}{dr} = \frac{[S]_o}{r_f} - \frac{[S]_o (k_1/D_{Sf})^{1/2}}{\tanh\left\{r_f(k_1/D_{Sf})^{1/2}\right\}} \tag{6-105c}$$

From these relationships it follows that

$$E_f = \frac{F_{obs}}{F} = \frac{3\coth\left\{r_f\left(k_1/D_{Sf}\right)^{1/2}\right\}}{r_f(k_1/D_{Sf})^{1/2}} - \frac{3D_{Sf}}{k_1 r_f^2} \tag{6-106}$$

When $r_f(k_1/D_{Sf})^{1/2} >> 1$, its $coth > 1$, and E_f approaches $3r_f^{-1}(k_1 D_{Sf})^{-1/2}$. This condition is produced by a large floc, high rate constant, and/or low diffusivity, and it leads

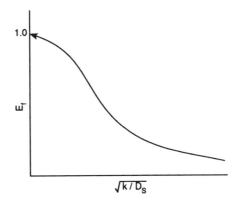

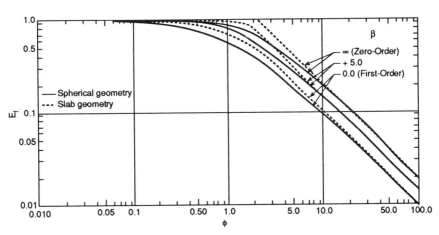

Figure 6-36. (a) General relationship between effectiveness factor and $\sqrt{k/D_s}$ for fixed r_f in first-order substrate uptake by diffusion-limited floc. [From Schroeder, E.D., *Water and Wastewater Treatment*, McGraw-Hill, New York, 1977. With permission.] (b) Effectiveness factor E_f vs. Φ (Equation 6-110) for zero-order and first-order reactions, expressing the range of conditions in Monod kinetics. $\beta = [S]_0/K_s$. First-order limit applies at $\beta \to 0$; zero order at $\beta \to \infty$. (From Bailey, J.E. and D.F. Ollis, *Biochemical Engineering Fundamentals*, McGraw-Hill, New York, 1977. With permission.)

to transport-limited growth. In contrast, when $r_f(k_1/D_{Sf})^{1/2} \ll 1$, $E_f \to 1$, and the system is reaction-rate controlled (see Figure 6-36a).

An analytical solution for Equation 6-96 is not available when R is a nonlinear Monod function. To simplify the analysis, the equation is transformed into a dimensionless form:

$$\frac{d^2[S^*]}{dr^{*2}} = \frac{2d[S^*]}{r^*dr^*} = \frac{Rr^2}{D_S[S]_o} = \frac{\phi[S^*]}{1+\beta[S^*]} \qquad (6\text{-}107)$$

The dimensionless boundary conditions are $[S^*] = 1$ at $r^* = 1$, and $d[S^*]/dr^* = 0$ at $r^* = 0$. The dimensionless variables are defined as follows: $r^* = r/r_c$; $[S^*] = [S]/[S]_o$; $\beta = [S]_o/K_S$; and

$$\phi = r_o \left[\frac{\rho_f \mu_{max}}{y K_S D_{Sf}} \right]^{1/2} = \frac{V_f}{A_f} \left[\frac{\rho_f k_{Tmax}}{K_S D_{Sf}} \right]^{1/2} \tag{6-108}$$

V_f and A_f are the volume and surface area of the floc particle, respectively. ϕ is a dimensionless parameter called the Thiele modulus.[*] ϕ^2 has the physical interpretation of a first-order rate divided by a diffusion rate:[59,83]

$$\phi^2 = \{r_c^3 (\rho_f k_{Tmax}/K_S)[S]_o\}/\{r_c D_{Sf}[S]_o\} \tag{6-109}$$

Both the numerator and denominator have units of g/s. As defined by Equation 6-109, ϕ^2 is the ratio of the rate in the absence of diffusion limitation to the rate of diffusion into the floc.

Because the intrinsic rate parameters k_{Tmax} ($= \mu_{max}/y$) and K_S may not be known, it is not always possible to determine ϕ. Bailey and Ollis[59] described an "observable modulus", Φ, which they related to E_f for substrate uptake by floc:

$$\Phi = Rr_f^2/D_{Sf}[S]_o \tag{6-110}$$

Figure 6-36b shows E_f vs. Φ for zero- and first-order conditions. The former applies to Monod kinetics when β ($= [S]_o/K_S$) $\rightarrow \infty$, and the latter applies when $\beta \rightarrow 0$. The curves thus encompass the range of conditions for Monod kinetics. It is apparent that E_f is ~1 when $\Phi < 0.3$ to 0.5, and $E_f \ll 1$ when $\Phi > ~1$ to 3. These values thus serve as criteria to define the rate-limiting process (mass-transport or biological reaction) in biological floc.

Diffusion limitation can distort linearized plots of kinetic data for microbial flocs.[159,160] This is particularly likely with large floc. Lineweaver-Burk plots of reciprocal specific uptake vs. 1/[S] show curvature at high [S] as diffusion limitation disappears. V_{max} (or μ_{max}) values obtained from such plots are not affected by diffusion limitation, but calculated K_S values are erroneously high if mass-transfer resistance is not eliminated.

Example 6-11. Kinetics of CO_2 Uptake by Free-Living Phytoplankton

This process is rate limiting for phytoplankton growth under some conditions; in quiescent water, CO_2 absorption by rapidly growing plankton may deplete the medium

---

[*] This is a widely used term in biochemical engineering. Thiele moduli were first used to analyze the effects of intraparticle diffusion on the efficiency of porous catalysts.[158] The mathematical form of ϕ depends on particle geometry and reaction order, but ϕ^2 generally represents the ratio of the reaction rate in the absence of diffusion limitation to the rate of diffusion. ϕ thus is related (nonlinearly) to E_f.

around the cells. Two sources may renew CO_2 at the cell surface: diffusion from the bulk solution and production from HCO_3^- within the boundary layer. Description of these processes requires a distributed diffusion-reaction model, even though the CO_2 is consumed at the boundary of the diffusion layer, rather than within it.

Although some algae can use HCO_3^- directly,[161] it is thought that most phytoplankton can assimilate only CO_2. Because $[CO_2]$ and $[HCO_3^-]$ cannot be varied independently without affecting pH, it is difficult to obtain unequivocal evidence for the use of HCO_3^- by algae. Gavis and Ferguson[162] derived a model for CO_2 uptake by algae, according to the concepts described in Section 6.10.5. Their model applies to nonmotile spherical cells in quiescent medium and assumes that only CO_2 can be assimilated. Equations for the model are given in Table 6-9; its conceptual basis is as follows.

If $[HCO_3^-]$ were high and rates of CO_2 formation were so rapid that chemical equilibrium always occurred (even at the cell surface), then the uptake rate would be controlled by C_T (total inorganic carbon concentration) and not by $[CO_2]$. Diffusion would be unimportant as long as $[HCO_3^-]$ remained high. However, rates of CO_2 formation are not infinitely fast, and equilibrium does not always prevail. Reactions 1–11 of Table 6-9 define the kinetics of CO_2 formation from HCO_3^-. At pH $> \sim$8, formation is controlled by reaction 1, direct dehydroxylation of HCO_3^-, but at pH $< \sim$8, formation is controlled by the ionization/recombination step followed by dehydration of H_2CO_3 (reactions 2 and 3). Reaction 2 is very rapid and always at equilibrium; reaction 3 thus is rate controlling. At pH near 8, both mechanisms contribute to CO_2 formation. As Equation 4 indicates, the rate of CO_2 formation depends on pH and concentrations of several carbonate species, all of which vary spatially in the boundary layer.

Several simplifying assumptions must be made to obtain a rate equation for CO_2 formation that can be integrated. Because $[CO_2]$ decreases toward the cell surface in the boundary layer, its formation rate from HCO_3^- increases as the cell is approached. Thus, $[HCO_3^-]$ must decrease with increasing proximity to the cell; this gradient allows diffusion of HCO_3^- into the CO_2-depleted zone. Because $[HCO_3^-]$ is much higher than $[CO_2]$, the relative change in $[HCO_3^-]$ is negligible, even when the fractional change in $[CO_2]$ is large. Thus, (**assumption 1**) we will consider $[HCO_3^-]$ spatially and temporally constant and at its equilibrium value (for the given pH and C_T). $[CO_2]$ is temporally, but not spatially, constant and lower than its equilibrium value in the boundary layer.

For every CO_2 formed from HCO_3^-, one OH^- is formed or one H^+ consumed (these are equivalent, since $H^+ + OH^- = H_2O$ is instantaneous). Because of the higher reaction rate near the cell surface, a gradient of OH^- is set up. Outward-diffusing OH^- balances inward-diffusing HCO_3^-, thus maintaining electroneutrality. At pH < 8, $[OH^-]$ is low ($< 1\ \mu M$), and a large fractional change in $[OH^-]$ and large pH gradient may be expected across the boundary layer, with higher pH near the cell. The magnitude of the gradient is difficult to predict. At low pH, $[CO_2]$ in the bulk solution is high, its formation is slow, and most of the CO_2 consumed by the cell is supplied by diffusion from the bulk solution. As a result, the gradient in OH^- is small compared with that for CO_2. At high pH, $[OH^-]$ is higher than $[CO_2]$, and the fractional change in $[OH^-]$ across the boundary layer is small; i.e., pH is nearly spatially uniform. Thus, (**assumption 2**) we will regard $[OH^-]$ and pH as spatially constant. These considerations lead to Equation 11 for CO_2 formation.

The steady-state concentration of CO_2 at cell receptor sites, $[CO_2]_r$, can be determined from the distributed diffusion/reaction equation in spherical coordinates (Equation 6-96), where **R** is given by Equation 11. Solution of this equation yields $[CO_2]_r$ as a quadratic function of $[CO_2]_c$ (the bulk solution value) and a dimensionless factor P^* (Equation 17). P^* is similar to P, the diffusion parameter of Example 6-10, and expresses the importance of diffusion limitation when the diffusing substance is replenished by chemical reaction in the

Table 6-9. Model Equations for Diffusion-Limited CO_2 Uptake by Algae

A. Formation of CO_2 from HCO_3^-

1. Two reactions are involved: direct formation and formation through H_2CO_3:

(1)
$$HCO_3^- \underset{k_2}{\overset{k_1}{\rightleftharpoons}} CO_2 + OH^-$$

(2)
$$H^+ + HCO_3^- \rightleftharpoons H_2CO_3 \quad \text{(instantaneous)}$$

(3)
$$H_2CO_3 \underset{k_4}{\overset{k_3}{\rightleftharpoons}} CO_2 + H_2O$$

2. Mass action (equilibrium) expressions can be written for reactions 1 to 3: (subscript e denotes an equilibrium concentration)

(4)
$$K_1 = k_1/k_2 = [OH^-]_e [CO_2]_e /[HCO_3^-]_e$$

(5)
$$K_2 = [H_2CO_3]_e /(10^{-pH})_e [HCO_3^-]_e \quad \text{(reaction 2)}$$

(6)
$$K_3 = k_3/k_4 = [CO_2]_e /[H_2CO_3]_e$$

3. The net rate of CO_2 formation is given by the following rate equation:

(7)
$$d[CO_2]/dt = k_1[HCO_3^-] - k_2[CO_2][OH^-] + k_3[H_2CO_3] - k_4[CO_2]$$

Substituting Equations 4 to 6 into Equation 7 yields:

(8)
$$d[CO_2]/dt = k_2 K_1 [HCO_3^-] - k_2[CO_2][OH^-] + $$
$$k_4 K_2 K_3 10^{-pH}[HCO_3^-] - k_4[CO_2]$$

From assumptions 1 and 2 (Example 6-11) that $[HCO_3^-]$ and $[OH^-]$ are spatially uniform and at their equilibrium concentrations in the depleted region, we can establish the following equalities:

(9)
$$k_2 K_1 [HCO_3^-] = k_2 [OH^-][CO_2]_e$$

(10)
$$k_4 K_2 K_3 10^{-pH}[HCO_3^-] = k_4[CO_2]_e$$

Substituting Equations 9 and 10 into Equation 8 and simplifying yields the following expression for the rate of CO_2 production:

(11)
$$d[CO_2]/dt = k'\{[CO_2]_e - [CO_2]\}$$

where $k' = k_2[OH^-] + k_4 \approx$ constant.

Table 6-9 (continued).

4. Values of the constants in Equations 1 to 11 are $k_2 = 8500\ M^{-1}\ s^{-1}$, $K_1 = 10^{-7.7}$, $K_2 = 10^{3.5}$, $K_3 = 10^{2.6}$

pH	7	8	9	10	11	12
$k'(s^{-1})$	0.0301	0.0385	0.115	0.88	8.53	85.0

B. Transport equation for CO_2 to algal cells

1. Substituting Equation 11 for $d[CO_2]/dt$ into the mass-transport equation for CO_2 in spherical coordinates (Equation 6-96) yields

(12)
$$\frac{D\,d}{r^2\,dr}\left[r^2\,\frac{d[CO_2]}{dr}\right] + k'\{[CO_2]_e - [CO_2]\} = 0$$

Boundary conditions are $[CO_2]_{r\to\infty} = [CO_2]_e$; $[CO_2]_{r=R} = [CO_2]_R$, where r = distance from cell center, R = cell radius, and ∞ = bulk solution. The solution of Equation 12 with these boundary conditions is

(13)
$$\frac{[CO_2] - [CO_2]_e}{[CO_2]_R - [CO_2]_e} = \frac{R}{r}\exp\{(R-r)\sqrt{k'/D}\}$$

2. The flux J_R ($\mu mol\ cm^{-2}\ h^{-1}$) to a cell is

(14) $$J_R = -3.6Dd[CO_2]/dr_{|R} = 3.6(D/R)\{R\sqrt{k'/D} + 1\}\{[CO_2]_e - [CO_2]\}$$

(3.6 is a units conversion factor ($3600\ s/h \div 1000\ cm^3\ L^{-1}$)), and the rate of CO_2 transfer (Q, $\mu mol\ h^{-1}$ to a spherical cell of surface area $4\pi R^2$ is $Q_R = 4\pi R^2 J_R$.

3. The rate of CO_2 uptake also is given by the Michaelis-Menten equation:

(15)
$$v = \frac{V[CO_2]_R}{\{K_S + [CO_2]_R\}}$$

Equating Equations 15 and Q_R and rearranging yields a quadratic equation:

(16) $$[CO_2]_R^2 + \{V/[14.4\pi RD(1 + \sqrt{k'/D})] + K_S + [CO_2]_e\}[CO_2]_R - K_S[CO_2]_e = 0$$

or (17) $$[CO_2]_R^{*2} + \{1/P^* + 1 - [CO_2]_e^*\}[CO_2]_e^* = 0$$

$[CO_2]_R^* = [CO_2]_R/K_S$ and $[CO_2]_e^* = [CO_2]_e/K_S$ are dimensionless concentrations, and $P^* = 14.4\pi RDK_S(1 + R\sqrt{k'/D})/V$ is a dimensionless diffusion limitation parameter.

Derived from Gavis, J. and J. F. Ferguson, *Limnol. Oceanogr.*, 20, 211 (1975).

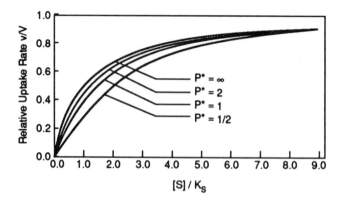

Figure 6-37. Transport-limited uptake of CO_2 by spherical plankton. P^*, defined in Example 6-10, is a dimensionless diffusion-reaction parameter analogous to P in Figure 6-29. [From Gavis, J. and J.F. Ferguson, *Limnol. Oceanogr.*, 20, 211 (1975). With permission.]

boundary layer. (P expresses the importance of diffusion limitation in the absence of chemical reaction.) In both cases diffusion is more important at low values of the parameter. Gavis and Ferguson[162] solved Equation 18 (Table 6-9) for $[CO_2]_i$ at several values of P^* and computed uptake as a function of $[CO_2]_c$ (Figure 6-37). Diffusion limitation causes significant deviations from Monod kinetics at $P^* < \sim 2.0$, which is analogous to the situation in Figure 6-31 for diffusion without replenishment by chemical reaction.

Because P^* includes the pH-dependent rate constant k' for the chemical formation of CO_2, its value depends on pH, as well as on the diffusivity of CO_2 and growth characteristics (K_S and V) of the algae. Because the assumption that $[OH^-]$ is spatially constant in the boundary layer may not be true at pH < ~9, the model should be restricted to cases above that pH.

It should be noted that pH is temporally constant only when the change in C_T causes $[OH^-]$ to change by less than ~10% at a given pH. For $C_T = 10^{-3} M$, this corresponds to consumption of $\sim 10^{-6} M$ inorganic carbon at pH 9 and $\sim 10^{-4} M$ at pH 10.9. At reasonable growth rates and plankton densities typical of bloom conditions (~10 mg [dry wt] L^{-1}), these amounts could be consumed in a period of minutes to hours.[162] Thus, the steady-state solutions may apply only to conditions at the time a set of measurements of pH, C_T, and algal density is made. Modeling the temporal dynamics of transport-limited CO_2 uptake by phytoplankton would require numerical solution of a nonlinear partial differential equation.

6.10.6 Diffusion-Reaction Models for Biological Films

Biofilms are common in natural and engineered systems, and diffusion-reaction processes in such films are of interest to aquatic scientists and engineers. Biofilms are found on ship hulls, piers, pipes, streambeds, macrophytes, and rocks — just about any submerged surface. Several treatment systems use biofilms: trickling filters, rotating biological contactors, and anaerobic sand filters (used for denitrification). The kinetics of substrate uptake by biofilms has received much attention, and many models are available.[144,145,163-168] The concepts used to describe biofilm kinetics are analogous to those used for bioflocs, but the equations differ because of the different geometry (flat surfaces and diffusion in one dimension). A general

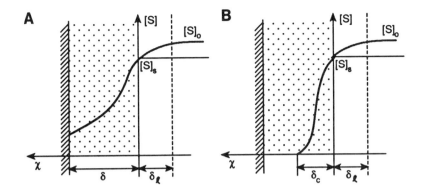

Figure 6-38. Substrate concentration profiles for (A) complete and (B) incomplete penetration of a biofilm (denoted by speckled area). $[S]_o$ = substrate concentration in bulk solution. Two profiles are shown in liquid phase for each case: solid line = well-mixed liquid with no stagnant boundary layer; dotted line = change in [S] through a stagnant boundary layer. Dashed line represents limit of boundary layer; $[S]_s$ = substrate concentration at biofilm surface. [Redrawn from LaMotta, E.J., *Environ. Sci. Technol.*, 10, 765 (1976).]

solution to the steady-state one-dimensional diffusion-reaction equation for nth-order reactions is available.[169] We will examine zero- and first-order cases.

Zero-Order Substrate Uptake

LaMotta[166] described this case for which Equation 6-95 becomes

$$d^2[S]/dx^2 = k_o/D_{Sf} \qquad (6\text{-}111)$$

k_o is the zero-order rate of substrate uptake (a constant with units of mass s^{-1} cm^{-3}). The boundary conditions depend on whether S penetrates the film completely or is depleted before it diffuses through the entire thickness δ of film (Figure 6-38). In the former case the boundary conditions are $[S] = [S]_o$ at $x = 0$ (the film surface) and $d[S]/dx = 0$ at $x = \delta$. (It is assumed that the wall is not penetrated.) When S is depleted within the film, the boundary conditions are $[S] = [S]_o$ at $x = 0$; $[S] = 0$ at $x = \delta_c$; and $d[S]/dx = 0$ at $x = \delta_c$ (no mass transport beyond $x = \delta_c$). For simplicity, diffusion from bulk solution to the film surface is assumed to be rapid, so that no concentration gradient builds up in the solution; this is not always a valid assumption in stagnant waters. Integration of Equation 6-111 yields equations for $d[S]/dx$ and $[S]$ with constants of integration that can be evaluated from the boundary conditions. For complete penetration by S, we obtain

$$d[S]/dx\big|_{x=0} = -\delta k_o/D_{Sf} \qquad (6\text{-}112)$$

The reaction rate with complete penetration, R_{cp}, is given by the flux of S to the film times its surface area. The former is given by Fick's first law; thus,

$$R_{pc} = A(F) = -AD_{Sf}\,d[S]/dx_{|x=0} = AD_{Sf}\,\delta k_o / D_{Sf} = A\delta k_o \qquad (6\text{-}113)$$

When diffusion is not limiting, the reaction rate thus is simply equal to the volume of the film times the zero-order rate constant.

Although the case of incomplete penetration yields a similar equation, the situation is more complicated than it first appears. For this case we obtain

$$R_{ip} = A\delta_c\,k_o \qquad (6\text{-}114)$$

In contrast to δ, however, δ_c is not a readily measurable variable. Moreover, it is not constant, but depends on D_{Sf}, $[S]_o$, and k_o. Thus, Equation 6-114 is not directly useful, and we need to replace δ_c with measurable quantities. From the boundary conditions it can be shown that

$$\delta_c = (2D_{Sf}[S]_o / k_o)^{1/2} \qquad (6\text{-}115)$$

and therefore Equation 6-114 becomes

$$R_{ip} = A(2k_o D_{Sf})^{1/2}([S]_o)^{1/2} \qquad (6\text{-}116)$$

According to this equation, diffusion control causes an intrinsically zero-order reaction to exhibit an apparent order of 0.5 with respect to [S]. In general, for an nth-order reaction, diffusion limitation leads to an observed reaction order of $(n+1)/2$.[166]

E_f for a zero-order biofilm is given by the ratio of Equations 6-113 and 6-114:

$$E_f = R_{ip}/R_{cp} = \delta_c/\delta \qquad (6\text{-}117)$$

As indicated by Equations 6-113 and 6-114, the zero-order rate of substrate removal from solution is simply proportional to the thickness of the film it penetrates. This can be seen in Figure 6-39 for glucose uptake by a laboratory analogue of biofilms used in wastewater treatment. The good fit indicates that the zero-order model is adequate to describe uptake kinetics in some biofilm reactors.

First-Order Substrate Uptake

The mathematics of the one-dimensional model with first-order kinetics is considerably more complicated, even though the starting equation looks simple:

$$d^2[S]/dx^2 = k_1[S]/D_{Sf} \qquad (6\text{-}118)$$

Boundary conditions are $[S] = [S]_o$ at $x = 0$, and $d[S]/dx = 0$ at $x = \delta$. Solutions to this equation are available in several texts:[83,145]

$$[S]_x/[S]_o = \cosh\{(k_1\delta^2/D_{Sf})^{1/2}(1 - x/\delta)\}/\cosh\{(k_1\delta^2/D_{Sf})^{1/2}\} \qquad (6\text{-}119)$$

and $$[S]_\delta/[S]_o = 1/\cosh\{(k_1\delta^2/D_{Sf})^{1/2}\} \qquad (6\text{-}120)$$

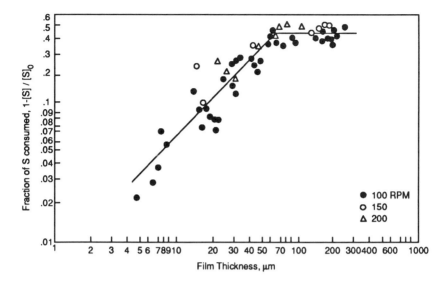

Figure 6-39. Plot of glucose removal by fixed biological film according to zero-order model; $[S]_o = 224$ mg L^{-1}. [From LaMotta, E.J., *Environ. Sci. Technol.*, 10, 765 (1976). With permission of the American Chemical Society.]

Steady-State Biofilm Model

The above models describe substrate uptake by biofilms, but do not consider film growth or decay. Film thickness thus is not determined by the model, but must be defined by the modeler. For short-term uptake studies, this is satisfactory because film thickness can be measured and does not change rapidly. A single-substrate biofilm model has been described[167,170] that couples the substrate flux into a biofilm at a given [S] to the biofilm thickness that this concentration will support at steady state. The model has four main equations:

1. Diffusion-reaction equation for [S] in the biofilm:

$$D_{Sf} \frac{d^2[S]_f}{dx^2} = \frac{k_{T\,max}[M]_f[S]_f}{K_S + [S]_f} \tag{6-121}$$

2. Flux of S from bulk liquid through boundary layer to biofilm surface:

$$F_S = -D_S d[S]/dx = D_S \{[S]_o - [S]_s\}/\delta_l \tag{6-122}$$

3. Net growth of bacterial mass in a differential section of biofilm:

$$A[M]_f \frac{dx}{dt} = \frac{k_{T\,max}[S]_f}{K_S + [S]_f} YA[M]fdx - bA[M]_f dx \tag{6-123}$$

4.   Steady-state biofilm thickness:

$$\delta_f = F_S Y / b[M]_f \qquad (6\text{-}124)$$

Symbols not previously defined are δ_l = thickness of liquid boundary layer, A = cross-sectional area of biofilm, and b = first order decay coefficient. Subscript f assigns a parameter to the biofilm.

Because Equation 6-121 contains a nonlinear Monod term, it cannot be solved analytically for $[S]_f$, and fluxes of S are computed numerically.[167,171] The equation for bacterial growth (Equation 6-123) includes a term for bacterial dieoff and/or maintenance/respiration; dieoff and respiration have equivalent effects on net biomass production. Because maintenance is considered explicitly in Equation 6-123, the correct coefficient is Y (the maximum yield coefficient, Equation 6-63). To solve the model, the steady-state assumption is made for Equation 6-123; i.e., the biofilm is assumed to reach a constant thickness. Equation 6-123 yields Equation 6-124 directly when this assumption is made. Steady state is achieved when the rate of energy capture by uptake of S and cell growth equals the rate of energy expenditure for cell maintenance (respiration and dieoff). The former is given by $F_S Y$ and the first term on the right side of Equation 6-123. The latter is given by the last term of Equation 6-123 integrated over the biofilm thickness: $bA[M]_f L_f$. Equating these terms leads to Equation 6-124.

Inclusion of a maintenance term in the growth equation leads to the finding that a threshold concentration of substrate, $[S]_{min}$, is needed to maintain a steady-state biofilm. Net growth is negative at sufficiently low [S], because the flux of S into the biofilm is insufficient to meet microbial maintenance needs, and the biofilm will gradually disappear. According to Rittmann and McCarty,[167,170] the existence of a threshold concentration (below which a steady-state biofilm cannot be maintained) implies that substrates may persist at residual levels in water, because the concentrations are too low to supply sufficient energy to sustain microorganisms. Although the persistence of residual levels of organic substrates is common, this explanation does not seem adequate for surface waters where biofilms in fact usually are present. Nonsteady-state conditions (periodic inputs of high [S]) may allow biofilms to persist in otherwise substrate-poor environments. On the other hand, steady-state conditions may be achieved in groundwater environments, and persistence of residual levels of organic substrates in them may reflect the disappearance of microorganisms (free-living or biofilms), because substrate concentrations are below the levels needed to maintain viable populations.

Rittmann and McCarty[167] computed F_S and δ_f at $[S] > [S]_{min}$ by solving dimensionless forms of Equations 6-121 to 6-124 simultaneously by computer. For computational reasons it is convenient to select a value of F_S and compute the corresponding steady-state values of δ_f and [S]. A given $[S] > [S]_{min}$ uniquely defines a steady-state flux F_S and film thickness δ_f according to the above equations (see Figure 6-40). F_S increases rapidly just above $[S]_{min}$; in fact, the apparent order of flux increase with respect to [S] initially is near infinity. The δ_f associated with a given flux (computed

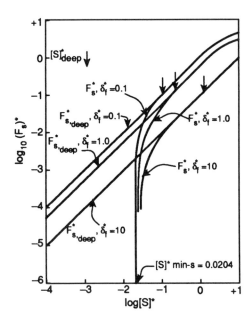

Figure 6-40. Dependence of steady-state flux F^*_s on substrate concentration $[S]^*$ for deep biofilms (F^*_s) at various δ^*_f ($\delta^*_f = J^*/(D^*_f b^*)$). Superscript * denotes dimensionless parameters. [From Rittman, B.E., and P.L. McCarthy, *Biotech. Bioeng.*, 22, 2343 (1980). With permission.]

from Equation 6-124) also increases rapidly as [S] increases just above $[S]_{min}$. At [S] near $[S]_{min}$, the film is thin, and S completely penetrates the "shallow" film. As [S] increases, both δ_f and F_S increase. Eventually, δ_f becomes so great that penetration by S is incomplete (a "deep" biofilm), and the reaction order becomes 0.5 to 1.0 (cf. Equation 6-116). The results in Figure 6-40 are analogous to those for the zero-order model (Figure 6-39 and Equations 6-112 to 6-117), which shows that substrate flux is proportional to film thickness in shallow films and independent of film thickness for deep films.

Biomass can be lost from films by shearing as well as dieoff or maintenance. Although shearing is a random (stochastic) event, both it and maintenance can be described as apparent first-order processes. Decay coefficients also can be written as a function of shear stress.[172] Shear stress is a minor factor for biofilm loss in fixed-bed reactors, but it is likely to be important in controlling the thickness of biofilms in turbulent natural streams (e.g., attached growths of filamentous algae and bacteria).

The usefulness of biofilm models to predict the fate of organic compounds in natural systems has been improved by several recent developments.[173-175] For example, effects of stream velocity on substrate mass transport to streambed biofilms was studied in laboratory simulations, and the results were analyzed by a one-dimensional stream water quality model with a biofilm substrate-uptake component.[173] Rates of substrate removal were more sensitive to short-term

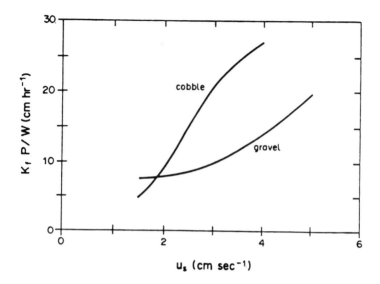

Figure 6-41. Comparison of calculated COD removal rate constants for biofilm model per unit projected streambed surface area ($K_f P/W$) for gravel and cobble streambeds as a function of shear velocity (u_s). A water temperature of 20°C was assumed. K_f = first-order flux constant (cm h^{-1}); P = amount of biofilm-covered surface area per stream length (cm); W = channel width (cm). [From Gantzer, C.J., B.E. Rittmann, and E.E. Herricks, *Water Res.*, 22, 709 (1988). With permission.]

changes in stream velocity for a cobble bed (mean diameter = 6 cm) than a gravel bed (mean diameter = 1.6 cm) (Figure 6-41). Streambed particle size evidently plays a more complicated role in determining substrate removal rates than can be explained on the basis of surface area alone. Particle size likely has a complex effect on hydraulic conditions within streambeds and influences removal rates via effects on interstitial void volume and porosity.

Uptake of inhibitory substrates by biofilms has been improved by applying the Haldane equation to a steady-state biofilm model.[174] For development of a steady-state biofilm from a monolayer, concentrations of inhibitory substrates like pentachlorophenol (PCP) must be between $[S]_{min}$ and $[S]_{max}$. The former is defined as above, and $[S]_{max}$ is the maximum concentration at which a cell can still produce enough energy to meet maintenance needs.[174] Two steady-state biofilm solutions occur for this model: the one associated with $[S]_{min}$ is stable, but the other, associated with $[S]_{max}$, is unstable to perturbations in $[S]$ around $[S]_{max}$. Interestingly, simulations showed that the $[S]_{max}$-associated steady-state value of $[S]_o$ (substrate concentration at the biofilm surface) is higher than $[S]_{max}$. This implies that biofilms can tolerate higher concentrations of inhibitory substrates than can dispersed organisms. This finding is explained by substrate use (at inhibited rates) in the near surface of the biofilm and by mass-transfer resistance within the biofilm, both of which result in the inner layers of biofilm being exposed to $[S] < [S]_{max}$.

6.11 MICROBIAL DIEOFF AND DISINFECTION KINETICS

6.11.1 Microbial Dieoff and Maintenance Losses

Microbial dieoff usually is described by first-order expressions such as

$$dN/dt = -k_{bd}N \qquad (6\text{-}125)$$

k_{bd} is a first-order dieoff coefficient, and N is microbial density (number per volume of medium). Net growth thus is defined as

$$dN/dt = (\mu - k_{bd})N \qquad (6\text{-}126)$$

Inclusion of a loss term for cell maintenance in Equation 6-65 is an analogous concept, and k_e (the maintenance coefficient) is analogous to a dieoff coefficient, as implied in the biofilm model of Section 6.10.6. Maintenance loss and dieoff are described by similar expressions, but mechanistically they are different. The former involves a decrease in biomass, but not necessarily a decrease in cell numbers. It is analogous to the weight loss of a dieting individual. Of course, individuals can die of starvation, but only after considerable weight loss has occurred. Similarly, a starved microbial culture gradually depletes its biomass because of maintenance requirements, but loss of cell numbers may be small until a substantial fraction of biomass is consumed. Dieoff thus refers to cell numbers, and maintenance loss refers to biomass.

Just as growth rate coefficients (μ) are functions of many environmental conditions, including temperature and [S], dieoff coefficients also vary widely, depending on the physiological state of the organisms and environmental conditions like temperature, sunlight, and the presence of toxicants. For example, k_{bd} is small compared with μ in exponentially growing cultures, but increases with mean cell age. In batch cultures $k_{bd} \approx \mu$ during the stationary phase, and $k_{bd} > \mu$ as the culture senesces. Table 6-10 summarizes dieoff coefficients for various microorganisms in aquatic systems. As discussed in Section 6.7.2, k_e increases with temperature according to an Arrhenius-like expression (in the normal temperature range of the microorganism). Dieoff rates (and k_{bd}) increase rapidly above a certain temperature (T_{opt}). Temperature-induced death rates of microorganisms follow the Arrhenius expression, and apparent activation energies have been reported for death of various organisms. Reflecting the high sensitivity of microorganisms to thermal death at $T > T_{opt}$, reported E_{act} values typically are high. For example, E. coli has E_{act} = 439 ± 38 kJ mol^{-1} for thermal death in the temperature range 50 to 60°C,[176] and the yeast Saccharomyces cerevisiae has E_{act} for thermal death of 355 kJ mol^{-1}.[177]

Dieoff rates of some bacteria are higher in sunlight than in the dark; Gameson and Gould[178] concluded that k_{bd} is directly proportional to light intensity: $k_{bd} = \alpha I_o$, where I_o is surface light intensity. When I_o is expressed in cal cm^{-2} h^{-1}, α is approximately 1. Light intensity decreases exponentially with depth in the water column, and the effect of light on microbial dieoff rates can be expressed approximately as[178]

Table 6-10. Dieoff Coefficients for Bacteria

Organism	Conditions	k_{bd} (day^{-1})	Ref.
Bacterial groups			
Total coliforms	Freshwater (summer or 20°C)	1.0–5.5	180
	Seawater (20°C)	0.7–3.0	181
Fecal coliforms	Seawater (sunlight)	37–110	182
Fecal streptococci	Seawater (sunlight)	18–55	182
Bacterial species			
Escherichia coli	Seawater (10–30%)	0.08–2.0	183
Streptococcus faecalis	Freshwater (20°C)	0.4–0.9	184
	Freshwater(4°C)	0.1–0.4	184
	Stormwater (20°C) days 0–3	0.3	185
	Stormwater days 3–14	0.1	185
	L. Ontario(18°C) days 0–10	1.0–3.0	186
	L. Ontario days 10–28	0.05–0.1	186
Salmonella typhimurium	Stormwater (20°C) days 0–3	1.1	185
	Stormwater days 3–14	0.1	185

Summarized from Thomann, R.V. and J.A. Mueller, *Principles of Surface Water Quality Modeling and Control*, Harper & Row, New York, 1987.

$$k_{bd} = \frac{\alpha I_o}{z k_e} \{1 - \exp(-k_e z)\} \tag{6-127}$$

where k_e is the vertical light extinction coefficient (m^{-1}), and z is depth (m).

The assumption that dieoff for a given population of microorganisms follows first-order kinetics described by a single k_{bd} is equivalent to assuming that the population is homogeneous, at least with regard to resistance to the environmental stress (e.g., a disinfecting agent or toxic substance) causing dieoff. It is well known that some microorganisms exhibit greater resistance to stress than do others, even within the same species. When this occurs, dieoff can be described by a series of first-order coefficients that apply to different time periods (Table 6-10). For example, dieoff of *Salmonella typhimurium* is relatively rapid during the first 3 days of incubation (apparent rate constant of ~1 day^{-1}), but after that only more resistant cells remain and the dieoff coefficient drops to ~0.1 day^{-1}.[185]

6.11.2 Kinetics of Disinfection

Elimination of potential pathogens is a primary objective of water and wastewater treatment, and several common disinfecting chemicals are used for this purpose. Chlorine (Cl_2) and its aqueous products (HOCl, OCl$^-$, chloramines) have been used most widely, but because of toxicity of residual chlorine compounds to higher organisms and undesirable side reactions of chlorine, alternative disinfectants are becoming more common. These include chlorine dioxide, iodine, and ozone, as well as UV light.[187] An understanding of the chemical reactions of disinfectants in solution is important in determining optimum disinfection conditions, and these reactions have received much attention in the past few decades (see Section 4.5). In this section we examine the dieoff kinetics of microbes exposed to disinfectants and describe some general mechanisms by which disinfectants cause microbial death.

When applied to microbial disinfection, the first-order rate equation is known as Chick's Law. The relationship was formulated in 1908.[188] For the law to apply, all conditions except microbial population density must remain constant; i.e., disinfectant concentration must not change over time. This often is difficult to achieve because of the chemical reactivity of disinfectants. Many examples are known of exponential dieoff of organisms exposed to disinfectants, but departures from Chick's Law also are common. Kill rates may increase or decrease with time (Figure 6-42). Increases with time (analogous to a lag phase) are attributed to the need for the disinfectant to build up to lethal concentrations at several centers in the organism before it dies.[189] This so-called multihit phenomenon has been reported for virus inactivation[190] and more commonly applies to multicellular organisms. Decreases with time may reflect a decrease in disinfectant concentration because of chemical reactions, but variable resistance among cells also may be responsible.

The effect of disinfectant concentration on microbial dieoff is expressed empirically by Watson's Law:[191]

$$C^n t = \text{Constant} = k' \qquad (6\text{-}128)$$

C is the concentration of disinfectant, n is the coefficient of dilution, t is the contact time required for a certain percentage kill, and k' is a species-dependent constant. Values of n < 1 indicate that contact time is more important than concentration in microbial dieoff; n > 1 implies that the efficiency of disinfection decreases rapidly with dilution. For disinfection of microbes, n usually is ~1.[189] Higher values were found when Watson's Law was used to describe dieoff of guppy fry exposed to NH_3 (n = 1.9) and KNO_3 (n = 1.4).[192]

Based on Chick's and Watson's Laws, Hom[189] proposed a general empirical expression for chemical disinfection:

$$dN/dt = -kNt^m C^n \qquad (6\text{-}129)$$

When m and n = 0, Equation 6-129 simplifies to a first-order equation (Chick's Law). When m = 0 and n ≠ 0, Equation 6-129 is n order in C. Substitution of $C^n = k'/t$ from Equation 6-128 into Equation 6-129 yields $dN/dt = -kNk'/t$, which can be integrated by separation of variables to

$$\ln(N/N_o) = -kk' \ln(t/t_o) = -k_n \ln(t/t_o) \qquad (6\text{-}130)$$

k_n is the rate constant for the n-order reaction. If a plot of $\ln N$ vs. $\ln t$ is linear, the reaction fits this model, and the rate constant can be obtained from the slope. When both n and m ≠ 0, substitution of k'/t for C^n yields $dN/dt = -kk'Nt^{m-1}$, which again can be integrated by separation of variables:

$$\ln(N/N_o) = k_{mn} t^m / m \qquad (6\text{-}131)$$

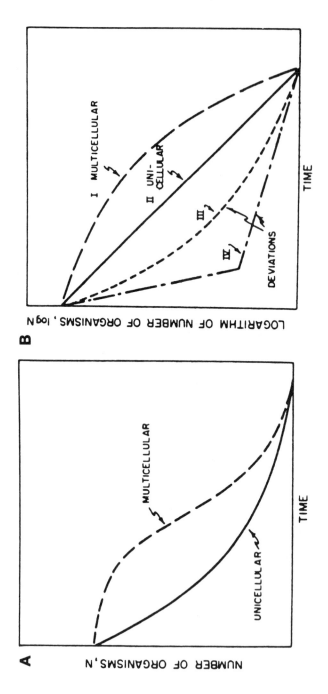

Figure 6-42.   Survival curves of organisms: (A) arithmetic plot showing typical exponential decay for unicellular organisms and lag phase for death of multicellular organisms; (B) semilog plots: curve II is for exponential decay (Chick's law); curves I, III, and IV are deviations from Chick's law. Curve I corresponds to multihit theory and also applies to multicellular organisms. Curves III and IV may result from presence of some resistant organisms in the population. [From Hom, L.W., J. Sanit. Eng. Div. ASCE, 98, 183 (1972). With permission.]

k_{mn} is the rate constant for the n- and m-order reaction. Reactions fitting this equation yield a linear relationship on a plot of $ln(ln(N))$ vs. $ln t$. The order m is obtained from the slope, and the intercept yields k_{mn}. Hom[189] reported that dieoff of coliforms in waste lagoons treated with Cl_2 followed Equation 6-130 (m = 0, n ≠ 0) at Cl_2 doses up to 1.0 mg L^{-1}. At higher doses dieoff followed Equation 6-131 (m and n ≠ 0). Coliform survival thus depended on reaction time, Cl_2 dose, and the number of surviving organisms. Hom reported m ≈ 0.8, which means the dieoff rate decreased with time. Values of m in the range 1.3 to 2.2 have been reported for disinfection when a lag phase occurred.[193,194] Fair et al.[194] proposed plotting *log* survival vs. t^2 to linearize such data.

Attempts have been made recently to develop mechanistic models of chlorine disinfection. Haas[193] proposed a two-step process: (1) reversible binding of disinfectant molecules (C) to receptor sites (S), and (2) inactivation of microorganisms at a rate determined by the amount of bound disinfectant:

$$C + S \underset{k_{-1}}{\overset{k_1}{\rightleftharpoons}} CS \overset{k_2}{\rightarrow} \text{inactive microbe} \qquad (6\text{-}132)$$

The number of binding sites per organism (β) is assumed to be constant, and both viable and killed microorganisms bind disinfectant in the same way. If the disinfectant concentration is constant and the inactivation step is first order in viable microorganisms (N), the following integrated rate law applies:

$$\ln(N/N_o) = \frac{-k_2 C \beta}{C + K_D} \left[t + \frac{\exp(-k_1 t(C + K_D)) - 1}{k_1(C + K_D)} \right] \qquad (6\text{-}133)$$

K_D is analogous to a Monod half-saturation constant. The model predicts a concentration-dependent lag in dieoff. The rate of inactivation determined from the linear portion of the curve follows a Monod-like function. Good fit was obtained when data on inactivation of a poliovirus by Cl_2 were applied to the model. Although Haas presented evidence for the existence of a "disinfectant-organism complex" (a key feature of the model), the chemistry of chlorine compounds suggests that the term complex should be construed loosely. It is questionable whether chlorine is bound chemically in a reversible fashion.

The effects of chemical disinfectants on microbial inactivation are temperature dependent, but apparent activation energies for chemically induced dieoff are much lower than those mentioned earlier for thermally induced dieoff. For example, E_{act} for ozone inactivation of *Mycobacterium fortuitum*[195] is 76.5 kJ mol^{-1}. This is high enough to rule out mass transfer of O_3 into cells as the rate-limiting step, but is only 20 to 25% of the E_{act} for thermally-induced microbial dieoff. E_{act} for SO_2-induced death of *S. cerevisiae* is 150 kJ mol^{-1};[177] in contrast, E_{act} = 355 kJ mol^{-1} for thermal death of the same organism.

Two types of behavior have been found on Arrhenius plots of dieoff rate constants for *E. coli* at different concentrations of disinfectant.[176] Sorbic acid (Figure 6-43A) yielded parallel lines (**type A**) at concentrations ranging from 0 to

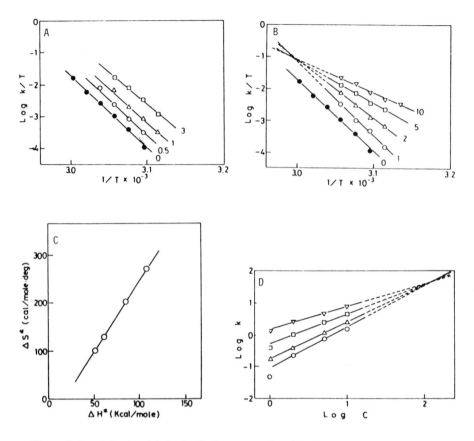

Figure 6-43. Arrhenius plots for dieoff rate constants (k_d) of *E. coli* at several concentrations of (A) sorbic acid and (B) an amphoteric surfactant. Disinfectant concentrations in % (A) and mg L^{-1} (B). (C) Linear relationship between $\Delta H^{\ne}$ and $\Delta S^{\ne}$ (the compensation effect [see Section 7.2]) for dieoff of *E. coli* in presence of amphoteric surfactant. Values obtained from plot in (B). (D) Effect of amphoteric surfactant concentration on k_d at several temperatures: (○) 48°C; (△) 50°C; (□) 52°C; (▽) 54°C. [From Tsuchido, T. and I. Shibasaki, *Biotech. Bioeng.*, 22, 107 (1980). With permission.]

3%, implying that E_{act} was constant (439 ± 38 kJ mol^{-1}) regardless of concentration. On the other hand, an amphoteric surfactant (Figure 6-43B) had decreasing slopes with increasing concentration (**type S**), and E_{act} decreased as concentration increased (from 451 to 213 kcal mol^{-1} over the range 1 to 10 µg/mL). The lines in Figure 6-43B intersect at a common point called the isokinetic temperature, θ. The dieoff rate is independent of disinfectant concentration at θ. The existence of an isokinetic point implies that the entropy and enthalpy of activation are linearly related (Figure 6-43C). This relationship, called the compensation effect, is discussed in Chapter 7.

Similarly, *log-log* plots of the dieoff rate constant vs. sorbic acid concentration yield parallel lines for various temperatures, while lines for the surfactant intersect at a common point (~114 µg/mL) at which the dieoff rate is independent of temperature. These findings suggest that the two disinfectants have different modes of action.[176] The surfactant (type S) is thought to enhance cellular injury and death during heating; the decrease in E_{act} with increasing concentration implies that it facilitates the action of heat on sensitive sites in the cells. Sorbic acid (type A) is thought to diminish the ability of cells to repair damage after heating. It had no effect on thermally induced death (E_{act} did not change as its concentration increased). Other examples of type A and S disinfectants have been reported.[176] These classes may have use in designing optimum conditions for disinfectants and also may be useful in evaluating mechanisms involved in chemically induced microbial death.

PROBLEMS

1. The reaction $A \rightarrow P$ is catalyzed by an enzyme. When $[A] = 3mM$. $v_o = 0.4$ units. When $[A] = 0.3$ mM, $v_o = 0.004$ units. What is v_o when $[A] = 1 \times 10\text{–}5\ M$?

2. The following initial velocities were obtained for an enzymatic reation at varying substrate concentrations. Plot the data acooring to the three linearized transformations of the Michaelis-Menten equation and determine the kinetic parameters form each plot. Which plot seems to give the most reliable results?

[S] (M)	0.002	0.005	0.010	0.020	0.040	0.100
v_o	0.0085	0.020	0.030	0.042	0.055	0.065

3. The maximum velocity, v_{max}, for a single-substrate enzymatic reaction was found to be 1.72 mM/min when the enzyme concentration was 1 mg/L. If the enzyme has one active site and its molecular weight is 50,000, what is the value (in min^{-1}) of k_3 (of Equation 6-2)?

4. The following data were obtained for an enzyme-catalyzed reaction in the abscence and presence of inhibitor I. Estimate K_S, V_{max}, and K_I. What kind of inhibition pattern does I exhibit?

[S] (μM)	0.1	0.15	0.2	0.3	0.5	1.0
v_o ($\mu M\ s^{-1}$) ($[I] = 0$)	0.029	0.039	0.044	0.054	0.067	0.081
v_o ($\mu M\ s^{-1}$) ($[I] = 0.1\ M$)	0.018	0.024	0.028	0.034	0.042	0.050

5. Given the values of K_S and μ_{max} for uptake of ammonium and growth of nitrifying bacteria in Table 6-4 and corresponding values for algae in Table 6-6, discuss the conditions under which nutrifiers may outcompete algae in rivers and lakes.

6. A bacterial culture is treated with mercuric chloride to determine its effectiveness as a bactericide. The following data are obtained under conditions where it can ba assumed that the $HgCl_2$ concentration remains constant:

Time (min)	Viable bacteria/mL	Time (min)	Viable bacteria/mL
2	549	6	82
3	323	7	42
4	236	8	19
5	138	10	10

Determine the order of the die-off reaction with respect to the bacteria and the rate constant for die-off. Calculate the initial number of viable bacteria in the sample and the time required to reduce the count to <1 per 100 mL.

7.   The effect of sodium ion on kinetics of ammonium oxidation by bacteria was studied by Visut (M.S. thesis, Asian Inst. Tech., Bangkok, 1985). Plot the following data according to the Lineweaver-Burk method and determine the type of enzyme inhibition pattern tha Na causes.

$[Na^+] = 0.137$ g/L

[S] (mg/L)	0.92	2.12	5.00	10.0	25
v (mg/L-h)	0.15	0.308	0.571	1.04	1.67

$[Na^+] = 1.052$ g/L

[S] (mg/L)	1.11	1.11	2.08	2.17	3.85	5.88	11.1	12.5	22.2	25
v (mg/L-h)	0.125	0.143	0.274	0.290	0.435	0.667	0.91	1.05	0.83	1.25

$[Na^+] = 5.26$ g/L

[S] (mg/L)	1.14	2.17	4.26	4.55	4.76	11.1	11.8	13.3	20	22.2	25
v (mg/L-h)	0.058	0.116	0.192	0.21	0.244	0.30	0.36	0.24	0.43	0.63	0.47

Data reported by Ohgaki and Wantawin, Chap. 13 in *Mathematical Submodels in Water Quality Systems,* S.E. Jørgensen and M.J. Gromiec (Eds.), Elsevier, Amsterdam, 1989.

REFERENCES

1.   Brown, A.J., *J. Chem Soc.* (Trans.), 61, 369 (1892); 81, 373 (1902).
2.   Henri, V., *Lois Générales de l'Action des Diastases*, Hermann et Cie, Paris, 1903.
3.   Cornish-Bowden, A., *Principles of Enzyme Kinetics*, Butterworths, London, 1976.
4.   Michaelis, L. and M.L. Menten, *Biochem. Z.*, 49, 333 (1913).
5.   Briggs, G.E. and J.B.S. Haldane, *Biochem. J.*, 19, 338 (1925).
6.   Endrenyi, L. (Ed.), *Kinetic Data Analysis*, Plenum Press, New York, 1981.
7.   Lineweaver, H. and D. Burk, *J. Am. Chem. Soc.*, 56, 658 (1934).
8.   Hanes, C.S., *Biochem. J.*, 26, 1406 (1942).
9.   Eadie, G.S., *J. Biol. Chem.*, 146, 85 (1942); Hofstee, B.H.J., *Nature*, 184, 1296 (1959).
10.   Dowd, J.E. and D.S. Riggs, *J. Biol. Chem.*, 210, 863 (1965).
11.   Cornish-Bowden, A., *Kinetic Data Analysis*, L. Endrenyi (Ed.), Plenum Press, New York, 1981, pp. 105–119.
12.   Lineweaver, H., D. Burk, and W.E. Deming, *J. Am. Chem. Soc.*, 56, 225 (1934); Burk, D., *Ergeb. Enzymforsch.*, 3, 23 (1934).
13.   Li, W.K.W., *Limnol. Oceanogr.*, 28, 185 (1983).
14.   Kendall, M.G. and A. Stuart, *The Advanced Theory of Statistics*, Vol. 2. Griffin, London, 1961; Yevjevich, V., *Probability and Statistics in Hydrology*, Water Resources Publishers, Fort Collins, CO, 1972.
15.   Kenney, B.C., *Water Resour. Res.*, 18, 1041 (1982).
16.   Eisenthal, R. and A. Cornish-Bowden, *Biochem J.*, 139, 715 (1974).
17.   Dugglesby, R.G., *Kinetic Data Analysis*, L. Endrenyi (Ed.), Plenum Press, New York, 1981, pp. 169–179.
18.   Marquardt, D.W., *J. Soc. Ind. Appl. Math.*, 11, 431 (1963).
19.   Schreiner, W.D., M. Kramer, S. Krischer, and Y. Langsam, *PC Tech. J.*, May, 170 (1985); Bevington, P.R., *Data Reduction and Error Analysis for the Physical Sciences*, McGraw-Hill, New York, 1969.
20.   King, E.L. and C. Altman, *J. Phys. Chem.*, 60, 1375 (1956).
21.   Duggleby, R.G. and J.F. Morrison, *Biochim. Biophys. Acta*, 481, 297 (1977).
22.   Barrio-Lage, G., F.Z. Parsons, and R.S. Nassar, *Environ. Sci. Technol.*, 21, 366 (1987).
23.   Mahler, H.R. and E.H. Cordes, *Biological Chemistry*, 2nd ed., Harper & Row, New York, 1971.
24.   Segel, I.H., *Enzyme Kinetics*, Wiley-Interscience, New York, 1975.
25.   Dixon, M. and E.C. Webb, *Enzymes*, 3rd ed., Academic Press, New York, 1979.
26.   Cleland, W.W., *Biochim. Biophys. Acta*, 67, 104 (1963).
27.   Cleland, W.W., *Biochim. Biophys. Acta*, 67, 173, 188 (1963).

28. Endrenyi, L., in *Kinetic Data Analysis*, L. Endrenyi (Ed.), Plenum Press, New York, 1981, pp. 137–167.
29. Hoffmann, M. and B. Lim, *Environ. Sci. Technol.*, 13, 1406 (1979).
30. Haldane, J.B.S., *Enzymes*, Longmans, London, 1930.
31. Megard, R.O., D.W. Tonkyn, and W.H. Senft, Jr., *J. Plankton Res.*, 6, 325 (1984).
32. Monod, J., J. Wyman, and J-P. Changeux, *J. Mol. Biol.*, 12, 88 (1965); Blangy, D., H. Buc, and J. Monod, *J. Mol. Biol.*, 31, 13 (1968).
33. Hill, A.J., *Biochem. J.*, 7, 471 (1913); Atkinson, D.E., *Ann. Rev. Biochem.* 35, 85 (1966).
34. Pauling, L., *Am. Sci.*, 36, 58 (1948).
35. Lienhard, G.E., *Science*, 180, 149 (1973).
36. Jencks, W.P., *Catalysis in Chemistry and Enzymology*, McGraw-Hill, New York, 1969.
37. Kraut, J., *Science*, 242, 533 (1988).
38. Bernhard, S.A., *The Structure and Function of Enzymes*, W.A.Benjamin, New York, 1968; Zeffren, E. and P.L. Hall, *The Study of Enzyme Mechanisms*, Wiley-Interscience, New York, 1973; Bender, M.L. and L. Brubacher, *Catalysis and Enzyme Action*, McGraw-Hill, New York, 1983; Ferscht, A.R., *Enzyme Structure and Mechanisms*, W.H. Freeman, San Francisco, 1985.
39. Schulz, P.G., *Science*, 240, 426 (1988); Wong, C.-H., *Science*, 244, 1145 (1989).
40. Raven, J.A., *Adv. Microb. Physiol.*, 21, 47 (1980); Cooney, C.L., *Biotechnology*, 1, 73 (1981); Daigger, G.T. and C.P. Grady, *Water Res.*, 16, 365 (1982).
41. Van Uden, N., *Ann. Rev. Microbiol.*, 24, 473 (1969).
42. Button, D.K., *Microb. Rev.*, 49, 270 (1985).
43. Monod, J., *Recherches sur la Croissance ds Cultures Bactèriennes*, Herman et Cie, Paris, 1942.
44. Levenspiel, O., *Biotech. Bioeng.*, 22, 1671 (1980).
45. Healey, F.P., *Microb. Ecol.*, 5, 281 (1980).
46. Tilman, D., M., Mattson, and S. Langer, *Limnol. Oceanogr.*, 26, 1020 (1981).
47. Droop, M.R., *J. Mar. Biol. Assoc. U.K.*, 48, 689 (1968).
48. Gerloff, G.C. and F. Skoog, *Ecology*, 35, 348 (1954); Mackereth, F.J., *J. Exp. Bot.*, 4, 296 (1953).
49. McCarthy, J.J., Physiological basis of phytoplankton ecology, T. Platt (Ed.), *Can. Bull. Fish. Aquat. Sci.*, 210, 211–233 (1981).
50. Droop, M.R., *Bot. Mar.*, 26, 99 (1983).
51. Rhee, G.-Y., *Adv. Microb. Ecol.*, 6, 33 (1982).
52. Okada, M., R. Sudo, and S. Aiba, *Biotech. Bioeng.*, 24, 143 (1982).
53. Tessier, G., *Rev. Sci. Extrait (Paris)*, 3208, 209 (1942).
54. Blackman, F.F., *Ann. Bot.*, 19, 281 (1905).
55. Esener, A.A., J.A. Roels, and N.W.F. Kossen, *Biotech. Bioeng.*, 25, 2803 (1983).
56. Lotka, A.J., *Elements of Mathematical Biology*, Williams & Wilkins, Baltimore, 1925; reprinted by Dover, New York, 1956. For details on use of the logistic equation in ecology, see Hutchinson, G.E., *Introduction to Population Ecology*, Yale University Press, New Haven, CT, 1978.
57. Webster, I.A., *Biotech. Bioeng.*, 25, 2981 (1983).
58. Powell, E.O., *Lab. Pract.*, 1145 (1965).
59. Bailey, J.E. and D.F. Ollis, *Biochemical Engineering Fundamentals*, McGraw-Hill, New York, 1977.
60. Herbert, D., *Symp. Int. Congr. Microbiol.*, 6, 381 (1958); *Cont. Cult.*, 6, 1 (1976).
61. Pirt, S.J., *Proc. R. Soc. London*, B 163, 224 (1965).
62. Smith, J.R., R.G. Luthy, A.C. Middleton, *J. Water Pollut. Contr. Fed.*, 60, 518 (1988).
63. Bierman, V.J., Jr., F.H. Verhoff, T.L. Poulsen, and M.W. Tenney, in *Modeling the Eutrophication Process*, E.J. Middlebrooks and T. Powers (Eds.), Ann Arbor Science, Ann Arbor, MI, 1973, pp. 89–109; DePinto, J.V., V.J. Bierman, Jr., and F.H. Verhoff, in *Modeling Biochemical Processes in Aquatic Ecosystems*, R.P. Canale (Ed.), Ann Arbor Science., Ann Arbor, MI, 1976, pp. 141–169; Grenney, W.J., D.A. Bella, and H.C. Curl, Jr., *Am. Nat.*, 107: 405 (1973); *J. Water Pollut. Contr. Fed.*, 46: 1751 (1974).
64. Dennis, R.W. and R.L. Irvine, *Water Res.*, 15, 1363 (1981).
65. Harder, A. and J.A. Roels, in *Advances in Biochemical Engineering*, Vol. 21, A. Fiechter (Ed.), Springer-Verlag, New York, 1982, pp. 55–107; Chiam, H.F. and I.J. Harris, *Biotech. Bioeng.*, 24, 37 (1982); Williams, F.M., *J. Theor. Biol.*, 15, 190 (1967).
66. Papageorgakopoulou, H. and W.J. Maier, *Biotech. Bioeng.*, Vol. 26, (1984).
67. Roels, J.A., *Biotech. Bioeng.*, 22, 33 (1980).

68. Frisch, H.K. and I.J. Gotham, *J. Theor. Biol.* 66, 665 (1977); *J. Math. Biol.*, 7, 149 (1979); Rhee, G.-Y., I.J. Gotham, and S.W. Chisholm, in *Continuous Culture of Cells*, Vol. II, P. Calcott (Ed.), CRC Press, Boca Raton, FL, 1981, pp. 159–186; Chisholm, S.W., pp. 150–181 in ref. 130; Zeuthen, E. (Ed.), *Synchrony in Cell Division and Growth*, Wiley-Interscience, New York, 1964.
69. Gotham, I.J. and G.-Y. Rhee, *J. Gen. Microbiol.*, 128, 199 (1982).
70. Van Uden, N., *Arch. Mikrobiol.*, 58, 155 (1967).
71. Hartmann, L. and G. J. Laubenberger, *J. Sanit. Eng. Div. ASCE*, 94, 247 (1968).
72. Rhee, G.-Y., *J. Phycol.*, 9, 495 (1973).
73. Sunda, W.G. and J.A.M. Lewis, *Limnol. Oceanogr.*, 23, 870 (1978); Anderson, D.M. and F.M.M. Morel, *Limnol. Oceanogr.*, 23, 283 (1978); McKnight, D.M. and F.M.M. Morel, *Limnol. Oceanogr.*, 24, 823 (1979).
74. Pawlowsky, U. and J.A. Howell, *Biotech. Bioeng.*, 15, 889 (1973).
75. Yang, R.D. and A.E. Humphrey, *Biotech. Bioeng.*, 17, 1211 (1975).
76. Andrews, J.F., *Biotech. Bioeng.*, 10, 707 (1968).
77. Sokol, W. and J.A. Howell, *Biotech. Bioeng.*, 23, 2039 (1981).
78. Papanastasiou, A.C. and W.J. Maier, *Biotech. Bioeng.*, 24, 2001 (1982).
79. Edwards, V.H. and C.R. Wilke, *Biotech. Bioeng.*, 10, 205 (1968).
80. Beltrame, P., P.L. Beltrame, and P. Carniti, *Biotech. Bioeng.*, 22, 2405 (1980).
81. Yoon, H., G. Klinzing, and H.W. Blanch, *Biotech. Bioeng.*, 19, 1193 (1977).
82. Reynolds, T.D., *Unit Operations and Processes in Environmental Engineering*, Brooks/Cole Engineering Division, Monterey, CA., 1982.
83. Hill, G.G., Jr., *An Introduction to Chemical Engineering Kinetics and Reactor Design*, John Wiley & Sons, New York, 1977.
84. Levenspiel, O. and K.B. Bischoff, *Ind. Eng. Chem.*, 51, 1431 (1959).
85. Eyring, H. and D.W. Urry, in *Theoretical and Mathematical Biology*, T.H. Waterman and H.J. Morowitz (Eds.), Blaisdell, New York, 1965, pp. 57–95.
86. Monod, J., J. Wyman, and J.P. Changeux, *J. Mol. Biol.*, 12, 88 (1965).
87. Esener, A.A., J.A. Roels, and N.W.F. Kossen, *Biotech. Bioeng.*, 23, 1401 (1981).
88. Esener, A.A., J.A. Roels, and N.W.F. Kossen, *Biotech Bioeng.*, 25, 2803 (1983).
89. Novak, J.T., *J. Water Pollut. Contr. Fed.*, 46, 1984 (1974).
90. Charley, R.C., D.G. Hooper, and A.G. McLee, *Water Res.*, 14, 1387 (1980).
91. Rhee, G.-Y. and I.J. Gotham, *Limnol. Oceanogr.*, 26, 635 (1981).
92. Goldman, J.C., *Limnol. Oceanogr.*, 22, 932 (1977).
93. Heijnen, J.J. and J.A. Roels, *Biotech. Bioeng.*, 23, 739 (1981).
94. Jorgensen, S.E., *Handbook of Environmental Data and Ecological Parameters*, Pergamon Press, New York, 1979.
95. Button, D.K., S.S. Dunker, and M.L. Morse, *J. Bacteriol.*, 113, 599 (1973).
96. Harrison, P.J., H.L. Conway, and R.C. Dugdale, *Mar. Biol.*, 35, 177 (1976).
97. Williams, P.J., *Limnol. Oceanogr.*, 18, 159 (1973).
98. Wright, R.T. and B.K. Burnison, in *Native Aquatic Bacteria: Enumeration, Activity and Ecology*, J.W. Costerton and R.R. Colwell (Eds.), ASTM, 1979, pp. 140–155; Krambeck, C., *Arch. Hydrobiol. Beih. Ergeb. Limnol.*, 12, 64 (1979).
99. Azam, F. and R.E. Hodson, *Mar. Ecol. Prog. Ser.*, 6, 213 (1981).
100. Bell, W.H., *Limnol. Oceanogr.*, 25, 1007 (1980).
101. Tarapchak, S.J. and L.R. Herche, *Can. J. Fish. Aquat. Sci.*, 43, 319 (1986).
102. Tarapchak, S.J. and L.R. Herche, *J. Environ. Qual.*, 18, 17 (1989).
103. Stumm-Zollinger, E., *Appl. Microbiol.*, 16, 133 (1968).
104. Papanastasiou, A.C. and W.J. Maier, *Biotech. Bioeng.*, 25, 2337 (1983).
105. Hutchinson, G.E., *Am. Nat.*, 95, 137 (1969).
106. Fredrickson, A.G. and G. Stephanopoulos, *Science*, 213, 972 (1981).
107. Aris, R. and A.E. Humphrey, *Biotech. Bioeng.*, 19, 1375 (1977).
108. Stephanopoulos, G., R. Aris, and A.G. Fredrickson, *Math. Biosci.*, 45, 63 (1977).
109. Kreyszig, E., *Advanced Engineering Mathematics*, 5th ed. John Wiley & Sons, New York, 1983.
110. Stewart, F.M. and B.R. Levin, *Am. Nat.*, 107, 171 (1973).
111. Lehman, J.T. and D. Scavia, *Proc. Nat. Acad. Sci. U.S.A.*, 79, 5001 (1982).

112. Lauffenberger, D. and B.P. Calcagno, *Biotech. Bioeng.*, 25, 2103 (1983).
113. MacLulich, D.A., Fluctuations in the numbers of the varying hare (Lepus americanus), University of Toronto Studies, Biol. Ser. 43, 1937; Elton, C., *Voles, Mice and Lemmings: Problems in Population Dynamics*, Oxford University Press, London, 1942.
114. Meers, J.L., in *Microbial Ecology*, A.I. Laskin and H. Lechvalier (Eds.), CRC Press, Boca Raton, FL, 1974, p. 151.
115. Lotka, A.J., *Elements of Physical Biology*, Williams and Wilkins, Baltimore, 1925; Volterra, V., in *Animal Ecology*, R.N. Chapman (Ed.), McGraw-Hill, New York, 1926, pp. 409–448.
116. Odum, E.P., *Fundamentals of Ecology*, 3rd ed. W.B. Saunders, Philadelphia, 1971; Krebs, C.J., *Ecology*, Harper & Row, New York, 1972.
117. Odum, H.T., *Systems Ecology: An Introduction*, Wiley-Interscience, New York, 1983.
118. Jost, J.L., J.F. Drake, A.G. Fredrickson, and H.M. Tsuchiya, *J. Bacteriol.*, 113, 834 (1973); *J. Theor. Biol.*, 41, 461 (1973); Bader, F.G., A.G. Fredrickson, and H.M. Tsuchiya, in *Modeling Biochemical Processes in Aquatic Ecosystems*, R.P. Canale (Ed.), Ann Arbor Science, Ann Arbor, MI, 1976, pp. 257–279.
119. Tsuchiya, H.M., J.F. Drake, J.L. Jost, and A.G. Fredrickson, *J. Bacteriol.*, 110, 1147 (1972).
120. May, R., *Science*, 156, 645 (1974).
121. Pool, R., *Science*, 243, 310 (1989).
122. Berryman, A. and J. Millstein, *Trends Ecol. Evol.*, 4, 23 (1989).
123. Gotham, I.J. and G.-Y. Rhee, *J. Phycol.*, 17, 309 (1981).
124. Gotham, I.J. and G.-Y. Rhee, *J. Phycol.*, 17, 257 (1981).
125. Dugdale, R.C., *Limnol. Oceanogr.*, 12, 685 (1967).
126. MacIsaac, J.J. and R.C. Dugdale, *Deep Sea Res.*, 16, 45 (1969).
127. Epply, R.W., J.N. Rogers, and J.J. McCarthy, *Limnol. Oceanogr.*, 14, 912 (1969).
128. Dugdale, R.C., B.H. Jones, Jr., J.J. MacIsaac, and J.J. Goering, Physiological basis of phytoplankton ecology, T. Platt (Ed.), *Can. Bull. Fish. Aquat. Sci.*, 210, 234–250 (1981).
129. Burmaster, D.E., *Amer. Nat.*, 113, 123 (1979).
130. Platt, T. (Ed.), Physiological basis of phytoplankton ecology, *Can. Bull. Fish. Aquat. Sci.*, 210 (1981).
131. O'Connor, D.J., R.V. Thomann, and D.M. DiToro, Dynamic water quality forecasting and management, U.S. EPA, Ecol. Res. Ser., EPA-660/3-73-009, 1973; Chen, C.W., *J. Sanit. Eng. Div. ASCE*, 96, 1085 (1970); Canale, R.P. (Ed.) *Modeling Biochemical Processes in Aquatic Ecosystems*, Ann Arbor Science, Ann Arbor, MI, 1976.
132. Rhee, G.-Y., *J. Phycol.*, 10, 470 (1974).
133. Rhee, G.-Y., *Limnol. Oceanogr.*, 23, 10 (1978).
134. Rhee, G.-Y. and I.J. Gotham, *J. Phycol.*, 16, 486 (1980).
135. Smith, V.H., *J. Phycol.*, 19, 306 (1983).
136. Nyholm, N., *Mitt. Int. Ver. Theor. Angew. Limnol.*, 21, 193 (1978).
137. Falkowski, P.G., *J. Theor. Biol.*, 64, 375 (1977).
138. Rhee, G.-Y. and I.J. Gotham, *Limnol. Oceanogr.*, 26, 649 (1981).
139. Maestrini, S.Y. and D.J. Bonin, Physiological basis of phytoplankton ecology, *Can. Bull. Fish. Aquat. Sci.*, Vol. 210, 264–278, 292–309 (1981).
140. Holm, N.P. and D.E. Armstrong, *Limnol. Oceanogr.*, 26, 622 (1981).
141. Tilman, D., *Resource Competition and Community Structure*, Princeton University Press, Princeton, NJ, 1982.
142. Tilman, D., *Science*, 192, 463 (1976); *Ecology*, 58, 338 (1977).
143. Sommer, U., *Verh. Int. Verein. Limnol.*, 24, (1989).
144. Atkinson, B., *Biochemical Reactors*, Pion Ltd., London, 1974.
145. Schroeder, E.D., *Water and Wastewater Treatment*, McGraw-Hill, New York, 1977.
146. Van Uden, N., *Z. Allg. Mikrobiol.*, 9, 385 (1969).
147. Johnson, M.J., *J. Bacteriol.*, 94, 101 (1967).
148. Powell, E.O., in *Microbial Physiology and Continuous Culture*, 3rd Int. Symp., E.O. Powell et al. (Eds.), Her Majesty's Stationery Office, London, 1967, pp. 34–55.
149. Pasciak, W.J. and J. Gavis, *Limnol. Oceanogr.*, 19, 881 (1974).
150. Pasciak, W.J. and J. Gavis, *Limnol. Oceanogr.*, 20, 211 (1975).

151. Stratton, J.A., *Electromagnetic Theory*, McGraw-Hill, New York, 1941.
152. Mierle, G., *J. Phycol.*, 21, 177 (1985).
153. Munk, W.H. and G.A. Riley, *J. Mar. Res.*, 11, 215 (1952).
154. Baillod, C.R. and W.C. Boyle, *J. Sanit. Eng. Div. ASCE*, 96, 525 (1970).
155. Matson, J.V. and W.G. Characklis, *Water Res.*, 10, 877 (1976).
156. Benefield, L. and F. Molz, *Biotech. Bioeng.*, 25, 2591 (1983).
157. Pipes, D.M., M.S. thesis, Rice University, Houston, TX, 1974.
158. Thiele, E.W., *Ind. Eng. Chem.*, 31, 916 (1939).
159. Ngian, K.F., S.H. Lin, and W.R.B. Martin, *Biotech. Bioeng.*, 19, 1773 (1977).
160. Shieh, W.K., *Water Res.*, 14, 695 (1980).
161. Raven, J.A., *Biol. Rev. (Camb.)*, 45, 167 (1970).
162. Gavis, J. and J.F. Ferguson, *Limnol. Oceanogr.*, 20, 211 (1975).
163. Atkinson, B., E.L. Swilley, A.W. Busch, and D.A. Williams, *Trans. Inst. Chem. Eng.*, 45, T257 (1967).
164. Young, J.C. and P.L. McCarty, *J. Water Pollut. Contr. Fed.*, 41, R160 (1969).
165. Williamson, K. and P.L. McCarty, *J. Water Pollut. Contr. Fed.*, 48, 9 (1976).
166. LaMotta, E.J., *Appl. Env. Microbiol.*, 31, 286 (1976); *Environ. Sci. Technol.*, 10, 765 (1976).
167. Rittmann, B.E. and P.L. McCarty, *Biotech. Bioeng.*, 22, 2343 (1980).
168. Ottengraf, S.P.P. and A.H.C. Van Den Oever, *Biotech. Bioeng.*, 25, 3089 (1983).
169. Frank-Kamenetskii, D.A., *Diffusion and Heat Transfer in Chemical Kinetics*, 2nd ed., J.P. Appleton (translator), Plenum Press, New York, 1969.
170. Rittmann, B.E. and P.L. McCarty, *Biotech. Bioeng.*, 22, 2359 (1980).
171. Rittmann, B.E. and P.L. McCarty, *J. Env. Eng. Div. ASCE*, 104, 889 (1978).
172. Rittmann, B.E., *Biotech. Bioeng.*, 24, 1341 (1982).
173. Gantzer, C.J., B.E. Rittmann, and E.E. Herricks, *Water Res.*, 22, 709 (1988).
174. Gantzer, C.J., *J. Env. Eng. Div. ASCE*, 115, 302 (1989).
175. Gantzer, C.J., H.P. Kollig, B.E. Rittmann, and D.L. Lewis, *Water Res.*, 22, 191 (1988).
176. Tsuchido, T. and I. Shibasaki, *Biotech. Bioeng.*, 22, 107 (1980).
177. Anacleto, J. and N. Van Uden, *Biotech. Bioeng.*, 24, 2477 (1982).
178. Gameson, A.L.H. and D.J. Gould, Paper No. 22, *Proc. Int. Symp. on Discharge of Sewage from Sae Outfalls*, Pergamon Press, London, 1974.
179. Thomann, R.V. and J.A. Mueller, *Principles of Surface Water Quality Modeling and Control*, Harper & Row, New York, 1987.
180. Mitchell, R. and C. Chamberlain, in *Indicators of Viruses in Water and Food*, G. Berg (Ed.), Ann Arbor Sci., Ann Arbor, MI, 1978, pp. 15–37.
181. Mancini, J.L., *J. Water Pollut. Contr. Fed.*, 50, 11 (1978).
182. Fujioka, R.S., H.H. Hashimoto, E.B. Siwak, and R.H.F. Young, *Appl. Environ. Microbiol.*, 37, 690 (1981).
183. Anderson, I.C., M. Rhodes, and H. Kator, *Appl. Environ. Microbiol.*, 35, 1147 (1979).
184. U.S. EPA, Analysis and Control of Thermal Pollution, Water Program Operations. EPA-430/1-74-010, Washington, D.C., 1974.
185. Geldreich, E.E. and B.A. Kenner, *J. Water Pollut. Contr. Fed.*, 41, R336 (1969).
186. Dutka, B.J. and K.K. Kwan, *Water Res.*, 14, 909 (1980).
187. Severin, B.F., *J. Water Pollut. Contr. Fed.*, 52, 2007 (1980).
188. Chick, H., *J. Hygiene*, 8, 92 (1908).
189. Hom, L.W., *J. Sanit. Eng. Div. ASCE*, 98, 183 (1972).
190. Young, D.C. and D.G. Sharp, *Appl. Environ. Microbiol.*, 33, 168 (1977).
191. Watson, H.E., *J. Hygiene*, 8, 536 (1908); Fair, G.M. et al., *J. Am. Water Works Assoc.*, 40, 1051 (1948).
192. Rubin, A.J. and G.A. Elmaraghy, *Water Res.*, 11, 927 (1977).
193. Haas, C.N., *Environ. Sci. Technol.*, 14, 339 (1980).
194. Fair, G.M., J.C. Geyer, and D.A. Okun, *Water and Waste Engineering*, John Wiley & Sons, New York, 1968.
195. Farooq, S., R.S. Englebrecht, and E.S. Chian, *Water Res.*, 11, 737 (1977).

"From its beginning the science of organic chemistry has depended on the empirical and qualitative rule that like substances react similarly and that similar changes in structure produce similar changes in reactivity... Linear free energy relationships constitute the quantitative specialization of this fundamental principle. "

L.P. Hammett[1]

CHAPTER 7

Prediction Methods for Reaction Rates and Compound Reactivity

OVERVIEW

The search for relationships among the equilibrium and kinetic properties of compounds has been a paradigm in chemistry for many years, and it reflects three major goals of chemists: (1) **categorize** information on the behavior of chemicals, (2) **explain** their behavior in terms of fundamental physical-chemical principles, and (3) **predict** unmeasured properties from measured properties of a chemical or from the behavior of related compounds. The discovery of unifying principles and predictive relationships is intellectually satisfying, but they also have practical benefits to water quality managers who need to predict the ecological and toxicological effects and fate of organic contaminants in natural waters.

Numerous relationships exist among the structural, physical-chemical, and/or biological attributes within classes of compounds. Simple examples include bivariate correlations between properties like aqueous solubility and octanol-water partition coefficients (K_{ow}) and correlations between equilibrium constants of related sets of compounds. Among the best-known relationships involving chemical attributes are the correlations between rate and equilibrium constants for related reactions, which are known as linear free-energy relationships, or LFERs. These correlations lead to the broader concepts of structure-activity and property-activity relationships, PARs and SARs, which seek to predict the environmental fate or bioactivity of related compounds based on correlations with their physico-chemical properties or structural features. Table 7-1 categorizes the "attribute relationships" used in chemical fate studies and defines some important terms.

Development of relationships to predict the fate and effects of aquatic contaminants from easily measured properties is useful for three reasons: (1) the large number of compounds potentially present in aquatic systems, (2) the variety and

Table 7-1. Characteristic Types of Attribute Relationships

Type of predictor variable[a]	Type of predicted variable[a]	Name	Example
Structure; molecular property	Activity	SAR	Bioaccumulation vs. connectivity index (e.g., $^1\chi^v$)
Property	Activity	PAR	Bioaccumulation vs. $log\,K_{ow}$
Property	Property	PPR	$log\,K_{ow}$ vs. $log\,S_w$ or vs. $log\,k'$ (chromatographic retention time)
Structure	Property	SPR	$log\,S_w$ vs. TSA (total molecular surface area)

[a] Definitions: Structure: geometric attribute of compound (based on size, shape, arrangement of atoms comprising molecule). Molecular property: attribute not measured directly, but calculated from molecular structure based on theoretical relationships like molecular orbital theory, or molecular mechanics; sometimes defined in terms of functional groups or molecular fragments; for purposes of this text, relationships based on molecular properties will be referred to as SARs. Property: physical-chemical attribute of compound that is measured directly or calculated from readily measurable variables. Activity: attribute related to a compound's biological or biochemical behavior.

complexity of physical-chemical processes involved in contaminant transport and transformation, and (3) the diversity of organisms that may be affected by contaminants or involved in their biodegradation. Most attention has focused on organic compounds, but some relationships have been reported for metal ions. Although attribute relationships are inherently empirical, most have an underlying theoretical basis. It is pertinent to note that the principles of thermodynamics, now regarded as firmly based in theory, originated from phenomenological observations. A goal of this type of research is to develop generalizations from empirical relationships so that a few fundamental principles can explain many observations.[2]

This chapter describes LFERs developed over the past 60 years and illustrates their applications to compounds of interest in natural aquatic systems. The underlying principles for property-activity and structure-activity relationships (PARs and SARs) are developed next, and their applications in predicting the behavior of aquatic contaminants are reviewed.

PART I. LINEAR FREE-ENERGY RELATIONSHIPS

7.1 INTRODUCTION

No global relationship exists between the thermodynamics of reactions (i.e., their energetics) and the rates at which they approach equilibrium. Many reactions have highly favorable thermodynamics ($\Delta G \ll 0$) but proceed imperceptibly, if at all. Such reactions are called kinetically hindered, and environmental examples abound. The absence of a general relationship is understandable from transition-state theory, which tells us that reaction rates are controlled by the energy difference between reactants and a transition-state rather than by the energy difference between reactants and products.

The generalization is misleading, however. Although no **single** relationship can predict reaction rates from thermodynamic data, kineticists have found that reaction rates are correlated with reaction energetics for many sets of similar compounds. Such relationships are called linear free-energy relationships, or LFERs, because they are linear correlations between logs of rate constants and logs of equilibrium constants for reactions of the compounds. Some empirical relationships between kinetic and equilibrium properties are nonlinear; the phrase **correlation analysis** sometimes is used to encompass these as well as linear relationships. For convenience we will use the term LFER.

Correlations among equilibrium constants of related sets of compounds also are LFERs. Stability constants for metal complexes with various ligands often are correlated with stability constants of another metal with the same ligands.[3] Kinetic LFERs are particularly useful, however, because they enable us to predict reaction rates from equilibrium properties that are more easily measured or more readily available. Kinetic LFERs also are valuable in improving our understanding of reaction mechanisms and rate-controlling steps.

7.1.1 Nature and Categories of LFERs

LFERs are widely used in organic chemistry, and effects of substituent groups on the kinetics and equilibria of reactions for homologous series of compounds have been studied since the 1920s. Substituents influence compound reactivity by altering electron density at the reaction site and by steric effects. The latter are more difficult to quantify, and the former have been studied primarily with meta- and parasubstituted aromatic compounds, where steric influences are minimized. Reactions favored by high electron density at the reaction site are accelerated by substituents like methyl groups that produce such an increase. Conversely, reactions favored by electron withdrawal are accelerated by electrophilic substituents like -Cl and -NO_2.

Electronic effects of substituents can be divided into two categories: (1) inductive, in which electron displacement is transmitted along a chain of atoms without reorganizing chemical bonds; and (2) resonance, which occurs in compounds with conjugated double bonds. The latter stabilizes certain resonance forms and changes electron densities at certain molecular positions. Inductive effects (sometimes called "field" effects) decrease rapidly with distance from a substituent group, but resonance effects can be felt at greater distances in conjugated molecules.

Four major categories of LFERs have been developed over past decades (Table 7-2). These relationships apply to many classes of organic and inorganic compounds and a wide range of reactions, including coordinative reactions (dissociation/association) of acids and metal complexes, hydrolysis, hydration, substitution, substituent group oxidations, and electron exchange between metal ions.

Table 7-2. Types of LFERs Applicable to Reactions in Aquatic Systems

Relationship	Types of reaction or reactants	Basis of LFER
Brønsted	Acid- and base-catalyzed reactions: hydrolysis, dissociation, association	Rate related to K_a or K_b of product or catalyst
Hammett (Sigma)	Numerous reactions of para- or meta-substituted aromatic compounds: hydrolysis, hydration of alkenes, substitution, oxidation, enzyme-catalyzed oxidations; some Type II photooxidations	Accounts for electron withdrawal/donation from/to reaction site by substituents on aromatic rings via resonance effects
Taft	Hydrolysis and other reactions of aliphatic organic compounds	Accounts for steric and polar effects of substituents on reactive site
Marcus	Outer-sphere electron exchange (redox) reactions of metal ions, chelated metals, and metal ion oxidation by organic oxidants like pyridines and quinones	Based on three components of energy needed to produce transition state; for series of related redox reactions, rate proportional to $E°$ or $pE°$

7.1.2 Theoretical Basis for LFERs

According to transition-state theory, rate constants are exponentially related to the free energy of activation (Equation 3-47). From thermodynamics we also know that equilibrium constants are exponentially related to the free energy of reaction. If two reactions exhibit a LFER, we can write

$$\ln k_2 - \ln k_1 = a\left\{\ln K_2 - \ln K_1\right\} \qquad (7\text{-}1a)$$

i.e.,

$$\left\{\Delta G_2^{\ddagger} - \Delta G_1^{\ddagger}\right\}/RT = a\left\{\Delta G_2° - \Delta G_1°\right\}/RT$$

For a series of reactants undergoing a given reaction,

$$\ln k_i = a \ln K_i + C, \qquad or \qquad \Delta G_i^{\ddagger} = a\Delta G_i° + C \qquad (7\text{-}1b)$$

The reactions on the kinetic and equilibrium sides of the equation do not have to be the same. For a series of i reactants and related reactions j and k,

$$\ln k_{ij} = a \ln K_{ik} + C, \qquad or \qquad \Delta G_{ij}^{\ddagger} = a\Delta G_{ik}° + C \qquad (7\text{-}1c)$$

Subscript i refers to the ith reactant in a series; the constant is a function of $\Delta G^{\neq}$ and $\Delta G°$ for a reference reaction.

As described in Chapter 3, $\Delta G^{\neq}$ consists of an entropy of activation and an enthalpy of activation. The former is associated with pre-exponential factor A of the Arrhenius equation, and the latter with E_{act}, which defines the sensitivity of the rate to temperature. In some reactions, substituent groups affect E_{act} (or $\Delta H^{\neq}$) and $\Delta H°$ primarily, while A (or $\Delta S^{\neq}$) and $\Delta S°$ change only slightly. For example, in the alkaline hydrolysis of benzoic acid esters, E_{act}, $\Delta H^{\neq}$, and $\Delta H°$ vary with substituent, but $\Delta S^{\neq}$ (hence, A) and $\Delta S°$ remain constant. Thus, $\Delta G^{\neq}$ and lnk vary in the same way as E_{act}, and lnk varies linearly with $\Delta G°$.

Frequently, however, $\Delta G^{\neq}$ (thus, lnk) and $\Delta G°$ (thus, lnK) are linearly related even though both A and E_{act} vary with substituent changes. In these cases $\Delta H^{\neq}$ and $\Delta S^{\neq}$ are correlated; this is called an isokinetic relationship (Figure 7-1a). Since $\Delta G^{\neq}$ $= \Delta H^{\neq} - T\Delta S^{\neq}$, the changes are compensating. Similar relationships are observed for overall changes in $\Delta H°$ and $\Delta S°$ of the reactions, and the net effect is a linear relationship between $\Delta G°$ and $\Delta G^{\neq}$. Although changes in substituents and solvent-reactant interactions may affect $\Delta H^{\neq}$ and $\Delta S^{\neq}$ in complicated ways, compensation between the two terms results in simple LFERs,[6] in which the change in k is less than if E_{act} or A changed alone. The compensation between $\Delta H^{\neq}$ and $\Delta S^{\neq}$ allows Equation 3-45 to be reformulated[7] as

$$k = A' \exp\left[-\left(\Delta H^{\neq}/R\right)\left(1/T - 1/\theta\right)\right] \tag{7-2}$$

A' is related to pre-exponential factor A, and θ is the isokinetic temperature, at which all k's for related series of reactions are the same.[2,7,8] At T < θ, reactions with smaller E_{act} occur more rapidly, but at T > θ, reactions with larger E_{act} are faster. θ can be obtained as the slope of a plot of ΔH vs. ΔS (or between $\Delta H^{\neq}$ and $\Delta S^{\neq}$): $\Delta H = \Delta H_o + \theta \Delta S$, but significant statistical problems may be encountered in evaluating θ.[9]

The existence of a LFER for a set of reactants is equivalent to saying that the free energy of activation is a constant fraction of the free energy of reaction for the reactants ($\Delta G^{\neq}/\Delta G° = a$). Although a quantitative theory is not available to explain why this should be, the basis for LFERs can be understood qualitatively from the energy diagrams in Figure 7-1b,c. Both diagrams show that if the energy curves for related reactions have similar shapes (a reasonable assumption), then by simple geometry the change in E_{act} must be proportional to the change in overall reaction energetics: $\Delta(\Delta E^{\neq}) \propto \Delta(\Delta E°)$ or $\Delta(\Delta H^{\neq}) \propto \Delta(\Delta H°)$. If the entropies ($\Delta S^{\neq}$ and $\Delta S°$) do not change between the two reactions, then we have $\Delta(\Delta G^{\neq}) \propto \Delta(\Delta G°)$ or lnk $\propto$ lnK. If the entropy terms are not constant, but ΔS and ΔH are correlated, the free-energy relationship still holds. Theoretical models for energy profiles along the reaction coordinate have been used to show that a unified theory is possible for many of the observations and relationships described in this chapter.[2] The broader philosophical and mathematical concepts that underpin LFERs have been described succinctly by Wold and Sjöström.[10]

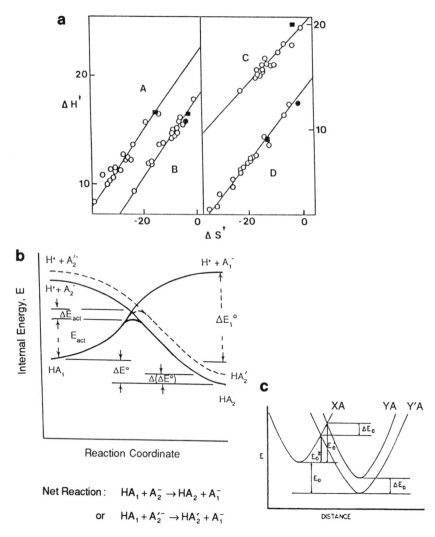

Figure 7-1. (a) Isokinetic plots (correlation of $\Delta S^{\neq}$ and $\Delta H^{\neq}$) for reaction of Fe^{III} with hydroxamates: line **A**, complex formation of $Fe(H_2O)_6^{3+}$ with monohydroxamates; line **B**, H^+-catalyzed hydrolysis of Fe-monohydroxamate complexes; line **C**, uncatalyzed hydrolysis of same complexes; line **D**, complex formation of $FeOH(H_2O)_5^{2+}$ with monohydroxamates. Closed circles: ferrioxamine B; closed squares: diferrioxamine B. [From Biruš, M., et al., *Croat. Chem. Acta*, 61, 33 (1988). With permission.] (b) Existence of LFERs can be understood from geometric similarity of energy profiles along the reaction coordinate for related reactants and reactions [modified from Adamson, A.W., *A Textbook of Physical Chemistry*, 2nd ed., Academic Press, New York, 1979]; or (c) from Morse diagrams for atom or ion transfer reactions of general type $X\text{-}A + Y \rightarrow (X \cdots A \cdots Y)$ $\rightarrow X + A\text{-}Y$ and $X\text{-}A + Y' \rightarrow (X \cdots A \cdots Y') \rightarrow A + A\text{-}Y'$. The parabolas represent reactant and product energy levels as a function of bond lengths. Similar diagrams apply to outer-sphere electron transfer reactions, as described in Section 7.4. In these cases the reaction coordinate represents the extent of solvent and bond length rearrangement from ground-state conditions for reactants and products.

7.2 THE BRØNSTED RELATIONSHIP

Brønsted and Pederson[11] were the first to describe a relationship between rate and equilibrium constants for a series of compounds. They found that $log k_B$ for base-catalyzed decomposition of nitramide, $H_2N_2O_2$, varies linearly with $log K_{HB}+$, the acidity constant of the conjugate acid of the catalyst. The reaction $(H_2N_2O_2 + B \rightarrow H_2O + N_2O + B)$ follows the rate equation

$$-d[H_2N_2O_2]/dt = (k + k_B[B])[H_2N_2O_2]$$ (7-3)

A plausible mechanism for the reaction is

$$B + HN=\overset{O}{\overset{\|}{N}}OH \rightarrow BH^+ + {}^-\overset{O}{\overset{\|}{N}}=NOH \qquad (slow)$$

$$^-N=\underset{\underset{O}{\|}}{N}OH \rightarrow N_2O + OH^- \quad (fast)$$

$$OH^- + BH^+ \rightarrow B + H_2O$$

Brønsted and co-workers found that other acid- or base-catalyzed reactions had rate constants that were log-linearly related to the acid or base dissociation constant of the catalyst and followed the general equations

$$\log k_{HB} = A + \alpha \log K_{HB}$$ (7-4a)

or

$$\log k_B = A + \alpha \log K_B$$ (7-4b)

These equations are known as the Brønsted Catalysis Law. Coefficient α may range between 0 and 1. The applicability of these relationships now has been established for numerous reactions subject to general acid-base catalysis (see reference 12 for a comprehensive review).

A commonly cited example of a Brønsted LFER is the correlation between k_f and K_a (Figure 7-2a) for Brønsted acids in the reaction

$$HA + H_2O \underset{k_r}{\overset{k_f}{\rightleftharpoons}} A^- + H_3O^+$$

which is linear over many orders of magnitude with slope $(\alpha) \approx 1.0$. This example is less profound than it may first appear, however. Rate constants for the reverse reaction (k_r) are nearly uniform because they are at the diffusion-controlled limit $(\sim 10^{11} \ M^{-1} \ s^{-1})$.[5] Because $K_a = k_f/k_r$, k_f must be proportional to K_a, and α_f must be unity. Conversely, for diffusion-controlled reactions, α_r must be zero. The linear

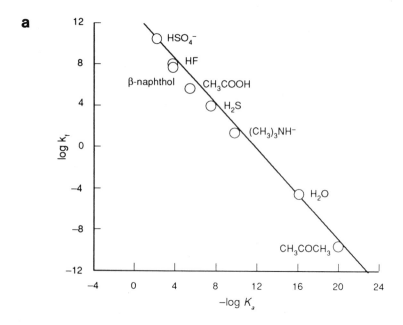

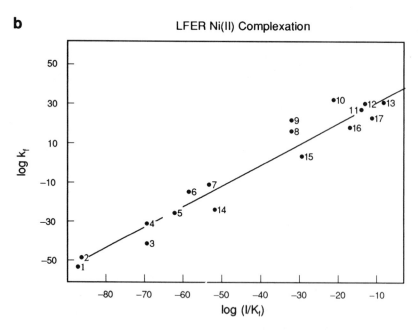

Figure 7-2. Brønsted LFERs: (a) acid dissociation rate constants vs. corresponding acid dissociation equilibrium constant. (From Adamson, A.W., *A Textbook of Physical Chemistry*, 2nd ed., Academic Press, New York, 1979. With permission.) (b) complex formation rate constant for reaction of $Ni(H_2O)_6^{2+}$ with various organic ligands vs. reciprocal of equilibrium stability constant, K_f. Note: $1/K_f$ = dissociation equilibrium constant for complexes. [From Hoffman, M.R., *Environ. Sci. Technol.*, 15, 345 (1981). With permission of the Amer. Chem. Soc.]

relationship between k_f and K_a thus must break down above some value of K_a. In the above case, when $K_a \gg 1$, k_f approaches the diffusion-controlled limit, and $\alpha_r \rightarrow 0$. A change in slope in Brønsted plots often is interpreted as signifying a change in the rate-limiting step from chemical to diffusion control, but it also could arise from changes in the relationship between $\Delta G°$ and $\Delta G^{\neq}$.[5,12] Proton transfer is the rate-limiting step in many acid-catalyzed reactions, and linear relationships between logk and logK, with $\alpha = 1$, thus are not surprising. Brønsted relationships with $\alpha < 1$ are very common, however; these can be explained only by semiquantitative arguments such as illustrated in Figure 7-1b,c. The value of α is useful in evaluating the nature of transition states. If an acid catalyst were completely dissociated before reacting, α would be 1. If $\alpha \approx 1$, the leaving group (conjugate base) is assumed to be loosely bound in the transition state.[13]

Similar LFERs also apply to complexation reactions of metal ions (Lewis acids). For example, dissociation rates of Ni^{II} complexes are correlated with corresponding dissociation equilibrium constants (Figure 7-2b). This suggests that the reactions occur by dissociative interchange in which breaking of the Ni^{II}-ligand bond predominates over formation of the Ni^{II}-water bond in the rate-determining step.[14] (For another example of a LFER involving metal ions, see Figure 4-2, which shows that the effectiveness of metal ions, in catalyzing the decarboxylation of malonic acid is correlated with the stability constants for the corresponding metal-malonate complexes.)

7.3 SIGMA RELATIONSHIPS FOR ORGANIC SUBSTITUENT EFFECTS

7.3.1 The Hammett Equation

A free-energy relationship with far-ranging implications was developed by Hammett[1,15] in the 1930s to explain substituent effects on rates of meta- or parasubstituted benzene compounds. Hammett found that hydrolysis rates of substituted ethyl benzoates, ionization constants, and esterification rates of substituted benzoic acids, and many other reactions, depend on the nature and position of the substituent on the ring, and the substituent effect is similar from one reaction to the next. Using rate and equilibrium constants for ionization of unsubstituted benzoic acid as a starting point, he defined the LFER:

$$\log(k/k_o) = \rho \log(K/K_o) \tag{7-5a}$$

where k is a rate constant for some reaction involving a series of benzoic acids; subscript o denotes unsubstituted benzoic acid; and ρ is a constant that depends on the reaction and solvent. Hammett defined the log of the ratio of equilibrium ionization constants (logK/K$_o$) as a substituent parameter σ, which is characteristic of a given substituent and ring position. (K_o is the ionization constant of benzoic acid in water at 25°C.) Thus,

$$\log(k/k_o) = \rho \sigma \tag{7-5b}$$

Equation 7-5b is the Hammett relationship; ρ for a given reaction is determined from this equation by least-squares fit of normalized rate data for substituted benzene compounds to the corresponding σ. Because ion recombination reactions $(H^+ + B^- \rightarrow HB)$ are diffusion controlled, k for ionization of benzoic acids is directly proportional to K_a, and therefore, $\rho = 1.00$ for this reaction. Table 7-3 lists ρ and σ values for common substituents and important reactions; note that the effects of m- and p-substituents (as expressed by σ) are thought to be additive. The information in Table 7-3 can be used to evaluate substituent effects on reactivity of many compounds. For example, $\sigma = -0.17$ for p-substituted methyl groups, and $\rho = 2.46$ for alkaline hydrolysis of methyl benzoates in 60% acetone. Thus, $log(k/k_o)$ for alkaline hydrolysis of p-methyl toluate is -0.42; i.e., hydrolysis of the p-methyl ester is about 0.38 times that of the unsubstituted ester. The Hammett relationship does not apply to ortho-substituents, because they exert steric as well as electronic effects.

Hammett[15] found good agreement of data with Equation 7-5b in 30 of 39 rate or equilibrium examples. Jaffe[16] later analyzed 371 sets of rate and equilibrium data on aromatic reactions statistically and found that about 70% had an excellent or satisfactory fit (r > 0.95) to the Hammett relationship. The predictive importance of the relationship is impressive.[9] It was found by various workers to apply not only to hydrolysis reactions, but also to aromatic substitution and oxidation reactions and enzyme-catalyzed reactions like the oxidation of phenols and aryl amines by peroxidase.[17] It also served as the starting point for the development of many related sigma-type relationships for reactions of both aromatic and aliphatic electrophiles and nucleophiles.

The fact that σ and its relatives (described below) represent the effects of substituents on a wide variety of reactions implies that it measures a fundamental property like electron density at the reactive site.[18] Equation 7-5 defines σ in terms of the relative ionization of benzoic acids. The extent of ionization (K_a) decreases as electron density increases at the O-H bond in a carboxylic acid group, and σ decreases with decreasing K_a. Thus, σ and electron density are inversely related, as shown by the σ values in Table 7-3. Electron-withdrawing groups like $-NO_2$ and -Cl have $\sigma > 0$; electron donors like $-NH_2$ have $\sigma < 0$. Similarly, ρ measures a reaction's sensitivity to electron density. Nucleophilic reactions that are hindered by high electron density have positive ρ values; electrophilic reactions that are accelerated by high electron density have negative values. Many books and review papers[16,18-22] describe the theoretical basis of the Hammett relationship in detail, along with its applications and limitations in predicting reaction rates and mechanisms.

7.3.2 Sigma Relationships for Electrophilic and Nucleophilic Reactions

In spite of its many successes, the basic Hammett relationship does not work well for many classes of aromatic reactions, especially those involving carbonium ion and carbanion intermediates. Substituent parameters have been developed to improve correlations for specific types of reactions. Among the most widely used are the σ^+ constants

Table 7-3. Values of Sigma and Related Constants for Some Common Substituents[a]

A. Substituent constants	σ_p	σ_m	σ_p+	$\sigma+_m$	$\overset{*}{\sigma}$	$E_s{}^b$	$E_s{}^c$
NH_2	−0.66	−0.15	–	–	+0.10		
OH	−0.35	+0.08	–	–	+0.25		
OCH_3	−0.26	+0.08	−0.76	+0.05	+0.25	0.99	
CH_3	−0.16	−0.07	−0.31	−0.06	−0.05	0.00	0.00
C_6H_5	−0.01	+0.06	−0.18	+0.11	+0.10		−2.55
H	0.00	0.00	0.00	0.00	0.00		1.24
F	+0.08	+0.35	−0.07	+0.35	+0.52	0.49	
Cl	+0.23	+0.37	+0.11	+0.40	+0.47	0.18	
Br	+0.23	+0.39	+0.15	+0.41	+0.45	0.00	
I	+0.28	+0.35	+0.14	+0.36	+0.39	−0.20	
CN	+0.68	+0.62	+0.66	+0.56	+0.58		
CH_3SO_2	+0.71	+0.65	–	–	+0.59		
NO_2	+0.79	+0.71	+0.79	+0.67	+0.63	−0.75	
ethyl							−0.38
n-propyl							−0.67
iso-propyl							−1.08
n-butyl							−0.70
iso-butyl							−1.24
t-butyl							−2.46

B. Reaction constants	ρ	$\overset{*}{\rho}$	δ^d
Ionization of benzoic acids	1.00		
OH⁻-catalyzed hydrolysis of ethylbenzoates	2.55		
Methylation of benzoic acids	−0.58		
Ionization of carboxylic acids, RCO_2H		1.72	
Alkaline hydrolysis of $[Co(NH_3)_5O_2CR]^{2+}$ in water		0.79	
Catalysis of nitramide decomposition by $RCOO^-$		−1.43	
Acid hydrolysis of formals, $CH_2(OR)_2$		−4.17	
Alkaline hydrolysis of primary amides[43]		1.60	
Ionization of orthobenzoic acids		1.79	
Hydrolysis of bromoalkanes[50]		−11.9	
Acid dissociation constants of aldehyde-bisulfites[49]		−1.29	
Alkaline hydrolysis of diphthalate esters[44]		4.59	1.52
Acid hydrolysis of orthobenzamides			0.81
Acid methanolysis of 2-naphthyl esters			1.38
Methyl iodide reaction with alkylpyridines			2.07

a   Sigma values from Weston, R.E., Jr., and H.A. Schwartz, *Chem. Kinetics,* Prentice-Hall, Englewood Cliffs, NJ, 1972; ρ^*, δ, and E_s from Shorter, J., pp. 71–117 in Chapman, N.B., and J. Shorter (Eds.), *Advances in Linear Free Energy Relationships,* Plenum Press, London, 1972; ρ values from Exner, O., pp. 1–69, *Ibid.,* except where references are cited; σ values for ionic substituents like −COO⁻ and −NH₃⁺ do not give accurate predictions of reaction rates[22] and are not included here.

b   Steric parameter for ortho substituents in benzoates (see Section 7.3.3).

c   Steric parameter for aliphatic system; values from ethyl to t-butyl are "corrected" (Hancock) constants (see Section 7.3.3).

d   Reaction constant for steric effects (see Section 7.3.3).

developed by Okamoto and Brown[23] for electrophilic reactions. These constants are based on hydrolysis rates of m- and p-substituted 2-chloro-2-phenylpropanes (cumyl chlorides) (**I**) in acetone, which proceed by electrophilic carbonium ion intermediates:

$$\text{(7-6)}$$

Formation of the intermediates is facilitated by a high electron density at the reactive carbon, i.e., by m- or p- electron donors. The electrophilic constant, σ^+, is defined by $-4.54\sigma^+ = log(k/k_o)$, where k_o is the hydrolysis rate constant for unsubstituted $\mathbf{I}$(X=H), k is the rate constant for a substituted form of $\mathbf{I}$, and -4.54 is the best estimate of ρ for reaction 7-6 based on Hammett's σ constants for meta substituents. Table 7-3 lists σ^+ values for some common substituents.

Similarly, a substituent constant σ^- was developed to fit situations in which the reacting center is especially electron rich, i.e., nucleophilic agents, including carbanions (see references 24,25 for reviews). Several other σ constants have been defined for specific classes of reactants in efforts to extend the applicability of the Hammett approach, and this proliferation has been criticized as artificial and ambiguous.[26] In large part the proliferation arose because, in extending the Hammett approach more widely, the σ constant began to represent other reactivity-influencing factors than those the original derivation intended. Multiparameter and multiterm extensions of the Hammett equation have been developed in an effort to account for these factors explicitly and thus eliminate the ambiguity in the σ's (and the need for so many different kinds of σ's). Several reviews describe these developments and their usefulness and limitations in assessing substituent effects on reaction rates.[24,27] Although σ relationships have obvious uses in predicting reaction rates, they are of even greater importance to organic chemists in unraveling mechanisms and developing reaction theories.

7.3.3 Separation of Resonance, Inductive and Steric Effects

The sigma constants, σ, σ^+, and σ^-, measure inductive and resonance effects on electron distributions. Relationships using these parameters do not work very well for ortho-substituents and for substituents on aliphatic compounds, which influence reaction rates by direct field (inductive) and steric effects. Separation of substituent constants into inductive, resonance, and steric terms has been the subject of many studies,[28-32] but the primary achievement was by Taft. He theorized that σ is the sum of inductive and resonance terms, $\sigma = \sigma_I + \sigma_R$, and evaluated σ_I (which he called σ^*) from rates of acid- and base-catalyzed hydrolysis of esters (RCOOR'), where R is the substituent being evaluated. This led to the equation

$$\sigma^* = \left[log(k/k_o)_b - log(k/k_o)_a\right]/2.48 = \left[log(k/k_o)_b - E_s\right]/2.48 \quad \text{(7-7)}$$

The factor 2.48 puts σ^* on the same scale as σ, and the k_o's are rate constants for acid (a) and base (b) hydrolysis of acetic acid esters (R is a methyl group in the reference compound). R′ usually is an ethyl or methyl group, but according to Shorter,[27] k does not depend on the nature of R′ (over the range in which it was varied). Equation 7-7 was based on the fact that acid hydrolysis rates of substituted benzoic acid esters are only slightly affected by the presence of substituents, but acid hydrolysis rates of aliphatic esters are strongly affected by substituents. These effects were assumed to be caused by steric factors; thus, $log(k/k_o)_a$ defines a steric substituent constant, E_s. It is reasonable to assume that steric factors affect base-catalyzed rates in the same way. Substituent effects on base hydrolysis of aliphatic compounds are composed of both inductive and steric effects, and subtraction of the steric effects (based on acid hydrolysis) thus yields a measure of the inductive effects. The inductive sigma, σ^*, is important because it allows one to evaluate substituent effects on rates of aliphatic reactions by a formula analogous to the Hammett equation or by the two-variable Taft-Pavelich equation:[30]

$$log(k/k_o) = \rho^*\sigma^* + \delta E_s \qquad (7\text{-}8)$$

where ρ^* is a reaction constant analogous to ρ, and δ is a reaction constant that measures the sensitivity of the reaction to steric effects, E_s. Table 7-3 lists σ^*, δ, and E_s for some common constituents and reactants.

Roberts and Moreland[32] estimated σ_I from substituent effects on ionization constants of 4-X-bicyclo[2.2.2]-octane-1-carboxylic acids (**II**):

$$X - C \overset{\displaystyle C - C}{\underset{\displaystyle C - C}{\overbrace{} - C - C - C}} - COOH \qquad\qquad \textbf{II}$$

$$\sigma_I = 0.683 \log(K_{HB}/K_o)$$

X is a variable substituent, K_o is the ionization constant of the unsubstituted compound, and the constant puts σ_I on the same scale as σ. Compound **II** is structurally similar to p-substituted benzene compounds, but because it is not aromatic, there can be no resonance contributions to substituent effects. A good correlation was found between this σ_I and Taft's σ^*.

7.4 LFERs FOR REDOX REACTIONS

7.4.1 Mechanisms of Redox Reactions

Oxidation-reduction reactions involving metal ions occur by two types of mechanisms: inner-sphere and outer-sphere (*os*) electron transfer. In the former case the oxidant and reductant approach to the point of sharing a common primary hydration sphere, and the activated complex involves a bridging ligand between the two metal ions (M-L-M′). Electron transfer often is achieved by ligand transfer, as described in Section 4.2.2. Inner-sphere redox reactions thus involve bond-formation and -breaking processes similar to other group transfer and substitution reactions, and activated complex theory (ACT) thus applies directly to these reactions. In *os* electron transfer the primary hydration spheres remain intact. The metal ions are separated by at least two water molecules (or other ligands), and only the electron moves between the ions. An *os* mechanism can be inferred if the redox reaction is much faster than the rates of ligand exchange of the metal ions involved. Because no bonds are made or broken in *os* transfers, conventional ACT requires a modification that involves the *Franck-Condon* principle.[5] This is described qualitatively as follows.

Electrons move much more rapidly than atoms or nuclei. For example, vibrational periods of atoms are on the order of 10^{-13} s, but electron transitions occur on a time scale of 10^{-15} s. Atoms thus can be regarded as fixed with regard to internuclear distances on the time scale of electron transitions. The question arises: if electronic transitions are so rapid, why are many redox reactions so slow? For aqueous metal ions the answer lies in the fact that the distribution of charge in the ions induces structural configurations in the ligands and surrounding solvent. Charge distributions and structural configurations differ between the reactants and products in an electron exchange reaction, and energy is required to distort the ions to a shape (the transition state) from which they can form the product ions spontaneously.

The time required for rearrangement of solvent to an equilibrium configuration is long compared with the time required for electronic transitions. Orientation of water molecules in an electric field requires $\sim 10^{-11}$ s (about the same as the lifetime of a solvent-reactant cage), which is about 10^4 times longer than that of an electronic transition. Reordering of metal-ligand bond lengths requires about 10^{-13} s. Because a transferred electron is in an unfavorable energy state as long as ligands and solvent are oriented in the reactant configuration, it will return to the donor (reductant), and no reaction will take place. For net electron transfer to occur, the ligands and solvent must rearrange to a nonequilibrium-state intermediate between the reactants and products before electron transfer occurs. This reorientation requires energy and is responsible for the slowness of redox processes.

The importance of these processes is illustrated by electron exchange reactions of Fe^{II}-Fe^{III} complexes.[33] Metal-water bond lengths in aquated Fe^{II} are longer than those in more highly charged Fe^{III}, and the rate constant for electron exchange is slow (4 M^{-1} s^{-1}). Metal-ligand bond lengths and molecular orbitals are similar in ferricyanide and ferrocyanide complexes, and these large, firmly coordinated ions have lower hydration energies, leading to a greatly reduced energy barrier and faster rate of electron exchange (3×10^2 M^{-1} s^{-1}).[18]

7.4.2 The Marcus Model for Outer-Sphere (*os*) Redox Reactions

Several theories have been developed to explain rates of electron transfer reactions.[33] The most widely used model was derived by Marcus[34,35] (see references

18,36,37 for more detailed descriptions). Marcus based his model on the idea that the activation energy for electron transfer reactions is the sum of three terms: (1) electrostatic work ω to bring two ions together; (2) energy needed to modify the solvent structure, $\Delta G^{\neq}_{solv}$; and (3) energy needed to distort the metal-ligand bond lengths, $\Delta G^{\neq}_{lig}$. The rate constant can be formulated in terms of the free energy required for these tasks:

$$k_{AB} = Z_{AB}\exp\left(-\Delta G^{\neq}_{AB}/RT\right) \qquad (7\text{-}9a)$$

where Z_{AB} is the collision frequency of A and B in solution and

$$\Delta G^{\neq}_{AB} = \omega_{elec} + \Delta G^{\neq}_{solv} + \Delta G^{\neq}_{lig} \qquad (7\text{-}9b)$$

Electrostatic work ω can be calculated readily from Coulomb's Law:

$$\omega_{elec} = z_A z_B e^2 / D_{H_2O} r_{AB} \qquad (7\text{-}10)$$

For reactions involving an ion and a neutral molecule such as O_2, ω_{elec} is zero.

The solvent and ligand energy terms were quantified by Marcus,[34] and his derivation led to the Marcus *cross-relation*:

$$k_{AB} = \left(k_{AA}k_{BB}K_{AB}f\right)^{0.5} \qquad (7\text{-}11a)$$

or $\ln k_{AB} \propto \Delta G^{\neq}_{AB} = 0.5\Delta G^{\neq}_{AA} + 0.5\Delta G^{\neq}_{BB} + 0.5\Delta G^{\circ}_{AB} - 0.5RT\ln f$ (7-11b)

where $\ln f = 0.25\left(\ln K_{AB}\right)^2 / \ln\left(k_{AA}k_{BB}/Z^2\right)$ (7-11c)

The factor 0.25 in Equation 7-11c derives from a geometric analysis of the activation barrier for such reactions, assuming the potential energy profiles along the reaction coordinate can be described by parabolas (as in Figure 7-1c).[36] Z is the collision frequency factor for uncharged molecules, originally estimated as $10^{11}\,M^{-1}\,s^{-1}$, but later revised by Marcus to $10^{12}\,M^{-1}\,s^{-1}$.[35] If $\Delta G^{\circ}_{AB} \approx 0$, the last term in Equation 7-11b can be ignored. k_{AA} and k_{BB} are rate constants for electron exchange between the oxidized and reduced states of an element ("self-exchange"):

$$ox_A + red_A^* \overset{k_{AA}}{\rightleftharpoons} red_A + ox_A^*; ox_B + red_B^* \overset{k_{BB}}{\rightleftharpoons} red_B + ox_B^* \qquad (7\text{-}12)$$

and K_{AB} is the equilibrium constant for the net redox reaction:

$$ox_A + red_B \rightleftharpoons red_A + ox_B \qquad (7\text{-}13)$$

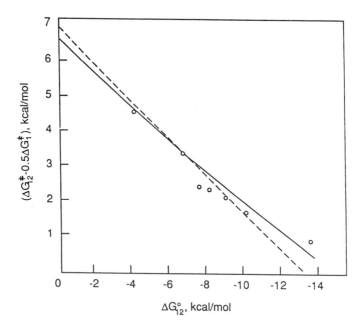

Figure 7-3. Marcus LFER for oxidation of various substituted phenanthroline Fe^{II} complexes by Ce^{IV}. [Data redrawn from Dulz, G. and N. Sutin, *Inorg. Chem.*, 2, 917 (1963).]

K_{AB} can be determined from the standard reduction potentials, $E°$, or the related $p\varepsilon°$s ($p\varepsilon° = E°/0.059$) of the half reactions. For one-electron changes:

$$\ln K_{AB} = \left(E_A° - E_B°\right)/0.059 \tag{7-14a}$$

or
$$\ln K_{AB} = p\varepsilon_A° - p\varepsilon_B° \tag{7-14b}$$

Rate constants for electron exchange reactions like Equation 7-12 are determined by isotopic exchange methods and are available for many metal ions.[18,38-41]

Under limiting conditions the Marcus theory leads to a simple LFER. In a plot of $\ln k$ vs. $\ln K$ (or $\Delta G_{AB}°$), the y-intercept value (where $\Delta G_{AB}° = 0$) is one-half the sum of exchange rate constants k_{AA} and k_{BB} (Equation 7-11b). For *os* redox reactions of two metal ions and a series of ligands complexing one metal ($A_{ox}L_i + B_{red} = A_{red}L_i + B_{ox}$), a plot of $\ln k$ vs. $\ln K$ (or $\Delta G_{AB}°$ or $p\varepsilon_{AB}$ for one-electron transfers) will have a slope of ~0.5 for small values of $\Delta G_{AB}°$. In such cases k_{BB} is constant, but K_{AB} varies. Exchange rate constants (k_{AA}) for the various complexes of A will vary somewhat, causing scatter about the line. An *average* value of k_{AA} can be estimated from the y-intercept (if k_{BB} is known). Figure 7-3 shows this for oxidation of $Fe(phen)_3^{2+}$ complexes by Ce^{IV}. An average k of $2 \times 10^3 \ M^{-1} \ s^{-1}$ was found for $Fe(phen)^{II-III}$ exchanges by this approach;[39] this compares with $k = 4 \ M^{-1} \ s^{-1}$ for the free ions (Table 7-4).

Table 7-4.   Rate and Equilibrium Constants for Some Outer-Sphere Electron-Transfer Reactions ($A_{ox} + B_{red} \rightleftharpoons A_{red} + B_{ox}$)

Reactants	k_A	k_B	K_{eq}	k_{calc}	k_{obs}
A B					
$Ce^{IV} + Fe(CN)_6^{4-}$	4.4	3×10^2	6×10^{12}	7×10^6	1.9×10^6
$Ce^{IV} + Mo(CN)_8^{4-}$	4.4	3×10^4	6×10^{10}	1.3×10^7	1.4×10^7
$Mo(CN)_8^{3-} + Fe(CN)_6^{4-}$	3×10^4	3×10^2	1×10^2	2.7×10^4	3.0×10^4
$Ce^{IV} + Fe^{2+}$	4.4	4.0	1×10^{12}	6×10^5	1.0×10^3
$Co^{III} + Fe^{2+}$	5	4.0	2×10^{18}	6×10^7	42
$Co(phen)_3^{3+} + Fe^{II}(cyt \mathbf{c})$	40	1.2×10^3	72	2.0×10^3	1.5×10^3
$Co(phen)_3^{3+} + V^{2+}$	40	3×10^{-3}	4×10^{10}	3.2×10^4	4.0×10^3
$Fe^{3+} + V^{2+}$	4.0	3×10^{-3}	8×10^{16}	1.7×10^6	1.8×10^4
$Fe^{3+} + Co(phen)_3^{2+}$	4.0	40	2×10^6	4.2×10^3	5.3×10^2

Tabulated from Weston, R.E., Jr., and H.A. Schwarz, *Chemical Kinetics,* Prentice-Hall, Englewood Cliffs, NJ, 1972, and Chou, M., C. Creutz, and N. Sutin, *J. Am. Chem. Soc.,* 99, 5615 (1977).

The slope in plots of *ln*k vs. *ln*K for *os* electron transfers is 0.5 only when ΔG_{AB}° is small, because at large values f is no longer ~1. The slope of a Marcus plot approaches unity for very endergonic reactions.[36] Linear plots of *ln*k vs. ΔG° also are found in some atom transfer redox reactions. Hammett and Brønsted LFERs for redox reactions of some organically complexed metal ions also yield linear plots of *log*k vs. E_{AB}° (Section 7.5.2). Thus, care needs to be taken in inferring redox mechanisms (inner sphere vs. outer sphere) from the existence of redox LFERs.

The Marcus theory has been verified for many metal-ion redox reactions,[42] as well as reactions between metal ions and bio-organic compounds.[43] Rate constants computed from the theory for electron exchange between +2- and +3-charged reactants generally agree with measured values within a factor of ~25, and k_{obs} usually is smaller than k_{pred} (Table 7-4).[38] Large deviations may indicate that a reaction occurs by an inner sphere mechanism. For example, k_{obs} is about 10^6 lower than k_{pred} for Fe^{II}-Co^{III} exchange, and the reaction is thought to involve an inner-sphere mechanism.[18] Reasons why the "theoretically" faster *os* mechanism does not occur are not always clear. In some cases (Mn^{II}-Ce^{IV} exchange), the slowness of *os* exchange reflects inaccessible orbitals, but this explanation does not apply to Co^{III}-Fe^{II} exchange. The Marcus theory assumes exchange reactions are adiabatic, and some have suggested that differences between predicted and measured values may be due to nonadiabatic effects. However, corrections for nonadiabadicity are insufficient to account for the differences,[41] and reasons for the discrepancies are still unresolved.

7.5 APPLICATIONS OF LFERs TO REACTIONS IN NATURAL WATERS

LFERs have been applied to reactions of many contaminants in natural waters. They are especially useful in applying fate models like EXAMS (see Chapter 5) to organic chemicals in aquatic systems. If experimental values of rate constants and transfer coefficients are not available and they can be estimated by LFERs, the applicability of the models can be greatly extended.

7.5.1 Hydrolysis Reactions

Hydrolysis reactions have received the most attention in developing predictive relationships for organic contaminants.[44-47] In most cases only base-catalyzed hydrolysis is important at natural water pH values (see Figure 4-6).

Brønsted relationships have been fit to many classes of compounds, including common classes of organic contaminants. For example, plots of second-order $\log k$ values for alkaline hydrolysis of carbamate pesticides vs. pK_a of the resulting alcohol showed good correlations within classes of carbamates,[44] but slopes varied among classes. The hydrolysis rates are very low, however, and degradation by hydrolysis is an important pathway for these pesticides only when the alcohol has a $pK_a < {\sim}12$. Table 7-5 lists a few Brønsted relationships for organic contaminants, and other examples are cited in Chapter 4.

Alkaline hydrolysis rates of organophosphate and organophosphorothionate esters also follow a Brønsted relationship (Equations 1–4, Table 7-5).[45] These classes of compounds include many widely used pesticides, as well as plasticizers and hydraulic fluid additives. Rate constants for O,O-dimethyl-O- and O,O-diethyl-O-(aryl)-phosphates and phosphothionates are correlated with the acidity of the conjugate acid of the aryl leaving group, which is preferentially hydrolyzed under alkaline conditions, leaving the dialkyl ester anions. When the leaving-group conjugate acid has a $pK_a \sim 10$ (e.g., phenol), $t_{1/2}$ for hydrolysis at pH 8 approaches 1 year. Alkaline hydrolysis of phosphate esters thus occurs readily in natural waters only when the leaving aryl group has strong electron-withdrawing substituents, which provide conjugate acid pK_a's $\ll 10$. The O,O,O-trialkyl esters are all very slow to hydrolyze ($t_{1/2}$ at pH 8 > 10^4 days) because the alcohols are very weak acids. The phosphorothionates are even less reactive, and hydrolysis of both alkyl and aryl phosphorothionates is quite slow under environmental conditions.

Hammett and related sigma relationships have been applied to aqueous reactions of several classes of organic contaminants. Alkaline hydrolysis of some triaryl phosphate esters follows a Hammett-like relationship (Equation 5 of Table 7-5). (Note: $\Sigma\sigma$ in Equation 5 is the sum of σ constants for the aromatic groups, and k_o is the hydrolysis rate constant for triphenyl phosphate [$k_o = 0.27 \, M^{-1} \, s^{-1}$; $t_{1/2} = 30$ days at pH 8].) Triaryl esters thus hydrolyze much more rapidly than trialkyl or dialkyl-monoaryl esters in alkaline waters. Photooxidation rates of deprotonated substituted 2-nitrophenols (2-NPs) by singlet oxygen ($^1O_2$) have been correlated with Hammett σ constants,[48] and substituent effects on pK_a values of these compounds also fit a σ relationship: $pK_a = pK_{aH} - \rho\Sigma\sigma_i$.[49] (This is not a kinetic LFER.) Nitrophenols are intermediates in the synthesis of dyes and pesticides and are used directly as herbicides and insecticides. 3-Methyl-2-NP did not fit the relationship, perhaps because of steric factors, but 4-phenyl-2-NP also did not fit the relationship, and an explanation is not apparent.[49] The reactivity of 9-substituted anthracenes toward aqueous nitrating

Table 7-5. Examples of LFERs for Aquatic Contaminants

I. Alkaline hydrolysis reactions		r^2	n	Ref.
Organophosphates (Brønsted):				42
1. O,O-diethyl-P	$logk = 0.28logK_a - 0.22$	0.93	4	
2. O,O-dimethyl-P	$logk = 0.28logK_a + 0.50$	0.97	4	
Organophosphorothionates (Brønsted):				42
3. O,O-diethyl-P	$logk = 0.21logK_a - 1.6$	0.95	4	
4. O,O-dimethyl-P	$logk = 0.25logK_a + 0.34$	0.97	5	
Triaryl phosphate esters (Hammett):				42
5.	$logk = 1.4\,\Sigma\sigma + logk_o$	0.99	4	
Aliphatic primary amides (Taft):				43
6.	$logk = 1.6\sigma^* - 1.37$	0.95	11	
Diphthalate esters (Taft-Pavelich):				44
7.	$logk = 4.59\sigma^* + 1.52E_s - 1.02$	0.97	5	

II. Other reactions		r^2	n	Ref.
pK$_a$s of substituted 2-nitrophenols (not a kinetic LFER):				46
8.	$pK_a = pK_{aH} - 2.59\Sigma\sigma_i$	0.98	17	
Nitration of 9-substituted anthracenes:				47
9.	$logk = -3.02\sigma_p^+ + 2.66$	0.90	6	
Acid hydrolysis of 2,2-substituted alkenes (Brown-Okamoto):				43
10.	$logk_H = -12.3\Sigma\sigma^+ - 8.5$	0.97	24	
pK$_a$s of aldehyde-bisulfite adducts (Taft):				49
11.	$pK_a = 12.1 - 1.29\Sigma\sigma^*$		6	
Hydrolysis of bromoalkanes (Taft):				50
12.	$log(k/k_o) = -11.9\sigma^*$	0.77	17	

species (N_2O_4, HNO_2, HNO_3) is correlated with electrophilic (σ_p^+) substituent values,[50] and the large negative ρ (–3.0; Equation 9; Table 7-5) indicates the transition state is electron deficient (i.e., the reaction is electrophilic).

Success in applying LFERs to hydrolysis reactions of aliphatic compounds has been mixed, reflecting the fact that steric and polar factors complicate the effects of substituent groups on rates of such reactions. Rates of acid-catalyzed hydration of 2,2-substituted alkenes to form secondary alcohols fit a σ^+ correlation (Figure 7-4a) for electrophilic reactions over almost 16 orders of magnitude in k:[46]

$$RR'C=CH_2 + H_2O \xrightarrow{H^+} RR'\overset{OH}{\underset{}{C}} - CH_3$$

This class of compounds includes vinyl chloride, propylene, styrene, and other compounds used as starting materials in polymer, pesticide, and dye manufacturing.

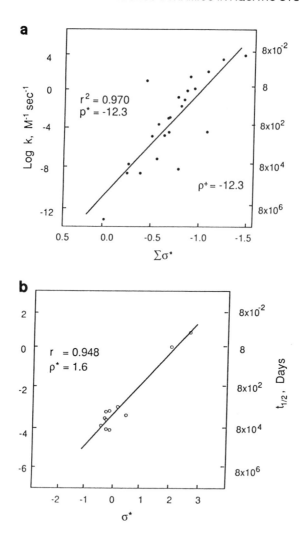

Figure 7-4. (a) Electrophilic σ⁺ LFER for hydration of 2,2-substituted alkenes, and (b) Taft σ*
LFER for alkaline hydrolysis of primary amides. [Redrawn from Wolfe, N.L., in
Dynamics, Exposure and Hazard Assessment of Toxic Chemicals, R. Haque
(Ed.), Ann Arbor Science, Ann Arbor, MI, 1980, p. 163. With permission.]

The correlation, though impressively wide, is not very precise; predicted and
measured values of rate constants agree only within about a factor of ten. Nonetheless, such LFERs can be useful in screening compounds for further study. Even
crude estimates (factor of ten) may be sufficient for compounds with (predicted)
rates that are very fast or very slow. Attention thus can be focused on obtaining more
reliable rates for compounds of intermediate reactivity (characteristic times of days
to a few years). It is interesting to note that σ⁺ constants were developed to describe
effects of aromatic substituents on electrophilic reactions. That they work at all for
aliphatic alkenes may be regarded as an unexpected bonus.

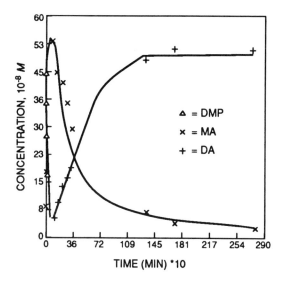

Figure 7-5. Concentration vs. time plot for consecutive reaction of dimethyl phthalate (DMP) hydrolysis to monomethyl intermediate (MA) and the diacid product (DA) at 30°C in 0.01 M NaOH; $k_1 = 6.9 \times 10^{-2}$ M^{-1} s^{-1}; $k_2 = 2.9 \times 10^{-2}$ M^{-1} s^{-1}. [From Wolfe, N.L., W.C. Steen, and L.A. Burns, *Chemosphere*, 9, 403 (1980). With permission.]

Base-catalyzed hydrolysis rates of aliphatic primary amides follow a sigma-type LFER with Taft σ^* constants for polar effects (Equation 6, Table 7-5; Figure 7-4b). Second-order rate constants for common primary amides, like acetamide, acrylamide, and propionamide, are in the range of about 10^{-3} M^{-1} s^{-1}, and half-lives at pH 8 are in the range ~300 to 1000 days.[46,51]

The alkaline hydrolysis of phthalate diesters has been fit to the Taft-Pavelich equation (7-8).[47] Dimethyl phthalate hydrolyzes to phthalic acid in a two-step process (Figure 7-5):

The first step is about 12 times faster than the second, and the diester is nearly all converted to the monoester before the diacid product is formed. Similar two-step sequences are presumed to exist for other phthalate diesters. An LFER was obtained from rate measurements on five phthalate diesters: dimethyl, diethyl, di-*n*-butyl, di-*iso*butyl, and di-(2-ethylhexyl) esters.[47] The constants ρ^* and δ were determined from multiple regression analysis of measured rate constants and values of σ^* and E_s for the alkyl substituents (Equation 7, Table 7-5). The fitted intercept (−1.02) compares closely with the measured rate constant ($logk_{OH} = -1.16 \pm 0.02$) for the dimethyl ester, for which σ^* and $E_s = 0$ by definition. Calculated half-lives under pseudo-first-order conditions of pH 8.0 and 30°C range from about 4 months for dimethyl phthalate to over 100 years for di-(2-ethylhexyl)phthalate.

Taft's σ^* parameter also has been shown to be linearly related to some thermodynamic and kinetic properties of aldehyde-bisulfite adducts.[52] These compounds, which include α-hydroxymethanesulfonate, may be important S^{IV} reservoirs in clouds, fog, and rain. Fairly good relationships were found between equilibrium properties like adduct acidity constants and $\Sigma\sigma^*$ values (Equation 11, Table 7-5), but rates constants for nucleophilic addition of SO_3^{2-} to aldehydes showed only a crude fit. Poor results also were found in applying σ^* to hydrolysis of volatile alkyl chlorides;[53] this appears to be a general characteristic of alkyl chloride reactions with nucleophiles.[54] However, a fair σ^* correlation has been reported for alkyl bromides (Equation 12, Table 7-5).

7.5.2 Reactions of Sulfur Nucleophiles

The lower oxidation states of sulfur (−II, sulfide, to +IV, sulfite) are electron rich and act as nucleophiles. These species react with alkyl halides to displace the more electronegative halide, and such reactions may be significant in anoxic aquatic environments such as sediments and groundwater under landfills. In the case of sulfide, the reaction produces a thiol: $RX + HS^- \rightarrow RSH + X^-$. Such reactions thus are analogous to hydrolysis reactions in which water acts as the nucleophile: $RX + H_2O \rightarrow ROH + H^+ + X^-$. Barbash and Reinhard[55] used two correlation equations from the literature for nucleophilic substitution reactions[25] to estimate rate constants of some sulfur nucleophiles with some alkyl halides likely to be found in contaminated groundwater. The first, attributed to Swain and Scott,[56] is similar in form to the Hammett equation:

$$\log\left(k_s / k_{H_2O}\right)_{SN} = sn \qquad (7\text{-}15)$$

k_S is the rate constant for nucleophilic substitution between reactant RX and a given sulfur nucleophile, $k_{H:O}$ is the rate constant for the corresponding hydrolysis reaction, *s* is an empirical constant characteristic of the substrate ($s \equiv 1$ for methyl bromide), and *n* defines the nucleophilicity of the compound (based on its reactivity with methyl bromide). The second is a two-term equation attributed to Edwards:[57]

$$\log\left(k_s / k_{H_2O}\right)_{SN} = \alpha E_n + \beta H \qquad (7\text{-}16)$$

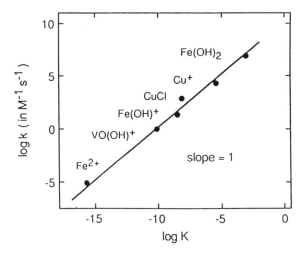

Figure 7-6. LFER for oxidation of Fe^{II} and other metal ion species. [From Wehrli, B., in *Aquatic Chemical Kinetics*, W. Stumm (Ed.), Wiley-Interscience, New York, 1990. With permission.]

E_n is an electron donor constant for the nucleophile related to the standard (oxidation) electrode potential ($E°$, American convention) for the oxidative dimerization of the nucleophile: $2X^- \rightleftharpoons X_2 + 2e^-$, $E_n = E° + 2.6$; H is the basicity of the nucleophile toward a proton (relative to water): $H = pK_a + 1.74$); and α and β are fitted substrate constants.

7.5.3 Redox Reactions

LFERs have been applied to some redox reactions of potential interest in aquatic systems, including reactions of dissolved metal ion oxidants and metal oxides with both organic and inorganic reductants. A Marcus LFER with slope = 1 fits the autoxidation rates of several metal ions, including Fe^{2+}, its hydroxide complexes, and Cu$^+$ (Figure 7-6). The rate-limiting step is assumed to be the endergonic reduction of O$_2$ to O$_2^{-}$,[36] and under these conditions the slope of a Marcus plot is 1. Wehrli[36] interpreted this finding to indicate that these autoxidation reactions proceed by *os* mechanisms. Oxidation by Co^{III} of some pyridine and bipyridyl derivatives, such as biologically important nicotinamide and the herbicides paraquat and diquat, also fit a Marcus LFER.[43] Oxidation rates of substituted phenols by soluble complexes of Fe^{III} (with cyanide, phenanthroline, or bipyridyl as ligands) are correlated with σ constants of the phenol substituents.[58] Similar trends were found for the reductive dissolution of Mn^{III,IV} oxides by substituted phenols.[59] Electron-donor groups ($\sigma < 0$) decrease phenolic reduction potentials, and electron-withdrawers ($\sigma > 0$) have the opposite effect. Correlation of oxidation rates with σ constants implies that the rates are correlated with reduction potentials of the phenols (thus with overall $E°$s and $\Delta G°$s of reaction, see Figure 7-7).

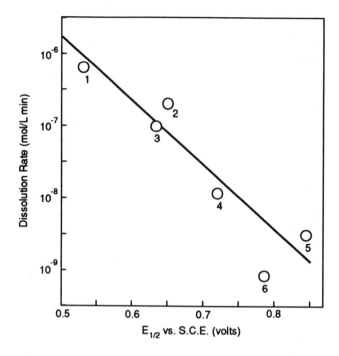

Figure 7-7. LFER for reductive dissolution of $Mn^{III/IV}$ oxides by substituted phenols: rate of Mn^{2+} formation vs. $E_{1/2}$ for phenols. [From Stone, A.T., *Environ. Sci. Technol.*, 21, 979 (1987). With permission of the Amer. Chem. Soc.]

7.5.4 Concluding Comments and Prospectus

The above discussion is not an exhaustive review of LFER applications in aquatic chemistry, but is illustrative of the published information on this topic. Several conclusions and comments are pertinent. First, the huge database regarding LFERs in the chemistry literature has hardly been tapped by existing aquatic studies. At least two factors may be responsible for this: (1) many compounds used by chemists in laboratory studies are not important environmental contaminants, and (2) reactions occurring in nonaqueous systems often are not relevant to the aquatic behavior of organic contaminants. Second, most reported LFERs for aquatic contaminants are based on correlations with very small sample sizes: $n = 4$ or 5 for many of the regression equations in Table 7-5. Consequently, a fair amount of uncertainty exists regarding the broad applicability of these relationships. Third, most of the reported correlations are not very precise. It is not clear whether this is a reflection primarily of measurement uncertainties or of an inherent lack of fit to the supposed linear relationships. Even when the correlation coefficient is high (e.g., $r > 0.95$), the *log-log* nature of the relationships, which often span many orders of magnitude, leads to large uncertainties in estimating k (easily ±5 times or ±10 times). As mentioned earlier, such LFERs still can be useful in screening groups of compounds for further study. Fourth, further developmental work is needed on some important classes of aquatic contaminants, like volatile chlorinated compounds, for which none of the existing LFERs seems to apply.

Finally, quantitative predictions based on a given LFER are limited to one class of compounds and a particular chemical reaction. Prediction of a compound's chemical reactivity strictly from its molecular structure is a much more ambitious task. This is a goal of computerized "expert systems" now under development. Such systems are of interest both to chemical manufacturers and to regulatory agencies that must be concerned about the potential environmental effects of an increasing number of anthropogenic chemicals. SPARC[60] is a prototype expert system that predicts UV-visible absorption spectra and pK_as for organic compounds from data on the compound's structure and reaction conditions (e.g., temperature, pH). Prediction of hydrolysis rate constants is under development, and the goal of SPARC's developers is to extend its capabilities to predict chemical reactivity and environmental properties on a broad scale.

In general, reactivity predictions are based on analysis of the energy differences between two states of a molecule — in broadest terms, an initial state and a "product" state. In the case of light absorption, the former is the ground state and the latter is some excited electronic state. For chemical equilibria the energy difference of interest is between a reactant and a product, and for reaction rates the relevant energy differences are between a reactant and a transition state. In theory such energy differences could be computed from first principles (*ab initio* calculations; see Section 3.5.1), but the energy differences are very small compared with total binding energies in most processes of interest. In spite of great improvements in computing power and algorithm efficiency, *ab initio* methods still are not feasible for large organic molecules. SPARC does not compute chemical properties from first principles, but uses a variety of methods developed by physical organic chemists; in particular, a blending of perturbed molecular orbital methods[61] and linear free-energy theory.

SPARC performs a reactivity assessment by first locating the potential sites within a molecule for a reaction of interest (e.g., light absorption, ionization, hydrolysis). These sites are called centers (C), and molecular structures attached to C are called perturbers (P). All reactions are analyzed as equilibria involving some energy change:

$$PC_o \overset{f(\Delta E)}{=} PC_f \qquad (7\text{-}17)$$

Perturbers are assumed to remain unchanged by the reaction, but the nature of P affects the reactivity of C; $f(\Delta E)$ is some function of the energy change of the reaction. For example, the ionization of phenol is described by

$$-OH \underset{}{\overset{K_a}{\rightleftharpoons}} -O^- + H^+$$

C_o, C_f, and P are $-OH$, O^-, and the phenyl group, respectively.[60] Energy changes are defined in terms of contributions of C and P, $\Delta E = \Delta E_C + \delta_P(\Delta E_C)$, where ΔE_C is the intrinsic value for C, and $\delta_P(\Delta E_C)$ represents a perturbation in energy change due to P. Both electrostatic and resonance contributions to δ_P are considered. Details of how these terms are computed or inferred from measured values are described elsewhere.[60] Predictions of absorption spectra and pK_as made by SPARC show good agreement with measured values, but much work remains before this method can be used as a general tool to predict the reactivity of compounds in the environment.

PART II. PROPERTY-ACTIVITY AND STRUCTURE-ACTIVITY RELATIONSHIPS IN ENVIRONMENTAL BIOLOGY AND CHEMISTRY

7.6 NATURE AND IMPORTANCE OF PARs AND SARs

The LFERs described in preceding sections are but a subset of the correlations and predictive relationships that have been developed to relate the behavior of compounds (especially their bio-activity) to their structural or equilibrium properties (see Table 7-1 for a classification of these relationships). The original impetus for research on PARs and SARs was the need to predict the effectiveness of new drugs and medicines, but this topic has become important in environmental chemistry and toxicology in recent years (see references 62–65 for reviews).

Attributes commonly used in PARs and SARs are listed in Table 7-6. Structural variables range from simple indices like carbon number and measures of molecular size (volume or surface area) to group additivity (fragment) parameters and complicated topological indices of molecular shape. Development of SARs and SPRs (structure-property relationships) for aquatic systems is described in Section 7.11.

In general, the physical-chemical properties listed in Table 7-6 are ratios of a property that are used to predict other properties. As such, they are manifestations of molecular structure rather than structural characteristics themselves.[66] Relationships based on these properties thus are property-activity relationships (PARs); the term structure-activity relationship (SAR) should be restricted to correlations based on structural or topological parameters. However, the literature is not consistent on this terminology, and the line between structural characteristics and properties resulting from structure is not always clearcut.

The physical-chemical properties used in PARs usually can be measured directly or calculated readily from measured quantities. However, as noted above, intercorrelations exist among several properties, as well as between some structural attributes and physical-chemical properties. Such correlations can be used for predictive purposes, thus giving rise to two additional categories of attribute relationships: structure-property and property-property relationships (SPRs and PPRs). Because analytical problems sometimes are encountered for physical-chemical properties like K_{ow} and S_w, SPRs and PPRs are useful, indirect ways of estimating the value of a property that serves as a predictor variable in a PAR (see Section 7.7.4). Some molecular parameters are not measured or computed from directly measured quantities, but instead are products of theoretical analysis, like

Table 7-6. Common Attributes Used in Environmental Correlations

Structural variables	Physical-chemical properties	Biological activity
Molar volume	Molecular weight	Bioaccumulation, BF
Surface area	Lipophilicity (hydrophobicity)	Acute toxicity, LC_{50}
Chain length	K_{ow} K_p (sediment-water partition coefficient)	Inhibition of a function or process like photosynthesis, respiration, or bioluminescence, EC_{50}
Topological properties of atom-bond arrangements (degree of branching): χ and χ^v molecular connectivity indices	K_{oc} (K_p normalized to organic content) Aqueous solubility, S_w	Biodegradation rate, k_1, (BOD_5/ThOD) ThOD = theoretical (ultimate) BOD
	Chromatographic retention time, k'	
Molecular fragment (group-additive) indices of fundamental atomic properties	Molar refractivity, R_m LFER parameters: σ, σ^+, σ^- LSER solvatochromic properties, π^*, β, σ_m Hard-soft acid parameters, σ_p, σ_k, for metals	

molecular mechanics theory. Parameters derived in this way (see Section 7.11.2 for examples) may be called **molecular properties** to distinguish them from empirical physical-chemical properties.

The physical-chemical properties used in PARs fall into two general categories: those related to a compound's availability or transportability, and those related to its chemical/biological reactivity. The most common property of compounds used in PARs is lipophilicity, as measured by the partition coefficient of a compound between water and *n*-octanol, i.e., K_{ow}. Many other properties used in PARs are correlated with K_{ow}: aqueous solubility (S_w), molar refractivity, sorption or partitioning onto solids, and chromatographic retention indices. In general, these properties reflect the aqueous thermodynamic activity of a compound or aqueous solvation free energy (see Section 7.8) rather than its chemical reactivity. The properties are defined and relationships among them are described in Sections 7.7 and 7.8. PARs resulting from them are discussed in Sections 7.9–7.10. Some LFER parameters like Hammett's σ have been correlated with the bio-activity of compounds and thus are components of PARs.

There are three principal processes of interest relative to the bio-activity of aquatic contaminants:

1. Bioaccumulation, i.e., uptake and retention of a compound by organisms, usually expressed in terms of the concentration of the contaminant in cellular biomass divided by the contaminant concentration in the growth medium

2.   Biodegradation, i.e., the complete or partial breakdown of a compound by organisms. The organism may obtain energy from the reaction or it may occur as a detoxifying mechanism or as a cometabolic process
3.   Acute toxicity: the rapid death of organisms, expressed as the median lethal concentration, LC_{50}, i.e., the contaminant concentration at which 50% of the organisms die within some prescribed test period, usually 24 or 48 hours

Inhibition of key biotic processes like respiration and photosynthesis also is used to indicate a compound's toxicity in PARs and SARs. The above processes are much more complicated than the chemical reactions for which LFERs and PARs are developed. Biotic processes consist of many bio-physico-chemical steps, and their elementary and rate-controlling steps are poorly defined compared with chemical reactions. Given this complexity, it is surprising that simple PARs and SARs exist, and more surprising that they often work well over a relatively wide range of compounds. The applicability of simple relationships over a range of organisms, processes, and compounds suggests that a few common steps are of critical importance: e.g., transport across cell membranes and chemical reaction with a key enzyme. Rates of transport across membranes depend on the lipophilicity of molecules, because membranes are partly made of lipids. Chemical reactions are subject to the same fundamental laws whether they occur *in vitro* or *in vivo*; consequently, cellular reactions often fit LFERs, like Brønsted or Hammett relationships, even if they are mediated by enzymes.

7.7 PHYSICAL-CHEMICAL MEASURES OF COMPOUND AVAILABILITY AND TRANSPORTABILITY

7.7.1 Definitions and Descriptions

K_{ow}. The relative lipophilic/hydrophilic character of a molecule has important implications for biological transport. Interest in this characteristic dates back to the classic studies of Meyer and Overton[67] near the turn of the 20th century, which showed a relationship between olive oil/water partition coefficients and the narcotic action of organic compounds. In general, a partition coefficient is the ratio of concentrations of a compound in two phases in equilibrium with each other. The partition coefficient of a compound between *n*-octanol (o) and water (w) is the standard measure of a compound's lipophilicity:

$$K_{ow} = C_o / C_w \qquad (7\text{-}18)$$

Aqueous solubility. S_w typically is expressed on a molar or a weight/volume basis, but some thermodynamic relationships express solubility on a mole fraction basis. In general, mole fraction solubilities, X_i, can be converted to molar solubilities, S_i, by dividing the former by the molar volume (V_{solv}) of the solvent; in the case of water, $V_w = 0.018$ L/mol.

The mole fraction solubility ($X_{i,aq}$) of a liquid nonelectrolyte i in water is inversely proportional to γ_w.[68]

$$X_{i,w} = a_{i,w}/\gamma_{i,w} = f_{i,w}/f_i^{ss}\gamma_{i,w} = 1/\gamma_{i,w} \qquad (7\text{-}19)$$

$a_{i,w}$ is the aqueous activity of i, and $f_{i,w}/f_i^{ss}\gamma_{i,w}$ is the ratio of the solute fugacity in solution and in its standard state (the pure liquid state). On a molar basis, Equation 7-19 becomes $S_i = 1/\gamma_{i,w}V_w$. Equilibrium is achieved when the solute's fugacity in pure form and in aqueous solution are equal. Activity coefficients of sparingly soluble liquid nonelectrolytes thus are much larger than unity. For solutes that are solids at the temperature of dissolution, a correction term must be added to Equation 7-19 to account for the energy required to convert the solid to a supercooled liquid at the temperature of the solution. This hypothetical state is the standard state for solids ($f_i^{ss} = f_i^{sL}$, where superscript sL denotes pure supercooled liquid). Two forms of the correction term exist, depending on assumptions about the change in solute heat capacity (ΔC_p) from solid to liquid.[68] Either ΔC_p is assumed to be small or zero, which leads to the following expression:

$$\ln X_{i,w} = \frac{-\Delta S_f}{R}\left[\frac{T_m}{T} - 1\right] - \ln \gamma_{i,w} \qquad (7\text{-}20a)$$

or ΔC_p is assumed equal to ΔS_f, which leads to:

$$\ln X_{i,w} = \left(-\Delta S_f/R\right)\ln\left(T_m/T\right) - \ln \gamma_{i,w} \qquad (7\text{-}20b)$$

ΔS_f is the entropy of fusion, T_m is the solute melting-point temperature, and T the temperature of interest. The latter assumption was found to apply to some PCBs.[69] In addition, although data are limited, aromatic molecules exhibit a narrow range of ΔS_f (~12 to 16 cal mol^{-1} K^{-1}). The mean value (13.2 cal mol^{-1} K^{-1}) is used to estimate of ΔS_f in the absence of data.[68]

Molar refractivity, R_m. This parameter is related to a compound's refractive index (n) and its specific molar volume (molecular weight M_w divided by density d), as defined by the Lorentz-Lorenz equation:[5]

$$R_m = \left[\frac{n^2 - 1}{n^2 + 2}\right]\frac{M_w}{d} \qquad (7\text{-}21)$$

R_m has units of cm^3 mol^{-1} and is an additive-constitutive property of molecules; i.e., values can be assigned to individual groups or fragments in compounds. R_m encodes information about the electronic configuration of molecules, in particular about the deformation of outer molecular orbitals caused by weak electric fields induced by other molecules, i.e., London dispersion forces. R_m is correlated with many physical-chemical properties of molecules and their activity in mixtures of molecules. It has been used in some PARs developed for medicinal and pharmacological studies, but only rarely in environmental applications.[63] The range of n for liquid organic compounds is small, typically between 1.36 and 1.60. The term within brackets in Equation 7-21 thus varies only over a range of ~0.21 to 0.34. Consequently, most of the variance in R_m reflects the variance in molar volume, and R_m really is a "corrected" form of molar volume.[22] Correlations of R_m with bio-activity parameters thus primarily reflect the correlation of bio-activity with molar volume. Nonetheless, in binding studies of organic ligands with the enzyme chymotrypsin, R_m provided slightly better correlations than M_W/d, suggesting that London dispersion forces are involved in the binding.[22]

Chromatographic retention indices. The primary mechanism of separation in many chromatographic procedures is the same as that in octanol-water partitioning: they involve partitioning of solutes between a stationary liquid phase coated onto a solid support and an immiscible mobile phase. Sorption also may be involved in the separations, but this is thought to be minor if the liquid stationary phase completely coats the solid support. Chromatographic retention indices of compounds commonly are expressed as k', defined as

$$k' = \left(R_t - R_{t,o}\right) / R_{t,o} \tag{7-22}$$

where R_t is the elution (retention) time of a compound, and $R_{t,o}$ is the elution time of an unretained solvent peak. $Log k'$ is linearly related to $log K_{ow}$,[70] and the former often is used to estimate the latter; this is an example of a PPR. $Log k'$ also has been used directly as a predictor variable in a few PARs.

Sediment-water and soil-water partition coefficients (K_p and K_{oc}). These are important parameters in determining the environmental fate of organic contaminants. K_p is defined analogously to K_{ow} as the ratio of the sorbed concentration (mg kg^{-1}) on a bulk sediment to the dissolved concentration (mg L^{-1}) of a contaminant when the two phases reach equilibrium. Thus, K_p has units of reciprocal density (L kg^{-1}). Because organic contaminants are associated primarily with the organic fraction of soils and sediments, it is useful to normalize soil- and sediment-water partition coefficients to their organic content, f_{oc} (the fraction of organic matter in the soil or sediment). Dividing K_p by f_{oc} yields the desired coefficient, K_{oc}.

7.7.2 Measurement of K_{ow}

Measurement of K_{ow} is conceptually straightforward; one simply measures the compound's concentration in both phases after a period of equilibration in a well-shaken flask. In practice the situation is much more complicated. K_{ow} values for aquatic contaminants range over many orders of magnitude, from $log K_{ow} \approx 1.1$ for aniline to more than 8 for highly chlorinated biphenyls (PCBs). Measurement of K_{ow} by shaker-flask methods is not practical for most PCBs, polynuclear aromatic hydrocarbons (PAHs), polychlorinated dibenzo-p-dioxins (PCDDs), and other highly lipophilic compounds ($log K_{ow} > 5$). Concentrations of such compounds in the aqueous phase of batch flasks are too low to measure accurately. A generator-column method[71] extends measurement to

Table 7-7. Methods to Determine K_{ow}

Method	Reference
1. Direct measurement	
a. conventional shaker flask method	
b. generator column method for highly lipophilic compounds	65
2. Hansch-Leo π approach for substituted aromatic compounds	69
$\pi_X = logK_{ow,X} - logK_{ow,H}$	
3. For chlorinated PCBs and PAHs	70
$logK_{ow} = (n + 1)^b logK_{ow}^{\circ}$	
n = number of Cl atoms	
$b \propto logK_{ow}^{\circ}$, ($K_{ow}$ of parent compound)	
4. Fragment constant approach	25,71,72
$logK_{ow} = \Sigma n_i f_i$	
n_i = number of fragments of type i in molecule	
f_i = contribution of ith fragment to $logK_{ow}$	
5. Estimation from UNIFAC	73
$K_{ow} = 0.115\gamma_w/\gamma_o$	
γ_w and γ_o calculated by UNIFAC	
6. Correlation with aqueous solubility, S_w	62,75,76
7. Correlation with $logk'$, HPLC retention coefficient	64,68,78–81
$k' = (t_r - t_o)/t_o$	

$logK_{ow} > 8$. Substantial variations (0.5 to ~2 log units) can be found among recent measured values of some common aquatic contaminants, even at $logK_{ow} < 5$:[72]

benzene	1.56-2.15	hexachlorobenzene	4.13-7.42	aldrin	5.52-7.40
toluene	2.11-2.73	pentachlorophenol	3.32-5.86	2-chlorobiphenyl	3.90-4.59
p,p'-DDT	3.98-6.36	trichloroethene	2.29-3.30	naphthalene	3.01-4.70

7.7.3 Calculation Methods for K_{ow}

Because direct measurement methods are tedious, time consuming, and often not very accurate, indirect methods (Table 7-7) often are used to estimate K_{ow} in environmental studies. These include group additivity methods in which the lipophilic contributions of molecular fragments have been estimated, as well as simple correlations (PPRs and SPRs) with other measured parameters (Figure 7-8). That so many methods exist to estimate or measure K_{ow} suggests there are serious problems in determining this important parameter. As a minimum, we can expect differences in the values of a compound's K_{ow} determined by different methods, and the resulting uncertainty needs to be taken into account when using K_{ow} to predict bio-activity. A large literature is available on this subject.[68,72,74]

π, **a substituent parameter for lipophilicity.** $LogK_{ow}$ is an additive property of molecular constituents. This means that it can be estimated from substituent

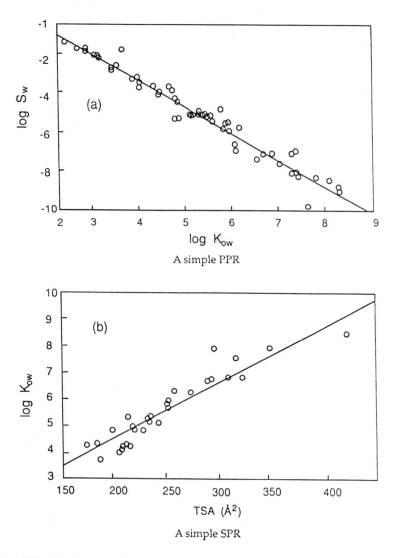

Figure 7-8.   Simple examples of structure-property and property-property relationships: (a) PPR of *log*K$_{ow}$ vs. *log* aqueous solubility for 55 aromatic hydrocarbons. [From Andren, A.W., W.J. Doucette, and R.M. Dickhut, in *Sources and Fates of Aquatic Pollutants*, R.A. Hites and S.J. Eisenreich (Eds.), Adv. Chem. Ser. 216, Am. Chem. Soc., Washington, D.C., 1987, p. 3.] (b) *log*K$_{ow}$ vs. total molecular surface area for 32 aromatic hydrocarbons. [From Douchette, W.J. and A.W. Andren, *Environ. Sci. Technol.*, 21, 821 (1987). With permission of the Amer. Chem. Soc.]

parameters without direct measurement. A substituent parameter π for lipophilicity of aromatic compounds has been defined:[22,75]

$$\pi_X = \log K_{ow.X} - \log K_{ow.H} \qquad (7\text{-}23)$$

Subscript H refers to a parent compound, and X to its derivative with substituent X. Values of π_X are constant for similar solvent systems;[75] octanol/water values for common substituents are listed in Table 7-8. For example, π for Cl is the difference in $logK_{ow}$ for chlorobenzene and benzene: $\pi_{Cl} = 2.84 - 2.13 = 0.71$. $LogK_{ow}$ for chlorophenol then is calculated as $logK_{ow}$ for phenol plus π_{Cl}. The resulting value, 2.17, compares with measured values of 2.15, 2.50, and 2.39 for 2-, 3-, and 4-chlorophenols. K_{ow} values for other substituted compounds can be calculated similarly, but errors increase with the number of substituents on the ring, and the method is inaccurate when $logK_{ow} > \sim6$.

A method analogous to the π approach was proposed to calculate K_{ow} for chlorosubstituted aromatic compounds (PCBs, polychlorinated PAHs) from the K_{ow} of the parent compound by the equation[76]

$$\log K_{ow} = (n+1)^b \log K_{ow}^o \qquad (7\text{-}24)$$

where n is the number of chlorine atoms in the molecule; superscript o refers to the parent compound; and exponent b is correlated with $logK_{ow}^o$.

Fragment methods. Several molecular fragment methods have been devised to calculate $logK_{ow}$ for parent compounds. The best known are those of Rekker[77,78] and Hansch and Leo.[22] Both methods are based on equations of the general form,

$$\log K_{ow} = \sum_i n_i f_i \left(+F_c\right) \qquad (7\text{-}25)$$

n_i is the number of fragments of type i in a molecule, and f_i is the contribution of the ith fragment to $logK_{ow}$. F_c is a factor that is added for certain kinds of molecules (hence, the parentheses) to account for lipophilic or lipophobic effects of structural features that aren't accounted for by the fragments themselves. Table 7-8 lists fragment constants and factors for both methods, and Example 7-1 illustrates their use to calculate $logK_{ow}$ values for several organic compounds.

Rekker used a deductive method and a database of over 1000 $logK_{ow}$ values to derive fragment constants by regression analysis. He concluded that a single factor or "magic constant" with a value of 0.289, or multiples of that constant, could account for the discrepancies in $logK_{ow}$ values calculated from fragment constants alone. In contrast, Hansch and Leo took a "constructive" approach to build a list of fragment constants from $logK_{ow}$ values for compounds of increasing complexity. For example, from $logK_{ow} = 0.45$ for H_2, they concluded that each H atom in a molecule contributes 0.45/2 to $logK_{ow}$, or $f_H = 0.225$ (rounded to 0.23). From $logK_{ow} = 1.09$ for CH_4 and 1.81 for CH_3CH_3, they calculated that f = 0.89 for CH_3, and $f_C = 0.20$ for C atoms in aliphatic compounds. f_C for C in aromatic rings was found to have a lower value (0.13). As Example 7-1 shows, the two methods give similar values for many compounds.

Example 7-1. Calculation of $logK_{ow}$ from Fragment Constants

The following calculations are based on the fragment and factor values of Rekker (R) and Hansch and Leo (HL) given in Table 7-8; M = measured value as reported in the reference cited in parentheses. It should be noted that agreement between the two methods for complicated and highly lipophilic compounds is not as good as shown below. Also, neither

Table 7-8. π and Fragment Values for Aromatic Substituents

A. π Values[a]

Substituent	π	Substituent	π	Substituent	π
H	0.00	F	0.14	NH_2	−1.23
C_6H_5	1.96	Cl	0.71	NO_2	−0.28
OC_6H_5	2.08	Br	0.86	CN	−0.57
$N=NC_6H_5$	1.65	OH	−0.67	SCN	0.41
CH_3	0.56	SH	0.39	CHO	−0.65
CF_3	0.88	OCH_3	−0.02	COO^-	−4.36

B. Hansch and Leo fragment constants (f) and factors (F)[a,b]

Group	f	f^{φ}	Group	f	f^{φ}	Group	f	f^{φ}
−Br	0.20	1.09	$-SO_2$	−2.67	−2.17	−CON=	−3.04	−2.80
−Cl	0.06	0.94	−H	0.23	0.23	−SCN	−0.48	−0.64
−F	−0.38	0.37	−NH−	−2.15	−1.03	−C(O)−	−1.90	−1.09
−N=	−2.18	−0.93	−OH	−1.64	−0.44	−COO⁻	−5.19	−4.13
−NO	—	0.11	−SH	−0.23	0.62	−COH	−1.10	−0.42
$-NO_2$	−1.16	−0.03	$-NH_2$	−1.54	−1.00	−CONH−	−2.71	−1.81
−O−	−1.82	−0.61	>C<	0.20	0.20	$-CONH_2$	−1.58	−0.82
−S−	−0.79	0.03	−CN	−1.27	−0.34	$-OCH_3$	−1.54	

Fragments fused in aromatic ring[c]

	f^{φ}		f^{φ}		f^{φ}
−N=	−1.12	−S−	0.36	−C(O)−	−0.59
−N<	−1.60	−NH−	−0.65	−CH=N−NH−	−0.47
−N=N−	−2.14	C	0.13	−N=CH−NH−	−.079
−O−	−0.08	CH	0.355		

Factors:

				F
1. involving bonds: unsaturation (normal)				−0.55
(conjugate to aromatic ring)				−0.42
length (n) of aliphatic chain				−0.12(n − 1)
size (n) of aliphatic ring				−0.09(n − 1)
2. multiple halogenation	(on same carbon)		X = 2	0.30
			X = 3	0.53
			X = 4	0.72
	(on adjacent carbons)			0.28(n − 1)
3. H-polar proximity: Chains	(one C separating polar groups)			$-0.42\Sigma f_1 + f_2$
	(two C separating polar groups)			$-0.26\Sigma f_1 + f_2$
Aromatic ring	(one C separating polar groups)			$-0.16\Sigma f_1 + f_2$
	(two C separating polar groups)			$-0.08\Sigma f_1 + f_2$

method gives very accurate estimates for polar compounds, such as substituted phenols, where $log K_{ow} < 2$.

Benzene (C_6H_6) R: $6 \times (0.155 + 0.182) = 6 \times 0.337 = \underline{2.02}$

HL: $6 \times 0.355 = \underline{2.13}$

M: $\underline{1.56} - 2.15$ (72)

Table 7-8 (continued).

C. Rekker fragment constants[b,d,e]

	f		f		f	f_φ		f	f_φ
C	0.155	C_6H_5	1.840	OH	−1.470	−0.314	NH_2	−1.420	−0.842
H	0.182	C_6H_4	1.658	Cl	0.057	0.924	NH	−1.814	−0.947
CH	0.337	C_6H_3	1.476	Br	0.249	1.116	SH	0.00	0.62
CH_2	0.519	CH_3	0.701	F	−0.476	0.391	CO	−1.643	−0.776

[a] From Hansch, C., and A. Leo, *Substituent Constants for Correlation Analysis in Chemistry and Biology*, J. Wiley, New York, 1979.
[b] f is for attachment to aliphatic structure; f[φ] is for attachment to aromatic ring. Attachment is from left for bivalent groups.
[c] Groups and f[φ] are underlined to indicate presence within an aromatic ring.
[d] Rekker's method has a single factor, $c_M = 0.289$. Multiples of c_M are added to Σf_i to obtain a proper estimate of $\log K_{ow}$. One c_M is added for each pair of aliphatic conjugated double bonds, each pair of conjugated or condensed aromatic rings, and other structural features involving electronegative groups.
[e] From Rekker, R.F., and H.M. DeKort, *Eur. J. Med. Chem.*, 14, 479 (1979).

Toluene $(C_6H_5CH_3)$ R: $5 \times 0.337 + 0.155 + 0.155 + 3 \times 0.182 = \underline{2.54}$

HL: $5 \times 0.355 + 0.13 + 0.20 + 3 \times 0.23 = \underline{2.80}$

M: $\underline{2.11 - 2.73}\ (72)$

2−Nitrophenol R: $4 \times 0.337 + 2 \times 0.155 - 0.314 - 0.053 = \underline{1.29}$

$\left(C_6H_4(OH)NO_2\right)$ HL: $4 \times 0.355 + 2 \times 0.13 - 0.44 - 0.03 = \underline{1.21}$

M: $\underline{1.89}\ (49)$

Biphenyl R: $2 \times (6 \times 0.155 + 5 \times 0.182) + c_m = 2 \times 1.84 + 0.289$

$\left(C_6H_5C_6H_5\right)$ $= \underline{3.97}\ \left(\text{Note: } c_m \text{ is Rekker's "magic constant"}\right)$

HL: $10 \times 0.355 + 2 \times 0.13 = \underline{3.81}$

M: $\underline{3.93}\ (70)$

$2,2',4,4',5,5'\,PCB$ R: $12 \times 0.155 + 4 \times 0.182 + 6 \times 0.924 + 0.289 = \underline{8.42}$

(See Figure 7-12c HL: $12 \times 0.13 + 4 \times 0.23 + 6 \times 0.94 - 0.16 \times 2(2 \times 0.94)$

for structure of PCBs) $-0.08 \times 2(2 \times 0.94) = \underline{7.22}$

(Note: the last two terms are F values for the effects

of polar groups on lipophilicity of adjacent H atoms)

M: $\underline{7.79}\ (70)$

Anthracene R: $10 \times 0.337 + 4 \times 0.155 + 2 \times 0.289 = \underline{4.57}$

$\left(C_6 H_4 {}^{CH}_{CH} C_6 H_4\right)$ HL: $10 \times 0.355 + 4 \times 0.225 = \underline{4.45}$

                                   M: $\underline{4.49}\,(70)$

DDT R: $2(4 \times 0.337 + 2 \times 0.155 + 0.924) + 2 \times 0.155 + 0.182$

$(p - Cl\phi)_2 CHCCl_3$ $+3 \times 0.057 = \underline{5.83}$

$(\phi = phenyl$ HL: $2(4 \times 0.355 + 2 \times 0.13 + 0.94) + 2 \times 0.20 + 0.23$

$group)$ $+3 \times 0.06 = \underline{6.05}$

                                   M: $\underline{6.07}\,(70)$

UNIFAC. Another fragment-based method for K_{ow} is based on UNIFAC (universal quasichemical functional groups activity coefficient). This procedure calculates activity coefficients of components in a mixture from the structures (i.e., functional groups) of each component (see Section 7.8). Arbuckle[79] showed that K_{ow} values of compounds can be estimated from the ratio of UNIFAC-calculated activity coefficients in octanol (o) and water (w):

$$K_{ow} = 0.115 \gamma_w / \gamma_o \tag{7-26}$$

The γ are calculated values for infinite dilution. The average error in K_{ow} calculated by this method for 27 priority pollutants was 0.41 *log* units; nearly all calculated values were greater than measured ones.

7.7.4 Correlation Methods to Estimate K_{ow}

Molecular weight (M_W). Overall, K_{ow} is only crudely correlated with M_W, but surprisingly good correlations have been obtained for some groups of compounds. For example, Rekker-based $log K_{ow}$ values were highly correlated ($r^2 = 0.998$) with M_W for 36 PAHs.[80] In contrast, a much poorer correlation ($r^2 = 0.67$) was found between experimentally determined $log K_{ow}$ and M_W for a more diverse group of 36 polycyclic aromatic compounds that included S-, N-, and O-containing heteroaromatic compounds.[74] Correlations with M_W thus typically have a limited range of applicability. This is not surprising, since M_W does not embody any structural (geometric or topologic) information about a molecule.

Aqueous solubility. Many inverse correlations of K_{ow} with S_w have been described for environmental contaminants since Hansch et al.[81] first reported a relatively good correlation ($r^2 = 0.87$) for 156 aliphatic and aromatic compounds. Most of the K_{ow} values were calculated from π values rather than measured. A review by Lyman et al.[82] mentioned 18 such correlations. Stronger correlations tend to occur with more restricted data sets, e.g., for 17 substituted nitrophenols, $\log K_{ow}$ $= -0.97 \log S_w - 0.04$ ($r^2 = 0.96$).[49] Because of the difficulties involved in measuring very small values of S_w, such correlations are not very helpful in estimating K_{ow} for highly lipophilic compounds.

Although S_w-K_{ow} correlations are empirical, the relationship has a thermodynamic basis.[83] As Equation 7-19 shows, S_w is inversely proportional to γ_w, and according to Equation 7-26, K_{ow} is proportional to the aqueous activity coefficient (γ_w) of a solute. Thus, we expect an inverse correlation between K_{ow} and S_w of organic compounds. K_{ow} essentially is a measure of the relative solubility of a compound in a nonpolar and a polar solvent, and hydrophobicity (low S_w) is a major driving force for solute partitioning into octanol. (Hydrophobicity and lipophilicity are not *exactly* the same phenomenon, however.) For liquid or supercooled liquid solutes, the following relationship applies:[68]

$$\log S = - \log K_{ow} - \log V_o^* - \log \gamma_o^* + \log\left(\gamma_w^* / \gamma_w\right) \qquad (7\text{-}27)$$

V_o^* is the molar volume of octanol saturated with water; γ_o^* and γ_w^* are activity coefficients of the solute in water-saturated octanol and octanol-saturated water, respectively; and γ_w is the value in pure water. For solid solutes, the melting point correction term of Equation 7-20 must be added to Equation 7-27, and solid solubility data need to be adjusted by this term if comparisons are to be made with data sets containing both liquid and solid solutes. If the solute forms an ideal solution in water-saturated octanol, $\gamma_o^* = 1$, and if the solute has the same solubility in pure octanol and water-saturated octanol, the last term in Equation 7-27 disappears. A plot of $\log S_w$ vs. $\log K_{ow}$ then will be linear with slope of -1 and y-intercept $= -\log V_o^*$. Regressions of $\log S_w$ vs. $\log K_{ow}$ thus should have slopes of unity if the above conditions are satisfied. As Figure 7-8a shows, a good correlation ($r^2 = 0.95$) exists over a wide range of the variables ($-\log S_w = 2$–13, $\log K_{ow} = 2$–8). Data were screened for accuracy, and solubilities of solids were corrected to the corresponding supercooled liquid solubility.[68] Nonetheless, the line of best fit still did not have a slope of unity.

Chromatographic k′. A widely-used technique to estimate K_{ow} involves its correlation with chromatographic retention indices. Chromatographic methods are attractive because of their speed and ability to separate large numbers of closely related compounds that are difficult to isolate (e.g., PCB congeners). A single chromatographic experiment can provide K_{ow} values over about 6 *log* units.[80] High-performance liquid chromatography (HPLC)[70,74,80,84-86] and, to a lesser extent, thin-layer chromatography[74,80,85] have been used for such studies, and reverse-phase (RP) procedures generally are employed. The mobile phase is a mixture of several polar solvents, commonly 70 to 95% methanol/water. The nonpolar stationary phase typically consists of long-chain (C_{18}) hydrocarbons bonded to an inert support. In RPHPLC, solutes elute from the column in order of lipophilicity — more lipophilic compounds have longer retention times. Compounds are detected by UV absorbance or mass spectrometry.[86] A calibration curve is constructed for a given set of experimental conditions by measuring k′ for a series of compounds with known K_{ow} values.

Example 7-2. Construction of a Calibration Curve to Predict $logK_{ow}$ from R_t

De Kock and Lord[70] obtained the data tabulated below for retention times of the compounds on a C_{18} reversed-phase HPLC column; $R_t - R_{t,o}$ is the compound elution time corrected for the solvent elution time. Compounds were selected to obtain a broad range of retention times and K_{ow}. $Logk'$ values were computed from Equation 7-18, and $logK_{ow}$ values were obtained from the literature or calculated from fragment constants. A plot of $logk'$ vs. average values from the literature (Figure 7-9) yielded the regression equation:

$$\log K_{ow} = 3.91(\log k') + 2.114; \qquad n = 8; \qquad r^2 = 0.90$$

As shown in Figure 7-9 and the following table, the relationship yielded reasonable estimates of $logK_{ow}$ for compounds not used to generate the regression.

Compound	$R_t - R_{t,o}$	$logk'$	$logK_{ow}$ Values[a] Calculated	Literature	Predicted
Calibration compounds					
Benzene	1.1	0.04	2.13	2.13	2.26
Toluene	1.7	0.23	2.79	2.74	3.00
Naphthalene	2.1	0.32	3.29	3.44	3.36
1,2,4 trichlorobenzene	3.5	0.54	4.07	4.19	4.22
p,p'DDD	4.5	0.65	5.80	5.90	5.00
p,p'DDT	8.0	0.90	6.19	6.07	5.63
Hexachlorobenzene	17.0	1.23	6.42	6.45	6.92
2,2',4,4',5,5'PCB	26.8	1.42	7.71	7.79	7.69
Other compounds					
Chlorobenzene		0.23	2.84	2.91	3.00
Biphenyl		0.39	4.00	3.93	3.63
Anthracene		0.69	4.45	4.49	4.80
Heptachlor		0.89	5.50	5.36	5.58
2,2',4,5,5'PCB		1.22	7.00	6.76	6.66

[a]   Calculated values based on fragment constants; literature values are means of 3 to 5 values from various reports; predicted values are from regression equation for $logk'$ vs. $logK_{ow}$ for these compounds.

Good correlations ($r^2 > 0.8$) can be obtained by HPLC and TLC methods; for selected groups of compounds, r^2 values as high as 0.99 have been reported, but accuracies in estimating K_{ow} still are only about ±0.5 log units.[85] For example, De Voogt and Govers[85] reported $r^2 = 0.984$ for a regression of $logk'$ against measured $logK_{ow}$ for ten polycyclic heteroaromatic compounds, but an 11th compound (carbazole) clearly did not fit the regression. A common and perhaps too facile explanation for nonfitting data is that the measured value is incorrect. In the case of K_{ow} values for lipophilic compounds this may be reasonable, given the difficulties in making the measurements. The reliability of HPLC and several other methods to predict $logK_{ow}$ has been evaluated for PAHs.[74,80] Molecular structure methods like the Rekker

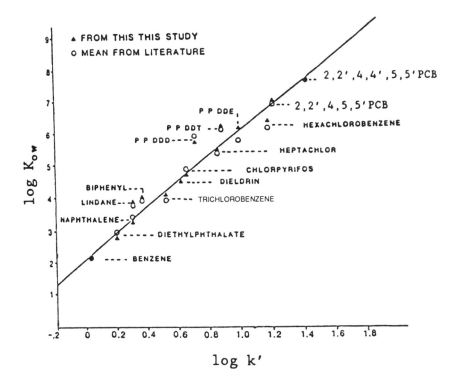

Figure 7-9. Calibration curve of $log K_{ow}$ vs. $log k'$ for compounds listed in Example 7-2. [From De Kock, A.C. and D.A. Lord, *Chemosphere*, 16, 133 (1987). With permission.]

fragment approach[77] and molecular connectivity indices (Section 7.11.3) were somewhat better predictors ($r^2 = 0.74$–0.998) than HPLC methods ($r^2 = 0.71$–0.984).

Finally, it is not a simple matter to say whether PARs based on parameters correlated with K_{ow} are surrogates for activity-K_{ow} relationships or in fact all these variables are surrogates for a more fundamental molecular property. Although K_{ow} has historical precedence as a measure of a compound's lipophilic character, it is not the only, or necessarily the best measure.[22] Given the difficulty of measuring K_{ow} for highly lipophilic compounds and the high correlation between $log K_{ow}$ and $log k'$, it may be appropriate to develop PARs directly from k' rather than from K_{ow} values estimated from k'. However, the importance of standardizing the measurement of k' needs to be emphasized so that PARs developed by different workers can be compared.

7.7.5 K_{ow} as a Predictor of K_p and K_{oc}

As the previous section showed, there are many intercorrelations among physical-chemical parameters related to solubility and chemical mobility. Not all are used to predict K_{ow}; in some, K_{ow} is the predictor variable, as in correlations with sediment-water and soil-water partition coefficients:[87-89]

$$\log K_p = a \log K_{ow} + \log f_{oc} + b \qquad (7\text{-}28)$$

a and b are fitted constants, and f_{oc} is the fractional organic content of the soil or sediment (or suspended particulates). K_p also has been correlated with S_w; an r^2 of 0.91 was found between $logK_p$ and $logS_w V$ for 14 aromatic hydrocarbons; V is the molar volume of the compound.[87] Given the high correlation between K_{ow} and S_w, such correlations are not surprising.

Equation 7-28 is based on the assumption that organic contaminants do not sorb onto mineral phases to a significant extent and their association with solid particles results from partitioning into organic components. Although this may not be exactly true, many studies have shown that sorption of lipophilic organic contaminants is very low on pure mineral sediments and increases with increasing sedimentary organic content. Dividing K_p by f_{oc} yields a partition coefficient K_{oc} that expresses the affinity of an organic solute for sediment organic carbon relative to that for water, and this leads to the equation

$$\log K_{oc} = a \log K_{ow} + b \qquad (7\text{-}29)$$

The slope in Equation 7-29 is a measure of the free-energy change of solute transfer from water to sediment organic matter compared with the free-energy change of solute transfer from water to octanol.[87] A slope of 1 implies that octanol and sediment organic matter have the same affinity for solutes, and this has been reported for various natural substrates and solutes.[88] However, other studies[87,89] have reported lower and/or higher values (0.72–1.15). Vowles and Mantoura[87] observed that it would be fortuitous for different classes of organic solutes to have the same affinities for octanol and natural organic matter. They found better correlations when organic solutes were placed into homologous classes, e.g.,

unsubstituted aromatics $\log K_{oc} = 1.20 \log K_{ow} - 1.13$ $r^2 = 0.998$

alkylbenzenes $\log K_{oc} = 0.90 \log K_{ow} - 0.46$ $r^2 = 0.996$

alkylnaphthalenes $\log K_{oc} = 0.77 \log K_{ow} + 0.37$ $r^2 = 0.992$

It is interesting to note that K_{oc} varies in a consistent way for homologous series of organic compounds such that substitutent or fragment constants can be generated. For example, the addition of another methylene group (CH_2) into alkylbenzene compounds produced a $\Delta logK_{oc}$ of 0.48 for estuarine sediments[87] and 0.39 ± 0.10 for soil organic matter.[89] Similarly, a $\Delta logK_d$ of 0.45 per methylene group was found for sorption of C_{10} to C_{14} linear alkylbenzene sulfonates by a river sediment.[90]

Considerable controversy exists about the constancy of the value of K_p (or K_{oc}) for suspended particles in aquatic systems. The so-called particle concentration phenomenon[91] results in a decrease in the apparent K_p for suspended sediment as sediment concentration increases, and this contradicts conventional equilibrium principles. Details of the controversy are beyond the scope of this book, but possible causes of the phenomenon have been critically assessed elsewhere.[92]

7.8 THEORETICAL APPROACHES TO PREDICTING SOLUBILITY: UNIFAC AND LSERs

Because of the difficulties in measuring S_w for nonpolar compounds, several empirical methods have been used to predict it from other variables. These techniques, which include correlations with K_{ow}, as described above, typically have problems of poor fit for some compounds, limited range of applicability, or difficulties in accurately determining the predictor variable (e.g., K_{ow}). The importance of S_w as a predictor variable for bioactivity of organic compounds has stimulated the development of more theoretical approaches, and considerable success has been achieved with two methods: UNIFAC and the more recent linear solvation energy relationships, LSERs. Both methods are based on an idea dating back at least to the 1930s that the energetics of solvation for nonelectrolytes could be accounted for by two terms: one relating the endergonic formation of a space in the solvent for the solute to occupy (a "solvent cavity"), and the other relating exergonic solute-solvent interactions.

UNIFAC is a fragment-based approach[93,94] in which activity coefficients (γ) are calculated as the sum of two components computed from the functional groups of a molecule:

$$\log \gamma = \log \gamma^C + \log \gamma^R \qquad (7\text{-}30)$$

γ^C, the combinatorial fraction, is based on functional-group surface areas and volumes. γ^R, the residual fraction, is based on interaction energies between functional groups and is calculated from vapor-liquid and liquid-liquid equilibria by correlation methods. Values of the group parameters needed to calculate γ^C and γ^R are available for 40 functional groups.[93,94] The original UNIFAC calculation method did not work well for long chain hydrocarbons, which tend to fold back on themselves to minimize contact with water. A revised calculation procedure[94,95] works better for such molecules, but does not perform as well as the original method for aromatic molecules.[96] The method has been used to calculate S_w, liquid-vapor equilibria, and liquid-liquid equilibria for many organic compounds (see references 68,96 for reviews and cautionary notes).

The LSER approach[97-107] explains the solubility of nonelectrolytes in terms of three processes: cavity-formation, nonspecific solvent-solute dipole interactions, and specific hydrogen-bonding interactions. The processes are modeled as linear combinations of the corresponding free-energy terms:

$$SP = SP_0 + \text{cavity term} + \text{dipolar term} + \text{hydrogen−bonding term} \quad (7\text{-}31)$$

SP is a solubility-related property. The cavity term is endergonic, since work must be done to separate the solvent molecules, but the dipolar and hydrogen-bonding terms are exergonic (to varying degrees). The solute characteristic that affects the cavity term is molar volume, V_2, and the solvent characteristic affecting this term is the square of the Hildebrand solubility parameter $(\delta_H^2)_1$;[108] (subscripts 1 and 2 refer to solvent and solute, respectively). δ_H is a measure of solvent-solvent interactions that are interupted by cavity formation.[97–99] In early LSER applications the liquid molar volume, V_1, (solute M_w divided by its liquid density) was used, but more recent studies have changed this to a computer-calculated intrinsic molar volume, V_1 (see Section 7.11.1).[107]

The exergonic effects of solute-solvent interactions are described in terms of "solvatochromic" parameters, so called because they are based on measurements of various spectroscopic shifts induced by solute-solvent interactions. For example, the solvatochromic dipolarity/polarizability parameter for solvents and solutes, π^*, is determined from bathochromic shifts in the frequency of maximum absorbance (ν_{max}) of nonprotonic solutes in a series of nonhydrogen-bonding acceptor (HBA) solvents.[98] For such systems the shift in ν_{max} is proportional to solvent polarity. "Bathochromic" refers to spectral shifts to lower frequencies (higher wavelengths). π^* describes the ability of a compound to stabilize a neighboring charge or dipole by nonspecific dielectric interactions and is proportional to the dipole moment of a molecule. Values of π^* tabulated in the literature[97,107] were obtained by averaging multiple (normalized) solvent effects on several spectroscopic properties.

The exergonic effects of hydrogen bonding (HB) for HB-acceptor (HBA) solutes in HB-donor (HBD) solvents like water are measured by two other solvatochromic parameters: α_1 and β_{m2}, the HBD acidity and HBA basicity, respectively.[104] Conversely, for HBD solutes in HBA solvents, the solvatochromic terms are β_1 and α_{m2}. Subscript m indicates that for compounds that can form hydrogen bonds with themselves, a value applicable to the nonself-associated solute is used.

A generalized equation encompassing these effects has the form[105–107]

$$SP = SP_0 + A\left(\delta_H^2\right)_1 V_{12} + B\pi_1^* \pi_2^* + C\alpha_1\beta_{m2} + D\beta_1\alpha_{m2} \qquad (7\text{-}32)$$

This can be simplified for specific applications. If one is interested in a single solute in a series of solvents, the factors related to the solute can be grouped into the coefficients of Equation 7-32, and the resulting equation relates specifically to solvent parameters:

$$SP = SP_0 + h\left(\delta_H^2\right)_1/100 + s\pi_1^* + a\alpha_1 + b\beta_1 \qquad (7\text{-}33)$$

If one is interested in a series of solutes in a single solvent or the distribution of a series of solutes between solvents, the solvent parameters become subsumed in the coefficients. The resulting equation relates to solute parameters; this is the typical situation of interest for aquatic contaminants:

$$SP = SP_0 + mV_{12}/100 + s\pi_2^* + a\alpha_{m2} + b\beta_{m2} \qquad (7\text{-}34)$$

The cavity term in Equations 7-33 and 7-34 is divided by 100 so that its scale is similar to those of the other terms. Equation 7-33 has been used in hundreds of correlations of solvent effects on numerous solute properties: uv/vis/ir/nmr/esr

Table 7-9. Predictive Equations Based on LSERs

	n	r^2	Ref.
HPLC retention coefficients:			
Polychlorinated benzenes and PCBs:			
1. $\log t_r' = -0.165 \pm 0.054 + 1.166 V_1/100 - 0.326\pi^* - 0.869\beta$	23	0.972	101
Mostly PAHs (65:35 acetonitrile:water)			
2. $\log k' = -0.49 + 1.207 V_1/100 - 0.110\pi^* - 0.764\beta$	32	0.989	101
(75:25 methanol:water)			
3. $\log k' = -0.76 + 1.596 V_1/100 - 0.135\pi^* - 0.783\beta$	32	0.989	101
Octanol-water partition coefficients:			
Aliphatic and aromatic compounds			
4. $\log K_{ow} = 0.45 + 5.15 V_1/100 - 1.29(\pi^* - 0.40\delta) - 3.60\beta$	103	0.983	178
Organic compound solubility in human blood:			
5. $\log S_b \approx \log\{S_g K_{b/g}\} = 1.35 - 3.05 V_2/100 - 0.22\pi^*_2 + 3.58\beta_{m2}$	27	0.974	100
(S_b = solubility in blood; S_g = solute concentration in its saturated vapor at 37°C; $K_{b/g}$ = Ostwald coefficient on blood; applies to restricted range of alkanes, alkylbenzenes, and chlorobenzenes)			
Binding to bovine serum albumin:			
6. $\log(1/C)(BSA) = 3.30 + 3.43 V_1/100 - 0.78\pi^* - 3.67\beta_m$			
$+ 0.37\alpha_m$	16	0.956	100
Nonreactive (narcotic) toxicity (golden orfe fish):			
7. $\log(1/LC_{50}) = -3.19 + 3.29 V_2/100 + 1.14\pi^*_2 - 4.60\beta_2 + 1.52\alpha_{m2}$	32	0.974	100
8. $\log(1/LC_{50}) = -3.73 + 3.16 V_2/100 + 1.65\pi^*_2 - 4.03\beta_2 + 2.09\alpha_{m2}$	28	0.962	100
(Equations 7 and 8 are based on separate data sets for toxicity (LC_{50}, mM) of similar sets of compounds to golden orfe)			
Inhibition of bioluminescence (Microtox test):			
9. $\log(EC_{50}) = 7.61 - 4.11 V_2/100 - 1.54\pi^*_2 + 3.94\beta_2 + 1.51\alpha_{m2}$	38	0.974	99
(EC_{50} in µM)			

absorption characteristics and rate and equilibrium constants.[102] Simple rules for estimating V_{12} and the solvatochromic parameters π^* and β for PCBs, PAHs, and polychlorinated benzenes by group additivity methods have been published[107] and make Equation 7-34 readily accessible for calculating solubility-related properties for these compounds. Equation 7-34 has yielded impressive results for several solubility-related properties (Table 7-9). Toxicity PARs based on the LSER model are described in Section 7.10.3.

7.9 A SIMPLE PROPERTY-ACTIVITY MODEL FOR BIOLOGICAL SYSTEMS

The concepts discussed in preceding sections can be used to develop simple models relating bioactivity and physico-chemical properties of organic compounds. Hansch and Fujita[109] were the first to develop such a model; the following discussion is based on their model, which sometimes is called the linear free-energy relationship model. It describes the bioactivity of compounds as an additive combination (*log-log* basis) of lipophilic and electronic effects; in theory, other factors like steric effects could be added to it.

The rate-limiting processes for biological response to organic compounds can be described in a general way as

$$
\begin{array}{l}
\text{compound} \\
\text{in extra-} \\
\text{cellular} \\
\text{phase}
\end{array}
\longrightarrow
\begin{array}{l}
\text{site of} \\
\text{action in} \\
\text{cellular} \\
\text{phase}
\end{array}
\quad
\begin{array}{c}
k_x \\
\rightarrow\rightarrow \\
\text{critical} \\
\text{reaction}
\end{array}
\quad
\begin{array}{l}
\\
\text{biological} \\
\text{response}
\end{array}
\qquad (7\text{-}35)
$$

The first step, transport by a random walk process from the external solution to the critical active site in the cell, is relatively slow, and the rate depends on molecular structure, especially on the lipophilic/hydrophilic character of the compound. For simplicity, we assume one rate-controlling reaction at the active site. The rate of biological response can be written:

$$
\text{Rate of bioresponse} = \text{d}(BC)/dt = Ak_x\left[C_x\right] \qquad (7\text{-}36)
$$

BC = biotic condition; A is the probability that a bioactive molecule will reach the site of action in a given time interval, and it depends in some way on compound lipophilicity (e.g., $log K_{ow}$). $[C_X]$ is the concentration of the compound in the external medium, and k_X is the rate constant for the critical step. The product $A[C_X]$ is the effective concentration at the site of action. Parameters A and k_X thus determine the compound's biological activity.

A linear relationship between $log K_{ow}$ and biological activity is not to be expected over an indefinite range of K_{ow}. At low values, activity should be low because transport across lipid membranes is hindered. A compound may have a greater bioactivity than otherwise expected, however, if it interacts with sites on cell walls, thus eliminating the need for transport across lipid layers. At very high K_{ow} values, activity may level off, yielding a "plateau model," or decrease, yielding an "optimum relationhip." Such trends can be explained by both equilibrium and kinetic models.[110] The former assume that low S_w at high K_{ow} limits the compound's cytoplasmic concentration, which is assumed to be in equilibrium with enzyme active sites. Kinetic models explain the falloff in activity in terms of a decreased probability of finding the site of action by random walk processes. At sufficiently high K_{ow} the rate is so slow that the intracellular (active site) concentration is insufficient to induce the biological response being studied during a test interval.

Hansch and Fujita[109] proposed that bioactivity follows a normal distribution with respect to $log K_{ow}$ or π (the substituent lipophilicity parameter) when other factors remain constant. They assumed that the maximum mobility of a molecule in a cell occurs when $K_{ow} = 1$. The probability of the compound finding a site of action by a random walk process then is maximized. For a series of compounds covering a range of π values, there will be an ideal value π_o, at which $K_{ow} = 1$. Any increase or decrease in π away from π_o slows the rate of movement of compounds toward the site of action, thus decreasing A. π_o is an idealized average value because cell structure is complex and transport to active sites entails many partitioning and sorption-desorption steps. Based on the argument that π exhibits an optimum, Hansch and Fujita proposed that A is related to π by a normal distribution:

$$A = a \exp\left(-\left[\pi - \pi_o\right]^2 / b\right) \tag{7-37}$$

a and b are fitted constants, and π_o is the idealized value (for a substituent with K_{ow} = 1). Other functions exhibiting a maximum or plateau also could be used to relate A and π. Several studies discussed later have reported parabolic relationships between $log K_{ow}$ and bioactivity.

The rate constant k_X for a compound with substituent X is assumed to be related to that of the parent compound by a Hammett-like relationship:

$$\log k_x = \rho\sigma + \log k_H \tag{7-38}$$

σ is a Hammett, Taft, or electrophilic sigma constant, as appropriate, ρ is a constant that expresses the sensitivity of the critical reaction to electronic effects, and k_H is the rate constant of the parent compound for the critical step. If the critical reaction involves hydrolysis, Equation 7-38 could be replaced by a Brønsted relationship. Substituting Equations 7-37 and 7-38 for A and k_X into Equation 7-36, we obtain

$$d(BC)/dt = ak_H[C]\exp\left(\rho\sigma - \left[\pi - \pi_o\right]^2 / b\right) \tag{7-39}$$

Bioassay results usually are reported in terms of the concentration of a compound required to give a certain response in a given measurement time (e.g., 96 h LC_{50} or % growth). Therefore, the term d(BC)/dt can be replaced by a constant. Dividing both sides of Equation 7-39 by [C], taking logs, and simplifying by collecting constants, we obtain

$$\log(1/C) = -c'\left(\pi^2 + \pi_o^2\right) + d'\pi\pi_o + e'\pi + \rho\sigma + f' + \log(k_H) \tag{7-40}$$

where c'–f' are coefficients. For a given parent molecule in a particular biological system, both k_H and π_o are constants, and Equation 7-40 simplifies to

$$\log(1/C) = -c\pi^2 + d\pi + \rho\sigma + e \tag{7-41}$$

Equations 7-40 and 7-41 are a general model that relates biological response to the physical-chemical properties π and σ. Under certain circumstances they can be simplified further.[109] We can consider four cases. When π_o is large compared with π, and σ is small or zero, Equation 7-41 reduces to

$$\textbf{Type I} \qquad \log(1/C) = a\pi + b \qquad\qquad (7\text{-}42)$$

This behavior is called nonspecific toxicity when the biological response is negative. Bactericidal action of phenols behaves this way.[109] This type of activity is limited in any series of compounds. As the lipophilic character of derivatives increases, π approaches or exceeds π_o, and if σ is still unimportant, Equation 7-41 takes a curvilinear form:

$$\textbf{Type II} \quad \log(1/C) = -a\pi^2 + b\pi + c \qquad\qquad (7\text{-}43)$$

If π is unimportant (transport across membranes is not rate limiting), Equation 7-41 reduces to a Hammett expression (**Type III**). Combination of **Types I** and **III** leads to **Type IV** behavior: bioactivity related linearly to both $\log K_{ow}$ and σ. Examples of these behavioral types are given in subsequent sections.

7.10 PROPERTY-ACTIVITY RELATIONSHIPS FOR AQUATIC ORGANISMS

The physical-chemical properties discussed in Sections 7.7 and 7.8 have been used to develop many PARs for aquatic organisms. This section describes PARs to predict three important types of bioactivity: bioaccumulation, biodegradation, and toxicity. Early work on these topics relied on empirical correlations with K_{ow} or S_w, but more complicated and mechanistic models using several predictor variables are becoming common (see reviews listed in reference 111).

7.10.1 Bioaccumulation of Organic Compounds

Bioaccumulation of organic compounds by aquatic organisms can be described by a simple first-order two-compartment model[112-115] (see Example 5-5), which yields the following equation:

$$C_b = \frac{k_1}{k_2} C_w \left(1 - \exp\left[-k_2 t\right]\right) \qquad\qquad (7\text{-}44)$$

k_1 and k_2 are uptake and clearance rate constants, respectively, and C_b and C_w are the concentrations of the compound in the organisms (biota) and water, respectively. Equation 7-44 applies if C_w is held constant and if the compound is not biodegradable. The bioaccumulation factor (BF) for a compound is usually defined as the ratio of its concentration in a test organism ($\mu g\ g^{-1}$ [wet wt]) to its concentration in water ($\mu g\ L^{-1}$): C_b/C_w. For a given compound and organism, BF depends on the time of exposure, and for meaningful comparisons, steady-state (quasiequilibrium) conditions are necessary. For these circumstances, BF is symbolized BF_∞. The bioconcentration factor is defined as the ratio of rate constants for

Table 7-10. Bioaccumulation PARs for Organic Compounds and Aquatic Organisms

Organism	Predictive equation[a]	r^2	n	Ref.
1. Mussels	$logBF = 0.86 logK_{ow} - 0.81$	0.91	16	114
2. "	$logBF = -0.68 logS_w + 4.94$	0.89	16	114
3. Fish	$logBF = 0.85 logK_{ow} - 0.70$	0.90	59	127
4. *Chlorella*	$logBF = 0.68 logK_{ow} + 0.16$	0.81	41	122
	Linear fit of most of same fish data used for Equation 3:			
5.	$logBF = logK_{ow} - 1.32$	0.95	51	129
or	$BF = 0.048 K_{ow}$			
	Parabolic BF-$logK_{ow}$ relationship:			
6. Chlorobenzenes	(BF expressed on lipid wt. basis)			
	$logBF = 3.41 logK_{ow} - 0.264(logK_{ow})^2 - 5.51$	—	6	109
7. PCBs	$RBF = 0.25(logK_{ow} \times SEC) - 0.75$	0.74	—	131
or	$RBF = 0.0014(t_r) logK_{ow} + 0.13$	0.94	45	131
	(RBF = relative bioaccumulation factor; SEC = empirical			
	steric effect coefficient; t_r = HPLC retention time)			
	Uptake and clearance rates in fish			
8. Uptake[b]	$logk_1 = 0.337 logK_{ow} - 0.373$	N.A.	—	128
9. Clearance	$-logk_2 = 0.663 logK_{ow} - 0.947$	0.95	—	128

[a] BF expressed on wet weight basis; when BF is expressed on dry weight basis, the constant in Equation 4 becomes 0.86.
[b] Derived from Equations 5 and 9, assuming $BF = k_1/k_2$.

uptake (k_1) and clearance (k_2): $K_{bf} = k_1/k_2$.[116] If equilibrium conditions apply, $BF = K_{bf} = C_b/C_w = k_1/k_2$. The literature is not consistent on the use of these terms, and the terms bioaccumulation factor and bioconcentration factor often are used interchangeably for equilibrium conditions.

Many studies have shown linear relationships between $logS_w$ and $logBF$ of organic compounds in aquatic organisms,[117] including bacteria,[118] algae,[119] mussels,[120] and fish,[121-123] as well as in sediments[124] and soils.[121,125] Linear relationships also have been reported between $logK_{ow}$ and $logBF$ in microorganisms,[126] algae,[127-129] zooplankton,[127,130] fish,[116,121,122,127,131-133] and mussels[120] (Table 7-10). Figure 7-10 shows linear relationships for bioaccumulation of organic compounds by the mussel *Mytilus edulis* vs. $logS_w$ and $logK_{ow}$ for the compounds. Most of the values in the figure are from lab studies, but some are based on environmental measurements where assumption of steady-state is more tenuous. Nonetheless, the correlations are reasonable ($r^2 = \sim 0.9$) over a wide range of K_{ow} and S_w. The equation of best fit between BF and K_{ow} for mussels is nearly the same as that for six fish species, but the slope and intercept of a $logBF$-$logK_{ow}$ relationship for *Chlorella* are considerably different (cf. Equations 1,3,4, Table 7-10). These results suggest that mussels may be good surrogates to estimate organic accumulation by fish, but algae probably are not. The slopes of correlations between $logBF$ and $logK_{ow}$ usually are close to one, and there is little theoretical basis for fractional values. Mackay[134] suggested that the true slopes are unity, which implies that BF and K_{ow} are linearly

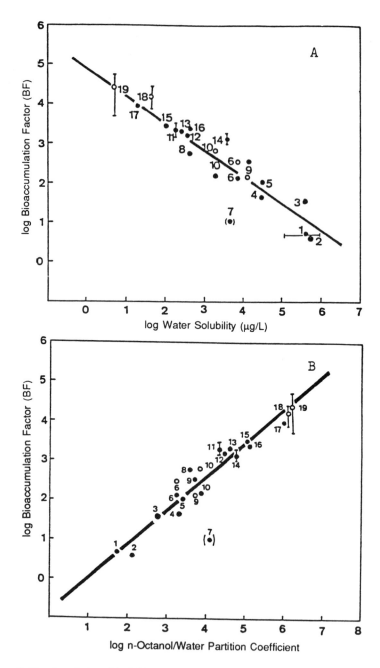

Figure 7-10. Correlation of *log*BF for mussels (*Mytilus edulis*) with (A) *log*S$_w$ and (B) *log*K$_{ow}$ for various organic compounds. [From Geyer, H.P., et al., *Chemosphere*, 11, 1121 (1982). With permission.]

related. He found that such a one-constant relationship (Equation 5, Table 7-10) provided a good fit for a data set that includes many of the values used to generate Equation 3 in that table.

Several studies[131,135] have shown that the uptake rate constant k_1 for fish is positively correlated with K_{ow}, and the clearance rate constant k_2 is negatively correlated (over the $log K_{ow}$ range 2.6 to 6.0) (Equations 8,9, Table 7-10). Uptake rates by fish thus increase with compound lipophilicity (the compounds are stored in fish lipids), and purging of compounds from fish lipids occurs more slowly for more lipophilic compounds. The net effect is to increase the time (t_{eq}) required to reach the equilibrium BF for more lipophilic compounds. BF_∞ is approached asymptotically; for practical purposes, the time required for C_b to reach 99% of its equilibrium value, t_{99}, is used to represent t_{eq} ($t_{99} = 4.61/k_2$) (Section 2.2.1). For $log K_{ow} = 6$, t_{eq} requires about 6 months. The equilibrium time rapidly increases beyond convenient experimental range for higher values of K_{ow}; t_{eq} was estimated to be 12 years for $log K_{ow} = 8$.[135] By manipulating the uptake-clearance equations, an empirical equation has been developed that relates the BF measured for fish under nonequilibrium conditions (exposure times $< t_{eq}$) to $log K_{ow}$ and the time of exposure.[131]

Log-linear relationships are not always obtained between BF and K_{ow}. BF values and uptake rates of chlorobenzenes by guppies[115] are described by a Type II parabolic relationship (Equation 6), with maximum uptake rate at $log K_{ow} = 5.4$ and maximum BF at $log K_{ow} = 6.5$. Other data[132] also show a leveling off or decrease in BF at $log K_{ow} > {\sim}5$ to 6. As discussed in Section 7.9, there are several reasons why nonlinear relationships can be expected between bioaccumulation and $log K_{ow}$. Figure 7-11 shows a hypothetical relationship between the transport rate of compounds through biological membranes and $log K_{ow}$. Transport rates are limited by lipid solubility at low K_{ow} and increase with increasing K_{ow}. A plateau occurs at intermediate-high K_{ow} values, as transport, rather than lipid solubility, becomes the controlling factor. Transport rates decline at very high K_{ow}, as the membrane gets clogged by the lipophilic compound or because of low availability of compound in the aqueous phase. Process models have been developed to explain the plateau and decrease in BF at high K_{ow}.[113,136] However, values of BF and K_{ow} for highly lipophilic compounds are uncertain, and existing data are not adequate to evaluate the accuracy of these models.

Other molecular characteristics besides K_{ow} (or related measures of lipophilicity) affect BF values for some classes of compounds. Bioaccumulation of PCB congeners by mullet (*Mugil cephalus*) and a polychaete (*Capitella capitata*) depends on congener shape, as well as lipophilicity. BF values are not linearly related to K_{ow} values of the congeners,[137] but display a bell-shaped relationship with much scatter (Figure 7-12a). Linear correlations were obtained when $log K_{ow}$ was multiplied by an empirical steric coefficient or by the HPLC elution time of the congener on an activated carbon stationary phase (Figure 7-12b). Planar isomers of PCBs have the longest HPLC retention times (are most strongly adsorbed) and the highest BF values. PCB bioaccumulation thus is affected by two main factors: lipophilicity and molecular shape. The importance of the three-dimensional structure of PCBs on bioactivity also was noted in a study on induction of mixed-function oxidase (MFO)

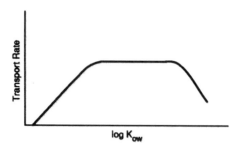

Figure 7-11. Conceptual model for transport of organic compounds into cells vs. K_{ow} of the compound.

by 50 different PCB congeners.[138] MFO is an important mammalian enzyme for detoxifying organic pollutants. The relative importance of Cl substitution in the ten possible positions of the biphenyl molecule on MFO induction was evaluated statistically by principal-component analysis. The first principal component, which explains the greatest proportion of the variance in MFO induction, weighted the 2, 2', 6, and 6' positions most heavily, i.e., positions ortho to the bond between the benzene rings. Presence or absence of Cl in these positions determines whether a congener is flat or not (Figure 7-12c).

Some poor correlations between BF and K_{ow} can be explained by biological factors. Poor relationships are found among compounds that are metabolized to varying degrees.[139,140] Compounds with high molecular weight and high K_{ow} have long half-lives in fish, and as mentioned above, equilibrium may not be reached within the duration of most lab tests. BF values obtained from lab tests on these compounds may be lower than those estimated from field observations. The major ambient source of such compounds in fish probably is contaminated food rather than direct uptake from the water.

7.10.2 Biodegradation Rates

The rate-controlling step in biodegradation may be either transport to the active site or binding/reaction with a key enzyme. Biodegradation rate constants thus may be related to lipophilic, electronic, and/or steric factors. Trends in biodegradation

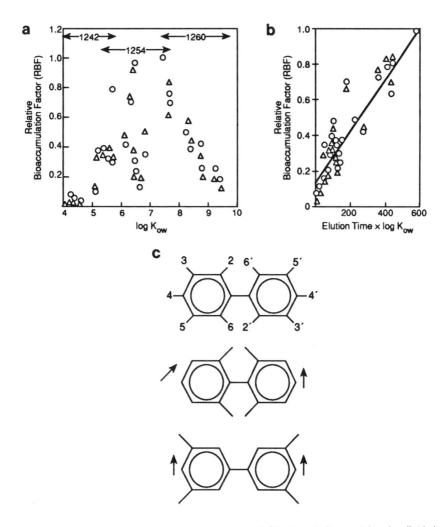

Figure 7-12. (a) Relative bioaccumulation factors of PCBs for polychaetes (o) and mullet (△) vs. $logK_{ow}$ for PCB congeners. Range of components in PCB mixtures, Arochlor 1242, 1254, and 1260, shown by arrows. (b) Relationship between relative bioaccumulation factor and product of chromatographic elution time times $logK_{ow}$ for same compounds shown in (a). [From Shaw, G.R. and D.W. Connell, *Environ. Sci. Technol.*, 18, 18 (1984). With permission of the Amer. Chem. Soc.] (c) Structure of biphenyl showing numbering system for carbon atoms. Presence of Cl substitution in 2, 2', 6, or 6' positions causes rings to rotate 90°; Cl substitution in 3 or 4 positions yields planar structures.

PARs and deviations from such PARs within a series of compounds can be useful in evaluating mechanisms and rate-limiting steps of degradation. For example, first-order biodegradation rate constants for alkyl esters of *p*-aminobenzoic acid (PAB) increase with alkyl chain length up to C_5 and then remain constant.[141] This parallels the trend in permeability of lipid membranes and aqueous diffusion layers for these compounds. Membrane permeability increases with K_{ow} up to some moderate value, but at higher values the rate-limiting step is K_{ow}-independent diffusion through an aqueous boundary layer around the cells. A model for boundary-layer-membrane transport[141] shows these trends, and the results suggest that the rate-determining step for biodegradation of these compounds is transported into the cells.

Biodegradation rate constants (k_b) have been correlated with $logK_{ow}$ for alcohols,[142,143] ketones,[142,144] phenols,[145] phthalate esters,[146] and 2,4-dichlorophenoxyacetic acids,[147] but not all the relationships are linear. For example, k_b values obtained from BOD experiments with 17 primary alcohols yielded a biphasic relationship,[142] with a large change in slope at $logK_{ow} \approx 3$ (Figure 7-13a). Degradation rates of C_1 to C_7 alcohols were affected only slightly by K_{ow}, but rates for C_8 to C_{12} alcohols decreased as K_{ow} increased (Equation 1, Table 7-11). Branched alcohols generally are more refractory than straight-chain alcohols,[142] and the differences in degradability are not explained by differences in K_{ow}, implying that enzymatic attack, rather than transport into cells, is the rate-limiting step.

The results for the alcohols shown in Figure 7-13a differ from an earlier study[143] that found a parabolic relationship among C_4-C_{14} straight-chain primary alcohols in activated sludge, with an increase in biodegradability from C_4 to C_8 and a decrease thereafter. C_{11} to C_{14} alcohols had much higher rate constants than predicted by the model. However, the earlier results are questionable because alcohol concentrations exceeded S_w values of the higher weight compounds.[142] Bacteria growing at the water/alcohol interface of the resulting micelles may have had greater access to the compounds than would be the case for a homogeneous solution. The increased degradation rates from C_4 to C_8 alcohols probably are not related to inhibitory effects (narcosis) of the shorter chain alcohols at high aqueous concentrations, because the toxicity of n-alcohols (at least in small fish and zooplankton) *increases* with carbon number up to C_{12}, and $log(1/LC_{50})$ is linearly correlated with $logK_{ow}$ (see Section 7.10.3).

Large differences also have been reported in biodegradability-K_{ow} trends between batch-bottle (BOD) experiments[142] and activated sludge systems[144] for aliphatic ketones (Figure 7-13b). The BOD study yielded a parabolic or bilinear relationship with $logK_{ow}$ over the range -0.24 (acetone) to 2.65 (2,6-dimethyl-4-heptanone), but activated sludge produced a linear correlation (Equations 2,3, Table 7-11). These differences have discouraging implications for the development of simple (batch-bottle) treatability assays.

Microbial degradation of phenols increases with decreasing lipophilicity and levels off at $K_{ow} \approx 2$ (Figure 7-14a). This was explained by a transport-degradation model[145] in which transport occurs by two parallel pathways:

$$S + B \underset{k_2}{\overset{k_1}{\rightleftharpoons}} S-B \overset{k_3(hydrophobic)}{\underset{k_4(hydrophilic)}{}}$$ (7-45)

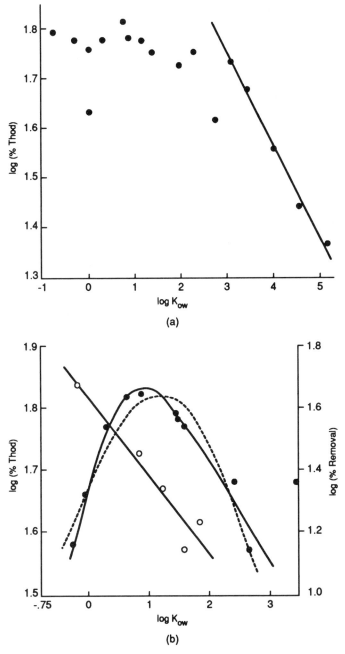

Figure 7-13. (a) Relationship between biodegradability (% ThOD) and *log* K_{ow} for 17 alcohols; % ThOD = 5-day BOD as % of theoretical BOD. (b) Similar relationship for aliphatic ketones. Solid straight line and open circles: data for activated-sludge systems. [From Ludzack, F.J. and M.B. Ettinger, *J. Water Pollut. Contr. Fed.*, 32, 1173 (1960). With permission.] Solid circles: BOD data; solid curved line: best fit of solid circles to bilinear equation; and dashed line: best fit of same data to parabolic equation. [Both (a) and (b) from Vaishnav, D.D., R.S. Boethling, and L. Babeu, *Chemosphere*, 16, 695 (1987). With permission.]

Table 7-11. Biodegradation PARs for Aquatic Organisms

Compounds	Predictive equation	r^2	n	Ref.
1. C_8–C_{12} linear alcohols (BOD assay)[a]				
	$log(\% \ ThOD) = -0.912 log K_{ow} + 2.338$	0.99	5	136
2. Acyclic ketones (BOD assay)				
	$log(\% \ ThOD) = 1.682 + 0.241 log K_{ow} - 0.106(log K_{ow})^2$	0.99	10	136
3. Acyclic ketones (activated sludge)			136,138	
	$log(\% \ removal) = 0.241 log K_{ow} + 1.617$	0.93	5	
4. Phthalates	$log k_b = -2.09 log t_r^2 + 1.19 log t_r - 1.15$	0.99	5	140
5. Phthalates	$log k_b = 2.1 log k_{OH} - 6$	0.93	4	142
6. Phthalates	$log k_b = 9.6\sigma^* + 3.2 E_s - 8.1$			
7. Anilines	$log k_b = -0.78\sigma_p + 1.04$	0.89	5	143
	k_b in mg/g [dry wt]·h			

[a]   Expressed as percent of theoretical BOD exerted in 5 days.

S adsorbs to, or desorbs reversibly from, cell surface B and either is transported through the lipid layer at a rate controlled by k_3 or diffuses through hydrophilic pores at a rate defined by k_4. If it is assumed that adsorption-desorption equilibrium is established rapidly, an equation similar to the Michaelis-Menten equation can be derived:

$$-d[S]/dt = \frac{k_1(k_3 + k_4)[B]_o[S]}{k_2 + k_3 + k_4 + k_1[S]}$$
(7-46)

$[B]_o$ is the biomass available for sorption. At high $[S]$, $k_1[S] > k_2 + k_3 + k_4$, and Equation 7-46 simplifies to

$$-d[S]/dt = [B]_o[S](k_3 + k_4)$$
(7-47)

For hydrophobic compounds, k_4 is negligible, and k_3 will be inversely proportional to K_{ow}. For hydrophilic compounds, k_3 will be negligibly small, k_4 will be dominant, and the rate will be independent of K_{ow}. These trends lead to the curve shown in Figure 7-14a. However, the data also fit a parabolic model, and the data are not sufficient to validate a mechanistic transport model. Development of simple expressions for low $[S]$ is complicated by uncertainties in the magnitudes of k_2, k_3, and k_4.

Values of k_b for degradation of di-n-alkyl phthalates by activated sludge fit a parabolic relationship with RP-HPLC retention times (R_t) of the compounds (Equation 4, Table 7-11). This represents Type II behavior of the bioactivity model.

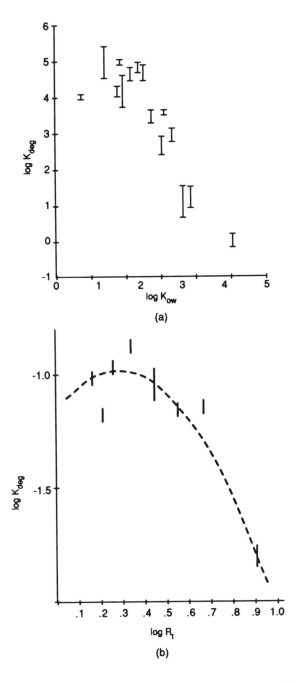

Figure 7-14. (a) Relationship between biodegradation rate constants and *log*K_ow for various phenols [from Banerjee, S., et al., *Environ. Sci. Technol.*, 18, 416 (1984). With permission]. (b) Relationship between biodegradation rate constants and *log* chromatographic retention times on reversed-phase hplc columns for di-*n*-alkyl phthalates. [From Urushigawa, Y. and Y. Yonezawa, *Chemosphere*, 5, 317 (1979). With permission.]

However, the data are rather scattered (Figure 7-14b), and the nature of the relationship is uncertain. $LogR_t$ is closely correlated ($r^2 = 0.999$) with $logK_{ow}$ for the compounds; membrane transport thus appears to be an important factor in the biodegradation of these compounds. However, correlation of k_b with K_{ow} does not necessarily mean that lipophilicity is the sole controlling factor; K_{ow} is correlated with other physical-chemical properties, including solubility and molecular surface area. Values of k_b for phthalate esters also are correlated with based-catalyzed hydrolysis rate constants (Equation 5, Table 7-11), which, in turn, fit a σ^* LFER (Equation 7, Table 7-5). Together these correlations yield a PAR that represents Type III behavior in the bioactivity model (Equation 6, Table 7-11). The correlation in Equation 5 (Table 7-11) was improved by including $logK_{ow}$ as a second independent variable, but because the sample size was small, this variable was not statistically significant. Nonetheless, a two-variable model (similar to the Type IV bioactivity model) is theoretically sound and reflects the two potentially rate-limiting steps: mass transport and chemical reaction.

Biodegradability is correlated with electronic factors, rather than lipophilicity, in some classes of compounds.[148-151] A high correlation ($r^2 = 0.97$) was found between k_b and alkaline hydrolysis rate constants of six organic compounds with diverse structures, even though no single LFER fit the compounds (an organophosphate pesticide, two chlorinated insecticides, a 2,4-D ester, and two methyl esters of aromatic acids).[148] Microbial degradation of phenols and anilines was described by a Hammett relationship (Equation 7, Table 7-11); the negative sign on σ_p in the equation indicates that electron-donor groups increase k_b; i.e., the reaction is electrophilic. Oxidation of phenols to catechols has been related to the size (van der Waals radius) of the substituent on the ring,[150] and half-saturation constants for some hydrolytic enzymes have been correlated with K_{ow}, σ, and R_M values of substrates.[151]

Fit of decomposition data to PARs like those in Table 7-11 does not necessarily mean that a process is biologically mediated. For example, degradation of the herbicide propyzamide (3,5-dichloro-N-(1,1-dimethylpropynyl)benzamide) and nine analogues in soil fit a two parameter model: $logk = -2.74 - 1.22\sigma + 0.58logK_{ow}$,[152] but other evidence (e.g., effect of temperature and sterilization) supported chemical hydrolysis, rather than microbial degradation, as the primary loss mechanism. Again, the negative sign on σ implies an electrophilic reaction. The correlation with K_{ow} suggests that chemical degradation of the compounds in soils occurs in the adsorbed phase, since K_{ow} is correlated with the strength of adsorption to soil particles (e.g., with K_p).

7.10.3 Toxicity Relationships

Narcosis

This toxic effect is defined as a nonspecific, reversible, physiological effect that is independent of chemical structure. Many classes of compounds behave this way, including aliphatic, aromatic, and chlorinated hydrocarbons, alcohols, ketones, aldehydes, and ethers, all of which are fairly unreactive molecules. The mechanism of narcotic action is thought to involve saturation of lipophilic membranes, leading to hindered electrolyte transport. Thermodynamic activities of various narcotics in blood are roughly the same for a given level of physiological effects.[153] Mammalian studies with volatile chemicals[154] have found that the activity needed to produce narcosis (a_{nar}) is about 0.02:

$$a_{nar} \approx 0.02 \approx P_{nar}/P^o \tag{7-48a}$$

P_{nar} is the partial pressure of the compound at the narcotic concentration; P^o is the partial pressure of the pure compound.

For nonvolatile chemicals in aquatic studies, Veith et al.[153] suggested that a_{nar} can be estimated by an analogous expression:

$$a_{nar} \approx C_{nar}/S_w \tag{7-48b}$$

C_{nar} is the aqueous concentration of chemical producing the narcotic effect (related to LC_{50}), and S_w is the water solubility of the compound. This relationship holds if the aqueous concentration of the chemical is related to its concentration in the fluid of the test organism (e.g., fish blood). If a_{nar} indeed is approximately constant, C_{nar} will be proportional to S_w. Because S_w and K_{ow} are inversely related, we would expect an inverse correlation between C_{nar} (or LC_{50}) and K_{ow} for compounds behaving as narcotics.

Toxicity PARs of some narcotic compounds for aquatic organisms are listed in Table 7-12. The acute toxicity (96 h LC_{50}) of C_1 to C_{13} aliphatic alcohols to a copepod and a small cyprinid fish (Equations 1 and 2) has been correlated with the number of carbon atoms (N) in the alcohol.[155] This simple structure-activity relationship works because structural parameter N is correlated with the physical-chemical property K_{ow}. For C_1 to C_6 alcohols, $logK_{ow} = 0.57N - 1.41$ ($r^2 = 0.999$). The ratio of K_{ow} for consecutive members of a homologous series can be written as $x = K_{ow(n+1)}/K_{ow(n)}$. Thus, $logx = logK_{ow}(n+1) - logK_{ow}(n)$ defines a substituent parameter (π) for lipophilicity of methylene groups in n-alcohols as the slope of above regression, i.e., 0.57. This is somewhat higher than the range of $\pi_{CH2} = 0.39$ to 0.48 calculated from K_{oc} values for homologous alkylbenzenes and K_d values for homologous alkylbenzene sulfonates.[87,89,90]

Many direct relationships have been reported between LC_{50} and K_{ow} for compounds that behave as narcotic toxicants.[153,156,157] For example, a good linear correlation was reported[157] between $logLC_{50}$ values for fathead minnows and $logK_{ow}$ for 50 organic compounds (including chloroalkanes, chlorobenzenes, alcohols, and ethers) over the range of –1 to 6 for $logK_{ow}$ (Equation 3, Table 7-12). On the other hand, a log-log plot of 96-h LC_{50} for the fathead minnow vs. K_{ow} for 10 alcohols[153] was linear only up to $logK_{ow} \approx 4$ (1-decanol) and then flattened out (Figure 7-15). a_{nar} was approximately constant (~0.025) for the lower homologs, in agreement with narcotic actions on mammals,[154] but increased above n-nonanol (e.g., 0.055 for n-decanol) until LC_{50} exceeded the water solubility (tridecanol) at which $a_{nar} > 1$.

Table 7-12. Toxicity PARs for Organic Compounds and Aquatic Organisms

Organism	Predictive equation	r^2	n	Ref.
Narcosis (unreactive compounds):				
1. *Nitocra spinipes* (copepod)	$logLC_{50} = -0.38N + 4.64$	0.97	3	149
2. Bleak (marine fish)	$logLC_{50} = -0.40N + 4.75$	0.97	13	149
3. Fathead minnows	$logLC_{50} = -0.87logK_{ow} + 4.87$	0.97	50	147
4. Fathead minnows; bilinear model:	$logLC_{50} = -0.94logK_{ow} + 0.94log(1 + 0.000068logK_{ow}) - 1.25$	—	100	147
Inhibition of primary production (Chlorobenzenes):				
5. *Ankistrodesmus* (green alga)	$-logEC_{50} = 0.985logK_{ow} - 2.626$	0.97	12	152
6. *Ankistrodesmus* (green alga)	$logEC_{50} = 0.587logS_w - 2.419$	—	12	152
Reactive compounds:				
7. Guppies (organic halides; k_{NBP} = rate constant for reaction with nucleophilic reagent 4-nitrobenzylpyridine)	$logLC_{50} = 1.30log(1604 + 1/k_{NBP}) - 4.35$	0.88	15	150
8. Shrimp (mostly phenols and anilines; LT = lethal threshold concentration extrapolated from 96-h bioassay)	$-logLT = 1.03logK_{ow} + 2.48$	0.61	45	155
9. Shrimp (phenols; $\Delta pH = pK_a(phenol) - pK_a(compound)$)	$-logLT = 0.48logK_{ow} + 0.54(\Delta pH) + 1.43$	0.92	23	155

Possible reasons for this include lower chemical activity of larger molecules and equilibration problems — long times are needed to achieve equilibrium between water and lipid phases for highly lipophilic compounds. Data for 50 other compounds (ketones, ethers, alkyl halides, substituted benzenes) fit the same relationship and are described by a bilinear equation (Equation 4, Table 7-12).

The slopes of the toxicity-K_{ow} relationships in Equations 3 and 4 of Table 7-12 are close to unity, as predicted by the narcosis model, in contrast to a slope of ~0.7 implicit in Equations 1 and 2 and the $logK_{ow}$ vs. N regression given above. The nonlinear nature of the relationship for compounds with high K_{ow} reflects the nonlinearity between $logS_w$ and $logK_{ow}$.[153] Although the results yield a good visual fit to the model, it must be emphasized that *log-log* plots spanning many orders of magnitude are deceptive in terms of the precision of predictions. Observed/predicted ratios ranged from about 0.12 to 4.22 for the compounds in Figure 7-15. In addition, caution must be used in applying such toxicity relationships to similar compounds not included in the original database, because minor changes in chemical structure can greatly change the chemical reactivity and biodegradability of compounds.[156]

Impressive predictive equations have been obtained with LSER formulations for toxicities of compounds that act by narcosis. For example, Equations 7 and 8 of Table 7-9 give LC_{50} predictions for a test species of fish and have r^2 values greater than 0.98. The equations are useful not just because they provide closer fits than single variable $logK_{ow}$ relationships, but also because they help explain why some

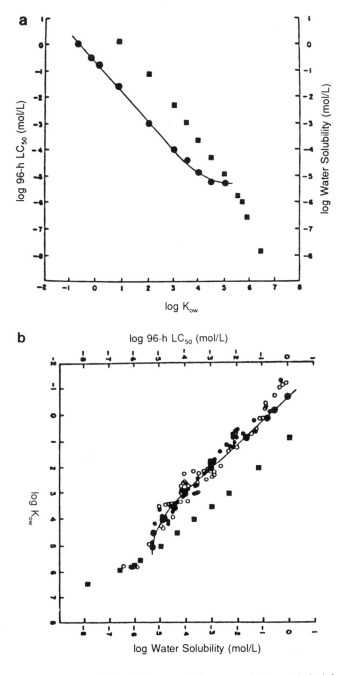

Figure 7-15. (a) Bilinear relationship between LC_{50} concentrations (circles) for fathead minnows and $log\,K_{ow}$ for C_1–C_{12} alcohols and between water solubility ($log\,S_w$) (squares) and $log\,K_{ow}$ for the same alcohols. (b) Same relationship for alcohols with data for 50 other narcotic compounds (ketones, ethers, alkyl halides, substituted benzenes included). Solid circles: fathead minnnows; open circles: guppies; squares: water solubility. [From Veith, G.D., D.J. Call, and L.T. Brooke, *Can. J. Fish. Aquat. Sci.*, 30, 743 (1983). With permission.]

compounds with similar K_{ow} values have different toxicities and why some compounds do not fit LSER models. For example, Kamlet et al.[106] concluded that carboxylic esters are more toxic than predicted by Equations 7 and 8 because of *in vivo* hydrolysis.

Similarly, correlation of Microtox EC_{50} values with LSER variables for 38 compounds yielded an r^2 of 0.974 (Equation 9, Table 7-9). The Microtox test is a rapid bioassay based on inhibition of bioluminescence in the bacterium *Photobacterium phosphoreum*. Bioluminescence results from a complicated series of energy-consuming enzymatic reactions. The test reflects chemical inhibition of enzymes, and its endpoint is the concentration (EC_{50}) that reduces light production by 50% compared with a toxicant-free solution. The test is attractive because of its speed (5 min) and simplicity, and it has been used as a screening tool in toxicity testing programs. Fair correlations are found between results of the Microtox test and 96-h LC_{50}s for fathead minnows.[150] Analysis of EC_{50} data by the LSERs indicated that alcohols are less toxic than phenols of the same K_{ow} because the phenols have greater polarity and HBD acidity. Such inferences provide incremental improvements in our understanding of the mechanisms of narcosis.

Inhibition of primary production in the green alga *Ankistrodesmus* is inversely correlated to $logK_{ow}$ and directly correlated to $logS_w$ for 12 chlorobenzenes (Equations 5 and 6, Table 7-12). The slope in Equation 5 is approximately one; thus, $1/EC_{50}$ and K_{ow} are linearly related. The slope of the correlation with S_w (Equation 6) is not near unity, but the general narcosis model may not apply to plants. The usefulness of Equations 5 and 6 in predicting the toxicity of other chlorobenzenes is marred by the fact that hexachlorobenzene was not toxic even at high concentrations, in spite of the fact that it has the highest K_{ow} and lowest S_w ($10^{-7.7}$ M) of the 12 compounds. The anomaly may reflect difficulties in keeping toxic doses of this compound in solution.[158]

Toxicity PARs for Reactive Compounds

Chemically reactive compounds generally are more toxic than unreactive compounds that induce narcosis (Figure 7-16). The toxicity of reactive compounds is affected by electronic structure, and lipophilicity is not always a major factor. Nucleophilic displacement (ND) reactions often are involved in the mechanisms by which they induce toxic effects.[156] These reactions are important because of the abundance of nucleophiles (-SH, -NH$_2$, -OH groups) on important biomacromolecules (enzymes, membranes, DNA). The ND reactivity of organic compounds is influenced by polar and resonance effects, and electronic substituent parameters should be useful in developing predictive PARs for the toxicity of such compounds. Hermens[156] found that 15 reactive organic halides (e.g., allyl chloride, benzyl chloride, chloroacetone) were much more toxic than predicted from a K_{ow}-toxicity relationship for unreactive compounds (Figure 7-16), and LC_{50} values (for guppies) were poorly correlated with K_{ow} values of the halides. A much better correlation was found between the LC_{50} values and the halide reactivity with a nucleophilic reagent, 4-nitrobenzylpyridine (Equation 7, Table 7-12). Similarly, chemically reactive

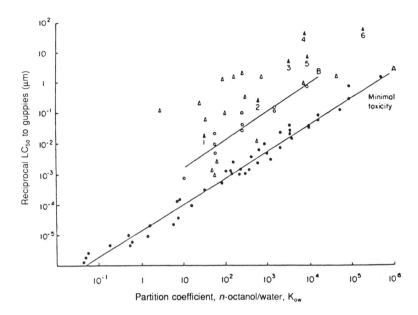

Figure 7-16. Correlation between *log* K_{ow} for various groups of organic compounds and reciprocal acute toxicity ($1/L_{50}$) to guppies. Closed circles: unreactive compounds (presumed mode of action is general narcosis); open circles: chloroanalines; open triangles: reactive organic halides; closed triangles: (1) 1,3-dinitrobenzene, (2) malathion, (3) lindane, (4) disulfiram, (5) rotenone, (6) dieldrin. [From Hermens, J.L.M., *Pest. Sci.*, 17, 287 (1986). With permission.]

chloroanilines are about ten times more toxic to guppies than are unreactive compounds with comparable K_{ow} values.[156,159] Addition of a Hammett term ($\Sigma\sigma$) to the *log*LC_{50}-*log*K_{ow} relationship improved the r^2 from 0.78 to 0.87, but the problem of multicollinearity (*log*K_{ow} and $\Sigma\sigma$ are themselves correlated; $r^2 = 0.75$) minimizes the usefulness of the two-variable model.

The inhibiting effects of chloro- and nitrophenols on degradation of simple phenol by activated sludge also is related to both *log*K_{ow} and $\Sigma\sigma$ for the compounds,[160] but different relationships apply to 2,6-disubstituted phenols than to compounds with no, or one, ortho substituent. The former compounds were less inhibitory and less sensitive to substituents, probably because of steric problems; ortho disubstituted phenols may not fit well in the active sites of enzyme(s) responsible for degradation of phenol.

Similarly, acute toxicities of 45 reactive aromatic compounds to the shrimp (*Crangon septemspinosa*) and soft-shelled clams (*Mya arenaria*) were not closely related to *log*K_{ow} alone, and inclusion of other properties improved the relationships[161] (Equations 8,9, Table 7-12). The best correlations for phenols were obtained when threshold toxicity was regressed against both K_{ow} and the difference between the pK_a's of the compound and phenol. Similar relationships were reported for 110 phenols with guppy LC_{50} values.[162] The general relationship of phenol toxicity to pK_a reflects the that fact cells absorb the ionized forms of these compounds much more slowly than the nonionized forms. Development of predictive relationships for clams was difficult because they can sense certain chemicals and avoid contact by closing. Lethal thresholds were obtained for only ten compounds. Some compounds that were toxic to shrimp caused no mortality in clams because the clams closed when exposed to the test solutions. Use of microorganisms to develop toxicity PARs (e.g., the Microtox test[105]) avoids such problems and simplifies interpretation of results, but this does not resolve the problem of predicting compound toxicity to higher organisms.

LC_{50} values for acute toxicity of eight triorganotin compounds to crab larvae[163] were poorly correlated with several measures of the compounds' electronic and steric features (σ and E_s constants), but a good correlation was found between LC_{50}s and K_{ow} values calculated from fragment constants. This suggests that uptake and toxicity of these hydrophobic compounds is determined by their aqueous activity and that they behave as narcotics. Further analysis[163] showed that toxicity was correlated with the molecular surface areas of the compounds. However, other studies with *Daphnia magna*[164,165] failed to find an adequate single-variable model to explain the toxicity of organotin compounds; multivariable equations including physical-chemical and topological properties were required for good predictive ability. (Other examples of PARs for toxicity of organic compounds are reviewed in reference 156.)

Metal Ion Toxicity

The bioavailability and toxicity of metals depends greatly on their "speciation", i.e., on their exact chemical forms: whether they are present as free (aquo) ions, complexes with simple inorganic or organic ligands, complexes with natural organic matter (aquatic humus), or sorbed to suspended particles. The extent to which metal ions exist in complexed forms depends on solution conditions such as pH, concentrations of carbonate, other inorganic ligands, organic chelating agents, and the nature of the metal ion of interest. Significant advances have been made in recent decades in classifying metal ions with regard to their tendency to form complexes with various ligands, and several reviews are available.[166-168] Metal-complexing trends with ligands have been correlated with metal ion electronegativity and with the square of the charge to the radius ratio (z^2/r), which is called the cation-hydrolyzing power.[166,169]

Pearson's hard and soft acid-base (HSAB) classification system[170] and the similar Ahrland-Chatt-Davies A, B, and borderline classes for cations[171] provide useful qualitative trends on metal-ligand binding preferences. Hard acids (roughly equivalent to class-A metal ions) have complete outer electron shells (inert-gas configuration). This class includes all alkali and alkaline earth ions, and trivalent Al, La, Co, Cr, Fe, and Mn. These ions have spherical, nonpolarizable shapes. In contrast, soft acids (roughly the same as class-B metals) have complete d shells (10 to 12 outer shell electrons) and are easily polarized. This class includes monovalent Ag, Au, Cu, Tl, divalent Cd, Hg, Sn, and trivalent Tl, Au, and In. Borderline metals in both systems include the first-row transition metals with partially filled d orbitals.

Among ligands, softness increases with increasing atomic weight in a given column of the Periodic Table; for example, F^-, OH^-, and N-containing ligands tend to be hard; Cl^-, HS^-, and PO_4^{3-} are soft. Hard and soft acids have well-characterized preferences for ligand binding: hard acids (A metals) tend to form complexes with hard bases, and soft acids (B metals) tend to form complexes with soft ligands. For hard acids, $N \gg P$, $O \gg S$, and $F \gg Cl$, and for soft acids, $P \gg N$, $S \gg O$, and $I \gg F$.[170,172] Moreover, hard cations tend to form weak complexes with electrostatic bonds, and the driving force for reaction is an increase in entropy caused by solvent loosening. Soft cations tend to form stronger covalent bonds, and the reactions tend to be enthalpy driven. Obviously, the scheme is not dichotomous, and the trends occur across a continuum.

Several parameters have been developed to quantify the concepts of hardness and softness. For example, σ_k is a parameter that decreases with increasing hardness and is determined from differences in outer-orbital and desolvation energies;[173] σ_p is a parameter that decreases with increasing softness and is defined in terms of metal-halide-bond energies:

$$\sigma_{p.M} = \frac{\left(\text{bond energy of } MF - \text{bond energy of } MI\right)}{\left(\text{bond energy of } MF\right)} \qquad (7\text{-}49a)$$

The HSAB parameters σ_p and σ_k are not related to Hammett's σ.

The term $\Delta\beta$ was developed similarly by Turner et al.[166] to quantify "A-ness" and "B-ness" of metal ions:

$$\Delta\beta = \log\beta_{MF} - \log B_{MCl} \qquad (7\text{-}49b)$$

Because B-type cations form weaker fluoride complexes than chloride complexes and the converse is true for A-type cations, a large negative $\Delta\beta$ is associated with B-type cations, and a large positive $\Delta\beta$ is associated with A-type cations. Turner et al.[166] divided the range of $\Delta\beta$ into four classes: A ($\Delta\beta > 2$), A' ($\Delta\beta = 0$ to 2), B' ($\Delta\beta = -2$ to 0), and B ($\Delta\beta < -2$). Nieboer and Richardson[174] used the product E_n^2r as a covalent index that groups the A, B, and borderline ions in a manner similar to $\Delta\beta$. The parameters z^2/r and $\Delta\beta$ can be used to develop two-dimensional classification schemes such as shown in Figure 7-17. The major categories defined by these dimensions each have similar ligand preferences and relative binding strengths, and they also show similarities in the types of bonds they form (electrostatic vs. covalent) and driving forces for bond formation (entropy changes vs. enthalpy changes).[166]

PARs relating the bioactivity of metals in aquatic organisms (e.g., bioaccumulation factors and/or acute toxicity) to atomic properties such as softness or extent of B character are relatively uncommon, despite much interest among scientists in metal contamination of surface waters.[175] Some early examples for PARs based on HSAB theory were developed by workers in mammalian toxicology. For example, mouse LD_{50} values for a large number of metal ions have been related[176] to the HSAB σ parameters:

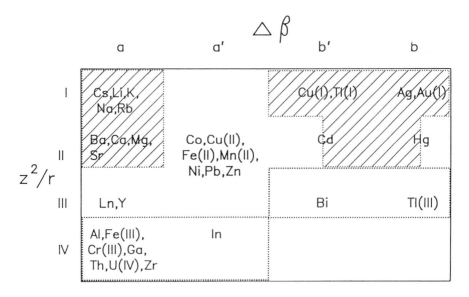

Figure 7-17. Bivariate metal classification scheme based on ionic and covalent bonding tendencies — cation polarizing power (z^2/r) and $\Delta\beta$, respectively. The scheme divides metals into four general classes with distinct ligand preferences and bonding strengths: very weakly complexed cations (IA, IIA); chloro-dominated cations (IBB', IIBB'); hydrolysis-dominated cations (IIIBB', IVAA'); and a residual group (IIA', IIIA) of variable speciation. [Redrawn from Turner, D.R., M. Whitfield, and A.G. Dickson, *Geochim. Cosmochim. Acta*, 45, 855 (1981); and reprinted from Brezonik, P.L., S. King, and C.E. Mach, in *Ecotoxicology of Metals: Current Concepts and Applications*, M. Newman and A. MacIntosh (Eds.), Lewis Publishers, Chelsea, MI, 1991.]

Soft acids: $\qquad$ $LD_{50} = 26.1\sigma_p + 0.076; \quad r^2 = 0.85$ $\qquad$ (7-50a)

Hard acids: $\qquad$ $LD_{50} = -2.85\sigma_k + 5.54; \quad r^2 = 0.99$ $\qquad$ (7-50b)

In addition, mouse LD_{50} values of divalent transition metal ions were found to follow the Irving-Williams order (Mn < Fe < Co < Ni < Cu > Zn).[176,177] From the regressions in Equation 7-50a,b, it is apparent that metal ion toxicity to mice increases with increasing softness within both the hard and soft acids.[178] Whether similar relationships apply to aquatic organisms is uncertain, but thus far, only limited success has been reported[179,180] in applying metal-softness relationships to other organisms. This perhaps is not surprising, since there likely is considerable variation among metals in regard to the chemical characteristic(s) key to their toxic mechanisms.[180]

Kaiser[181] was successful in relating the toxicity of metal ions to aquatic organisms by empirical regression equations involving some fundamental atomic properties. The equations have the general form

$$pT = a_o + a_1 \log\left(AN/\Delta IP\right) + a_2\Delta E_o \qquad (7\text{-}51)$$

pT is the negative log of metal ion concentration (M) with a certain toxicity, AN is the atomic number of the metal, ΔIP is the difference in the ion's ionization potential (eV) and the ionization potential of the next lower oxidation state of the element, and ΔE_o is the absolute value of the electrochemical potential between the ion and the first stable reduced state. Coefficients a_o, a_1, and a_2 depend on the group of metals, biota, and type of toxic effect being determined. The terms ΔIP and ΔE_o are related to outer-orbital electronic properties of atoms, and AN is related at least crudely to ionic size. Inclusion of AN in the regressions allows successful predictions for ions having similar ionization and electrochemical potentials, but different ionic radii (e.g., Na and Sr). Correlation coefficients as high as 0.96 were reported for toxicity correlations involving various biota and sets of metals, but the extent to which a regression equation can be applied beyond the set of data from which it was generated remains to be determined.

A relatively good predictive relationship has been reported between the *log* of metal bioaccumulation, *log*VCF (volume concentration factor), by marine algae and the negative *log* of the solubility product, $-log K_{so}$(MOH), for 21 metal hydroxides.[182] In turn, the *log* concentration of metal ions at which the growth of marine diatoms was reduced by 50% (EC_{50}) is related exponentially to *log*VCF, and EC_{50} also is linearly correlated with $log K_{so}$(MOH). Algae are relatively simple organisms that lack the variety of detoxifying and metal elimination mechanisms found in higher organisms. To this extent, relationships between the aqueous chemistry or atomic properties of metals and their behavior in aquatic flora should be simpler to derive and more straightforward than relationships involving aquatic fauna.

7.11 STRUCTURE-ACTIVITY RELATIONSHIPS (SARs)

Numerous structural characteristics are used as predictor variables in SARs. The approaches used to develop SARs range widely in complexity and sophistication and can be grouped into four categories:

I.   Numerical indices such as carbon number or chain length
II. Basic measures of molecular size such as total molecular volume (TMV) and total molecular surface area (TSA)
III. Group additivity (fragment) parameters based on fundamental atomic properties
IV. Topological parameters based on molecular shape (including degree of branching)

Type I relationships are limited in their range of application; Equations 1 and 2 (Table 7-12), which apply to straight-chain primary alcohols, are examples. This section focuses on categories II to IV and gives examples of such SARs and aquatic applications. Development of molecular structure indices that are related to bioactivity is of great interest in medicinal and drug chemistry. Environmental scientists interested in this topic should be aware of the large body of information in the medicinal and pharmaceutical chemistry literature.

7.11.1 Estimation of Molecular Volume and Surface Area

Volume and surface area are perhaps the simplest integrative measures of molecular structure (although they are not related to molecular conformation in any simple way). They have been used in SPRs to predict physical-chemical properties such as K_{ow} and S_w and in SARs to predict the bioactivity of organic compounds. As described in Section 7.8, molar volume is an important parameter in the LSER model; the energy required to create a cavity in the solvent for a solute is related to its molar volume (or surface area).

Although molar volumes of many liquid compounds can be determined simply from their molecular weight and liquid density, this approach is not available for compounds not obtainable in pure form (e.g., polychlorinated and brominated isomers of some aromatic compounds). The *liquid* molar volumes of substances that are solids at environmental temperatures also cannot be determined this way. Several computer programs have been developed to estimate total molecular (or molar) volume (TMV) from interatomic distances (bond lengths) and van der Waals radii.[183,184] A group additivity method not requiring computer calculations has been described that applies to many compounds of environmental interest,[107] and a simple atomic additivity scheme also is available.[185] Example 7-3 lists the rules and volume data needed for these methods and compares estimates from them for several compounds.

Example 7-3. Simple Methods to Estimate Molecular Volumes

The method of Kamlet et al.[107] is based on computer algorithms developed by Pearlman[183] and Leahy.[184] It applies to chlorinated, brominated, and nitrated benzenes, biphenyls, and PAHs.

1. Estimated parent compound volumes (V_1, cm³ mol⁻¹) are benzene = 49.1; biphenyl = 92.0; fluorene = 96.
2. For addition of a fused aromatic ring to a compound, add 6.55 for each additional CH: e.g., to form naphthalene from benzene, add 4(6.55) = 26.2; to form pyrene from naphthalene, add 6(6.55) = 39.3 to V_1.
3. To replace an aromatic H by CH_3 or add a side chain CH_2, add 9.8 to V_1.
4. To replace an aromatic H by Cl, Br, or NO_2, add 9.0, 13.3, or 14.0, respectively, to V_1.

McGowan's method[185] is based on an atomic additive scheme. The following atomic volumes (cm³ mol⁻¹) are used: C = 16.35, H = 8.71, Cl = 20.95, O = 12.43, N = 14.39 , and Br = 26.21.

Molar volumes are calculated by summing the contributions from all atoms and subtracting 6.56 for each bond (whether single, double, or triple). This yields a value V_x, which is related to molar volume V_1 by a regression equation derived by Abraham and McGowan:[186]

$$V_I = 0.597 + 0.6823 V_x; \quad n = 209, \quad r^2 = 0.998, \quad s.d. = 1.2 \ cm^3 \ mol^{-1}$$

Sample calculations for hexachlorobenzene are

modified Leahy method: $V_i = 49.1 + 6(9) = 103.1$;

McGowan method: $V_x = 6V_c + 6V_{cl} - 12(6.56) = 145.1, V_I(calc) = 99.6$

Kamlet et al.[107] found that the two methods gave similar values of V_I for a variety of compounds:

Compound	V_I (Leahy)	V_I (McGowan)
biphenyl	92.0	91.0
4,4′-dichlorobiphenyl	110.0	107.6
decachlorobiphenyl	182.0	174.5
triphenylene	127.7	125.0
1,2,3-trichlorobenzene	76.1	74.5

No simple method is available to directly measure the total surface areas (TSA) of molecules in solution, but TSAs can be calculated by several computer programs[187-190] from data on bond distances and angles and van der Waal's radii. Hermann[187] described TSA as the area of the solvent (water) cavity surface, as defined by the centers of the water molecules in the first layer around a solute. More recent "shorthand" methods ignore the solvent radius and compute the surface area of the pure solute.[190]

TSA can account for effects of branching, cyclization, and positional isomerization on the solubility of molecules. Several SPRs have been reported between aqueous solubility and TSA for PCBs, PBBs, PCDDs, and PAHs, e.g.,

$$\log S_w = -0.0313 TSA + 1.588 \quad n = 66; \quad r^2 = 0.94 \quad (ref.68) \quad (7\text{-}52)$$

$$\log X_w = -0.0469 TSA - 13.15 + 3.30 T_m; \quad n = 45; \quad r^2 = 0.935 \quad (ref.69) \quad (7\text{-}53)$$

Equation 7-52 applies to the molar aqueous solubility (S_w) of some chlorinated aromatic compounds (cf. Figure 7-8), and Equation 7-53 to the mole fraction aqueous solubility (X_w) of PCBs; T_m is the melting-point temperature of the congener.

Within a given series of related compounds, several molecular-size parameters have been found to correlate well with compound biodegradability (as measured by 5-day BOD values). These include substituent van der Waals radii,[150] a length factor, and accessible molecular surface area.[191]

7.11.2 Group Additivity Approaches

Some group additivity or molecular fragment methods used to estimate physical-chemical properties (e.g., K_{ow}) were described earlier. This general approach also has been used to develop formulas to predict bioactivity. Fragment constants have been defined on the basis of atomic or group properties derived from principles of molecular mechanics or indirectly from spectroscopic data or physical properties of molecules (as in the "i/o" concept described below). Such computed parameters, which are not quite structural nor physical-chemical attributes, as defined in Table 7-1, may be called molecular attributes. There is no precise way to decide whether bioactivity relationships based on these group additivity methods should be called PARs or SARs; for convenience, relationships based on these methods are discussed in this section.

One of the earliest efforts to quantify relationships between organic structure and physical-chemical-biological attributes is the i/o (inorganic/organic) concept of Fujita.[192] According to this concept the organic character of a molecule is derived from nonpolar, covalently bonded carbon atoms. Pure hydrocarbons have only nonpolar van der Waals interactions and thus have a purely "organic" character. Addition of methylene groups to a homologous series of hydrocarbons causes physical properties, such as boiling point and heat of atomization, to change in a regular fashion. Each methylene group (CH_2) in a molecule arbitrarily is assigned an "o" value of 20.

The inorganic character of a molecule is related to the presence of other elements or "residues of inorganic compounds", such as NH_3, SH_2, and SO_3H_2, which impart polar character to a molecule and cause dipole-dipole and hydrogen-bonding interactions. Addition of these substituents to a hydrocarbon causes regular changes in physical properties compared with a pure hydrocarbon. Values were assigned to inorganic substituents based on their effects on boiling points; the hydroxyl group served as the reference substituent and was given a value of 100. Values for other groups were calculated from the ratio of the distance between the boiling-point curve for the methane series of hydrocarbons and the curve for the substituted homologs to the distance between the methane hydrocarbon and alcohol boiling-point curves. Table 7-13 lists i values for common substituents. Some substituents exhibit both inorganic and organic character; i.e., large deviations occur between actual values of physical constants, and those calculated from carbon-based o values and boiling point-based i values. Fujita interpreted these deviations as a reflection of covalent bond character in the inorganic groups, which weakens their polarity. Corrections for these effects (i.e., organic character values; Table 7-13) were determined by considering solubilities and/or partition coefficients.

According to Fujita, physiologically active compounds should have approximately equal i and o character because they must diffuse through both lipid and aqueous media, and he stated that most effective drugs have $\Sigma i/\Sigma o \approx 1.0$. On the other hand, bioaccumulation should be related to the organic character. Matsuo[193] found the following correlations based on accumulation of seven organophosphate pesticides by mosquito fish (*Gambusia affinis*):

$$\log BF = -0.016\Sigma i + 0.014\Sigma o; \quad n = 7, \quad r^2 = 0.94 \qquad (7\text{-}54a)$$

$$\log BF = -6.3\Sigma i/\Sigma o + 5.4; \quad n = 7, \quad r^2 = 0.96 \qquad (7\text{-}54b)$$

Table 7-13. Inorganic/Organic (i/o) Character Values

Substituent	i	Substituent	o	i
-SO$_2$-NH-CO-	260	>SO$_2$	40	110
≡N-OH, -SO$_3$H	250	-CSSH	120	80
-CO-NH-CO-NH-	240	-SCN	90	80
≡N-OH, -NH-CO-NH-	220	-CSOH	80	80
≡N-NH- -CO-NH-NH-	210	-NCS	90	75
-CO-NH-	200	-NO$_2$	70	70
-COOH	150	>P-	20	70
-OH	100	-NO	50	50
>Hg	95	=S	50	10
>N-	70	-I	80	10
>CO	65	-Br	60	10
-COOR	60	-Cl	40	10
>C=NH	50	-F	5	5
-N=N-	30	Iso >-	-10	0
-O-	20	Tert. ->-	-20	0
benzene (nucleus)	15			

Summarized from Fujita, A., *Pharm. Bull.* (Japan), 2, 163 (1954).

Similar correlations were reported between $\Sigma i/\Sigma o$ ratios and BF of chloronaphthalenes and chlorobenzenes in carp:[194]

Chloronaphthalenes:

$$\log BF = -(97.3 \pm 25.6)\Sigma i / \Sigma o + (30.9 \pm 7.3); \quad n = 10, \quad r^2 = 0.91 \qquad (7\text{-}54c)$$

Chlorobenzenes:

$$\log BF = (55.3 \pm 11.4)\Sigma i / \Sigma o - (7.4 \pm 2.2); \qquad n = 12, \quad r^2 = 0.92 \qquad (7\text{-}54d)$$

The first three regressions have the "correct" signs, increasing BF with increasing organic character, but chlorobenzenes exhibit the opposite relationship. This can be explained by the fact that $\Sigma i/\Sigma o$ of the parent compound, benzene, is less than the i/o character of the Cl atom (0.25). Consequently, $\Sigma i/\Sigma o$ increases with the number of Cl atoms in chlorobenzenes. Nonetheless, the lipophilic character (i.e., K_{ow}) increases with increasing Cl content. The *posteriori* nature of this explanation is troublesome and implies that generalizations made from i/o trends must be made cautiously.

A fragment-based SAR derived from more fundamental molecular characteristics recently was described to predict biodegradability of organic compounds.[191,195] Five-day BOD data on several series of compounds were correlated with the differences in absolute atomic charge, $\Delta|\delta|_{x\text{-}y}$, across a key bond common to a class of molecules. $\Delta|\delta|_{x\text{-}y}$ was calculated by obtaining the minimum energy configuration of a molecule by a molecular mechanics program.[196] Provided that the correct bond is chosen for the charge calculation, impressive correlations ($r^2 > 0.99$) can be achieved (Table 7-14). This can be seen by comparing the correlation based on $\Delta|\delta|_{C\text{-}O}$ ($r^2 > 0.99$) with that for $\Delta|\delta|_{C\text{-}N}$ bonds ($r^2 = 0.15$) for

Table 7-14. Predictive Equations for Biodegradation Rates of Organic Compounds Based on Calculated Difference in Atomic Charge Across Key Bonds[a]

Class	Equation	r^2	n^b	Ref.
1. Phenols				
	$BOD = (0.998 \times 10^3)\Delta|\delta|_{C\text{-}O} + 2.108$	0.983	11	189
2. Carboxylic acids				
	$BOD = (0.996 \times 10^3)\Delta|\delta|_{C\text{-}O} + 3.234$	0.974	40	189
3. Alcohols				
	$BOD = (1.023 \times 10^3)\Delta|\delta|_{C\text{-}O} + 1.504$	0.990	20	189
4. Sulfonates				
	$BOD = (1.037 \times 10^3)\Delta|\delta|_{S\text{-}O} + 1.453$	0.943	20	189
5. Amino acids				
	$BOD = (1.087 \times 10^3)\Delta|\delta|_{C\text{-}O} - 2.986$	0.995	8	185
	$BOD = -(0.069 \times 10^3)\Delta|\delta|_{C\text{-}N} + 76.97$	0.154	8	185
6. Aromatic and aliphatic amines				
	$BOD = (1.004 \times 10^3)\Delta|\delta|_{C\text{-}N} - 0.106$	0.999	15	185
7. Halogenated hydrocarbons				
	$BOD = (1.009 \times 10^3)\Delta|\delta|_{C\text{-}x} + 0.204$	0.999	9	185
8. Aromatic and aliphatic aldehydes				
	$BOD = (1.008 \times 10^3)\Delta|\delta|_{C=O} + 0.497$	0.991	6	185
9. Groups 1–4 above plus 7 ketones and 14 ethers				
	$BOD = (1.015 \times 10^3)\Delta|\delta|_{x\text{-}y} + 1.906$	0.978	112	189

[a] Rates expressed as fraction of theoretical oxygen demand exerted in conventional 5-day BOD test.

[b] n = number of compounds in regression equation.

amino acids. Amino acid biodegradation thus is controlled by the charge difference at the C-O bond (and presumably some reaction involving that bond and not the C-N bond). It is interesting to note that poor correlations were obtained for sulfonates when $\Delta|\delta|_{S\text{-}O}$ was computed for the anionic forms, but good correlations were obtained when undissociated (acid) forms were used to compute atomic charges. It also is significant to note that the slopes and intercepts of the regression equations are similar for the six classes of compounds in Table 7-14. All the data thus can be combined into a single predictive equation that has remarkable precision, considering the diversity of compounds and its use of only a single parameter.

The correlations in Table 7-14 were obtained by ignoring the signs of charges on the atoms, suggesting that some property related to the square of the charge is the controlling factor. More recent studies[195] have suggested that this property is the "electrophilic superdelocalizability" (S_E), which is a measure of the reactivity of atomic centers. For 19 alcohols, the following correlation was obtained:

$$BOD = 0.093 S_E - 3.163 \qquad r^2 = 0.96 \qquad (7\text{-}55)$$

S_E is for the carbon atom to which the -OH group is attached.

Fragment-based approaches also have been used to develop predictive relationships for gas-phase reactions, e.g., reaction of ·OH radicals with organic compounds.[197] Four types of processes are considered in calculating a compound's overall reactivity with ·OH: hydrogen atom abstraction from CH and OH bonds in saturated compounds, ·OH addition to >C=C< and -C≡C- bonds, ·OH addition to aromatic rings, and ·OH reactions with N and S groups. The total reaction rate constant for a molecule is the sum of the constants for these four processes. The H-abstraction rate constant was based on -CH$_3$, -CH$_2$-, >CH-, and -OH group rate constants, with values for each depending on the identity of the neighboring substituents (e.g., -CH$_2$-, >CH- groups). Group rate constants and substituent factors were derived directly from kinetic data on alkanes by nonlinear regression. Once these factors were obtained, they were used to obtain substituent factors for haloalkanes, and this sequential process was continued for more complicated compounds. Other SARs and PARs for ·OH gas-phase reactions also have been reported[198] (see Mill[199] for a recent review of SARs and PARs applicable to both gas- and liquid-phase photochemical reactions).

7.11.3 Molecular Connectivity Indices

The idea that the geometric structure of a molecule encodes information on its chemical properties and biological activity has led to the analysis of molecular data by topographical methods and the development of topological indices such as the Weiner Path Number,[200] Branching Index of Randić[201] and several lesser known indices.[63,195] The Branching Index forms the basis for "molecular connectivity" (mc or χ) indices[66] that are widely used as predictor variables in SARs. The mc approach is a method of bond counting that describes the "connectedness" of nonhydrogen atoms in organic compounds. It considers four types of bonding patterns: paths, clusters, chains, and path/clusters (Figure 7-18). Two series of "χ indices" (simple, χ, and valence, χ^v) can be calculated from the reduced representation (carbon skeleton) of a molecule. The former are based on the number of sigma bonds an atom has with nonhydrogen atoms; the latter are based on an atom's valence electrons minus the number of hydrogen atoms bonded to it. For valence-based indices, carbon atom valences are calculated from the formula

$$\delta_i = 4 - n_{Hi} \qquad (7\text{-}56)$$

Connectivity indices are defined from the reciprocal square roots of the δ_i. The zero-order index, $^0\chi^v$, is the sum of the reciprocal square root valences over all carbon atoms in a molecule:

$$^0\chi^v = \sum_{i=1}^{n} \left(\delta_i \right)^{-1/2} \qquad (7\text{-}57)$$

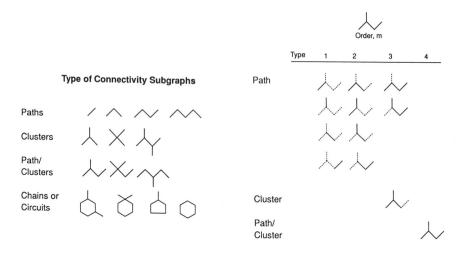

Figure 7-18. (a) Four types of molecular connectivity subgraphs; (b) subgraphs for isopentane according to type and order; order m = number of contiguous bonds in subgraph. (Modified from Kier, L.B. and L.H. Hall, *Molecular Connectivity in Chemistry and Drug Research*, Academic Press, New York, 1976.)

The first order index, $^1\chi^v$, is the sum over all bonded pairs of carbon atoms in a molecule of the pairwise products of the reciprocal square root valences:

$$^1\chi^v = \sum_{s=1}^{N_e} \left(\delta_i \delta_j\right)_s^{-1/2} \tag{7-58}$$

N_e is the number of edges in the reduced carbon skeleton (i.e., the number of C-C pairs). Second- and third-order indices are defined similarly:

$$^2\chi^v = \sum_{s=1}^{n_{nl}} \left(\delta_i \delta_j \delta_k\right)_s^{-1/2} \qquad ^3\chi^v = \sum_{s=1}^{n_{nl}} \left(\delta_i \delta_j \delta_k \delta_l\right)_s^{-1/2} \tag{7-59}$$

Second-order indices are sums over all pairs of adjacent edges (connected C-C pairs) in a molecule, and third-order indices are sums of reciprocal square root valence products over four adjacent atoms. In general, order is equal to the number of bonds in the substructure being considered. Whereas the lower-order indices each have only one type of bonding arrangement (paths in $^1\chi^v$ and $^2\chi^v$), $^3\chi^v$ may contain path, cluster, and chain terms (see Figure 7-18). Each type is evaluated separately; a compound thus may have three different $^3\chi^v$ indices ($^3\chi^v_p$, $^3\chi^v_{cl}$, $^3\chi^v_{ch}$). Fourth-order and higher indices may have all four bonding types. Higher-order indices are defined in ways analogous to Equation 7-59.[66] χ^v indices up to tenth-

Table 7-15. Heteroatom Valence δ Values for χ^v Indices

Group	δ	Group	δ	Group	δ
NH_4^+	1	H_3O^+	3	>S-	0.944
NH_3	2	H_2O	4	|	
-NH_2	3	-OH	5	=S=	3.58
>NH	4	-O-	6	|	
=NH	4	=O	6	-F	-20^a
>N-	5	O (both		-Cl	0.690
≡N	5	nitro)	6	-Br	0.254
=N-	5	O (both		-I	0.085
>N= (nitro)	6	carboxyl)	6		

a This value leads to negative edge terms, corresponding to the negative root of $(\delta_i\delta_F)^{-1/2}$; this contribution to $^1\chi^v$ is subtracted.[66]

order summations have been used to develop SARs. Although computation of χ^v is conceptually simple, the procedure is time-consuming when all types and orders of indices are calculated for a group of compounds. For example, 68 different χ indices can be calculated up to order ten, including both simple and valence types. Other ways of modifying the basic approach extends the possible number of mc indices even further. Computer programs are available to simplify the task.[203-205]

Procedures also have been developed to calculate χ^v indices for molecules with O, S, N, or halogen atoms and unsaturated compounds. Presence of O and N atoms in molecules is accommodated into χ^v indices in a nonempirical way, with δ^v values obtained by substituting the nominal valence of the atom for that of carbon in Equation 7-56 (Table 7-15). Thus, δ^v for O in hydroxyl groups is $6 - 1 = 5$ and in carbonyl groups is $6 - 0 = 6$. Similarly, δ^v for primary amines is $5 - 2 = 3$; δ^v for secondary amines is 4, and so on. A nitro N is formed from a nitroso group by adding an edge to give $\delta^v = 5 - 0 + 1 = 6$.

The nominal valence procedure does not work for fluorine and atoms beyond the first row of the periodic table. The valence approach implies that $\delta^v = 7$ in fluorine and all other halogens. However, use of this δ^v value for F leads to mc indices that do not correlate well with physical properties of fluorocompounds.[66] Moreover, it is well known that physical and chemical properties of halogenated compounds, RX, vary with the identity of X. To circumvent this problem, δ^v values for halogens and sulfur groups, for which similar problems were encountered, were estimated empirically from mc-molar refraction relationships for a series of substituted benzene compounds (Table 7-15). Unsaturated bonds are handled readily by Equation 7-56. Example 7-4 shows that this approach yields different χ^v values for various C_4 compounds.

Example 7-4. Calculation of χ Indices for Simple Compounds

Reduced skeletons for some four-carbon compounds are shown below. The numbers above or below the atoms indicate the valences of the nonhydrogen atoms (that for Cl is from Table 7-15). χ^v indices are calculated from Equations 7-57 to 7-59.

$$
\begin{array}{ccccc}
1 & 2 & 2 & 2 & 5 \\
\text{C–} & \text{C–} & \text{C–} & \text{C–} & \text{OH}
\end{array}
\quad n\text{–butanol}
$$

$$^{0}\chi^{v} = 1/\sqrt{1} + 3/\sqrt{2} + 1/\sqrt{5} = \underline{3.568}$$

$$^{1}\chi^{v} = 1/\sqrt{1} \times 2 + 2/\sqrt{2} \times 2$$

$$+ 1/\sqrt{2} \times 5 = \underline{1.523}$$

$$^{2}\chi^{v} = 1\sqrt{1 \times 2 \times 2} + 1/\sqrt{2 \times 2 \times 2}$$

$$+ 1/\sqrt{2 \times 2 \times 5} = \underline{1.077}$$

$$
\begin{array}{ccccc}
1 & 2 & 2 & 2 & 0.69 \\
\text{C–} & \text{C–} & \text{C–} & \text{C–} & \text{Cl}
\end{array}
\quad n\text{–butyl chloride}
$$

$$^{0}\chi^{v} = 1 + 3/\sqrt{2} + 1/\sqrt{0.69} = \underline{4.324}$$

$$^{1}\chi^{v} = 1/\sqrt{2} + 2/\sqrt{4} + 1/\sqrt{1.38}$$

$$= \underline{2.558}$$

$$^{2}\chi^{v} = 1/\sqrt{4} + 1/\sqrt{8} + 1/\sqrt{2.76} = \underline{1.455}$$

$$
\begin{array}{ccccc}
1 & 2 & 2 & 3 & 6 \\
\text{C–} & \text{C–} & \text{C–} & \text{C=} & \text{O}
\end{array}
\quad n\text{–butanone}
$$

$$^{0}\chi^{v} = 1 + 2/\sqrt{2} + 1/\sqrt{3} + 1/\sqrt{6}$$

$$= \underline{3.400}$$

$$^{1}\chi^{v} = 1/\sqrt{2} + 1/\sqrt{4} + 1/\sqrt{6} + 1/\sqrt{18}$$

$$= \underline{1.851}$$

$$^{2}\chi^{v} = 1/\sqrt{4} + 1/\sqrt{12} + 1/\sqrt{36} = \underline{0.955}$$

$$
\begin{array}{ccccc}
1 & 2 & 2 & 2 & 3 \\
\text{C–} & \text{C–} & \text{C–} & \text{C–} & \text{N}
\end{array}
\quad n\text{–butylamine}
$$

$$^{0}\chi^{v} = 1 + 3/\sqrt{2} + 1/\sqrt{3} = \underline{3.699}$$

$$^{1}\chi^{v} = 1/\sqrt{2} + 2/\sqrt{4} + 1/\sqrt{6} = \underline{2.115}$$

$$^{2}\chi^{v} = 1/\sqrt{4} + 1/\sqrt{8} + 1/\sqrt{12} = \underline{1.142}$$

$$
\begin{array}{ccc}
 & 1 & \\
1 & \text{C} & 2 \quad 5 \\
\text{C–} & \text{C–} & \text{C–O} \\
 & 3 &
\end{array}
\quad iso\text{–butanol}
$$

$$^{0}\chi^{v} = 2/\sqrt{1} + 1/\sqrt{3} + 1/\sqrt{2} + 1/\sqrt{5}$$

$$= \underline{3.731}$$

$$^{1}\chi^{v} = 2/\sqrt{1 \times 3} + 1/\sqrt{3 \times 2}$$

$$+ 1/\sqrt{2 \times 5} = \underline{1.879}$$

$$^{2}\chi^{v} = 1/\sqrt{1 \times 3 \times 1} + 2/\sqrt{1 \times 3 \times 2}$$

$$+ 1/\sqrt{3 \times 2 \times 5} = \underline{1.576}$$

$$
\begin{array}{c}
 1 \quad\quad 4\\
 \text{C} \quad 5 \\
1\text{C–} \; \text{C–O} \quad tert\text{–butanol}\\
 \text{C} \\
 1
\end{array}
$$

$$^{0}\chi^{v} = 3 + 1/\sqrt{4} + 1/\sqrt{5} = \underline{3.947}$$

$$^{1}\chi^{v} = 3/\sqrt{4} + 1/\sqrt{20} = \underline{1.724}$$

$$^{2}\chi^{v} = 3/\sqrt{4} + 3/\sqrt{20} = \underline{2.171}$$

Developers of the mc concept showed that mc indices for many classes of compounds are correlated with physical properties such as boiling point and density. A few examples relevant to aquatic contaminants are given in Table 7-16, and a comprehensive review is available.[206] $^{1}\chi^{v}$ (but not $^{1}\chi$) is closely related to R_{m}

Table 7-16. SARs AND SPRs Based on Connectivity Indices

Predicted variable	Predictor equation	r^2	n	Ref.
SPRs				
1. Molar refractivity	$R_M = 4.46 + 5.23\,^1\chi^v + 3.66\,^1\chi$	0.98	65	200
2. Aqueous solubility:				
(alkylbenzenes)	$\ln S_w = -1.045\,^1\chi + 2.129\,^1\chi^v + 0.671$	0.98	13	60
(aliphatic alcohols)	$\ln S_w = 6.702 - 2.666\,^1\chi$	0.96	51	60
3. $log\,K_{ow}$	$log\,K_{ow} = 1.48 + 0.95\,^1\chi$	0.97	138	60
(diverse O- or N-containing compounds)				
4. Molecular surface area	$TSA = 24.6\,^1\chi + 57.7$	0.91	72	66
5. Soil sorption coefficient	$log\,K_{om} = 0.53\,^1\chi + 0.54$	0.95	72	66
6. PAH HPLC retention times	$log\,k' = 0.328\,^2\chi^v - 0.762$	0.97	26	204
7. Henry's constant				
	$log\,H = 1.29 + 1.005\Phi - 0.468\,^1\chi^v - 1.258I$	0.99	180	209
(Φ = polarizability; I = indicator variable to differentiate hydrogen-bonding compounds, HBCs (I = 1), from nonHBCs (I = 0).				

Toxicity SARs

8. Inhibition of *Aspergillus niger* by substituted benzyl alcohols:				
(a)	$log\,(1/C) = 0.99\,^1\chi^v + 0.656\sigma - 1.768$	0.88	19	201
(b)	$log\,(1/C) = 1.987\,^4\chi^v_p + 0.507\sigma + 0.365$	0.93	19	201
(C = minimum inhibitory concentration; σ = Hammett constant)				
9. Chlorophenols to guppies:	$log\,LC_{50} = 3.26 - 0.67\,^1\chi^v$	0.96	10	203
10. Diverse aromatic and aliphatic compounds to *Daphnia magna*:				
11.	$log\,LC_{50} = 4.86 - 0.90\,^1\chi^v$	0.70	13	203
Chlorophenols to protozoan *Entosiphon*:				
	$log\,TLC = 4.07 - 0.99\,^1\chi^v$	0.79	10	203
12.	$log\,TLC = 4.20 - 1.38\,^2\chi^v$	0.82	10	203
(TLC = toxic limit concentration)				
13. Biosorption	$log\,BS = 0.445 + 0.673\,^1\chi^v$	0.949	18	206
14. Bioaccumulation	$log\,BF = 0.147 + 0.789\,^1\chi^v$	0.916	21	206
15. Toxicity	$log\,LC_{50} = 5.582 - 1.192\,^1\chi^v$	0.815	31	206
Chlorosis induced in *Lemna minor* by substituted phenols:				60
16.	$log\,EC_{50} = 0.55 - 1.116\,^1\chi^v - 0.603\Sigma\sigma$	0.92	25	
17.	$-log\,EC_{50} = 2.12 + 1.114\,^1\chi^v - 0.273pK_a$	0.93	25	
Acute toxicity of organotin compounds to *Daphnia magna*:				158
18.	$-log\,LC_{50} = 0.518\,log\,K_{ow} + 2.59$	0.57	12	
19.	$-log\,LC_{50} = 0.41\,log\,K_{ow} + 0.52pK_a + 0.10$	0.96	12	
20.	$-log\,LC_{50} = 0.21\,log\,K_{ow} + 0.51pK_a$	0.98	12	
	$+ 0.21\,^1\chi^v - 0.82$			
21.	$-log\,LC_{50} = 0.75\,^1\chi^v - 5.63$	0.86	12	
22. Acute toxicity of 123 diverse compounds to red killifish:				205
	$log\,LC_{50} = 3.72 - 0.33\,^3\chi_p - 0.28\,log\,K_{ow}$	0.77	123	
($^3\chi_p$ = simple, 3rd-order, path-type index)				

and S_w values of a diverse group of saturated and unsaturated aliphatic and halogenated aromatic compounds, and the simple first-order index $^1\chi$ is strongly correlated with $log\,K_{ow}$ for a diverse group of compounds that includes aliphatic and aromatic esters, acids, and alcohols.[66] The fact that $^1\chi$ and $^1\chi^v$ are correlated with so many physico-chemical properties of organic compounds suggests that they act as a measure of some underlying fundamental molecular property. Most probably this is molecular surface area. Sabljic[72] found a moderately good correlation between TSA and $^1\chi$ (Equation 4, Table 7-16); compounds included in the correlation include hydrocarbons, halogenated hydrocarbons, PAHs, and halogenated phenols. For the same compounds he found a strong correlation between soil sorption coefficients (K_{om}) and $^1\chi$ (Equation 5). The average difference between predicted and observed K_{om} values for a smaller data set (n = 37) was 0.24 *log* units.[209] As Figure 7-19 shows, polar and ionic compounds like anilines, carbamates, nitrobenzenes, organic acids, organophosphates, and ureas all had lower $log K_{om}$ values than predicted from their $^1\chi$ values and Equation 5 of Table 7-16. Empirical group correction factors were derived based on the finding that a given polar group causes a given amount of deviation from the K_{om}-$^1\chi$ relationship.

Early SAR applications of mc indices related χ values to enzyme inhibition, microbial and mammalian toxicities, and narcotic and insecticidal properties. More recent reports have correlated mc indices with uptake rates of organic compounds by algae,[129] bioaccumulation factors, and toxicity of pollutants to various aquatic organisms (Table 7-16). Moderate to good correlations (r^2 = 0.70–0.96) were reported between LC_{50} values and $^1\chi^v$ or $^2\chi^v$ indices for several species of fish, *Daphnia*, and a protozoan, *Entosiphon*,[209] but the accuracy of the relationship depends on the test organism and the similarity of structure for the compounds being compared.

In general, the compounds used to develop mc SARs should be sufficiently similar in structure that their metabolic fates and mechanisms of attack are the same or closely related. For example, good bivariate correlations (r^2 = 0.89–0.99) were found[211] for a data set of unsubstituted PAHs among all the following variables: molecular weight, $^1\chi^v$ to $^6\chi^v$ indices, HPLC retention times, thin-layer chromatographic R_f values, $log K_{ow}$, sediment/water partition coefficients, bioconcentration factors, and LC_{50} to *Daphnia pulex*. For this class of compounds, physico-chemical and bioactivity measures thus are interrelated and also correlated with mc indices. However, for a diverse set of 123 chemicals (including insecticides, fungicides, herbicides, and other compounds) poorer bivariate correlations (r^2 = 0.54–0.69) were found between toxicity to *Orizias lapites* (red killifish) and a variety of structure/property indices, including $^3\chi^v$, K_{ow}, i/o, and molecular weight.[212] A given mc index can be used to predict several types of bioactivity for a given series of compounds,[213] e.g., biosorption, bioaccumulation, and acute toxicity (Equations 13–15, Table 7-16). Not surprisingly, the precision of the relationship decreases with the complexity of the biological process being modeled.

Although SARs based on mc indices have impressive predictive abilities thus far, few insights have been obtained concerning the underlying basis for them. As mentioned earlier, $^1\chi$ is a steric characteristic closely related to molecular surface area, and it also is assumed to reflect the "branchedness" of a compound. Specific structural characteristics that may be encoded in other χ indices are not well defined. The nonmechanistic nature of the SARs based on χ indices is troubling to some scientists, but the indices do have many attractive features, including simplicity of calculation (by computer) and applicability to a wide variety of compounds, organisms, and biological activities. In addition, χ indices are deterministic; i.e., there is no uncertainty in the value of a given index for a particular compound. In contrast, considerable uncertainties may exist in values of other physico-chemical or structural properties (K_{ow}, S_w, TSA) that are measured directly or calculated by methods described elsewhere in this chapter. An important assumption of conventional regression analysis is that the dependent variable contains all the uncertainty in each data pair, but that clearly is not the case in PARs and SARs involving most properties and structural metrics. Violation of this assumption can have major effects on the slopes of regression equations.[72]

7.12 USE OF STATISTICAL TECHNIQUES TO DEVELOP AND ANALYZE PARs AND SARs

Statistical procedures have several important uses in studies on PARs and SARs (see reference 65 for a discussion of statistical considerations in developing such relationships). First, regression analysis is the standard technique to relate predictor and response variables to each other and form quantitative, predictive equations. Second, a variety of descriptive statistics and statistical techniques are used to evaluate the significance and reliability of such equations and compare sets of equations. Third, several exploratory statistical methods are used to simplify complicated data sets and extract the "unique" information from sets of intercorrelated variables. The last two applications can be helpful in inferring the underlying basis for PARs and SARs, establishing mechanisms of toxicity for a set of compounds, and evaluating factors affecting their toxicity.

Several studies have shown that SARs involving the steric parameter $^1\chi^v$ are improved by including explicit measures of electronic and lipophilic character, such as σ, i/o, and K_{ow}, in multiple regression models. For example, chlorosis induced in *Lemna minor* (duckweed) by substituted phenols is correlated with $^1\chi^v$ ($r^2 = 0.84$). Addition of either pK_a for the phenol or a Hammett $\Sigma\sigma$ term[66] increased the R^2 to 0.92 and reduced the standard error by 40% (Equations 16 and 17, Table 7-16). ($\Sigma\sigma$ is the sum of σ values for substituents on a phenol.) As the two equations suggest, $\Sigma\sigma$ and pK_a are closely correlated ($r^2 = 0.95$).

Similarly, good correlations were found between LC_{50} values for *Daphnia magna* and K_{ow} within a given subclass of alkyltin chlorides ($RSnCl_3$, R_2SnCl_2, etc.), but a poor correlation was found for all compounds together[164] (Equation 18, Table 7-16). Conversely, correlations were *not* significant between LC_{50} and pK_a in a given subclass, but a moderate correlation was found in the combined set. A multiple regression involving both K_{ow} and pK_a for the combined set yielded a high correlation, and inclusion of $^1\chi$ improved the R^2 slightly. Within a given subclass, toxicity of alkyltin chlorides to *Daphnia* thus depends primarily on compound lipohilicity. pK_a explains much of the difference in toxicity between subclasses, but little of the variation within a subclass. This is because pK_a changes stepwise with the number of Cl atoms attached to the tin atom and has little relationship to the nature of the alkyl groups. Addition of $^1\chi$ improved the regression only slightly, probably because $^1\chi$ and $logK_{ow}$ are closely correlated. Caution must be observed

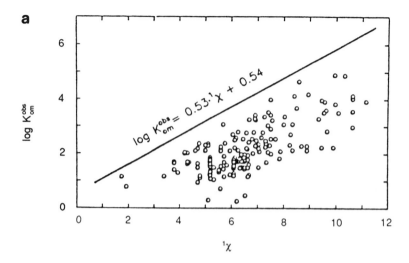

b

Chemical Group	No. of compds.	$log K_{om}^{calcd} - log K_{om}^{obs}$	average deviation
substituted benzenes and pyridines	6	1.00	0.20
organic phosphates (group I)	9	1.03	0.28
carbamates	5	1.05	0.11
anilines	12	1.08	0.17
nitrobenzenes	6	1.16	0.22
phenylureas	15	1.88	0.30
triazines	11	1.88	0.15
acetanilides	17	1.97	0.26
uracils	3	1.99	0.24
alkyl N-phenylcarbamates	10	2.01	0.16
3-phenyl-1-methylureas	5	2.07	0.28
3-phenyl-1-methyl-1-methoxyureas	4	2.13	0.10
dinitrobenzenes	7	2.28	0.29
3-phenyl-1,1-dimethylureas	16	2.36	0.23
organic acids	8	2.39	0.22
3-phenyl-1-cycloalkylureas	4	2.76	0.14
organic phosphates (group II)	5	3.19	0.22

Figure 7-19.   (a) Ionic and polar compounds do not fit the $^1\chi$ predictive relationship for soil sorption coefficients (K_{om}) derived for nonpolar compounds; all values are lower than predicted. (b) Differences between observed soil sorption coefficients and those predicted from χ indices are relatively constant within groups of polar and ionic compounds; these differences are used as polarity correction factors. (c) A two-variable regression using both $^1\chi$ and the polarity correction factor as predictor variables yields a good fit between measured and predicted K_{om} values. [From Sabljic, A., *Environ. Sci. Technol.*, 21, 358 (1987). With permission of the Amer. Chem. Soc.]

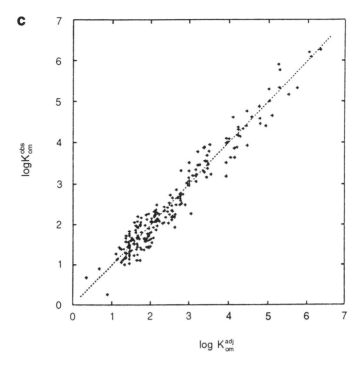

Figure 7-19 (continued).

in interpreting multiple regression models because multicollinearity among predictor variables like $^1\chi$ and K_{ow} is common.

Some progress in developing and interpreting multivariable SARs has been made by principal component analysis (PCA) and factor analysis (FA). These multivariate statistical methods are used to simplify the structure of complex data sets.[138,213-215] The goal of PCA is to reduce such data sets from many dimensions (variables) to a few new component variables.[216] The first-principal component (PC_1) is the linear combination of original variables that explains the maximum variance in the data. PC_2 is the linear combination of variables that explains as much of the remaining variance as possible, with the constraint that it be uncorrelated to PC_1; higher-order components are defined similarly. Factor analysis extends the inferential power of PCA one step further. By axis-rotation procedures, FA produces component variables in which only one or few of the original variables have high loadings. (Unrotated PCs may have high loadings of many variables, making interpretation of the PCs difficult.) Factor analysis thus simplifies interpretation of the component variables in terms of the original variables. PCA and FA produce two useful results: (1) a few new variables, the PCs, usually explain much of the variability in the original data; and (2) the component variables are orthogonal (i.e., uncorrelated), thus eliminating problems arising from multicollinearity among the original variables.

Burkhard et al.[215] used PCA to analyze different mc indices computed for a series of n-alkanes and all PCB congeners. Over 98% of the variance in 108 different mc indices they computed was explained by the first four PCs. Moreover, each PC seemed to have a simple physical interpretation. PC_1 was thought to measure branching. The n-alkanes, which have no branching, had essentially the same PC_1 values, while biphenyl and decachlorobiphenyl, which have the minimum and maximum branching among the PCB congeners, were end members in a ranking of congeners by PC_1 (Figure 7-20). PC_2 seems to represent molecular size or bulkiness; it increased directly with the number of methylene groups in n-alkanes. PC_3 is thought to reflect relative structural flexibility of molecules, and PC_4, which explained less than 1% of the variance in the data, is thought to describe the symmetry of PCB congeners. Information was not reported on the relative loadings of the mc indices in the PCs, but it would be interesting to relate the physical interpretations of the PCs to the mc indices themselves.

7.13 SUMMARY

Numerous relationships have been developed between chemical reactivity and biological activity, on the one hand, and equilibrium properties and/or structural characteristics of compounds, on the other. Linear correlations between *log*s of rate constants and *log*s of equilibrium constants for related reactions (linear free-energy relationships or LFERs) are available for many chemical reactions, ranging from acid-base dissociations to redox reactions, and for a wide variety of compounds. A comprehensive theory explaining why a few types of LFERs apply to such a variety of reactions and compounds remains elusive, but the existence of LFERs can be understood qualitatively on the basis of geometric similarities in reaction profiles for related compounds. In addition, the LFERs can be explained in terms of a few general influences of substituent groups, such as steric factors and electron-donating or -withdrawing properties.

LFERs have been developed to predict reaction rates of organic pollutants in natural waters. These relationships are useful in assessments of a compound's reactivity by a given pathway (e.g., hydrolysis). Because moderate scatter exists in most LFERs, predicted rates should not be used, however, if accurate results are needed, and experimentally based rate constants should be obtained.

Simple correlations can be used to predict transport properties of organic compounds in aquatic systems. For example, sediment/water partition coefficients have been correlated with S_w and with K_{ow} values of numerous organic compounds; these essentially are LFERs. S_w and K_{ow} also can be predicted with good accuracy from basic molecular properties, like molecular surface area and volume, and from properties relating to the energetics of solution by linear solvation energy relationships (LSERs).

Simple correlations also exist between biological activities of organic compounds and their structural/equilibrium properties. Property-activity and structure-activity relationships are useful in assessing the hazards posed by organic pollutants in aquatic environments; through such relationships several important biological

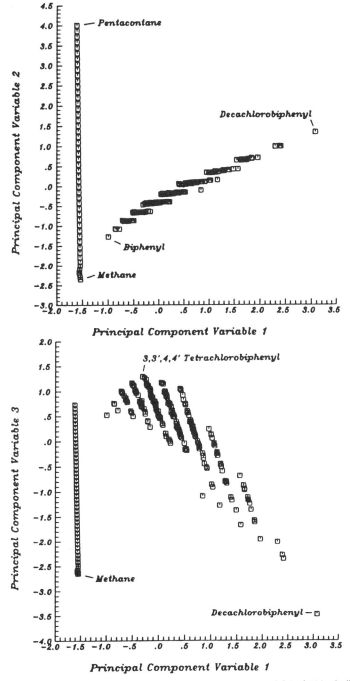

Figure 7-20.   Plots of principal component variables obtained from PCA of 108 χ indices for the C_1–C_{55} alkanes and all PCB congeners. PC1 describes the degree of branching; PC2 describes molecular size; PC3 describes molecular flexibility. Together, PC1–PC3 explain 98% of the total variance in the data set. [From Burkhard, L.P., A.W. Andren, and D.E. Armstrong, *Chemosphere*, 12, 935 (1983). With permission.]

processes can be predicted from easily measured or calculated properties. Biodegradation rate constants, bioaccumulation factors, and acute toxicity values of organic compounds for some aquatic organisms are related to transport/availability parameters like K_{ow} and S_w and/or to reactivity parameters like σ. Structural characteristics like molecular surface area and connectivity indices also are useful in predicting the same bioactivity properties. Many other groups of organic compounds are candidates for PARs, and there are many opportunities for further research in this field.

Finally, analysis of the fit, or lack of fit, of different classes of compounds to various PARs and SARs has led to improved understanding of the mechanisms whereby compounds produce a given physiological response. Indeed, mechanistic inferences may be a more important consequence of the relational analyses described in this chapter than are the predictions of contaminant behavior derived thus far from PARs and SARs. The subject of property- and structure-activity relationships is evolving rapidly, and the relationships described in this chapter undoubtedly will be refined as work continues.

REFERENCES

1. Hammett, L.P., *Physical Organic Chemistry*, 2nd ed., McGraw-Hill, New York, 1970; foreword to *Advances in Linear Free Energy Relationships*, N.B. Chapman and J. Shorter (Eds.), Plenum Press, London, 1972.

2. Agmon, N., *Int. J. Chem. Kinetics*, 13, 333 (1981).

3. Chipperfield, J. R., in *Advances in Linear Free Energy Relationships*, N.B. Chapman and J. Shorter (Eds.), Plenum Press, London, 1972, pp. 321–368; Langmuir, D., in *Chemical Models*, E.A. Jenne (Ed.), ACS Symp. Ser. 93, Am. Chem. Soc., Washington, D.C., 1979, pp. 353–387.

4. Biruš, M., G. Krznarić, N. Kujundžić, and M. Pribanić, *Croat. Chem. Acta,* 61, 33 (1988).

5. Adamson, A.W., *A Textbook of Physical Chemistry*, 2nd ed., Academic Press, New York, 1979.

6. Laidler, K.J., *Chemical Kinetics*, McGraw-Hill, New York, 1965; *Reaction Kinetics, Vol. 2, Reactions in Solution*, Pergamon Press, Oxford, 1963.

7. Boudart, M., *Kinetics of Chemical Processes*, Prentice-Hall, Englewood Cliffs, NJ, 1968.

8. Leffler, J.E., *J. Org. Chem.*, 20, 1202 (1955).

9. Exner, O., in *Advances in Linear Free Energy Relationships*, N.B. Chapman and J. Shorter (Eds.), Plenum Press, London, 1972, pp. 1–69.

10. Wold, S. and M. Sjöström, in *Correlation Analysis in Chemistry*, N.B. Chapman and J. Shorter (Eds.), Plenum Press, London, 1978, pp. 1–54.

11. Brønsted, J.N. and K. Pederson, *Z. Phys. Chem.*, 108, 185 (1923); Brønsted, J.N., *Trans. Faraday Soc.*, 25, 630 (1928).

12. Bell, R.P., in *Correlation Analysis in Chemistry*, N.B. Chapman and J. Shorter (Eds.), Plenum Press, London, 1978, pp. 55–84.

13. Pearson, R.G., in *Advances in Linear Free Energy Relationships*, N.B. Chapman and J. Shorter (Eds.), Plenum Press, London, 1972, pp. 281–319.

14. Hoffmann. M.R., *Environ. Sci. Technol.*, 15, 345 (1981).

15.   Hammett, L.P., *J. Am. Chem. Soc.*, 59, 96 (1937).

16.   Jaffe, H.H., *Chem. Rev.*, 53, 191 (1953).

17.   Job, D. and H.B. Dunford, *Eur. J. Biochem.*, 66, 607 (1976).

18.   Weston, R.E., Jr. and H.A. Schwarz, *Chemical Kinetics*, Prentice-Hall, Englewood Cliffs, NJ, 1972.

19.   Wells, P.R., *Linear Free Energy Relationships*, Academic Press, New York, 1968.

20.   Chapman, N.B. and J. Shorter (Eds.), *Advances in Linear Free Energy Relationships*, Plenum Press, London, 1972.

21.   Chapman, N.B. and J. Shorter (Eds.), *Correlation Analysis in Chemistry*, Plenum Press, London, 1978.

22.   Hansch, C. and A. Leo, *Substituent Constants for Correlation Analysis in Chemistry and Biology*, John Wiley & Sons, New York, 1979.

23.   Okamota, Y. and H.C. Brown, *J. Org. Chem.*, 22, 485 (1957); Brown, H.C. and Y. Okamoto, *J. Am. Chem. Soc.*, 80, 4979 (1958).

24.   Shorter, J., *Correlation Analysis in Chemistry*, N.B. Chapman and J. Shorter (Eds.), Plenum Press, London, 1978, pp. 119–173.

25.   Duboc, C., *Correlation Analysis in Chemistry*, N.B. Chapman and J. Shorter (Eds.), Plenum Press, London, 1978, pp. 313–355.

26.   van Bekkum, H., P.E. Verkade, and B.M. Wepster, *Rec. Trav. Chim.*, 78, 815 (1959).

27.   Shorter, J., in *Advances in Linear Free Energy Relationships*, N.B. Chapman and J. Shorter (Eds.), Plenum Press, London, pp. 71–117.

28.   Taft, R.W., Jr., *J. Am. Chem. Soc.*, 74, 2729,3120 (1952); *J. Am. Chem. Soc.*, 75, 4231,4534, 4538 (1953); *J. Phys. Chem.*, 64, 1805 (1960).

29.   Taft, R.W., Jr., in *Steric Effect in Organic Chemistry*, M. S. Newman (Ed.), Wiley, New York, 1956, pp. 556–675.

30.   Taft, R.W., Jr. and Pavelich, W.A., *J. Am. Chem. Soc.*, 79, 4935 (1957).

31.   Hancock, C.K., E.A. Myers, and B.J. Yager, *J. Am. Chem. Soc.*, 83, 4211 (1961); Idoux, J.P., P.T.R. Hwang, and C.K. Hancock, *J. Org. Chem.* 38, 4239 (1973).

32.   Roberts, J.D. and W.T. Moreland, *J. Am. Chem. Soc.*, 75, 2167 (1953).

33.   Amis, E.S., *Solvent Effects on Reaction Rates and Mechanisms*, Academic Press, New York, 1966.

34.   Marcus, R.A., *J. Phys. Chem.*, 67, 853 (1963); *J. Chem. Phys.*, 38, 1858 (1963).

35.   Marcus, R.A., *Int. J. Chem. Kinetics*, 13, 865 (1981).

36.   Wehrli, B., in *Aquatic Chemical Kinetics*, W. Stumm (Ed.), Wiley-Interscience, New York, 1990, pp. 311–336.

37.   Sutin, N., in *Inorganic Reactions and Methods*, J.J. Zuckerman (Ed.), VCH, Weinheim, 1986, pp. 16–46.

38.   Campion, R.J., N. Purdie, and N. Sutin, *Inorg. Chem.*, 3, 1091 (1964).

39.   Dulz, G. and N. Sutin, *Inorg. Chem.*, 2, 917 (1963).

40.   Proll, P.J., in *Comprehensive Chemical Kinetics*, Vol. 7, C.H. Bamford and C.F.H. Tipper (Eds.), Elsevier, Amsterdam, New York, 1972, pp. 56–152.

41.   Chou, M., C. Creutz, and N. Sutin, *J. Am. Chem. Soc.*, 99, 5615 (1977).

42.   Kalin, K.D. and J.K. Yandell, *Inorg. Chem.*, 23, 1184 (1984); Anast, J.M., A.W. Hamburg, and D.W. Margerum, *Inorg. Chem.*, 22, 2139 (1983).

43.   Fanchiang, Y.-T., *Int. J. Chem. Kinetics*, 14, 1305 (1982).

44.   Wolfe, N.L., R.G. Zepp, and D.F. Paris, *Water Res.*, 12, 565 (1978).

45.   Wolfe, N.L., *Chemosphere*, 9, 571 (1980).

46.   Wolfe, N.L., in *Dynamics, Exposure and Hazard Assessment of Toxic Chemicals*, R. Haque (Ed.), Ann Arbor Science, Ann Arbor, MI, 1980, pp. 163–178.

47.   Wolfe, N.L., W.C. Steen, and L.A. Burns, *Chemosphere*, 9, 403 (1980).

48. Sculley, F.E. and J. Hoigne, *Chemosphere*, 16, 681 (1987).

49. Schwarzenbach, R., R. Stierli, B.R. Folsom, and J. Zeyer, *Environ. Sci. Technol.*, 22, 83 (1988).

50. Nielsen, T., *Environ. Sci. Technol.*, 18, 157 (1984).

51. Charton, M., *J. Org. Chem.*, 41, 2906 (1976).

52. Betterton, E.A., Y. Erel, and M.R. Hoffmann, *Environ. Sci. Technol.*, 22, 92 (1988).

53. Vogel, T. and M. Reinhard, *Environ. Sci. Technol.*, 20, 992 (1986).

54. Okamoto, K., I. Nitta, T. Imoto, and H. Shingu, *Bull. Chem. Soc. Jpn.*, 40, 1905 (1967); T. Vogel, personal communication, 1989.

55. Barbash, J.E. and M. Reinhard, in *Biogenic Sulfur in the Environment*, E. S. Saltzman and W.J. Cooper (Eds.), ACS Symp. Ser. 393, Am. Chem. Soc., Washington, D.C., pp. 101–138.

56. Swain, C.G. and C.B. Scott, *J. Am. Chem. Soc.*, 75, 141 (1953).

57. Edwards, J.O., *J. Am. Chem. Soc.*, 76, 1540 (1954).

58. Murty, P.S.R. and R.K. Panda, *Indian J. Chem.*, 11, 1003 (1973); Rao, P.V. S., B.A.N. Murty, R.V.S. Murty, and K.S. Murty, *J. Indian Chem. Soc.*, 55, 207 (1978); Kimura, M. and Y. Kaneko, *J. Chem. Soc. (Dalton Trans.)*, 341 (1984); and Rao, P.V.S., K.V. Subbaiah, P.S.N. Murty, and R.V.S. Murty, *Indian J. Chem. Sec. A*, 19A, 257 (1980).

59. Stone, A.T., *Environ. Sci. Technol.*, 21, 979 (1987).

60. Karickhoff, S.W., V.K. McDaniel, C. Melton, A.N Vellino, D.E. Nute, and L.A. Carriera, *Environ. Toxicol. Chem.*, 10, 1405 (1991); unpublished manuscript, 1989.

61. Dewar, M.J.S. and R.C. Dougherty, *The PMO Theory of Organic Chemistry*, Plenum Press, New York.

62. Haque, R. (Ed.), *Dynamics, Exposure and Hazard Assessment of Toxic Chemicals*, Ann Arbor Science, Ann Arbor, MI, 1980.

63. Nirmalakhandan, N. and R.E. Speece, *Environ. Sci. Technol.*, 22, 606 (1988).

64. Brezonik, P.L., in *Aquatic Chemical Kinetics*, W. Stumm (Ed.), Wiley-Interscience, New York, 1990, pp. 113–143.

65. Tosato, M.L., S. Marchini, L. Passerini, A. Pino, L. Eriksson, F. Lindgren, S. Hellberg, J. Jonsson, M. Sjöström, B. Skagerberg, and S. Wold, *Environ. Tox. Chem.*, 9, 265 (1990).

66. Kier, L.B. and L.H. Hall, *Molecular Connectivity in Chemistry and Drug Research*, Academic Press, New York, 1976.

67. Meyer, H., *Arch. Exp. Pathol. Pharmakol.*, 42, 110 (1899); Overton, E., *Studien uber die Narkose*, Fischer Publications, Jena, Germany, 1901.

68. Andren, A.W., W.J. Doucette, and R.M. Dickhut, in *Sources and Fates of Aquatic Pollutants*, R.A. Hites and S.J. Eisenreich (Eds.), Adv. Chem. Ser. 216, Am. Chem. Soc., Washington, D.C., 1987, pp. 3–26.

69. Opperhuizen, A., F.A. Gobas, J.M. Van der Steen, and O. Hutzinger, *Environ. Sci. Technol.*, 22, 638 (1988).

70. De Kock, A.C. and D.A. Lord, *Chemosphere*, 16, 133 (1987).

71. Woodburn, K.B., W.J. Doucette, and A.W. Andren, *Environ. Sci. Technol.*, 18, 457 (1984).

72. Sabljic, A., *Environ. Sci. Technol.*, 21, 358 (1987).

73. Doucette, W.J. and A.W. Andren, *Environ. Sci. Technol.*, 21, 821 (1987).

74. De Voogt, P., J.W.M. Wegener, J.C. Klamer, G.A. Van Zijl, and H. Govers, *Biomed. Environ. Sci.*, 1, 194 (1988).

75. Fujita, T., J. Isawa, and C. Hansch, *J. Am. Chem. Soc.*, 86, 5175 (1964); Leo, A., C. Hansch, and D. Elkins, *Chem. Rev.*, 71, 525 (1971); Hansch, C., A. Leo, S.H. Unger, K.H. Kim, D. Nikaitani, and E.J. Lien, *J. Med. Chem.*, 16, 1207 (1973); Hansch, C., in *Dynamics, Exposure and Hazard Assessment of Toxic Chemicals*, R. Haque (Ed.), Ann Arbor Sci., Ann Arbor, MI, 1980, pp. 273–286.

76. Kaiser, K.L.E., *Chemosphere*, 12, 1159 (1983).

77. Rekker, R.F., *The Hydrophobic Fragmental Constant*, Elsevier, Amsterdam, 1977.

78. Rekker, R.F. and H.M. De Kort, *Eur. J. Med. Chem.*, 14, 479 (1979).

79. Arbuckle, W.B., *Env. Sci. Technol.*, 17, 537 (1983).

80. Ruepert, C., A. Grinwis, and H. Govers, *Chemosphere*, 14, 279 (1985).

81. Hansch, C., J.E. Quinlan, and G.L. Lawrence, *J. Org. Chem.*, 33, 347 (1968).

82. Lyman, W.J., W.F. Reehl, and D.H. Rosenblatt, *Handbook of Chemical Property Estimation Methods. Environmental Behavior of Organic Compounds*, McGraw-Hill, New York, 1982.

83. Mackay, D., R. Mascarenhas, W.Y. Shiu, S.C. Valvani, and S.H. Yalkowsky, *Chemosphere*, 9, 257 (1980); Chiou, C.T., D.W. Schmedding, and M. Maines, *Environ. Sci. Technol.*, 16, 4 (1982); Miller, M.M., S.P. Wasik, G. Huang, W.Y. Shiu, and D. Mackay, *Environ. Sci. Technol.*, 19, 529 (1985).

84. Rapaport, R.A. and S.J. Eisenreich, *Environ. Sci. Technol.*, 18, 163 (1984).

85. De Voogt, P. and H. Govers, *Chemosphere*, 15, 1467 (1986).

86. Burkhard, L.P. and D.W. Kuehl, *Chemosphere*, 15, 163 (1986); Burkhard, L.P., D.W. Kuehl, and G.D. Veith, *Chemosphere*, 14, 1551 (1985); Eadsforth, C.V., *Pest. Sci.*, 17, 311 (1986).

87. Vowles, P.D. and R.F.C. Mantoura, *Chemosphere*, 16, 109 (1987).

88. Karickhoff, S.W., *Chemosphere*, 10, 833 (1980); Means, J.C., S.G. Wood, J.L. Hassett, and W.L. Banwart, *Environ. Sci. Technol.*, 14, 1524 (1980).

89. Schwarzenbach, R. and J. Westall, *Environ. Sci. Technol.*, 15, 1360 (1981).

90. Hand, V.C. and G.K. Williams, *Environ. Sci. Technol.*, 21, 370 (1987).

91. O'Connor, D.J. and J.P. Connolly, *Water Res.*, 14, 1517 (1980).

92. DiToro, D.M., *Chemosphere*, 14, 1503 (1985); DiToro, D.M., J.D. Mahony, P.R. Kirchgraben, A.L. O'Byrne, L.R. Pasquale, and D.C. Picirilli, *Environ. Sci. Technol.*, 20, 55 (1986); Mackay, D. and B. Powers, *Chemosphere*, 16, 745 (1987).

93. Fredenslund, A., R.L. Jones, and J.M. Prausnitz, *Am. Inst. Chem. Eng. J.*, 21, 1086 (1975); Fredenslund, A., J. Gmehling, and P. Rasmussen, *Vapor- Liquid Equilibria Using UNIFAC, a Group-Contribution Method*, Elsevier, New York, 1977.

94. Gmehling, J., P. Rasmussen, and A. Fredenslund, *Ind. Eng. Chem. Process Des. Dev.*, 21, 118 (1982). Note: the interaction parameters tabulated in this paper give much better estimates of S_w than the original parameter values tabulated in reference 93.

95. Kikic, I., P. Alessi, P. Rasmussen, and A. Fredenslund, *Can. J. Chem. Eng.*, 58, 253 (1980).

96. Arbuckle, W.B., *Environ. Sci. Technol.*, 20, 1060 (1986).

97. Kamlet, M.J., P.W. Carr, R.W. Taft, and M.H. Abraham, *J. Am. Chem. Soc.*, 103, 6062 (1981).

98. Kamlet, M.J., J.-L.M. Abboud, M.H. Abraham, and R.W. Taft, *J. Org. Chem.*, 48, 2877 (1983).

99. Taft, R.W., J.-L.M. Abboud, M.J. Kamlet, and M.H. Abraham, *J. Solution Chem.*, 14, 153 (1985).

100. Kamlet, M.J., M.H. Abraham, R.M. Doherty, and R.W. Taft, *J. Am. Chem. Soc.*, 106, 464 (1984).

101. Taft, R.W., M.H. Abraham, R.M. Doherty, and M.J. Kamlet, *Nature*, 313, 384 (1985).

102. Kamlet, M.J., R.M. Doherty, M.H. Abraham, J.-L.M. Abboud, and R.W. Taft, *CHEMTECH*, 16, 566 (1985).

103. Kamlet, M.J., R.M. Doherty, M.H. Abraham, and R.W. Taft, *Carbon*, 23, 549 (1985).

104. Taft, R.W., M.H. Abraham, G.R. Famini, R.M. Doherty, and M.J. Kamlet, *J. Pharm. Sci.*, 74, 807 (1985); Kamlet, M.J., R.M. Doherty, J.-L.M. Abboud, M.H. Abraham, and R.W. Taft, *J. Pharm. Sci.*, 75, 338 (1986); Abboud, J.-L.M., K. Sraidi, G. Guiheneuf, A. Negro, M.J. Kamlet, and R.W. Taft, *J. Org. Chem.*, 50, 2879 (1985).

105. Kamlet, M.J., R.M. Doherty, G.D. Veith, R.W. Taft, and M.H. Abraham, *Environ. Sci. Technol.*, 20, 690 (1986).

106. Kamlet, M.J., R.M. Doherty, R.W. Taft, M.H. Abraham, G.D. Veith, and D.J. Abraham, *Environ. Sci. Technol.*, 21, 149 (1987).

107. Kamlet, M.J., R.M. Doherty, P.W. Carr, D. Mackay, M.H. Abraham, and R.W. Taft, *Environ. Sci. Technol.*, 22, 503 (1988).

108. Hildebrand, J.H. and R.L. Scott, *The Solubility of Nonelectrolytes*, 3rd ed., Dover, New York, 1964.

109. Hansch, C. and T. Fujita, *J. Am. Chem. Soc.*, 86, 1616 (1964).

110. Hyde, R. and E. Lord, *Eur. J. Med. Chem.*, 14, 199 (1979).

111. Many reviews of PARs for aquatic organisms have been published, including *QSAR in Environmental Toxicology*, K.L. Kaiser, (Ed.), D. Reidel, Boston, 1984; a series of seven papers from a symposium: Physicochemical Properties and their Role in Environmental Hazard Assessment, in *Pest. Sci.*, 17, 256–325 (1986); Veith, G.D. and D.E. Konasewich, (Eds.), *Structure-Activity Correlations in Studies of Toxicity and Bioconcentration with Aquatic Organisms*, Int. Joint Comm. Great Lakes Adv. Bd., Windsor, Ont., 1975; Eaton, J.G., P.R.P. Parrish, and A.C. Hendricks (Eds.), *Aquatic Toxicology*, ASTM STP 707, American Society of Testing and Materials, 1980; also see reference 64.

112. Branson, D.R., G.E. Blau, H.C. Alexander, and W.B. Neely, *Trans. Am. Fish. Soc.*, 4, 785 (1975).

113. Mackay, D. and A.I. Hughes, *Environ. Technol.*, 18, 439 (1984).

114. Bruggeman, W.A., L.B.J.M. Martron, D. Kooiman, and O. Hutzinger, *Chemosphere*, 10, 811 (1981).

115. Konemann, H. and K. van Leeuwen, *Chemosphere*, 9, 3 (1980).

116. Neely, W.B., D.R. Branson, and G.E. Blau, *Environ. Sci. Technol.*, 8, 1113 (1974).

117. Esser, H.O., *Pest. Sci.*, 17, 265 (1986).

118. Grimes, D.J. and S.M. Morrison, *Microb. Ecol.*, 2, 43 (1975).

119. Geyer, H., R. Viswanathan, D. Freitag, and F. Korte, *Chemosphere*, 10, 1307 (1981).

120. Geyer, H., P. Sheehan, D. Kotzias, D. Freitag, and F. Korte, *Chemosphere*, 11, 1121 (1982).

121. Kenaga, E.E. and C.A.I. Goring, pp. 78–115 in *Aquatic Toxicology*, J.G. Eaton, P.R.P. Parrish, and A.C. Hendricks (Eds.), ASTM, STP 707, American Society of Testing and Materials, 1980.

122. Veith, G.D., K.J. Macek, S.R. Petrocelli, and J. Carrol, in *Aquatic Toxicology*, pp. 116–129.

123. Chiou, C.T., V.H. Freed, D.W. Schmedding, and R.L. Kohnert, *Environ. Sci. Technol.*, 11, 475 (1977).

124. Karickhoff, S.W., D.S. Brown, and T.A. Scott, *Water Res.*, 13, 241 (1979).

125. Chiou, C.T., L.J. Peters, and V.H. Freed, *Science*, 206, 831 (1979).

126. Steen, W.C. and S.W. Karickhoff, *Chemosphere*, 10, 27 (1981).

127. Ellgehausen, H., J.A. Guth, and H.O. Esser, *Ecotoxicol. Environ. Safety*, 4, 134 (1980).
128. Geyer, H., G. Politzki, and D. Freitag, *Chemosphere*, 13, 269 (1984).
129. Mailot, H., *Environ. Sci. Technol.*, 21, 1009 (1987).
130. Southworth, G.R., J.J. Beauchamp, and P.K. Smieder, *Water Res.*, 12, 973 (1978).
131. Hawker, D.W. and D.W. Connell, *Chemosphere*, 14, 1835 (1985).
132. Opperhuizen, A., E.W. v.d. Velde, F.A. Gobas, D.A. Liem, J.M. v.d. Steen, and O. Hutzinger, *Chemosphere*, 14, 1871 (1985).
133. Veith, G.D., D.L. DeFoe, and B.V. Bergstedt, *J. Fish. Res. Board Can.*, 36, 1040 (1979).
134. Mackay, D., *Environ. Sci. Technol.*, 16, 274 (1982); Mackay, D. and A.I. Hughes, *Environ. Sci. Technol.*, 18, 439 (1984).
135. Hawker, D.W. and D.W. Connell, *Chemosphere*, 14, 1205 (1985).
136. Thomann, R., *Ecol. Model.*, 22, 145 (1984).
137. Shaw, G.R. and D.W. Connell, *Environ. Sci. Technol.*, 18, 18 (1984).
138. Clarke, J.U., *Chemosphere*, 15, 275 (1986).
139. Oliver, B.G. and A.J. Niimi, *Environ. Sci. Technol.*, 19, 842 (1985).
140. Astles, D.J., R. Pearce, D. Griller, H.M. Schwartz, and D.C. Villeneuve, *Chemosphere*, 16, 803 (1987).
141. Parsons, A.R., A. Opperhuizen, and O. Hutzinger, *Chemosphere*, 6, 1361 (1987).
142. Vaishnav, D.D., R.S. Boethling, and L. Babeu, *Chemosphere,* 16, 695 (1987).
143. Yonezawa, Y. and Y. Urushigawa, *Chemosphere*, 8, 139 (1979); Yonezawa, Y., et al., *Kogai Shigen Kenkyusho Iho*, 11, 77 (1981).
144. Ludzack, F.J. and M.B. Ettinger, *J. Water Pollut Contr. Fed.*, 32, 1173 (1960).
145. Banerjee, S., P.H. Howard, A.M. Rosenberg, A.E. Dombrowski, H. Sikka, and D.L. Tullis, *Environ. Sci. Technol.*, 18, 416 (1984).
146. Urushigawa, Y. and Y. Yonezawa, *Chemosphere*, 5, 317 (1979).
147. Paris, D.F., N.L. Wolfe, and W.C. Steen, *Appl. Environ. Microbiol.*, 47, 7 (1984).
148. Wolfe, N.L., D.F. Paris, W.C. Steen, and G.L. Baughman, *Environ. Sci. Technol.*, 14, 1143 (1980).
149. Pitter, P., *Coll. Czech. Chem. Commun.*, 49, 2891 (1984).
150. Paris, D.F., N.L. Wolfe, and W.C. Steen, *Appl. Environ. Microbiol.*, 45, 1153 (1983).
151. Carotti, A., C. Raguseo, and C. Hansch, *Chem.-Biol. Interact.,* 52, 279 (1985).
152. Cantier, J.M., J. Bastide, and C. Coste, *Pest. Sci.*, 17, 235 (1986).
153. Veith, G.D., D.J. Call, and L.T. Brooke, *Can. J. Fish. Aquat. Sci.*, 30, 743 (1983).
154. Mullins, L.J., *Chem. Rev.*, 54, 289 (1954).
155. Bengtsson, B.-E., L. Renberg, and M. Tarkpea, *Chemosphere*, 13, 613 (1984).
156. Hermens, J.L.M., *Pest. Sci.*, 17, 287 (1986).
157. Konemann, H., *Toxicology*, 19, 209 (1981).
158. Wong, P.T.S., Y.K. Chau, J.S. Rhamey, and M. Docker, *Chemosphere*, 13, 991 (1984).
159. Hermens, J.L.M., P. Leeuwangh, and A. Musch, *Ecotoxicol. Environ. Safety*, 8, 388 (1984).
160. Beltrame, P., P.L. Beltrame, and P. Carniti, *Chemosphere*, 13, 3 (1984).
161. McLeese, D.W., V. Zitko, and M.R. Peterson, *Chemosphere*, 8, 53 (1979).
162. Lipnick, R.L., et al., *ASTM Spec. Tech. Publ.,* 891, 153 (1986).
163. Laughlin, R.B., Jr., W. French, R.B. Johannesen, H.E. Guard, and F.E. Brinckman, *Chemosphere*, 13, 575 (1984).
164. Vighi, M. and D. Calamari, *Chemosphere*, 14, 1925 (1985).

165. Vighi, M. and D. Calamari, *Chemosphere*, 16, 1043 (1987).

166. Turner, D.R., M. Whitfield, and A.G. Dickson, *Geochim. Cosmochim. Acta*, 45, 855 (1981).

167. Brezonik, P.L., S. King, and C.E. Mach, in *Ecotoxicology of Metals: Current Concepts and Applications*, M. Newman and A. MacIntosh (Eds.), Lewis Publishers, Chelsea, MI, p. 1–29.

168. Morgan, J.J. and W. Stumm, in *Metals and their Compounds in the Environment*, E. Merian (Ed.), VCH, Weinheim, 1991, pp. 67–103.

169. Phillips, C.S.G. and R.J.P. Williams, *Inorganic Chemistry*, Clarendon Press, London, 1965.

170. Pearson, R.G., *J. Am. Chem. Soc.*, 25, 3533 (1963); Pearson, R.G. (Ed.), *Hard and Soft Acids and Bases*, Dowden, Hutchinson, & Ross, Stroudsburg, PA, 1973.

171. Arhland, S., J. Chatt, and N.R. Davies, *Q. Rev. Chem. Soc.*, 12, 265 (1958).

172. Stumm, W. and J.J. Morgan, *Aquatic Chemistry*, 2nd ed., Wiley-Interscience, New York, 1981.

173. Ahrland, S., *Struct. Bonding*, 5, 144 (1968).

174. Nieboer, E. and D.H.S. Richardson, *Environ. Pollut.*, B1, 3 (1980).

175. Wood, J.M. and H.-K. Wang, *Environ. Sci. Technol.*, 17, 582A (1983); Campbell, P.G.C. and P.M. Stokes, *Can. J. Fish. Aquat. Sci.*, 42, 2034 (1985).

176. Jones, M.M. and W.K. Vaughn, *J. Inorg. Nucl. Chem.*, 40, 2081 (1978).

177. Shaw, W.H.R., *Nature*, 192, 754 (1961).

178. Williams, M.W. and J.E. Turner, *J. Inorg. Nucl. Chem.*, 43, 1689 (1981); Williams, R.J.P., *Phil. Trans. R. Soc. Lond.*, Ser.B 57, 294 (1981).

179. Turner, J.E., M.W. Williams, K.B. Jacobson, and B.E. Higerty, in *Quantitative Structure Activity Relationships in Toxicology and Xenobiochemistry*, M. Tichy (Ed.), Elsevier, New York, 1985, pp. 1–8.

180. McDonald, D.G., J.P. Reader, and T.R.K. Dalziel, in *Acid Toxicity and Aquatic Animals*, R. Morris, E.W. Taylor, D.J.A. Brown, and J.A. Brown (Eds.), Soc. Exper. Biol., Seminar Ser. 34, Cambridge University Press, Cambridge, 1989, pp. 221–242.

181. Kaiser, K.L.E., *Can.. J. Fish. Aquat. Sci.*, 37, 211 (1980).

182. Fisher, N.S., *Limnol. Oceanogr.*, 31, 443 (1986).

183. Pearlman, R.S., in *Partition Coefficient Determination and Estimation*, W.J. Dunn, J.H. Block, and R.S. Pearlman (Eds.), Pergamon Press, New York, 1986, pp. 3–20.

184. Leahy, D.J., *J. Pharm. Sci.*, 75, 629 (1986).

185. McGowan, J.C., *J. Appl. Chem. Biotechnol.*, 28, 599 (1978); *J. Appl. Chem. Biotechnol.*, 34A, 38 (1984).

186. Abraham, M.H. and J.C. McGowan, *Chromatographia*, 23, 243 (1987).

187. Hermann, R.B., *J. Phys. Chem.*, 76, 2754 (1972).

188. Valvani, S.C., S.H. Yalkowsky, and G.L. Amidon, *J. Phys. Chem.*, 80, 829 (1976).

189. Pearlman, R.S., *Quant. Chem. Prog. Exch. (QCPE) Bull.*, 1, Prog. 413 (1981); in *Physical Chemical Properties of Drugs*, S.H. Yalkowsky, A.A. Sinkula, and S.C. Valvani (Eds.), Marcel Dekker, New York, 1980, pp. 321–347.

190. Gavezzotti, A., *J. Am. Chem. Soc.*, 107, 962 (1985).

191. Dearden, J.C. and R.M. Nicholson, *Pest. Sci.*, 17, 305 (1986).

192. Fujita, A., *Pharm. Bull. (Jpn.)*, 2, 163 (1954).

193. Matsuo, M., *Chemosphere*, 8, 477 (1979).

194. Matsuo, M., *Chemosphere*, 10, 1073 (1981).

195. Dearden, J.C. and R.M. Nicholson, in *QSAR in Environmental Toxicology-II*, K.L.E. Kaiser (Ed.), Reidel Publishers, Amsterdam, pp. 83–89.

196. Such programs are available from the Quantum Chemical Program Exchange at Indiana University, Bloomington, IN.

197. Atkinson, R., *Int. J. Chem. Kinetics*, 19, 799 (1987).

198. Darnall, K.R., R. Atkinson, and J.N. Pitts, Jr., *J. Phys. Chem.*, 82, 1581 (1978); Gaffney, J.S. and S.Z. Levine, *Int. J. Chem. Kinetics*, 11, 1197 (1979); Sanhueza, E. and E. Lissi, *Int. J. Chem. Kinetics*, 13, 317 (1981); Nielsen, T., *Environ. Sci. Technol.*, 18, 157 (1984); Gusten, H., L. Klasinc, and D. Maric, *J. Atmos. Chem.*, 2, 83 (1984).

199. Mill, T., *Environ. Toxicol. Chem.*, 8, 31 (1989).

200. Weiner, H., *J. Am. Chem. Soc.*, 17, 2636 (1957).

201. Randić, M., *J. Am. Chem. Soc.*, 97, 6609 (1975).

202. Mercier, C. and J.-E. Dubois, *Eur. J. Med. Chem.*, 14, 415 (1979); Basak, S.C., D.P. Gieschen, and V.R. Magnuson, *Environ. Toxicol. Chem.*, 3, 191 (1984); Devillers, J., P. Chambon, D. Zakarya, M. Chastrette, and R. Chambon, *Chemosphere*, 16, 1149 (1987).

203. Hall, L.H. and L.B. Kier, *Eur. J. Med. Chem.*, 16, 399 (1981).

204. Veith, G., State-of-the-Art of Structure Activity Methods Development, U.S. EPA-600/S3-81-029, Duluth, MN, 1981.

205. Sabljic, A. and N. Trinajstic, *Acta Pharm. Jugosl.*, 31, 189 (1981); a microcomputer version called GRAPH III is available from R. McDiarmid, Laboratory of Chemical Physics, NIH, Bdg.2, Rm B1-07, Bethesda, MD 20892.

206. Kier, L.B. and L.H. Hall, *Molecular Connectivity in Structure-Activity Analysis*, Research Studies, London, 1986.

207. Kier, L.B. and L.H. Hall, *J. Pharm. Sci.*, 65, 1806 (1976).

208. Hall, L.H. and L.B. Kier, *J. Pharm. Sci.*, 66, 642 (1977).

209. Sabljic, A., *J. Agric. Food Chem.*, 32, 243 (1984).

210. Koch, R., *Chemosphere*, 11, 925 (1982).

211. Govers, H., C. Ruepert, and H. Aiking, *Chemosphere*, 13, 227 (1984).

212. Yoshioka, Y., T. Mizuno, Y. Ose, and T. Sato, *Chemosphere*, 15, 195 (1986).

213. Koch, R. *Toxicol. Environ. Chem.*, 6, 87 (1983).

214. Moulton, M.P. and T.W. Schultz, *Chemosphere*, 15, 59 (1986).

215. Burkhard, L.P., A.W. Andren and D.E. Armstrong, *Chemosphere*, 12, 935 (1983).

216. Harmon, H.H., *Modern Factor Analysis*, 3rd ed., 487 p., Univ. of Chicago Press, 1976.

217. Nirmalakhandran, N.N. and R.E. Speece, *Environ. Sci. Technol.*, 22, 1349 (1988).

> "Praise to thee, my Lord, for all thy creatures, Above all Brother Sun, Who brings us the day and lends us his light."
>
> Canticle of Brother Sun, St. Francis of Assisi

CHAPTER 8

Photochemical Reactions in Natural Waters

8.1 INTRODUCTION

8.1.1 Importance of Photochemical Processes

The most important light-related process in the biosphere is photosynthesis of organic matter by plants. This provides the reduced, high-energy compounds that drive other ecosystem processes, and life as we know it would not be possible without photosynthesis. In the short term, humans are concerned primarily with terrestrial primary production because it provides most of our food and all our natural fiber and wood. Over geological time and global scale, aquatic primary production is responsible for maintaining the O_2 balance of the atmosphere. Aquatic systems (primarily the oceans) account for 2 to 3 times as much of the global primary production as do terrestrial systems. Although primary production is fundamentally a photochemical process, the subject is properly in the domain of biology rather than chemistry. Numerous aquatic biology books discuss primary production and its controlling factors, and except for a brief example in Chapter 6, photosynthesis is not treated in this book.

Natural water photochemistry is a relatively new field of study, and the importance of photochemical reactions in aquatic systems is not yet completely understood. Nonetheless, we do know that many aquatic solutes are affected directly or indirectly by light. Many biorefractory organic contaminants are degraded in aquatic systems by photochemical reactions, and these processes are important in the breakdown of petroleum and crude oil present in surface waters from spills and leaks. Natural organic matter, especially aquatic humus, also is degraded photochemically. The mechanisms of these reactions involve reactive inorganic intermediates such as singlet oxygen and hydroxyl radicals, as well as organic radicals and radical anions. Photochemical reactions also affect inorganic substances; for example, nitric oxide is produced by photoreduction of nitrate and nitrite, and the redox cycles of Fe and Mn are partially mediated by photochemical reactions. Photochemical transformations in aquatic systems are not limited to dissolved substances; recent studies have shown that oxide particles that act as semiconductors photocatalyze reactions of organic compounds adsorbed to their surfaces. This chapter describes the major mechanisms of photochemical transformations, rate equations and predictive models for such reactions, the chemistry of reactive inorganic intermediates in natural waters, and the importance of these processes in controlling the aquatic fate of natural and synthetic organic compounds.

8.1.2 Photophysical Processes

The primary process in all photochemical reactions is the absorption of a photon of light energy. Consequently, a necessary (but not sufficient) condition for a photochemical reaction is that at least one reactant (which we will call S) must be able to absorb light in the UV-visible range. The amount of energy in a photon of light is not trivial relative to activation energies for chemical reactions. For example, one photon at 300 nm is equivalent to 6.65×10^{-19} joules or $\sim 400 \, kJ \, mol^{-1}$ of photons, i.e., per Einstein (Ei). At 400 nm, a mole of photons is equivalent to 300 kJ (see Table 2-7). For midday, midsummer conditions at 40° latitude, solar radiation at the Earth's surface is 8.9×10^{19} photons $m^{-2} \, s^{-1}$ ($\sim 0.54 \, Ei \, m^{-2} \, h^{-1}$) between 300 and 400 nm and 4.2×10^{20} photons $m^{-2} \, s^{-1}$ ($2.5 \, Ei \, m^{-2} \, h^{-1}$) between 300 and 500 nm. This is equivalent to about $0.025 \, Ei \, L^{-1} \, h^{-1}$ if all the light is absorbed in the top meter.

Photon absorption raises an orbital electron from its ground state into a higher-energy state, producing a more reactive species, S^*. Several fates may befall S^*, and only some of these involve chemical reactions (Figure 8-1). Electronic transitions among excited states that leave the molecules intact are called photophysical processes, in contrast to photochemical processes that produce molecular transformations. Important photophysical processes that S^* may undergo are listed in Table 8-1.

Most ground-state molecules have an even number of electrons, i.e., they are not free radicals. The electron spins normally are paired in the ground state, which thus is a singlet state (S_0). Absorption of a photon promotes an electron to an excited singlet state (e.g., $^1S^*_1$, $^1S^*_2$). Both the ground and excited states have a range of vibrational and rotational states, and for molecules in solution, a continuum of energies is associated with each "level." Excited states lose their excess vibrational or rotational energy rapidly by radiationless thermal equilibration and drop to the state's lowest energy level (e.g., $^1S_2^*$), which in turn can pass into a vibrationally excited level of a lower excited state ($^1S_1^{*v}$) by "internal conversion". Excited singlet states generally are too short-lived ($\tau < 10^{-9}$ s) to react with solutes in natural waters. The process of "intersystem crossing" inverts the spin of an electron so that two electrons have parallel spins, and the molecule has a net spin of 1. Such triplet states ($^3S^*$) have much longer lifetimes than do singlet states, and most photochemical reactions involve excited triplet states.

The excited singlet state may emit a photon of a higher wavelength (lower

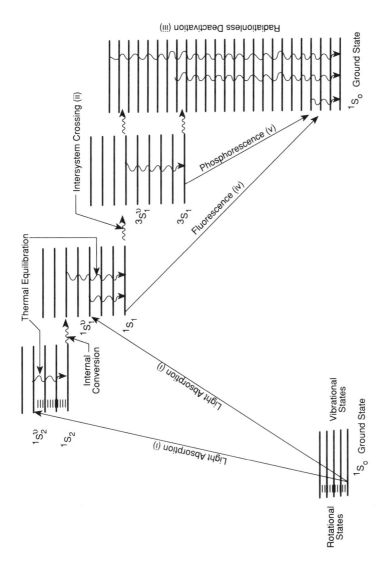

Figure 8-1. Photoactivation of ground-state molecule (assumed to be in singlet state) and return to ground state by various photophysical processes. (Redrawn from Adamson, A.W., *Textbook of Physical Chemistry*, 2nd ed., Academic Press, New York, 1979.)

Table 8-1. General Photophysical and Photochemical Reactions

Photophysical processes

$S_o + h\nu \rightarrow {}^1S^*$	light absorption; excitation	(i)
${}^1S^* \rightarrow {}^3S^*$	intersystem crossing	(ii)
${}^1S^*$ or ${}^3S^* \rightarrow S_o$	radiationless deactivation	(iii)
${}^1S^* \rightarrow S_o^v + h\nu$	fluorescence	(iv)
${}^3S^* \rightarrow S_o^v + h\nu$	phosphorescence	(v)
${}^3S^* + P \rightarrow S_o + {}^3P^*$	sensitization	(vi)
${}^3S^* + Q \rightarrow S_o\ (+ Q^*)$ ${}^\cdot Q \rightarrow Q$	quenching of S^* and (radia-tionless) deactivation of quenching agent	(vii)
${}^3S^* + O_2 \rightarrow S_o + {}^1O_2^*$	formation of singlet oxygen	(viii)
${}^1O_2 + Q \rightarrow O_2 + {}^3Q$ ${}^3Q \rightarrow Q$	quenching of singlet oxygen and (radiationless) deactivation of quenching agent	(ix)

Photochemical reactions

$S^* \rightarrow$ products, or $S^* + R \rightarrow\ \rightarrow$ products	direct photolysis; may involve many steps and other ground-state reactants	(x)
${}^3S^* + HR \rightarrow SH \cdot + R\cdot$	hydrogen transfer, leading to "photosensitized" oxidation of HR by free radical mechanism	(xi)
${}^1O_2 + R \rightarrow RO_2$ or $R'O + R''O$	addition of singlet oxygen to olefinic organic molecule	(xii)
${}^3S^* + O_2 \rightarrow S_{ox} + O_2^{\cdot -}$	oxidation of light-absorbing molecule and reduction of O_2 to superoxide ion	(xiii)
${}^3S^* + R \rightarrow S + R^*$ $R^* \rightarrow\ \rightarrow$ products	energy transfer from sensitizer to reactant, producing excited reactant state that subsequently forms product(s)	(xiv)

energy) than was absorbed and return to the ground state by the process of fluorescence. The balance of the excess energy is converted to thermal energy. Emission of a photon by a triplet state (returning the molecule to the ground singlet state) is called phosphorescence. Excited states also may return to the ground state by radiationless deactivation in which the excess energy is converted to vibrational energy of the molecule or adjacent solvent molecules.

Finally, a molecule in a singlet or triplet excited state may transfer its excess energy to a receptor molecule (R), which may react chemically or return to the

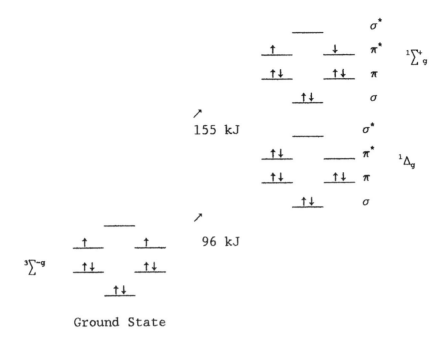

Figure 8-2. Molecular orbital diagram of ground-state molecular oxygen ($^3\Sigma^-_g$ state) and the first two excited states: $^1\Delta_g$ (singlet oxygen, $^1O_2$) and $^1\Sigma^+_g$. The $^1\Delta_g$ state is particularly reactive because it has an unfilled π^* orbital (lowest unoccupied molecular orbital, LUMO) and can accept a pair of electrons; the other π^* orbital is filled (the highest occupied molecular orbital, HOMO).

ground state by the avenues mentioned previously. The most important example of energy transfer from excited molecules is the formation of singlet oxygen $^1O_2$, the first excited state of O_2. Molecular O_2 is unusual in that its ground state has two unpaired electrons (i.e., a triplet state), and thus it is highly reactive. In $^1O_2$, the unpaired electrons become paired (Figure 8-2). Ground state O_2 is a very efficient quencher of triplet excited states, because $^1O_2$ is only about 96 kJ mol^{-1} above the ground state and thus is readily formed. In many respects this important photochemical intermediate is less reactive than ground state O_2, but it is an exceptional Lewis acid and has a special affinity to add to electron-rich organic molecules like olefins.

8.1.3 Photochemical Processes

Alternatively, S* may undergo chemical reaction and form other compounds by a variety of bond rearrangements that may involve other solutes and/or solvent molecules (Equations x–xiv, Table 8-1). Chemical reaction of a light-absorbing molecule (S) is called direct photolysis. Chemical reaction of a receptor molecule (R) or of other molecules that react with R is called indirect photolysis or photosensitized reaction. The latter can be induced by at least four types of interactions with sensitizer molecules (Equations xii–xiv, Table 8-1). Organic compounds that are not susceptible to direct photolysis because they do not absorb light at wavelengths available in the ambient environment (>295 nm) still can be photodegraded by indirect mechanisms.

Although some organic contaminants react directly with excited sensitizer molecules, the reactions commonly proceed through a more complicated sequence in which excited sensitizer molecules react to form highly reactive transient intermediates, and these in turn react with the organic contaminant, as described in the schematic diagram in Figure 8-3. By far the most important photosensitizing agent in natural waters is dissolved natural organic matter, DOM, especially aquatic humic material — humic and fulvic acids (symbolized in this chapter as AH). Recent studies have shown that nitrate and iron can be significant photosensitizers in some aquatic systems. The transient intermediates are mostly free radicals or radical ions ($^1O_2$ is an exception) that are formed from common inorganic species (H_2O, O_2, nitrate) or from DOM. A variety of scavenging or quenching agents aside from organic contaminants may react with the transient intermediates. Because these intermediates have very low concentrations, they are difficult to measure directly. Reagents that exhibit a special affinity for a given intermediate and react in a predictable manner — called "probes" — are used to measure production rates and estimate steady-state concentrations of transient intermediates (see Section 8.3.2).

Direct photolysis can be described by straightforward kinetic models, at least when rates are measured in terms of reactant loss. The variety of radicals that may be present in water and the myriad of reactions they can undergo with each other and with organic compounds and metal ions greatly complicate development of predictive kinetic models for photosensitized reactions. Essential data on rate constants are unknown for many reactions, and probably many reactions are themselves unknown. Reactions with aquatic humus are especially problematic. Consequently, chemists are able to derive simulation models for only a few fairly simple indirect photochemical processes occurring in natural waters. In contrast, highly complicated reaction sequences have been identified and simulated for important atmospheric reaction sequences such as smog formation in the troposphere and ozone formation and loss in the stratosphere.

8.2 KINETICS OF PHOTOCHEMICAL REACTIONS

8.2.1 Direct Photolysis

Kinetic models of photochemical reactions are based on two fundamental laws: the **Grotthus-Draper Law** (first law of photochemistry), which states that only light that is absorbed can effect chemical change, and the **Stark-Einstein Law** (second law of photochemistry), which states that one molecule is activated for each light photon absorbed by a system. In simplest terms, direct photolysis rates are proportional to the number of photons absorbed per unit time,[3,4] $I_a(\lambda)$, times the fraction of absorbed light that produces chemical change, i.e., the quantum yield ϕ_d.

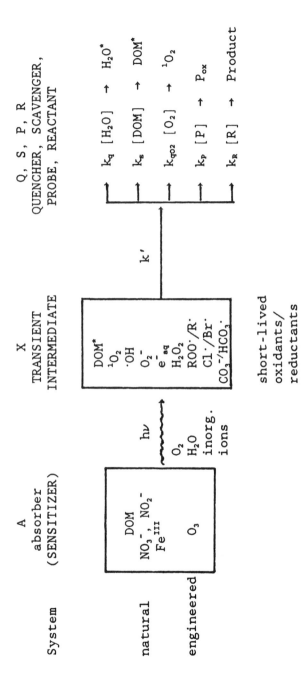

Figure 8-3. Schematic diagram of photosensitized reactions in aquatic systems, illustrating role of photosensitizing agents and alternative reaction pathways for transient intermediates.

Light absorption is proportional to [S] at low concentrations of absorbing compounds: $I_a^s(\lambda) = k_{a\lambda}[S]$. Direct photolysis thus is a first-order process, and the rate constant at a given λ, $k_{d\lambda}$ is $k_{a\lambda}\phi_d$ (t^{-1}):

$$\text{Rate}_\lambda = I_a^s(\lambda)\phi_d = k_{a\lambda}\phi_d[S] = k_{d\lambda}[S] \tag{8-1}$$

As noted in Section 8.1.2, not all absorbed photons produce chemical change; competing photophysical processes may return unstable excited molecules to the stable ground state. The fraction of absorbed photons that results in a given photochemical or photophysical process is called the primary quantum yield, ϕ_i (dimensionless). The sum of the ϕ_i for all processes that deactivate an excited molecule, by definition, is unity. Of principal interest is the primary quantum yield for chemical reaction of the excited molecule, i.e., direct photolysis, ϕ_d. This coefficient usually is not wavelength dependent.[3,4] The measured quantum yield for a direct photochemical reaction, ϕ_m, may differ from ϕ_d because secondary thermal reactions may occur. For example, the photoexcited molecule may be a free radical that initiates a chain reaction, in which case ϕ_d may be much greater than 1. Zepp and Cline[4] argued that chain reactions induced by direct photolysis of pollutant organics are rare because their concentrations are so low in natural waters and because phenolic humic materials present in many waters inhibit chain reactions. Measured and primary quantum yields (ϕ_m and ϕ_d) for direct photolysis usually are equal to each other (and less than 1).

The specific light absorption rate, $k_{a\lambda}$, varies with wavelength and depends on the molar absorptivity of S at λ (ε_λ) and the available light at λ. For reactions in natural waters, the latter is defined in terms of the "scalar irradiance," $E_o(\lambda,z)$, a variable that depends on λ, depth, and solar angle. $E_o(\lambda,z) = I(\lambda,z)D$, where $I(\lambda,z)$ is the radiant flux at λ on a horizontal surface at depth z, and D is a distribution coefficient — the ratio of the mean pathlength of light in a thin vertical layer divided by the thickness of the layer. E_o and D are derived in Section 8.2.4. For near surface conditions in natural waters, Equation 8-2 applies:

$$k_{a\lambda} = 2.3E_o(\lambda,0)\varepsilon_\lambda \tag{8-2}$$

$E_o(\lambda,0)$ is the scalar irradiance just below the water surface. Molar absorptivity $(M^{-1}\ cm^{-1})$ is obtained from **Beer's Law** ($A_\lambda = \varepsilon_\lambda lc$), where A_λ is the absorbance measured by spectrophotometer, l is the cell pathlength (cm), and c is molar concentration of S. The observed rate constant for direct photolysis is obtained by integrating over the wavelength range of light absorbed by S:

$$k_d = 2.3\phi_d \int_{\lambda 1}^{\lambda 2} E_o(\lambda,0)\varepsilon_\lambda\, d\lambda \tag{8-3}$$

Plots of $\phi_d\varepsilon_\lambda$ vs. λ are called action spectra and are a common way of portraying effects of wavelength on photolysis. Because scalar irradiance varies substantially with λ in the ambient environment, a more useful plot involves $k_{a\lambda}$ (from Equation

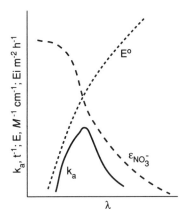

Figure 8-4. Sketch of NO_3^- absorptivity (ε_λ, dashed line), intensity of solar radiation at water surface (scalar irradiance, $E°$, dotted line), and resulting $k_{a\lambda}$ (solid line) for photolysis of nitrate as a function of wavelength (see Equation 8-2). λ_{max} for $k_a = 320$ nm.

8-2) vs. λ (Figure 8-4). The overlap of the absorptivity curve for a reactant with the irradiance curve readily shows whether the reactant has a high or low rate of photon absorptions in a natural water. Determining photolysis rate constants by Equation 8-3 is tedious because of the necessity to evaluate the integral containing E_o and ε. Some workers[5,6] calibrate photolysis rate constants in terms of total or visible solar radiation at the Earth's surface:

$$\text{Rate} = \left(k_{obs} / I_t\right)[S] = k'_d[S] \qquad (8\text{-}4)$$

$k'_d = k_{obs}/I_t$ has units of $s^{-1}/\mu Ei\ cm^{-2}\ s^{-1}$, or $cm^2/\mu Ei$. This works well for compounds such as nitrosamines, polynuclear aromatic hydrocarbons, and Fe cyanide complexes (Figure 8-5a), which absorb in the visible and uv-A regions (>320 nm). The intensity of incident light at these wavelengths does not vary much with solar angle or cloud cover.[1] This approach also is valid for photosensitized reactions induced by uv-A and visible light, but not for compounds like dioxans and most pesticides that absorb primarily in the uv-B region (280 to 320 nm), because the spectral quality of this light varies with solar angle (Figure 8-5b). The spectral character of uv and visible light varies with depth in natural waters, because of selective absorbance of light by water and natural solutes. As a result, the depth dependence of photolysis rates is not necessarily the same as that for total or visible-light intensity.

The effect of quenching (reaction vii, Table 8-1) on quantum yields of direct photolysis reactions is given by the **Stern-Volmer equation**:[7,8]

$$\phi_d^o / \phi_d^q = 1 + k_{vii}\tau_o[Q] \qquad (8\text{-}5)$$

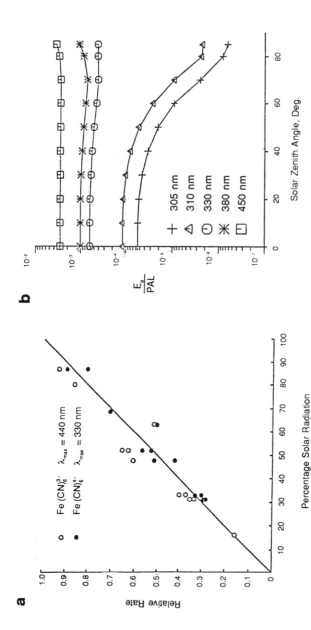

Figure 8-5. (a) Photolysis rate for iron cyanide complexes, which absorb primarily in the visible range, is a linear function of integrated visible radiation (400–700 nm) over a broad range of light conditions; axes are normalized to clear sky values. (b) Ratio of scalar irradiance, E_o, at various λ wavelengths to integrated irradiance over 350–700 nm (range of photosynthetically active light, PAL) vs. solar zenith angle. Ratios computed for just below the water surface. [From Zepp, R.G., in *Dynamics, Exposure, and Hazard Assessment of Toxic Chemicals*, R. Haque (Ed.), Ann Arbor Science, Ann Arbor, MI, 1980, p. 69. With permission.]

$\phi_d{}^o$ and $\phi_d{}^q$ are quantum yields of a direct photolysis reaction in the absence and presence of quenching agent Q, respectively, k_{vii} is the rate constant for the quenching of S^* by Q, and τ_o is the lifetime of S^* in the absence of Q. Dissolved oxygen is a very good quenching agent,[9] and in this case the product of quenching ($^1O_2$) (reaction viii, Table 8-1) itself may induce a photosensitized reaction (reaction xii, Table 8-1). As Equation 8-5 indicates, quenching is not significant if τ_o is sufficiently small. For most compounds k_{viii} (for quenching by O_2) $= 2 \times 10^9\, M^{-1}\, s^{-1}$.[9] Under O_2-saturated conditions (0.25 mM), the product $k_{viii}[O_2] \approx 5 \times 10^5\, s^{-1}$, and quenching by O_2 thus affects ϕ for direct photolysis significantly when $\tau_o > 1$–$2 \times 10^{-7}\, s$.

8.2.2 Indirect Photolysis

As mentioned previously, photosensitized reactions are mechanistically complex. The rate-limiting step to form products may be a nonphotochemical reaction of ground-state reactants with intermediates produced from receptor molecules, and the rate of product formation may not be directly related to the kinetics of the primary photochemical process. A model that is both general and precise in describing reaction rates for complicated mechanisms thus is not feasible.

A simple kinetic model can describe the fate of primary receptor molecules, however.[3] The model includes six processes:

$$
\begin{array}{ll}
S + h\nu \;\rightarrow^1 S^* & (i) \\[4pt]
^1S^* \;\rightarrow^3 S^* & (ii) \\[4pt]
^3S^* + R \;\rightarrow S +\, ^3R^* & (vi) \\[4pt]
^3S^* + O_2 \;\rightarrow S +\, ^1O_2 & (viii) \\[4pt]
^3S^* \;\rightarrow S & (iii) \\[4pt]
^3R^* \;\rightarrow\rightarrow \text{products} &
\end{array}
\qquad (8\text{-}6)
$$

The first five equations are numbered as they appear in Table 8-1. The overall reaction rate of receptor molecules is the product of four terms:

$$
\text{Rate} = I_a^s \phi_{ls} \phi_{et} \phi_r \qquad (8\text{-}7)
$$

I_a^s is the rate of light absorption by sensitizer S over all λ, ϕ_{ls} is the fraction of absorbed light resulting in triplet formation, ϕ_{et} is the fraction of triplet energy transferred from $^3S^*$ to R, and ϕ_r is the quantum yield for reaction of R. In turn, ϕ_{et} is given by the ratio of the rate of Equation vi to the sum of the rates of all loss processes for $^3S^*$:

$$
\phi_{et} = \frac{k_{vi}\left[^3S^*\right]\left[R\right]}{k_{iii}\left[^3S^*\right] + k_{viii}\left[^3S^*\right]\left[O_2\right] + k_{vi}\left[^3S^*\right]\left[R\right]} \qquad (8\text{-}8a)
$$

$[^3S^*]$ appears in all terms on the right side of Equation 8-8a and can be eliminated. Because $[O_2] \gg [R]$ in natural waters, reaction vi usually is small compared with reaction viii. Thus, ϕ_{et} becomes

$$\phi_{et} = \frac{k_{vi}[R]}{k_{iii} + k_{viii}[O_2]} \qquad (8\text{-}8b)$$

The rate of light absorption by S at a given λ, $I_{a\lambda}^s$, is obtained from Equation 8-2:

$$I_{a\lambda}^s = k_{a\lambda}[S] = 2.3\varepsilon_{\lambda S}E_0(\lambda,0)[S] \qquad (8\text{-}9a)$$

$\varepsilon_{\lambda S}$ = molar absorptivity of S at λ, and $E_0(\lambda,0)$ is the scalar irradiance just below the water surface. The total rate of light absorption by S is obtained by integrating over the range in which S absorbs light:

$$I_a^s = 2.3[S]\int_{\lambda 1}^{\lambda 2} \varepsilon_{\lambda s}E_o(\lambda,0)d\lambda \qquad (8\text{-}9b)$$

Substituting Equations 8-8b and 8-9b into Equation 8-7 and collecting terms yields the desired expression for near surface rates:

$$\text{Rate} = \frac{k_s[S][R]}{k_{iii} + k_{viii}[O_2]} \qquad (8\text{-}10)$$

$$k_s = \left\{ 2.3 \int \varepsilon_\lambda E_o(\lambda,0)d\lambda \right\} k_{vi}\phi_{ls}\phi_r$$

A wavelength-independent expression (Equation 8-4) can be used to approximate k_s for some photosensitized reactions under near-surface conditions.

8.2.3 Evaluation of Light Intensity in Natural Waters

The principal difference between the rate equations for photolysis and rate expressions for thermal reactions is the dependency of the former on the rate of light absorption. This is a function of both environmental and reactant parameters ($k_{a\lambda} = 2.3E_0(\lambda)\varepsilon_\lambda$). ε_λ varies with compound and wavelength, but is readily measured by spectrophotometry. Scalar irradiance, $E_0(\lambda)$, the measure of available light, depends on many factors and is difficult to predict exactly, especially as a function of depth in aquatic systems (see Section 8.2.4). Both the spectral quality and intensity of light vary with solar angle, cloud cover, depth in the water column, and composition of material in the water, especially in the uv-B range (280 to 320 nm), which is responsible for direct photolysis of many organic pollutants.[2] Surface irradiance below 320 nm decreases rapidly because of absorbance by stratospheric ozone; for the same reason, essentially no solar radiation less than 295 nm reaches the Earth's surface. As mentioned earlier, light intensity in the uv-B region decreases much more rapidly with increasing solar angle than it does in the uv-A (320 to 400 nm) or visible regions. Seasonal and latitudinal variations in stratospheric and tropospheric ozone concentrations also affect uv-B intensity at the Earth's surface.

As solar radiation penetrates the sky, some is scattered by aerosols, some is absorbed by molecules, and some is transmitted to Earth. Scattering in the atmosphere increases with decreasing wavelength; the scattered light, which is predominantly short wavelengths, illuminates the sky and causes its blue color. Light at the Earth's surface thus has two components: that directly from the sun (I_d), and that scattered from the sky (I_s). Factors affecting incident irradiance are understood sufficiently that $G(\lambda)$, global irradiance (sun + sky) at the Earth's surface, and $E_o(\lambda, 0)$, total scalar irradiance $(I_d + I_s)D$ just below the water surface, can be modeled as functions of season, latitude, and time of day. Computer codes are available to compute $G(\lambda)$ and $E_o(\lambda)$.[4,10-12]

Effects of cloud cover on the spectral quality and intensity of surface light are somewhat uncertain, but probably less important than most people would predict. Contrary to intuition, clouds transmit uv radiation to a greater extent than visible light, but the difference is small, except for thick cloud layers and large solar angles.[4,13] Consequently, the reduction in solar uv radiation at the Earth's surface can be estimated by the reduction in total radiation caused by clouds. An empirical relationship for the effects of cloud cover on fractional reduction uv intensity F_c is[14]

$$F_c = 1 - 0.056 C_f \qquad (8\text{-}11)$$

C_f is the fraction (in tenths) of the sky covered by clouds. According to Mo and Green,[11] the annual average reduction in uv intensity caused by cloud cover is less than a factor of two in several U.S. urban areas.

Factors affecting the spectral quality of light as a function of depth in the water column include the nature and concentrations of light-scattering particles and light-absorbing natural dissolved organic matter (DOM), especially aquatic humus, as well as absorbance by water itself. (Concentrations of organic pollutants usually are too low to significantly affect light absorption in the water column.) These factors are relatively well understood,[3,4,15-17] but their quantitative importance in a given water body is difficult to predict, especially over time. Thus, it is much simpler to predict "near-surface" photolysis rates than depth-integrated rates.

Light attenuation in pure water varies with wavelength, and water is less transparent at both extremes of the uv-visible spectrum than at midrange (blue light). Nonetheless, the transparency of pure water to uv light is much greater than commonly supposed, and the thickness of the zone in which uv-mediated photochemical reactions may occur (the so-called photic zone) is relatively large. It is convenient to define the bottom of the photic zone as the depth at which light intensity is 1% of incident light (i.e., the depth at which 99% of the incident light has been attenuated). From **Lambert's Law**,

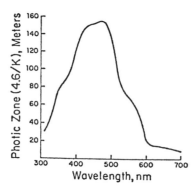

Figure 8-6. Approximate photic zone of open ocean as function of wavelength. [From Smith, R.C. and J.E. Tyler, *Photochem. Photobiol. Rev.*, 1, 117 (1976). With permission.]

$$E(\lambda, z) = E_o(\lambda)\exp(-K_\lambda z) \tag{8-12}$$

The photic zone depth is equal to $4.6/K_\lambda$, where K_λ is the diffuse attenuation coefficient of the water at λ. K_λ varies with λ and depends on the nature and concentration of light-scattering and light-absorbing material in the water (assumed constant over depth in a well-mixed system).

The photic zone of oligotrophic midocean water, which is very low in uv-absorbing DOC and light-scattering particles, ranges from about 30 m at the high end of the uv-B range (~320 nm) to about 160 m near 500 nm and again to much smaller values (<20 m) above 600 nm (Figure 8-6). Because of light scattering by particles and absorbance by organic matter, the photic zone of most freshwaters is much less than that in pristine ocean waters. Light attenuation by particle scattering varies with wavelength, but to a smaller extent than does absorbance by DOM. The absorptivity of humic materials is low above ~450 nm and increases exponentially with decreasing λ in the low visible and uv regions (Figure 8-7; see Section 8.4.3). Photosynthetic pigments (chlorophylls and carotenoids) from phytoplankton absorb strongly in the uv region and have two broad peaks of absorbance in the visible region: from about 400 to 500 nm and above 600 nm. The wavelength dependency of light absorbance in natural waters thus depends on the nature and source of light-absorbing DOM, i.e., on relative concentrations of aquatic humus and plant pigments[18] (Figure 8-7). However, both sources cause rapidly decreasing light transmittance with decreasing λ below 500 nm, the range with sufficient energy to induce photochemical reactions.

The relative importance of turbidity and DOM on the spectral quality of light varies widely among water bodies and can be evaluated by the ratio $4.6/K_R : 2Z_{SD}$.[3] K_R is the light attenuation coefficient determined by a Robertson light meter, which measures uv-B light (~312 nm), and Z_{SD} is the Secchi disk transparency (in m), which is a measure of visible light transmittance.[19] This ratio is as high as 0.5 in

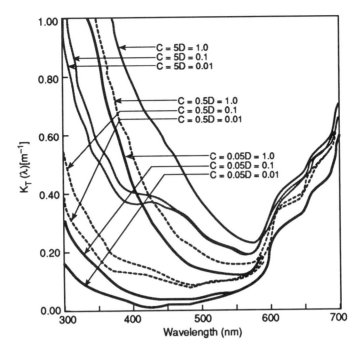

Figure 8-7. Effects of varying concentrations and ratios of dissolved organic matter (DOM) and phytoplankton chlorophyll on diffuse light attenuation coefficient K as a function of λ. D = DOM, mg L^{-1}; C = chlorophyll, μg L^{-1}. [From Zafiriou, O. C., et al., *Environ. Sci. Technol.*, 18, 358A (1984). With permission of the Amer. Chem. Soc.]

turbid waters where light attenuation is dominated by silt and clay, and as low as 0.05 in waters where DOM causes most of the light attenuation. $2Z_{SD}$ often is accepted as the limit of the photic zone for for net primary production; $4.6K_R$ is a measure of the uv-B photic zone.

Measurement of light attenuation in natural waters is more complicated than it might seem, even if we ignore complications arising from wavelength variations. Two principal approaches are used, but they do not measure the same phenomenon and thus yield different attenuation coefficients. *Beam attenuation coefficients* (α) measure the transmittance of a collimated beam of light through water, usually with a conventional spectrophotometer. Beam attenuation is caused by two phenomena: molecular absorption and particle scattering. Most scattering occurs in a forward direction in natural waters, but not exactly in the same direction as the collimated beam. Consequently, most of the forward scattered light does not reach the detector in a spectrophotometer, and

$$\alpha = a + s \qquad (8\text{-}13)$$

a is the absorption coefficient, and s the total scattering coefficient. In contrast, *diffuse attenuation coefficients* (K) are determined by measuring solar irradiance directly at various depths in a natural water. K is obtained from the slope of a plot of ln(irradiance) vs. depth: $I(z) = I(0)\exp(-Kz)$. K, I, a, a, and s are all understood to be wavelength dependent. K is smaller than a in natural waters, because the irradiance measured with a light meter includes forward-scattered light.[3,20] Several approximate relationships have been derived for K:[20,21]

$$K \approx \alpha - s_f \qquad\qquad (8\text{-}14a)$$

$$K \approx Da + s_b \qquad\qquad (8\text{-}14b)$$

where s_f and s_b are forward and back components of the scattering coefficient and D is the distribution function (ratio of mean pathlength of light in a layer to the thickness of the layer). In particle-free water, α theoretically is equal to K,[3] but even in very clear ocean waters, $\alpha > K$ in the uv range. Because forward-scattered light can induce photochemical reactions, diffuse attenuation coefficents (K) are the best measure of light penetration in natural waters, especially those with significant turbidity.

8.2.4 Calculation Model for Direct Photolysis in Natural Waters

Rates of photochemical reactions depend on light intensity, which varies in known, but somewhat complicated ways as functions of time of day, season, latitude, elevation of the water body, depth in the water, and (in many cases) the thickness of the ozone layer. Consequently, calculation of ambient photolysis rates from lab-measured rate constants is time consuming and tedious. The above factors also need to be taken into account in comparing ambient rate measurements made at different times and/or places. GCSOLAR[12] is a calculation model for photochemical reactions in natural waters, which computes direct photolysis rates and half-lives of pollutants as a function of $E_o(\lambda,z)$ in waters with low-suspended solids, where light attenuation is caused primarily by absorption. The model accounts for the above factors and is available as an interactive program for personal computers. Knowledge of programming languages is not needed to run GCSOLAR. This section describes the physical basis for the model[4] and some things that can be done with it.

When light beams in the atmosphere encounter a water surface, a fraction is reflected at an angle equal to that of incident radiation, and refraction changes the direction of light that enters the water (Figure 8-8). The fraction of direct sunlight reflected by water surfaces is less than 0.10, except at very high solar zenith angles, Z_A. If the sky is uniformly bright, the reflected fraction of sky radiation is ~0.07 .[4] Refraction changes the angle of light beams according to **Snell's Law**: $n = \sin Z_A / \sin\theta$, where n = refractive index (n = 1.34 for water), and θ is the angle of refracted light (from the vertical) in water. θ increases with increasing Z_A and reaches a maximum of 48° at $Z_A > 85°$.

The intensity of sun and sky radiation on a horizontal plane below the surface of a water body, symbolized by W_λ, can be computed from (sun + sky) light intensity

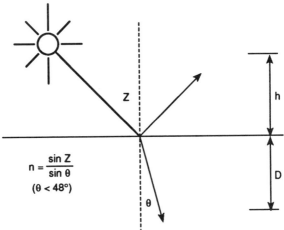

Z = angle of incidence (solar zenith angle in case of direct sunlight);
θ = angle of refraction

Intensities and Pathlengths of Solar Radiation in Atmosphere and Water Bodies

	Atmosphere[a]	Water Body
Direct intensity on horizontal plane	S cos z	S cos z (1-RFD) = I_d
Sky intensity on horizontal plane	H	H(1-RFS) = I_s
Pathlength of direct radiation	h sec z	D sec θ = ℓ_d
Average pathlength of sky radiation	2 h	1.2 D = ℓ_s

Figure 8-8. Schematic of sunlight passage into a water body and formulas for pathlengths and intensities of direct and sky radiation in atmosphere and water body. [From Zepp, R.G. and D.M. Cline, *Environ. Sci. Technol.*, 11, 359 (1977). With permission of the Amer. Chem. Soc.]

at the Earth's surface (G_λ) by taking reflection into account (Figure 8-8). The amount of light absorbed within a given depth of water depends on the attenuation coefficient and the pathlength taken by light beams. Pathlengths of light in water differ for sun and sky radiation. The former can be derived by simple trigonometric relationships (Figure 8-8). The latter depends on the average pathlength of sky radiation in the atmosphere[22] and has been estimated to be ~1.2(water depth).[4] The intensities and pathlengths of sun and sky light are fundamental parameters of the model, which is derived as follows.

The amount of light absorbed per unit time ($I_{a\lambda}$) over some pathlength ℓ (cm) in a completely mixed water body is

$$I_{a\lambda} = W_\lambda - I_{\ell\lambda} = W_\lambda\left(1 - 10^{-\alpha_\lambda \ell}\right) \qquad (8\text{-}15)$$

W_λ = incident light intensity, $I_{\ell\lambda}$ = light intensity at pathlength ℓ from incident, and α_λ = decadic light attenuation coefficient. The rate of absorption, $I'_{a\lambda}$ per unit volume for underwater light in a layer of thickness z is

$$I'_{a\lambda} = \frac{W_\lambda}{z}\left(1 - 10^{-\alpha_\lambda \ell}\right) = \frac{1}{z}\left\{I_{d\lambda}\left(1 - 10^{-\alpha_\lambda \ell_d}\right) + I_{s\lambda}\left(1 - 10^{-\alpha_\lambda \ell_s}\right)\right\} \quad (8\text{-}16)$$

I_d, I_s, ℓ_d, and ℓ_s are defined in Figure 8-8.

Addition of a photoreactive compound to water increases the absorption coefficient to $\alpha_\lambda + \varepsilon_\lambda[S]$, where ε_λ is molar absorptivity. Generally, $\varepsilon_\lambda[S] \ll \alpha_\lambda$, and the fraction of the absorbed light absorbed by S $\approx \varepsilon_\lambda[S]/\alpha_\lambda$. The average rate of light absorption at a given concentration thus is a constant fraction of the average rate of light absorption. This may be expressed as

$$I''_{a\lambda} = I_{a\lambda}\varepsilon_\lambda[S]/j\alpha_\lambda = ka_\lambda[S] \quad (8\text{-}17)$$

$k_{a\lambda} = I_{a\lambda}\varepsilon_\lambda/j\alpha_\lambda$, and j is a constant that converts light intensity into units compatible with [S]; $j = 6.02 \times 10^{20}$ when [S] is in M and light intensity is in photons cm^{-2} s^{-1}.[4] Two limiting conditions can be considered to simplify the determination of $k_{a\lambda}$.

1. If $\alpha_\lambda \ell_d$ and $\alpha_\lambda \ell_s$ both are greater than 2, the system has absorbed nearly all the solar radiation available for direct photolysis, and the exponential terms in Equation 8-16 can be ignored. From Equations 8-15 to 8-17, $k_a\lambda$ thus becomes

$$k_{a\lambda} = \frac{\left(I_{d\lambda} + I_{s\lambda}\right)\varepsilon_\lambda}{jz\alpha_\lambda} = \frac{W_\lambda \varepsilon_\lambda}{jz\alpha_\lambda} \quad (8\text{-}18)$$

Values of W_λ applicable to midday and midseason clear sky conditions at 40°N latitude are listed in Table 8-2. In effect, Equation 8-18 defines rate constants for direct photolysis in optically opaque systems — water bodies deep enough to absorb all incident light. Under these conditions, wavelength-specific rate constants are directly proportional to the total irradiance entering the water and the molar absorptivity of the compound and are inversely proportional to the light attenuation coefficient and depth of the water column over which photolysis is occurring.

2. If $\alpha_\lambda \ell_d$ and $\alpha_\lambda \ell_s$ both are less than ~0.02, the numerator terms in parentheses in Equation 8-16 are approximately equal to 2.3 times the value of the exponent. Thus, $k_{a\lambda}$ becomes independent of α_λ and can be expressed as

$$k_{a\lambda} = 2.3\left(I_{d\lambda}\ell_d + I_{s\lambda}\ell_s\right)\varepsilon_\lambda/jz \quad (8\text{-}19)$$

Substituting expressions for ℓ_d and ℓ_s from Figure 8-8 into Equation 8-19 yields

$$k_{a\lambda} = 2.3\varepsilon_\lambda Z_\lambda/j \quad (8\text{-}20a)$$

where

$$Z_\lambda = I_{d\lambda} \sec\theta + 1.2 I_{s\lambda} \quad (8\text{-}20b)$$

Values of Z_λ applicable to midday and midseason clear sky conditions at 40°N latitude are given in Table 8-2. Equation 8-20 is applicable to near-surface

Table 8-2. Seasonal Values of W_λ and Z_λ for Latitude 40°N

Wavelength (nm)	W_λ				Z_λ			
	Spr	Sum	Fall	Wtr	Spr	Sum	Fall	Wtr
	$(10^{13}$ photons cm^{-2} s^{-1} 2.5 nm$^{-1})$							
297.5	0.024	0.065	0.008	0.000	0.027	0.072	0.009	0.000
300.0	0.105	0.219	0.043	0.006	0.120	0.240	0.052	0.007
302.5	0.369	0.657	0.185	0.030	0.419	0.723	0.223	0.037
305.0	1.06	1.63	0.555	0.039	1.21	1.81	0.670	0.170
307.5	0.95	2.74	1.12	0.369	2.23	3.05	1.35	0.450
310.0	3.25	4.44	1.73	0.698	3.72	4.95	2.08	0.854
312.5	5.10	6.43	3.08	1.45	5.84	7.17	3.71	1.77
3l5.0	6.83	8.36	4.10	2.22	7.80	9.33	4.94	2.71
3l7.5	8.67	10.3	5.32	2.96	9.92	11.5	6.41	3.62
320.0	10.3	12.1	6.63	4.08	11.7	13.5	8.08	4.98
	$(10^{13}$ photons cm^{-2} s^{-1} 3.75 nm^{-1}							
323.1	19.3	22.6	11.9	7.40	22.1	25.2	14.4	9.06
	$(10^{13}$ photons cm^{-2} s^{-1} 10 nm$^{-1})$							
330	66.9	76.2	42.1	27.9	76.1	84.6	50.8	34.2
340	77.8	87.5	50.0	34.1	88.0	96.3	60.4	42.0
350	83.5	93.8	53.3	36.3	94.2	103	64.5	44.9
360	89.5	100	56.8	38.3	101	110	68.7	47.9
370	99.7	112	62.3	41.8	112	122	75.4	52.0
380	110	124	67.9	45.0	124	135	82.2	56.2
390	133	148	89.5	64.6	149	161	108	80.5
400	191	212	129	93.1	213	231	156	116
410	251	279	170	123	280	302	206	154
420	258	287	175	127	288	310	212	159
430	249	277	170	123	277	298	205	154
440	295	327	201	146	327	351	244	184
450	332	368	227	164	368	394	275	208
460	335	372	230	167	371	398	279	211
470	347	384	238	172	384	411	289	219
480	355	394	244	177	392	420	296	225
490	336	372	231	168	371	396	281	213
500	343	380	236	171	378	404	287	218

Note: W_λ and Z_λ are for λ intervals centered about the value given in column 1; e.g., W_{330} is for λ range 325 to 335 nm.

Tabulated from Zepp, R.G. and D.M. Cline, *Environ. Sci. Technol.*, 11, 359 (1977).

conditions where less than ~5% of the incident light is absorbed (including contributions of $\varepsilon_\lambda[S]$ to absorption). According to Equation 8-20, $k_{a\lambda}$ is independent of depth (but the depth to which the equation applies obviously depends on the rate of light attenuation).

Comparison of Equation 8-20a with Equation 8-2 shows that Z_λ/j is the same as the near-surface scalar irradiance, $E_o(\lambda,0)$. It also can be shown that $E_o(\lambda) = W_\lambda D_\lambda/j$, where D_λ is the distribution function

$$D_\lambda = I_d \sec\theta + 1.2 I_s / (I_d + I_s) \qquad (8\text{-}21)$$

Algebraic manipulation of Equation 8-21 shows that D_λ is the ratio of the mean pathlength of light, ℓ, in a given layer to the thickness, z, of the layer. The depth-averaged irradiance in a well-mixed system can be calculated from Lambert's Law:[3]

$$E_o^{av}(\lambda,z) = E_o(\lambda,0)\{1 - \exp(-K_\lambda z)\}/K_\lambda z = E_o(\lambda,0)F_\lambda/K_\lambda z \quad (8\text{-}22a)$$

or

$$I_\lambda^{av} = W_\lambda\{1 - \exp(K_\lambda z)\}/K_\lambda z = W_\lambda F_\lambda/K_\lambda z \quad (8\text{-}22b)$$

F_λ is the fraction of light attenuated at depth z. When $K_\lambda z = 0.1$, the depth-averaged irradiance is 95% of the incident irradiance. Near-surface rates (computed from incident irradiance) thus apply to the region where $K_\lambda z < \sim 0.1$.

The computer program GCSOLAR determines $W(\lambda)$, $Z(\lambda)$, and k_a, and it models direct photolysis rates at specified latitudes, longitudes, elevations, seasons, and times of day. Figure 8-9 shows that computed rates compare favorably with measured direct photolysis rates for several compounds. Example 8-1 illustrates use of the program to compare photolysis rates for differing geographic locations and seasons of the year.

The model described above applies to cases where light attenuation by scattering is less important than that by absorption. In these cases, the distribution function (D_λ) is less than 1.2.[20] In waters where scattering is the dominant factor in attenuation, D_λ increases with depth as light becomes increasingly diffuse. Because direct photolysis rate constants are proportional to D_λ, this should increase photolysis rates. Miller and Zepp[20] evaluated the effects of scattering by suspended sediments on D_λ in laboratory photolysis experiments with a phenyl ketone that absorbs in the mid-uv. They computed the ratio D_s/D_{dw}, (subscripts s and dw refer to water with suspended sediments and distilled water, respectively) and found that the ratio varied from 1.06 to 1.69 for six rivers and ponds. In contrast, D_s/D_{dw} was 1.03 (near the expected value of unity) for a nonscattering solution of humic acid. If $D_{dw} = 1.2$, the range of D_s for the suspensions was 1.3 to 2.0 (ave = 1.6 ± 0.2). This implies that, on average, the mean photolysis rate was about a third higher in the suspensions than in clear water.

Example 8-1. Use of GCSOLAR to Compare Photolysis Rates

In its current configuration, GCSOLAR does not have a library of light-attenuation coefficients (k_λ) for natural waters or absorbance data for organic contaminants of interest relative to loss by photolysis in aquatic systems. This information needs to be supplied by the user. However, the following data are supplied with GCSOLAR for k_λ of pure water and ε_λ, the molar extinction coefficient of methoxychlor, an organochlorine insecticide:

λ	297.5	300	302.5	305	307.5	310	312.5	315	317.5	320	323.1	330
ε	11.1	4.67	1.90	1.10	0.80	0.53	0.33	0.27	0.16	0.10	0.06	0.02
k_λ	69	61	57	53	49	45	43	41	39	37	35	29

The units of ε are $M^{-1} cm^{-1}$; k_λ is in $10^{-5} cm^{-1}$. The quantum yield, ϕ_d, for methoxychlor is 0.32.

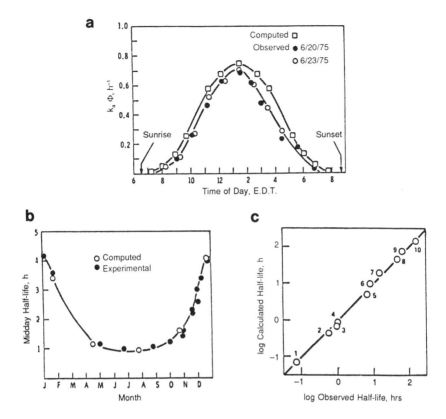

Figure 8-9. Comparison of measured rates of direct photolysis vs. rates computed by model of Zepp and Cline:[4] (a) Photolysis of DMDE, an analog of DDT, vs. time of day; (b) photolysis of DMDE vs. month of year at Athens, GA; (c) calculated vs. observed half-lives for various pesticides and their derivatives: (1) N-nitrosoatrazine; (2) trifluralin; (3) DMDE; (4) DMDE in hexadecane (HD); (5) DDE in HD; (6) diphenyl mercury; (7) phenylmercuric acetate; (8) 2,4-D-butoxyethyl ester in HD; (9) carbaryl; (10) 2,4-D butoxyethyl ester. All compounds in water, except as noted. [From Zepp, R.G. and D.M. Cline, *Environ. Sci. Technol.*, 11, 359 (1977). With permission of the Amer. Chem. Soc.]

The program is run by inserting simple one-word commands and parameter values. The italicized words below indicate commands needed to obtain near-surface photolysis rate constants for methoxychlor at midseason and several latitudes:

	Time of day					
	Midday k_d, s^{-1}			Integrated full day k_d, day^{-1}		
	Latitude			*Latitude*		
Season	30°	40°	50°	30°	40°	50°
Spring	3.79E-7	2.76E-7	1.84E-7	8.11E-3	6.04E-3	4.13E-3
Summer	4.56E-7	3.77E-7	2.94E-7	1.03E-2	8.87E-3	7.27E-3
Fall	2.81E-7	1.76E-7	8.89E-8	5.28E-3	3.21E-3	1.55E-3
Winter	1.72E-7	8.24E-8	2.86E-8	3.01E-3	1.32E-3	4.14E-4

These commands also produce rate constants and corresponding half-lives at various times of day corresponding to preset solar altitudes. For example, the midsummer results for methoxychlor at 40°N and 90° longitude are

Solar altitude	0	5	10	20	30	40	50	60	70	70.09
Morning time	4.74	5.29	5.76	6.65	7.52	8.39	9.28	10.24	11.84	11.99
Afternoon time	19.24	18.69	18.22	17.33	16.46	15.59	14.70	13.74	12.14	11.99
k_d, 10^{-6} s^{-1}	0.0	0.24	0.70	3.28	8.42	15.6	23.6	31.4	37.7	37.7
$t_{1/2}$, 10^3 h	—	81.0	27.4	5.9	2.3	1.2	0.81	0.61	0.51	0.51

8.3 INORGANIC PHOTOCHEMISTRY OF NATURAL WATERS

8.3.1 Inorganic Reactants and Intermediates Produced by Photolysis Reactions

A few inorganic ions undergo direct photolysis in natural waters, but this process is not an important sink for any major or minor inorganic species. It *is* important, however, as a source of reactive intermediates that can induce indirect photolysis reactions. Nitrate, bicarbonate, bromide (in seawater), and especially dissolved O_2, and water itself, all undergo such reactions. In addition, some transition metals, notably iron and manganese, participate in redox cycles that involve direct photolysis of metal-organic acid complexes by charge transfer mechanisms (see Section 8.4.4).

Important free radicals, radical anions, and other reactive photointermediates produced from water or dissolved oxygen include superoxide anion ($\cdot O_2^-$), hydroxyl radical ($\cdot OH$), hydroperoxyl radical ($HO_2\cdot$), hydrogen peroxide (H_2O_2), hydroperoxide ion (HO_2^-), singlet oxygen ($^1O_2$), and the hydrated electron (e^-_{aq}). Reactive intermediates such as carbonate ion radical (CO_3^-) and nitric oxide (NO) are formed from common inorganic ions (carbonate and nitrate/nitrite, respectively) in fresh and saltwater; Br_2^- and I are produced photochemically in seawater. Table 8-3 lists reactions that form these species in water. Various organic radicals also are formed directly or indirectly by photoreactions (see Section 8.5.3).

8.3.2 Scavenging Reagents: Probes for Reactive Intermediates

Except for NO, which is relatively stable, and H_2O_2, which is only moderately reactive, the products of reactions listed in Table 8-3 are highly reactive and thus are important inducers of indirect photolysis. Because of their reactivity, steady-state concentrations are low ($<10^{-6}$ M) to extremely low (10^{-14} to 10^{-18} M) and difficult to measure directly. Recent improvements in pulsed laser spectroscopy[23,24] permit direct observation of some phototransients, including triplet-state photoexcited humic material and hydrated electrons. However, the most common method of measuring these intermediates is indirect, through their reaction with "scavenging reagents" or probes (Table 8-4). To be useful as a probe, a molecule should have the following characteristics:

- It **should not** undergo direct photolysis, quench sensitizer triplets, or react with radical intermediates other than the one of interest.
- It **should** have a high rate constant for reaction with the species of interest, and yield a single, stable product (or a few products) that can be measured easily.

If the rate constant k_p is known for reaction of probe P with a photointermediate (symbolized as X_{IP}), and if the rates of production and loss of X_{IP} can be assumed constant, the steady-state concentration of X_{IP} can be estimated as follows:[25]

$$-d[P]/dt = k_p [P][X_{IP}]_{ss} = k_{exp}[P]; \qquad [X_{IP}]_{ss} = k_{exp}/k_p \qquad (8\text{-}23)$$

For Equation 8-23 to apply, [P] must be sufficiently low that it does not affect the the lifetime of X_{IP} or $[X_{IP}]_{ss}$ in the water being studied. Steady-state production and loss of X_{IP} usually can be assumed for reasonable incubation times (hours) of natural samples if light conditions are held constant.

8.3.3 Peroxides, Superoxide, and Hydroxyl Radicals

Hydrogen peroxide is a common photointermediate and/or photoproduct in aquatic systems. It has been found in seawater, rain, lakes, and sewage lagoons,[26] often in surprisingly high concentrations (up to $10^{-5} M$; Table 8-5). Concentrations in seawater tend to be lower than in freshwater, but data are sparse. Concentrations of H_2O_2 increase in many waters (even groundwaters) when they are illuminated and decrease fairly rapidly in the dark, thus supporting the idea that the species is formed photochemically. First-order appearance rates on the order of $0.5\ h^{-1}$ were measured in Jacks Lake, Ontario (zero-order rates up to $\sim 10^{-6} M\ h^{-1}$), and first-order loss rates measured in the dark were $\sim 0.2\ h^{-1}$ (~ 0.2–$0.9 \times 10^{-7} M\ h^{-1}$).[37] These rates are ~ 5 to 10 times faster than those reported for marine environments. H_2O_2 is a fairly weak oxidant (especially at low concentrations), but it reacts with many organic compounds, including phenols, unsaturated fatty acids, and organic N and S compounds.[38] However, the actual fate of H_2O_2 in surface waters is not well known. H_2O_2 photolyzes ($\lambda < 300$ nm) to highly reactive ·OH radicals (reaction 9, Table 8-3), but the half-life for this reaction (~ 150 h) is too long for it to be an important source of ·OH in natural waters.

H_2O_2 in rainfall may arise from absorption of preformed compound from the atmosphere by cloud or rain droplets or by aqueous reactions involving O_3 and other reactive species in the droplets. The most likely mechanism for H_2O_2 formation in surface waters is photochemical reduction of dissolved O_2 by natural DOM to produce O_2^{-}, which disproportionates to form H_2O_2 and O_2 (Table 8-3, reactions 5 to 7). Evidence for this mechanism includes a linear dependency of H_2O_2 formation rate on [DOC] and the finding that the enzyme superoxide dismutase, which catalyzes the disproportionation reaction of O_2^{-}, accelerates the rate of H_2O_2 production.[27] Not all the H_2O_2 in surface waters is necessarily the result of abiotic photochemical reactions. A marine coccolithophore (phytoplankter), *Hymenomonas carterae*, produces H_2O_2 in the dark extracellularly.[39] Rates as high as 1 to 2×10^{-14} mol cell^{-1} h^{-1} were measured, and this could account for much or all of the H_2O_2 in

Table 8-3. Reactions Producing Inorganic Photointermediates in Natural Waters

Formation of O_2^- (superoxide radical ion)
- (1) Charge transfer within photoexcited sensitizer-O_2 complexes:
  $$S + O_2 + hv \rightarrow [S^- \cdot O_2] \rightarrow S_{ox} + O_2^-$$
  (S = photosensitizer (reducing agent), e.g., DOM, Fe^{2+})
- (2) Photoionization:
  $$DOM + hv \rightarrow DOM^+ + e^-_{aq}; \quad e^-_{aq} + O_2 \rightarrow O_2^-$$
- (3) Semiconductor pathways:
  $$TiO_2 + hv \rightarrow TiO_2^+ + e^- \xrightarrow{O_2} O_2^-$$
  ("hole")

Formation of hydroperoxyl by photoreduced carbonyl compounds
- (4) $>C=O + hv \rightarrow [>\overset{\cdot}{C}\text{-O}^\cdot] \xrightarrow{ROH} >\overset{\cdot}{C}\text{-OH} \xrightarrow{O_2} >C=O + \cdot OOH$
  (triplet)

Formation of H_2O_2
- (5) $S + O_2 + hv \rightarrow S_{ox} + O_2^-$ (i.e.,reaction 1, 2, or 3)
- (6) $O_2^- + H^+ \rightarrow HO_2^\cdot$
- (7) $2HO_2^\cdot \rightarrow H_2O_2 + O_2$ (catalyzed by superoxide dismutase)
- or (8) $HO_2^\cdot + R + H^+ \rightarrow H_2O_2 + R_{ox}$ (R = reducing agent)

Photolysis of H_2O_2:
- (9) $H_2O_2 + hv \rightarrow 2OH^\cdot$ ($\lambda < 300$ nm)

Nitrate and nitrite photolysis
- (10) Nitrate: $NO_3^- + hv \rightarrow NO_2^- + O(^3P)$ (λ_{max} 302 nm)
- (11) $O(^3P) + O_2 \rightarrow O_3$
- or (12) $NO_3^- + hv \rightarrow NO_2 + \cdot O^-$ (λ_{max} 302 nm)
- (13) $\cdot O^- + H^+ \rightarrow OH^\cdot$
- (14) Nitrite: $NO_2^- + hv \rightarrow NO + \cdot O^-$
- (15) $\cdot O^- + H_2O \rightarrow OH^\cdot + OH^-$
- (Net 14+15: $NO_2^- + H_2O + hv \rightarrow NO + OH^\cdot + OH^-$) ($\lambda \approx 300$–370 nm)
- or (16) Nitrous acid: $HNO_2 + hv \rightarrow NO + OH^\cdot$ } (a null cycle)
- (17) $NO + OH^\cdot \rightarrow HNO_2$
- (18) $NO + HO_2^\cdot \rightarrow NO_2 + OH^\cdot$
- (19) $NO_2^- + O_3 \rightarrow NO_3^- + O_2$
- (20) $2NO + 0.5O_2 + H_2O \rightarrow 2H^+ + 2NO_2^-$

Production of radicals by OH·
- (21) $OH^\cdot + Br^- \rightarrow Br^\cdot + OH^-$
- (22) $Br^\cdot + Br^- \rightarrow Br_2^-$
- (23) $OH^\cdot + HCO_3^- \rightarrow HCO_3^\cdot + OH^-$
- (24) $OH^\cdot + DOM \rightarrow OH^- +$ organocations $\rightarrow\rightarrow$ oxidation products

Production of aquated electron from organic matter
- (25) $DOM + hv \rightarrow DOM^+ + e^-_{aq}$

Iron photolysis
- (26) $FeOH^{2+} + hv \rightarrow Fe^{2+} + OH^\cdot$
- (27) $Fe^{2+} + H_2O_2 \rightarrow FeOH^{2+} + OH^\cdot$ (Fenton's reaction)

Ozone photolysis
- (28) $O_3 + H_2O + hv \rightarrow H_2O_2 + O_2$ (see Figure 8-12a for further details)

Table 8-4. Scavenging Reagents used as Probes for Photochemical Intermediates

Intermediate	Probe reagent	Measured product	Reference
·OH	benzene	phenol	41
	benzoic acid	hydroxybenzoic acids, primarily salicylic acid	171
	cumene (isopropylbenzene)	side chain and ring oxidation products	25,40
	bromide (Br−)	Br_2^-	63
	thiocyanate (SCN−)	$(SCN)_2^-$	63
	butyl chloride	loss of probe	25
	pyridine	hydroxypyridine, other polar compounds	40
ROO·	pyridine	pyridine N-oxide	40
	cumene	side-chain oxidation products	40
O_2^-	NH_2OH	NO_2^-	58
$^1O_2$	2,5-dimethylfuran (DMF)	diacetylethylene, H_2O_2	46,48, 49,51
	furfuryl alcohol	6-hydroxy(2H)pyran-3(6H)-one, H_2O_2	51
	disulfoton	corresponding sulfoxide	48
	histidine	loss of probe[a]	52
	tryptophan	loss of probe[a]	53
	2,3-dimethyl-2-butene		54
	1,4-diazabicyclo-[2.2.2]octane (DABCO)	none (quenches $^1O_2$)	9
	β-carotene	none (quenches $^1O_2$)	9
e^-_{aq}	2-chloroethanol	chloride	57
	paraquat		23
	NO_3^-	NO_2^-	58

[a]   Probe loss measured because complex mixture of products is formed.

seawater. At a reasonable concentration of 10^5 cells/L, these algae could produce the 1 to 20×10^{-8} M of H_2O_2 typically found in marine surface waters in 5 to 200 h. Physiological reasons for the process are still unknown.

Superoxide anion (O_2^-) can be formed by several mechanisms (Table 8-3, reactions 1–3), but their relative importance in natural waters is unknown. Concentrations of O_2^- in natural waters can be estimated kinetically from the loss rate of H_2O_2, if one assumes that disproportionation of O_2^- is the only source of H_2O_2 and that both species are at steady state. An upper limit of $\sim 10^{-8}$ M for O_2^- was calculated for seawater.[18] Although O_2^- decays mainly by further reduction to form H_2O_2, it is important to realize that O_2^- also is a reducing agent (returning to O_2). Formation of Cu^+ in seawater is thought to occur by reaction of Cu^{II} with O_2^- (see Section 4.6).

The hydroxyl radical (·OH) is a very important atmospheric intermediate in photochemical smog formation and the stratospheric ozone cycle, but its importance in aquatic systems is less certain. It is formed in many ways (Table 8-3, reactions 9,13,15,16,18,26–28): from the photolysis of H_2O_2, NO_3^-, and NO_2^-;

Table 8-5. Concentrations of H_2O_2 Measured in Natural Waters

Water body	Location	$[H_2O_2]$, 10^{-7} M Initial	After irradiation	Reference
Surface freshwater				
Five rivers	SE U.S.	0.9–3.2	6.4–46	27
Spring	Pennsylvania	0	3.1	27
Reservoir	Russia	6–15	18–26	28
Volga River area	Russia	13–32	–	29
Agricultural waters	California			
Irrigation		–	18–68	30
Runoff		–	23–53	30
Sewage				
Lagoon	California	–	125–325	30
	Russia	12–85	–	28
Groundwater				
	Texas	0	6–48	27
	Arizona	0	0.6	27
Estuarine	Chesapeake Bay	0.03–17	–	31
Seawater				
Coastal water	Texas	0.14–1.7	–	32
Coastal water	Gulf of Mexico	1.0–2.4	–	33
Offshore	Gulf of Mexico	0.9–1.4	–	33
Open ocean	N. Atlantic	–	0.35–1.5	34
Rainwater	California	10–30	–	35
	Florida, Bahamas	9–75	–	36

– indicates data not reported.

Adapted from Cooper, W.J., et al., *Environ. Sci. Technol.*, 22, 1156 (1988) with other references added.

decomposition of O_3; reaction of NO with ·OOH; photoreduction of $FeOH^{2+}$; reaction of Fe^{2+} and H_2O_2 (Fenton's reagent); and possibly from photolysis of humic compounds by unknown mechanisms. The relative importance of these mechanisms is still uncertain, but photoreduction of nitrate and reactions involving iron are thought to be major sources in freshwater. Typical production rates for ·OH also are uncertain, and whether it is formed at significant rates (compared with other photointermediates) in aquatic systems is controversial.

Haag and Hoigne[25] estimated production rates of ·OH in Greifensee, a eutrophic Swiss lake, to be about 10^{-11} M/s, which is 10^{-3} times the production rate of $^1O_2$ in the same lake, and the mean lifetime of ·OH radicals was about 10^{-5} s (comparable to that of $^1O_2$). However, ·OH is a much more reactive, and less selective, species than $^1O_2$. ·OH concentrations in eutrophic surface waters were estimated to be about 10^{-17} M, based on the rate of photolysis of the probe cumene (isopropylbenzene),[40] and about 1 to 3×10^{-16} M in Greifensee, based on scavenging experiments with butyl chloride.[25] Based on low production rates and low $[\cdot OH]_{ss}$ values, Haag and Hoigne concluded that $\cdot OH$ is not an important sink for organic contaminants in surface waters, but for reasons discussed below, this conclusion may not apply to all natural waters.

The potential importance of nitrate photolysis as a source of $\cdot OH$ was first suggested by studies with $10^{-5}\ M$ benzene as an $\cdot OH$ probe.[41] Benzene reacts with $\cdot OH$ to form phenol, and phenol production in water samples from several German lakes was directly correlated with $[NO_3^-]$, but inversely correlated with [DOC]. On this basis the authors concluded that photolysis of NO_3^- was a more important source of $\cdot OH$ than was DOM. The $[\cdot OH]_{ss}$ in the water was estimated to be $5 \times 10^{-16}\ M$. More recent and detailed studies on nitrate photolysis (see Section 8.3.6) indicate that $\cdot OH$ production rates from nitrate photolysis range from $\sim 2 \times 10^{-12}$ to $2 \times 10^{-11}\ M\ s^{-1}$ ($\sim 7 \times 10^{-9}$ to $10^{-8}\ M\ h^{-1}$), which is sufficient to make $\cdot OH$ important in the photodegradation of organic contaminants.

Oxidation of Fe^{2+} by O_2 via the Haber-Weiss mechanism produces H_2O_2. This may be significant regarding $\cdot OH$ production, because Fe^{2+} and H_2O_2 (this combination is known as Fenton's reagent) react rapidly to form $\cdot OH$ and Fe^{3+} (reaction 27, Table 8-3). High concentrations of both reactants could be produced in colored lakes, which often have high Fe concentrations. Hypolimnetic waters of low-alkalinity, oligotrophic lakes have Fe concentrations as high as several mg L^{-1} because of reductive dissolution of ferric oxides at the sediment-water interface.[42] Because water clarity is high in such lakes, light may penetrate to depths where Fe concentrations are high. Fenton's reagent is a highly effective oxidant. Consequently, the conclusion of Haag and Hoigne[25] that $\cdot OH$ is not an important photointermediate may not apply to waters with higher concentrations of Fe^{III} and/ or nitrate than found in Greifensee.

$\cdot OH$ is highly reactive both in air and water (Table 8-6a). Rate constants for reactions of $\cdot OH$ with many organic compounds and inorganic ions are known, and many of the reactions are diffusion limited. Few rate constants are available for reaction of $\cdot OH$ with aquatic contaminants, but in some cases they can be estimated from other compounds. The principal aquatic sinks for $\cdot OH$ are dissolved organic matter and HCO_3^- (with which it reacts to form CO_3^-). A major sink for $\cdot OH$ in seawater is Br^-, which forms $Br\cdot$ and reacts rapidly with another Br^- to form the ion-radical Br_2^-.[43] This species decays by reaction with CO_3^{2-}, HCO_3^-, and the ion pairs $CaCO_3^0$, $MgCO_3^0$, and $NaCO_3^-$.[44] The pseudo-first-order rate constant is $\sim 2.5 \times 10^3$ s^{-1} (carbonate species are approximately constant and in great excess in seawater).

8.3.4 Singlet Oxygen

Singlet oxygen is the best studied inorganic photochemical intermediate. It is formed by collision of ground-state O_2 with excited triplet states of sensitizer molecules (reaction viii, Table 8-1). Its role in environmental photochemistry was first proposed in 1970,[45] and the first measurements of $^1O_2$ concentrations in natural waters were reported in 1977.[46] Energy transfer is rapid when the triplet energy of the sensitizer exceeds that of the energy acceptor.[7] The energy level of $^1O_2$ is only

Table 8-6. Reactivities of Important Photointermediates

A. Hydroxyl radical (·OH)[b]

Compound	k_2[a]	Compound	k_2[a]	Compound	k_2[a]
Aromatics and unsaturated compounds: (addition reactions)		chloroacetate	5.5E7	**Sulfur compounds**	
benzene	7.5E9	butyric acid	1.9E9	H_2S	1.8E10
chlorobenzene	4.5E9	oxalic acid	8.0E6	HS^-	9.0E9
phenol	1.4E10	citric acid	5.0E7	DMS	5.2E9
phenate ion	9.1E9	lactic acid	4.3E8	DMSO	7.0E9
m-chlorophenol	7.2E9	5-chlorouracil	5.2E9	cysteine → cystine	
o-chlorophenol	8.2E9	succinic acid	1.2E8	(via H abstraction)	1.3E10
o-cresol	1.1E10	C_2–C_4 ethyl esters	2–8E8		
p-cresol	1.3E10	diethyl ether	2–4E9		
o-hydroxyphenol	1.1E10	dioxane	2.4E9	**Inorganic ions (electron transfer)[c]**	
salicylate	5.6E9	formaldehyde	2.0E9	Cr^{3+}	3.2E8
toluene	3.0E9	acetone	7.0E7	Mn^{2+}	>1.4E8
acetophenone	6.5E9	methylethylketone	9.0E8	Fe^{2+}	3–5E8
benzophenone	9.0E9	methane	2.4E8	$Fe(CN)_6^{4-}$	1.1E10
1,1-dichloroethene	4.1E9	chloroform	1.4E7	CO_3^{2-}	4.2E8
1,2-dichloroethene	4.7E9	CH_3CN	3.5E6	HCO_3^-	1.0E7
		CH_3NO_2	3.1E8	CN^-	4.5E9
H abstraction reactions		methylamine	1.9E7	NO_2^-	1–7E9
methanol	8.0E8	acrylamide	3.3E9	HPO_4^{2-}	<1.0E7
ethanol	1.8E9	ethylenediamine	1.0E8	H_2O_2	1.2E7
propanol	2.7E9	acetonitrile	3.5E6	Cl^-	1.5E10
n-butanol	3.7E9			Br^-	5.0E9
n-pentanol	4.8E9	**Deamination reactions**		I^-	3.4E10
n-octanol	6.1E9	alanine	4.4E7		
formic acid	2.8E9	aspartic acid	2.1E7	**Radical reactions**	
acetic acid	9.0E6	histidine (addition reaction)	2–5E9	OH + OH → H_2O_2	5.3E9
acetate ion	8.0E7	lysine	4.0E8	OH + O^- → HO_2^-	<2.0E10
bromoacetate	4.4E7	simple peptides	E8–E9	OH + H → H_2O	7.0E9
		enzymes	~E11	OH + HO_2^- → $H_2O + O_2$	6.0E9

B. Singlet oxygen ($^1O_2$)[d]

Compound	pH	k_2[a]	Compound	pH	k_2[a]	Compound	k_2[a]
p-cresol	11.5	3.5E8	p-nitrophenol	8.8	2–3E6	histidine	5E7
(pK_a = 10.2)	8.3	1.1E7	(pK_a = 7.2)			tryptophan	4E7
phenol	11.5	1.8E8	2,5-dimethylfuran		1.4E8	methionine	3E7
(pK_a = 9.9)	8.0	2–3E6	furfuryl alcohol		1.2E8	cyclohexene	3E3
2,4-dichlorophenol			2,3-dimethyl-2-butene		1.1E8	organic solvents	
(pK_a = 7.8)	9.6	1.2E8	2-methyl-2-pentene		1E6	(e.g., C_6H_6, CH_3OH)	
	5.5	7E5	diethylsulfide		1.8E7	(quenching)	~3E3
2,4,6-trichlorophenol			1-methylcyclohexene		2E5	(reaction)	<1E0
(pK_a = 6.1)	9.0	1.2E8					
	4.2	2E6					

[a] Values in M^{-1} s^{-1}.

[b] Values tabulated from Cooper, W.J., et al., in *Aquatic Humic Substances; Influences on Fate and Treatment of Pollutants*, P. MacCarthy and I.H. Suffet (Eds.), Adv. Chem. Ser. 219, Am. Chem. Soc., Washington, D.C., 1989, p. 333; and Dorfman, L.M. and G.E. Adams, Natl. Stand. Ref. Data Ser. No. 46, U.S. Nat. Bureau of Standards, Washington, D.C., 1973.

[c] $M^{n+} + OH· \rightarrow M^{n+1} + OH^-$, or $M^{n-} + OH· \rightarrow M^{(n-1)-} + OH^-$.

[d] Values tabulated from Foote, C.S., in *Free Radicals in Biology*, Vol. II, W. Pryor (Eds.), Academic Press, New York, 1976, p. 85; and Scully, F.E. and J. Hoigne, *Chemosphere*, 16, 681 (1987).

96 kJ mol^{-1} above that of ground-state O_2, whereas triplet state energies of synthetic organic compounds typically are much higher:[7] DDT, 330; parathion, 242; hexachlorobenzene, 293; benzene, 355; phenol, 343; biphenyl, 272; pyrene, 205; benzo[a]pyrene, 176 (all in kJ mol^{-1}). As a result, energy transfer to O_2 is very efficient; γ_{O_2}, the fraction of encounters of triplet sensitizers with ground-state O_2 that yield $^1O_2$ has a value near 1 . The rate constant for formation of $^1O_2$ (k_{viii}) has a value of ~2 × 10^9 M^{-1} s^{-1} for most triplet sensitizers.[9]

Only a small fraction of the photoproduced $^1O_2$ reacts chemically; most of it is quenched to ground-state O_2 by water.[18] The rate constant for quenching of $^1O_2$ by H_2O is k_d = 2.5 × 10^5 s^{-1}, and the lifetime of $^1O_2$ in water is short; τ_o ($1/k_d$) = 4 × 10^{-6} s.[47] Because rate constants for reactions of $^1O_2$ even with highly reactive organic compounds are at most 10^7 to 10^8 M^{-1} s^{-1} (Table 8-6b), quenching is the predominant fate of $^1O_2$, except at very high organic concentrations. Nonetheless, the small fraction of $^1O_2$ that does react represents a significant sink for some compounds, including reduced sulfur compounds and some amino acids, e.g., tryptophan and histidine (see Section 8.5.4).

Concentrations of $^1O_2$ in natural waters are estimated by adding probes to the water. Early studies[46,48,49] used 2,5-dimethylfuran (DMF), which reacts with $^1O_2$ in a characteristic way to yield diacetylethylene and H_2O_2 as the principal oxidation products (Figure 8-10). Wolff et al.[50] suggested that $^1O_2$ concentrations estimated with DMF are too high because secondary reactions of DMF with H_2O_2 yield the same products as does $^1O_2$, but other workers[51] have not confirmed these results. The anomalous results of Wolff et al. may be attributable to the high DMF concentrations (mM) they used; this may have led to high concentrations of H_2O_2 and thus to secondary oxidation of DMF. More recently, furfuryl alcohol (FFA) has been used as a trapping agent for $^1O_2$.[51] Its principal reaction products with $^1O_2$ are 6hydroxy(2H)pyran-3(6H)-one and H_2O_2 (Figure 8-10), and H_2O_2 does not react with FFA. Both DMF and FFA react rapidly with $^1O_2$; rate constants are 6.3 × 10^8 and 1.2 × 10^8 M^{-1} s^{-1}, respectively. Use of furans to trap $^1O_2$ has been criticized because some are susceptible to oxidation by species other than $^1O_2$ (e.g., H_2O_2). Several other probes have been used as probes for $^1O_2$: disulfoton (a dialkyl sulfide insecticide);[48] histidine and α-chymotrypsin;[52] tryptophan;[53] and 2,3-dimethyl-2 butene (DMB).[54] However, these acceptors are relatively untested and also may lack specificity for $^1O_2$.

Computation of $[^1O_2]_{ss}$ from data on oxidation rates of probes is done by kinetic analysis of the production and loss mechanisms for $^1O_2$.[51,55] The resulting equation is

$$\left[^1O_2 \right] = I_a \phi_s / k_d \tag{8-24}$$

I_a is the rate of light absorption by the water sample, and ϕ_s is the quantum yield for production of singlet oxygen production. Like Equation 8-23, Equation 8-24 assumes that probe concentration [P] is sufficiently low that it does not affect the lifetime of $^1O_2$. The quantum yield for production of $^1O_2$ must be computed from the measured quantum yield ϕ_{PO_2} for the reaction P + $^1O_2 \rightarrow PO_2$:

$$\phi_s = \phi_{PO_2} / \gamma_P \lambda_r \tag{8-25}$$

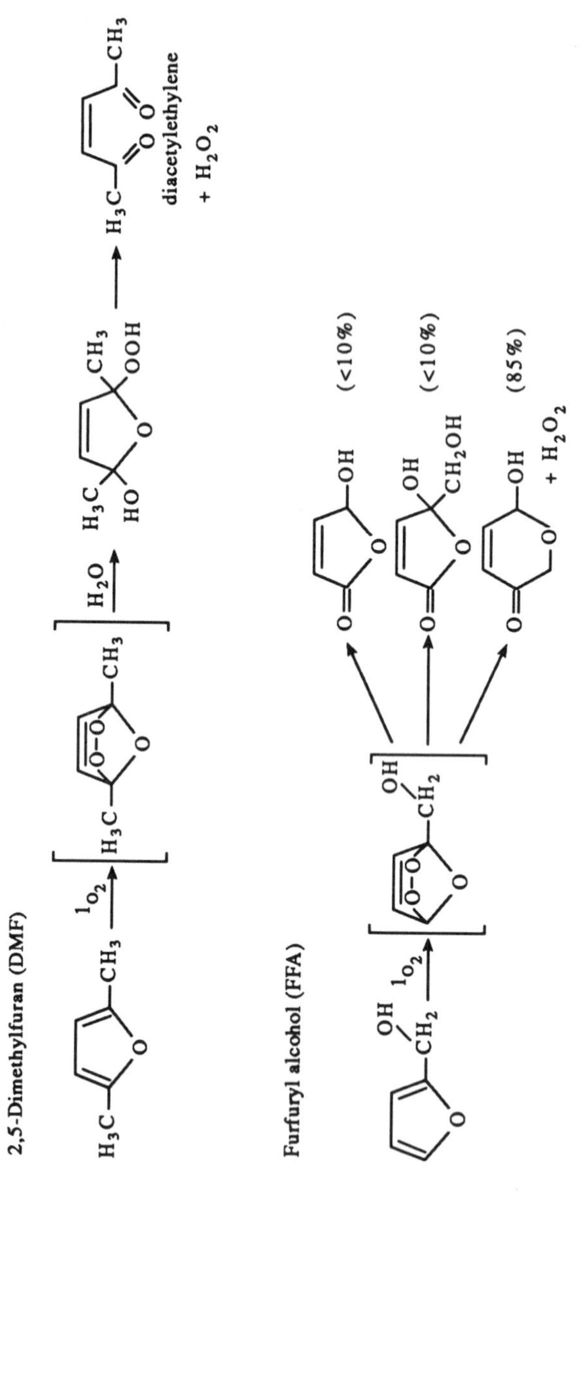

Figure 8-10.   Reactions of trapping agents with singlet oxygen.

ϕ_{PO_2} can be measured either from the loss of dissolved O_2 in a water sample (or the production of PO_2) divided by the total amount of light absorbed by the sample in some period of measurement. The former method is more accurate in that oxidation of P may yield a variety of products, some of which may not be measured. Loss of O_2 thus is the method of choice for calibration purposes. Because this is a more tedious analysis, measurement of PO_2 is done routinely once the ratio of O_2 lost to PO_2 formed is determined. γ_r is the fraction of collisions of $^1O_2$ with P that result in reaction. For DMF and FFA, it is assumed that $\gamma_r = 1$ and there is no physical quenching of $^1O_2$ by P. The "interception efficiency", γ_P, is the fraction of $^1O_2$ that reacts with P:

$$\gamma_A = k_A[A] / \{k_A[A] + k_d\} = [A] / \{[A] + \beta\} \qquad (8\text{-}26)$$

$\beta = k_d/k_A$ is the reactivity index.[9] β has units of concentration (k_d is a first-order rate constant, and k_A is a second-order constant) and represents the concentration of P at which quenching of $^1O_2$ by P and by water are equal. $\beta = 1.2 \times 10^{-3}$ M for DMF and 2.3×10^{-3} M for FFA. Methods for determining β were described by Haag et al.[51] It should be noted that quantum yields for production of $^1O_2$ vary with λ (Figure 8-11), whereas quantum yields for direct photolysis reactions usually are λ independent. This complicates the analysis and necessitates replacement of the λ-integrated I_a and λ-independent ϕ, and ϕ_{PO_2} in Equations 8-24 and 8-25 with λ-dependent values and integrating over λ.[51,55]

Concentrations of $^1O_2$ in the range 2 to 20×10^{-13} M were reported for various surface waters in the southeastern U.S.,[46] and highest concentrations occurred in highly colored waters. $^1O_2$ concentrations of 0.3 to 5×10^{-14} M per mg L^{-1} of DOC were found in Swiss waters under midday and spring-through-fall sunlight conditions.[51,56] Although humic material is thought to be the source (photosensitizing agent) of $^1O_2$ in surface waters, more highly colored waters actually produced less $^1O_2$ per mg DOC per liter for a given amount of absorbed light. The concentrations of $^1O_2$ mentioned above are sufficient to oxidize some especially reactive organic contaminants in a few hours, but half-lives of most contaminants are orders of magnitude longer.[56]

8.3.5 Hydrated Electrons

The hydrated electron (e^-_{aq}), a highly reactive and very short-lived species, has been identified as a photoactivation product of dissolved organic matter in natural waters[23,24] (Equation 25, Table 8-3). Primary quantum yields of 0.005 to 0.008 have been measured for electron ejection from DOM of natural waters,[57] and steady-state concentrations have been estimated as $\sim 10^{-17}$ M per mg DOC per liter.[57,58] The lifetime of the average hydrated electron is estimated to be less than 1 ns,[58] and the distance it can travel between formation and reaction thus is very short. This can be estimated from the Einstein-Smoluchowski equation (Equation 3-23a) as $\sim(2Dt)^{1/2}$. If we assume t = 1 ns and $D = 10^{-5}$ cm^2 s^{-1}, a typical value for small ions, the distance before reaction for e^-_{aq} is about 1.4 nm. This implies that most of

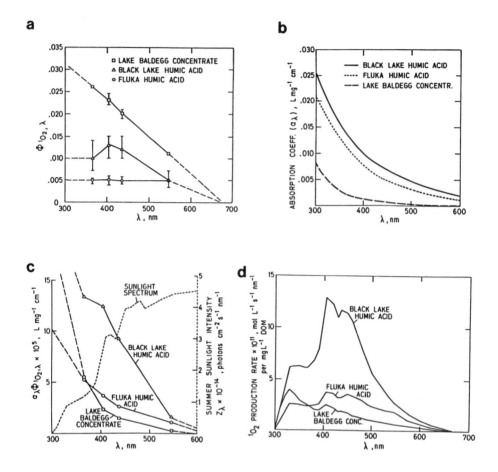

Figure 8-11. Effect of wavelength on photosensitized production of $^1O_2$ from humic material: (a) variation in quantum efficiency, $\phi_{^1O_2}$ with λ; (b) absorption spectra α vs. λ, of humic acids; (c) action spectra, $\alpha_\lambda\phi_{O_2}$ (solid lines) and summer sunlight spectrum (near-surface) at 40° N, midday; and (d) calculated $^1O_2$ production rates (midday, summer, 40° N) in near-surface water. [From Haag, W.R., et al., *Chemosphere*, 13, 641 (1984). With permission.]

the electrons produced by photoactivation do not even leave the organic macromolecule from which they were produced, but recombine with the DOM. This conclusion was verified[57] by the finding that quantum yields for formation of Cl⁻ from 2-chloroethanol (a probe reagent for e^-_{aq}) are about two orders of magnitude lower than primary quantum yields. Much longer lifetimes (~1 ms) were estimated for e^-_{aq} based on laser pulse photolysis.[57] Reasons for this are not certain, but it appears that the high concentrations of e^-_{aq} produced by laser pulses result in a more efficient escape of electrons from DOM, in effect lowering the importance of recombination in controlling the lifetime of e^-_{aq}.

The fate of e^-_{aq} that manage to escape the DOM from whence they were produced is still uncertain. The species reacts with O_2 to produce superoxide (O_2^-) (Equation

2, Table 8-3) at a diffusion-controlled rate ($k = 2 \times 10^{10} \, M \, s^{-1}$), and it reacts with nitrate nearly as fast ($8–11 \times 10^9 \, M^{-1} \, s^{-1}$) to produce nitrite ions. Rapid reactions also occur with organic compounds, especially cations like the herbicide paraquat[23] and electrophilic chlorinated compounds. Concentrations of O_2 generally are much higher than those of nitrate and organic contaminants, and thus the principal fate of e^-_{aq} in bulk solution probably is formation of superoxide ion, which quickly reacts to produce hydrogen peroxide. However important this reaction may be as a sink for e^-_{aq}, it apparently accounts for only a small fraction of the H_2O_2 produced photochemically in surface waters.[57]

8.3.6 Photochemistry of Inorganic Nitrogen Species

The role of photochemical reactions in the inorganic nitrogen cycle has been postulated for many years, but there was little definitive evidence for them until recently. Rao and Dhar[59] first suggested that ammonia is oxidized photochemically in tropical soils in 1931. Based on an inability to culture nitrifying organisms from seawater, ZoBell[60] suggested that NH_4^+ is oxidized photochemically in the sea, but this is not strong evidence. In 1936, Rakestraw and Hollaender[61] reported oxidation of NH_4^+ in seawater (but not distilled water) under intense UV irradiation. Nearly 30 years later, Hamilton[62] found no evidence for this process, but his experimental techniques were unsophisticated. More recent laboratory studies[45] demonstrated that $^1O_2$ can oxidize ammonia to nitrite and nitrate, and the authors suggested that aromatic compounds may sensitize this reaction in seawater. However, environmental data still have not been reported to verify this.

The photochemical activity of oxidized nitrogen forms is better documented. Based on relatively crude experiments, Hamilton[62] concluded that nitrite was not photoactive in seawater, but that nitrate was reduced at measurable rates. More recently, Zafiriou and co-workers[62a] demonstrated that both ions are photoactive. For nitrite the principal reaction (Equations 14,15, Table 8-3) results in the formation of nitric oxide (NO) and ·OH. The quantum yield of ·OH from NO_2^- varies with λ, from 0.015 to 0.08 over the range 298 to 371 nm.[63] NO and ·OH radicals recombine in distilled water to form a null cycle, but ·OH in seawater is scavenged by Br^-, HCO_3^-, or DOM (Equations 21–24, Table 8-3). The net effect is to allow NO to accumulate at rates more than $10^{-12} \, M \, h^{-1}$ during daylight, and NO levels to reach $10^{-11} \, M$ or higher. These values represent significant (5 times) supersaturation compared with the partial pressure of NO in pristine marine air. Photolysis of NO_2^- thus may make the sea a *source* of atmospheric NO. Rapid thermal reactions consume NO at night, but the exact mechanism is not well understood. The reaction of NO with O_2 to produce NO_2^- (Equation 20, Table 8-3) is too slow to account for the loss of NO in the dark, and reaction of NO with other radicals is suspected. Net photolysis of NO_2^- in surficial seawater for midsummer, midlatitude conditions averages about 10% per day (range 2 to 27%).[64]

For nitrate the principal photochemical reaction upon absorption of a photon ($\lambda_{max} = 302$ nm) is production of NO_2^- and either ·O$^-$ or O(^{3}P), ground-state atomic oxygen (Equations 10,12, Table 8-3).[65] ·O$^-$ reacts rapidly with H^+ to yield ·OH.

$O(^3P)$, which is much less reactive than $\cdot O^-$, reacts with O_2 to form O_3. In turn, O_3 reacts rapidly with NO_2^- or decomposes to $\cdot OH$ (see next section).

The photoproduction rate of $\cdot OH$ from nitrate via Equations 12,13 of Table 8-3 is given by Equation 8-1 as $k_d[NO_3^-] = k_a\phi_d[NO_3^-]$, where k_a is the wavelength-integrated light absorption coefficient, and ϕ_d is the quantum efficiency for $\cdot OH$ production. According to Zepp et al.,[65] $\phi_d = 0.013$–0.017 over the range 20 to 30°C. k_a is obtained from the uv absorbance spectrum of NO_3^- and seasonally varying solar irradiance (Table 8-2). It has a midday, midsummer value of 1.78×10^{-5} s^{-1} at 40°N. For the same conditions, the near-surface production rate of $\cdot OH$ from NO_3^- thus is $2.5 \times 10^{-7}[NO_3^-]$. A reasonable range of $[NO_3^-]$ in freshwaters is 10^{-5} to 10^{-4} M. The rate of nitrate-induced $\cdot OH$ production under these conditions is about 2.5×10^{-12} to 2.5×10^{-11} M s^{-1}. This calculation agrees (order of magnitude) with the estimate of Haag and Hoigne[25] for $\cdot OH$ production in Greifensee and suggests that nitrate is a major source of $\cdot OH$ in lakes.

Example 8-2. Rates of Nitrate-Induced $\cdot OH$ Oxidation of Organic Compounds

Rate constants for reactions of $\cdot OH$ with many organic compounds are available (Table 8-6a), and it is possible to estimate the half-lives of such compounds in surface waters based on NO_3^--induced $\cdot OH$ photooxidation.

Zepp et al.[65] reported the following midseason and annual mean values of k_d for nitrate-induced $\cdot OH$ production:

| | | | k_d, 10^{-3} day^{-1} | | |
latitude (°N)	spring	summer	fall	winter	annual mean
20	7.7	8.4	6.1	4.7	6.7
30	7.0	8.2	4.7	3.0	5.7
40	5.9	7.7	3.2	1.6	4.6
50	4.6	6.8	1.8	0.6	3.4

These values can be used to estimate seasonal and annual average rates near surface $\cdot OH$ production rates in various lakes or rivers. For example, $[NO_3^-]$ has increased in Lake Superior from about 70 µg N per liter in the early 1900s to ~300 µg N per liter (2.15×10^{-5} M) at present.[66] The latitude of the lake is ~47°N, which yields a mean annual k_d of ~3.6 $\times 10^{-3}$ day^{-1}. The daily rate of $\cdot OH$ production at the lake surface (mean annual basis) is

$$v_o = k_d[NO_3^-] = 3.6 \times 10^{-3} \times 2.15 \times 10^{-5} = 7.7 \times 10^{-8} \text{ mol/L-day}$$

or 2.8×10^{-2} mol/m^3-year. Assuming no net buildup of $\cdot OH$, the average loss rate for $\cdot OH$ is equal to the above production rate. The main scavenging agents for $\cdot OH$ in the lake water probably are bicarbonate and natural organic matter, and the time-averaged surface water $\cdot OH$ concentration, $[\cdot OH]_o$, thus is

$$\text{Rate of } \cdot OH \text{ production} = \text{Rate of } \cdot OH \text{ loss}$$

$$= \{k_{bc}[HCO_3^-] + k_{oc}[DOC]\}[\cdot OH]_o$$

k_{bc} is the rate constant ($10^7 M^{-1} s^{-1}$) for reaction of ·OH with HCO_3^- ($\sim 10^{-3} M$); k_{OC} is an average rate constant for reaction of ·OH with natural organic matter (expressed as dissolved organic carbon, DOC). Much of the lake's DOC probably is fairly unreactive; we will assume $k_{OC} \approx 10^8 M^{-1} s^{-1}$. The lake's [DOC] is low, ~ 1 mg L^{-1}, perhaps equivalent to $10^{-5} M$ in organic structural units. Thus,

$$7.7 \times 10^{-8} = \left\{ 10^7 \times 10^{-3} + 10^8 \times 10^{-5} \right\} \times 8.64 \times 10^4 [\cdot OH]_o$$

or

$$[\cdot OH]_o = 8.08 \times 10^{-17} M$$

where 8.64×10^4 is a units conversion factor, s day^{-1}.

The loss rate at the water surface of a contaminant C that reacts with ·OH at a typical rate ($k_C \approx 5 \times 10^9 M^{-1} s^{-1}$) is

$$-d[C]/dt = k_c [\cdot OH]_o [C] = k'[C]$$

where k' ($= 0.35$ day^{-1}) is a pseudo-first-order rate constant for contaminant loss at the water surface for the estimated $[\cdot OH]_o$.

If mixing in the photic zone is more rapid than photolysis (which is reasonable), it is pertinent to compute photic-zone average rates, $v(z)$, and areal fluxes (F_A) for ·OH. The diffuse light attenuation coefficient (K_R) for Lake Superior is 1.4 m^{-1},[3] and the photic zone is $\sim 4.6 K_R$, or 6.4 m. $v(z)$ is approximately

$$v(z) = v_o \left\{ 1 - \exp(-K_R z) \right\} / z K_R = 8.6 \times 10^{-9} mol\ L^{-1}\ day^{-1}$$

or 3.1×10^{-3} mol m^{-3} year^{-1}. $F_A \approx v_o/K_R$;[65] thus, $F_A = 2.2 \times 10^{-3}$ mol m^{-2} year^{-1}. The photic-zone average [·OH] thus becomes $9.0 \times 10^{-18} M$, and k' for contaminant C becomes 3.9×10^{-3} day^{-1}, or 1.4 year^{-1}. Photolysis within the photic zone thus is fairly rapid. However, in Lake Superior that zone includes only a small fraction of the lake's volume ($z_{av} = 145$ m).

Net photolysis of nitrate to nitrite has been observed in seawater spiked with nitrate,[67] but at ambient nitrate levels no net loss occurred. It was hypothesized that a null cycle occurs at low (ambient) concentrations. Overall, the photochemical reactions of nitrate and ammonium in aquatic systems probably are of greater significance for the radical intermediates they produce than for the net consumption or transformation of the inorganic N ions.

8.3.7 Ozone

Although ozone is a major product of photochemical reactions in the atmosphere, it apparently is not formed by photochemical reactions in surface waters, except as a minor product in nitrate photolysis (Equation 11, Table 8-3). Its main source in natural waters is thought to be absorption from the atmosphere, but atmospheric levels are low. Polluted urban air contains ~ 100 to 200 ppb O_3, and rural air normally is several orders of magnitude lower. The solubility of O_3 in water is substantially greater than that of O_2 (12.2 vs. 1.35 mmol L^{-1} atm^{-1}; Henry's law constants (H) at 25°C are 0.082 and 0.74 atm m^3 mol^{-1}, respectively). The deposition velocity of O_3 to natural water surfaces is not known precisely, but a range of ~0.1 to 0.5 cm s^{-1} can be estimated by analogy to other moderately unreactive gases for which we can ignore enhancement of mass transfer by chemical reaction in the boundary layer. These values can be used to obtain order-of-magnitude estimates of O_3 transfer from the atmosphere into water bodies: 0.01 to 0.1 mol m^{-2} $year^{-1}$. Thus, atmospheric inputs are quite small (but perhaps not negligible), especially in remote areas.

Little is known about the importance of ozone as a reactant in natural waters, but detailed studies have been conducted on the reactions of O_3 added to drinking water and wastewater as a disinfectant and oxidant.[68-76] The decomposition of O_3 is mechanistically complicated and involves free radicals (including O_2^- and ·OH), a chain mechanism, and hydroxide-ion initiation (Figure 8-12a). The high reactivity of ozone with organic compounds under alkaline conditions results from the formation of reactive intermediates like ·OH. In addition, O_3 can react directly with organic compounds, but it is highly selective in this regard. Direct ozonation reactions tend to be favored under acidic conditions because the lifetime of O_3 is longer at low hydroxide concentrations. Direct reaction of O_3 with organic solutes generally is described as a second-order process (first order in each reactant), and rate constants have been determined for a wide variety of organic compounds (Table 8-7). Reactions with various alkenes are around 10^5 M^{-1} s^{-1}, but rate constants for aromatic compounds are five to seven orders of magnitude lower. In general, rate constants for organic acids increase with pH, and the deprotonated forms react more quickly.

Several studies have shown that O_3 decomposition in water is second order with respect to $[O_3]$ at low pH,[74] and the loss rate increases rapidly with pH at values more than ~5 (Figure 8-12b). There is less agreement about the reaction order under neutral-to-alkaline conditions; some authors[75] report rates first order in $[O_3]$, and others report data that fit second-order plots.[74,76] Alkaline decomposition of O_3 is believed to follow a chain mechanism,[77] and ozonation of cyanide at high pH also is thought to occur by a chain mechanism involving the O_3 decomposition products (Figure 8-12a).[78] The mechanism gives rise to a rate equation that is fractional order in $[CN^-]$ (Example 2-6). At pH 12, the rate is so rapid that it is mass-tranport limited.

Recent studies have shown that the combination of ozone and UV light is highly effective in oxidizing organic contaminants,[68,79] and reactor systems have been developed[80] to take advantage of this fact. The effectiveness of photolytic ozonation apparently results from the photolysis of O_3 at 254 nm to yield H_2O_2, which further reacts with O_3 to form highly reactive ·OH radicals[81] (Figure 8-12a). The conjugate base of H_2O_2, HO_2^-, ($pK_a = 11.6$), accelerates decomposition of O_3 and H_2O_2 at high pH. ·OH is the species primarily responsible for organic compound decomposition in such reactors.

8.4 NATURE AND PHOTOCHEMISTRY OF NATURAL DOM

8.4.1 Natural Organic Chromophores: Structure and Origin of Aquatic Humus

As mentioned several times in this chapter, dissolved organic matter (DOM) of natural origin is of central importance in understanding the photochemistry of natural waters. Natural DOM contains a rich variety of chromophores (light absorbing substances) of both allochthonous (external or terrestrial) origin and autochthonous (internal) origin, e.g., algal exudates. The structures of these chromophores are poorly known, although we can speculate about them based, for example, on knowledge of microbial metabolites. The term UC (unknown chromophore) is used sometimes to describe DOM components that absorb light, and UPC (unknown photoreactive chromophore) is used for DOM components that absorb light and induce photochemical reactions.[18] These terms exemplify our level of ignorance on this subject.

Aquatic humus (AH)[*] is an important component of the UC and UPC content of surface waters — indeed, it is the main component in brown-stained waters of dystrophic lakes, swamps, bogs, and streams draining swamps and bogs. The structure of AH has been the subject of great controversy and numerous studies (for recent reviews on the chemistry of humic materials see references 82,83). Early work by Shapiro[84] indicated that AH consists primarily of aliphatic acids of relatively low molecular weight (several hundred daltons). Subsequent studies by Black and Christman[85] suggested that AH is much larger (colloidal) and mostly aromatic. The idea that AH is macromolecular (size in thousands of daltons) was supported in many later measurements by Sephadex size-exclusion chromatography.[86] The inference of aromaticity was supported by studies that identified hydroxy- and methoxy-substituted aromatic compounds nearly exclusively as the products of mild oxidative degradation of AH.[87] Models of AH (Figure 8-13) thus emphasized these components as structural units.

More recent studies have cast doubt on these views. Inferring the size of complicated molecules from chromatographic elution patterns with size-exclusion gels is very difficult[88] because AH molecules interact with size-exclusion gels, and elution depends on several factors besides molecular size. For example, ionic sites on the gels tend to repel ionic solutes, thus decreasing their ability to penetrate gel interstices and accelerating their movement through the column. This produces erroneously high estimates of molecular size. Conversely, aromatic compounds tend to adsorb onto gel surfaces. This retards the movement of such compounds through Sephadex columns and leads to erroneously low estimates of molecular

---

[*]Humic material consists primarily of humic acids (HA) and fulvic acids (FA), but the latter predominate in AH. HA and FA are operationally defined by soil chemists, based on solubility in acids and bases. HA is the fraction of organic material extracted from soil under alkaline conditions that is insoluble in strong acid. FA is the acid-soluble fraction of alkali-extracted organic matter. The classes are broad and structurally ill defined, but FAs generally are smaller and more highly charged (more ionic functional groups) than HAs.

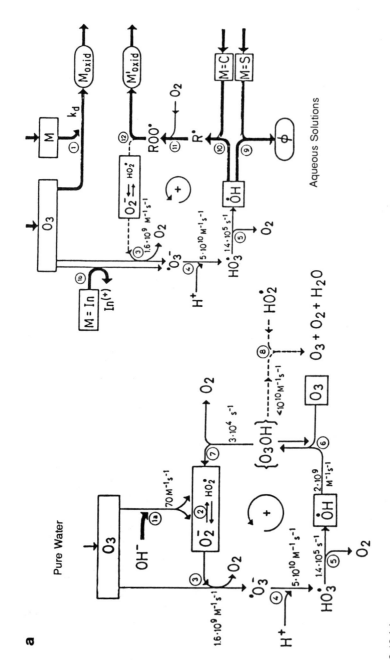

Figure 8-12 (a).

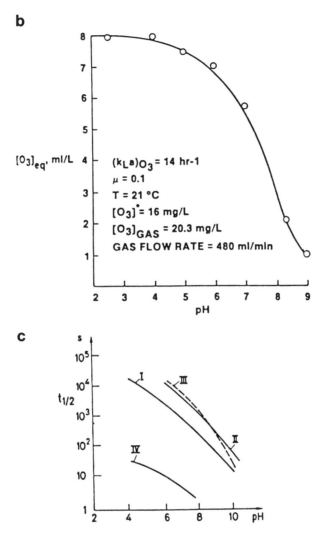

Figure 8-12. (a) Reactions of O_3 in pure water and aqueous solutions of solutes (M) that react directly with O_3 {1}, initiate radical chains {1b}, or interact with ·OH radicals produced from O_3 {9–12}. [From Staehelin, J. and J. Hoigne, *Environ. Sci. Technol.*, 19, 1206 (1985). With permission of the Amer. Chem. Soc.] (b) Steady-state aqueous concentration of O_3 decreases with pH, showing O_3 decomposition is more rapid at high pH. [From Kalmaz, E.E. and N.M. Trieff, *Chemosphere*, 15, 183 (1986). With permisison.] (c) Half-life of O_3 in various solutions decreases with pH: I, 0.05 M phosphate buffer; II, 2 mM sodium bicarbonate; III, 0.3mM methyl mercury hydroxide; IV, 0.08 mM benzene. [From Hoigne, J. and H. Bader, *Water Res.*, 17, 173 (1983). With permission.]

size. The effects of electrostatic repulsion and aromatic adsorption can be minimized (but not eliminated) by proper selection of eluants and gels,[88,89] but most estimates of AH molecular size published in the 1970s must be regarded with suspicion.

Table 8-7. Rate Constants for Direct Ozonation of Various Organic Compounds[a]

Compound	k_2[b]	Compound	k_2[b]	Compound	k_2[b]

A. Nondissociating compounds

Substituted benzenes		**Substituted alkanes**		**Halomethanes**	
nitrobenzene	0.9E-1	t-butanol	3.0E-3	carbon	
chlorobenzene	7.5E-1	ethanol	3.7E-1	tetrachloride	<5E-3
methyl benzoate	1.1E0	methanol	2.4E-2	chloroform	<1E-1
benzene	2E0	butylamine	<2E-2	methylene	
benzaldehyde	2.5E0	I-octanol	8E-1	chloride	<1E-1
toluene	1.4E1	acetone	3E-2	bromoform	<2E-2
o-xylene	9.0E1	formaldehyde	1E-1		
anisole	2.9E2	acetaldehyde	1.5E0	**Miscellaneous**	
naphthalene	3.0E3	n-octanal	8E0	urea	5E-2
				dioxane	3E-1
PAHs		**Substituted ethylene**		glucose	4.5E-1
pyrene	4.0E4	tetrachloroethylene	<1E-1	diethylether	1.1E0
phenanthrene	1.5E4	trichloroethylene	1.7E1	dipropylsulfide	>2E5
benzo[a]pyrene	0.6E4	1,1-dichloroethylene	1.1E2	ethylmercaptan	>2E5
		allylbenzene	1.2E5		
		styrene	3E5		

Compound	k_2(HB)	k_2(B⁻)	Compound	k_2(HB)	k_2(B⁻)

B. Dissociating compounds

acetic acid	<3E-5	<3E-5	ammomia	0.0	2E1
oxalic acid	—	<4E-2	methylamine	—	1.4E5
malonic acid	<4E0	7E0	dimethylamine	<1.3E-1	1.9E7
formic acid	5E0	1E2	aniline	—	9E7
benzoic acid	—	1.2E0	glycine	—	1.3E5
4-chlorophenol	6.0E2	6.0E8	alanine	—	6.4E4
2,4-dichlorophenol	<1.5E3	~8E9	imidazole	2.2E1	4E5
2,4,5-trichlorophenol	<3E3	>1E9	pyridin	1E-2	3E0
phenol	1.3E3	1.4E9			
salicylic acid	<5.0E2	3.0E4			

[a] Compounds in (A) summarized from Hoigne, J. and H. Bader, *Water Res.*, 10, 377 (1976) except for PAHs, which are summarized from Butkovik, V., et al., *Environ. Sci. Technol.*, 17, 546 (1983). Compounds in (B) summarized from Hoigne, J. and H. Bader, *Water Res.*, 17, 173 (1983). Rate constants determined from absolute rate of O_3 consumption or from relative rates of disappearance of two reactants in presence of O_3 (where reaction rate is known for one compound). ·OH radical scavengers (e.g., t-butanol, bicarbonate) were added to solutions to minimize interferences from decomposition products of O_3.
[b] k_2 in M^{-1} s^{-1}.

Several lines of experimental evidence[90] now indicate that AH is relatively small — average molecular weight of 800 to 1000 for aquatic FA and ~3000 for aquatic HA.[91] In addition, recent structural data from high-resolution ^{13}C NMR[92] indicate that both aquatic and soil humic materials contain a much larger aliphatic component than previous degradation studies suggest. Only 16 to 20% of the carbon atoms in aquatic FA are aromatic, and the carbon content of HA is ~30% aromatic.[91] The dominant core structure of AH thus appears to be aliphatic. How

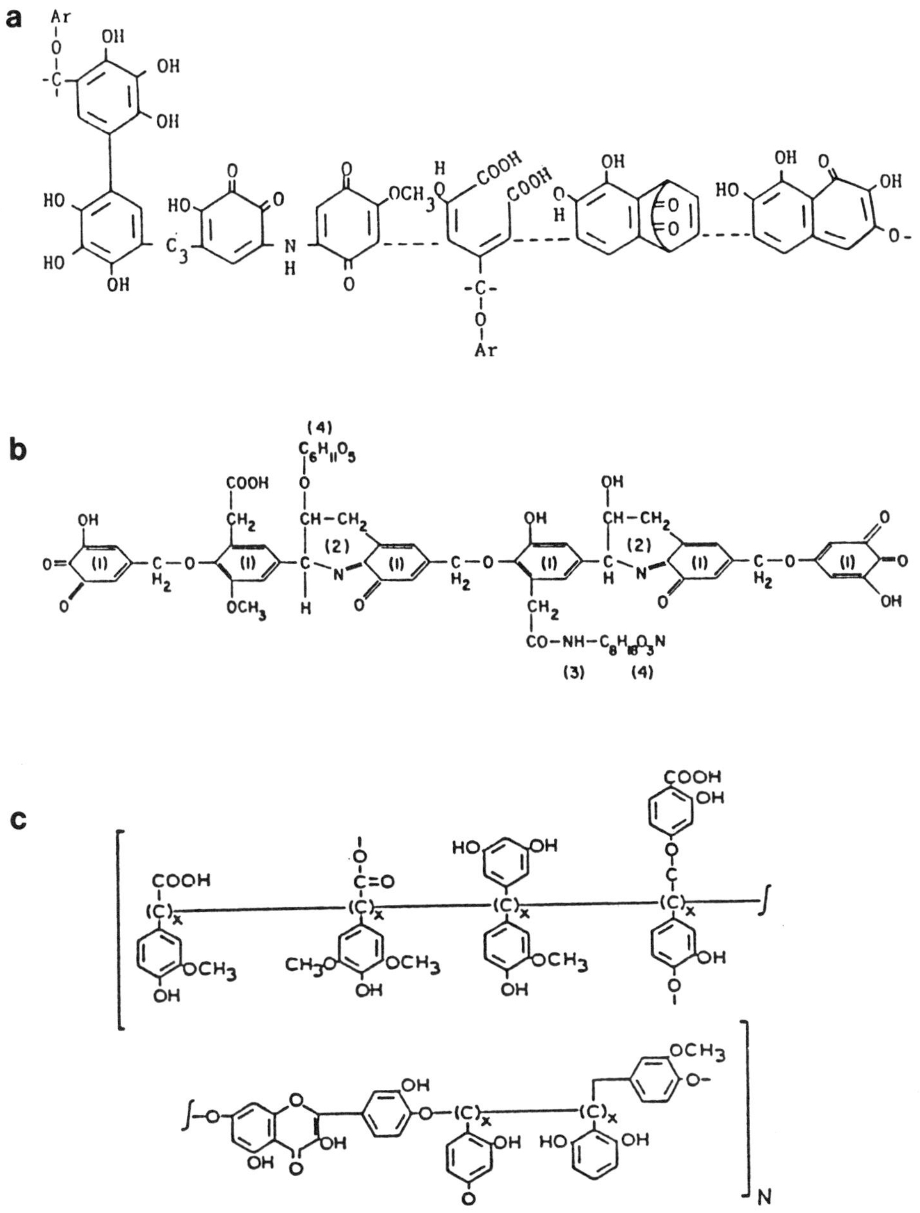

Figure 8-13. Structural models of humic substances: (a) humic acid according to Flaig; (b) Dragunov's structure of humic acid, as given by Kononova; (c) aquatic humus according to Christman, R.F. and M. Ghassemi, *J. Am. Water Works Assoc.*, 58, 723 (1966). (d) soil humic matter according to Schnitzer, M., in *Soil Organic Matter*, M. Schnitzer and S. U. Khan (Eds.), Elsevier, New York, 1978. (e) fulvic acid according to Buffle; and (f) humic acid according to Stevenson. [(a), (b), (e), and (f) redrawn from Aiken, G.R., et al., *Humic Substances in Soil, Sediment, and Water,* Aiken, G.R., et al. (Eds.), John Wiley & Sons, New York, 1985; original references cited therein.]

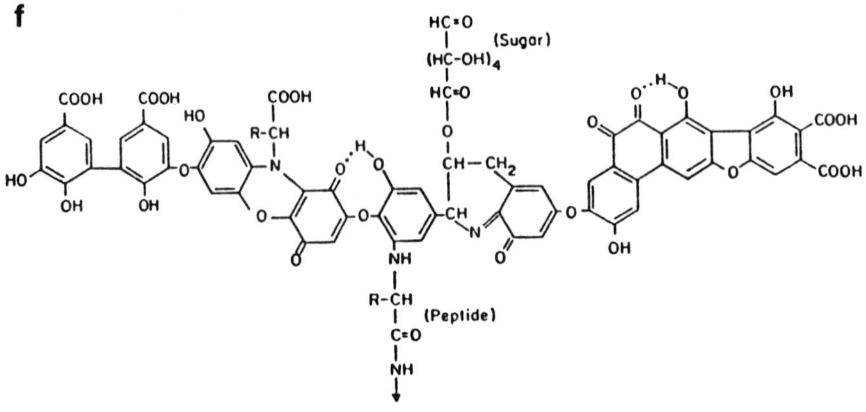

Figure 8-13 (continued).

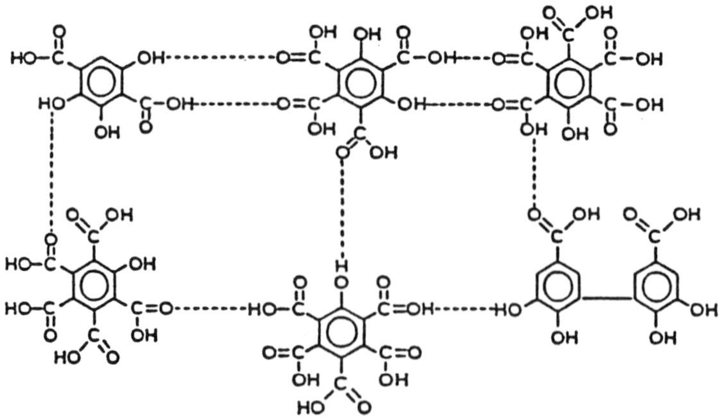

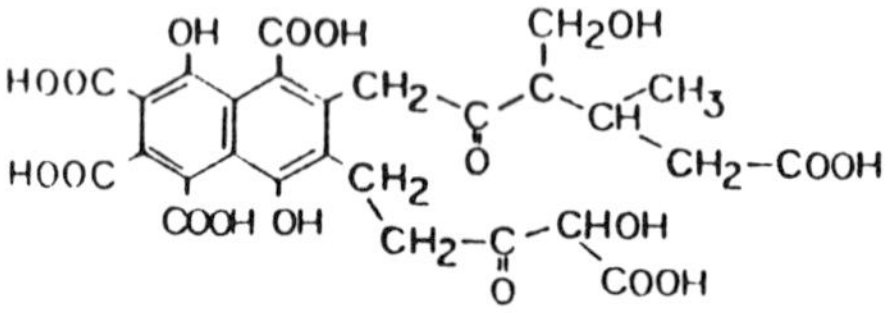

can the NMR and degradation data be reconciled? It is likely that the latter data are correct but incomplete. Although aromatic compounds predominate in the products of oxidative degradation, yields from such experiments are only a few percent of the starting material. Even "mild" oxidants like alkaline permanganate apparently destroy aliphatic portions of AH molecules, but leave the aromatic rings intact.

8.4.2 Formation Mechanisms for AH

The classical model for the origin of freshwater AH emphasizes allothchonous (terrestrial) sources. Some have considered freshwater AH simply as solubilized soil humic materials, but this is too simple. A broader, less explicit view is that soil and aquatic humic materials have a common precursor — woody vegetation — but formation can take place in soil, organic litter, natural waters, and sediments. In fact, there is as much uncertainty about the origin of AH as there is about its structure! Two general mechanisms have been proposed for formation of humic materials in terrestrial environments: degradation of macromolecules, especially lignin from wood, and condensation into new macromolecules of the monomers from these degradation reactions plus tannins and flavanoids extracted from leaves and bark.

At least three mechanisms have been proposed for autochthonous production of AH: (1) transition metal-catalyzed polymerization of phenolic compounds like catechol;[93] (2) enzyme-mediated oxidation of DOM, e.g., by phenolases[94] and laccases,[95] which are capable of coupling (polymerizing) phenols; and (3) the so-called "browning" reaction, which forms Schiff bases between amino groups of amino acids and aldehyde or ketone groups of sugars.[96] The Schiff bases are themselves reactive and undergo rearrangements, cyclizations, and decarboxylations to form complex brown mixtures known as melanoidins.[97] None of these reactions is sufficient to account for the aliphatic content of AH, however, and although they can be demonstrated in laboratory systems, their importance in the environment remains unknown. It also is possible that AH in lakes is partly derived from photoinduced free radical reactions of algal lipid exudates, similar to the mechanism described below for formation of marine AH. Autochthonous formation is likely to be important in lakes that do not receive swamp drainage and have relatively low color, but high DOC levels.

Enzymatic coupling of phenols by extracellular fungal laccases has been studied with model compounds, including phenolic derivatives of lignin and several aromatic xenobiotic pollutants. Oligomeric products have been identified,[95] and this process may be important in binding organic pollutants to AH. Laboratory experiments have shown that ^{14}C-labeled pesticides are incorporated into humic matter produced by fungal cultures.[98] The photoinduced formation of an estrogenic conjugate by methoxychlor and hydroquinone[99] suggests that photochemical reactions may play a role in binding pollutants to AH.

Hypothesized structures for autochthonous marine AH (Figure 8-14) are different from the "classical" pictures of aquatic and soil humic materials. Marine AH isolated from oligotrophic waters like the Sargasso Sea is largely aliphatic and is thought to be formed *in situ* by free radical autoxidative cross-linking of unsaturated lipids excreted by algae.[100,101] The condensation reaction is accelerated by light and transition metals (Figure 8-14). This mechanism is supported by experiments in which fulvic acid-like material was formed when model triglyceride esters of unsaturated fatty acids (e.g., trilinolein) were added to seawater.[102] Natural humic material was extracted from seawater prior to the experiments, and the solutions were incubated in sunlight for up to 7 days. The ir spectrum of

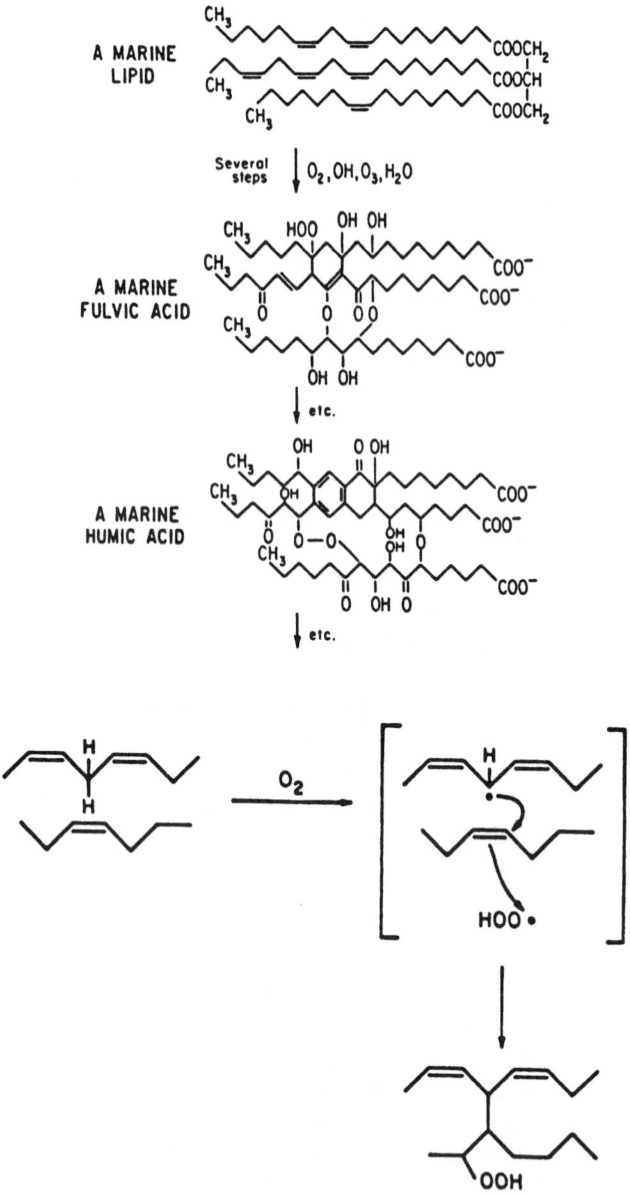

Figure 8-14.   Hypothetical structures of marine fulvic and humic acids thought to be derived from marine lipids by autoxidation mechanism catalyzed by light and transition metals. (From Harvey, G.R. and D.A. Boran, in *Humic Substances in Soil, Sediment, and Water*, John Wiley & Sons, New York, 1985. With permission.)

synthetic products is similar to that of authentic marine FA (Figure 8-15). Marine FA is thought to consist of 2 to 4 cross-linked fatty acid chains and has a molecular weight similar to that of freshwater FA (900–1200). Continued autoxidation of

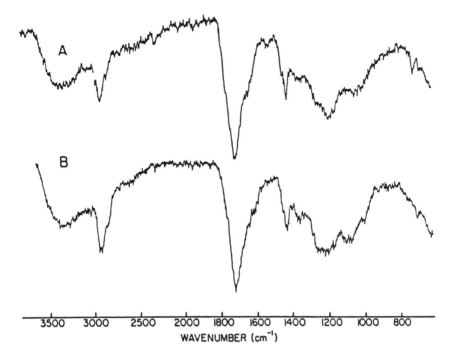

WAVENUMBER (cm⁻¹)

Figure 8-15. Infrared spectrum of natural marine fulvic acid (A) isolated from Gulf of Mexico and synthetic fulvic acid (B) prepared by autoxidation of trilinolein. (From Harvey, G.R. and D.A. Boran, in *Humic Substances in Soil, Sediment, and Water*, John Wiley & Sons, New York, 1985. With permission.)

marine FA may cause further intramolecular cross-linking and lead to formation of less soluble HAs, but the favored reaction pathway apparently is cleavage by sunlight-produced $^1O_2$ to low-molecular-weight acids and aldehydes.[100] Of course, not all AH in seawater is autochthonous; that found in coastal areas probably is primarily of riverine or terrestrial origin and probably similar in structure to freshwater AH.

8.4.3 Spectral and Acidic Properties of Aquatic Humus

AH is not a single compound or small group of compounds, but a heterogeneous mixture of innumerable structural entities produced by several mechanisms. Nonetheless, AH, FA, and HA derived from various sources have fairly constant physicochemical properties and must have similar structural features. A property of obvious relevance to photochemical activity is the uv-visible absorbance spectrum. Soil humic isolates and AH all have simple spectra characterized by gradually increasing absorbance as λ decreases (Figure 8-11b). Absorbance spectra of humic material obtained from various commercial and freshwater sources are similar and can be described by an exponential equation:[103]

$$e_\lambda = a \exp\big(b(450 - \lambda)\big)$$ (8-27)

a and b are fitted coefficients, and e_λ is a λ-dependent absorption coefficient. $e_\lambda = A_\lambda/C_o\ell$, where A_λ is absorbance at λ, C_o is the concentration of DOC (mg L^{-1}), and ℓ is lightpath (m). Means (± 1 s.d.) of a and b for 11 natural waters were 0.60 ± 0.28 and 0.0145 ± 0.0006, respectively. It must be emphasized, however, that $A_\lambda/C_o\ell$ is not constant among all waters, although it may be approximately so for waters that are visibly brown stained.

Spectral data for marine humic material were fit to a similar equation:[104]

$$e_\lambda = e_{\lambda o} \exp\big\{-0.140(\lambda - \lambda_o)\big\}$$ (8-28)

$e_{\lambda o}$ is the light absorption coefficient (m^{-1}) for marine AH at any fixed wavelength (λ_o) between 280 and 450 nm. The exponential coefficient in Equation 8-28 has an uncertainty (s.d.) of ± 0.0025 and agrees closely with the value of coefficient b in Equation 8-27. According to Nyquist,[105] an AH concentration of 1 mg L^{-1} yields a value of e = 0.212 m^{-1} at 450 nm; the concentration of marine AH thus can be estimated from absorbance data by the following equation:[104]

$$[AH] = 4.72 e_\lambda \exp\big\{0.140(\lambda - 450)\big\}$$ (8-29)

The carboxylate content of AH is of interest for charge-transfer photoredox reactions (see next section). Although some differences exist among reports in the literature, carboxylate concentrations of AH are relatively constant (within a factor of about 2). For example, Oliver et al.[106] isolated AH from 19 sources, including streams, lakes, wetlands, and groundwater, in diverse areas of North America and found a carboxylate content of 10.5 (± 1.7) μeq mg^{-1} organic carbon. Eshelman and Hemond[107] reported a value of 7.5 μeq mg^{-1} DOC in surface and groundwater samples from Massachusetts, but a lower value (5.5 μeq mg^{-1}) has been reported for some surface waters in Norway and Scotland.[108] A similar range of carboxylate concentrations have been reported for DOC in bog water (4.8 to 9.4 μeq mg^{-1}).[99-111]

8.4.4 Photolysis of Organic Acid-Transition Metal Complexes

Carboxylate groups (-COO$^-$) are the most common organic anions in natural waters and also the most common organic ligands. They are present in a wide range of compounds: simple organic acids like acetic and citric acid, which play prominent roles in metabolic processes; complicated molecules like aquatic humus (humic and fulvic acids); polymeric biomolecules, including proteins and polysaccharides; and synthetic molecules like aminopolycarboxylates, e.g., EDTA and NTA, which enter natural waters through municipal and industrial wastes. The carboxylate groups of some of these substances are subject to direct photolysis that causes decarboxylation by a ligand-to-metal charge-transfer (LMCT) mechanism involving transition-metal complexes.

Decarboxylation of polyaminocarboxylic acids by this mechanism has been known for many years; photodecarboxylation of the ferric-EDTA complex was described in 1952,[112] and the photodecomposition of Fe^{III}-NTA complexes was reported 20 years later,[113] when NTA was being considered as a replacement for phosphate in detergents. The net reaction is

$$N \begin{array}{c} \diagup CH_2 - COO \diagdown \\ - CH_2 - COO - Fe^{III} \\ \diagdown CH_2 - COO \diagup \end{array} \xrightarrow[H_2O]{hv} IDA + CH_2O + CO_2 + Fe^{II} \tag{8-30}$$

IDA stands for iminodiacetic acid; CH_2O is formaldehyde. Fe^{II} is reoxidized to Fe^{III} by dissolved O_2, a rapid reaction at neutral and alkaline pH. The net reaction thus involves oxidation of NTA by O_2, with Fe^{III} as a cyclic catalyst that is photoreduced (Equation 8-30) and thermally oxidized.

The charge-transfer process involves absorption of light by a metal-ligand complex. This promotes an electron into an excited state — from an occupied orbital of the organic ligand to an unoccupied orbital of the metal ion. Figure 8-16 illustrates the process from a molecular orbital perspective.* Electronic transitions from ligand to metal orbitals occur with high probability because the orbitals are unlike in spin characteristics,[115] and molar absorptivity is high. Iron is an ideal metal for charge-transfer reactions in natural water, because it is fairly abundant, has two oxidation states, forms strong complexes (especially Fe^{III}), and has a charge-transfer absorption region at near uv wavelengths that penetrate water reasonably well. Several other transition metal ions have suitable properties to catalyze decarboxylation reactions by LMCT mechanisms: Cu^{II}-Cu^{I}, Co^{III}-Co^{II}, $Mn^{III,IV}$-Mn^{II}. Significant rates of NTA photooxidation have been reported[116] for Cu^{II}-NTA complexes irradiated at 350 nm.

Photolysis of Fe^{III}-EDTA by the LMCT mechanism proceeds by a series of steps (Figure 8-17a), yielding several intermediates and products,[117] including IDA, glycine, and compounds with one and two acetate groups removed from EDTA. The time course for production and disappearance of the intermediates and products varies with pH (Figure 8-17b) and is most rapid at low pH. At a light intensity of 4000 ft-candles (6.9×10^{-3} watts cm^{-2}) and an initial Fe^{III}-EDTA concentration of 1.6 mM, EDTA removal was complete in 24 h at pH 4.5 to 6.9. Direct evidence also exists[118] for photoreduction of Fe^{III} by citric acid (Figure 8-18); tannic acid also reduces Fe^{III}, but this reaction does not require light. Photoreduction of Fe^{III} by citric acid was greatest at pH 3.2, probably because of increased OH$^-$ competition for Fe^{III} at higher pH.

Indirect evidence exists[118] for Fe^{III}-mediated photooxidation of many simple mono-, di-, and triprotic carboxylic acids (Table 8-8); O_2 consumption rates by

---

*Charge-transfer interactions (electron donor-acceptor complexes) also occur between dissolved humic materials and organic compounds like chlorinated benzoquinones that are strong electron acceptors.[114]

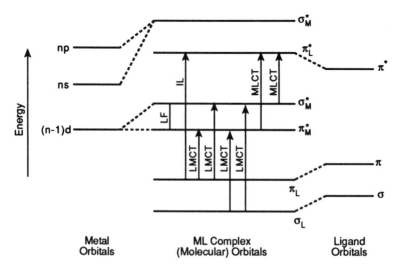

Figure 8-16. Molecular orbital (MO) energy diagram showing possible electronic transitions for octahedral ML complexes. Lines connect atomic orbitals to MOs in which that have the greatest participation. Key: n = electron shell number (principal quantum number); p,s,d = second (angular) quantum number; σ and π represent sigma- and pi-type bonding MOs; subscripts M and L represent MOs derived from metal and ligand atomic orbitals, respectively; LF = ligand-field transition, representing the splitting energy Δ between e_g and t_{2g} d-orbitals of the metal ion in an octahedral field; IL = internal ligand transition; LMCT = ligand-to-metal charge-transfer; MLCT = metal-to-ligand charge-transfer. (Modified from Balzani, V. and V. Carassiti, *Photochemistry of Coordination Compounds*, Academic Press, New York, 1970.)

solutions of the acids in the presence of Fe^{III} generally were higher in the light than in the dark, probably because photoinduced LMCT reduced Fe^{III} to Fe^{II}, which then was reoxidized to Fe^{III} by O_2. As expected, oxalic acid had a high O_2 consumption rate in the light; reduction of Fe^{III} by oxalic acid, $2Fe^{III} + C_2O_4^{2-} \rightarrow 2Fe^{II} + 2CO_2$, is used in chemical actinometry[119,120] to measure quantum efficiencies (Section 2.6.4). Modest O_2 consumption rates were found for several amino acids (Table 8-8), and photooxidation of glycine has been reported.[34] Photodecomposition may help explain the paucity of free amino acids in surface waters, but planktonic uptake probably is a more important loss mechanism. Other mechanisms besides photo-LMCT reactions must be responsible for O_2 consumption in solutions of reactive aromatic species like pyrogallol (1,2,3-trihydroxybenzene) and hydroquinone, which lack carboxylate groups, and tannic acid, which had comparable rates in the light and dark and consumed O_2 in the absence of Fe^{III} (Table 8-8).

8.4.5 Photodegradation of Aquatic Humus

Humic substances are biologically refractory and often are considered to be chemically stable as well. As indicated in the previous section, however, they are photochemically reactive, and thus are not the inert, long-term sinks for organic

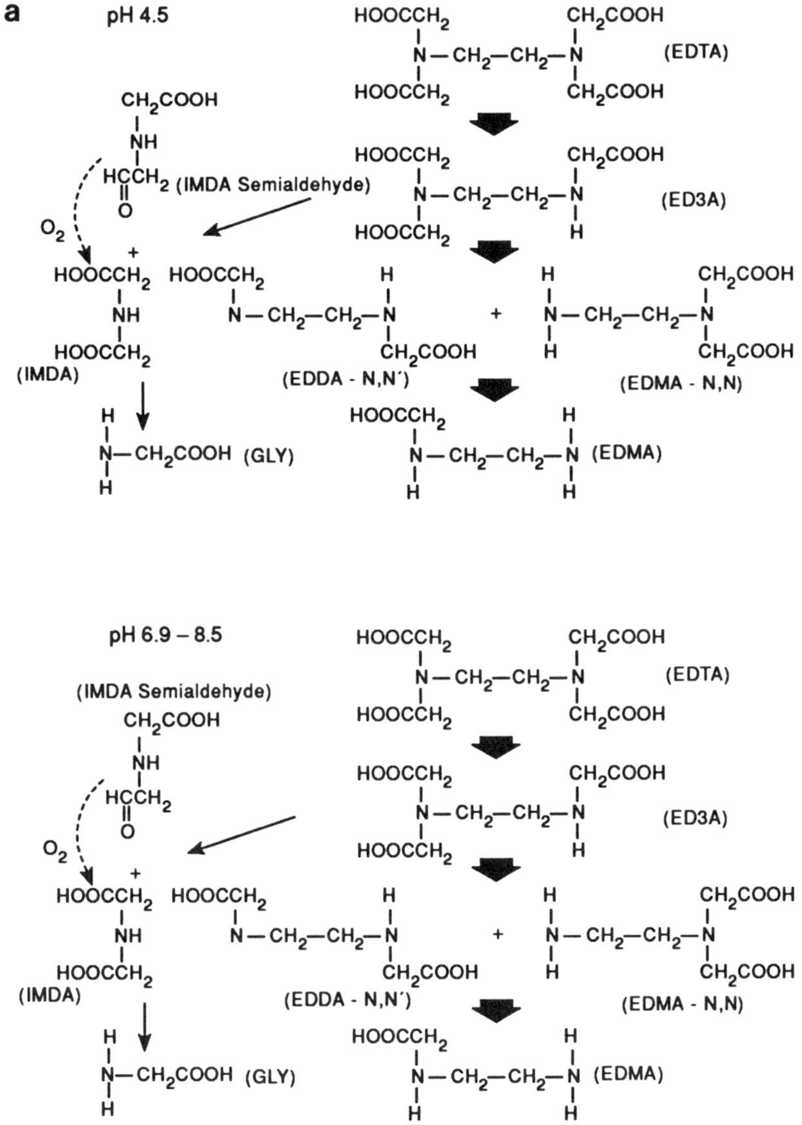

Figure 8-17. (a) Proposed degradation schemes for Fe^{III}-EDTA at pH 4.5 and 6.9-8.5, based on (b) concentrations of EDTA and its photodegradation products in presence of Fe^{III} vs. time at three pH values. Note: IMDA semialdehyde in proposed schemes was not detected experimentally. [From Lockhart, H.B., Jr. and R.V. Blakely, *Environ. Sci. Technol.*, 9, 1035 (1975). With permission of the Amer. Chem. Soc.]

carbon they once were thought to be. Radiocarbon dates of soil humus vary widely, but the average age is about 1000 years.[121] In contrast, carbon dating of AH from the Okefenokee Swamp yielded an age of only 20 years.[122] In addition, dissolved FA isolated from the Amazon River recently was found to have a minimum of 40 to 47%

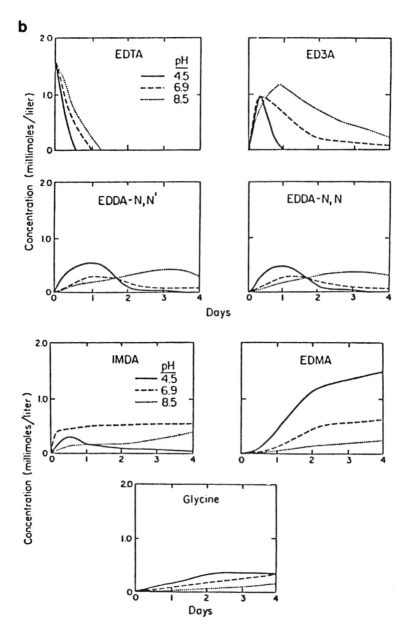

Figure 8-17 (continued).

"bomb" carbon by accelerator mass-spectrometric analysis of its ^{14}C content.[123] Bomb carbon is defined as carbon derived from the atmosphere since 1954, when testing of thermonuclear weapons began to increase the ^{14}C content of atmospheric CO_2. In contrast, HA from the same locations had only 21 to 26% bomb carbon;

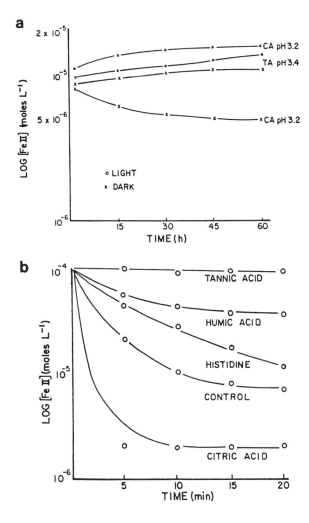

Figure 8-18. Effect of light and organic acids on Fe^{II}-Fe^{III} redox chemistry. (a) Reduction of Fe^{III} to Fe^{II} occurs only in light (500 μEi m^{-2} s^{-1}) with citric acid (CA), but in light and dark with tannic acid (TA). Decrease in [Fe^{II}] with CA in dark represents net oxidation of Fe^{II}. Initial conditions: Fe^{III} = 2×10^{-4} M; CA, TA = 100 mg L^{-1}. (b) Effect of various organic acids on rate of Fe^{II} oxidation under dark conditions. Initial conditions: pH 6.3; 0.1 M phosphate buffer; 25°C; pO_2 = 0.21 atm; organic acids = 100 mg L^{-1}; Fe^{II} = 1×10^{-4} M. [From Miles, C.J. and P.L. Brezonik, *Environ. Sci. Technol.*, 15, 1089 (1981). With permission of the Amer. Chem. Soc.]

which suggests that this fraction is somewhat older than the FA fraction (although still comparatively young). Precursor compounds for aquatic humic matter may reside in terrestrial vegetation several years before being released to soil/water systems. HA, which is more hydrophobic than FA, probably is released from soils more slowly than FA. If the ages of the Okefenokee and Amazon AH are

Table 8-8. Fe^{III}-Mediated Oxidation Rates of Some Organic Compounds[a]

Compound[b]	pH	O_2 consumption rate, μM h^{-1}	
		light[c]	dark
Tannic acid	1.9	18.1	16.6
Pyrogallol (1,2,3-trihydroxybenzene)	2.0	11.6	17.5
Propyl gallate	—		
Hydroquinone	2.8	<0.3	<0.3
Diphenylamine	—	2.8	1.9
1,2,4,5-benzene-carboxylic acid	—	<0.3	<0.3
Citric acid	1.8	25.9	<0.3
Malic acid	1.9	15.9	<0.3
Malonic acid	2.7	0.3	0.3
Oxalic acid	1.8	20.6	0.3
Glycine	2.9	0.9	<0.3
Proline	2.5	<0.3	<0.3
Lysine	2.8	1.3	<0.3
NTA	2.8	0.9	<0.3

[a] Expressed in terms of O_2 consumption rate; modified from Miles, C.J. and P.L. Brezonik, *Environ. Sci. Technol.*, 15, 1089 (1981).

[b] All at 100 mg L^{-1} in presence of 6 mg L^{-1} of Fe^{III}; all compounds except pyrogallol, hydroquinone, and diphenylamine have at least one carboxylic acid group.

[c] 200 μEi m^{-2} s^{-1}.

representative, it is apparent that AH is rather young. Evidence presented in this section also indicates it is chemically reactive in aquatic environments and that a large fraction of the AH degrades on the time scales of weeks to months.

The bleaching effect of sunlight on dissolved organic color has been known for many decades; Hutchinson[124] reported studies dating back to the late 19th century that demonstrated substantial loss of color in lakewater incubated in bottles near the lake surface. Aberg and Rodhe[125] found that sterilized and unsterilized water lost color at comparable rates when incubated in the light, thus implying that the process was photochemical rather than microbial. More recent studies indicate that the presence of dissolved O_2 is essential for the bleaching reaction[46] and that the loss of color and DOC from lakewater incubated in the light follows first-order kinetics.[89] Half-lives of color and DOC were about 35 to 45 days (equivalent to roughly 500 h of sunlight).

Photoreduction of Fe^{III} by AH and subsequent reoxidation of Fe^{II} by dissolved O_2 has been measured in highly colored lakewaters, and photoinduced O_2 consumption rates as high as 0.12 mg L^{-1} h^{-1} have been reported.[118] Charge-transfer decarboxylation also occurs when polycarboxylates (citric acid, humic material) adsorbed onto metal oxide surfaces are irradiated in the near UV (see Section 8.6.1). These processes may be important in removing carboxylate ligands from DOM, in chemical decomposition of AH, in solubilizing Fe, and as a sink for dissolved O_2. Humic-colored water often are undersaturated in O_2; concentration of 3 to 4 mg L^{-1} (~30 to 50% saturation) are common in surface waters of swamps and colored lakes that have no apparent anthropogenic source of biodegradable organic matter. An iron cycle like that discussed earlier could be responsible for the unsaturated conditions in some cases.

As Figure 8-19 shows, rates of O_2 consumption were much faster in AH-rich swamp water in the light than in the dark and were approximately first order in dissolved O_2; they increased in direct proportion to the amount of iron (added as Fe^{III} salt) and the concentration of DOC; they increased hyperbolically with increasing incident light intensity; and they increased slowly with increasing pH. The relative insensitivity to pH probably represents net effects on separate Fe^{III}-reduction and Fe^{II}-oxidation steps. As mentioned above, photoreduction of Fe^{III} by polycarboxylic acids decreases with increasing pH; Fe^{II} oxidation increases rapidly with increasing pH (with $(OH^-)^2$ at pH > ~5). Two other types of evidence support the hypothesis that O_2 consumption in colored waters occurs by a charge-transfer iron-cycle mechanism. The molar ratio of CO_2 produced to O_2 consumed was near the theoretical value of 2.0 obtained by adding the Fe^{III}-reduction and Fe^{II}-oxidation steps. In addition, methylation of carboxyl groups on AH isolates to form esters that do not form complexes with Fe caused a substantial decrease in the rate of O_2 consumption.[118]

If charge-transfer decarboxylation is important, one would expect the carboxyl content of AH to decrease with increasing exposure to sunlight. Miles[89] found slightly lower carboxyl acidity in AH from a highly colored lake than in AH from the lake's inlet stream (8.9 vs. 10.0 µeq mg^{-1}). The lake isolate had higher carbon and hydrogen and lower oxygen contents (49.7, 7.92, and 39.1%, respectively) than the stream isolate (48.4, 5.67, and 44.0%). Taken together, these data suggest that the lake AH, which was exposed to sunlight for a longer period, was more humified* than the stream isolate. Decarboxylation also is a major mechanism in the uv degradation of soil fulvic acid.[127]

Little is known about the photodegradation products of AH. Several studies have reported no dramatic changes in various indicators of organic structure (uv, ir, nmr spectra, gc-ms)[89,91] when AH samples were partially oxidized by uv light, but photolysis has been reported to cause a gradual decrease in the average molecular size of AH.[128,129] In addition, a decrease in the intensity of ESR signals was found with increasing uv irradiation of HA.[130] This was interpreted to indicate lower production of long-lived free radicals by aged HA. Solutions with added cumene and benzene (scavengers for ·OH and ROO·, respectively) showed decreased production of these radicals when HA was "aged" by preirradiation. Photochemical aging thus seems to produce humic material that is less photoreactive. On the other hand, there is some evidence that bacteria are able to use photolytic products of AH for growth and that light acts as a "priming agent" for microbial degradation of AH.[97,128]

---

* Humification is a diagenetic process that converts soluble FA and HA to large (insoluble) molecules (humins) with fewer carboxylic and phenolic groups. The extent of humification of organic matter can be described on a plot of elemental ratios: H:C vs. O:C — a van Krevelen diagram.[126] More humified matter has lower ratios of both H:C and O:C.

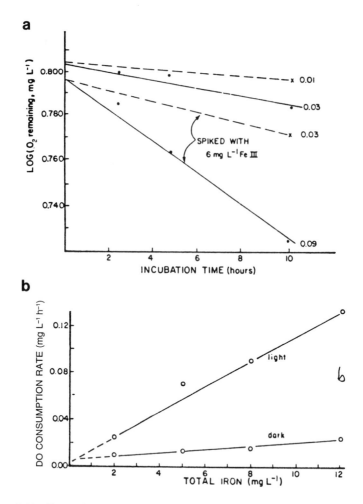

Figure 8-19. Fe-catalyzed photochemical O_2 consumption in humic-rich swamp water from northern Florida (color = 410 chloroplatinate units; DOC = 68 mg L^{-1}; Fe_T = 2.0 mg L^{-1}. (a) First-order plot of O_2 consumption vs. time in light (o) and dark (x) with and without added Fe^{III}; (b) effect of added Fe^{III} on O_2 consumption rates in light and dark; (c) effect of added aquatic humus on O_2 consumption rate in light; (d) effect of incident light intensity of O_2 consumption rate; and (e) pH dependence of O_2 consumption rates in light and dark. [From Miles, C.J. and P.L. Brezonik, *Environ. Sci. Technol.*, 15, 1089 (1981). With permission of the Amer. Chem. Soc.]

8.5 PHOTOCHEMISTRY OF ORGANIC CONTAMINANTS IN NATURAL WATERS

8.5.1 Direct Photolysis Reactions

A variety of biologically refractory organic pollutants can be decomposed by direct photochemical reactions; many others are subject to indirect photolysis. Table 8-9 lists rate constants and half-lives for direct photolysis of some important

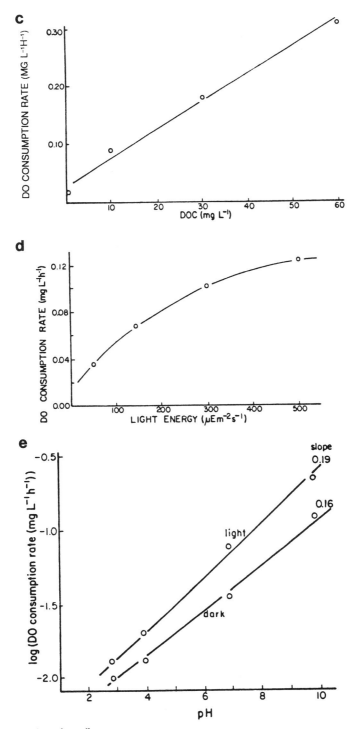

Figure 8-19 (continued).

Table 8-9. Direct Photolysis Rates and Quantum Yields for Some Organic Contaminants[a]

Compound	ϕ_d	k_s	$t_{1/2}$	Ref.
Anthracene	0.003	8.5	0.081	3
Benz[a]anthracene	0.0033	10.5	0.066	7
	0.032	12.7	0.054	133
Benzo[a]pyrene	0.00089	11.7	0.059	7
	0.00089	13.0	0.053	133
Benzo[f]quinoline	0.014	11.5	0.06	7
	0.014	16.0	0.043	133
Benzo[b]thiophene	0.1	0.049	14.1	133
Carbaryl	0.006	0.105	6.6	7
9H-Carbazole	0.0076	5.75	0.12	7
		6.63	0.10	133
7H-Dibenzo[c,g]carbazole	0.0028	16.7	0.041	133
Dibenzo[e,g]carbazole	0.0033	20.9	0.033	7
Dibenzothiophene	0.0005	0.12	5.7	7
	0.0005	0.13	5.3	133
3,3'-Dichlorobenzidine	—	25.6	0.0027 (1.5 min)	3
DDE	0.3	0.75	0.92	7
2,4-D-butoxyethyl ester	0.17	0.058	12.0	7
DMDE	0.3	7.3	0.094	7
Diphenyl mercury	0.056	0.86	0.8	7
Methyl parathion	0.00017	0.077	9.0	7
Naphthacene	0.013	230.0	0.003 (1.8 min)	3
Naphthacene	0.0015	0.09	7.7	3
N-Nitrosoatrazine	0.3	75.0	0.0092 (4.8 min)	7
Parathion	0.00015	0.069	10.0	7
Phenyl mercuric acetate	0.25	2.76	0.25	7
Pyrene	0.0022	9.3	0.074	3
Quinoline	0.00033	0.028	25.0	7
	0.00033	0.031	22.3	133
Trifluralin	0.002	17.7	0.039	7

[a] ϕ_d measured at 313 nm or 366 nm; k_s in day^{-1} (average 24-h rate for midsummer 40° near-surface conditions); $t_{1/2}$ in days for same conditions.

organic pollutants. Where possible, constants are expressed in terms of midsummer sunlight conditions (24 h average) at 40° latitude.[4] Some organic compounds normally considered to be environmentally persistent photodegrade surprisingly fast, e.g., various dioxins,[131] some PCBs,[132] benzo[a]pyrene and some other PAHs,[133] and DDE (a degradation product of DDT).[134] This seemingly contradictory situation is explained by the fact that these compounds are highly hydrophobic and are found primarily in organisms and sediments where little opportunity exists for photodegradation.

The relative importance of direct photolysis vs. photosensitized degradation has been assessed for many organic compounds. Photosensitized reactions are not important for most of the compounds listed in Table 8-9, because in most cases direct photolysis is so rapid. The relative importance of the two processes can be determined by comparing light-induced loss rates of a compound in distilled water (dw) (or photochemically inert organic solvent) and in a natural water (nw) under the same light conditions. Natural waters used in such comparisons usually are rich in aquatic humus (AH), which is effective as a photosensitizing agent by generating e^-_{aq}, free radicals and singlet oxygen. The only loss mechanism in distilled water is direct photolysis. If k_{dw}/k_{nw} is greater than or equal to 1.0, photosensitized reactions are not important. For example, photolysis of hexachlorocyclopentadiene (hex) was slightly lower in colored river water than in distilled water.[135] Direct photolysis of this compound is so rapid ($k_x > 4$ h^{-1}; $t_{1/2} = 10$ min) that indirect mechanisms are unimportant. Nonetheless, a simulation of the fate of hex by the model EXAMS[135] showed that the recovery time of a pond receiving a spill of hex would be 3 to 4 months because hex is absorbed rapidly by sediments, where it is protected from photolysis. It desorbs back into the overlying water at a slow rate.

Similarly, photolysis rates of seven out of eight polycyclic aromatic hydrocarbons (PAHs) studied by Mill et al.[133] were lower in colored river water than in distilled water. The lower rates were explained by quenching of excited PAH molecules by AH. Dissolved O_2 had little effect on photolysis rates, leading the authors to suggest that the reaction proceeds by singlet states. Triplet states transfer their energy rapidly to O_2 (i.e., they are quenched by O_2). Strong absorption bands at $\lambda > 300$ nm offset small quantum efficiencies for these PAHs, and half-lives are in the range of minutes to hours.

Direct photolysis of some organic pollutants involves complicated mechanisms in which various intermediates and products may be produced by competing photoreactions and dark reactions. Example 8-3 illustrates reaction sequences of several pesticides, but mechanisms and products have not been determined for many compounds. Photolysis of the chlorinated insecticide methoxychlor in the presence of hydroquinone (HQ) results in a larger compound, rather than degradation of the parent material;[99] direct radical coupling of methoxychlor with HQ forms an estrogenic conjugate, 4,4'-dimethoxy-$\alpha\alpha'$-dichlorostilbene, through a long sequence of reactions. Although HQ is not found in natural waters, quinones are common in humic material, and this reaction may be a model for the incorporation of pesticides into AH. Similarly, photoproducts of DDE undergo photocyclization reactions to produce complicated polycyclic structures.[134]

Example 8-3. Mechanisms for Direct Photolysis of Organic Contaminants

A. 3,4-Dichloroaniline (DCA). This compound photolyzes directly ($\lambda > 290$ nm) to 2-chloro-5-aminophenol (~80% yield) with $\phi_d = 0.05$ at 313 nm.[136] ϕ_d is not affected by pH in the range 4 to 12 or by CN$^-$ up to 0.35 M. A molecular orbital calculation indicated a large shift in electron density from the N atom to the aromatic ring in the excited singlet state, with the meta carbon receiving the largest increase. The rate of photolysis was not affected by O_2 or sorbic acid, which are effective triplet quenchers. These findings suggest the reaction proceeds by formation of an aryl cation intermediate from the excited singlet state (i.e., an S_N1 mechanism); the absence of any effect of nucleophiles on ϕ_d rules out an S_N2 (bimolecular) reaction.

B. Pentachlorophenol (PCP). Direct photolysis mechanisms are not necessarily short or simple. PCP photodegrades through a series of competing pathways[137] that involve initial photonucleophilic substitution of -OH groups for -Cl to produce ortho-, meta-, and paradiols. The orthodiol is oxidized to dicarboxylic acid **II** with concomitant ring cleavage. The meta and para compounds are oxidized in the presence of O_2 (dark and photoreactions) to yield quinones **III** that subsequently undergo ring cleavage and further oxidation to small molecules, ultimately producing CO_2 and Cl^-. Intermediates in the sequence were identified by conventional organic separation and analysis procedures.

C. 2,4-Dichlorophenoxyacetic acid (2,4-D). Direct photolysis of this herbicide proceeds by at least three pathways.[138] Oxidation of the side chain involves photogeneration of e^-_{aq} from the carboxylate ion, followed by loss of CO_2. The resulting radical (**II**) reacts with O_2 to produce a formate ester that is hydrolyzed to 2,4-dichlorophenol. In turn, this compound hydrolyzes to a diol (**III**) and triol (**IV**), and these compounds undergo coupling reactions to produce polycyclic structures (**V**). Substitution of -OH for the ortho -Cl on 2,4-D probably occurs by an aryl cation intermediate similar to the mechanism for DCA. The side chain of the product (**VI**) can undergo photooxidation in the same way as the parent compound, ultimately leading to the diol (**III**). Reductive replacement of the para -Cl by -H probably is a free radical process, with the H extracted from organic solutes. The product (**VII**) continues to react by -OH substitution for -Cl and side-chain oxidation to produce the triol (**IV**) and polycyclic structures.

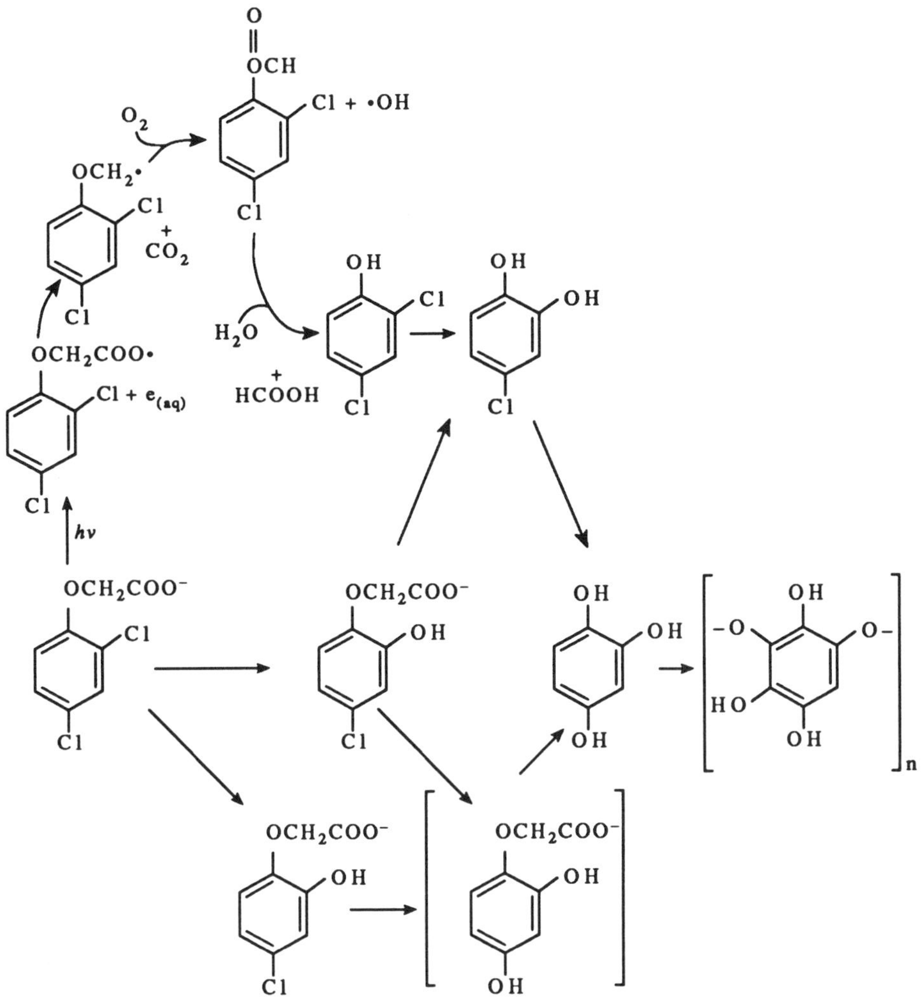

D. DDE. The major degradation route for DDT in the environment is dehydrochlorination to DDE (1,1-*bis*(p-chlorophenyl)-2,2-dichloroethylene), which is subject to microbial, chemical, and photochemical degradation. The latter involves photoisomerization that proceeds by exchange of a vinyl -Cl and ortho aromatic H.[134] Photolysis of the photoisomers is much slower than that of DDE itself. The mechanism is thought to involve cleavage of a vinyl C-Cl bond to yield a pair of free radicals. Because of the solvent cage phenomenon, the radicals do not diffuse apart immediately. Some of the radicals recombine (a null reaction), others form the photoisomer shown below, and still others react with O_2, H_2O, or other solutes to form other products. Photoisomerization does not occur in the vapor phase, which suggests that the solvent cage is essential for this mechanism.

8.5.2 Photosensitized Reactions of Organic Contaminants in Water: General Mechanisms and Photosensitizing Agents

In photosensitized reactions, organic reactants need not absorb light directly. For example, the pesticide aldrin, which does not absorb light at $\lambda > 250$ nm, is unreactive in light-irradiated distilled water, but is rapidly photooxidized to dieldrin, photoaldrin, and other products in light-irradiated water containing μM levels of H_2O_2.[139] H_2O_2 does not oxidize aldrin in the dark. Accelerated photolysis of organic compounds in natural waters (compared with distilled water) is common and usually attributed to the presence of dissolved organic matter, especially AH, which acts as a photosensitizer and provides alternative pathways to direct photolysis. In addition, the products of such reactions may be amenable to microbial degradation even if the reactants are not, as has been reported for several polychlorinated phenols.[140]

Excited (triplet)-state AH can induce photosensitized reactions of organic contaminants by three of the major mechanisms described in Table 8-1: (1) free radical formation via H atom transfer from contaminants to excited state AH; (2) quenching of excited AH by O_2 to form $^1O_2$, which adds directly to certain contaminants; and (3) direct energy transfer from excited AH to contaminant molecules. Zepp and co-workers[48,49] demonstrated this by adding reagents known to react by one of the mechanisms to solutions of humic substances. For example, aniline transfers H atoms from its nitrogen group to photosensitizer molecules, producing anilino free radicals. Azobenzene (C_6H_5-N=N-C_6H_5) is a product of these free radicals. Similarly, 2,5-dimethylfuran (DMF) is oxidized mainly to diacetylethylene by $^1O_2$ (Figure 8-10), and the insecticide disulfoton is oxidized to corresponding sulfoxide by $^1O_2$. Cis-1,3-pentadiene is a model for reactions induced by direct energy transfer from a photosensitizer. This compound isomerizes to the trans form by accepting triplet energy from photosensitizers,[141] but it does not isomerize when exposed to light in distilled water. Zepp et al.[48] found that efficiencies of these reactions were similar for eight commercial sources of HA and FA and four AH-rich natural waters. For example, DMF oxygenation rates for the 12 samples were within a factor of 3 and pH-independent in the range 5 to 10. The similarity in photosensitizing abilities of different humic samples corresponds to a narrow range of light absorption coefficients for them and suggests that the behavior of humic substances is fairly constant regardless of origin.

Reactions that result from hydrogen atom transfer are called **Type I** photosensitized oxidations; reactions involving $^1O_2$ are called **Type II** processes. When production and reaction of the various inorganic photointermediates are also included, the possibilities for photosensitized organic reactions become quite complicated, as the scheme in Figure 8-3 suggests.

Although DOM is thought to be the most important photosensitizing agent in most natural waters, it is not the only one. As described in Section 8.3.6, NO_3^- and NO_2^- absorb UV light and produce reactive radicals, including $\cdot OH$, NO, and NO_2. Recent studies have shown that incubation of biphenyl in the light in the presence of NO_3^- or NO_2^- results in nitration and hydroxylation of biphenyl to form nitrophenylphenols.[142] This finding is of interest because aromatic nitro compounds are mutagens. Rather than simply degrading organic matter, photochemical reactions may enhance the toxicity of some organic contaminants.

Iron species also may be important photosensitizers for oxidation of aquatic organic contaminants. As mentioned earlier, iron is a source of $\cdot OH$ radicals. Fenton's reagent, $Fe^{2+} + H_2O_2$,[143] is an effective oxidant of organic compounds like chlorophenols in laboratory systems,[144] because Fe^{2+} and H_2O_2 react rapidly to form $\cdot OH$ radicals and Fe^{3+} (reaction 27, Table 8-3), both of which are good oxidants. Photoreduction of Fe^{III}-AH complexes and reductive dissolution of ferric oxide by AH produces Fe^{2+}, and oxidation of Fe^{2+} by O_2 via the Haber-Weiss mechanism produces H_2O_2 (see Section 4.6.5). Further work is needed to define the role of inorganic species like nitrate and iron in the photochemical behavior of organic matter in lakes.

8.5.3 Nature of Organic Radicals Formed from DOM

The general nature of the organic radical intermediates produced from dissolved organic matter (DOM) by photoactivation can be predicted, even though exact structures of DOM are not known. Photochemical activation and photochemically produced inorganic radicals generate at least four types of organic radicals from DOM: alkyl, alkoxyl, alkylperoxyl, and carbonyl triplets (Table 8-10). As discussed earlier, $\cdot OH$ concentrations are very low in natural waters; consequently, there is some question whether reactions ii and iii are important, even though they have high rate constants — 10^8 to 10^{10} M^{-1} s^{-1}.[145,146] Alkyl radicals react rapidly with O_2 to produce alkylperoxyl radicals ($RO_2\cdot$).[146a] Reaction of $RO_2\cdot$ with NO to produce alkoxyl radicals ($RO\cdot$) and NO_2 is important in atmospheric photochemistry, but it is not important in water at [NO] < 10^{-10} M.[62] Concentrations of NO produced from

Table 8-10. Classes of Reactions that Generate Organic Radicals

(1) H atom transfer to excited triplet state organic compound

$$RH \xrightarrow{hv} {}^3RH$$
$${}^3RH + R'H \rightarrow HRH\cdot + \cdot R' \quad \text{(alkyl radical)}$$

(2) H atom transfer to hydroxyl radical
$$HO\cdot + RH \rightarrow H_2O + R\cdot$$

(3) Hydroxyl addition to double bond

$$HO\cdot + {>}C{=}C{<} \rightarrow HO\text{-}\overset{|}{\underset{|}{C}}\text{-}C\overset{/}{\diagdown}$$

(4) Addition of O_2 to alkyl radical to form peroxyl radical
$$R\cdot + O_2 \rightarrow RO_2\cdot$$

(5) Reaction of peroxyl radical with NO to form alkoxy radical plus NO_2
$$RO_2\cdot + NO \rightarrow RO\cdot + NO_2$$

(6) H atom transfer to alkoxy radical to form alcohol plus alkyl radical
$$RO\cdot + R'H \rightarrow ROH + R'\cdot$$

(7) Photochemical formation of n-π excited state from carbonyl group

$$\underset{R'}{\overset{R}{\diagdown}}C{=}O \xrightarrow{hv} \underset{R'}{\overset{R}{\diagdown}}C\text{-}O\cdot$$

photolysis of NO_2^- generally are lower than this in the open ocean, but higher concentrations could be produced in some freshwaters where $[NO_2^-]$ is higher. Carbonyl groups, including quinones, which are thought to be abundant in AH, are photoexcited directly to an n-π triplet state (reaction vii) that is similar to alkoxyl radicals in reactivity.[145] The principal organic radical intermediates resulting from photolysis of DOM thus are thought to be peroxyls ($RO_2\cdot$) and carbonyl triplets ($>$C-O·), which behave like alkoxyl radicals.

The relative importance of various types of organic radicals can be estimated from their widely differing reactivities toward different classes of organic compounds (Table 8-11). Peroxyl radicals react with phenols, amines, and hydroperoxides about as rapidly as alkoxyl radicals, but alkoxyl radicals are much more reactive (by 10^5) than peroxyl radicals toward alkanes, olefins, and alcohols. Model compounds like cumene (isopropylbenzene) and pyridine have been used in laboratory studies to evaluate the relative importance of alkoxyl and peroxyl radicals formed from DOM.[40] These compounds form different oxidation products, depending on the free radical generated in distilled water systems. For example, ·OH radicals formed by photolysis of H_2O_2 oxidize cumene and form side-chain and ring oxidation products in a 30:70 ratio, but RO_2· radicals produced thermally from azo-*bis*(2-carbomethoxypropane) (MAB) yield only side-chain oxidation products. Similarly, pyridine reacts with RO_2· to yield only pyridine N-oxide (C_5H_5NO), whereas reaction with OH· yields hydroxypyridines and more polar products, but no N-oxide.

Table 8-11.   Reactivities of Alkylperoxyl and Alkoxyl Radicals Toward Oxidation of Various Classes of Organic Compounds

Class	k_{ox}, M^{-1} s^{-1}	
	ROO·	RO·
Alkanes	10^{-2}	7×10^{-4}
Alcohols	10^{-2}	10^5
Amines	10^4	3×10^6
Hydroperoxides	10^5	2×10^8
Olefins	9×10^{-2}	5×10^5
Phenols	10^4	3×10^6

From Buxton, G., et al., *J. Phys. Chem. Ref. Data*, 17, 513 (1988).

Mill et al.[40] added cumene to several natural water samples and found both side-chain and ring oxidation products. The ratio of product types suggested that both OH· and RO$_2$· were responsible. Addition of pyridine to the water samples yielded several unidentified polar products plus pyridine N-oxide; this also suggested that both OH· and RO$_2$· are involved. The authors assumed that peroxyl radicals were responsible for the cumene side-chain products and pyridine N-oxide, and they estimated steady state [RO$_2$·] and [OH·] in the waters by kinetic analysis (Example 8-4). Estimates of [RO$_2$·] ranged from about 10^{-9} to 10^{-8} M under near-surface, daylight conditions, but [·OH]$_{ss}$ was only 10^{-18} to 10^{-17} M (somewhat lower than more recent estimates). Estimated half-lives for oxidation of the most reactive organic pollutants by ·OH were more than 80 days, whereas the estimated RO$_2$· concentrations would degrade the most reactive pollutants much more rapidly (t$_{1/2}$ of a few days or less).

Example 8-4. Kinetic Analysis by Probes for ROO· and ·OH Radicals

Given: ·OH reacts with cumene (CuH) to give primarily ring oxidation products, R, (mainly 2- and 4-isopropylphenols), and ROO· reacts with CuH to give only side-chain oxidation products, S.
Assume: [ROO·] and [·OH] maintain steady-state concentrations during experiments at constant illumination.
Derive: kinetic expressions for [ROO·]$_{ss}$ and [·OH]$_{ss}$ and calculate values from data reported by Mill et al.[40]
    The total loss rate for CuH is

(1) $$-d[CuH]/dt = k_T[CuH]$$

(2) $$= d[S]/dt + d[R]/dt = k_H[ROO·]_{ss}[CuH]$$

where (3) $$k_T = k_H[ROO·]_{ss} + k_a[·OH]_{ss}$$

k_H is the rate constant for H atom transfer from the side chain to ROO·, and k_a is the rate constant for ·OH addition to the ring. Integration of Equation 1 yields

(4)
$$[CuH]_t = [CuH_o] \exp(-k_T t)$$

and (5) $\Sigma P = [S]_t + [R]_t = [CuH]_o - [CuH]_t = [CuH]_o \{1 - \exp(-k_T t)\}$

where ΣP is the sum of all products (side chain and ring) of CuH photooxidation. From Equation 3, the fraction of ΣP represented by S is $k_H[ROO\cdot]/k_T$, and

(6)
$$[S]_t = \frac{k_H}{k_T}[ROO\cdot]_{ss}[CuH]_o \{1 - \exp(-k_T t)\}$$

or (7)
$$[ROO\cdot]_{ss} = \frac{[S]_t k_T}{k_H[CuH]_o - \{1 - \exp(-k_T t)\}}$$

Similarly, (8)
$$[R]_t = \frac{k_a}{k_T}[\cdot OH]_{ss}[CuH]_o \{1 - \exp(-k_T t)\}$$

or (9)
$$[\cdot OH]_{ss} = \frac{[R]_t k_T}{k_a[CuH]_o \{1 - \exp(-k_T t)\}}$$

From a separate experiment, k_H was found to be 10 M^{-1} s^{-1}, and k_a was estimated to be 3 $\times 10^9$ M^{-1} s^{-1}. When 274 μM CuH was added to Aucilla River water (a highly colored river in north Florida) and the sample was incubated at constant light intensity for 5 days, the following amounts of products were formed: $[S]_t = 13$ μM; $[R]_t = 6.9$ μM; i.e., $\Sigma P = 19.9$ μM. From Equation 5, k_T is found to be 1.745×10^{-7} s^{-1}. From Equation 7, $[ROO\cdot] = 1.1 \times 10^{-8}$ M, and from Equation 9 $[\cdot OH] = 2 \times 10^{-17}$ M.

The nature of the radicals produced from DOM is still uncertain. Mill suggested that a significant proportion of the radicals produced from DOM are alkoxyl-like carbonyl triplets and concluded that estimates of $RO_2\cdot$ concentrations like those in Example 8-4 are too high. Faust and Hoigne[147] reported that photosensitized oxidation of alkylphenols is controlled by a transient oxidant derived from DOM. They hypothesized that the oxidant was an organic peroxyl radical, but it could be a mixture of radicals, including carbonyl triplets. Production of these intermediates from DOM was sufficient to degrade alkylphenols in a Swiss lake at half-lives of 1 day to several months. These rates were much faster than calculated rates for alkylphenol degradation by $\cdot OH$ and $^1O_2$ in the lake and are consistent with $ROO\cdot$ concentrations near 1×10^{-10} M or carbonyl triplet concentrations near 1×10^{-15} M.

8.5.4 Comparison of Type I and Type II Photosensitized Mechanisms

The molecules most amenable to Type I (free radical) mechanisms are those that are easily reduced (quinones) or oxidized (phenols, amines, and, to a lesser extent, alcohols, ethers, and carboxylic acids). Table 8-12 describes typical Type I mechanisms. Type II mechanisms, in which $^1O_2$ adds directly to organic molecules, are favored by compounds that are less readily oxidized or reduced (olefins, aromatic compounds); $^1O_2$ is an electrophile and reacts only with electron-rich compounds. At least five categories of Type II reactions are known, and examples are listed in Table 8-12. Reaction rates of $^1O_2$ with nonionizable organic compounds are pH independent, but rates for organic acids and bases are not. In all cases the less protonated form is much more reactive. For example, rates increase in proportion to the fraction of phenols present in ionized form,[54] and the protonated compounds are at least two orders of magnitude less reactive than the anions. Rate constants for reactions of $^1O_2$ with substituted phenols are correlated with both Hammett σ values and $E_{1/2}$ (the half-wave potential for oxidation of the phenols).[148]

Some compounds (phenols, some olefins) can be oxidized by both Type I and Type II mechanisms. The relative importance of the two mechanisms in these cases depends on two types of competition: (1) between reactant and ground-state O_2 for the triplet sensitizer, and (2) between the decay rate (k_d) of $^1O_2$ and the rate ($k_A[A]$) at which $^1O_2$ reacts with the organic compound. O_2 concentrations in natural waters ($2–3 \times 10^{-4}$ M) are much higher than those of organic reactants ($10^{-9}–10^{-6}$ M). As a result, most of the triplet sensitizer energy is trapped by O_2, and Type II mechanisms are favored. Nonetheless, $^1O_2$ is a selective reagent and does not account for all photosensitized oxidations in surface waters.

In general, rates of both Type I and Type II oxidations depend on the concentration of sensitizer molecules, but Type II reactions are relatively independent of the nature of these molecules, because rate constants for production of $^1O_2$ are about 2×10^9 M^{-1} s^{-1} for most sensitizers.[9] Effects of O_2 concentration on rates of the two reaction types are difficult to predict. O_2 is a reactant in Type I mechanisms, and increasing $[O_2]$ thus should increase oxidation rates. However, reaction with O_2 may not be the rate-limiting step. Increasing $[O_2]$ actually may decrease Type I rates, because of competition for sensitizer. It would seem that production of $^1O_2$ should depend on the concentration of ground-state O_2; however, O_2 has a high efficiency for trapping triplet sensitizer energy, and all the energy at typical sensitizer concentrations is trapped at $[O_2] > {\sim}10^{-6}$ M. Type II reaction rates thus are independent of $[O_2]$ at levels found in surface waters.

The mean lifetime of $^1O_2$ in water is short because quenching by H_2O is rapid ($k_d = 2.5 \times 10^5$ s^{-1}). If $k_d > k_A[A]$, most of the $^1O_2$ will return to ground state without inducing a reaction. If $k_d = k_A[A]$, half the $^1O_2$ is quenched, and half reacts with the organic reactant. The ratio $\beta = k_d/k_A$ gives the concentration of a reactant required to trap half the $^1O_2$. β ranges from ${\sim}10^{-3}$ M for 2,5-dimethylfuran (DMF) and substituted phenolate ions, which react rapidly with $^1O_2$, to more than 1 M for unreactive solvents. Table 8-6b lists k_A values for reaction of $^1O_2$ with organic compounds.

Several techniques can be used to determine whether a reaction involves a Type II mechanism. Comparison of products and kinetics of a photooxidation reaction with those of chemically produced $^1O_2$ has been used in a few cases,[149] but the reagents used to produce $^1O_2$ chemically (H_2O_2 and NaOCl) are strong oxidants and may directly oxidize the substrate of interest.[9] Microwave generation of $^1O_2$ also

Table 8-12. Mechanisms of Type I and II Photosensitized Reactions

Type I reactions

(1)   Oxidation of alcohols

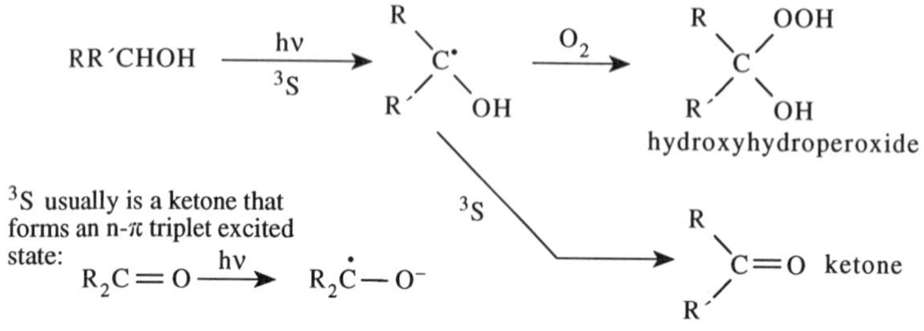

$$RR'CHOH \xrightarrow[^3S]{h\nu} \underset{R'}{\overset{R}{>}}\!\!C\!\cdot\!\!-OH \xrightarrow{O_2} \underset{R'}{\overset{R}{>}}\!\!C\!\!\underset{OH}{\overset{OOH}{<}}$$

hydroxyhydroperoxide

^3S usually is a ketone that forms an n-π triplet excited state:

$$R_2C=O \xrightarrow{h\nu} R_2\dot{C}-O^-$$

$^3S \longrightarrow \underset{R'}{\overset{R}{>}}C=O$ ketone

(2)   Oxidation of amines

$$R_2N-CHR'_2 \xrightarrow{^3S} R_2N-\dot{C}R'_2 \xrightarrow{O_2} \longrightarrow R_2NH + R'_2C=O$$

(3)   Oxidation and dealkylation of amines

$$RNR'_2 \xrightarrow[RR'C=O,\ O_2]{h\nu} \longrightarrow \underset{H}{\overset{R'}{RNC=O}} + RNH + RNH_2$$

(4)   Decarboxylation of substituted carboxylic acids, $RXCH_2COOH$

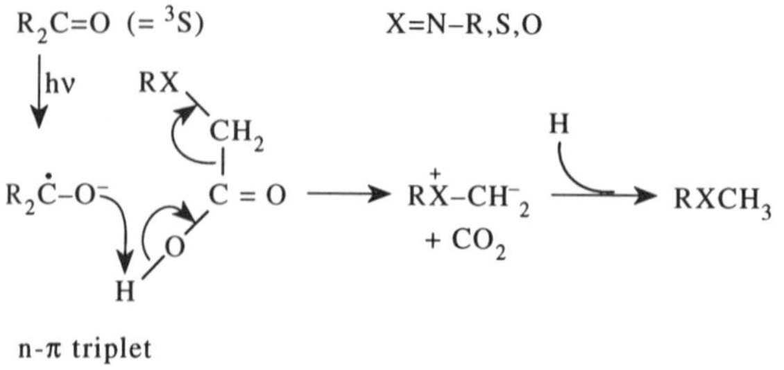

$$R_2C=O\ (= {}^3S)   \qquad X=N-R,S,O$$

$$\downarrow h\nu \quad RX$$

$$R_2\dot{C}-O^- \qquad \overset{CH_2}{\underset{C=O}{|}} \longrightarrow \underset{+CO_2}{\overset{+}{RX-CH_2}} \overset{H}{\longrightarrow} RXCH_3$$

n-π triplet

(5)   Autoxidation of olefins

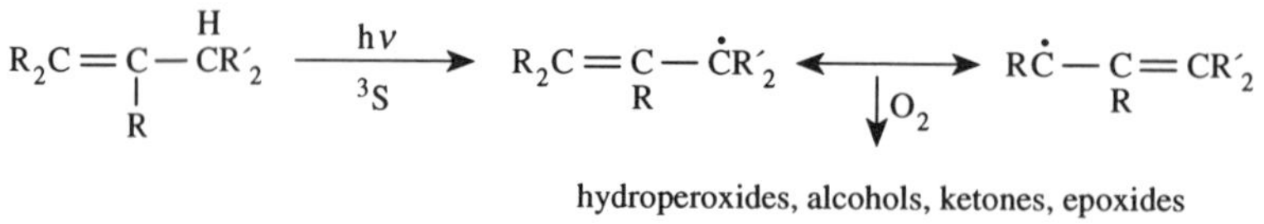

$$R_2C=\underset{R}{\overset{H}{C}}-CR'_2 \xrightarrow{h\nu \atop ^3S} R_2C=C-\dot{C}R'_2 \underset{R}{\longleftrightarrow} R\dot{C}-C=CR'_2 \atop R$$

$$\downarrow O_2$$

hydroperoxides, alcohols, ketones, epoxides

Table 8-12 (continued).

Type II reactions

(1)   1,4 additions to dienes and heterocycles

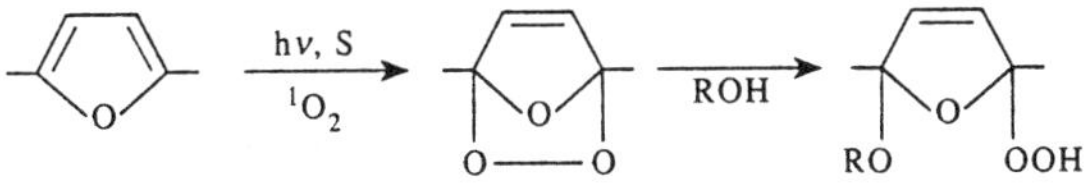

(2)   Oxidation of sterols

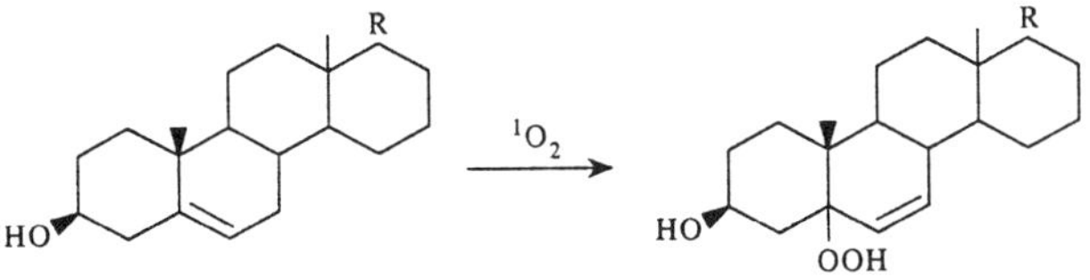

(3)   1,2 additions to electron-rich olefins

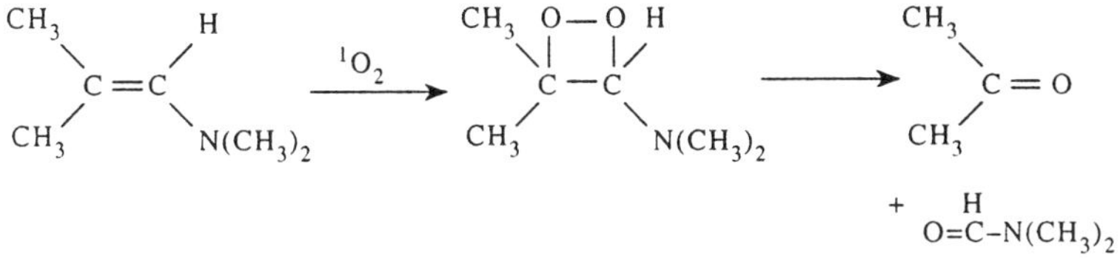

(4)   Oxidation of sulfides to sulfoxides

$$RR'S \xrightarrow{\ ^1O_2\ } [RR'\overset{+}{S}\text{-}OO^-] \xrightarrow{\ RR'S\ } 2RR'SO$$

(5)   Oxidation of phenols

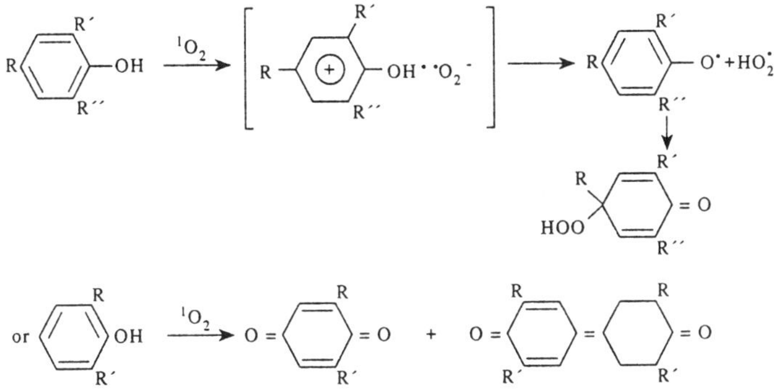

From Foote, C.S., in *Free Radicals in Biology*, Vol. II, W. Pryor (Ed.), Academic Press, New York, 1976, p. 85.

produces highly reactive atomic oxygen and ozone.[150] Dyes like methylene blue and rose bengal are efficient sensitizers of $^1O_2$. An increase in reaction rate in the presence of these dyes is evidence for a Type II mechanism. However, these dyes photodegrade within a matter of hours, and radical intermediates produced from the dyes could influence reaction rates. A more useful approach involves competitive inhibition by adding a known $^1O_2$ acceptor (probe) to the reacting system. Dimethylfuran and 1,4-diazabicyclo[2.2.2]octane (DABCO) are common probes. β-carotene does not react with $^1O_2$, but is a powerful quencher of $^1O_2$ and thus a good probe. Competitive substrates must not be good Type I substrates; they should have a high affinity for $^1O_2$ and an uncomplicated chemistry. Another method involves comparing reaction rates in H_2O and D_2O.[151] The lifetime of $^1O_2$ is much longer in D_2O than in H_2O, and Type II reactions are faster in D_2O when the reaction is first order in [A], i.e., when $k_A[A] < k_d$. At sufficiently high [A], all $^1O_2$ reacts with A before decaying, and the rate is independent of both [A] and the lifetime of $^1O_2$.

The significance of Type I mechanisms in photochemical transformations could be evaluated if the total flux of free radicals could be measured. Nitric oxide, NO, a relatively stable free radical, reacts rapidly with organic alkyl and peroxyl radicals, as well as major inorganic radicals like ·OH and superoxide. Zafiriou et al.[152] proposed using the difference between NO loss rates under light and dark conditions, ΔNO_{pho}, as a measure of total free radical flux. Using this technique he measured fluxes in the range 0.2 to 3 nM min^{-1} in Atlantic Ocean waters. Similarly, Blough[153] used a series of organic nitroxide compounds, which are stable radicals, to determine radical production by humic acid. The nitroxides react rapidly with radicals produced from HA to yield diamagnetic products, and the loss of ESR signal can be used to measure the rate of nitroxide disappearance (and thus radical production by HA). Radical production rates measured by this method were on the order of 6 to 10 nmol min^{-1} (mg HA)$^{-1}$ at a light intensity of 20 mW cm^{-2}.

8.5.5 Photochemistry of Oil and Petroleum Products in Aquatic Systems

Crude oil and refined petroleum products are released into surface waters by accidental spills during transit, from well "blowouts", and sometimes from illegal dumping and acts of terrorism. These releases comprise highly complex mixtures of aliphatic and aromatic hydrocarbons that vary widely in molecular weight and include some heterocyclic nitrogenous and organosulfur compounds. Many processes are involved in weathering and transformation of these mixtures in aquatic environments: evaporation of low-weight hydrocarbons, dissolution, emulsification, sorption onto suspended particles, microbial degradation, and photooxidation. The processes occur both concurrently and consecutively (i.e., the products of one process may be reactants for another). As a result, it is difficult to sort out the importance of a single process like photooxidation on the overall weathering/degradation of oil and petroleum products in natural waters. Several reviews of the subject are available;[154,155] this section reviews photochemical aspects of petroleum weathering.

Two types of experiments have been conducted to assess the role of photochemistry in oil weathering: (1) controlled laboratory experiments with individual compounds or simple mixtures, and (2) experiments with authentic "spill" material. The former provide information on rates, products, and transformation mechanisms for specific compounds, but it is difficult to extrapolate such results to the processes that occur in complex, nonaqueous petroleum mixtures. For example, photooxidation of dimethylnaphthenes was much slower in the absence of oil and took place exclusively in benzylic positions, whereas oxidation of naphthalene moiety itself predominated in the presence of oil.[156] Studies on individual compounds are useful primarily to assess photooxidation rates of petroleum compounds once they are dispersed and solubilized. Experiments with actual spill materials are potentially more realistic, but they are more difficult to interpret. Typically they yield qualitative or at best semi-quantitative information. Rather than looking at disappearance of specific compounds and formation of new ones, such studies generally follow the weathering process by measuring changes in aggregate variables such as viscosity and total mass of organic residue or functional group analysis by ir spectroscopy. Some experiments of this type described in the literature were done under conditions not realistic for photooxidation in the environment, e.g., light sources with $\lambda < 300$ nm and/or the absence of an aqueous phase to remove water-soluble oxidation products from the organic phase.

Extrapolation of measured photodegradation rates from short-term experiments (days to weeks) with crude oil suggests that complete photooxidation of oil-spill residues would require quite long times — years to many decades. This suggests that photooxidation is not very important in the overall disappearance of oil residues, which often are gone from spill sites in a matter of months. This conclusion is erroneous. Although photooxidation may account for only a small percentage of the total loss, it is significant, especially in the initial stages of weathering, for at least five reasons:

1. Photooxidation introduces oxygen into highly reduced hydrocarbons, yielding alcohols, aldehydes, and fatty acids, which are much more susceptible to biodegradation than the parent hydrocarbons.
2. Fatty acid products emulsify and disperse oil films, and the dispersed emulsions are degraded more readily by photolysis/biolysis.
3. Oxygenated photoproducts often are more toxic than parent materials, although they may pose less a hazard to large organisms like aquatic birds than do undispersed spills.
4. Photooxidation rapidly changes the viscosity of crude and refined oil residues. In general, viscous oils contract and become more viscous with slight exposure to photooxidizing conditions (within 2 h of irradiation);[157,158] nonviscous oils become less viscous and spread more readily over the water surface with exposure to a few hours of light. The decrease in viscosity is thought to reflect emulsification and solubilization induced by introduction of oxygenated groups into the hydrocarbons.
5. The contraction process mentioned above is thought to arise from polymerization reactions, possibly by free radical mechanisms, that ultimately are responsible for formation of tarry residues in seawater.

Both free radical and $^1O_2$-photosensitized mechanisms are likely to be involved in photooxidation of petroleum residues. Induction of free radical reactions was demonstrated with 1-naphthol as a model sensitizing agent.[159] Naphthol extracts H atoms from hydrocarbons to produce alkyl radicals that react with ground-state O_2 to produce organic peroxides. These react further to produce hydroperoxides, alcohols, phenols, aldehydes, lactones, and other oxygenated compounds[154,160] (Figure 8-20). The potential role of $^1O_2$ in photooxidation of oil spills was demonstrated by adding methylene blue, an efficient sensitizer of $^1O_2$, to the PAH

a

$$X + h\nu \longrightarrow X^{\bullet}$$
$$X^{\bullet} + RH \longrightarrow XH\cdot + R\cdot$$
$$XH\cdot + O_2 \longrightarrow X + HO_2\cdot$$
$$R\cdot + O_2 \longrightarrow RO_2\cdot$$
$$RO_2\cdot + RH \longrightarrow RO_2H + R\cdot$$
$$RO_2\cdot + XH\cdot \longrightarrow RO_2H + X$$
$$RO_2H \longrightarrow RO\cdot + \cdot OH$$
$$RO\cdot + RH \longrightarrow ROH + R\cdot$$
$$RO_2H + R\cdot \longrightarrow RO\cdot + ROH$$

X = xanthone; $X^{\bullet}$ = xanthone triplet;
RH = n-hexadecane.

b

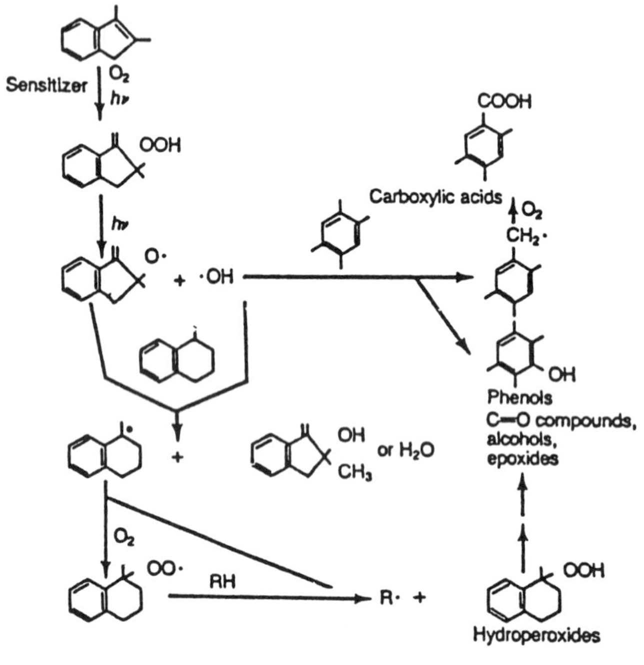

Figure 8-20. Some proposed mechanisms for photodegradation of petroleum: (a) photosensitized oxidation of hexadecane; (b) fuel oil oxidation; (c) crude oil and thiocyclane oxidation. [Redrawn from Payne, J.R. and C.R. Phillips, *Environ. Sci. Technol.*, 19, 569 (1985); original sources: (a) Gesser, H.P., T.A. Widman, and Y.B. Tewori, *Environ. Sci. Technol.*, 11, 605 (1977); (b) Larson, R.A. et al., *Environ. Sci. Technol.*, 13, 965 (1979); (c) Burwood, R. and G.C. Speers, *Estuarine Coast Mar. Sci.*, 2. 117 (1974).]

phenanthrene, which was dissolved in hexane and floated on water.[161] A variety of oxidation products were identified from irradiated samples. Some PAHs found in crude oil (e.g., perylene) could act as photosensitizers of $^1O_2$ in spills.

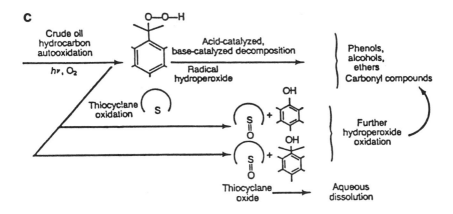

Figure 8-20 (continued).

The occurrence of free radical and $^1O_2$ mechanisms in early stages of petroleum photolysis also has been demonstrated by adding mechanism-specific inhibitors to spill materials. β-carotene, a quencher of $^1O_2$, inhibited formation of photooxidation products of refined oils almost completely.[162] Indirect evidence suggests that organosulfur compounds, such as thiols and thiocyclanes, which occur in many crude oils, are temporary inhibitors of free radical mechanisms. They quench such reactions by reacting with ·OOH and other radicals to produce sulfoxides such as thiocyclane oxide.[163] When the reduced sulfur compounds are exhausted, free radical mechanisms can proceed to oxidize hydrocarbons in oil spills.

Functional group analysis by IR spectroscopy has confirmed the formation of oxygenated compounds as early products of petroleum photooxidation. Groups identified in this way include: -OH (aliphatic alcohols, phenols, naphthols), -C=O (aldehydes and ketones), C-O-C (alkyl and aryl ethers), −S=O (sulfoxides), and -COOH (aliphatic and aromatic acids). In addition, photooxidation products have been identified in monitoring of oil and water at sites of major spills like the Amoco Cadiz accident[164] off the coast of France in 1977 and the Ixtoc I well blowout[165] off the coast of Mexico in 1979. Alkyl-substituted dibenzothiophene sulfoxides were identified in samples from the Amoco Cadiz spill. Numerous C_9 to C_{11} fatty acid methyl esters were found in extracts from sunlight-exposed aquaria to which Ixtoc crude was introduced. The parent material had $n\text{-}C_{11}$ to $n\text{-}C_{32}$ alkanes, but the photooxidized material had no alkanes below $n\text{-}C_{15}$. Various benzoic, naphthanoic, phenanthroic, and other PAH carboxylic acids also were found in these experiments.

In summary, photooxidation is important in petroleum weathering and accounts for physical (viscosity) changes in spill residues, as well as many chemical transformations: in particular, the introduction of oxygen into hydrocarbons. The products of photooxidation are more water soluble and more biodegradable, but in some cases more toxic to microorganisms. Additional work is needed to characterize the products of photooxidation and determine their fate. Only sketchy information is available on the mechanisms of petroleum photooxidation, and almost no information is available on the rate-controlling factors.

8.6 PHOTOCHEMISTRY ON SUSPENDED OXIDE PARTICLES

Photochemical processes in aquatic systems are not limited to substances in solution; at least two kinds of photochemical reactions involve particles: (1) the reductive dissolution of iron and manganese oxides is photoassisted by a LMCT process involving surface complexes of organic acids; and (2) various metal oxides behave as semiconductors and photocatalyze redox reactions of organic solutes. In the former class of reactions, the organic acid sorbed to the oxide surface acts as the chromophore, absorbs light energy, and transfers an electron to the lattice metal ion to which it is complexed. In the latter case the oxide particles absorb light energy directly, and the resulting excited states produce a charge separation (electrons and holes, e^-/h^+), which form oxidizing or reducing sites on the particle surface.

8.6.1 Charge-Transfer Photoreductive Dissolution Reactions

The finding that photoredox reactions of iron-organic acid complexes can involve insoluble iron (hydr)oxides[166,167] as well as soluble iron complexes[118] greatly expands the range of water bodies for the reaction. Ferric iron exists primarily as (hydr)oxide particles in surface waters (waters that are highly stained with humic material are exceptions). Waite and Morel[167] were the first to observe the photoreductive dissolution phenomenon. They found elevated levels of dissolved Fe^{2+} when water samples containing iron (hydr)oxide particles and either citrate or AH were illuminated in the near uv ($\lambda_{max} = 360$ nm). Insoluble $Mn^{III,IV}$ oxides also are reduced to soluble Mn^{2+} by marine humic acids[168] and freshwater fulvic acid.[169] Dissolution rates increase linearly with concentration of Mn oxide particles and hyperbolically with increasing [AH] and are substantially higher in sunlight.

Example 8-5. Kinetics of Photoassisted Mn-Oxide Dissolution by Fulvic Acid

The dissolution of $MnO(OH)_2$ can be explained[169] as a four-step sequence involving rapid sorption of humic molecules on the (hydr)oxide surface to form a surface complex and subsequent electron-transfer (the LMCT mechanism):

(1) Adsorption $>Mn^{IV}OH + AH \rightleftharpoons >Mn^{IV}A + H_2O$

(2) Electron transfer $>Mn^{IV}A \rightarrow >Mn^{IV}OMn^{II}A_{ox}$

(3) Ligand exchange $>Mn^{IV}OMn^{II}A_{ox} + H_2O \rightarrow >Mn^{IV}OMn^{II}OH_2^+ + A_{ox}^{2-}$

(4) Detachment $>Mn^{IV}OMn^{II}OH_2^+ + H^+ \rightleftharpoons >Mn^{IV}OH + Mn^{2+} + H_2O$

The rate-limiting step is electron transfer; steps 1 and 4 are rapid equilibrium processes, and step 3 is a fairly rapid ligand exchange (hydrolysis) reaction (see Section 4.4.2). Not all the Mn^{II} is released in step 4; a significant fraction remains sorbed onto the oxide surface, especially at high pH and concentrations of high solids. The reductive dissolution process increases slightly with pH, but only by a factor of ~2 between pH 4 and 7, and increases in hyperbolic fashion with concentration of fulvic acid (Figure 8-21).

The reaction rate for Mn^{II} production is first order in the concentration of surface complex:

(5)
$$d\left[MN^{II}\right]_T / dt = k_T\left[S{>}Mn^{IV}A\right]$$

Subscript T stands for total Mn^{II}, both free and adsorbed species; $[S{>}Mn^{IV}A]$ is the concentration of Mn^{IV}-humic surface complex. If adsorption is treated as a pre-equilibrium step, we can express the rate as:

(6)
$$d\left[Mn^{II}\right]_T / dt = k_T K^c \left[{>}Mn\right]_T \left[\frac{[AH]_T}{1 + K^c[AH]_T}\right]$$

$[{>}Mn]_T$ is the total concentration of Mn^{IV}-oxide surface sites, K^c is the conditional formation constant for surface complexes ($K^c = [{>}Mn^{IV}A]/[Mn^{IV}OH][AH]$), and $[AH]_T$ is the total concentration of humic material. It is assumed that the amount of AH adsorbed onto oxide surfaces is a small fraction of the total AH. Approximate parameter values for Equation 6 are given below.

	pH 4.0		pH 7.1	
	k_r min^{-1}	K_c $(mg/L)^{-1}$	k_r min^{-1}	K_c $(mg/L)^{-1}$
light	0.55	$10^{-0.67}$	1.23	$10^{-1.44}$
dark	0.31	$10^{-0.79}$	0.53	$10^{-1.64}$

These values were obtained at $[AH]_T = 0$ to 40 mg L^{-1} and $MnO_2 = 1$ to 10 μM and should not be used under greatly different conditions.

The initial studies of Waite and Morel showed that photodissolution occurred with both lepidocrocite (γ-FeOOH) and amorphous ferric (hydr)oxide, but the latter was much more (~50 times) photoactive. More recent studies[170] have shown that iron-oxide polymorphs vary widely with respect to photoelectrochemical properties. For example, rates of oxalate and sulfite photooxidation varied more than 100 times among six polymorphs (α and γ-Fe$_2$O$_3$, α, β, δ, and γ-FeOOH), and rate constants for direct electron transfer from the particles to an electrode varied by more than 10^4. The variations appeared to reflect intrinsic differences in crystal and surface structure rather than differences in surface area or band gap (this term is defined in Section 8.6.2). Because of the large differences among Fe polymorphs (and especially because amorphous FeOOH is many times more photoactive than the crystalline forms), the former must be carefully removed from synthetically prepared suspensions of a polymorph if one is interested in studying its photochemical behavior.[171]

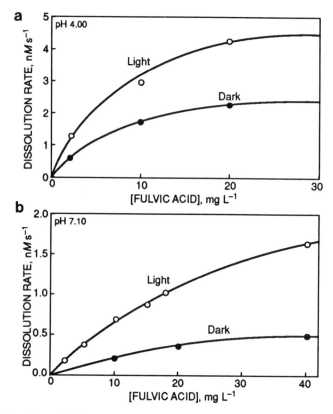

Figure 8-21. Effect of fulvic acid concentration on Mn^{IV} dissolution rate at (a) pH 4.0 and (b) pH 7.0. [From Waite, T.D., I.C. Wrigley, and R. Szymczak, *Environ. Sci. Technol.*, 22, 778 (1988). With permission of the Amer. Chem. Soc.]

Early photochemical studies with Fe and Mn oxides focused on the importance of the process as a means of solubilizing Fe and Mn rather than on the fate of the organic ligand. However, Waite and Morel did interpret their findings in terms of a photoinduced LMCT process in which an electron is transferred from the adsorbed organic ligand to >Fe^{III} on the oxide surface. As described in Section 8.4.4, this probably leads to decarboxylation of organic acid ligands. Subsequent studies have demonstrated that the reaction occurs in iron-rich streams,[172] confirmed the process as a charge transfer reaction, and determined the nature of some organic products. In addition, the list of organic reactants has been expanded to include other carboxylic acids, amino acids, and mercapto acids, all of which are believed to form inner-sphere complexes with Fe^{III} on the oxide surfaces, as well as adsorbed amines, alcohols, and glycols, which probably do not (see reference 171).

For example, uv photolysis (300 to 400 nm) of ethylene glycol adsorbed onto goethite at pH 6.5 was found to produce formaldehyde and glycoaldehyde. The latter required the presence of oxygen.[173] The proposed mechanism involves electron transfer from the adsorbed glycol to an excited Fe^{III} center to produce Fe^{2+}

plus an organic cation, both of which subsequently react in solution to yield the aldehydes. O_2 and ·OH are involved in the solution-phase reactions. Although ethylene glycol is not a common contaminant in aquatic systems, its functional groups are similar to those of natural polysaccharides (from microbial secretions), which commonly coat oxide surfaces in natural waters and sediments. Whether photocatalyzed degradation of natural organic coatings is a significant process or not is yet to be determined.

Extrapolation from results for one model compound (like ethylene glycol) to a whole class of organic compounds (like polysaccharides) must be done cautiously. The range of photoactivity within a class can be large, as was shown for Fe^{III}-carboxylic acid photoreactions in solution (Table 8-8) and with Fe (hydr)oxides. Quantum yields of Fe^{2+} from goethite with carboxylic anions $RCOO^-$ vary over three orders of magnitude depending on the nature of R.[171]

Ferric-(hydr)oxides are photoactive even in the absence of organic ligands. Waite and Morel found that irradiation of organic-free suspensions of γ-FeOOH with near-UV light led to increased concentrations of Fe^{2+} in solution.[167] Irradiation of organic-free suspensions of hematite (α-Fe_2O_3) also led to increased Fe^{2+} and O_2.[174] Two explanations are possible.[171] (1) Charge transfer could occur from hydroxide bound to an excited surface lattice $>Fe^{III}$:

$$>Fe^{III}\left(:OH^-\right)+h\nu \rightarrow >Fe^{III*}\left(:OH^-\right)\rightarrow >Fe^{II}+\cdot OH \qquad (8\text{-}31)$$

Fe^{II} rapidly dissociates from the lattice, and further solution-phase reactions of ·OH lead to O_2. (2) Alternatively, the product could be formed by the semiconductor photoredox model described in Section 8.6.2. According to this model, electrons and holes photogenerated within the Fe-oxide semiconductor lattice are scavenged by surface sites to produce (respectively) $>Fe^{II}+OH^-$ and $>Fe^{III}+\cdot OH$. This model cannot be distinguished from the first mechanism by data obtained to date, but the short diffusion lengths of electrons and holes in Fe-oxide semiconductors argue against this mechanism.[171,175] Similarly, Faust and Hoffmann[176] analyzed the energetics and quantum yields for reductive dissolution of α-Fe_2O_3 by bisulfite (HSO_3^-), but were unable to obtain conclusive evidence to decide whether the mechanism involved excitation of the O^{2-} to Fe^{3+} charge-transfer band in the bulk solid (a semiconductor model) or a photoassisted charge transfer involving Fe^{III}-S^{IV} surface complexes. Variations in quantum yields over the range 350 to 676 nm were consistent with both models.

The ·OH produced in iron-oxide photolysis could be generated from the surface reaction (Equation 8-31) or could arise from solution-phase oxidation of Fe^{2+} by O_2 via the Haber-Weiss mechanism (Section 4.2.2). Using probe analyses with benzoate (which reacts with ·OH to form hydroxybenzoates, primarily salicylate), Cunningham et al.[171] concluded that most of the ·OH is produced indirectly by thermal oxidation of Fe^{2+} in solution. This conclusion is based on comparison of quantum yields for Fe^{2+} and salicylate in systems containing only goethite plus probe and goethite plus oxalate or succinate plus probe.

8.6.2 Semiconductor Photochemistry

The photocatalytic activity of metal oxides has been a subject of speculation for many years, but significant particle-catalyzed photochemical reactions have been demonstrated only recently.[177-179] Becquerel demonstrated a photovoltaic effect with a semiconductor/liquid junction over 150 years ago.[180] In 1931 Rao and Dhar suggested that ammonia is oxidized photochemically on soil particles,[59] but strong evidence and a reasonable mechanism were lacking. Production of H_2O_2 by illuminated suspensions of ZnO was reported in the 1950s, and simple organic compounds like phenol, formate, and oxalate were found to increase the rate of H_2O_2 production.[181] A mechanism for this process that involves electrons and holes (see next paragraph) was described in 1967.[182] Finally in 1972, Fujishima and Honda[183] reported sustained oxidation of H_2O on illuminated TiO_2 in a photoelectrochemical cell. Although related reduction of water to produce H_2 could not be achieved without supplying additional energy via a 0.2 V bias, this report, coming just before the energy crisis of the mid-1970s, stimulated much research on the photocatalytic activity of metal-oxide semiconductor/liquid junction systems. Much of this work was aimed at converting light energy into useful chemical energy by splitting water to generate H_2. Such endothermic reductions properly are called photosynthetic or photoelectrosynthetic. Although progress has been achieved in finding stable catalysts for the water-splitting reaction,[184] economic fuel generation by such processes is yet to be realized. However, research toward this goal has led to the finding that semiconductor oxides are effective photocatalysts for thermodynamically feasible, but kinetically slow, chemical transformations.

The mechanism of photoredox catalysis by semiconductors is explained by the model of Gerischer and Willig[185] and is illustrated in Figure 8-22. Photon absorption by the solid promotes an electron from the ground state to an excited state and leads to charge separation, i.e., the formation of a free electron and corresponding positive hole (e^-/h^+). The ground state is referred to as the *valence band* (VB), and the excited state as the *conduction band* (CB); the bands correspond respectively to the highest occupied (ground state) molecular orbitals and lowest unoccupied orbitals. The energy difference, E_{bg}, between the bands, expressed in electron volts or equivalent wavelength of light energy, is called the *band-gap energy*.

When a semiconductor is placed in a solution containing a redox couple, charge transfer occurs across the interface to equilibrate potentials. The resulting electrical field causes band bending (Figure 8-22) from the lattice interior to the interface, and this facilitates charge separation in n-type semiconductors, which are electron-rich as the result of the presence of electron donors. When an e^-/h^+ pair forms as the result of photon absorption in the space-charge region of the semiconductor, the electron migrates toward the interior of the crystal in the almost-empty conduction band. The hole migrates to a surface site (by inward migration of valence-band electrons), where it acts as a strong oxidizing agent, accepting an electron from an adsorbed solute. The buildup of electrons in the lattice makes particles electrophoretically mobile,[186] but electrons do not build up indefinitely. Conduction band electrons have three possible fates: (1) they may migrate to other surface sites that act as

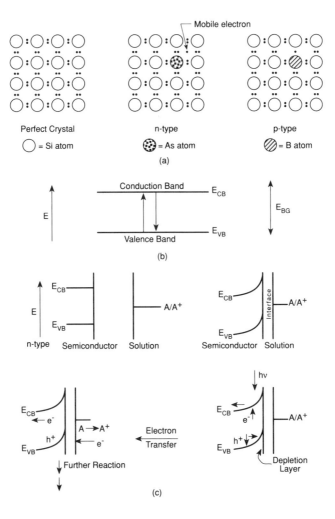

Figure 8-22. Schematic of photocatalysis by semiconductors according to model of Gersicher and Willig. [After Nozik, A.J., *Ann. Rev. Phys. Chem.*, 29, 189 (1978).]

reducing agents for sorbed species, (2) they can be discharged at a cathode in a photoelectrochemical cell, or (3) they may cause photoreductive dissolution of the semiconductor solid.

The effectiveness of semiconductors as photoredox catalysts depends on at least five system attributes: (1) the CB and VB band positions, i.e., potentials of the e^- and h^+ (reducing and oxidizing) sites, expressed in volts; (2) E_{bg}, the band-gap energy, which determines the maximum wavelength of light capable of exciting the semiconductor; (3) interfacial charge-transfer kinetics; (4) stability of the semiconductor toward photoreductive or photooxidative dissolution; and (5) sorption characteristics of the surface and possible reactants, e.g., extent of surface coverage, proximity of sorbate reactive sites to oxidizing or reducing sites at the interface, etc.

Table 8-13.   Band Positions and Band-Gap Energies for Some Semiconductor Photocatalysts and Oxidation Potentials of Some Organic Groups

Semi-conductor	Valence band V vs. NHE	Conduction band V vs. NHE	Band-gap energy		pH	Ref.
			eV	nm[a]		
TiO_2 (anatase)	+3.0	0.0	3.0	413	1.0	179
SnO_2	+3.8	+0.3	3.5	354	1.0	179
Fe oxides[c]			1.94–2.12	640–585	12.0	170
α-Fe_2O_3 (hematite)[b]			2.34	530	—	176
α-FeOOH (goethite)[b]			2.64	470	—	176
ZnO	+2.9	−0.1	3.0	413	1.0	179
WO_3	+2.9	+0.2	2.7	459	1.0	179
CdS	+2.0	−0.4	2.4	517	1.0	179
CdSe	+1.4	−0.3	1.7	729	1.0	179
GaAs	+0.7	−0.7	1.4	886	1.0	179
GaP	+1.1	−1.2	2.3	539	1.0	179
SiC	+1.6	−1.4	3.0	413	1.0	179

Oxidation potentials for some organic compounds

Compound	Oxid. pot. (V vs. SCE)[d]
Acetic acid	1.6
N,N-dimethylacetamide	1.02
4-Aminoaniline	0.18
Anthracene	1.09
1,3-Butadiene	2.31
Triethylamine	0.66
4-Methyl-heptane-2,6-dione	1.56
2,6-Di-tert-butyl-4-methylphenol	1.21
Pyridine (C_5H_5N)	1.82
Pyrrole (C_4H_4N)	0.76
Thiophene (C_4H_4S)	1.84

[a]   Maximum wavelength with sufficient energy to allow e^-/h^+ formation.
[b]   Reported values for dry hematite and goethite (α-FeOOH).
[c]   Range for α- and γ-Fe_2O_3 and α-, β-, γ-, and δ-FeOOH.
[d]   Measured in acetonitrile; from Fox, M.A., *Acct. Chem. Res.*, 16, 314 (1983).

These attributes vary with solution conditions such as pH and ionic strength. Table 8-13 lists VB and CB band positions and E_{bg} for common semiconductor photocatalysts, and oxidation potentials for organic compounds with various functional groups. Comparison of the band positions for the semiconductors with the redox potentials for the compounds suggests that most of the semiconductors potentially can catalyze a wide variety of organic redox reactions. Most of the listed semiconductors are not found in natural waters, but some of the oxides are.

Semiconductors that are not naturally occurring nonetheless offer potential as photoredox catalysts in engineered reactors to treat organic-contaminated wastewaters. This is an active research topic, and applications to treat industrial wastes and contaminated groundwater are under development. Early work on this topic showed that cyanide and sulfite could be oxidized photocatalytically by TiO_2.[187] Complete photodegradation of C_1 volatile organochlorine compounds (VOCs) — dichloromethane, chloroform, and carbon tetrachloride — on illuminated TiO_2 was demonstrated in 1983;[188] other VOCs such as dichloroethane and perchloroethylene also are photodegradable on TiO_2.[189,190] Photodegradation of pentachlorophenol (PCP),[191] benzene,[192] biphenyl,[192] dioxins,[192,193] PCBs,[193] aromatic hydrocarbons,[194] and the surfactant dodecylbenzene sulfonate,[195] has been demonstrated with TiO_2. Several other semiconductor suspensions, including ZnO, CdS, SnO_2, and WO_3, have been studied with some of the above compounds, but TiO_2 generally is the most efficient catalyst for the oxidation reactions. The half-life of PCP (initial concentration = 4.5×10^{-5} M) is about 8 min in the presence of 2 g L^{-1} of TiO_2.[191] Similarly, the herbicide 2,4,5-T (at 1×10^{-4} M) was mineralized to CO_2 and HCl within 180 min in the presence of 2 g L^{-1} of TiO_2.[196] In all cases O_2 was the ultimate oxidant; it or some other oxidant must be present to trap the electrons produced by photoactivation and minimize e^-/h^+ recombination.

One drawback to the use of TiO_2 as a photocatalyst is its large band-gap energy (~ 3.2 eV). This means that it is photoactivated primarily by uv light ($\lambda_{max} = 387$ nm). Other factors aside, semiconductors with smaller band gaps and higher λ_{max} would be preferred for processes using natural sunlight. For example, CdS ($\lambda_{max} = 517$ nm) is highly effective in oxidizing sulfur-containing compounds, including H_2S, cysteine,[197] and thiols,[198] and it also is effective in oxidizing phenols[199] and EDTA.[197] Unfortunately, illumination of CdS in photocatalytic reactions results in oxidative dissolution of the catalyst. The factors favoring oxidation of organic compounds also favor CdS corrosion,[198] giving rise to unacceptable (toxic) concentrations of cadmium in the reacting solutions, especially at low pH, and this problem must be resolved before CdS can be used in full-scale treatment reactors.

Rates of photocatalytic oxidation generally increase in a nonlinear (hyperbolic) fashion with increasing concentration of the oxidized compound. These results suggest that oxidation rates are proportional to the surface area of the catalyst covered by the organic compound and that compound-surface interactions can be described by the Langmuir adsorption isotherm. Based on these findings, several authors[188,200] have derived kinetic models leading to Langmuir-Hinshelwood-type rate expressions. However, addition of e^-/h^+ recombination to the mechanism also produces a rate expression of the same form.[198] Photocatalytic oxidation rates with TiO_2 and CdS are first order in O_2 concentration[199,201] and light intensity[199,202] and vary with pH in a manner consistent with the extent of compound adsorption to the CdS surface.[198,199] Direct reaction between holes at the CdS surface and adsorbed compounds is thought to occur. In contrast, adsorption and direct reaction between holes and organic compounds are not always necessary with TiO_2, and indirect photocatalytic oxidation mechanisms have been postulated. For example, mechanisms involving ·OH radicals have been proposed for photooxidation of 2,4,5-T[196] and phenol[202] on TiO_2.

Photocatalytic production of H_2O_2 and organic peroxides has been observed in illuminated suspensions of ZnO, TiO_2, and desert sand in the presence of O_2 and organic electron donors.[203] Steady-state concentrations of H_2O_2 as high as 10^{-4} M were measured in ZnO suspensions. Appreciable yields of H_2O_2 were obtained only in the presence of electron donors that adsorbed to the particle surfaces. The adsorbed donors (D) react with a valence-band hole:

$$D + h_{vb}^+ \rightarrow D^+ \tag{8-32a}$$

This inhibits e⁻/h⁺ recombination and permits conduction band electrons to react with O_2:

$$O_2 + 2e_{cb}^- + 2H_2O \rightarrow 2OH^- + H_2O_2 \tag{8-32b}$$

Organic peroxides constituted more than 40% of the total peroxide in ZnO suspensions containing acetate.[203] Further indirect photolysis reactions undoubtedly are induced by the peroxides and other photointermediates like ·OH.

Photoproduction of H_2O_2 has been demonstrated by a novel hybrid catalyst consisting of cobalt tetrasulfophthalocyanine ($Co^{II}TSP$) covalently bonded to the surface of TiO_2.[204] The $Co^{II}TSP$ is reduced to Co^ITSP under anoxic conditions when the hybrid catalyst is illuminated ($\lambda < 380$ nm), but under oxic conditions in the presence of electron donors, the bound Co complex serves as an electron relay and produces H_2O_2 at quantum yields of 0.16 to 0.49. TiO_2-bound $Co^{II}TSP$ also is an effective photocatalyst for the oxidation of aqueous sulfur dioxide to sulfate.[205] The oxidation occurs at holes on the TiO_2 surface, and conduction-band electrons are directed to the bound $Co^{II}TSP$, producing H_2O_2. Although unmodified TiO_2 also catalyzes the photooxidation of SO_2, the hybrid catalyst is more effective when free radical inhibitors are present.

Finally, solid catalysts can be used to accelerate thermal reactions of transient intermediates generated photochemically from dissolved reactants. For example, the reductive dechlorination of carbon tetrachloride and trichloroethylene (TCE) has been achieved with the following system:[206]

$$Ru(bp)_3^{2+} \xrightarrow{hv} Ru(bp)_3^{2+*}$$

$$Ru(bp)_3^{2+*} + MV^{2+} \rightarrow Ru(bp)_3^{3+} + MV^+$$

$$MV^+ + H_2O \xrightarrow{Pt} MV^{2+} + H + OH^- \tag{8-33}$$

$$8H + C_2HCl_3 \xrightarrow{Pt} C_2H_6 + 3HCl$$

where bp = bipyridyl, and MV = methylviologen. $Ru(bp)_3^{2+}$ acts as the sensitizing agent and transfers an electron to methylviologen in sunlight. The reduced MV reacts with water on a colloidal platinum catalyst and produces hydrogen atoms that act as reducing agents for chlorinated pollutants. TCE is completely dechlorinated to ethylene and then to ethane. Reduction of 8.97 μmol of TCE in a 25 mL reactor vessel was 89% complete in 10 h of irradiation in sunlight.

8.6.3 Summary

Aquatic photochemistry is a very rapidly growing field with significant practical applications. Photochemical reactions are responsible for the decomposition of many aquatic organic contaminants, including petroleum-related materials and synthetic compounds of industrial origin. Humic substances are important sensitizers of photochemical reactions, especially in freshwater systems, but many organic contaminants absorb light and photolyze directly. A variety of metal oxides and sulfides that behave as semiconductors can act as photocatalysts for the decomposition of organic contaminants; some of these minerals are found in natural systems (e.g., iron oxides), and others are of interest for possible use in engineered treatment systems. Both photocatalysis and UV-ozonation processes offer the potential for rapid and economical treatment of industrial waste streams and contaminated groundwaters.

REFERENCES

1. Adamson, A.W., *Textbook of Physical Chemistry,* 2nd ed., Academic Press, New York, 1979.
2. Hoigne, J., in *Aquatic Chemical Kinetics,* W. Stumm (Ed.), Wiley-Interscience, New York, 1990, pp. 43–70.
3. Zepp, R.G., in *Dynamics, Exposure, and Hazard Assessment of Toxic Chemicals,* R. Haque (Ed.), Ann Arbor Science, Ann Arbor, MI, 1980, pp. 69–110.
4. Zepp, R.G. and D.M. Cline, *Environ. Sci. Technol.,* 11, 359 (1977).
5. Mancini, J.L., *Environ. Sci. Technol.,* 12, 1274 (1978).
6. Broderius, S.J. and L.L. Smith, U.S. EPA Report, Duluth, MN, 1979.
7. Zepp, R.G. and G.L. Baughman, in *Aquatic Pollutants: Transformation and Biological Effects,* O. Hutzinger, I.H. van Lelyveld and B.C.J.Zoeteman (Eds.), Pergamon Press, New York, 1978, pp. 237–263.
8. Turro, N.J., *Molecular Photochemistry,* W.A. Benjamin, New York, 1965.
9. Foote, C.S., in *Free Radicals in Biology,* Vol. II, W. Pryor (Ed.), Academic Press, New York, 1976, pp. 85–133.
10. Braslau, N. and J.V. Dave, *J. Appl. Meteorol.,* 12, 601 (1973).
11. Mo, T. and A.E.S. Green, *Photochem. Photobiol.,* 20, 483 (1974).
12. Center for Exposure Assessment Modeling, GCSOLAR Version 1.1, U.S. EPA, Athens, GA, 1988.
13. Spinhirne, J.D. and A.E.S. Green, *Atmos. Environ.,* 12, 2449 (1978).
14. Buttner, K., *Physik. Bioklimat,* Leipzig, 1938.
15. Smith, R.C. and J.E. Tyler, *Photochem. Photobiol. Rev.,* 1, 117 (1976).
16. Calkins, J. and J.R.V. Zaneveld, in *Impacts of Climatic Change on the Biosphere,* CIAP Monograph 5, U.S. Dept. Transport., Report DOT-TST-75-55, 1975, pp. 2–267, (available from NTIS, Springfield, VA).
17. Smith, R.C. and K.S. Baker, *Limnol. Oceanogr.,* 23, 260 (1978); Smith, R.C., J.E. Tyler, and C.R. Goldman, *Ibid.,* 18, 189 (1973); Smith, R.C. and K.S. Baker, *Ibid.,* 27, 500 (1982).
18. Zafiriou, O.C., J. Joussot-Dubien, R.G. Zepp, and R.G. Zika, *Environ. Sci. Technol.,* 18, 358A (1984).
19. Preisendorfer, R.W., *Limnol. Oceanogr.,* 31, 909 (1986); Tyler, J.E., *Limnol. Oceanogr.,* 13, 1 (1968).
20. Miller, G.C. and R.G. Zepp, *Water Res.,* 13, 453 (1979).
21. Tyler, J.E. and R.W. Preisendorfer, in *The Sea,* Vol. I, Wiley-Interscience, New York, 1962, pp. 313–325.
22. Leighton, P.A., *Photochemistry of Air Pollution,* Academic Press, New York, 1961.

23. Hoigne, J., B.C. Faust, W.R. Haag, F.E. Scully, Jr., and R.G. Zepp, in *Aquatic Humic Substances; Influence on Fate and Treatment of Pollutants,* P. MacCarthy and I.H. Suffet (Eds.), Adv. Chem. Ser. 219, Am. Chem. Soc., Washington, D.C., 1989, pp. 363–381.

24. Fischer, A.M., D.S. Kliger, J.S. Winterle, and T. Mill, *Chemosphere,* 14, 1229 (1985); Frimmel, F.H., H. Bauer, J. Putzien, P. Murasecco, and A.M. Braun, *Environ. Sci. Technol.,* 21, 541 (1987).

25. Haag, W.R. and J. Hoigne, *Chemosphere,* 14, 1659 (1985).

26. Cooper, W.J., R.G. Zika, R.G. Petasne, and J.M.C. Plane, *Environ. Sci. Technol.,* 22, 1156 (1988).

27. Cooper, W.J. and R.G. Zika, *Science,* 220, 711 (1983).

28. Sinelnikov, V.E. and A.S. Liberman, *Trans. Inst. Biol. Vnutr. Vod., Akad Nauk SSSR,* 29, 27 (1974).

29. Sinelnikov, V.E., *Gidrobiol. Zh.,* 7, 115 (1971).

30. Draper, W.M. and D.G. Crosby, *Arch. Environ. Contam. Toxicol.,* 12, 121 (1983).

31. Helz, G.R. and R.J. Kieber, in *Water Chlorination: Chemistry, Environmental Impact, Health Effects,* R.L. Jolley et al. (Eds.), Lewis Publishers, Chelsea, MI, 1985, pp. 1033–1040.

32. Van Baalen, C. and J.E. Marler, *Nature,* 211, 951 (1986).

33. Zika, R.G., J.W. Moffett, R.G. Petasne, W.J. Cooper, and E.S. Saltzman, *Geochim. Cosmochim. Acta,* 49, 1173 (1985).

34. Zika, R.G., Ph.D. thesis, Dalhousie University, Halifax, N.S., 1978.

35. Kok, G.L., *Atmos. Environ.,* 14, 653 (1980).

36. Zika, R.G., E.S. Saltzman, W.L. Chameides, and D.D. Davis, *J. Geophys. Res.,* 87, 5015 (1982).

37. Cooper, W.J. and D.R.S. Lean, *Environ. Sci. Technol.,* 23, 1425 (1989).

38. Cooper, W.J., R.G. Zika, R.G. Petasne, and A.M. Fischer, in *Aquatic Humic Substances; Influence on Fate and Treatment of Pollutants,* P. MacCarthy and I.H. Suffet (Eds.), Adv. Chem. Ser. 219, Am. Chem. Soc., Washington, D.C., 1989, pp. 333–362.

39. Palenik, B., O.C. Zafiriou, and F.M.M. Morel, *Limnol. Oceanogr.,* 32, 1365 (1987).

40. Mill, T., D.G. Hendry, and H. Richardson, *Science,* 207, 886 (1980).

41. Russi, H., D. Kotzias, and F. Korte, *Chemosphere,* 11, 1041 (1982).

42. Mach, C.E. and P.L. Brezonik, *Sci. Tot. Environ.,* 87/88, 269 (1989).

43. Dorfman, L.M. and G.E. Adams, Natl. Stand. Ref. Data Ser. No. 46 (U.S. Nat. Bur. Standards), Washington, D.C., 1973.

44. Zafiriou, O.C., M.B. True, and E. Hayon, in *Photochemistry of Environmental Aquatic Systems,* R.G. Zika and W.J. Cooper (Eds.), ACS Symp. Ser. 327, Am. Chem. Soc., Washington, D.C., 1987, pp. 89–105; True, M.B. and O.C. Zafiriou, *Ibid.,* pp. 106–115.

45. Joussot-Dubien, J. and A. Kadiri, *Nature,* 227, 700 (1970).

46. Zepp, R.G., N.L. Wolfe, G.L. Baughman, and R.C. Hollis, *Nature,* 267, 421 (1977).

47. Rogers, M.A.J. and P.J. Snowden, *J. Am. Chem. Soc.,* 104, 5541 (1982).

48. Zepp, R.G., G.L. Baughman, and P.F. Schlotzhauer, *Chemosphere,* 10, 109 (1981).

49. Zepp, R.G., G.L. Baughman, and P.F. Schlotzhauer, *Chemosphere,* 10, 119 (1981).

50. Wolff, C.J.M., M.T.H. Halmans, and H.B. van der Heijde, *Chemosphere,* 10, 59 (1981).

51. Haag, W.R., J. Hoigne, E. Gassmann, and A. M. Braun, *Chemosphere,* 13, 631 (1984).

52. Baxter, R.M. and J.H. Carey, *Freshwater Biol.,* 12, 285 (1982).

53. Momzikoff, A., R. Santus, and M. Giraud, *Mar. Chem.,* 12, 1 (1983).
54. Scully, F.E. and J. Hoigne, *Chemosphere,* 16, 681 (1987).
55. Haag, W.R., J. Hoigne, E. Gassmann, and A. M. Braun, *Chemosphere,* 13, 641 (1984).
56. Haag, W.R. and J. Hoigne, *Environ. Sci. Technol.,* 20, 341 (1986).
57. Zepp, R.G., A.M. Braun, J. Hoigne, and J.A. Leenheer, *Environ. Sci. Technol.,* 21, 485 (1987).
58. Breugem, P., P. van Noort, S. Velberg, E. Wondergem, and J. Zijlstra, *Chemosphere,* 15, 717 (1986).
59. Rao, G.G. and N.R. Dhar, *Soil Sci.,* 31, 379 (1931).
60. ZoBell, C.E., *Science,* 77, 27 (1933).
61. Rakestraw, N.W. and A. Hollaender, *Science,* 84, 442 (1936).
62. Hamilton, R.D., *Limnol. Oceanogr.,* 9, 107 (1964).
62a. Zafiriou, O.C., M. McFarland, and R.H. Bromund, *Science,* 207, 637 (1980); Zafiriou, O.C., *J. Geophys. Res.,* 79, 4491 (1974).
63. Zafiriou, O.C. and R. Bonneau, *Photochem. Photobiol.,* 45, 723 (1987).
64. Zafiriou, O.C., in *Chemical Oceanography,* Vol. 8, J.P. Riley (Ed.), Academic Press, New York, pp. 339–379.
65. Zepp, R.G., J. Hoigne, and H. Bader, *Environ. Sci. Technol.,* 21, 443 (1987).
66. Weiler, J., *J. Great Lakes Res.,* 4, 370 (1978).
67. Zafiriou, O.C. and M.B. True, *Mar. Chem.,* 8, 9, 33 (1979).
68. Glaze, W.H., *Environ. Sci. Technol.,* 21, 224 (1987).
69. Hoigne, J. and H. Bader, in *Organometals and Organometalloids Occurrence and Fate in the Environment,* F.E. Brinckman and J.M. Bellama (Eds.), ACS Symp. Ser. 82, Am. Chem. Soc., Washington, D.C., 1978, pp. 292–313.
70. Hoigne, J. and H. Bader, *Water Res.,* 10, 377 (1976).
71. Hoigne, J. and H. Bader, *Water Res.,* 17, 173,185 (1983).
72. Butkovic, V., L. Klasinc, M. Orhanovic, and J. Turk, *Environ. Sci. Technol.,* 17, 546 (1983).
73. Hoigne, J., in *Handbook of Ozone Technology and Applications,* Vol. 1, R.G. Rice and A. Netzer (Eds.), Ann Arbor Science, Ann Arbor, MI, 1982, pp. 341–377.
74. Kalmaz, E.E. and N.M. Trieff, *Chemosphere,* 15, 183 (1986).
75. Staehelin, J. and J. Hoigne, *Environ. Sci. Technol.,* 16, 676 (1982).
76. Gural, M.D. and P.C. Singer, *Environ. Sci. Technol.,* 16, 377 (1982).
77. Gorbenko-Germanov, D.S. and I.V. Kozlova, *Dokl. Acad. Nauk SSSR,* 210, 851 (1973); *Russ. J. Phys. Chem.,* 48, 93 (1974); Czapski, G.A., *Rev. Phys. Chem.,* 22, 171 (1971); Peleg, M., *Water Res.,* 10, 361 (1976).
78. Zeevalkink, J.A., D.C. Visser, P. Arnoldy, and C. Boelhouwer, *Water Res.,* 14, 1375 (1980).
79. Peyton, G.R., F.Y. Huang, J.L. Burleson, and W.H. Glaze, *Water Res.,* 16, 454 (1982); Glaze, W.H., G.R. Peyton, S. Lin, F.Y. Huang, and J.L. Burleson, *Water Res.,* 16, 448 (1982).
80. Peyton, G.R., M.A. Smith, and B.M. Peyton, Research Report 206, Water Resources Center, University of Illinois, Urbana, 1987.
81. Peyton, G.R. and W.H. Glaze, *Environ. Sci. Technol.,* 22, 761 (1988); in *Photochemistry of Environmental Aquatic Systems,* R.G. Zika and W.J. Cooper (Eds.), ACS Symp. Ser. 327, Am. Chem. Soc., Washington, D.C., 1987, pp. 76–88.
82. Christman, R.F. and E.T. Gjessing (Eds.), *Aquatic and Terrestrial Humic Materials,* Ann Arbor Science, Ann Arbor, MI, 1983.
83. Aiken, G.R., D.M. McKnight, R.L. Wershaw, and P. MacCarthy (Eds.), Humic *Substances in Soil, Sediment, and Water,* John Wiley & Sons, New York, 1985.

84. Shapiro, J., *Limnol. Oceanogr.*, 2, 161 (1957); *J. Am. Water Works Assoc.*, 56, 1062 (1964).
85. Black, A.P. and R.F. Christman, *J. Am. Water Works Assoc.*, 55, 753, 897 (1963).
86. Gjessing, E.T., *Nature*, 208, 1091 (1965); *Environ. Sci. Technol.*, 4, 437 (1970).
87. Christman, R.F. and M. Ghassemi, *J. Am. Water Works Assoc.*, 58, 723 (1966); Ghassemi, M. and R.F. Christman, *Limnol. Oceanogr.* 13, 583 (1968).
88. Swift, R.S. and A.M. Posner, *J. Soil Sci.*, 22, 237 (1971).
89. Miles, C.J., Ph.D. thesis, University of Florida, Gainesville, 1983.
90. Wershaw, R.L. and G.R. Aiken, in *Humic Substances in Soil, Sediment, and Water*, John Wiley & Sons, New York, 1985, pp. 477–492.
91. Malcolm, R.L., in *Humic Substances in Soil, Sediment, and Water*, John Wiley & Sons, New York, 1985, pp. 181–209.
92. Wershaw, R.L., in *Humic Substances in Soil, Sediment, and Water*, John Wiley & Sons, New York, 1985, pp. 561–582; Wilson, M.A., *J. Soil Sci.*, 32, 167 (1981); Hatcher, P.G., I.A. Breger, L.W. Dennis, and G.E. Maciel, in *Aquatic and Terrestrial Humic Materials*, Ann Arbor Science, Ann Arbor, MI, 1983, pp. 37–82.
93. Larson, R.A. and J.M. Hufnal, Jr., *Limnol. Oceanogr.*, 25, 505 (1980).
94. Steinberg, C., *Arch. Hydrobiol. Suppl.*, 53, 48 (1977); De Haan, H., G. Halma, T. De Boer, and J. Haverkamp, *Hydrobiologia*, 78, 87 (1981).
95. Bollag, J.-M., in *Aquatic and Terrestrial Humic Materials*, Ann Arbor Science, Ann Arbor, MI, 1983, pp. 127–141.
96. Stuermer, D.H., Ph.D. thesis, Woods Hole Oceanography Institute, Woods Hole, MA, 1975.
97. Steinberg, C. and U. Muenster, in *Humic Substances in Soil, Sediment, and Water*, John Wiley & Sons, New York, 1985, pp. 105–145.
98. Wolf, D.C. and J.P. Martin, *Soil Sci. Soc. Am. Proc.*, 40, 700 (1976).
99. Chaudhary, S.K., R.H. Mitchell, P.R. West and M.J. Ashwood-Smith, *Chemosphere*, 14, 27 (1985).
100. Harvey, G.R. and D.A. Boran, in *Humic Substances in Soil, Sediment, and Water*, John Wiley & Sons, New York, 1985, pp. 233–247.
101. Harvey, G.R., D.A. Boran, L.A. Cheasl, and J.M. Tokar, *Mar. Chem.*, 12, 119 (1983).
102. Harvey, G.R. and D.A. Boran, *Trans. Am. Geophys. Union*, 63, 62 (1982).
103. Zepp, R.G. and P.F. Schlotzhauer, *Chemosphere*, 10, 479 (1981).
104. Højerslev, N.K., Abstract, NATO-ARI workshop on photochemistry of natural waters, Woods Hole, MA, 1983.
105. Nyquist, G., Thesis, Dept. Anal. Mar. Chem., University of Gothenburg, Sweden, 1979.
106. Oliver, B.G., E.M. Thurman, and R.L. Malcolm, *Geochim. Cosmochim. Acta*, 47, 2031 (1983).
107. Eshelman, K.N. and H.F. Hemond, *Water Resour. Res.*, 21, 1503 (1985).
108. Henriksen, A. and H.M. Seip, *Water Res.*, 14, 809 (1980).
109. Gorham, E., S.J. Eisenreich, J. Ford, and M.V. Santlemann, in *Chemical Processes in Lakes*, W.Stumm (Ed.), Wiley-Interscience, New York, 1985, pp. 339–363.
110. McKnight, D.M., E.M. Thurman, R.L. Wershaw, and H.F. Hemond, *Ecology*, 66, 1339 (1985).
111. Urban, N.R. and S.J. Eisenreich, Tech. Report No. 130, Water Resource Research Center, University Minnesota, St. Paul, 1989.
112. Jones, S.S. and F.A. Long, *J. Phys. Chem.*, 56, 25 (1952).

113. Trott, T., R.W. Henwood, and C.H. Langford, *Environ. Sci. Technol.,* 6, 367 (1972).
114. Melcer, M.E., M. Zalewski, M.A. Brisk, and J.P. Hassett, *Chemosphere,* 16, 1115 (1987).
115. Purcell, K. and J. Kotz, *Inorganic Chemistry,* W.B. Saunders, Philadelphia, 1977.
116. Langford, C.H., M. Wingham, and V.S. Sastri, *Environ. Sci. Technol.,* 7, 820 (1973).
117. Lockhart, H.B., Jr. and R.V. Blakely, *Environ. Sci. Technol.,* 9, 1035 (1975).
118. Miles, C.J. and P.L. Brezonik, *Environ. Sci. Technol.,* 15, 1089 (1981).
119. Hatchard, C.G. and C.A. Parker, *Proc. R. Soc. London,* Ser. A235, 518 (1956).
120. Baker, A.D., A. Casadavell, H.D. Gafney, and M. Gellender, *J. Chem. Educ.,* 57, 314 (1980).
121. Schnitzer, M., in *Soil Organic Matter,* M. Schnitzer, and S.U. Khan (Eds.), Elsevier, New York, 1978, pp. 1–64.
122. Thurman, E.M. and R.L. Malcolm, in *Aquatic and Terrestrial Humic Materials,* Ann Arbor Science, Ann Arbor, MI, 1983, pp. 1–23.
123. Hedges, J.I., J.R. Ertel, P.D. Quay, P.M. Grootes, J.E. Richey, A.H. Devol, G.W. Farwell, F.W. Schmidt, and E. Salati, *Science,* 231, 1129 (1986).
124. Hutchinson, G.E., *Treatise on Limnology,* Vol. I, Wiley-Interscience, New York, 1957, and citations therein; Spring, W., *Bull. Acad. Belg. Cl. Sci.,* 34, 578 (1898); and Whipple, G.C., *The Microscopy of Drinking Water,* John Wiley & Sons, New York, 1899.
125. Aberg, B. and W. Rodhe, *Symb. Bot. Upsaliens,* 5, No.3, (1942).
126. Visser, S.A., *Environ. Sci. Technol.,* 17, 412 (1983).
127. Chen, Y., S.U. Khan, and M. Schnitzer, *Soil Sci. Soc. Am. J.,* 42, 292 (1978).
128. Strome, D.J. and M.C. Miller, *Verh. Int. Verein Limnol.,* 20, 1248 (1978).
129. Gilbert, E., *Vom Wasser,* 55, 1 (1980).
130. Kotzias, D., M. Herrmann, A. Zsolnay, R. Beyerle-Pfnur, H. Parlar, and F. Korte, *Chemosphere,* 16, 1463 (1987).
131. Choudhry, G.G. and G.R.B. Webster, *Chemosphere,* 15, 1935 (1986).
132. Dulin, D., H. Drossman, and T. Mill, *Environ. Sci. Technol.,* 20, 72 (1986).
133. Mill, T., W.R. Mabey, B.Y. Lan, and A. Baraze, *Chemosphere,* 10, 1281 (1981).
134. Zepp, R.G., N.L. Wolfe, L.V. Azarraga, R.H. Cox, and C.W. Pape, *Arch. Environ. Contam. Toxicol.,* 6, 305 (1977).
135. Wolfe, N.L., R.G. Zepp, P. Schlotzhauer, and M. Sink, *Chemosphere,* 11, 91 (1982).
136. Miller, G.C., M.J. Miille, D.G. Crosby, S. Sontum, and R.G. Zepp, *Tetrahedron,* 35, 1797 (1979).
137. Wong, A.S. and D.G. Crosby, *J. Agric. Food Chem.,* 29, 125 (1979).
138. Crosby, D.G. and A.S. Wong, *J. Agric. Food Chem.,* 21, 1049 (1973); Zepp, R.G., N.L. Wolfe, J.A. Gordon, and G.L. Baughman, *Environ. Sci. Technol.,* 9, 1144 (1975).
139. Ross, R.D. and D.G. Crosby, *Chemosphere,* 5, 277 (1975); Draper, W.M. and D.G. Crosby, *J. Agric. Food Chem.,* 32, 231 (1984).
140. Hwang, H.-M., R.E. Hodson, and R.F. Lee, *Environ. Sci. Technol.,* 20, 1002 (1986); Miller, R.M., G.M. Singer, J.D. Rosen, and R. Bartha, *Environ. Sci. Technol.,* 22, 1215 (1988).
141. Lamola, A.A. and G.S. Hammond, *J. Chem. Phys.,* 43, 2129 (1965).
142. Suzuki, J., T. Sato, A. Ito, and S. Suzuki, *Chemosphere,* 16, 1289 (1987).
143. Fenton, H.J.H., *Chem. News,* 33, 190 (1876).
144. Barbeni, M., C. Minero, E. Pelizzetti, E. Borgarello, and N. Serpone, *Chemosphere,* 16, 2225 (1987).

145. Buxton, G., C.L. Greenstock, W.P. Helman, and A.B. Ross, *J. Phys. Chem. Ref. Data,* 17, 513 (1988).

146. Mill, T. and D.G. Hendry, in *Comprehensive Chemical Kinetics,* Vol. 16, C.H. Bamford and C.F.H. Tipper (Eds.), Elsevier, Amsterdam, 1980, pp. 1–87.

146a. Walling, C., *Free Radicals in Solution,* John Wiley & Sons, New York, 1958.

147. Faust, B. and J. Hoigne, *Environ. Sci. Technol.,* 21, 957 (1987).

148. Tratnyek, P.G. and J. Hoigne, *Environ. Sci. Technol.,* 25, 1596 (1991).

149. Foote, C.S., *Acct. Chem. Res.,* 1, 104 (1968).

150. Herron, J.T. and R.E. Huie, *Ann. N.Y. Acad. Sci.,* 171, 229 (1970); *Environ. Sci. Technol.,* 4, 685 (1970).

151. Nilsson, R., P.B. Merkel, and D.R. Kearns, *Photochem. Photobiol.,* 16, 117 (1972).

152. Zafiriou, O.C., N.V. Blough, E. Micinski, B. Dister, D. Kieber, and J. Moffett, *Mar. Chem.,* 30, 45 (1990); Zafiriou, O.C., pers. comm., 1988.

153. Blough, N.V., *Environ. Sci. Technol.,* 22, 77 (1988).

154. Payne, J.R. and C.R. Phillips, *Environ. Sci. Technol.,* 19, 569 (1985).

155. Jordan, R.E. and J.R. Payne, *Fate and Weathering of Petroleum Spills in the Marine Environment: A Literature Review and Synopsis,* Ann Arbor Science, Ann Arbor, MI, 1980.

156. Sydnes, L.K., S.H. Hansen, and I.C. Burkow, *Chemosphere,* 14, 1043 (1985).

157. Berridge, S.A., et al., *J. Inst. Petrol.,* 54, 300 (1968).

158. Klein, A.E. and N. Pilpel, *Water Res.,* 8, 79 (1974).

159. Klein, A.E. and N. Pilpel, *J. Chem. Soc. Faraday Trans.,* 70, 1250 (1974).

160. Hansen, H.P., *Mar. Chem.,* 3, 183 (1975).

161. Patel, J.R., et al., in *Symposium on Carcinogenic Polynuclear Aromatic Hydrocarbons in the Marine Environment,* U.S. EPA, Washington, D.C., 1978, pp. 1–32.

162. Larson, R.A. and L.L. Hunt, *Photochem. Photobiol.,* 28, 553 (1978).

163. Larson, R.A., L.L. Hunt, and D.W. Blankenship, *Environ. Sci. Technol.,* 11, 492 (1977).

164. Patel, J.R., E.B. Overton, and J.L. Laseter, *Chemosphere,* 8, 557 (1979).

165. Overton, E.B., et al., in: Proc. Symp. Prelim. Results from September 1979 Researcher/Pierce IXTOC 1 Cruise, Off. Mar. Assess., Nat. Ocean. Atmos. Admin., Key Biscayne, FL, 1980, pp. 341–386.

166. Waite, T.D., Ph.D. thesis, Massachusetts Institute of Technology, Cambridge, 1983.

167. Waite, T.D. and F.M.M. Morel, *J. Colloid Interface Sci.,* 102, 121 (1984).

168. Sunda, W.G., S.A. Huntsman, and G.R. Harvey, *Nature,* 301, 234 (1983).

169. Waite, T.D., I.C. Wrigley, and R. Szymczak, *Environ. Sci. Technol.,* 22, 778 (1988).

170. Leland, J.K. and A.J. Bard, *J. Phys. Chem.,* 91, 5076 (1987).

171. Cunningham, K.M., M.C. Goldberg, and E.R. Weiner, *Environ. Sci. Technol.,* 22, 1090 (1988).

172. McKnight, D.M., B.A. Kimball, and K.E. Bencala, *Science,* 240, 637 (1988).

173. Cunningham, K.M., M.C. Goldberg, and E.R. Weiner, *Photochem. Photobiol.,* 41, 409 (1985).

174. Haupt, J., J. Peretti, and R. Van Steenwinkel, *Nouv. J. Chim.,* 8, 633 (1984).

175. Kiwi, J. and M. Graetzel, *J. Chem. Soc. Faraday Trans.* 1, 83, 1101 (1987).

176. Faust, B.C. and M.R. Hoffmann, *Environ. Sci. Technol.,* 20, 943 (1986).

177. Fox, M.A., *Acct. Chem. Res.,* 16, 314 (1983).

178. Wrighton, M.S., *Acct. Chem. Res.,* 12, 303 (1979).

179. Nozik, A.J., *Ann. Rev. Phys. Chem.,* 29, 189 (1978).

180. Becquerel, E., *C.R. Hebd. Seances Acad. Sci.,* 9, 561 (1839).

181. Markin, M.C. and K.J. Laidler, *J. Phys. Chem.*, 57, 363 (1953); Rubin, T.R., J.G. Calvert, G.T. Rankin, and W.M. MacNevin, *J. Am. Chem. Soc.*, 75, 2850 (1953); Calvert, J.G., K. Theuer, G.T. Rankin, and W.M. MacNevin, *J. Am. Chem. Soc.*, 76, 2575 (1954).

182. Morrison, S.R. and T. Freund, *J. Chem. Phys.*, 47, 1543 (1967).

183. Fujishima, A. and K. Honda, *Nature,* 37, 238 (1972).

184. Harriman, A., M.-C. Richoux, P.A. Christensen, S. Mosseri, and P. Neta, *J. Chem. Soc. Faraday Trans.* 1, 83, 3001 (1987).

185. Gerischer, H. and F. Willig, *Top. Curr. Chem.*, 61, 33 (1976).

186. Dunn, W.W., Y. Aikawa, and A.J. Bard, *J. Am. Chem. Soc.*, 103, 3456 (1981).

187. Frank, S.N. and A.J. Bard, *J. Phys. Chem.*, 81, 1484 (1977); *J. Am. Chem. Soc.*, 99, 303 (1977).

188. Ollis, D.F. and A.L. Pruden, *Environ. Sci. Technol.*, 17, 628 (1983); Pruden, A.L. and D.F. Ollis, *J. Catalysis,* 82, 404 (1983); Hsiao, C.Y., C.L. Lee, and D.F. Ollis, *J. Catalysis,* 82, 418 (1983).

189. Ahmead, S. and D.F. Ollis, *Solar Energy,* 32, 597 (1984); Ollis, D.F., C.Y. Hsiao, L. Budiman, and C.L. Lee, *J. Catalysis,* 88, 89 (1984).

190. Ollis, D.F., *Environ. Sci. Technol.*, 19, 480 (1985).

191. Barbeni, M., E. Pramauro, E. Pelizzetti, E. Borgarello, and N. Serpone, *Chemosphere,* 14, 195 (1985).

192. Barbeni, M., E. Pramauro, E. Pelizzetti, E. Borgarello, N. Serpone, and M.J. Jamieson, *Chemosphere,* 15, 1913 (1986).

193. Pelizzetti, E., M. Borgarello, C. Minero, E. Pramauro, E. Borgarello, and N. Serpone, *Chemosphere,* 17, 499 (1988).

194. Matthews, R.W., *Water Res.*, 20, 569 (1986).

195. Hidaka, H., H. Kubota, M. Gratzel, E. Pelizzetti, and N. Serpone, *J. Photochem.*, 35, 219 (1986).

196. Barbeni, M., M. Morello, E. Pramauro, E. Pelizzetti, M. Vicenti, E. Borgarello, and N. Serpone, *Chemosphere,* 16, 1165 (1987).

197. Spikes, J.D., *Photochem. Photobiol.*, 34, 549 (1981).

198. Davis, A.P. and C.P. Huang, *Water Res.*, 25, 1273 (1991).

199. Davis, A.P., J.M. Tseng, and C.P. Huang, Proc., Oak Ridge Model Conf., Vol. I, Pt. 3, (1987); Davis, A.P. and C.P. Huang, *Water Sci. Technol.*, 21, 455 (1989); *Water Res.*, 24, 543 (1990).

200. Al-Ekabi, H. and N. Serpone, *J. Phys. Chem.*, 92, 5726 (1988).

201. Tseng, J. and C.P. Huang, in *Emerging Technologies in Hazardous Waste Management,* D.W. Tedder and F.G. Pohland (Eds.), ACS Symp. Ser. 422, Am. Chem. Soc., Washington, D.C., 1990, pp. 12–39.

202. Tunesi, S. and M.A. Anderson, *Chemosphere,* 16, 1447 (1987).

203. Kormann, C., D.W. Bahnemann, and M.R. Hoffmann, *Environ. Sci. Technol.*, 22, 798 (1988).

204. Hong, A.P., D.W. Bahnemann, and M.R. Hoffmann, *J. Phys. Chem.*, 91, 2109 (1987).

205. Hong, A.P., D.W. Bahnemann, and M.R. Hoffmann, *J. Phys. Chem.*, 91, 6245 (1987).

206. Wang, T.C. and C.K. Tan, *Environ. Sci. Technol.*, 22, 916 (1988).

Index

B

Backward differencing, 46
Bacterial respiratory enzymes, 315, see also
    specific types
Base catalysis, 170, 195
Base-catalyzed hydrolysis, 167, 172, 564
Batch cultures, 451–452, 490
Batch reactors, 340, 451
Beer's law, 650
Benzenes, 583, 590, 618, 625, 721
Benzoate, 322
Benzophenone-*cis*-1,3-pentadiene, 99
Bicarbonate, 261, 316
Bidentate ligands, 321
Bimolecular collisions, 35, 78
Bimolecular reactions, 135, 169
Binding-site energy, 76
Bioaccumulation, 380, 579, 617, 628
    of organic compounds, 598–602
    reactors and, 378
Biochemical oxygen demand (BOD), 68, 69,
    604, 619, 621
Biochemical processes, 5, 13, 115, see also
    specific types
    enzyme-catalyzed reactions and, see
        Enzyme-catalyzed reactions
Biodegradation, 580
    of herbicides, 490
    rates of, 602–608
    reactors and, 378, 383, 393
Biological films, 534–540
Biological processes, 5, 21, 117, see also
    specific types
    chaos and, 499–500
    property-activity relationships for, 595–598
    rate equations for, 25
    reaction rates and, 110
Biomass, 4, 451, 483
Biospheric biomass, 4
Biphenyls, 721
Birnessite, 268
Bisubstrate reactions, 436–439
Blackman model, 457
BOD, see Biochemical oxygen demand
Boiling point of water, 117
Boltzmann's constant, 126, 129, 135, 139
Boltzmann's theory, 139
Bond dissociation, 52
Bonding ligands, 177
Boundary conditions, 282, 342, 349, 352, 530
Boundary layers, 278, 516
Breakpoint chlorination, 48, 233–235
Bromoform, 241

Bronsted acid-base catalysis, 170–177
Bronsted acids, 177, 192
Bronsted catalysis, 170–177, 559
Bronsted equation, 156, 157, 247, 267
Bronsted relationship, 556, 559–561, 569,
    570, 580, 597
Buffer catalysis, 190, 224–226

C

Calcite, 114, 149, 300, 304, 308, 316
Calcium, 311
Calcium carbonate, 294, 299–309
Calcium carbonate compensation depth
    (CCD), 300
Carbamates, 220, 570, see also specific types
Carbon, 4, 101, 273, 290
Carbonate minerals, 292–293, see also
    specific types
Carbon cycle, 397, 404, 406, 408
Carbon dioxide, 195, 273, 290, 300, 303, 305
    absorption of, 290
    dissolution and, 308
    hydration of, 145, 158, 194
    transfer of, 291
    transport of, 533
    uptake of, 530, 532
Carbonic acids, 194, see also specific types
Carbon monoxide, 50
Carbon tetrachloride, 722
Catalysis, 34, 36, see also specific types
    acid, 170, 192, 206, 230, 233
    acid-base, see Acid-base catalysis
    anion, 184–185
    Arrhenius, 170–177
    auto-, see Autocatalysis
    base, 170, 195
    Bronsted, 170–177, 559
    buffer, 190, 224–226
    enzyme, see Enzyme-catalyzed reactions
    homogeneous, see Homogeneous catalysis
    by ions, 180–181
    of Lewis acids, 177–179
    mechanisms of, 169–179
    photo-, 269
    rate expressions for, 25, 185–193
    as reaction mechanism change, 168–169
Catalysts, 35, 36, see also specific types
    auto-, 168
    characteristics of, 168
    defined, 168
    effectiveness of, 524
    recycled, 186–187
    redox reactions and, 258, 260

E

sulfur cycle and, 7, 9
Nonionic reactions, 65, see also specific types
Nonlinear regression, 57, 59, 62–64, 68, 74, 116
  direct-fitting, 425
  enzyme-catalyzed reactions and, 421, 425–427, 434
  gas transfer and, 282
  microbial processes and, 479
Nonrandom errors, 58
Nonstiff differential equations, 46
Normal mode oscillation, 142
Nuclear magnetic resonance (NMR), 196, 682, 684
Nucleophiles, 574–575, see also specific types
Nucleophilic attack, 249
Nucleophilic reactions, 562–564, see also specific types
Nucleophilic substitution, 223–224, 574
Nutrient models, 363–377, see also specific types
  mass-balance, 363–365
  residence times in, 365–366
  trophic conditions and, 370–373
Nutrients, 339, see also specific types
  loading of, 363, 364
  micro-, 456–457
  microbial growth and, 451–458
  models of, see Nutrient models
  multiple, 507–509
  uptake of, 505–507

O

Ockham's razor, 85
Octanol-water partition coefficients, 383, 553
Oil, 710–714
Olefins, 183, see also specific types
Oligoclase, 324
One-dimensional diffusion-reaction equations, 349
One-dimensional mass-transport equations, 350, 351–352
One-electron redox cycles, 181
One-electron transfers, 266
One-step reactions, 80, 88–89, 148
Open systems, 339, 361–363, see also specific types
Orbital energy, 196
Order, 26–27
Ordered sequential mechanisms, 437, 439
Organic acids, 293, 311–312, 628, see also specific types
Organic acid-transition metal complexes, 688–690

Organic compounds, 311, see also specific types
  bioaccumulation of, 598–602
  in CFSTRs, 378–380
  chlorination of, 236–245
  color-causing, 228
  compartment models for, 382–383
  fate of, 378–383
  hydrolysis of, 218–223
  photochemical reactions of, 696–714
    photolysis and, 696–701
  photosensitized reactions of, 702–703
  transport of, 378–383
  volatile, 720, 721
Organic ligands, 261, see also specific types
Organic matter, dissolved, see Dissolved organic matter (DOM)
Organic radicals, 703–706, see also specific types
Organic sulfides, 183
Organochlorines, 229, see also specific types
Organofluorophosphonate, 170
Organophosphates, 620, 628
Orthoclase, 322
Orthophosphates, 48, 217, see also specific types
Oscillation, 142
Othmer-Thakar relationship, 283
Outer-sphere complexes, 198, 201, 205, 311
Outer-sphere electron transfer, 565
Outer-sphere exchange, 184
Outer-sphere mechanisms, 185, 257, 312
Outer-sphere redox reactions, 566–569
Overflow rate, 341
Oxalate, 312
Oxidation, 4, 19, 48, 181, see also Oxygen; Oxygenation
  of ammonium, 69
  auto-, see Autooxidation
  by chain mechanisms, 50
  of chlorine, 236
  complete, 69
  of copper, 272–273
  of dissolved organic matter, 685
  enzyme-mediated, 695
  homogeneous catalysis and, 179
  intermediate, 169, 180–181
  of iron, 64, 190, 245, 267, 669
  of iron sulfides, 19
  of manganese, 64, 190
  metal-catalyzed, 167
  of metal ions, 5
  of nitrogen, 48
  nonenzymatic, 439